Photoreceptors and Calcium

ADVANCES IN EXPERIMENTAL MEDICINE AND BIOLOGY

Recent Volumes in this Series

Volume 506
LACRIMAL GLAND, TEAR FILM, AND DRY EYE SYNDROME 3: BASIC SCIENCE AND CLINICAL RELEVANCE
Edited by David A. Sullivan, Michael E. Stern, Kazuo Tsubota, Darlene A. Dartt, Rose M. Sullivan, and B. Britt Bromberg

Volume 507
EICOSANOIDS AND OTHER BIOACTIVE LIPIDS IN CANCER, INFLAMMATION, AND RADIATION INJURY 5
Edited by Kenneth V. Honn, Lawrence J. Marnett, Santosh Nigam, and Charles Serhan

Volume 508
SENSORIMOTOR CONTROL OF MOVEMENT AND POSTURE
Edited by Simon C. Gandevia, Uwe Proske and Douglas G. Stuart

Volume 509
IRON CHELATION THERAPY
Edited by Chiam Hershko

Volume 510
OXYGEN TRANSPORT TO TISSUE XXIII: OXYGEN MEASUREMENTS IN THE 21ST CENTURY: BASIC TECHNIQUES AND CLINICAL RELEVANCE
Edited by David F. Wilson, John Biaglow and Anna Pastuszko

Volume 511
PEDIATRIC GENDER ASSIGNMENT: A CRITICAL REAPPRAISAL
Edited by Stephen A. Zderic, Douglas A. Canning, Michael C. Carr and Howard McC. Snyder III

Volume 512
LYMPHOCYTE ACTIVATION AND IMMUNE REGULATION IX: LYMPHOCYTE TRAFFIC AND HOMEOSTASIS
Edited by Sudhir Gupta, Eugene Butcher and William Paul

Volume 513
MOLECULAR AND CELLULAR BIOLOGY OF NEUROPROTECTION IN THE CNS
Edited by Christian Alzheimer

Volume 514
PHOTORECEPTORS AND CALCIUM
Edited by Wolfgang Baehr and Krzysztof Palczewski

Volume 515
NEUROPILIN: FROM NERVOUS SYSTEM TO VASCULAR AND TUMOR BIOLOGY
Edited by Dominique Bagnard

Volume 516
TRIPLE REPEAT DISORDERS OF THE NERVOUS SYSTEM
Edited by Lubov T. Timchenko

Volume 517
DOPAMINERGIC NEURON TRANSPLANTATION IN THE WEAVER MOUSE MODEL OF PARKINSON'S DISEASE
Edited by Lazaros C. Triarhou

Volume 518
ADVANCES IN MALE MEDIATED DEVELOPMENTAL TOXICITY
Edited by Bernard Robaire and Barbara F. Hales

A Continuation Order Plan is available for this series. A continuation order will bring delivery of each new volume immediately upon publication. Volumes are billed only upon actual shipment. For further information please contact the publisher.

Photoreceptors and Calcium

Edited by

Wolfgang Baehr, Ph.D.
Moran Eye Center/EIHG
University of Utah Health Science Center
15 North/ 2030 East
Salt Lake City, UT 84112-5330
USA
email: wbaehr@hsc.utah.edu

Krzysztof Palczewski, Ph.D.
Department of Ophthalmology
University of Washington
1959 NE Pacific St.
Seattle, WA 98195
USA
email: palczews@washington.edu

Springer Science+Business Media, LLC

Library of Congress Cataloging-in-Publication Data

CIP applied for but not received at time of publication.

Photoreceptors and Calcium
Edited by Wolfgang Baehr and Krzysztof Palczewski
ISBN 978-0-306-47415-6 ISBN 978-1-4615-0121-3 (eBook)
DOI 10.1007/978-1-4615-0121-3
AEMB volume number: 514

Originally published by Kluwer Academic / Plenum Publishers, New York in 2002

http://www.wkap.nl

Landes Bioscience
810 S. Church Street, Georgetown, TX 78626
http://www.landesbioscience.com; http://www.eurekah.com
Landes tracking number: 1-58706-139-2

10 9 8 7 6 5 4 3 2 1

A C.I.P. record for this book is available from the Library of Congress.

PREFACE

The role of Ca^{2+} as an internal messenger in visual transduction of vertebrate and invertebrate organisms has been explored intensely in the recent past. Since the early 1970s, calcium ions and cyclic GMP (whose levels are controlled by Ca^{2+} in vertebrates) have been recognized as important second messengers. Particularly in the last decade, however, the role of Ca^{2+} in visual transduction has been re-evaluated and a proliferation of research has documented a multiplicity of roles.

It is now evident that Ca^{2+} modulates phototransduction by acting at several sites through a host of small Ca^{2+}-binding proteins. For example, in phototransduction of vertebrates, Ca^{2+}-free forms of guanylate cyclase activating proteins (GCAPs) activate guanylate cyclase, modulating levels of cGMP, a key event in the return of photoreceptors to pre-bleach conditions. Defects in genes encoding guanylate cyclase or guanylate cyclase activating proteins lead to severe diseases of the retina (e.g., Leber congenital amaurosis, rod/cone dystrophy, or cone dystrophy), thus emphasizing the important role of these proteins in phototransduction. Similarly, mutant genes encoding cation or Ca^{2+} channels (cyclic nucleotide-gated cation channels located in the cell membrane and L-type voltage-gated Ca^{2+} channels located at the synapse of photoreceptors) lead to retinitis pigmentosa or congenital stationary night blindness. In phototransduction of invertebrate organisms (e.g., Drosophila and Limulus), the role of Ca^{2+} is similarly central, but distinct, from that of vertebrates.

The book, "Photoreceptors and Calcium," reviews our understanding of the role of Ca^{2+} in phototransduction, dark- and light-adaptation, recovery from bleaching and return to the dark state, as well as in synaptic signaling of photoreceptors and their second-order neurons (rod and cone bipolar cells), and its potential role in the neighboring retinal pigment epithelium. By no means is this review designed to be comprehensive, but most topics under intense scrutiny during the last 10 years are discussed.

It is our pleasure to thank the contributors, reviewers of individual chapters, and publisher (Ron Landes) who made this endeavor possible. In particular, we would like to thank members of our laboratories for their patience and support during the hectic period (1.5 years) of soliciting contributions, reviewing and editing this book.

Wolfgang Baehr, Ph.D.
Moran Eye Center/EIHG
University of Utah Health Science Center
Salt Lake City, UT, USA

Krzysztof Palczewski, Ph.D.
Department of Ophthalmology
University of Washington
Seattle, WA, USA

PARTICIPANTS

Mohammad Akhtar
Division of Biochemistry and Molecular Biology
School of Biological Sciences
University of Southampton
Basset Crescent East
Southampton S 016 7PX
United Kingdom

James B. Ames
Center for Advance Research in Biotechnology
University of Maryland Biotechnology Institute
9600 Gudelsky Drive
Rockville, MD 20850
USA
email: james@carb.nist.gov

Gautam B. Awatramani
Oregon Hearing Research Center
Vollum Institute
3181 SW Sam Jackson Park Road
Portland, OR 97201
USA

Wolfgang Baehr
Moran Eye Center/EIHG
University of Utah Health Science Center
15 North/ 2030 East
Salt Lake City, UT 84112-5330
USA
email: wbaehr@hsc.utah.edu

Sherry Ball
Cleveland V.A. Hospital
Research Service 151W
10701 East Boulevard
Cleveland, OH 44106
USA
email: slb25@po.cwru.edu

Steven A. Barnes
Departments of Physiology & Biophysics & Ophthalmology
Dalhousie University
Sir Charles Tupper Building
Halifax, B3H 4H7
Nova Scotia
email: sbarnes@is.dal.ca

Paul Joerg Bauer
Institute for Biological Information Processing
P.O.B. 1913
52425 Jülich
Germany
email: p.j.bauer@fz-juelich.de.

Hanno Bolz
Institute of Human Genetics
University Hospital Eppendorf
Butenfeld 42
D-22529 Hamburg
Germany

Peter D. Calvert
Department of Ophthalmology
Harvard Medical School
243 Charles Street
Boston, MA 02114
USA
email pdcalvert@meei.harvard.edu

Ching-Kang Jason Chen
Departments of Ophthalmology and Visual Sciences & Human Genetics
University of Utah
Salt Lake City, UT 84112-5330
USA
email: jason@hmbg.utah.edu

Chunhe Chen
Department of Physiology and Biophysics
University of Colorado Health Science Center
Denver, CO 80262
USA

Jeannie Chen
University of Southern California
1333 San Pablo St.
BMT 401
Los Angeles, CA 90033
USA
email: Jeannie@hsc.usc.edu

Conan B. Cooper
Department of Physiology and Biophysics
University of Calgary
3330 Hospital Drive NW
Calgary ALBERTA T2N 4N1
Canada

Evelyne C. Deery
School of Biological Sciences
Queen Mary, University of London
Mile End Road
London E1 4NS
United Kingdom

Douglas J. Demetrick
Department of Pathology
University of Calgary
3330 Hospital Drive NW
Calgary ALBERTA T2N 4N1
Canada

Alexander M. Dizhoor
Department of Ophthalmology
Kresge Eye Institute
Wayne State University
4717 St. Antoine
Detroit, MI. 48201
USA
email: adizhoor@med.wayne.edu

Andreas Gal
Institute of Human Genetics
University Hospital Eppendorf
Butenfeld 42
D-22529 Hamburg
Germany
email: gal@uke.uni-hamburg.de

Chaoxian Geng
Department of Biological Sciences
Purdue University
Lilly Hall of Life Sciences
West Lafayette IN 47907-1392
USA
email: cgeng@bilbo.bio.purdue.edu

Andreas Gießl
Institute of Zoology
Johannes Gutenberg Universität Mainz
Mullerweg 6
Mainz D-55099
Germany

Wojciech A. Gorczyca
Polish Academy of Sciences
Institute of Immunology & Experimental Therapy
Department of Microbiology
R. Weigla 12, 53-114 Wroclaw, Poland
email: gorczyca@immuno.iitd.pan.wroc.pl

Ronald G. Gregg
Department of Biochemistry & Molecular Biology and Ophthalmology & Visual Sciences
University of Louisville
Louisville, KY 40202
USA

Françoise Haeseleer
Department of Ophthalmology
University of Washington
School of Medicine
Box 356485
Seattle, WA. 98195-6485
USA
email: fanfan@u.washington.edu

David M. Hunt
Institute of Ophthalmology
University College London
11-43 Bath Street
London, EC1V 9EL
United Kingdom

Mitsuhiko Ikura
Division of Molecular and Structural Biology
Ontario Cancer Institute and Department of Medical Biophysics
University of Toronto
610 University Avenue
Toronto, Ontario M5G 2M9
Canada

Satoru Kawamura
Department of Biology
Graduate School of Science
Osaka University
1-1 Machikaneyama
Toyonaka, Osaka 560-0043
Japan
email: kawamura@bio.sci.osaka-u.ac.jp

Melanie E.M. Kelly
Departments of Pharmacology & Ophthalmology
Dalhousie University
Sir Charles Tupper Building
Halifax, B3H 4H7
Nova Scotia

Karl-Wilhelm Koch
Institut fur Biologische Informationsverarbeitung 1,
Forschungszentrum Jh lich
D-52425 Juelich
Germany
email: k.w.koch@fz-juelich.de

Juan I. Korenbrot
Department of Physiology
School of Medicine
University of California at San Francisco
San Francisco, CA 94143
USA

Yiannis Koutalos
Department of Physiology and Biophysics
University of Colorado Health Sciences Center
4200 E 9th Ave
Denver, CO 80262
USA
email: Yiannis.koutalos@uchsc.edu

Michael R. Kreutz
Department Neurochemistry/Molecular Biology
Leibnitz Institute for Neurobiology,
D-39108 Magdeburg
Brenneckestr. 6
Germany

Eric Lasater
Department of Ophthalmology and Visual Sciences
John Moran Eye Center
University of Utah
50 N. Medical Dr.
Salt Lake City, UT 84132
USA
email eric.lasater@hsc.utah.edu

K. Nicholas Leibovic
Department of Biophysics and Physiology
State University of N.Y. at Buffalo
105 High Park Blvd
Buffalo, N.Y.
USA
email: bphknl@ubvms.cc.buffalo.edu

John E. Lisman
Department of Biology and Volen Center for Complex Systems
Brandeis University
Waltham, MA 02454-9110
USA
email: lisman@brandeis.edu

Clint L. Makino
Department of Ophthalmology
Harvard Medical School
243 Charles Street
Boston, MA 02114
USA
email: cmakino@meei.harvard.edu

Ana Mendez
The Mary D. Allen Laboratory for Vision Research
Doheny Eye Institute
Los Angeles, CA 90033
USA

Baruch Minke
Department of Physiology
Hadassah Medical School
The Hebrew University
P.O.Box 12272
Jerusalem 91120
Israel
email: minke@md.huji.ac.il

Robert S. Molday
Department of Biochemistry/Molecular Biology
University of British Columbia
2146 Health Sciences Mall
Vancouver BC V6T 1Z3
Canada
email: molday@unixg.ubc.ca

Oleg Moskvin
Department of Ophthalmology
Kresge Eye Institute
Wayne State University
School of Medicine
Detroit, MI 48201
USA

Sabita K. Murthy
Department of Pathology
University of Calgary
3330 Hospital Drive NW
Calgary ALBERTA T2N 4N1
Canada

Kei Nakatani
Institute of Biological Sciences
University of Tsukuba
Japan

Mitsuru Nakazawa
Department of Ophthalmology
Hirosaki University School of Medicine
5 Zaifu-cho
Hirosaki 036-8562 Aomori
Japan

Richard J. Newbold
School of Biological Sciences
Queen Mary, University of London
Mile End Road
London E1 4NS
United Kingdom

Johannes Oberwinkler
Department of Neurobiophysics
University of Groningen
Nijenborgh 4
9747 AG Groningen
The Netherlands
email: j.oberwinkler@phys.rug.nl

Hiroshi Ohguro
Department of Ophthalmology
Hirosaki University School of Medicine
5 Zaifu-cho
Hirosaki 036-8562 Aomori
Japan
email: ooguro@cc.hirosaki-u.ac.jp

Joseph E. O'Tousa
Department of Biological Science
University of Notre Dame
Notre Dame IN 46556
USA
email: jotousa@nd.edu.

William Pak
Department of Biological Sciences
Purdue University
Lilly Hall of Life Sciences
West Lafayette IN 47907-1392
USA
email: wpak@bilbo.bio.purdue.edu

Krzysztof Palczewski
Department of Ophthalmology
University of Washington
1959 NE Pacific St.
Seattle, WA 98195
USA
email: palczews@washington.edu

Annette M. Payne
Department of Biological Sciences
Institute of Cancer Genetics and Pharmacogenomics
Brunel University
Uxbridge, Middlesex UB8 3PH
United Kingdom

Richard Payne
Department of Biology
University of Maryland
College Park, MD 20742
USA

Pavel P. Philippov
A.N. Belozersky Institute of Physico-Chemical Biology
M.V.Lomonosov Moscow State University
119899 Moscow
Russia
email: ppph@belozersky.msu.ru

Clemens F.M. Prinsen
Department of Physiology and Biophysics
University of Calgary
3330 Hospital Drive NW
Calgary ALBERTA T2N 4N1
Canada

Alexander Pulvermüller
Institut für Medizinische Physik und Biophysik
Humboldt-Universität zu Berlin
Universitätsklinikum Charité
10098 Berlin
Germany

Sridhar Raghavachari
Department of Biology and Volen Center for Complex Systems
Brandeis University
Waltham, MA 02454-9110
USA

Tatiana I. Rebrik
Department of Physiology
School of Medicine
University of California at San Francisco
San Francisco, CA 94143
USA

Jan Reiners
Institute of Zoology
Johannes Gutenberg Universität Mainz
Mullerweg 6
Mainz D-55099
Germany

C. Reissner
Department Neurochemistry/ Molecular Biology
Leibnitz Institute for Neurobiology,
D-39108 Magdeburg
Brenneckestr. 6
Germany

Edwin A. Richard
Department of Biology and Volen Center for Complex Systems
Brandeis University
Waltham, MA 02454-9110
USA
email: richard@brandeis.edu

Rita Rosenthal
Institut fuer Klinische Chemie
Universitaetsklinikum Benjamin Franklin
Hindenburgdamm 30
Berlin 12200
Germany

Paul P. M. Schnetkamp
Department of Physiology and Biophysics
University of Calgary
3330 Hospital Drive NW
Calgary ALBERTA T2N 4N1
Canada
email: pschnetk@acs.ucalgary.ca

Constanze Seidenbecher
Department Neurochemistry/ Molecular Biology
Leibnitz Institute for Neurobiology,
D-39108 Magdeburg
Brenneckestr. 6
Germany
email: seidenc@ifn-magdeburg.de

Ivan I. Senin
A.N. Belozersky Institute of Physico-Chemical Biology
M.V.Lomonosov Moscow State University
119899 Moscow
Russia

Ari Sitaramayya
Eye Research Institute
Oakland University
423 Dodge Hall
Rochester, MI 48309-4401
USA
email: ari@oakland.edu

Malcolm Slaughter
Department of Physiology and Biophysics
University at Buffalo School of Medicine
124 Sherman Hall
Buffalo, NY 14214
USA
email: mslaught@buffalo.edu

Izabela Sokal
Department of Ophthalmology
School of Medicine
University of Washington
Seattle, WA
USA

Eduardo Solessio
Center for Vision Research
SUNY
Upstate Medical University
Syracuse, NY
USA

Olaf Strauß
Institut fuer Klinische Chemie
Universitaetsklinikum Benjamin Franklin
Hindenburgdamm 30
Berlin 12200
Germany
email strauss@ukbf.fu-berlin.de

Robert T. Szerencsei
Department of Physiology and Biophysics
University of Calgary
3330 Hospital Drive NW
Calgary ALBERTA T2N 4N1
Canada

Shuji Tachibanaki
Department of Biology
Graduate School of Science
Osaka University
1-1 Machikaneyama
Toyonaka, Osaka 560-0043
Japan

Martin James Warren
School of Biological Sciences
Queen Mary, University of London
Mile End Road
London E1 4NS
United Kingdom
email: m.j.warren@qmw.ac.uk

René Warren
Department of Biochemistry/Molecular Biology
University of British Columbia
2146 Health Sciences Mall
Vancouver BC V6T 1Z3
Canada

Theodore G. Wensel
Baylor College of Medicine
One Baylor Plaza
Houston, TX 77030
USA
email: twensel@bcm.tmc.edu

Susan E. Wilkie
Institute of Ophthalmology
University College London
11-43 Bath Street
London, EC1V 9EL
United Kingdom

Uwe Wolfrum
Institute of Zoology
Johannes Gutenberg Universität Mainz
Mullerweg 6
Mainz D-55099
Germany
email: wolfrum@mail.uni-mainz.de

Akio Yamazaki
Departments of Ophthalmology and Pharmacology
Kresge Eye Institute
Wayne State University
School of Medicine
Detroit, MI 48201
USA
email: ayamazak@med.wayne.edu

Russell K. Yamazaki
Department of Pharmacology
Wayne State University
School of Medicine
Detroit, MI 48201
USA

King-Wai Yau
Departments of Neuroscience and Ophthalmology and the Howard Hughes Medical Institute
The Johns Hopkins University School of Medicine
Baltimore, MD 21205
USA

CONTENTS

COLOR PLATES

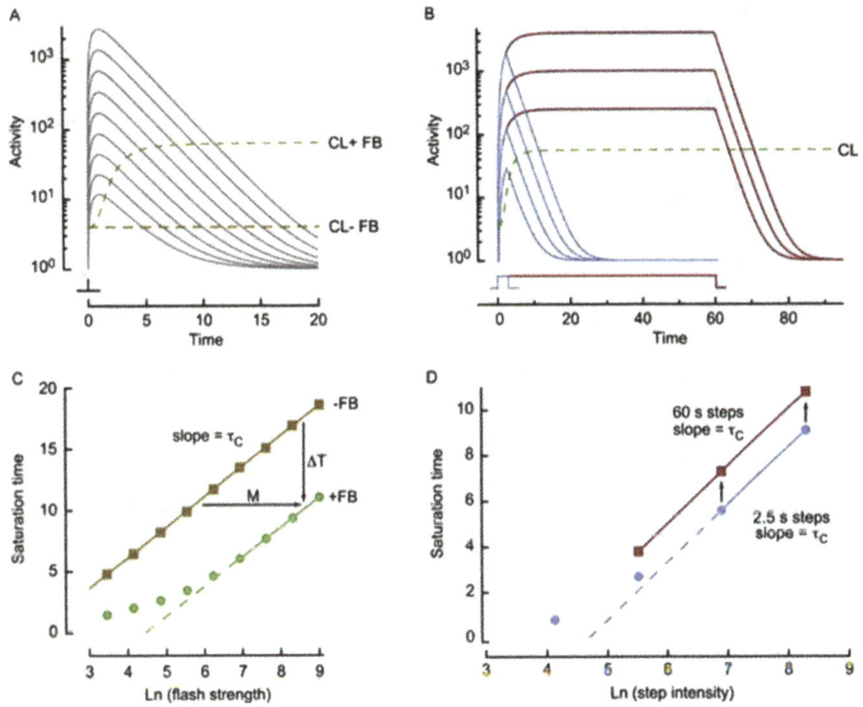

Figure C.1. Models for the saturation behavior of the rod response to bright flashes and steps as shown in Chapter 3, Figure 3. A. Waves of PDE activity (continuous lines, in arbitrary units) after flashes of nine photon densities. Activity declines exponentially with a time constant, τ_c. The CL rises as a function of time after the flash due to Ca^{2+} feedback regulation (dashed, green line). The horizontal, dashed, orange line represents an unchanging CL as might be encountered in the absence of feedback. B. Time course of PDE activity produced by 2.5 or 60 s steps of light in the absence of any slow phase of adaptation (Eq. 3). At the end of the step, activity decreases exponentially with the same time constant as after a flash, τ_c. C. Dependence of saturation time on the flash strength. Saturation time is given as the time to the crossing of CL in the presence (+FB) or absence (–FB) of feedback in A. The relationship is linear with slope τ_c for flash strengths in which the PDE activity intersects a constant CL. The reduction in the saturation time at a given flash strength, ΔT, due to feedback onto CL corresponds to M, the increase in flash strength required to produce a particular saturation time. M expresses the degree of adaptation. D. Dependence of saturation time on step intensity. Step response saturation time is defined as the interval between the end of the step and the intersection of the PDE activity with CL in B. The time in saturation after a 60 s step of a given intensity (squares) exceeds that for a 2.5 s step (circles) since 2.5 s was insufficient for PDE activity to attain steady state.

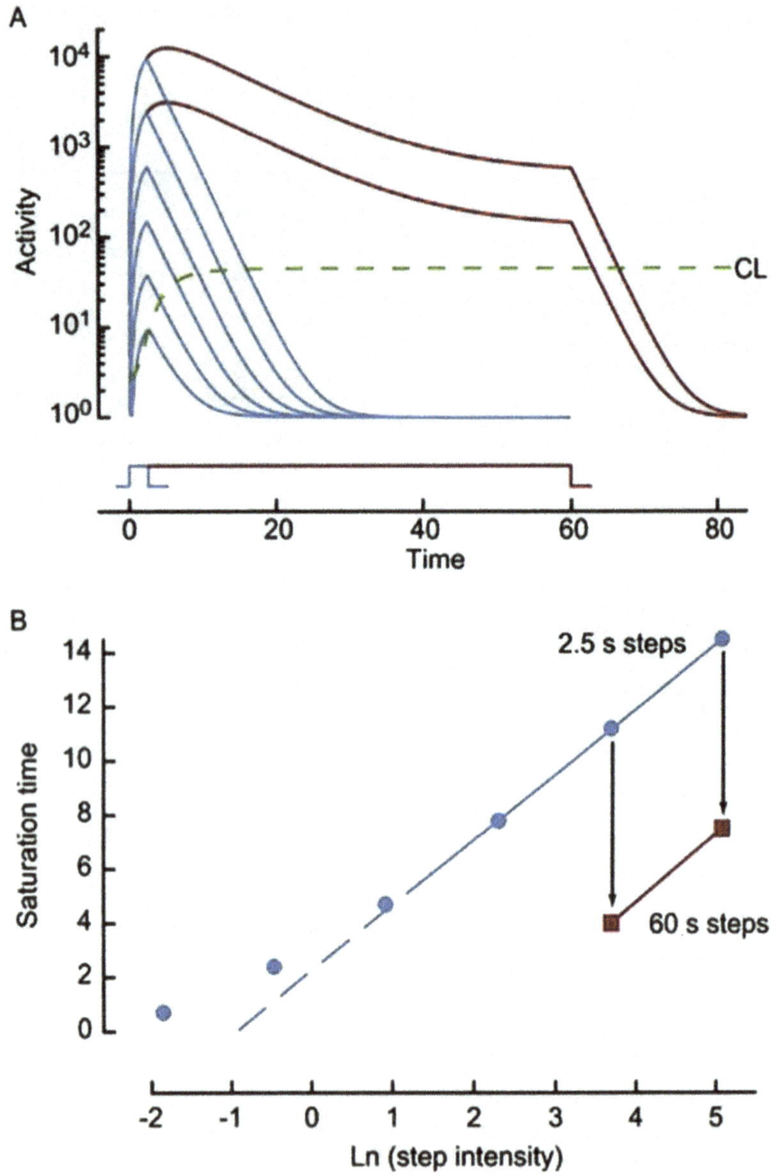

Figure C.2. Revised model of adaptation to continuous light as shown in Chapter 3, Figure 6. A. Time courses of PDE activity (continuous lines) in response to 2.5 and 60 s of illumination. Six illumination levels are shown for 2.5 s steps and two for 60 s steps. The change in CL is the same as that during a flash response (see Fig. 3). The PDE activity gradually declines with prolonged, 60 s illumination due to the slow phase of adaptation. Nevertheless, the kinetics of PDE decline are the same after the extinction of 2.5 and 60 s steps. B. Relation between saturation time after step shutoff and the natural logarithm of the step intensity. The slopes of the relations are the same for short and for long steps but the saturation time at a given intensity is shorter for the long step.

CALCIUM AND PHOTOTRANSDUCTION

Kei Nakatani,[1] Chunhe Chen,[2] King-Wai Yau,[3] and Yiannis Koutalos[2]

ABSTRACT

Visual phototransduction, the conversion of incoming light to an electrical signal, takes place in the outer segments of the rod and cone photoreceptor cells. Light reduces the concentration of cGMP, which, in darkness, keeps open cationic channels present in the plasma membrane of the outer segment. Ca^{2+} plays an important role in phototransduction by modulating the cGMP-gated channels as well as cGMP synthesis and breakdown. Ca^{2+} is involved in a negative feedback that is essential for photoreceptor adaptation to background illumination. The effects of Ca^{2+} on the different components of rod phototransduction have been characterized and can quantitatively account for the steady state responses of the rod cell to background illumination. The propagation of the Ca^{2+} feedback signal from the periphery toward the center of the outer segment depends on the Ca^{2+} diffusion coefficient, which has a value of $15\pm1\ \mu m^2\ s^{-1}$. This value shows that diffusion of Ca^{2+} in the radial direction is quite slow providing a significant barrier in the propagation of the feedback signal. Also, because the diffusion coefficient of Ca^{2+} is much smaller than that of cGMP, the decline of Ca^{2+} in the longitudinal direction lags behind the propagation of excitation by the decline of cGMP.

INTRODUCTION

Vertebrate rod and cone photoreceptors respond to light with a membrane hyperpolarization. Visual transduction, the conversion of light stimulus to the electrical response, takes place in the outer segments of the photoreceptors. In the dark,

[1]Institute of Biological Sciences, University of Tsukuba, Japan; [2]Department of Physiology and Biophysics, University of Colorado School of Medicine, U.S.A; [3]Departments of Neuroscience and Ophthalmology and the Howard Hughes Medical Institute, The Johns Hopkins University School of Medicine, U.S.A.

cations enter the photoreceptor outer segments through channels located in the plasma membrane, keeping the cell partially depolarized. This depolarization maintains the release of neurotransmitter from the photoreceptor synaptic terminal. Light leads to closure of these channels, resulting in membrane hyperpolarization and a decrease in neurotransmitter release from the synaptic terminal (Fig. 1A).

Calcium ions, Ca^{2+}, have been known for a long time to exert a profound effect on the light transduction process by photoreceptors. In fact, Ca^{2+} was observed to mimic the effects of light, an observation that led to the "Ca^{2+} hypothesis", the proposal that Ca^{2+} is the intracellular messenger of phototransduction acting by blocking the light-sensitive conductance.[1,2]

THE EFFECTS AND TARGETS OF CALCIUM

Subsequent experiments were inconsistent with the Ca^{2+} hypothesis of phototransduction[3]. Among other observations, Ca^{2+} was shown to pass through the light-sensitive conductance[4,5] and be extruded by a $Na^+/Ca^{2+},K^+$ exchanger[5,6]. As a result, the Ca^{2+} concentration in photoreceptor outer segments actually declines in the light,[7-10] as the light-sensitive channels close and the Ca^{2+} influx stops while the efflux continues. Furthermore, cyclic GMP (cGMP) was shown to gate the only cationic channels located on the outer segment plasma membrane.[11] It was subsequently demonstrated that the cGMP-gated channels are identical to the light-sensitive conductance.[12] Thus, the intracellular messenger of phototransduction is cGMP, while Ca^{2+} plays an important modulating role in the process.

The rod outer segment is packed with hundreds of membrane sacs, the disks, which are regularly stacked on top of each other. The initial biochemical reactions of visual transduction take place on the surface of these disks. The current scheme for visual phototransduction, summarizing the work from several laboratories over many years (for recent reviews, see refs. 13-15), is shown in Figure 1B. This scheme describes the process that takes place within the outer segments of the rod photoreceptors. Cones are thought to function in a broadly similar way. cGMP is produced by a guanylate cyclase and hydrolyzed by a phosphodiesterase. In the dark, the two enzymatic activities are in a balance that maintains a steady-state concentration of cGMP. cGMP binds to and keeps open the cationic channels located in the plasma membrane of the outer segment, allowing the steady influx of cations in the dark. Incoming photons are absorbed by the light-sensitive pigment rhodopsin, leading to the formation of an enzymatically active form, metarhodopsin II. Metarhodopsin II activates the trimeric G-protein transducin, by catalyzing the exchange of GTP for GDP on the transducin α-subunit. The activated transducin α-subunit then stimulates the hydrolytic activity of the phosphodiesterase. As a result, the cytosolic cGMP concentration falls, and the cGMP-gated channels close, stopping the influx of cations and hyperpolarizing the cell membrane. The recovery from the excitation by light involves the deactivation of each of the active intermediates of the light-activated cGMP cascade: the activated rhodopsin molecules are deactivated through phosphorylation by rhodopsin kinase and subsequent capping by a 48 kDalton protein

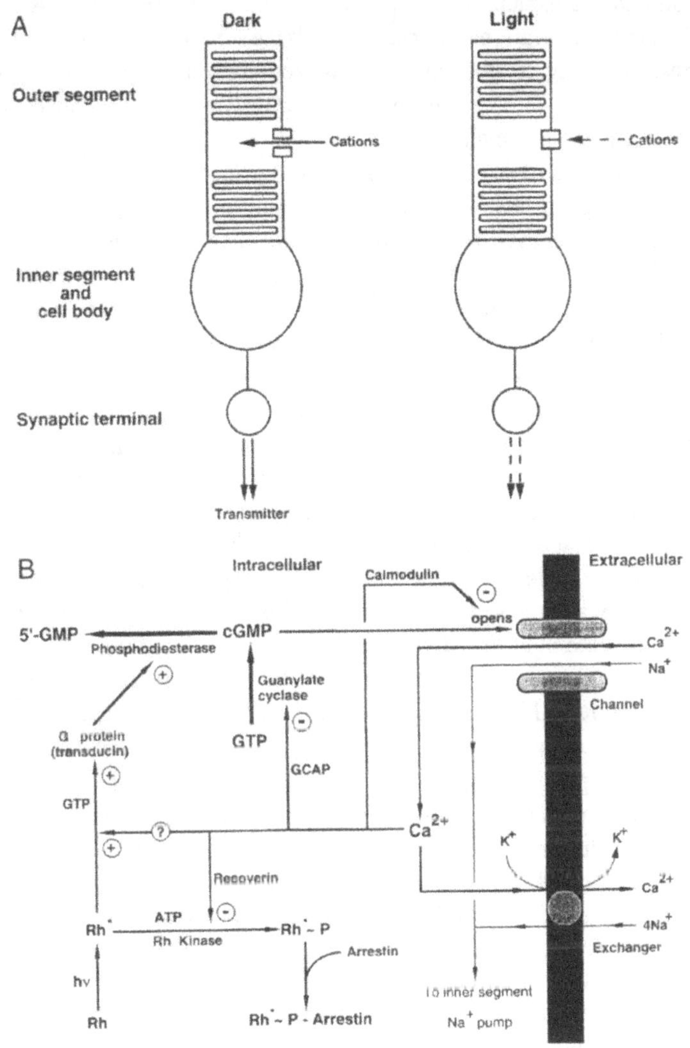

Figure 1. The transduction of light stimuli in rod photoreceptors. (A) Schematic diagram of the rod photoreceptor showing the overall response of the cell to light. (B) Phototransduction scheme in rods. cGMP is produced by a guanylate cyclase and hydrolyzed by a phosphodiesterase. In the dark, cGMP binds to and keeps open cationic channels located in the plasma membrane of the outer segment, allowing the steady influx of cations. Light (hv) activates the visual pigment, rhodopsin (Rh), initiating a cascade of reactions leading to the stimulation of cGMP hydrolysis and closure of the channels. The recovery from light involves the deactivation of the light-activated rhodopsin (Rh*) through phosphorylation by rhodopsin kinase and subsequent capping by a 48 kDalton protein called arrestin. Transducin deactivates through the hydrolysis of bound GTP to GDP, ending the stimulation of the cGMP phosphodiesterase. Ca^{2+} enters the outer segment through the cGMP-gated channels and is extruded by a $Na^+/Ca^{2+},K^+$ exchanger. Ca^{2+} inhibits the guanylate cyclase through Guanylate Cyclase Activating Proteins (GCAPs), reduces the apparent affinity of the cGMP-gated channels for cGMP through calmodulin, and inhibits the deactivation of photoactivated rhodopsin through recoverin. There are perhaps additional targets of Ca^{2+}, but none has been identified unequivocally. Adapted from ref. 52.

called arrestin; the transducin α-subunit deactivates through the hydrolysis of bound GTP to GDP; the deactivation of transducin ends the stimulation of the cGMP phosphodiesterase, bringing the phosphodiesterase activity and the hydrolysis of cGMP down to the original level. The concentration of cGMP recovers through continuing synthesis by the guanylate cyclase.

Ca^{2+} plays a modulating role in this light transduction process by acting at several sites through a host of small Ca^{2+}-binding proteins.[16] Ca^{2+} inhibits the guanylate cyclase through two Guanylate Cyclase Activating Proteins (GCAPs)[17-21] and this inhibition of cGMP synthesis is responsible for the old observation that Ca^{2+} can mimic the effect of light. Ca^{2+} also reduces the apparent affinity of the cGMP-gated channels for cGMP, probably through the mediation of calmodulin.[22] Finally, Ca^{2+} inhibits the deactivation of photoactivated rhodopsin through recoverin,[23,24] a Ca^{2+}-binding protein. Recoverin has been suggested to function by inhibiting rhodopsin kinase, and hence the phosphorylation and deactivation of photoactivated rhodopsin.[25-28] However, its function remains controversial.[29] At the whole rod photoreceptor cell level, the effects of Ca^{2+} manifest themselves at an early and at a late stage during the light response,[30,31] consistent with the Ca^{2+} effects on rhodopsin deactivation and guanylate cyclase, respectively. There are perhaps additional targets of Ca^{2+},[32] but none has been identified unequivocally.

A lot of progress in understanding phototransduction has come about through an important experimental technique that has allowed the extensive study of the process at the single cell level. This technique is suction pipette recording[33]. In its basic form the technique involves the drawing of a photoreceptor inner segment inside a recording pipette (Fig. 2A). This allows recording of the light-sensitive current by the electrode, while allowing the exposure of the outer segment to different extracellular solutions. A modification of the technique is the truncated rod outer segment preparation[12] (Fig. 2B), which consists of drawing the rod outer segment inside the recording pipette and then shearing off the remainder of the cell with a glass probe. The resulting truncated outer segment, open at one end, allows the recording of current flowing through the cGMP-gated channels and the dialysis of the outer segment cytoplasm through the bath. The preparation can be used as a micro-test-tube to study phototransduction and its components in situ.

These experimental techniques allow the detailed examination of how the different effects of Ca^{2+} influence the response of the cell to light. As mentioned before, Ca^{2+} enters the outer segment through the cGMP-gated channels and is extruded by the $Na^+/Ca^{2+},K^+$ exchanger. In the dark, the balance of these two activities results in a stable Ca^{2+} concentration of 500-800 nM.[9,34] Upon stimulation by light, cGMP-gated channels close and the Ca^{2+} influx decreases or stops altogether. The Ca^{2+} extrusion continues and as a result, the Ca^{2+} concentration declines. The decline in the Ca^{2+} concentration has at least three consequences: it relieves the inhibition on the rhodopsin kinase, allowing the faster deactivation of rhodopsin; it removes the inhibition on the guanylate cyclase, bringing about an increase in cGMP synthesis; it increases the apparent affinity of the cGMP-gated channels for cGMP, allowing the channels to remain open at lower cGMP concentrations. All three effects of Ca^{2+} counteract the original effect of light, and constitute part of a negative

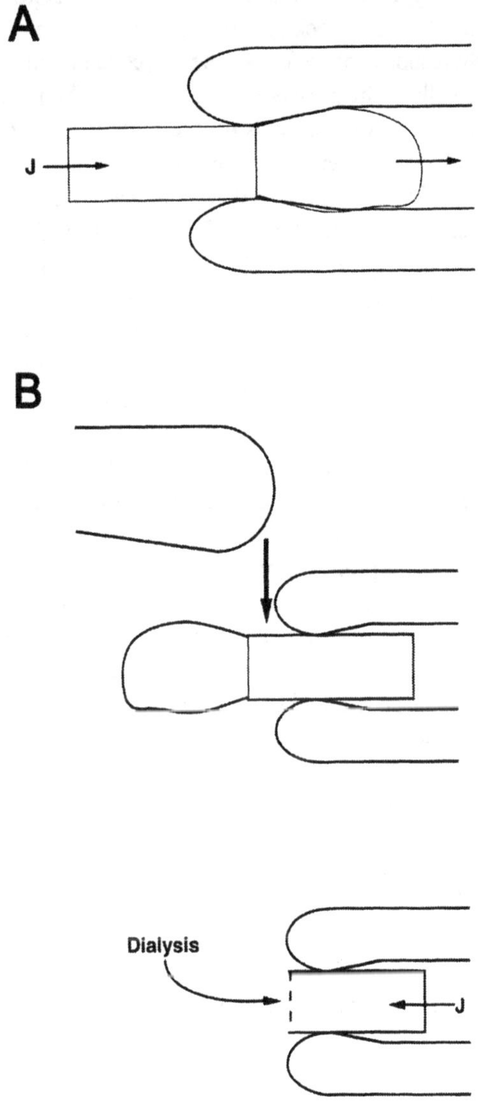

Figure 2. (A) Suction pipette recording technique. The inner segment of a photoreceptor (rod in this case) is drawn inside a recording pipette. This permits recording of the light-sensitive current by the electrode, while allowing the exposure of the outer segment to different extracellular solutions. Alternatively, the outer segment may be drawn inside the recording pipette. (B) The truncated rod outer segment preparation. A rod photoreceptor outer segment is drawn inside a recording pipette and the remainder of the cell is sheared off with a glass probe. This allows the recording of current flowing through the cGMP-gated channels and the dialysis of the outer segment cytoplasm through the bath. The preparation can be used as a micro-test-tube to study the phototransduction components.

feedback loop that is triggered by the initial closure of the cGMP-gated channels. The importance of this Ca^{2+} feedback in phototransduction can be demonstrated by its effect on the light response of the cells. The effect can be judged by measuring the responses of the cells in the presence and absence of the feedback. In order to remove the feedback without compromising the ability of the cell to respond to light, an experimental trick was employed by Matthews et al.[35] and Nakatani and Yau:[36] extracellular Ca^{2+} was removed so that there was no Ca^{2+} influx through the cGMP-gated channels, and extracellular Na^+ was substituted by guanidinium or Li^+. These cations go through the cGMP-gated channels, but do not support the extrusion of Ca^{2+} by the $Na^+/Ca^{2+},K^+$ exchanger, thereby eliminating Ca^{2+} efflux. In this way, light-sensitive currents and light responses can be recorded, but Ca^{2+} fluxes across the outer segment plasma membrane are minimized, thereby suppressing the feedback. Figure 3 shows the responses of a rod photoreceptor cell in the presence and the absence of the Ca^{2+} feedback.[37] These responses are the changes in the light-sensitive current elicited by light stimuli. In the dark, there is a steady current of ca. 32 pA flowing into the rod outer segment. Because the current is inward, it is plotted as negative. Light stimulation reduces the circulating current, giving rise to the upward deflections in the plots. The response to a dim flash of light, shown as an inset, is larger and slower in the absence of the Ca^{2+} feedback. That is, the reduction in the circulating light-sensitive current is more pronounced, and it takes longer for the current to recover. The feedback also dramatically affects the responses to steps of light, shown in the main panels. In the presence of the feedback, the light-sensitive current declines immediately after the switching on of the light, but after a while it slowly recovers to a new steady state. This relaxation reflects the time it takes for the effects of the Ca^{2+} feedback to set in. After removal of the feedback, the responses are larger and the relaxation is abolished. As a result, the response of the cell reaches saturation at lower light intensities, the dynamic range of the cell becomes significantly limited, and the cell fails to adapt to the presence of the background light.

MEASURING THE EFFECTS OF CALCIUM

The experiment shown in Figure 3 exemplifies one of many manifestations of the effects of Ca^{2+} feedback on the light response of the rod photoreceptor. It also provides a relatively simple means of evaluating the roles of the different Ca^{2+} targets in bringing about the dramatic increase in the operating range of the rod. This approach focuses on the steady-state response of the rod to light and avoids the consideration of kinetic effects that are still incompletely understood due to lack of sufficient quantitative information.

The question then can be formulated as follows: can the effects of Ca^{2+} on the different components of phototransduction account for the behavior of the whole rod cell as exemplified in the experiment of Figure 3? In order to answer this question, the effect of Ca^{2+} on each one of the targets shown in Figure 1B has to be separately quantified. A large body of experimental information is available on these

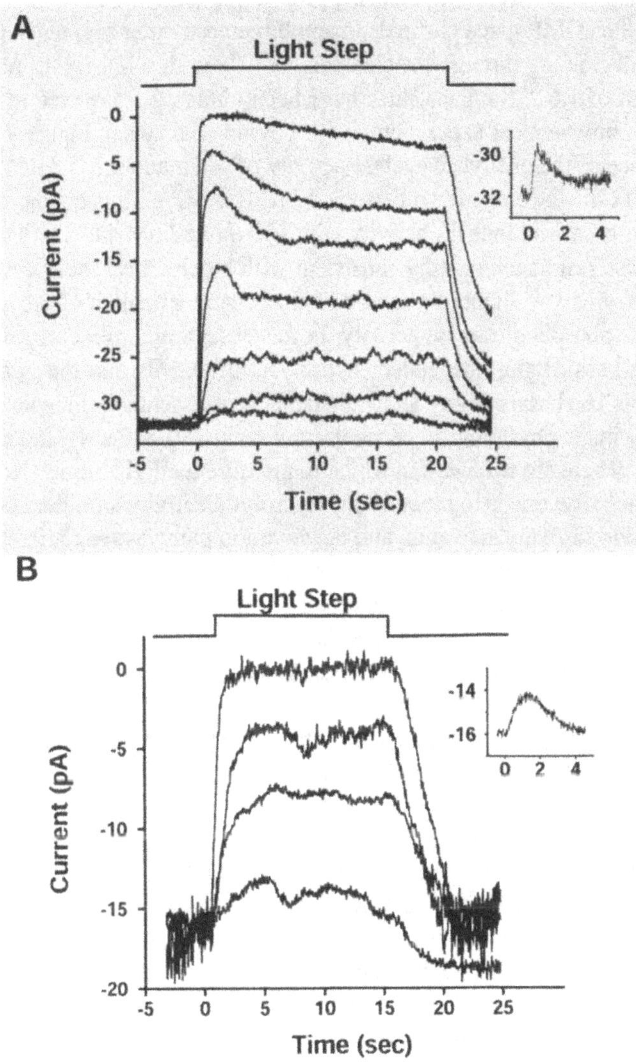

Figure 3. Responses of two salamander rod cells to steps of light. As the light intensity increased, the current suppression increased. (A) In normal Ringer's solution, in the presence of the Ca^{2+} feedback. The light intensities were 0.7, 2.5, 9.0, 28.5, 103.3, 220.9 and 1300.6 hν μm^{-2} s^{-1}. (B) With the Ca^{2+} feedback removed. The light intensities were 0.7, 2.5, 9.0 and 28.5 hν μm^{-2} s^{-1}. Insets: Responses of the cells to dim flashes in the presence and the absence of the Ca^{2+} feedback. The flash was delivered at time 0; the intensity in Ringer's was 1.3 hν μm^{-2}, while in the absence of the Ca^{2+} feedback was 0.7 hν μm^{-2}. Reproduced from ref. 37 by copyright permission of the Rockefeller University Press.

Ca^{2+} effects, including the characterization of these effects in single rod outer segments using an electrophysiological approach.[38] Figure 4A shows the Ca^{2+} dependence of the cGMP-gated channels in amphibian rod outer segments. Ca^{2+} reduces the current flowing through the channels to 40% with a $K_{1/2}$ of 48 nM and a Hill coefficient of 1.6.[39] Similar values have been obtained from excised patches[40] as well as in biochemical experiments with bovine channels.[22] Figure 4B shows the dependence of the guanylate cyclase activity of salamander rod outer segments on Ca^{2+}. In 0 Ca^{2+} the enzyme synthesizes 13 μM cGMP s^{-1} in a single rod outer segment, while Ca^{2+} inhibits the activity with a $K_{1/2}$ of 87 nM and a Hill coefficient of 2.1.[41] These results are broadly consistent with biochemical measurements for the bovine enzyme.[17,18] Figure 5 shows a similar characterization of the Ca^{2+} modulation of the phosphodiesterase activity. In this case, there are two activities to consider: the basal, light-independent hydrolysis of cGMP, and the light-stimulated hydrolysis. The basal activity, shown in panel A, is essentially independent of Ca^{2+}, especially in the physiological concentration range (50-800 nM), and has a value of ~0.3 s^{-1}.[37] There are two ways in which to measure the light-stimulated phosphodiesterase activity: one is to measure the hydrolytic activity stimulated by a flash of light; in this case, the activation and deactivation pathways are reflected in the kinetics of the hydrolytic activity, which initially rises and subsequently falls. A complication in this situation is that the effect of Ca^{2+} on the hydrolytic activity would reflect not only the effect on the specific targets, but also the kinetics of the effect, along with the kinetics of the Ca^{2+} change. Another way is to measure the steady-state hydrolytic activity elicited by a step of light; in this case the activation and deactivation pathways are lumped together, but the kinetics of the Ca^{2+} change are not relevant. So, there is only a single parameter, the steady-state light-stimulated activity that needs be measured as a function of Ca^{2+}. Figure 5, panel B, shows the dependence of the salamander steady-state light-stimulated phosphodiesterase activity on Ca^{2+}.[37] The solid line describes the enhancement of the light-stimulated phosphodiesterase activity by Ca^{2+} and has a $K_{1/2}$ of 400 nM, and a Hill coefficient of 1. The estimated maximal activity is 0.16 $s^{-1}/(h\nu\ \mu m^{-2}\ s^{-1})$ (units of rate of hydrolysis per unit light intensity). It should be emphasized that this measured Ca^{2+} dependence of the light-stimulated phosphodiesterase activity does not rely on the mechanism for the Ca^{2+} modulation and the role of recoverin. It is a direct empirical measure of the cGMP hydrolytic activity and its modulation by Ca^{2+}.

In order to relate these effects of Ca^{2+} to the response of the cell, we also need the relation between Ca^{2+} concentration and light intensity. This has been directly measured using Ca^{2+}-sensitive fluorescent dyes in isolated rod outer segments, and was found that the steady-state Ca^{2+} concentration is approximately proportional to the circulating current.[9] The reason behind this relation is that the steady-state Ca^{2+} concentration at different light intensities is determined by the balance between influx through the cGMP-gated channels and efflux through the $Na^+/Ca^{2+},K^+$ exchanger. The exchanger has a K_m for Ca^{2+} of 1-2 μM,[9,42] and so, for the physiological Ca^{2+} concentrations, operates in the linear range. Therefore, at steady state, the influx of Ca^{2+} is proportional to the light-sensitive current, while the efflux is

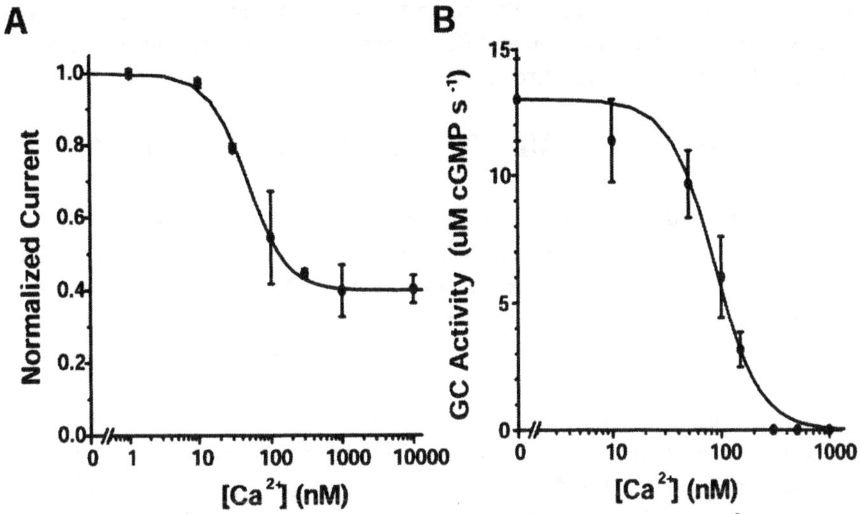

Figure 4. Effects of Ca^{2+} on the cGMP-gated channels and on the guanylate cyclase. (A) Ca^{2+} dependence of the cGMP-gated current. Data were obtained from truncated rod outer segments of bullfrog. The current was elicited with 20 μM cGMP. The currents have been normalized over the value of the current measured in the absence of Ca^{2+}. Reproduced from ref. 39 with permission. (B) Ca^{2+} dependence of the guanylate cyclase activity. Data were obtained from truncated rod outer segments of the salamander. Guanylate cyclase activity was elicited with 2 mM GTP in the presence of 0.5 mM free Mg^{2+}. Reproduced from ref. 41 by copyright permission of the Rockefeller University Press.

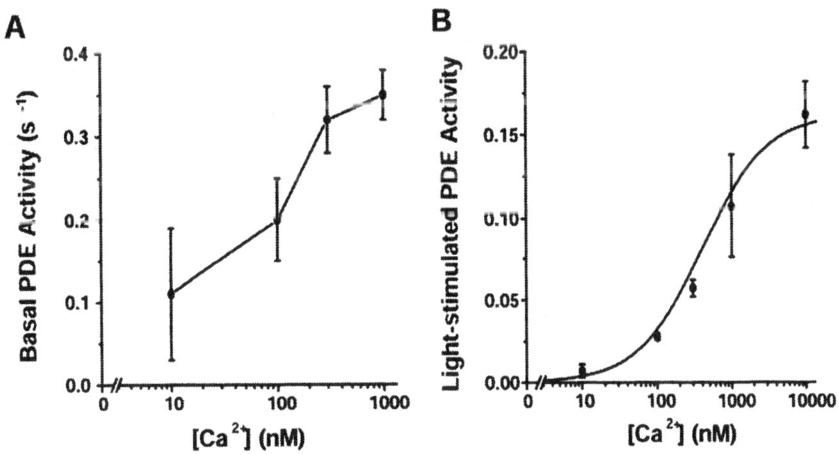

Figure 5. Effects of Ca^{2+} on the phosphodiesterase activity. (A) Lack of significant Ca^{2+} dependence of the basal, light-independent activity. Data were obtained from truncated salamander rods in the presence of 0.1 mM GTP, 0.2 mM ATP and 0.5 mM free Mg^{2+}. (B) Ca^{2+} dependence of the light-stimulated phosphodiesterase activity. Data were obtained from truncated salamander rods in the presence of 0.1 mM GTP, 0.2 mM ATP and 0.5 mM free Mg^{2+} to ensure the activation of transducin and rhodopsin kinase. The units of the phosphodiesterase activity are $s^{-1}/(h\nu\ \mu m^{-2}\ s^{-1})$ (hydrolytic rate per unit light intensity). Reproduced from ref. 37 by copyright permission of the Rockefeller University Press.

proportional to the Ca^{2+} concentration in the rod outer segment. As a result, the steady-state Ca^{2+} concentration at different background light intensities is proportional to the light-sensitive current.

CALCIUM AND REGULATION OF LIGHT SENSITIVITY

Figures 4 and 5 provide the characterizations of the Ca^{2+} effects on the different components of the phototransduction process. Given the phototransduction model in Figure 1B, these effects can be used to predict the changes in the light-sensitive current elicited by steps of light. The actual measurements of the current, like those shown in Figure 3, can be used for comparison. The whole procedure will give us a sense of how well we understand the role of Ca^{2+} in phototransduction.

Figure 6A shows the predicted responses as solid lines. For comparison, the responses of cells, obtained from experiments like the one shown in Figure 3, are shown in the presence (filled triangles) and absence (open circles) of the Ca^{2+} feedback. There is good agreement between the experimental measurements of intact cell responses and the responses predicted from the measurements of the individual components. Moreover, the curves correctly predict the effect of the Ca^{2+} feedback on the dynamic range of the cell. Since we know the effects of Ca^{2+} on each of the components, we can also evaluate the contribution of each component on the dynamic range of the cell. Figure 6B shows the different contributions for the Ca^{2+} effects on the guanylate cyclase and the phosphodiesterase. The Ca^{2+} modulation of the channel does not provide much change and is not shown. In the absence of any Ca^{2+} modulation, we obtain the curve on the left (curve 1), which is the same as the curve in the absence of the Ca^{2+} feedback in Figure 6A. The cell has high light sensitivity, but the response saturates at relatively low light intensities, limiting the operating range of the cell. Adding the Ca^{2+} modulation of the cyclase lowers the sensitivity and shifts the response-intensity relation to the right (curve 3). These predictions have been experimentally realized with genetically modified mice lacking both GCAPs and the Ca^{2+} modulation of the cyclase.[43] Adding the Ca^{2+} modulation of the phosphodiesterase only has no effect on the light sensitivity at low light intensities, but changes the slope of the response-intensity relation, allowing the cell to operate over a larger range of intensities (curve 2). Adding both Ca^{2+} modulations shifts the curve to the right, and reduces the slope, allowing the cell to operate over a large dynamic range (curve 4).

Another way of looking at the relevance of the different Ca^{2+} modulations is to examine the contribution of each modulation to the light threshold at different background light intensities. Light threshold is the intensity required to produce a criterion response. Since the Ca^{2+} feedback resists in a sense the effect of light, the contributions of each Ca^{2+} modulation will be reflected in the light intensity needed to produce the criterion response. The threshold increases with background light intensity but the relative contributions of different processes also change. Figure 7 shows these contributions at backgrounds of different light intensities. The term 'cGMP' refers to the need to reduce the cGMP concentration by the appropriate

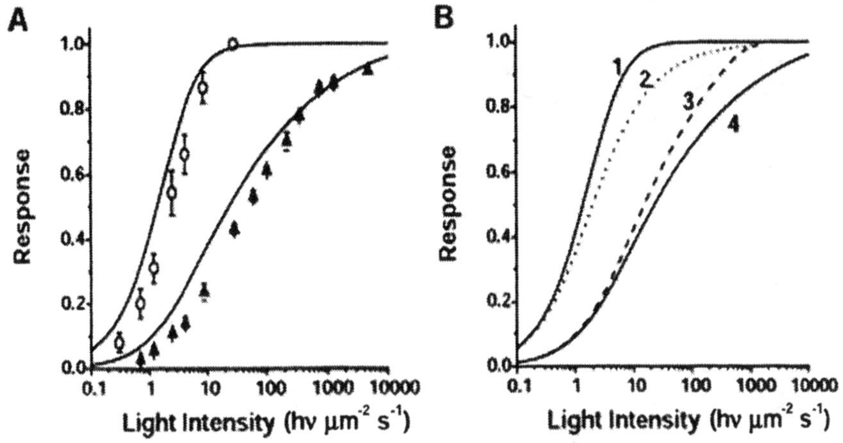

Figure 6. The characterized components of phototransduction can predict the light responses of the intact cell. (A) Comparisons between predicted and experimental responses to steps of light. Responses are measured as the fractional suppressions of light-sensitive current. The filled triangles are data from intact salamander rods in Ringer's and in the presence of the Ca^{2+} feedback, while the open circles represent measurements from cells for which the feedback was removed. The solid lines are the responses predicted from the characterized effects of Ca^{2+}, with and without feedback, assuming a Ca^{2+} concentration in the dark of 500 nM. (B) Contribution of the different Ca^{2+} modulations to the overall response of the cell. Reproduced from ref. 37 by copyright permission of the Rockefeller University Press.

amount to produce the criterion response. As the background light intensity increases, this reduction has to occur at lower cGMP concentrations and at higher background phosphodiesterase activities, since the hydrolytic activity increases with background light. The 'cGMP' term contributes an approximately constant proportion to the overall threshold and in the absence of the Ca^{2+} feedback, this term would be the only one contributing. In the presence of the Ca^{2+} feedback, light has to overcome the reduction in Ca^{2+} as well. There are three 'Ca^{2+}' terms, corresponding to the three Ca^{2+}-modulated enzymatic processes: the guanylate cyclase, 'GC', the phosphodiesterase, 'PDE', and the channel, 'Ch'. As the background light intensity increases, the Ca^{2+} concentration decreases, and the relative contribution of each of the Ca^{2+} contributions to the overall threshold changes. The channel term is generally very small and does not contribute much to the threshold over the full dynamic range of the cell. At low light intensities, most of the threshold is contributed by the Ca^{2+} modulation of the guanylate cyclase. The Ca^{2+} modulation of the light-stimulated phosphodiesterase becomes the dominant term at higher light intensities.

Pugh et al[13] have provided a different perspective by examining the ways in which each of the Ca^{2+} modulations analyzed above affects the adaptation of the rod to background light. They make the important distinction between effects on the operating range and on the light sensitivity of the rod. From this perspective, there are two interesting aspects of the Ca^{2+} modulation of the cyclase. One, during a flash response, the transient decrease in Ca^{2+} leads to a dynamic increase in cGMP

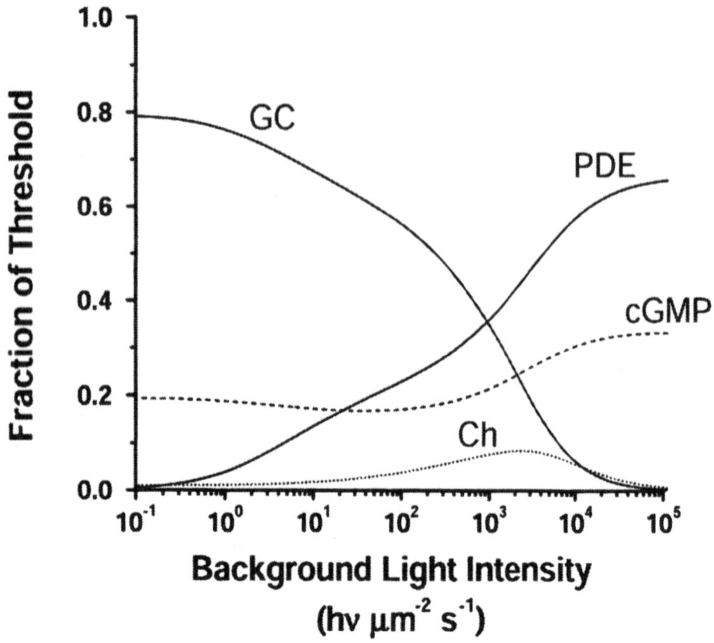

Figure 7. Plot of calculated threshold contributions by each pathway as a function of background ligh intensity. The contributions to threshold have been calculated for a Ca^{2+} concentration in the dark of 50(nM. Reproduced from ref. 37 by copyright permission of the Rockefeller University Press.

synthesis, which reduces the light sensitivity of the rod. Two, in the presence of ε background light, the lower Ca^{2+} concentration leads to an increased rate of cGMF synthesis and the maintenance of a higher cGMP concentration. As a result, more channels remain open extending the operating range of the cell. However, the maintenance of more open channels also increases the light sensitivity of the rod. Witl this in mind, Figure 6B can be seen as analyzing the effect of the different Ca^{2+} modulations on the operating range of the rod, while Figure 7 analyzes the effect or light sensitivity.

When considering the importance of the different modulations to light adaptation an important clarification is necessary. Although the Ca^{2+} feedback is essentia for the light adaptation of the rod to background light, it is not the only factor responsible,[44] as also hinted by the presence of the 'cGMP' term in Figure 7. Keeping in mind the distinction between effects on operating range and sensitivity, it is important to note that the major factor responsible for the desensitization of the roc during light adaptation is the increase in the steady state rate of cGMP hydrolysis by the phosphodiesterase.[45]

Calcium Diffusion

As mentioned before, the dynamic effects of Ca^{2+} concentration changes on the light response are still incompletely understood due to lack of sufficient quantitative information. One of the factors mediating the dynamic effects of Ca^{2+} is the Ca^{2+} diffusion coefficient. Upon closure of the cGMP-gated channels, the Ca^{2+} concentration will begin to decrease next to the plasma membrane of the rod outer segment. Subsequently, and through diffusion, the decline in concentration will propagate toward the center of the outer segment (Fig. 8A). The rate at which the decline in Ca^{2+} concentration propagates from the periphery toward the center of the outer segment is also the rate at which the Ca^{2+} signal propagates radially to be sensed by the different Ca^{2+} targets. This rate depends on the pumping activity of the exchanger and on the apparent Ca^{2+} diffusion coefficient. The apparent Ca^{2+} diffusion coefficient will be significantly affected by the binding of Ca^{2+} to intracellular components. Immobile components would slow down Ca^{2+} diffusion, while highly mobile components would tend to speed up Ca^{2+} diffusion. The protein mediators of the Ca^{2+} effects on phototransduction components described above represent one class of Ca^{2+}-binding components that can affect Ca^{2+} diffusion.

The apparent Ca^{2+} diffusion coefficient can be estimated by measuring the radial profile of the Ca^{2+} concentration after stimulation by light.[46] For this, a fluorescent Ca^{2+} indicator dye, fluo-3, is used to monitor the Ca^{2+} concentration using a laser scanning confocal microscope. The confocal microscope collects fluorescence from only a thin slice of cytoplasm, allowing the measurement of the profile of the Ca^{2+} concentration along a diameter of the rod outer segment. These experiments are carried out with salamander rod photoreceptors, which are quite large, 10-12 μm in diameter, allowing the measurement of the fluorescence intensity profile with sufficient resolution.

For an experiment, salamander rod photoreceptors loaded with the Ca^{2+}-sensitive dye fluo-3 and placed in an experimental chamber, are brought on the stage of the confocal microscope. A rod photoreceptor is focused under infrared light and then positioned so that the laser scan that will excite the dye fluorescence passes along the diameter of the outer segment and is perpendicular to the rod outer segment axis (Fig. 8B). The first laser scan activates the phototransduction cascade and leads to closure of the cGMP-gated channels, initiating the reduction in the Ca^{2+} concentration. However, because a single scan takes less than 4 ms to complete, the channels do not have time to close during the first scan, so the initial fluorescence profile is flat and reflects the resting Ca^{2+} concentration in the dark (Fig. 9A, filled triangles). At the end of the experiment, the incorporated fluo-3 is saturated with Ca^{2+}, providing a measure of the total dye concentration in the outer segment (Fig. 9A, open circles). All the measured fluorescence profiles are normalized over the fluorescence of the Ca^{2+}-saturated fluo-3. This allows the estimation of Ca^{2+} concentrations, and for this cell the Ca^{2+} concentration in the dark was 770 nM. After closure of the channels, the Ca^{2+} concentration begins to drop. Figure 9B shows the fluorescence profiles 0.5 and 1.0 s after the initial scan, while Figure 9C shows the profiles

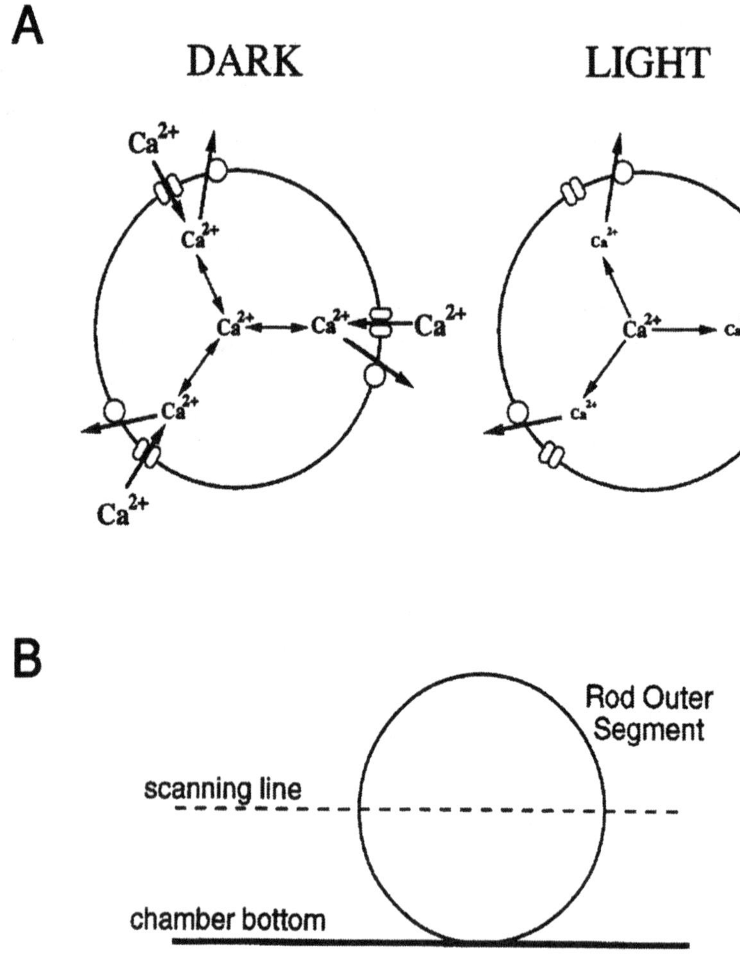

Figure 8. (A) Schematic diagram of a cross-section of a rod outer segment, showing the creation of a Ca^{2+} concentration gradient upon stimulation of the cell by light. In the dark, Ca^{2+} enters though the light-sensitive channels and is extruded by the $Na^+/Ca^{2+},K^+$ exchanger. At steady state, Ca^{2+} at different distances from the plasma membrane of the outer segment is at equilibrium with Ca^{2+} next to the plasma membrane, so that the Ca^{2+} concentration is uniform throughout. Light stimulation closes the light-sensitive channels, so that the Ca^{2+} influx stops while the efflux continues; this will result in the reduction of the Ca^{2+} concentration next to the plasma membrane and the reduction will propagate toward the center of the outer segment. The ensuing gradient of Ca^{2+} concentration between periphery and center will depend on the apparent Ca^{2+} diffusion coefficient. (B) Experimental geometry for measuring the gradient of the Ca^{2+} concentration with a laser scanning confocal microscope. Reproduced from ref. 46 with permission.

1.5 and 2.5 s after the initial scan. The solid lines in the panels of Figure 9 are fits to the data points based on a diffusion model that allows the estimation of the apparent Ca^{2+} diffusion coefficient that can account for the observed fluorescence profiles. A significant gradient of Ca^{2+} concentration between the edge and the center of the

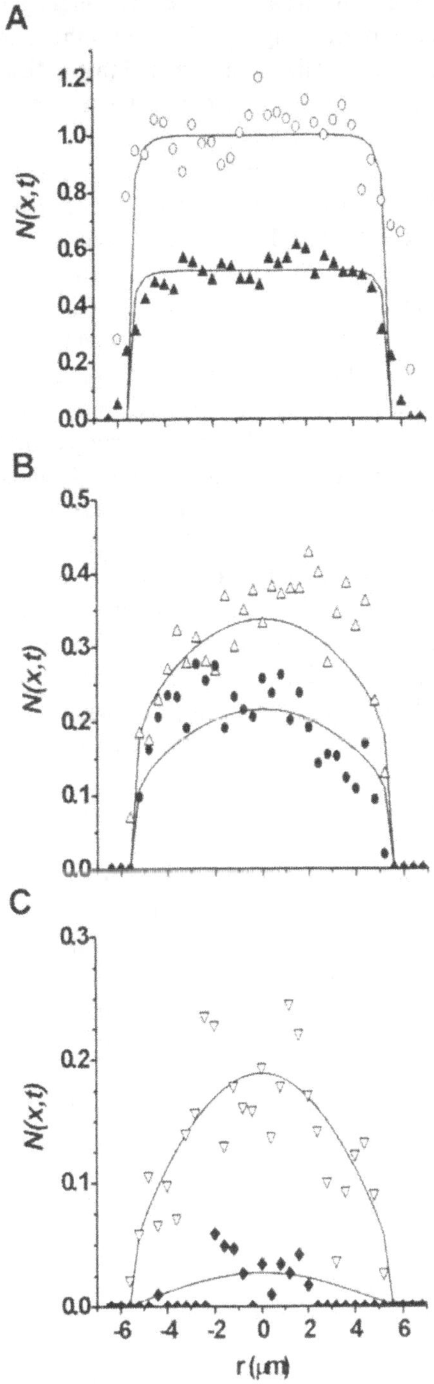

Figure 9. Normalized fluorescence intensity profiles, obtained with confocal laser line scans from a tiger salamander rod photoreceptor loaded with fluo-3. (A) Profiles of the first (triangles) and last (circles) scans; the first scan is from the dark-adapted photoreceptor while the last is under conditions which saturate the internalized fluo-3 with Ca^{2+}. (B) Profiles of the scans obtained 0.5 s (triangles) and 1.0 s (circles) after the first one. (C) Profiles of the scans obtained 1.5 s (inverted triangles) and 2.5 s (diamonds) after the initial scan. Reproduced from ref. 46 with permission.

outer segment appears 1.5 s after the initial scan, and a dome is evident in the fluorescence intensity record (Fig. 9C, inverse open triangles). This shows that diffusion cannot keep up with the extrusion of Ca^{2+} by the exchanger and that diffusion is limiting for the rate at which the Ca^{2+} concentration declines in the outer segment after channel closure.

From experiments like the one shown in Figure 9, a value $D = 15\pm1\ \mu m^2\ s^{-1}$ was obtained for the radial diffusion coefficient of Ca^{2+} [46]. For a salamander rod outer segment with a diameter $d = 12\ \mu m$, the rate at which the signal of the decline in the Ca^{2+} concentration will be spreading from the periphery toward the center of the disks will be given by $\sim \pi^2 \cdot D / d^2 = 1.0\ s^{-1}$, corresponding to a time constant of ~1.0 s, longer than the time-to-peak of the light response to dim flashes, which is ~0.5 s. [47] This suggests that diffusion provides a physiologically significant barrier for the propagation of the Ca^{2+} concentration decline signal, and the distribution of the Ca^{2+}-sensitive enzymes on the surface of the disks may be an important parameter in the determination of the kinetics of the recovery of the light response. Assuming that the Ca^{2+} diffusion coefficient is the same in the rod outer segments of other species, we can estimate the rate of propagation of the Ca^{2+} decline signal on the basis of the radii of the outer segments. So, in mammalian rod photoreceptors, with an outer segment diameter $1/10^{th}$ that of the salamander, the rate of propagation would be about 100 times faster, resulting in a time constant of ~7 ms, much shorter than the time-to-peak of the dim flash response, which is ~150-250 ms in mammalian rods.[48] Thus, it appears that Ca^{2+} diffusion provides an important barrier in the radial propagation of the Ca^{2+} decline signal in the large amphibian photoreceptors, but not in the much smaller mammalian ones.

The signal of the decline in Ca^{2+} concentration also spreads longitudinally, along the length of the rod outer segment. This spread is particularly relevant in the case of the single-photon response, when a single rhodopsin molecule has been excited. The longitudinal Ca^{2+} diffusion coefficient will be 6-7 times lower than the radial coefficient because of baffling by the disks.[49,50] Thus, the longitudinal Ca^{2+} diffusion coefficient will be $\sim 2.3\ \mu m^2\ s^{-1}$, many times smaller than the respective diffusion coefficient for cGMP, which is 30-60 $\mu m^2\ s^{-1}$.[49,50] In the case of the single-photon response, or, more generally, in the case of the response to localized stimulation by light, cGMP is diffusing along the length of the outer segment down its concentration gradient toward the site of excitation where it is hydrolyzed. As the decline in the cGMP concentration is spreading beyond the site of excitation, cGMP channels are closing, initiating a reduction in Ca^{2+} concentration. Because the Ca^{2+} mobility is so much lower than that of cGMP, the changes in Ca^{2+} concentration will follow the closure of the channels. Therefore, the longitudinal spread of the Ca^{2+} decline signal will follow the spread of the cGMP decline signal. Indeed, this is what is observed when the spread of cGMP decline is measured electrically from the closure of channels and the Ca^{2+} decline is measured with a Ca^{2+}-sensitive dye.[51] In fact, because of the tight coupling between cGMP and Ca^{2+}, there is a single length constant that describes the spread of both cGMP and Ca^{2+}. This experimentally measured length constant can be quantitatively accounted for by the values of the cGMP and Ca^{2+} diffusion coefficients.[51]

CONCLUSION

cGMP is the intracellular messenger of the phototransduction process in retinal rod photoreceptors, but Ca^{2+} plays an important modulatory role. Ca^{2+} affects several of the phototransduction enzymes and pathways and is involved in a negative feedback that is essential for the adaptation of the rod cell to background illumination. The steady state responses of the cell to background illumination can be quantitatively accounted for on the basis of the characterized effects of Ca^{2+} on the different components.

After the closure of the light-sensitive channels upon stimulation by light, the diffusion of Ca^{2+} in the radial direction is quite slow providing a significant barrier in the radial propagation of the feedback signal. Also, because the diffusion coefficient of Ca^{2+} is much smaller than that of cGMP, the decline of Ca^{2+} in the longitudinal direction lags behind the propagation of excitation by the decline of cGMP.

Although we have a detailed understanding of the role of Ca^{2+} in rod phototransduction, several important questions remain unanswered. Not all Ca^{2+} targets have been unequivocally established, thereby limiting our grasp of the effect of Ca^{2+} on the kinetics of the light response. In the same vein, the role of Ca^{2+} buffering and its effects on the kinetics of the light response also remain unclear. Unlike rod phototransduction, the role of Ca^{2+} in cone phototransduction has barely been studied. Cones are generally less sensitive and have faster light responses than rods. But there are differences in kinetics and sensitivity between different types of cones as well. It is not clear what the factors responsible for such differences might be. Could it be differences in the transduction enzymes? Or perhaps differences in the concentrations of the enzymes, the topology of the photoreceptor outer segments, or Ca^{2+} homeostasis are important? Promising approaches for answers to such questions may be extending the techniques for measuring single cell enzymatic activities from rods to cones, or the use of transgenic mice with changes targeted to the cone photoreceptors.

REFERENCES

1. Hagins WA. The visual process: Excitatory mechanisms in the primary receptor cells. Ann Rev Biophys Bioeng 1972; 1:131-158.
2. Yoshikami S, Hagins WA. Control of dark current in vertebrate rods and cones. In: Langer H, ed. Biochemistry and Physiology of Visual Pigments. New York: Springer 1973:245-255.
3. Pugh EN Jr. The nature and identity of the internal excitational transmitter of vertebrate phototransduction. Ann Rev Physiol 1987; 49:715-741.
4. Hodgkin AL, McNaughton PA, Nunn BJ. The ionic selectivity and calcium dependence of the light-sensitive pathway in toad rods. J Physiol 1985; 358:447-468.
5. Yau K-W, Nakatani K. Electrogenic Na-Ca exchange in retinal rod outer segment. Nature 1984; 311:661-663.
6. Cervetto L et al. Extrusion of calcium from rod outer segments is driven by both sodium and potassium gradients. Nature 1989; 337:740-743.
7. Yau K-W, Nakatani K. Light-induced reduction of cytoplasmic free calcium in retinal rod outer segment. Nature 1985; 313:579-582.
8. McNaughton PA, Cervetto L, Nunn BJ. Measurement of the intracellular free calcium concentration in salamander rods. Nature 1986; 322:261-263.
9. Gray-Keller MP, Detwiler PB. The calcium feedback signal in in the phototransduction cascade of vertebrate rods. Neuron 1994; 13:849-861.
10. McCarthy ST, Younger JP, Owen WG. Free calcium concentrations in bullfrog rods determined in the presence of multiple forms of Fura-2. Biophys J 1994; 67:2076-2089.
11. Fesenko EE, Kolesnikov SS, Lyubarsky AL. Induction by cyclic GMP of cationic conductance in plasma membrane of retinal rod outer segment. Nature 1985; 313:310-313.
12. Nakatani K, Yau K-W. Light suppressible, cyclic-GMP-sensitive conductance in the plasma membrane of a truncated rod outer segment. Nature 1985; 317:252-255.
13. Pugh EN Jr, Nikonov S, Lamb TD. Molecular mechanisms of vertebrate photoreceptor light adaptation. Curr Opin Neurobiol 1999; 9:410-418.
14. Fain GL et al. Adaptation in vertebrate photoreceptors. Physiol Rev 2001; 81:117-151.
15. Ebrey TG, Koutalos Y. Vertebrate Photoreceptors. Prog Retin Eye Res 2001; 20:49-94.
16. Polans A, Baehr W, Palczewski K. Turned on by Ca^{2+}! The physiology and pathology of Ca^{2+}-binding proteins in the retina. Trends Neurosci. 1996; 19:547-554.
17. Dizhoor AM et al. The human photoreceptor membrane guanylyl cyclase, RetGC, is present in outer segments and is regulated by calcium and a soluble activator. Neuron 1994; 12:1345-1352.
18. Gorczyca WA et al. Purification and physiological evaluation of a guanylate cyclase activating protein from retinal rods. Proc Natl Acad Sci USA 1994; 91:4014-4018.
19. Palczewski K et al. Molecular cloning and characterization of retinal photoreceptor guanylyl cyclase-activating protein. Neuron 1994; 13:395-404.
20. Frins S et al. Functional characterization of a guanylyl cyclase-activating protein from vertebrate rods. Cloning, heterologous expression and localization. J Biol Chem 1996; 271:8022-8027.
21. Kachi S et al. Detailed localization of photoreceptor guanylate cyclase activating protein-1 and –2 in mammalian retinas using light and electron microscopy. Exp Eye Res 1999; 68:465-473.
22. Hsu YT, Molday RS. Modulation of the cGMP-gated channel of rod photoreceptor cells by calmodulin. Nature 1993; 361:76-79.
23. Dodd RL. The role of arrestin and recoverin in signal transduction by retinal rod photoreceptors [PhD Thesis]. Palo Alto: Stanford University 1998.
24. Erickson MA et al. The effect of recombinant recoverin on the photoresponse of truncated rod photoreceptors. Proc Natl Acad Sci USA 1998; 95:6474-6479.

25. Kawamura S. Rhodopsin phosphorylation as a mechanism of cyclic GMP phosphodiesterase regulation by S-modulin. Nature 1993; 362:855-857.
26. Gorodovikova EN et al. Recoverin mediates the calcium effect upon rhodopsin phosphorylation and cGMP hydrolysis in bovine retina rod cells. FEBS Lett. 1994; 349:187-190.
27. Chen CK et al. Ca^{2+}-dependent interaction of recoverin with rhodopsin kinase. J Biol Chem 1995; 270:18060-18066.
28. Klenchin VA, Calvert PD, Bownds MD. Inhibition of rhodopsin kinase by recoverin. Further evidence for a negative feedback system in phototransduction. J Biol Chem 1995; 270:16147-16152.
29. Otto-Bruc AE et al. Phosphorylation of photolyzed rhodopsin is calcium-insensitive in retina permeabilized by alpha-toxin. Proc Natl Acad Sci USA 1998; 95:15014-15019.
30. Matthews HR. Static and dynamic actions of cytoplasmic Ca^{2+} in the adaptation of responses to saturating flashes in salamander rods. J Physiol 1996; 490:1-15.
31. Matthews HR. Actions of Ca^{2+} on an early stage in phototransduction revealed by the dynamic fall in Ca^{2+} concentration during the bright flash response. J Gen Physiol 1997; 109:141-146.
32. Lagnado L, Baylor DA. Calcium controls light-triggered formation of catalytically active rhodopsin. Nature 1994; 367:273-277.
33. Baylor DA, Lamb TD, Yau K-W. The membrane current of single rod outer segments. J Physiol 1979; 288:589-611.
34. Sampath AP et al. Bleached pigment produces a maintained decrease in outer segment Ca^{2+} in salamander rods. J Gen Physiol 1998; 111:53-64.
35. Matthews HR et al. Photoreceptor adaptation is mediated by cytoplasmic calcium concentration. Nature 1998; 334:67-69.
36. Nakatani K, Yau K-W. Calcium and light adaptation in retinal rods and cones. Nature 1998; 334:69-71.
37. Koutalos Y, Nakatani K, Yau K-W. The cGMP-phosphodiesterase and its contribution to sensitivity regulation in retinal rods. J Gen Physiol 1995; 106:891-921.
38. Koutalos Y, Yau K-W. Characterization of guanylyl cyclase and phosphodiesterase activities in single rod outer segments. Methods Enzymol 2000; 315:742-752.
39. Nakatani K, Koutalos Y, Yau K-W. Ca^{2+}-modulation of the cGMP-gated channel of bullfrog retinal rod photoreceptors. J Physiol 1995; 484:69-76.
40. Gordon SE, Downing-Park J, Zimmerman AL. Modulation of the cGMP-gated ion channel in frog rods by calmodulin and an endogenous inhibitory factor. J Physiol 1995; 486:533-546.
41. Koutalos Y et al. Characterization of guanylate cyclase activity in single rod outer segments. J Gen Physiol 1995; 106:863-890.
42. Lagnado L, Cervetto L, McNaughton PA. Calcium Homeostasis in the Outer Segments of Retinal Rods from the Tiger Salamander. J Physiol 1992; 455:111-142.
43. Mendez A et al. Role of guanylate cyclase-activating proteins (GCAPs) in setting the flash sensitivity of rod photoreceptors. Proc Natl Acad Sci USA 2001; 98:9948-9953.
44. Gray-Keller MP, Detwiler PB. Ca^{2+} dependence of dark- and light-adapted flash responses in rod photoreceptors. Neuron 1996; 17:323-331.
45. Pugh EN Jr, Lamb TD. Phototransduction in vertebrate rods and cones: Molecular mechanisms of amplification, recovery and light adaptation. In: Stavenga DG, DeGrip WJ, Pugh EN Jr, eds. Handbook of Biological Physics, Volume 3. Elsevier Science B.V. 2000:183-255.
46. Nakatani K, Chen C, Koutalos Y. Calcium diffusion coefficient in rod photoreceptor outer segments. Biophys J 2002; 82:728-739.
47. Baylor DA, Nunn BJ. Electrical properties of the light-sensitive conductance of rods of the salamander Ambystoma tigrinum. J Physiol 1986; 371:115-145.
48. Nakatani K, Tamura T, Yau K-W. Light adaptation in retinal rods of the rabbit and two other nonprimate mammals. J Gen Physiol 1991; 97:413-435.
49. Koutalos Y, Nakatani K, Yau K-W. Cyclic-GMP diffusion coefficient in rod photoreceptor outer segments. Biophys J 1995; 68:373-382.

50. Koutalos Y et al. Diffusion coefficient of the cyclic GMP analog, 8-(fluoresceinyl) thioguanosine 3',5'-cyclic monophosphate in the salamander rod outer segment. Biophys J 1995; 69:2163-2167.
51. Gray-Keller M et al. Longitudinal spread of second messenger signals in isolated rod outer segments of lizards. J Physiol 1999; 519:679-692.
52. Koutalos Y, Yau K-W. Regulation of sensitivity in vertebrate rod photoreceptors by calcium. Trends Neurosci. 1996; 19:73-81.

THE CALCIUM GRADIENT ALONG THE ROD OUTER SEGMENT

K. Nicholas Leibovic

ABSTRACT

Vertebrate photoreceptor outer segments renew themselves by growing new membrane near the base and shedding old membrane at the tip. Parallel to the resulting age gradient along the outer segment there have also been observed gradients of membrane composition, rhodopsin phosphorylation, cGMP regeneration, responsiveness to light and others. This chapter describes the calcium gradient which has been found to exist along the outer segment. The concentration of calcium which increases towards the tip is due to an increase in buffered calcium. Since calcium is involved in a network of regulatory processes this gradient has implications for the transduction cascade as it affects the light response, as well as on disc shedding and other functions of the outer segments.

INTRODUCTION

Vertebrate rod and cone outer segments continually renew themselves by growing new membrane out from the inner segment and shedding excess length at the distal end. As in all cells, calcium is also involved in a network of regulatory activities in photoreceptors.

Calcium is essential in the control of the transduction cascade which amplifies the absorbed photon energy to generate a membrane response, and it also plays an important role in light and dark adaptation. In particular, calcium participates in the modulation of the cGMP sensitive conductance, rhodopsin phosphorylation and cGMP generation in the rod outer segment (ROS), as well as in a number of metabolic

---

Department of Physiology and Biophysics, State University of N.Y., at Buffalo, N.Y., and Medical College of Virginia, Virginia Commonwealth University, Richmond, VA.

functions in the rod inner segment (RIS) and the synaptic release of glutamate, the photoreceptor transmitter.

The purpose of this chapter is to demonstrate that there is a gradient of internal calcium in rod outer segments and to consider some functional implications of this finding.

Gradients along the ROS

The vertebrate rod outer segment consists of a plasma membrane enclosing a stack of membraneous discs. As it renews itself at the inner to outer segment junction, new membrane is formed, invaginates and is pinched off to form discs. Into these are incorporated proteins which have been assembled in the inner segment. The discs migrate to the tip of the outer segment where they fuse and form packets which are shed from the ROS and ingested and phagocytosed in the pigment epithelium (PE). This is illustrated in Figure1. Under normal physiological conditions the balance between disc formation and shedding keeps the ROS length constant. Due to these processes there is an age gradient from base to tip along the ROS (e.g., for reviews see refs. 1-4).

Other gradients have also been observed along the ROS. Thus there is a gradient of disc membrane composition with cholesterol in higher concentration at the base than the tip[5] a gradient of rhodopsin phosphorylation which is more pronounced near the base;[6] a gradient of light responses which are larger and faster at the base;[7] a gradient of recovery from bleaching[8] and a gradient of cGMP regeneration after a flash of light[9] which are both faster at the base; and there is a gradient of calcium exchange between base and tip across the plasma membrane.[10]

It has been speculated that all these gradients are the result of aging due to the renewal of the ROS. Indeed, we have shown that this is the case with the response gradient.[11] Moreover, it is worth recalling in this context that many processes, including transduction, occur locally - in the case of the latter, near the site of photon absorption.[9;12;13] Therefore, the ROS can be considered as a stack of progressively aging modules.

The Electrical Response and Transduction

Photoreceptors respond to light with a membrane hyperpolarization and reduction of inward current. The responses are graded with light intensity. Flashes of increasing intensity produce responses of increasing amplitude and duration up to a saturating intensity after which no further increase in amplitude occurs. More intense flashes merely increase the duration of the response (Fig. 2). Analysis of the response waveform can help in unraveling the underlying mechanisms which produce them.[14] Specifically, the saturated response has been used to study calcium exchange across the ROS membrane, the deactivation kinetics of rhodopsin, the local regeneration of cGMP and age related changes along the ROS.[9;11;15-17]

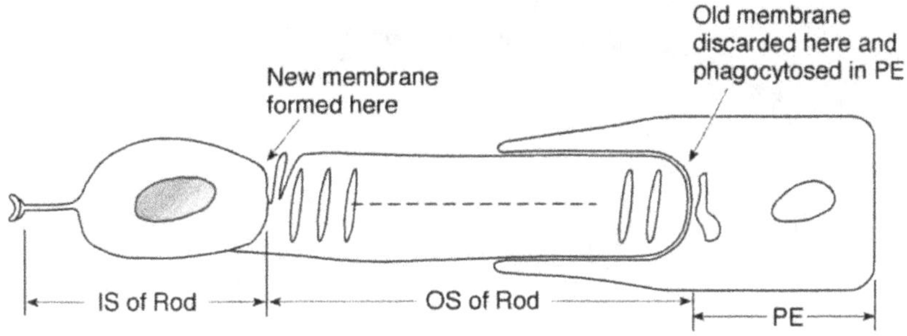

Figure 1. Illustration of rod outer segment renewal and shedding.

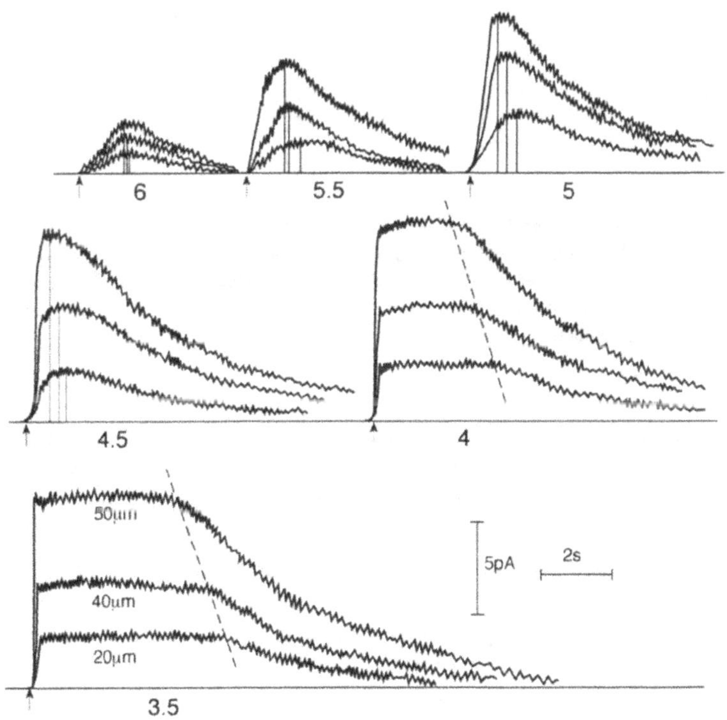

Figure 2. Responses of Bufo rods to flashes of light. The numbers below each set of three superimposed records are the 100 msec flash intensities in nominal ND (neutral density) units ranging from 6 ND (dim) to 3.5 ND (bright) flashes. The arrow at the beginning of each record marks the timing of the flash. At each intensity the currents were recorded from three different lengths of ROS in the recording electrode (28, 40 and 50 μm, respectively) As the intensity increases, the amplitude and the duration of the response increases up to a saturating amplitude, after which only the duration continues to increase. As the recorded ROS length decreases the response amplitude decreases and the duration increases. The full vertical lines indicate progressively later peak responses with decreasing ROS length. The dashed lines indicate a lengthening of the saturation time both with increasing flash intensity and decreasing ROS length.

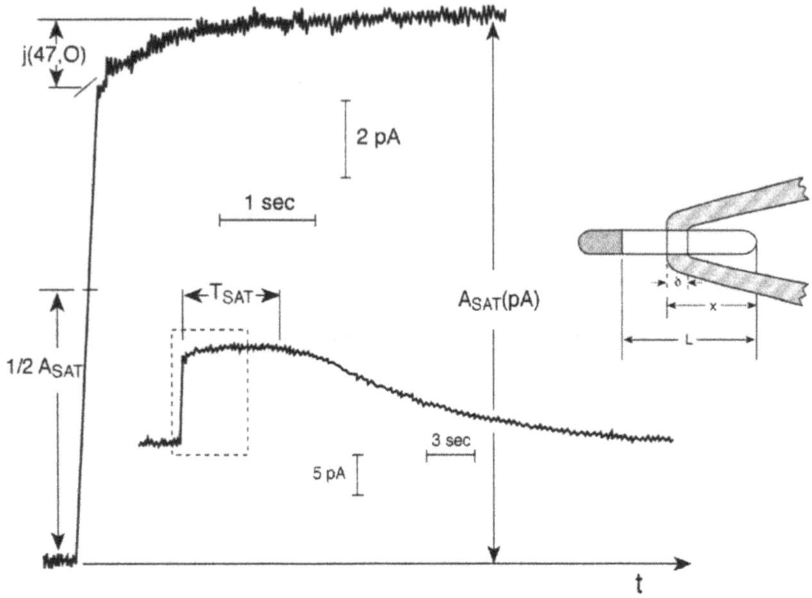

Figure 3. The calcium exchange current. The inset on the right shows the recording electrode with a length x of ROS inside it. δ is the dead space at the tip of the electrode. The current is recorded from the length (x - δ). Part of the original waveform in the dashed rectangle in the lower inset is enlarged as shown. T_{SAT} is the saturation time, A_{SAT} is the saturated response amplitude, j (47,0) is the exchange current for a 47 μm length of ROS at time 0.

Response saturation is due to the closure of all the cGMP gated channels in the ROS. The current response (Fig. 3) is characterized by a rapid upstroke during which the current is shut off, a slow relaxation to a plateau which is due to the calcium exchanger removing $1Ca^{2+} + 1K^{+}$ from the ROS in exchange for $4Na^{+}$,[18] the plateau itself and a slow return of the current to baseline as new cGMP is being regenerated.

In brief, the transduction cascade underlying the electrical response starts with the absorption of a photon by rhodopsin R and its conversion to its active form R^*. The latter activates a G-protein (transducin) T to T^* which in turn activates a phosphodiesterase PDE to PDE^* in which form it hydrolyzes cGMP. Since the latter opens the channels in the plasma membrane which allow Na^+ and Ca^{2+} to flow into the ROS, the hydrolysis of cGMP leads to channel closure and membrane hyperpolarization. These events underlie response initiation. The response is shut off by the inactivation of R^* through phosphorylation, the decay of T^* and PDE^* to their inactive forms and the regeneration of cGMP to reopen the membrane channels. This regeneration is facilitated by low internal free calcium $[Ca^{2+}]_i$ which increases the activity of a guanylate cyclase. The lowering of $[Ca^{2+}]_i$ occurs when the flow of calcium into the cell is prevented by channel closure while the calcium exchanger continues to extrude $[Ca^{2+}]_i$ from the ROS. The low $[Ca^{2+}]_i$ also facilitates R^*

phosphorylation through its action on a rhodopsin kinase which participates in the process. Rhodopsin spans the disc membrane with transducin and phosphodiesterase in association with the latter. The rate-limiting step of response activation is the encounter through diffusion of R^* with T. The transduction cycle has been reviewed extensively (e.g., see refs. 19-25).

THE CALCIUM GRADIENT

In view of the important role of calcium in phototransduction and the presence of a light response gradient it is of some interest to investigate the internal calcium gradient along the ROS. We did this by using the fact that the relaxation to the plateau after the initial current shut-off in the saturated response is due to the continuing extrusion of $[Ca^{2+}]_i$ through the calcium exchanger after the cGMP gated channels have been closed.[26] In the dark some 10% to 15% of the ROS inward current is carried by calcium ions[15] and the calcium exchanger current is directly proportional to the number of Ca^{++} ions removed.

Materials and Methods

The following analysis is based on experiments by ourselves and others using suction electrode recordings on isolated, single retinal rods.[27] In our laboratory the experimental animals were Bufo marinus and Xenopus laevis. Dissociation of the retina was as described elsewhere.[28] Details of the electrophysiological recordings, optical stimulation and data acquisition were as published and will not be repeated here.[9;10;17;29;30] In this study, currents were recorded from different lengths of the ROS in the suction electrode in response to saturating flashes of light.

Data Analysis

The experimental data from Bufo and Xenopus were quite comparable. There was the usual variability between individual cells which was dealt with by normalization.[10]

The calcium exchange current is illustrated in Figure 3. It has been shown that its time course is independent of flash intensity and that it can be represented by a single exponential, although in some cases two exponentials give a better fit.[10;15;31]

In general, the exchange current j depends on time t as well as the length x of ROS inside the electrode. Note that this length includes the electrode dead space as shown in the inset to Figure 3. So, for example, j(50,0) is the current recorded from a length (50 – the length of the dead space)μm. We found that the exponential decay time constant τ also depends on x and that it increases as the length x of ROS inside the suction electrode decreases. Therefore

$$j(x,t) = j(x,0) \exp[-t/\tau(x)] \quad (1)$$

The charge transfer by the exchange current over a length x of ROS is

$$q(x) = \int_0^\infty j(x,t)\,dt$$

$$= j(x,0)\,\tau(x) \qquad (2)$$

Let $i(x,t)$ = current density along the ROS.

Then $j(x,0) = \int_d^x i(z,0)\,dz$ (3)

where Δ = dead space length at the electrode tip (see Fig. 3).

If we know i(x,t) then j(x,0) can be found from (3). Experimentally, however, we only know j(x,0) and so i(x,0) can only be inferred. We consider three cases:

$$i = i_0 = \text{constant; then } j(x,0) = i_0\,(x - d) \qquad (4A)$$

$$i = i_0 \pm i_1 x\text{; then } j(x,0) = \pm 1/2 i_1(x^2 - d^2) + i_0(x - d) \qquad (4B)$$

$$i = i_0 \exp(i_1 x)\text{; then } j(x,0) = (i_0/i_1)[\exp(i_1 x) - \exp(i_1 d)] \qquad (4C)$$

where i_0 and i_1 are constants.

Results

The Calcium Exchange Currents

The mean values of the combined Bufo and Xenopus data sets before normalization were

$j(50,0) = 1.7$ pA $\qquad \tau(50) = 0.55$ sec.

The normalized values, distinguished by the suffix N, of j(x,0) and τ(x) are plotted on Figures 4 and 5 respectively. The normalized values of q(x), from eq.(2), are plotted on Figure 6.

A regression analysis was performed on the data to find the best fitting equations and parameters. From the data of Figure 4 the best fit for the exchange current was given by the linear equation 4(A), resulting in

$$j_N(x,0) = 0.038x - 0.871 \qquad (5)$$

Accordingly the normalized exchange current density is constant and equal to

$i_0 = 0.038$ pA/μm

with δ= 23μm approx.

Similarly, the best fit for the normalized time parameter after a regression analysis of linear, quadratic and exponential forms was

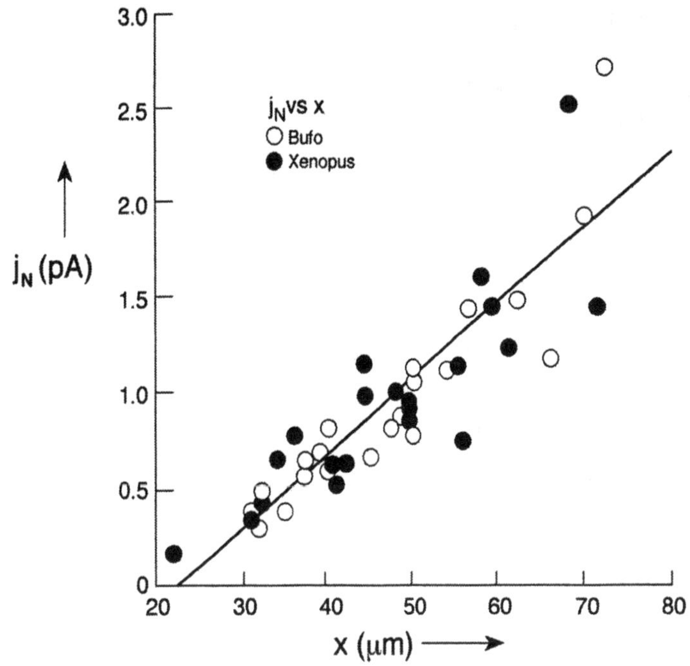

Figure 4. The normalized calcium exchange current j_N in pA as a function of recorded ROS length x in μm. Open circles are the Xenopus and closed circles the Bufo data points. The line through the points is the best fitting regression line.

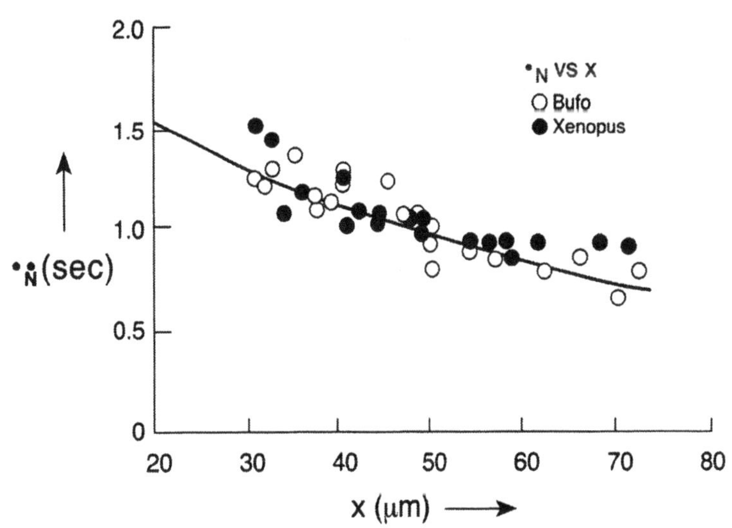

Figure 5. The normalized time constant τ_N in sec. as a function of recorded ROS length. Data points as in Figure 3. The curve through the points is the best fitting exponential

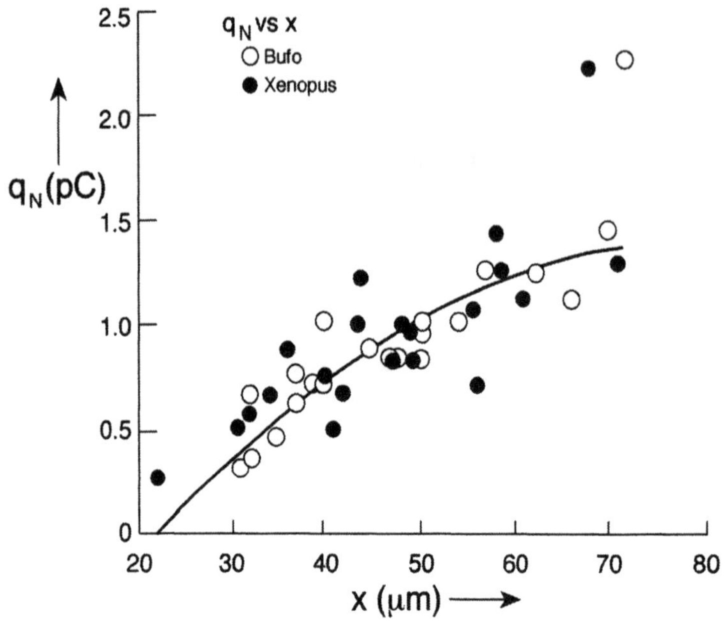

Figure 6. The normalized charge transfer q_N in pC vs. the recorded ROS length. Data points as shown in Figure 3. The curve is derived from $j_N(x,0)\tau_N(x)$ as in the text eq. (2).

$$\tau_N(x) = 0.73 + 2.98\exp(-0.049x) \qquad (6)$$

From equations (2), (4), (5) we can calculate the normalized charge transfer after a saturating flash. Correcting for the normalizations, the mean charge transfer over a $(50 - \delta)\mu m$ length of ROS is approximately

$$q(50) = 0.9 \text{ pC}$$

Over a 1μm length near the base of the ROS:

$$q(50) - q(49) = 0.02 \text{ pC}$$

and over a 1μm length near the tip of the ROS:

$$q(\delta + 1) - q(\delta) = 0.06 \text{ pC}$$

In terms of the number of calcium ions extruded these figures translate to:

$5.9*10^6$ Ca^{2+} over a $(50 - \delta)\mu m$ long ROS (7A)

$1.3*10^5$ Ca^{2+} over the 49 to 50 μm near the base of the ROS (7B)

$3.9*10^5$ Ca^{2+} over 23 to 24 μm near the tip of the ROS (7C)

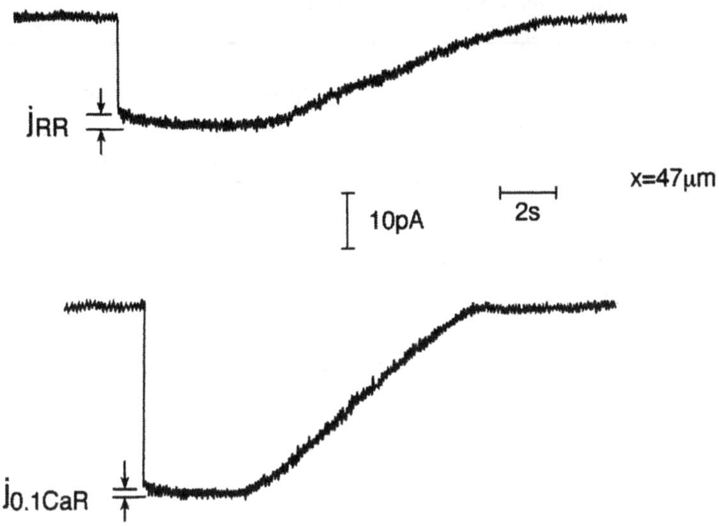

Figure 7. Comparison of responses in Ringer solution (RR) vs. responses in a solution (0.1CaR) in which calcium is at 0.1 the concentration in (RR). (A): response in (RR); (B): response in (0.1CaR). Compared to (A) the response in (B) is larger, the saturation time T is shorter and the return to baseline is faster. Note, however, that the exchange current j in (B) is smaller than in (A).

The free $[Ca^{2+}]_i$ inside the ROS has been estimated to be around 400nM.[32-34] For a $(50 - \delta)\mu m$ long ROS of diameter 6μm and with half the internal space occupied by discs, 400 nM is equivalent to $9.19*10^4$ Ca^{2+}. But during response saturation there is no influx of calcium into the ROS. It therefore follows from (7A) that most of the calcium extruded by the exchanger must come from internal buffer stores. Moreover, from (7B) and (7C) it is also apparent that more calcium is extruded near the tip than the base of the ROS.

Effects of Lowered Calcium

The calcium gradient between base and tip of the ROS which is accompanied by a response gradient, has parallels in the responses of the rod as a whole when it is exposed to different concentrations of calcium. To investigate this further, the rod inner segment was drawn into the suction electrode and the outer segment was either immersed in a regular Ringer solution (RR) or superfused with a solution containing 0.1 of the regular calcium (0.1CaR). In the low calcium solution the light responses were larger and exhibited faster kinetics, but the exchange currents were smaller. This is shown in Figure 7. The mean values for the exchange current j(x,0) and saturated response amplitude A from five experiments are given in Table 1.

The calcium exchange current is proportional to the internal free calcium concentration in the steady state[35] and it has been shown to obey the equation

$$j/j_{SAT} = [Ca]_i^{2+} / ([Ca]_i^{2+} + K_{Ca}) \quad (8)$$

Table 1. Comparisons of response amplitude and Ca^{2+} exchange current in RR and in 0.1CaR. Lengths are in μm, currents in pA.

	ROS Length	A(RR)	j(RR)	A(0.1CaR)	j(0.1CaR)
Mean ± S.D	38 ± 5.5	25.9 ± 5.1	1.49 ± 0.28	42 ± 7.4	0.89 ± 0.14

over a wide range of conditions,[15;33] where j_{SAT} is the maximum j at a saturated $[Ca^{2+}]_i$, the free internal calcium concentration, and K_{Ca} is a constant. The reduction of j in low calcium solution implies, according to (8), a reduction of free internal calcium, since j_{SAT} and K_{Ca} are constant for a given length of ROS.

DISCUSSION

The Distributions of Free and Buffered Calcium

Equation (8) can be rearranged to give

$$[Ca^{2+}]_i = K_{Ca}\, j / (j_{SAT} - j) \tag{9}$$

We can take it that (9) holds for any length x of ROS. But, from (4A) j is linear in x. Then if we make the reasonable assumption that j_{SAT} is also linear in x, i.e.,

$$j_{SAT}(x,0) = i_{0SAT}\,(x - \delta)$$

then

$$[Ca^{2+}]_i = K_{Ca}\, i_0(x - \delta) / (i_{0SAT} - i_0)(x - \delta)$$

$$= K_{Ca}\, i_0/(i_{0SAT} - i_0)$$

$$= \text{constant.}$$

Now, the numbers in eq.(7) show that the amount of calcium extruded over a length (50 - δ)mm, i.e., 27μm, of ROS is almost 64 times the amount of free calcium which is consistent with Lagnado et al[33] who found a ratio of free to exchangeable bound calcium of 74. Then taking account of the gradient along the ROS, this ratio is even larger near the tip and smaller at the base. This implies that there is a larger store of internal calcium near the tip. It could come about, in principle, either by a change in the affinity of the buffer for calcium or else by the presence of more buffer. This is discussed further below.

Two kinds of buffer, a high and a low affinity one respectively, have been identified in the ROS,[33] with the total exchangeable calcium $[Ca]_T$ (free + bound) being given by

$$[Ca]_T = C[Ca]_i / \{[Ca]_i + K_{buff}\} + (B + 1)[Ca]_i \tag{10}$$

where $[Ca]_T$ = total exchangeable calcium

C = capacity of high affinity buffer

K_{buff} = binding constant of high affinity buffer

B = calcium bound to low affinity buffer / $[Ca]_i$ (see ref. 33)

suggested that the low affinity buffer may be associated with the phospholipid headgroups of the disc membrane. But calcium is known to affect membrane behavior. Therefore the calcium gradient could be expected to influence interactions between membrane associated molecules of the transduction cascade.

Consequences of the Calcium Gradient for the Light Response

The following expression has been derived for the time "constant" τ of the calcium exchanger:[33]

$$\tau = FV\{(C / j_{SAT})(K_{Ca} / K_{buff}) + (B + 1)K_{Ca} / K_{buff}\} \tag{11}$$

where F = Faraday constant

V = internal ROS volume.

From our results the most likely candidates for increasing τ towards the tip in the above equation are C and B, the capacities of the high and low affinity buffers respectively. The other parameters are less likely to be significant because of the linearity or near linearity of the exchange current and the constancy or near constancy of internal free calcium along the length of the ROS. One can evaluate the contributions which K_{Ca} and K_{buff} would make to the calcium gradient if C and B were constant. As regards K_{Ca}, from eq.(8), with $[Ca]_i$ constant and with j and j_{SAT} linear, K_{Ca} is constant. As regards K_{buff}, let $\Delta[Ca]_T$ be the difference between $[Ca]_T$ in the dark and after a saturating flash. Assume that (10) holds at all points along the ROS and $[Ca]_i$ goes from 400nM before to 40nM after the flash. Then with B, C constant it can be shown that $\Delta[Ca]_T$ has a maximum for $K_{buff} = 0.1265\mu M$ approximately. Suppose, for purposes of illustration, that K_{buff} has this value near the tip. At physiological $[Ca]_i$ the low affinity buffer makes a relatively small contribution to equation (10). Then neglecting this term, it can be shown that for a ratio $\Delta[Ca]_{T(tip)} / \Delta[Ca]_{T(base)} = 3$, as in eqs. (7B) and (7C), $K_{buff} = 0.01\mu M$ near the base. This is more than a twelvefold change along the ROS, which would imply a correspondingly large change in the properties of the buffer. This is an extreme situation, but if one assumes values for K_{buff} found experimentally,[33] the results are even more questionable. It therefore appears that a change in K_{buff} alone cannot account for the calcium gradient.

It has long been known from recordings of whole rods that lowering the external calcium speeds up the light response and increases its amplitude.[36;37] This is also the case here as shown in Figure 7. As we have seen from eq. (8), a reduction of external calcium implies a reduction of internal calcium as well and this is consistent with the smaller exchange current in Figure 7b. The changes on lowering external calcium are due to effects on the plasma membrane and the cell interior. These effects can include an increased affinity of the cGMP gated channels for cGMP[38] activation of the guanylate cyclase,[39] facilitation of R^* phosphorylation, and a possible modification, due to changes in buffered calcium, of the disc membrane affecting the diffusional encounters of R^* with T.

Could the above effects of calcium explain why the responses are smaller and slower at the tip than the base of the ROS? We have to remember that we have only found a gradient of bound, not free calcium, along the ROS. As a result of this gradient and the lengthening value of τ, $[Ca^{2+}]_i$ stays higher, longer at the tip than the base during the light response.

Considering first the response kinetics, we have to account for the slower rise to peak and slower return to baseline as illustrated in Figure 2. The former has to do with response initiation and the latter with its shut-off. The rate limiting step in response initiation is the diffusional encounter of R^* with T (e.g., see ref. 25) In this connection it is worth recalling three relevant findings: first, the gradient of membrane composition along the ROS, in particular the cholesterol gradient;[40] cholesterol which tends to stabilize the membrane decreases toward the tip; second, the proposal[33] that the disc membrane acts as a buffer for internal calcium which, we have shown, increases toward the tip; and third, the modulation of membrane properties by calcium; for example, calcium has been shown to induce phase separation and fusion, as well as changes in packing density and hydrophobicity of vesicle and bilayer membranes.[41-44] If the encounters of R^* with T in the disc membrane are slowed by such gradients it would account for the slower response initiation.

As regards shutting off the response, we know that calcium is involved in the regeneration of cGMP and the phosphorylation of R^*. Both these processes are accelerated in low calcium. The persistence of $[Ca]_i$ at the tip can therefore cause a slower return to baseline of the response.

It is more difficult to explain the gradient of response amplitude depending, as it does, on the number of open cGMP gated channels in the plasma membrane. But there may be a combination of factors at work. First, there could be a slight increase of $[Ca^{2+}]_i$ towards the tip which is not detectable with the resolution of our methods. Second, as already mentioned, there could be a decreased affinity of the cGMP gated channels for cGMP at the tip. These channels can also be considered as a buffer for calcium in equilibrium with other buffers as well as the free calcium inside the ROS. It is known that an increase in calcium decreases the affinity of the channels for cGMP.[38] In that case the current density would be expected to decrease towards the tip. There may not be any strong evidence for that from our results, in agreement with the data of ref. 45 who measured current density directly. However, an examination of their Figure 5B suggests the possibility that, in fact, some decrease of current density toward the tip may occur.

Possible Functional Implications of the Calcium Gradient

It remains an open question whether the $[Ca^{2+}]_i$ gradient, like other gradients along the ROS, is an inevitable concomitant of aging or whether it plays any functionally useful role. One possibility may have to do with disc shedding. Old disc packet formation and shedding at the tip, like new disc closure on separation from the plasma membrane, require membrane fusion which depends on calcium. This has been shown to be true for vertebrate photoreceptors.[46] Disc shedding is initiated preferentially by the onset of light after darkness.[47] As we have seen, calcium remains relatively elevated, especially at the tip, following light stimulation. This suggests that the processes involved in disc shedding may be facilitated by the higher calcium content and the persistence of increased calcium after light onset near the tip of the ROS.

REFERENCES

1. Young RW. The renewal of photoreceptor cell outer segments. J Cell Biol 1967; 33:61-72.
2. Young RW. Visual cells and the concept of renewal. Invest Ophthalmol Vis Sci 1976; 15:700-25.
3. Besharse JC. The daily light-dark cycle and rhythmic metabolism in the photorecptor-pigment epithelial complex. Progr in Retinal Res 1982; 1982:81-124.
4. Bok D. Retinal photoreceptor-pigment epithelium interactions. Friedenwald lecture. Invest Ophthalmol Vis Sci 1985; 26:1659-94.
5. Boesze-Battaglia K, Albert AD. Cholesterol modulation of photoreceptor function in bovine retinal rod outer segments. J Biol Chem 1990; 265:20727-30.
6. Shichi H, Williams TC. Rhodopsin phosphorylation suggests biochemical heterogeneities of retinal rod disks. J Supramol Struct 1979; 12:419-24.
7. Schnapf JL. Dependence of the single photon response on longitudinal position of absorption in toad rod outer segments. J Physiol 1983; 343:147-59.
8. Baylor DA, Lamb TD. Local effects of bleaching in retinal rods of the toad. J Physiol 1982; 328:49-71.
9. Leibovic KN, Pan KY. The saturated response of vertebrate rods and its relation to cGMP metabolism. Brain Res 1994; 653:325-9.
10. Leibovic KN, Bandarchi J. Phototransduction and calcium exchange along the length of the retinal rod outer segment. Neuroreport 1997; 8:1295-300.
11. Leibovic KN, Bandarchi J. Effects of light and temperature on the response gradient of retinal rod outer segments. Brain Res 1997; 750:321-4.
12. Cornwall MC, Fein A, MacNichol Jr EF. Spatial localization of bleaching adaptation in isolated vertebrate rod photoreceptors. Proc Natl Acad Sci USA 1983; 80:2785-8.
13. Matthews G. Spread of the light response along the rod outer segment: an estimate from patch-clamp recordings. Vision Res 1986; 26:535-41.
14. Leibovic KN. Response waveforms of vertebrate photoreceptors: what are the underlying mechanisms? Biol Cybern 1978; 31:125-35.
15. Nakatani K, Yau KW. Calcium and magnesium fluxes across the plasma membrane of the toad rod outer segment. J Physiol 1988; 395:695-729.
16. Pepperberg DR, Cornwall MC, Kahlert M et al. Light-dependent delay in the falling phase of the retinal rod photoresponse. Vis Neurosci 1992; 8:9-18.
17. Bandarchi J, Leibovic KN. Effects of animal age on the responses along the outer segment of retinal rod photoreceptors. Neuroreport 1997; 8:581-5.

18. Cervetto L, Lagnado L, Perry RJ et al. Extrusion of calcium from rod outer segments is driven by both sodium and potassium gradients. Nature 1989; 337:740-3.
19. Stryer L. Cyclic GMP cascade of vision. Ann Rev Neurosci 1986; 9:87-119.
20. Pugh EN, Cobbs WH. Visual transduction in vertebrate rods and cones: a tale of two transmitters, calcium and cGMP. Vision REs 1986; 26:1613-43.
21. Yau KW, Baylor DA. Cyclic GMP-activated conductance of retinal photoreceptor cells. Annu Rev Neurosci 1989; 12:289-327.
22. Leibovic KN. Vertebrate Photoreceptors. In: Leibovic KN, ed. Science of Vision. New York: 1990.
23. McNaughton PA. Light response of vertebrate photoreceptors. Physiol Rev 1990; 70:847-84.
24. Yarfitz S, Hurley JB. Transduction mechanisms of vertebrate and invertebrate photoreceptors. J Biol Chem 1994; 269(20):14329-14332
25. Baehr W, Liebman PA. Visual Cascade. Encyclopedia of Life Sciences, MacMillan, 2000.
26. Yau KW, Nakatani K. Light-induced reduction of cytoplasmic free calcium in retinal rod outer segment. Nature 1985; 313:579-82.
27. Baylor DA, Lamb TD, Yau KW. The membrane current of single rod outer segments. J Physiol 1979; 288:589-611.
28. Leibovic KN. A new method of non-enzymatic dissociation of the Bufo retina. J Neurosci Methods 1986; 15:301-6.
29. Leibovic KN, Dowling JE, Kim YY. Background and bleaching equivalence in steady-state adaptation of vertebrate rods. J Neurosci 1987; 7:1056-63.
30. Leibovic KN, Bandarchi J. Recovery from bleaching in photoreceptors promoted by biotin, pyruvate, and glucose. Vis Neurosci 1990; 4:489-92.
31. Gray-Keller MP, Detwiler PB. The calcium feed back signal in the phototransduction cascade of vertebrate rods. Neuron 1994; 13:849-61.
32. Korenbrot JI, Miller DL. Cytoplasmic free calcium concentration in dark-adapted retinal rod outer segments. Vision REs. 1989; 29:939-48.
33. Lagnado L, Cervetto L, McNaughton PA. Calcium homeostasis in the outer segments of retinal rods from the tiger salamander. J Physiol 1992; 455:111-42.
34. McCarthy ST, Younger JP, Owen WG. Free calcium concentrations in bullfrog rods determined in the presence of multiple forms of Fura-2. Biophys J 1994; 67:2076-89.
35. Younger JP, McCarthy ST, Owen WG. Light-dependent control of calcium in intact rods of the bullfrog Rana catesbeiana. J Neurophysiol 1996; 75:354-66.
36. Hagins WA, Yoshikami S. Proceedings: A role for Ca2+ in excitation of retinal rods and cones. Exp Eye Res 1974; 18:299-305.
37. Waloga G. Effects of calcium and guanosine-3',5'-cyclic-monophosphoric acid on receptor potentials of toad rods. J Physiol 1983; 341:341-57.
38. Hsu Y-T, Molday RS. Modulation of the cGMP-gated channel of rod photoreceptor cells by calmodulin. Nature 1993; 361:76-9.
39. Koch K-W, Stryer L. Highly cooperative feedback control of retinal rod guanylate cyclase by calcium ions. Nature 1988; 334:64-6.
40. Boesze-Battaglia K, Fliesler SJ, Albert AD. Relationship of cholesterol content to spatial distribution and age of disc membranes in rod outer segments. J Biol Chem 1990; 265(31):18867-18870.
41. Papahadjopoulos D, Vail WJ, Newton C et al. Studies on membrane fusion. III. The role of calcium-induced phase changes. Biochim Biophys Acta 1977; 465:579-98.
42. Koter M, de Kruijff B, van Deenen LL. Calcium-induced aggregation and fusion of mixed phosphatidylcholine- phosphatidic acid vesicles as studied by 31P NMR. Biochim Biophys Acta·1978; 514:255-63.
43. Zidovetzki R, Atiya AW, De Boeck H. Effect of divalent cations on the structure of dipalmitoylphosphatidylcholine and phosphatidylcholine/phosphatidylglycerol bilayers: an 2H-NMR study. Membr Biochem 1989; 8:177-86.

44. Ohki S, Zschornig O. Ion-induced fusion of phosphatidic acid vesicles and correlation between surface hydrophobicity and membrane fusion. Chem Phys Lipids 1993; 65:193-204.
45. Karpen JW, Loney DA, Baylor DA. Cyclic GMP-activated channels of salamander retinal rods: spatial distribution and variation of responsiveness. J Physiol 1992; 448:257-74.
46. Greenberger LM, Besharse JC. Photoreceptor disc shedding in eye cups. Inhibition by deletion of extracellular divalent cations. Invest Ophthalmol Vis Sci 1983; 24:1456-64.
47. LaVail MM. Rod outer segment disk shedding in rat retina: relationship to cyclic lighting. Science 1976; 194:1071-4.

THE TIME COURSE OF LIGHT ADAPTATION IN VERTEBRATE RETINAL RODS

Peter D. Calvert and Clint L. Makino

ABSTRACT

The photoresponse of a rod wanes over time in steady illumination, as light loses its efficacy in generating the response. Such desensitization is adaptive because it extends the range of ambient light levels over which the rod signals changes in light intensity by several orders of magnitude. Adaptation begins to unfold rapidly after the onset of light with a time constant of ~1 s, causing the rod's sensitivity to steady light to decrease by nearly two log units. Thereafter, a much slower phase of adaptation evolves with a time constant of 9 s. During this phase the rod's sensitivity decreases by an additional log unit. Both phases are dependent upon the light-induced fall in intracellular Ca^{2+}. The fast phase of light adaptation can be attributed to Ca^{2+} feedback processes regulating the lifetime of photoactivated rhodopsin, cGMP synthesis and sensitivity of the cGMP-gated channel to cGMP. Although the mechanism(s) of the slow phase is not yet known, it appears to include further regulation of the lifetime of photoactivated rhodopsin.

INTRODUCTION

Retinal rods mediate vision under dim light by signaling the presence of single photons, yet they are also capable of signaling changes in light intensity over a wide range of ambient lighting conditions. Rods from cold-blooded vertebrates are especially proficient, boasting a dynamic range that exceeds four log units. The conflicting requirements of high absolute sensitivity and wide dynamic range are reconciled

Department of Ophthalmology, Harvard Medical School and the Massachusetts Eye and Ear Infirmary, Boston, MA 02114.

by the rod's ability to adapt to light. In this Chapter, we examine the time course of the transition of the rod from the dark to the light adapted state.

Rods transduce light into an electrical signal in their outer segments, where the capacity to signal the presence of a single photon is achieved through a highly amplified biochemical cascade (Fig. 1) (reviewed in refs. 1-4). Photon capture by a rhodopsin molecule causes it to assume a conformation (R*) that promotes the exchange of a GTP for a GDP on transducin (T). Catalytic activation of more than a hundred transducin molecules by the R* marks the first amplifying stage of the cascade.[5,6] With GTP bound, each transducin relieves an inhibition on a single phosphodiesterase (PDE). In a second stage of amplification, each PDE hydrolyzes several hundred cGMP molecules per second.[5] The ensuing fall in cGMP leads to the cooperative closure of cGMP-gated channels on the plasma membrane in a third amplifying stage of the cascade. Channel closure prevents the entry of a million Na^+ ions into the rod outer segment. This interruption of the rod's circulating current constitutes the photoresponse. While the large amplification in the phototransduction cascade is essential for signaling single photons, it will tend to drive the rod into saturation at rather low intensities. To forestall saturation, the rod engages a set of negative feedback reactions that reduce the overall impact of photon absorptions.

Ca^{2+} serves as the messenger for these feedbacks (Fig. 1; reviewed in refs. 1-4,7-9). In darkness, Ca^{2+} enters through the cGMP-gated channel and is removed by a $Na^+/K^+,Ca^{2+}$ exchanger. Closure of the channels in response to light blocks the entry of Ca^{2+}, whereupon the continued extrusion of Ca^{2+} by the exchanger drops internal Ca^{2+} down to a low level. One action of the decline in Ca^{2+} is to reduce the PDE activity. PDE remains active as long as transducin retains GTP and as long as R* continues to activate transducins. Transducin deactivates after it hydrolyzes its bound GTP, a process that is accelerated by its PDE effector and the RGS9-Gβ5 complex. R* shuts off after it is phosphorylated by rhodopsin kinase (RK) and subsequently binds arrestin (Arr). Ca^{2+} decline removes an inhibition of rhodopsin kinase by the Ca^{2+}-binding protein recoverin (Rec) thus shortening the lifetime of R*. The result is that fewer transducin-activated PDE molecules are produced by each R* and there is a reduction in the overall rate of cGMP hydrolysis. Another action of Ca^{2+} decline, mediated by guanylate cyclase activating proteins (GCAPs), is to stimulate cGMP synthesis by guanylate cyclase (GC). In a third action, dissociation of Ca^{2+} from calmodulin or a related protein raises the affinity of the channel for cGMP, allowing it to open at lower cGMP concentrations. There is physiological evidence for additional feedback onto PDE activity, but neither the mechanism nor the target is known. The net effect of these Ca^{2+} feedbacks is to attenuate the response to steady light by increasing the fraction of open channels at a given light intensity. While this loss of step sensitivity cannot prevent saturation entirely, it greatly extends the range of ambient light intensities over which the rod remains differentially sensitive. To avoid confusion between step sensitivity and differential sensitivity, we will restrict our use of the term "sensitivity" to the description of the relationship between response amplitude and light intensity during steady illumination.

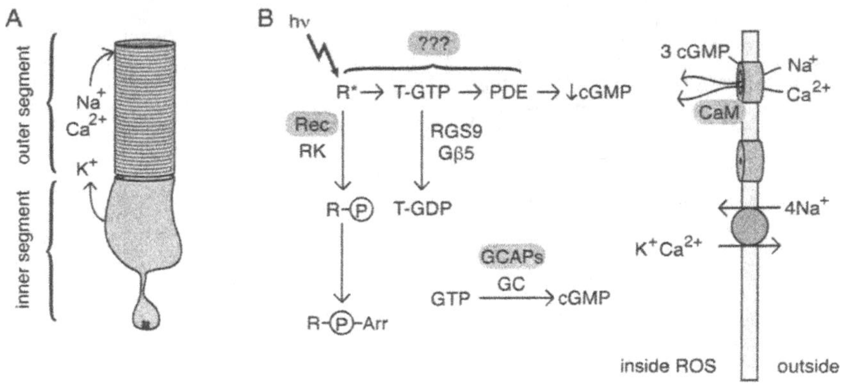

Figure 1. A. Circulating current of the rod. B. Phototransduction cascade. Sites of Ca^{2+} feedback are highlighted.

HOW IS LIGHT ADAPTATION MANIFESTED?

Figure 2A shows photoresponses of a frog rod to steps of light. In extremely dim light, the response was the summation of individual single photon responses. In single trials responses fluctuated erratically (e.g., gray trace), due to the stochastic absorption of photons. Responses to higher intensities became smoother and more reproducible. They rose to a peak and then sagged to an intermediate level within 5-10 s. For responses in which the peak surpassed the half-maximal level, the sag showed a second, slower phase that continued to develop over many tens of seconds of illumination. The responses to the brightest steps no longer exhibited the fast phase since the response saturated transiently. Nevertheless, the slow phase still managed to rescue the rod from saturation.

When the response amplitudes were plotted as a function of the light intensity at various times after step onset, it became immediately apparent that the rod progressively lost sensitivity with time in the light and that the loss was most pronounced at higher intensities. Figure 2B shows amplitudes at 2.5 s, near the peak of the response, at 10 s, near the end of the fast sag and at 60 s after light onset, as indicated by the corresponding symbols in Figure 2A. The total magnitude of desensitization after 60 s of illumination was gauged by the light intensity required to produce a given response amplitude relative to the intensity required in the absence of adaptation. The relation between response amplitude and light intensity in the absence of light adaptation is given by:[10-13]

$$\frac{r}{r_{max}} = 1-(1+IF)^{n} \qquad (1)$$

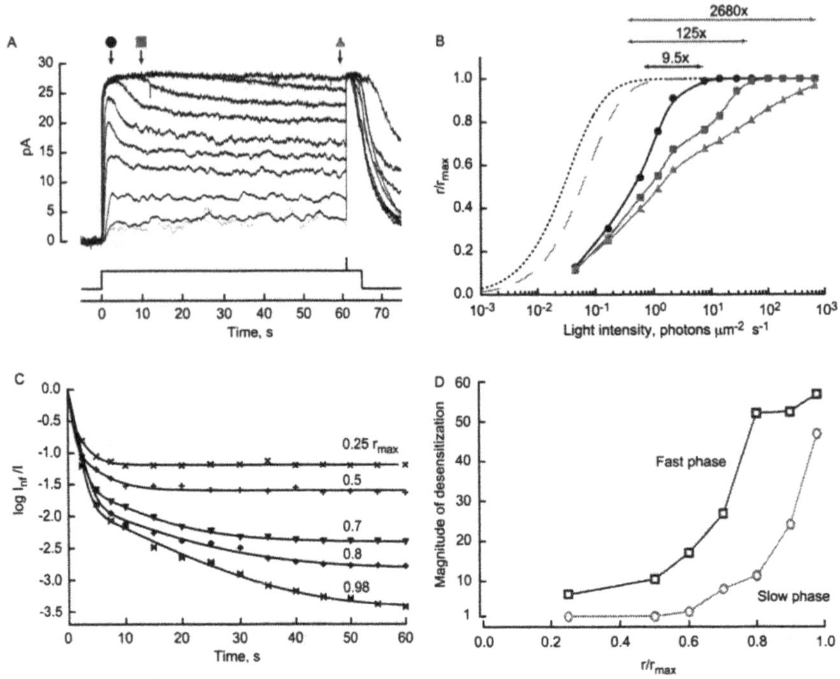

Figure 2. A. Averaged photocurrent responses of a frog rod to steps of light. For the dimmest intensity, the response to a single trial is shown in gray. A bright flash (56-7500 photons μm^{-2}), presented 61 s after step onset, demonstrated that the recovered current was suppressible by light. B. Response-intensity relations for the cell in A at 2.5 s (●), 10 s (■) and 60 s (▲) after light onset, including some intensities omitted in A for clarity. Lines connecting the points were drawn by eye. The broken lines predict the response in the absence of light adaptation (Eq. 1). The dotted line applies for responses measured at t > 9s after light onset. The dashed line is for responses measured at 2.5 s after light onset; it corrects for the pre-steady state PDE activation (Eq. 3), based on $k_R = 2\ s^{-1}$ (Ref. 18) and the k_E for this rod, 0.3 s^{-1}. The magnitudes of desensitization, given at the top of the figure, were found from the ratios of the intensities required to elicit a response of 0.98 r_{max}. C. The time course of desensitization for different response levels. Continuous lines are fittings with Eq. 4. The τ_n and A_n values varied with response level. For the 0.98 r_{max} response level, 57-fold desensitization was achieved with a time constant of 0.9 s. An additional 47-fold sensitivity loss evolved with a time constant of 9.8 s. D. The increase in magnitude of each phase of adaptation as a function of response amplitude, from C.

where r is the response amplitude, r_{max} is the maximal response, I is the intensity, n is the Hill coefficient for gating of the cGMP-gated channel by cGMP, and F is a sensitivity factor relating the fractional response amplitude to the rate of cGMP hydrolysis by light-activated PDE.[13] F may be approximated from the integral of the fractional single photon response in the absence of Ca^{2+} feedback, the rod's light collecting area (a)[14] and the Hill coefficient for the cGMP-gated channel (n)[13]:

$$F = \frac{a}{n} \cdot \int_0^t \frac{r(t)}{r_{max}} dt \tag{2}$$

F can be found from the product of the integral of the normal single photon response and a "feedback factor" equal to the ratio of single photon response areas in the absence and presence of Ca^{2+} feedback. Experiments on toad and salamander rods in which the exchanger was disabled and on mouse rods lacking GCAPs indicate that Ca^{2+} feedback normally reduces the single photon response integral by a factor of 5.7-11.[15-19] For the analysis of frog rods, we used a feedback factor of 5.7 obtained from toad rods[19] since their response kinetics in normal Ringer's most closely resemble those of frog rods. The relation between fractional response and step intensity predicted in the absence of adaptation is shown by the dotted line in Figure 2B. Thus at the 0.98 r_{max} response level, the total magnitude of desensitization was 2680-fold. On average, frog rods desensitized 3130-fold over 60 s of continuous illumination.[13]

Equation 2 describes light-stimulated PDE activity at the steady state level. However, at early times after the onset of light F is smaller because PDE activity has not yet reached a steady state. So the position of the relation given by Eq. 1 shifts to higher intensities. The rate at which PDE reaches a steady state depends on the shutoff time constants for R* and transducin-activated PDE. Provided that the activities of R* and transducin-PDE shut off after a flash with exponential time courses,[20] the value for F will change as a function of time after light onset according to:[13]

$$\frac{F(t)}{F(\infty)} = \frac{k_R k_E}{k_R - k_E} \int_0^t \left(\exp(-k_E t) - \exp(-k_R t)\right) dt = 1 - \frac{k_R \exp(-k_E t) - k_E \exp(-k_R t)}{k_R - k_E} \tag{3}$$

where k_R and k_E are the rate constants for rhodopsin and transducin shutoff, respectively. k_R, measured in salamander, was ~2 s^{-1} (ref. 18), while k_E for amphibian rods ranges from 0.33-0.5 s^{-1} (refs. 13,18,20,21). k_E dominates the overall cascade shutoff, so PDE activity will reach 95% of its steady state, light-stimulated activity in $3/k_E$, or ~ 9 s. The response-intensity relation predicted for 2.5 s exposures in the absence of light adaptation is plotted in Figure 2B as a dashed line. Thus, for the 0.98 r_{max} response level, a 9.5-fold loss in sensitivity was already in place at 2.5 s. By 10 s, the loss in sensitivity reached 125-fold, while an additional 21-fold loss occurred over the next 50 s.

The time course of the change in sensitivity was obtained by extending the response-intensity analyses to additional times after light onset. The logarithm of the intensity required to elicit a given response amplitude in the absence of light adaptation relative to that with adaptive mechanisms in place is mapped out as a function of exposure duration in Figure 2C. On the assumption that both phases initiated at light onset and ignoring any effect of the partial recovery of the rod's circulating current during the step response on the intracellular level of Ca^{2+}, the change in sensitivity was described as the product of two exponentials:[13]

$$\frac{I_{nf}}{I(t)} = \frac{(1+A_1\exp(-t/\tau_1))(1+A_2\exp(-t/\tau_2))}{(1+A_1)(1+A_2)} \quad (4)$$

where I_{nf} is the intensity required for the rod to reach a certain response amplitude in the absence of light adaptation, $1+A_1$ and $1+A_2$ are the full magnitudes of desensitization and $I_{nf}/I(\infty) = (1+A_1)^{-1}(1+A_2)^{-1}$. On average the time constants of the fast and slow phases taken at the 0.98 r_{max} level were 0.8 and 9.2 s,[13] respectively, although it should be noted that the kinetics of the fast phase were difficult to resolve using this method. The average magnitude attributed to the fast phase was 78-fold, while that attributed to the slow phase was about 40-fold[13]. The relative magnitudes may differ slightly if the slow phase began after a delay rather than at the onset of light.

The two temporal phases of light adaptation operate at distinct levels of current suppression. Figure 2D shows the amplitudes of the fast and slow phases derived from the curve fittings in Figure 2C for different response levels. The fast phase impacted the entire range of responses while the slow phase was not detected until more than 50% of the circulating current was suppressed. Since the slow phase did not appear to saturate at 0.98 r_{max}, it may provide further "desensitization" after complete suppression of the current by supersaturating light steps.

HOW CAN ADAPTATION NEAR SATURATION BE STUDIED?

Adaptation was greatest at intensities that saturated the rod initially. This property posed a technical problem because the small amount of circulating current that recovered upon adaptation was difficult to measure reliably. Furthermore, any recovery of the circulating current would prevent internal Ca^{2+} from reaching its minimal level and would complicate Ca^{2+} dynamics. In order to circumvent these problems and determine the full effect of light adaptation, we modified an approach developed by Pepperberg et al[21] and Lyubarsky et al.[20] Their approach was to use saturating flashes to measure the rate constant of cascade shutoff and to gauge the magnitude of desensitization imparted by Ca^{2+} feedback reactions during flash responses. We substituted saturating steps for flashes to probe the adaptation that occurs during responses to continuous illumination. A brief description is given below.

Bright flashes quickly close all of the cGMP-gated channels and elicit a maximal, saturating response. With increasing flash strength, the response remains saturated for a longer period, before recovering. The basis for this behavior is outlined in Figure 3A,C. Bright flashes produce waves of PDE activity that surpass a critical level (CL) at which all of the cGMP-gated channels close (Fig. 3A). Brighter flashes

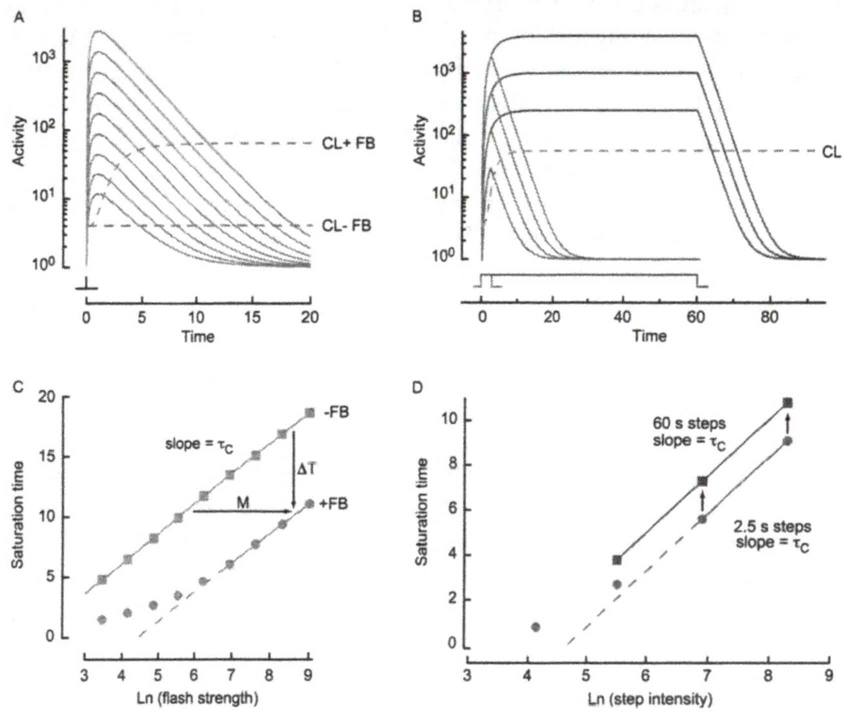

Figure 3. Models for the saturation behavior of the rod response to bright flashes and steps. A. Waves of PDE activity (continuous lines, in arbitrary units) after flashes of nine photon densities. Activity declines exponentially with a time constant, τ_c. The CL rises as a function of time after the flash due to Ca^{2+} feedback regulation (dashed, green line). The horizontal, dashed, orange line represents an unchanging CL as might be encountered in the absence of feedback. B. Time course of PDE activity produced by 2.5 or 60 s steps of light in the absence of any slow phase of adaptation (Eq. 3). At the end of the step, activity decreases exponentially with the same time constant as after a flash, τ_c. C. Dependence of saturation time on the flash strength. Saturation time is given as the time to the crossing of CL in the presence (+FB) or absence (-FB) of feedback in A. The relationship is linear with slope τ_c for flash strengths in which the PDE activity intersects a constant CL. The reduction in the saturation time at a given flash strength, ΔT, due to feedback onto CL corresponds to M, the increase in flash strength required to produce a particular saturation time. M expresses the degree of adaptation. D. Dependence of saturation time on step intensity. Step response saturation time is defined as the interval between the end of the step and the intersection of the PDE activity with CL in B. The time in saturation after a 60 s step of a given intensity (squares) exceeds that for a 2.5 s step (circles) since 2.5 s was insufficient for PDE activity to attain steady state. A color version of this figure is included in the insert that precedes Chapter 1.

produce higher activity. As the activity declines exponentially it eventually crosses the CL a second time, after which the circulating current begins to recover. For a constant CL (horizontal, dashed, orange line, Fig. 3A) saturation time rises linearly with the natural logarithm of the flash strength (orange squares, Fig. 3C) and the slope of the relation gives the time constant of the shutoff of PDE activity, τ_c.[21] However, the CL is not static after a light impulse. The CL is largely set by the rate of cGMP synthesis by guanylate cyclase and the sensitivity of the channel for cGMP, both of which are targets for Ca^{2+} feedback. As a result, the CL increases dynamically in response to the light-induced Ca^{2+} decline (dashed green line, Fig. 3A) such that the PDE activity meets the CL sooner than if the CL had remained at the dark-adapted level. Therefore, when feedback is operational, saturation time is shorter for a given flash strength than it is in the absence of feedback (Fig. 3C). Furthermore, at flash strengths where the PDE activity meets the CL before it has stabilized, the relation between saturation time and ln(flash strength) is shallower. Once internal Ca^{2+} reaches a minimum or falls outside the range over which guanylate cyclase activity and channel cGMP sensitivity are affected, the CL stabilizes and saturation time increases linearly with ln(flash strength) with slope τ_c. An example of this behavior is shown in Figure 4A,B. A steady CL was established at a ln(flash strength) of about 7 photons μm^{-2} as indicated by the saturation time at which the relation "peeled away" from the linear region. On average, the CL stabilized after saturation times of 6.1 ± 0.5 s (mean ± SEM, n=9)[13] (cf. refs. 20-22).

The magnitude, M, of Ca^{2+}-dependent rod desensitization after a bright flash is defined by the rightward shift of the saturation time on the ln(flash strength) axis caused by Ca^{2+} feedback (Fig. 3C). For an invariant τ_c, M is related to the difference in saturation time, ΔT, for a given flash presented in the presence and absence of Ca^{2+}-feedback according to:[20]

$$M = \exp(\Delta T/\tau_c). \tag{5}$$

It is notable that M after flashes is ~14-fold,[20] a far cry from the ~3100-fold observed in rods during continuous illumination.[13]

The saturation behavior elicited by intense steps of light may be interpreted similarly (Fig. 3B,D). A step of light produces PDE activity that rises according to Eq. 3. Given enough time, it will reach a steady level. In the absence of the slow phase of adaptation, this steady level would be maintained for the duration of the step. After the light is extinguished the PDE activity declines exponentially. Here we define saturation time as the time required for PDE activity to cross the CL after the step is turned off. The slope of the relation between saturation time and ln(step intensity) again reflects the time constant of the decline in PDE activity, τ_c (Fig. 3D), provided that the PDE activity crosses the CL after the latter has stabilized at a constant value. Saturation time would rise with step duration until the PDE activity reaches a steady state and would be constant thereafter. For example, with $k_R = 2\ s^{-1}$ and $k_E = 0.4\ s^{-1}$ the PDE would reach 95% of steady state in 7.9 s whereas after 2.5 s, it would reach only 56% of the steady state level (Eq. 3). Thus, the time in saturation after a light step of 2.5 s duration would be less than after a 60 s step of the same intensity.

Experiments show the opposite to be true! The recovery phases from 2.5 s and 60 s steps of increasing intensity are illustrated in Figure 5. The slopes of the relations between saturation time and the natural logarithm of the step intensity appeared to be similar for 2.5 and 60 s steps, but for a given step intensity, saturation

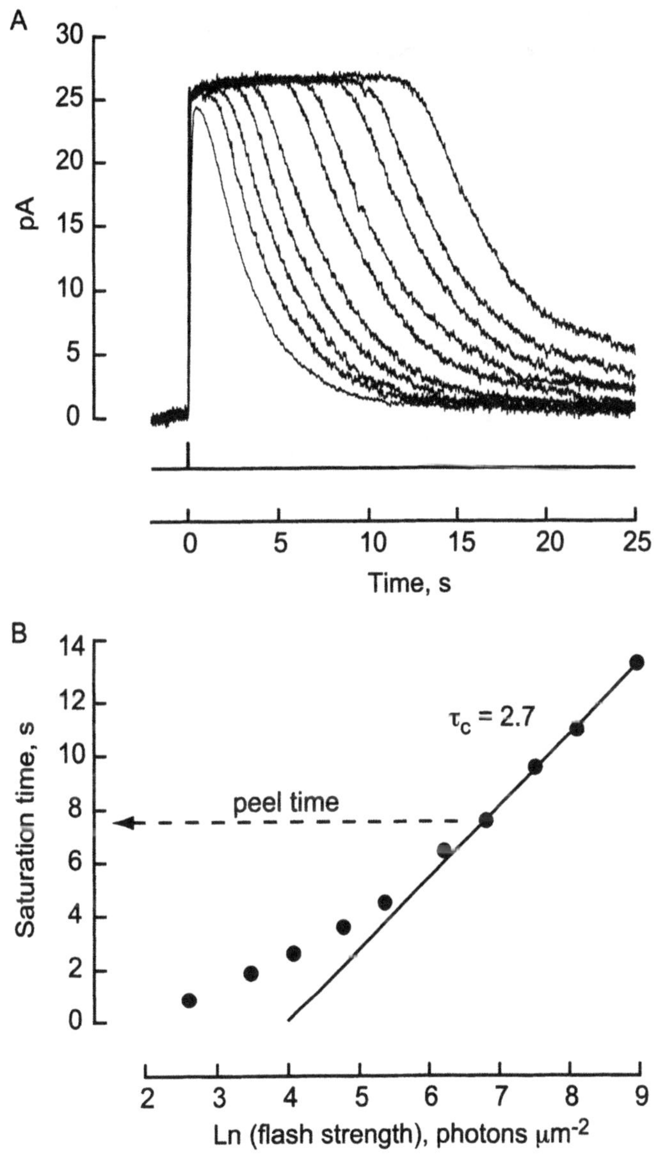

Figure 4. A. Averaged responses of a bullfrog rod to bright flashes. B. Saturation times plotted against the natural logarithm of the flash strength. In practice, it is difficult to determine the moment that the response emerges from saturation, so saturation time was measured as the interval from the flash to 10% recovery of the circulating current. The solid line is a linear regression of the saturation times elicited by the three brightest flashes. The dashed arrow indicates the "peel time", defined as the shortest saturation time that lies on the linear regression.

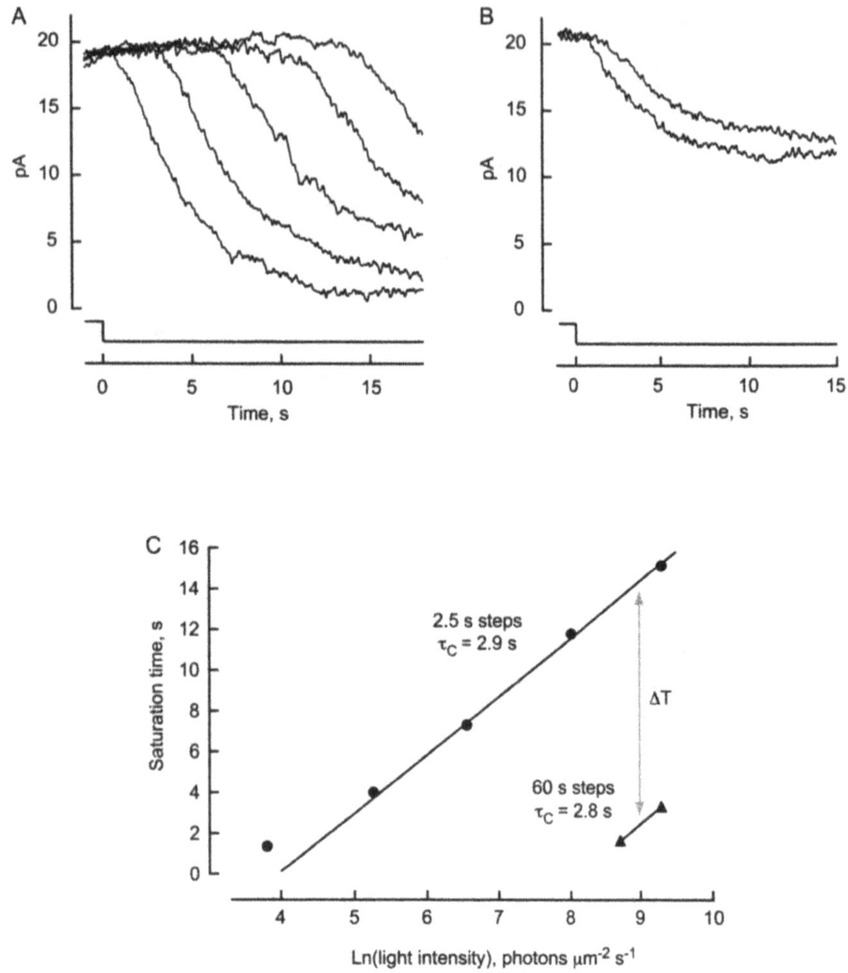

Figure 5. Saturation behavior of a frog rod to short, 2.5 s (A) and long, 60 s (B) steps of bright light. Only the final 2 s of the light exposure and the initial response recoveries are shown. C. Relations between saturation time and ln(light intensity) for 2.5 s and 60 s steps, from A and B. The difference in saturation time after the two sets of steps over the linear regions (gray vertical line) was 11.8 s. Then from Eq. 5, the magnitude of adaptation occurring between 2.5 and 60 s of illumination was 68-fold. After correction for the pre-steady state PDE activity attained during the 2.5 s step (Eq. 3), the magnitude of adaptation operating between 2.5 s and 60 s was 140-fold.

time was much greater after the short step than after the long step (Fig. 5C). This strongly suggests that a molecular mechanism(s) underlying light adaptation engaged between 2.5 and 60 s of continuous illumination. A simple interpretation of these observations is that PDE activity was not maintained at a steady state level during long steps, but instead slowly declined with time in the light (Fig. 6).

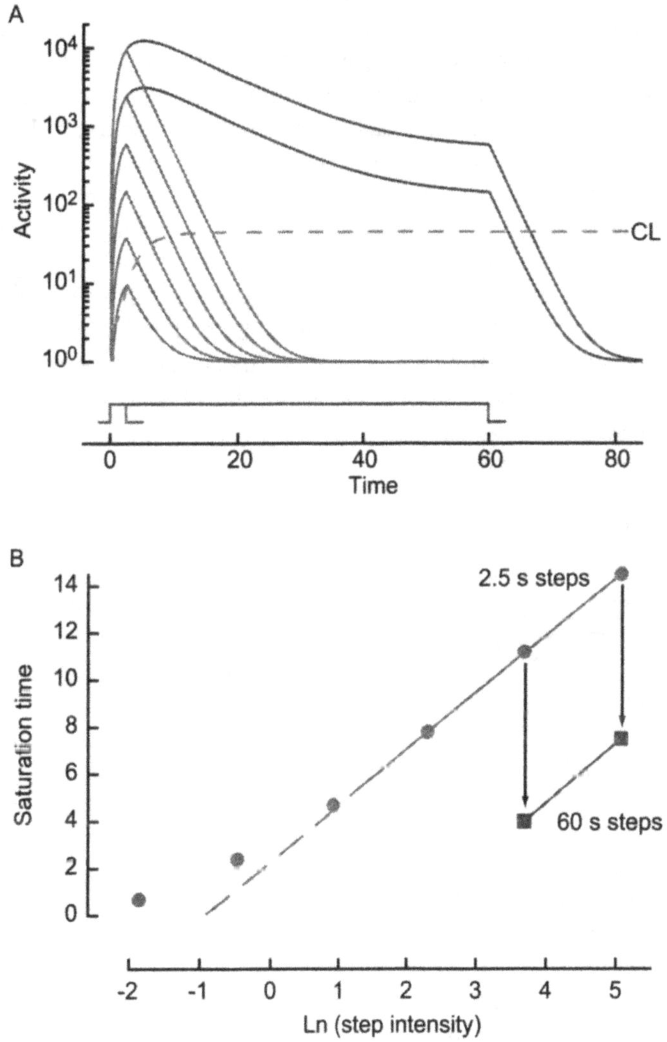

Figure 6. Revised model of adaptation to continuous light. A. Time courses of PDE activity (continuous lines) in response to 2.5 and 60 s of illumination. Six illumination levels are shown for 2.5 s steps and two for 60 s steps. The change in CL is the same as that during a flash response (see Fig. 3). The PDE activity gradually declines with prolonged, 60 s illumination due to the slow phase of adaptation. Nevertheless, the kinetics of PDE decline are the same after the extinction of 2.5 and 60 s steps. B. Relation between saturation time after step shutoff and the natural logarithm of the step intensity. The slopes of the relations are the same for short and for long steps but the saturation time at a given intensity is shorter for the long step. A color version of this figure is included in the insert that precedes Chapter 1.

HOW RAPIDLY CAN ADAPTATION OPERATE?

The peel time observed in the relation between saturation time and ln (flash strength) (Fig. 4) suggests that adaptive changes in CL after bright flashes are complete within 5-8 s. Therefore the underlying fast phase process(es) must have a time constant of approximately 1-2 s. The change in saturation time with step duration for steps of fixed intensity provided a means for probing the time course of the slow phase (Fig. 7). Step intensities were selected so that the total time in saturation (step duration plus saturation time) would exceed ~9 s in order to isolate slow phase mechanisms from any changes in CL produced by flashes. The saturation time after the longest step was subtracted from that after each of the shorter steps and the result was used to calculate M, the magnitude of desensitization (Eq. 5). After correcting for pre-steady state PDE activity according to Eq. 3 (Fig. 7B, inset), sensitivity fell exponentially with step duration (Fig. 7B). The average time constant was 8.7 s, similar to the value obtained using the approach in Fig. 2.[13] Thus, during continuous illumination the slow phase of adaptation develops ~8-fold more slowly than the process(es) responsible for desensitization after bright flashes.

WHAT IS THE TOTAL DESENSITIZATION?

Since the τ_c appears unchanged by step duration, the magnitude of desensitization that operates between steps of 2.5 and 60 s may be calculated according to Eq. 5. In this case ΔT is the difference in saturation time between the 2.5 and 60 s steps at a given intensity (Fig. 5C) and M must be corrected for the pre-steady state PDE activity attained during the 2.5 s step by a factor of $F(\infty)/F(2.5)$. For the cell in Fig. 5C, M was ~140-fold. The average magnitude of adaptation operating during continuous illumination was 77-fold,[13] comparable to the magnitude of the fast phase given above.

The magnitude of 77-fold for the slow phase may be an underestimate, because the method was designed to exclude *any* adaptation that operates during the first 2.5 s of illumination. If the slow phase initiated at light onset with a time constant of ~9 s, then its full magnitude would be approximately 1.3 times larger than its magnitude at 2.5 s, i.e., 100-fold. Hence the total desensitization due to both phases may be about 7800-fold. While the rod clearly does not apply the full desensitization towards partial recovery from saturation and extension of its operating range, the desensitization occurring after saturation may help the rod to recover its circulating current more rapidly following a prolonged exposure to very bright light.

WHAT ARE THE MECHANISMS OF FAST ADAPTATION?

Ca^{2+} feedback is essential for light adaptation. If intracellular Ca^{2+} is prevented from changing in a rod, neither the fast nor the slow phase of light adaptation appears to develop (Fig. 8).[16,23,24] The molecular details of Ca^{2+} regulation of three

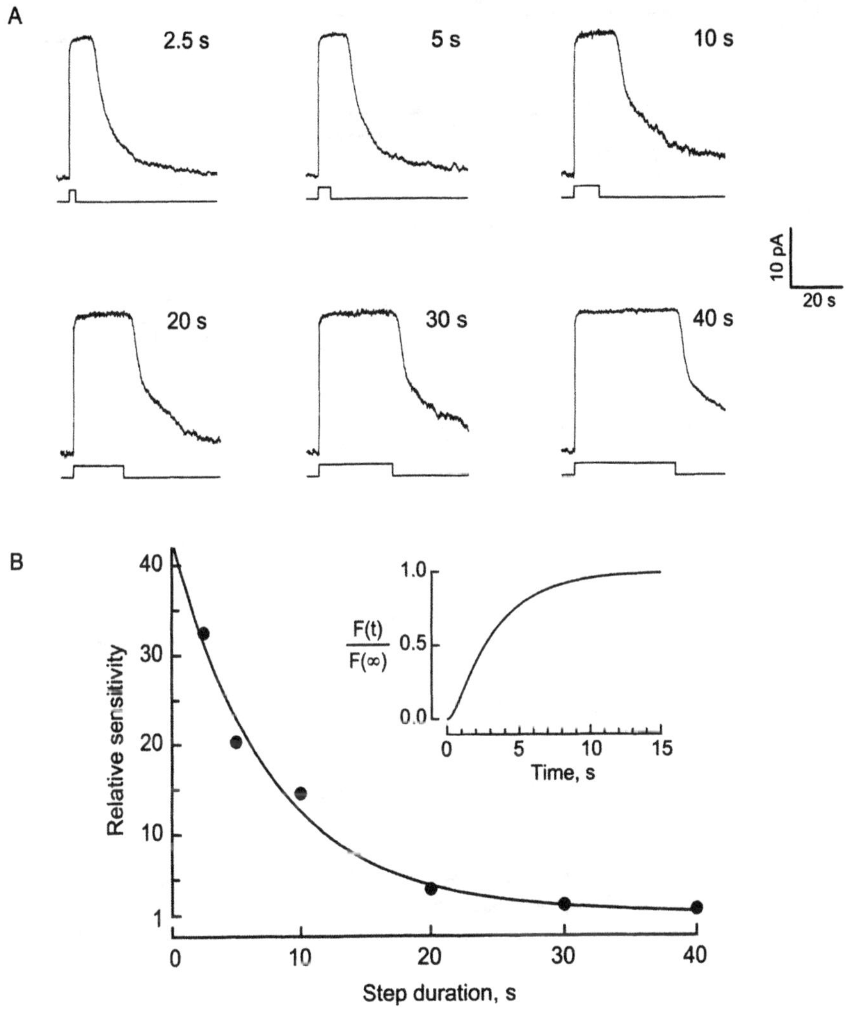

Figure 7. Time course of the slow phase of light adaptation in a frog rod. A. Averaged responses to bright steps of varying duration at a constant intensity. Step durations are listed in the upper right corner of each panel. B. Exponential fall in sensitivity with step duration. The overall magnitude was 42-fold and the time constant was 8 s. Inset: Time course for the light induced change in PDE activity relative to that at steady state, as a function of time after light onset (Eq. 3). $k_R = 2\ s^{-1}$, τ_c determined for this rod was 2.7 s so $k_E = 1/\tau_c = 0.37\ s^{-1}$.

feedback targets: PDE, GC and cGMP-gated channels are described in other Chapters. The magnitude of Ca^{2+}-dependent regulation of light-stimulated PDE activity is thought to be 2-10-fold.[12,25,26] The rate of cGMP synthesis was estimated to change 5-30-fold over the physiological range of Ca^{2+} (refs. 26-28) while regulation of the sensitivity of the channel to cGMP was estimated to change 2-7-fold.[29-31] The exact magnitude of each mechanism under physiological conditions is not known, in part,

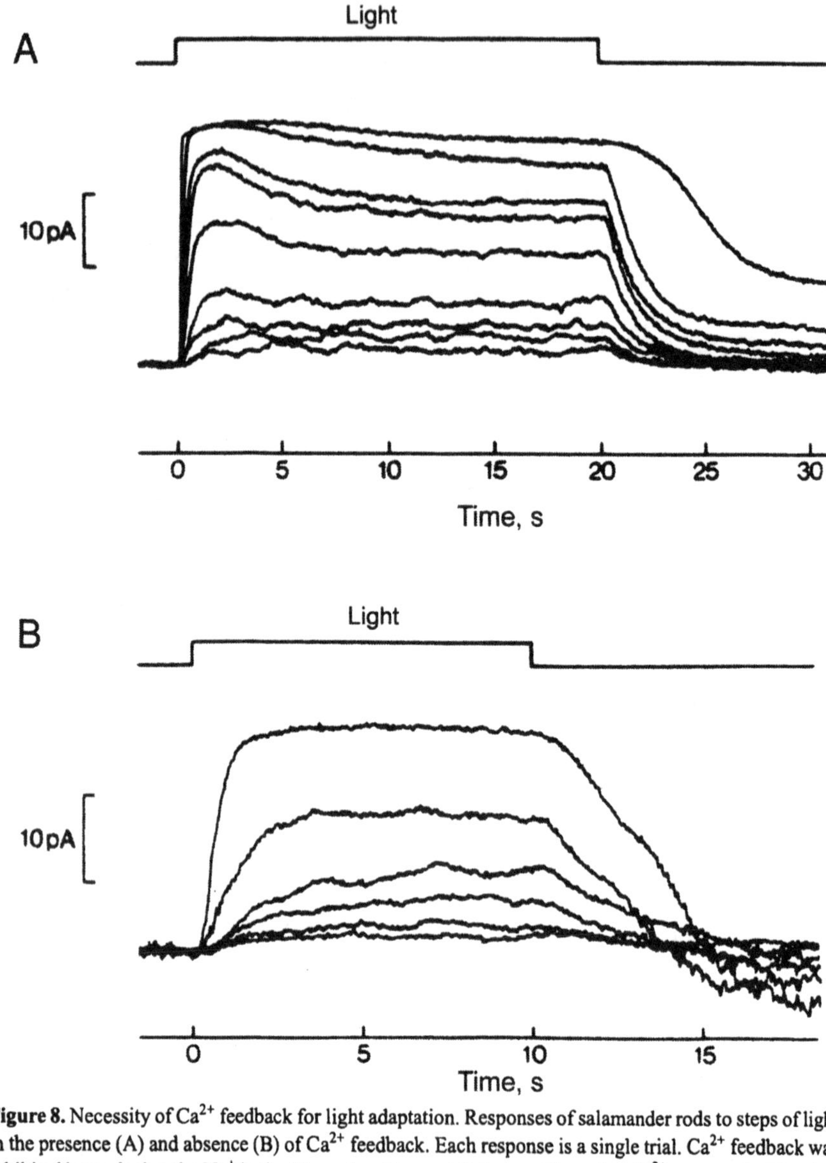

Figure 8. Necessity of Ca^{2+} feedback for light adaptation. Responses of salamander rods to steps of light in the presence (A) and absence (B) of Ca^{2+} feedback. Each response is a single trial. Ca^{2+} feedback was inhibited by replacing the Na^+ in the Ringer's with guanidinium and lowering Ca^{2+} to a nominal level of 0-5 μM. In this solution, the rod failed to light adapt; it was more sensitive to light and the response did not sag over time. Adapted with permission from: Nakatani K, Yau K-W. Nature 1988; 334:69-71. © 1988 MacMillan Magazines Limited.

because there are uncertainties in the concentrations of Ca^{2+} in the rod outer segment in darkness and in saturating light. But taking the product[12,13] of the largest values for the magnitudes of each mechanism, these Ca^{2+} feedbacks could at best, provide for an ensemble ~2100-fold desensitization, falling somewhat short of the total ~7800-fold required. In the discussion to follow, we present arguments for why these Ca^{2+} feedbacks are likely to mediate only the fast phase of light adaptation.

The rate at which a Ca^{2+} feedback reaction engages after the onset of light depends on how fast Ca^{2+} changes, on how fast the feedback target senses that change and on the kinetic parameters of the regulation of the target by Ca^{2+}. The kinetics of Ca^{2+} removal from the rod are dictated in part by the $Na^+/K^+,Ca^{2+}$ exchanger. Four Na^+ enter and one K^+ exits, each time a Ca^{2+} is extruded, so there is a net inward current carried by the exchanger that allows it to be monitored during suction electrode recording of the rod's circulating current. The exchange current can be isolated from the current passing through the cGMP-gated channels after the rapid closure of the channels by presentation of bright light[32]. In frog rods we observed a decline in the exchanger current along a double exponential time course; on average ~70% decayed with a time constant of 0.6 s and ~30% with a time constant of 6 s[13] (e.g., Fig. 9). McCarthy et al[33] were able to distinguish three temporal phases of Ca^{2+} decline in frog rods. In their experiments ~48% of the exchanger current decayed with a time constant of 0.25 s, ~28% decayed with a time constant of 1.35 s and the remainder decayed with a time constant of 6.75 s. Simultaneous measurement of cytoplasmic Ca^{2+} decline using Ca^{2+}-indicator dyes showed that the kinetics of the decay of the exchange current provided a reasonable approximation of the average kinetics of Ca^{2+} decline in the rod.[33,34]

A fall in Ca^{2+} is sensed by GCAP quite rapidly. Using a stopped flow apparatus, Sokal et al[35] estimated a k_{-1} for Ca^{2+} of 37.8 s^{-1} at 25° C by monitoring the fluorescence changes in a mutant GCAP1 containing tryptophan reporter groups. The times required for recoverin/rhodopsin kinase and GCAPs/guanylate cyclase to respond to changes in Ca^{2+} were tested biochemically by measuring rhodopsin phosphorylation and cGMP synthesis in high Ca^{2+} (10 μM), in low Ca^{2+} (10 nM) and after a stepped change from high to low Ca^{2+} (ref. 28). The rates of rhodopsin phosphorylation and cGMP synthesis increased after Ca^{2+} was lowered with subsecond time constants (Fig. 10). Thus recoverin and GCAPs react very quickly to a fall in Ca^{2+}. The kinetics of Ca^{2+} regulation of the cGMP-gated channel have not been measured for physiological levels of Ca^{2+}. However, in internally perfused, truncated rod outer segments, the channel's sensitivity to cGMP appeared to change with a $T_{1/2}$ of 7 s after a step change in internal Ca^{2+} from 1 μM to nominally zero Ca^{2+} (ref. 36). The time course of Ca^{2+} regulation of the channel did not change when the starting Ca^{2+} concentration was increased from 1 to 50 μM, suggesting that the measurement was not hindered by dialyzation time. The apparent lack of delay in the response of the channel suggests that Ca^{2+} feedback on the channel initiates rapidly. Furthermore, in the intact rod, the close proximity of the exchanger to the channel[37,38] causes the Ca^{2+} concentration localized around the channel to fall even more quickly than in the bulk cytoplasm (see also Ref. 33). Thus channel regulation is likely to be faster than expected from the $T_{1/2}$ of 7 s observed in truncated rod outer segments.

The speed with which a Ca^{2+} feedback target responds to changes in Ca^{2+} in a living cell also depends on its $K_{1/2}$ for Ca^{2+} and the resting level of Ca^{2+} in darkness. The more disparate the two values, the longer it will take for feedback to operate.

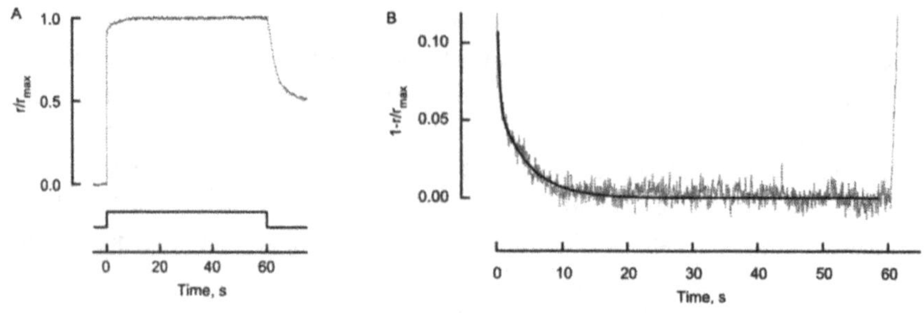

Figure 9. Kinetics of $Na^+/K^+,Ca^{2+}$ exchange in a frog rod. A. The response to a 60 s, saturating step. B. The peak of the response was inverted and enlarged (noisy trace) to show the decline in Ca^{2+} effected by the $Na^+/K^+,Ca^{2+}$ exchanger. The exchange current declined as the sum of two exponentials (smooth line) with time constants of 0.4 and 5 s.

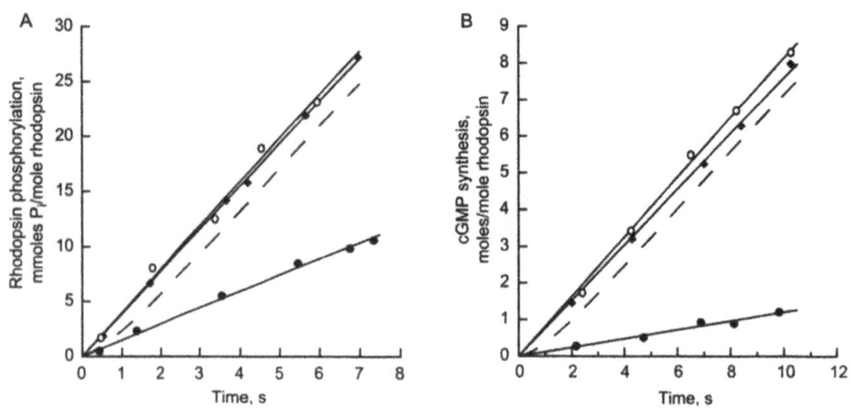

Figure 10. The rapidity of Ca^{2+} regulation on rhodopsin phosphorylation (A) and cGMP synthesis (B) in rod outer segment homogenates from frog. Rhodopsin phosphorylation was initiated after a full bleach with the addition of [γ-^{32}P] ATP. Formation of cGMP was initiated with the addition of [α-^{33}P]GTP. Enzymatic activities were measured at high Ca^{2+} (●), low Ca^{2+} (○) and as Ca^{2+} was switched from high to low levels by the simultaneous addition of 1 mM EGTA (◆). The reactions at high and low Ca^{2+} progressed linearly with time. The dashed lines were generated according to:

$$y(t) = \alpha_{high} t + (\alpha_{low} - \alpha_{high}) \int_0^t (1 - e^{-t/\tau}) dt, \quad (6)$$

where α_{high} and α_{low} are the enzymatic rates at high Ca^{2+} and low Ca^{2+}, respectively, and $\tau = 1$ s. Thus the transitions of both enzymatic activities were significantly faster than a second. Reproduced with permission from: Calvert PD, Ho TW, LeFebvre YM, Arshavsky VY. J Gen Physiol 1998; 111: 39-51. © 1998 The Rockefeller University Press.

However, physiological determinations of the onset rates of Ca^{2+} feedbacks do not require any knowledge of the Ca^{2+} levels and have been carried out in intact rods. Light adaptation commences most rapidly upon exposure to bright, saturating light, where the removal of Ca^{2+} by the $Na^+/K^+,Ca^{2+}$ exchanger is unopposed by Ca^{2+}

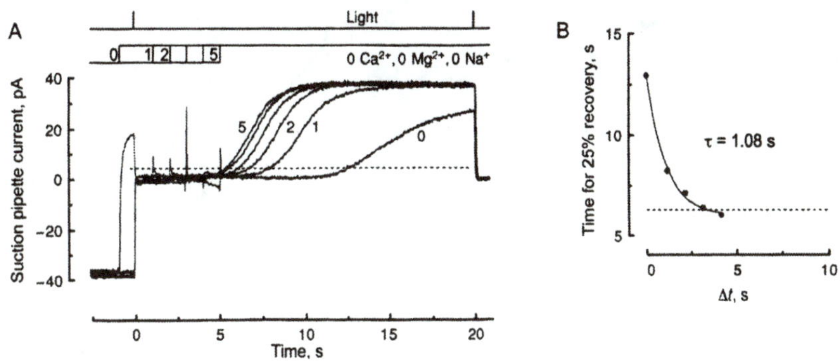

Figure 11. Rapidity of CL change in a salamander rod during light adaptation. A. At various times relative to a bright flash, the medium bathing the rod was rapidly switched to one lacking Na^+, Mg^{2+} and Ca^{2+} in order to inhibit $Na^+/K^+,Ca^{2+}$ exchange and essentially "clamp" internal Ca^{2+} after different levels of light-induced decline. The photoresponses inverted because K^+ became the major permeant ion, exiting through the cGMP-gated channels. Delaying the switch to 0 Ca^{2+}, 0 Mg^{2+}, 0 Na^+ solution permitted increasing Ca^{2+} feedback onto CL and shortened saturation time. A second flash presented after 20 s confirmed that recovery from the first flash was due to the reopening of light-sensitive channels. The dashed line demarcates 25% recovery of the circulating current. B. The exponential decline in saturation time with Δt, the delay between the first test flash and the switch in bathing medium, mirrors the time course of feedback onto CL. The dashed line shows the time required for 25% recovery in Ringer's solution. Adapted from: Matthews HR. J Physiol 1996; 490.1: 1-15. © 1996 The Physiological Society.

influx. The time course of the change in CL was tracked by effectively halting Ca^{2+} feedback onto guanylate cyclase and the cGMP-gated channel at various intervals after a saturating flash and then measuring the saturation time of the response to that flash[25] (Fig. 11). While this procedure also halts feedback onto R* lifetime, the magnitude of this effect is relatively small (see below). These experiments suggest that CL changes with a time constant of 1.1 s, in good agreement with the peel time analysis discussed above. The time course for regulation of light-induced PDE activity was assessed from the change in saturation time in the response to a test flash, delivered at different times after a bright conditioning flash[22] (Fig. 12A,B). Adaptation proceeds faster than the change in saturation time because the magnitude of desensitization is an exponential function of the difference in saturation time (Eq. 5). After replotting the results from Figure 12B, the time constant for the onset of light adaptation due to the reduction in light-induced PDE activity is 1.2 s (Fig. 12C). Taken together it is apparent that Ca^{2+} feedbacks onto PDE activity and CL respond promptly to the onset of light and participate in the fast phase of light adaptation.

The nature of the feedbacks onto PDE activity and the CL are fundamentally different. The target for feedback onto PDE activity disappears with a time constant < 0.5 s after a brief flash[39] (Fig. 13). This period is short relative to the kinetics of Ca^{2+} decline. Consequently regulation over PDE activity contributes little "feedback" onto the response to a flash. Rather, it "feeds forward" to impact responses to subsequent photon absorptions during repetitive flash stimulation or stimulation

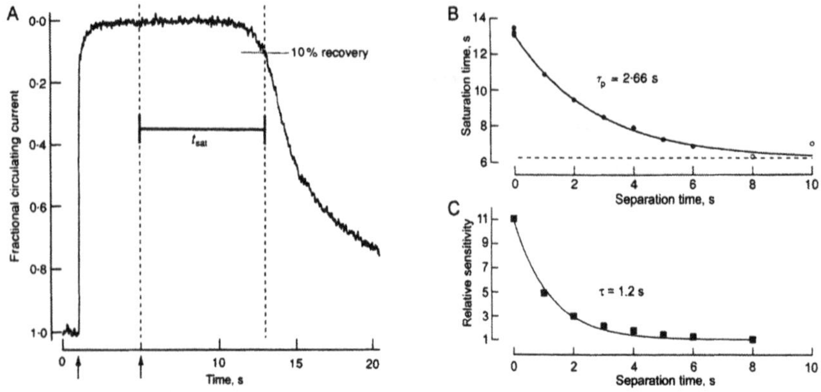

Figure 12. Rapid onset of regulation of light-stimulated PDE activity in a toad rod. A. The rod was conditioned by a bright flash (arrow on the left), and then probed after various intervals with a test flash (arrow on the right). Saturation time was determined as the time between the second flash and 10% recovery of the circulating current. Even the shortest saturation time allowed complete feedback onto CL, so the method isolated the effects of feedback onto light-stimulated PDE activity. B. Saturation time declined exponentially with the increase in time separating the conditioning and test flashes with a time constant of 2.66 s. Open circles were not included in the fit. C. Saturation time was converted to relative sensitivity according to Eq. 5 using a τ_c of 2.9 s. Sensitivity declined exponentially with a time constant of 1.2 s. Panels A and B adapted with permission from: Murnick JG, Lamb TD. J Physiol 1996; 495.1:1-13. © 1996 The Physiological Society.

with continuous light. In contrast, regulation of CL is based strictly on intracellular Ca^{2+} concentration and is independent of illumination. This dichotomy explains why the magnitude of the fast phase of light adaptation during steps is 4 to 5-fold greater than adaptation after a flash.

WHAT UNDERLIES THE SLOW PHASE OF ADAPTATION?

An important clue to the mechanism behind the slow phase of adaptation is that it is present in the responses to subsaturating steps but is absent from the response to bright flashes, even when the flash produces much greater PDE activity and drives Ca^{2+} to a much lower level. The requirement of continuous illumination for the slow adaptation phase implicates a mechanism that regulates a target within the primary transduction cascade.

Actions on the primary cascade that would lead to desensitization include a reduction in the ability of light to form R* or reductions in the catalytic activities or lifetimes of R* or PDE*. Nikonov et al[26] found that for intensities that produce a steady response of less than 0.8 r_{max}, the initial rising phase of the flash response was unchanged. This observation places a powerful constraint on the site of action of the mechanism underlying the slow phase by ruling out a decrement in R* formation or in R* or PDE* catalytic activities. PDE* lifetime is governed by the shutoff of light-activated transducin. A mechanism exists for the regulation of transducin

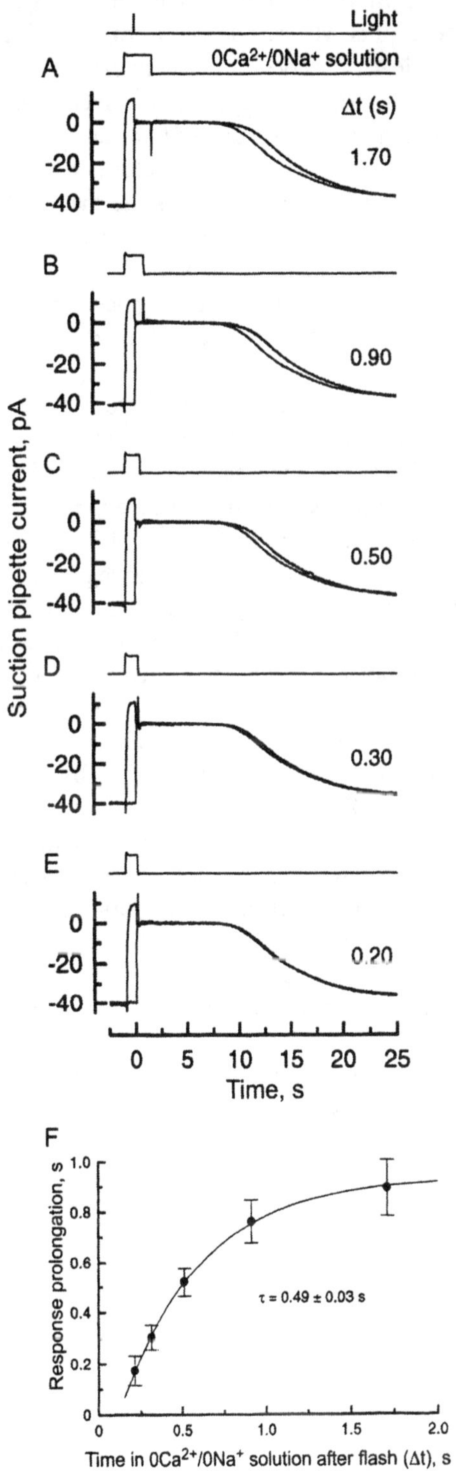

Figure 13. Time span for Ca^{2+} regulation of flash-induced PDE activity. In panels A-E, the response of a salamander rod in normal Ringer's to a saturating flash is superimposed on the response to a flash delivered with the ROS bathed in a modified solution lacking Na^+ and Ca^{2+}. Exposure to the modified solution continued after the flash for the period, Δt, listed on the right side of each panel. Removal of Na^+ and Ca^{2+} disabled the $Na^+/K^+,Ca^{2+}$ exchanger reversibly, largely preventing internal Ca^{2+} from changing at the beginning of the flash response. Switching back to normal Ringer's while the rod was still saturated allowed for full Ca^{2+} feedback onto the CL, so prolongation of the saturation time arose from varying degrees of Ca^{2+} feedback onto PDE activity. F. The increase in saturation time became proportionately shorter as the perfusion time in 0 Na^+, 0 Ca^{2+} lengthened. The continuous line is an exponential fit to the averaged results from 8 rods. Reprinted with permission from: Matthews HR. J Gen Physiol 1997; 109: 141-146. © 1997 The Rockefeller University Press.

shutoff; it is accelerated by the release of cGMP from noncatalytic binding sites on PDE.[28,40,41] However, insignificant amounts of cGMP are released from these sites at light intensities where the slow phase is clearly observed.[13] Furthermore, a shortened transducin lifetime would reduce τ_c, but τ_c was unchanged by Ca^{2+} feedback[20] or by light adaptation[42] in flash response experiments and appeared to be unchanged for short steps and long steps[13] (Figs. 4 and 5). By exclusion, the mechanism mediating the slow phase must target the lifetime of R*, at least for intensities that result in responses $< 0.8\ r_{max}$. Only one Ca^{2+}-dependent molecular mechanism capable of regulating R* lifetime has been described to date, the Ca^{2+}-recoverin inhibition of rhodopsin kinase.[43-45] Yet this mechanism is invoked during the fast phase and the magnitude of this regulation, 3 to 5-fold,[26,28] appears insufficient to explain the magnitude of the slow adaptation phase.

More than one mechanism could be at play during the slow phase. For high intensities producing steady responses greater than $0.8\ r_{max}$, there is no experimental evidence ruling out regulation of R* and PDE* catalytic activities. A gain reduction in light-dependent activation of PDE was reported in physiological experiments on truncated rods internally dialyzed with lower than normal Ca^{2+} (ref. 46). Protein kinase C could have been cast in a leading role. It is present in rod outer segments[47-49] and has been shown to phosphorylate primary excitatory cascade components: rhodopsin,[50,51] arrestin,[52] the α subunit of transducin[53] and PDE.[54] However, flash and step sensitivities are unaffected by perfusion of phorbol ester activators of protein kinase C or a protein kinase C inhibitor.[55]

Biochemical studies suggest several other possibilities. The steady state PDE activity may be lowered by an interaction of PDE with GARP2[56] or by ADP ribosylation[57] and/or phosphorylation of the PDE inhibitory subunit.[58,59] Another intriguing possibility involves RGS9, one of the proteins that set the lifetime of T*. RGS9 undergoes Ca^{2+}-dependent phosphorylation and becomes dephosphorylated in response to light[60] (see the Chapter by Wensel). The effect of phosphorylation has not been determined. Conceivably, it could also lower the overall PDE activity. It is not yet known whether any of these mechanisms are regulated by Ca^{2+}in the physiological range nor is it known how powerful an influence they exert over PDE activity in the intact rod.

The magnitude of the slow phase is greatest for responses that saturate transiently. Longer initial saturation periods lead to greater desensitization as Ca^{2+} drops to lower and lower levels. This interpretation is supported by the correspondence between the time constant for the onset of the slow phase of adaptation and the slower of the two time constants for the Ca^{2+} exchange current. Yet interestingly, the slow phase of adaptation appears to be unperturbed by the eventual reopening of channels which allow the internal Ca^{2+} to partially recover (Fig. 2). Such behavior cannot be explained in terms of a slowly equilibrating process. One explanation may be that the activation and reversal of the process are set in motion by very different conditions. For example, a regulatory site may become phosphorylated at very low Ca^{2+} but may be dephosphorylated only at high Ca^{2+}, after complete recovery of the circulating current. Alternatively, reversal of the slow phase mechanism could be independent of Ca^{2+} altogether.

MODEL OF LIGHT ADAPTATION DURING CONTINUOUS ILLUMINATION

Based on the evidence presented we now provide a model for the time course of light adaptation (refer to Fig. 1). Rapid adaptation is provided by Ca^{2+}-dependent feedback onto mechanisms that underlie the CL, the guanylate cyclase and the cGMP-gated channel. These mechanisms dominate the desensitization at lower response amplitudes and operate during bright flashes as well as steps. After a flash a modest portion of the total adaptation is contributed by regulation of R* lifetime. Adaptation to continuous light also involves regulation of the primary cascade. One mechanism, operating on the lifetime of rhodopsin engages quickly and boosts the magnitude of the fast phase observed after a flash 4 to 5-fold. A second mechanism, likely with the same target, evolves nearly 10-fold more slowly, providing a further 40 to 70-fold adaptation. The result is a slow decline in the activity of light-activated PDE during continuous illumination while the rate of PDE shutoff after termination of the light is unaffected (Fig. 6).

WHY SO MANY MECHANISMS?

Ca^{2+} serves as a messenger for light adaptation, recruiting a corps of feedback mechanisms operating on numerous targets. It is interesting to speculate as to why there are so many mechanisms. Adaptation lowers the rod's step sensitivity by nearly four orders of magnitude, wherein each mechanism contributes roughly an order of magnitude or less. Perhaps it is difficult to gain such an enormous amount of desensitization with any single mechanism. A second consideration is that it may be advantageous to spread out light adaptation over time, particularly if the effects are not rapidly reversible. One wonders whether Ca^{2+} regulates additional phases of light adaptation that operate on even longer time scales. Finally, distributing adaptation over multiple mechanisms provides a margin of safety. Should any one mechanism of adaptation become disabled, e.g., due to a genetic mutation, the other mechanisms would remain intact. Instead of a total abolition of adaptation, there would only be a relatively slight loss in dynamic range. Furthermore, it may be easier to compensate for an expanded set of congenitally defective phototransduction components if regulation exists at multiple sites.

ACKNOWLEDGMENTS

We thank Dr. V. I. Govardovskii for many stimulating and insightful discussions. This work was supported by a Career Development Award from Research to Prevent Blindness (CLM), E. Mathilda Ziegler Foundation for the Blind, Inc. (CLM) and the National Eye Institute: EY12944 (CLM), EY06857 (PDC).

REFERENCES

1. Rodieck RW. The First Steps in Seeing. Sunderland, Massachusetts: Sinauer; 1998.
2. Pugh EN Jr, Lamb TD. Phototransduction in vertebrate rods and cones: Molecular mechanisms of amplification, recovery and light adaptation. In: Stavenga DG, DeGrip WJ, Pugh EN Jr, eds. Handbook of Biological Physics: Molecular Mechanisms in Visual Transduction. Amsterdam: Elsevier Science B. V., 2000:183-255.
3. Roof DJ, Makino CL. The structure and function of retinal photoreceptors. In: Albert DM, Jakobiec FA, eds. Principles and Practice of Ophthalmology. second ed. Philadelphia: W. B. Saunders Co, 2000:1624-1673.
4. Fain GL, Matthews HR, Cornwall MC et al. Adaptation in vertebrate photoreceptors. Physiol Rev 2001; 81:117-151.
5. Leskov IB, Klenchin VA, Handy JW et al. The gain of rod phototransduction: reconciliation of biochemical and electrophysiological measurements. Neuron 2000; 27:525-537.
6. Heck M, Hofmann KP. Maximal rate and nucleotide dependence of rhodopsin-catalyzed transducin activation: initial rate analysis based on a double displacement mechanism. J Biol Chem 2001; 276:10000-10009.
7. Bownds MD, Arshavsky VY. What are the mechanisms of photoreceptor adaptation? Behav Brain Sci 1995; 18:415-424.
8. Pugh EN Jr, Nikonov S, Lamb TD. Molecular mechanisms of vertebrate photoreceptor light adaptation. Curr Opin Neurobiol 1999; 9:410-418.
9. Burns ME, Baylor DA. Activation, deactivation, and adaptation in vertebrate photoreceptor cells. Annu Rev Neurosci 2001; 24:779-805.
10. Forti S, Menini A, Rispoli G et al. Kinetics of phototransduction in retinal rods of the newt Triturus cristatus. J Physiol (Lond) 1989; 419:265-295.
11. Matthews HR, Fain GL, Murphy RLW et al. Light adaptation in cone photoreceptors of the salamander: A role for cytoplasmic calcium. J Physiol (Lond) 1990; 420:447-469.
12. Koutalos Y, Nakatani K, Yau K-W. The cGMP-phosphodiesterase and its contribution to sensitivity regulation in retinal rods. J Gen Physiol 1995; 106:891-921.
13. Calvert PD, Govardovskii VI, Arshavsky VY et al. Two temporal phases of light adaptation in retinal rods. J Gen Physiol 2002; 119:129-145.
14. Baylor DA, Lamb TD, Yau K-W. Responses of retinal rods to single photons. J Physiol (Lond) 1979; 288:613-634.
15. Fain GL, Lamb TD, Matthews HR et al. Cytoplasmic calcium as the messenger for light adaptation in salamander rods. J Physiol (Lond) 1989; 416:215-243.
16. Matthews HR, Murphy RLW, Fain GL et al. Photoreceptor light adaptation is mediated by cytoplasmic calcium concentration. Nature 1988; 334:67-69.
17. Mendez A, Burns ME, Sokal I et al. Role of guanylate cyclase-activating proteins (GCAPs) in setting the flash sensitivity of rod photoreceptors. Proc Natl Acad Sci USA 2001; 98:9948-9953.
18. Nikonov S, Engheta N, Pugh EN Jr. Kinetics of recovery of the dark-adapted salamander rod photoresponse. J Gen Physiol 1998; 111:7-37.
19. Rieke F, Baylor DA. Molecular origin of continuous dark noise in rod photoreceptors. Biophys J 1996; 71:2553-2572.
20. Lyubarsky A, Nikonov S, Pugh EN Jr. The kinetics of inactivation of the rod phototransduction cascade with constant $Ca^{2+}{}_i$. J Gen Physiol 1996; 107:19-34.
21. Pepperberg DR, Cornwall MC, Kahlert M et al. Light-dependent delay in the falling phase of the retinal rod photoresponse. Vis Neurosci 1992; 8:9-18.
22. Murnick JG, Lamb TD. Kinetics of desensitization induced by saturating flashes in toad and salamander rods. J Physiol (Lond) 1996; 495:1-13.
23. Nakatani K, Yau K-W. Calcium and light adaptation in retinal rods and cones. Nature 1988; 334:69-71.

24. Calvert PD, Govardovskii VI, Makino CL. Potent modulation of phototransduction amplification dominates adaptation of vertebrate rods to prolonged bright illumination. Invest Ophthalmol Vis Sci 2000; 41:S112.
25. Matthews HR. Static and dynamic actions of cytoplasmic Ca^{2+} in the adaptation of responses to saturating flashes in salamander rods. J Physiol (Lond) 1996; 490:1-15.
26. Nikonov S, Lamb TD, Pugh EN Jr. The role of steady phosphodiesterase activity in the kinetics and sensitivity of the light-adapted salamander rod photoresponse. J Gen Physiol 2000; 116:795-824.
27. Koutalos Y, Nakatani K, Tamura T et al. Characterization of guanylate cyclase activity in single retinal rod outer segments. J Gen Physiol 1995; 106:863-890.
28. Calvert PD, Ho TW, LeFebvre YM et al. Onset of feedback reactions underlying vertebrate rod photoreceptor light adaptation. J Gen Physiol 1998; 111:39-51.
29. Gordon SE, Downing-Park J, Zimmerman AL. Modulation of the cGMP-gated ion channel in frog rods by calmodulin and an endogenous inhibitory factor. J Physiol (Lond) 1995; 486:533-546.
30. Hsu Y-T, Molday RS. Modulation of the cGMP-gated channel of rod photoreceptor cells by calmodulin. Nature 1993; 361:76-79.
31. Nakatani K, Koutalos Y, Yau K-W. Ca^{2+} modulation of the cGMP-gated channel of bullfrog retinal rod photoreceptors. J Physiol (Lond) 1995; 484:69-76.
32. Yau K-W, Nakatani K. Electrogenic Na-Ca exchange in retinal rod outer segment. Nature 1984; 311:661-663.
33. McCarthy ST, Younger JP, Owen WG. Dynamic, spatially nonuniform calcium regulation in frog rods exposed to light. J Neurophysiol 1996; 76:1991-2004.
34. Gray-Keller MP, Detwiler PB. The calcium feedback signal in the phototransduction cascade of vertebrate rods. Neuron 1994; 13:849-861.
35. Sokal I, Otto-Bruc AE, Surgucheva I et al. Conformational changes in guanylyl cyclase-activating protein 1 (GCAP1) and its tryptophan mutants as a function of calcium concentration. J Biol Chem 1999; 274:19829-19837.
36. Sagoo MS, Lagnado L. The action of cytoplasmic calcium on the cGMP-activated channel in salamander rod photoreceptors. J Physiol (Lond) 1996; 497:309-319.
37. Bauer PJ, Drechsler M. Association of cyclic GMP-gated channels and Na^+-Ca^{2+}-K^+ exchangers in bovine retinal rod outer segment plasma membranes. J Physiol (Lond) 1992; 451:109-131.
38. Schwarzer A, Schauf H, Bauer PJ. Binding of the cGMP-gated channel to the Na/Ca-K exchanger in rod photoreceptors. J Biol Chem 2000; 275:13448-13454.
39. Matthews HR. Actions of Ca^{2+} on an early stage in phototransduction revealed by the dynamic fall in Ca^{2+} concentration during the bright flash response. J Gen Physiol 1997; 109:141-146.
40. Arshavsky VY, Gray-Keller MP, Bownds MD. cGMP suppresses GTPase activity of a portion of transducin equimolar to phosphodiesterase in frog rod outer segments. Light-induced cGMP decreases as a putative feedback mechanism of the photoresponse. J Biol Chem 1991; 266:18530-18537.
41. Arshavsky VY, Bownds MD. Regulation of deactivation of photoreceptor G protein by its target enzyme and cGMP. Nature 1992; 357:416-417.
42. Pepperberg DR, Jin J, Jones GJ. Modulation of transduction gain in light adaptation of retinal rods. Vis Neurosci 1994; 11:53-62.
43. Chen C-K, Inglese J, Lefkowitz RJ et al. Ca^{2+}-dependent interaction of recoverin with rhodopsin kinase. J Biol Chem 1995; 270:18060-18066.
44. Kawamura S. Rhodopsin phosphorylation as a mechanism of cyclic GMP phosphodiesterase regulation by S-modulin. Nature 1993; 362:855-857.
45. Klenchin VA, Calvert PD, Bownds MD. Inhibition of rhodopsin kinase by recoverin. Further evidence for a negative feedback-system in phototransduction. J Biol Chem 1995; 270:16147-16152.

46. Lagnado L, Baylor DA. Calcium controls light-triggered formation of catalytically active rhodopsin. Nature 1994; 367:273-277.
47. Inoue M, Isayama Y. Calcium ion and phospholipid-dependent protein kinase in rod outer segment. Jpn J Ophthalmol 1984; 28:47-56.
48. Kapoor CL, Chader GJ. Endogenous phosphorylation of retinal photoreceptor outer segment proteins by calcium phospholipid-dependent protein kinase. Biochem Biophys Res Com 1984; 122:1397-1403.
49. Wood JG, Hart CE, Mazzei GJ et al. Distribution of protein kinase C immunoreactivity in rat retina. Histochem J 1988; 20:63-68.
50. Kelleher DJ, Johnson GL. Purification of protein kinase C from bovine rod outer segments. J Cyclic Nucleotide Protein Phosphorylation Res 1985; 10:579-591.
51. Newton AC, Williams DS. Involvement of protein kinase C in the phosphorylation of rhodopsin. J Biol Chem 1991; 266:17725-17728.
52. Weyand I, Kuhn H. Subspecies of arrestin from bovine retina. Equal functional binding to photoexcited rhodopsin but various isoelectric focusing phenotypes in individuals. Eur J Biochem 1990; 193:459-467.
53. Zick Y, Sagi-Eisenberg R, Pines M et al. Multisite phosphorylation of the alpha subunit of transducin by the insulin receptor kinase and protein kinase C. Proc Natl Acad Sci USA 1986; 83:9294-9297.
54. Udovichenko IP, Cunnick J, Gonzales K et al. Phosphorylation of bovine rod photoreceptor cyclic GMP phosphodiesterase. Biochem J 1993; 295:49-55.
55. Xiong W-H, Nakatani K, Ye B et al. Protein kinase C activity and light sensitivity of single amphibian rods. J Gen Physiol 1997; 110:441-452.
56. Korschen HG, Beyermann M, Muller F et al. Interaction of glutamic-acid-rich proteins with the cGMP signalling pathway in rod photoreceptors. Nature 1999; 400:761-766.
57. Bondarenko VA, Yamazaki M, Hayashi F et al. Suppression of GTP/Tα-dependent activation of cGMP phosphodiesterase by ADP-ribosylation by its γ subunit in amphibian rod photoreceptor membranes. Biochemistry 1999; 38:7755-7763.
58. Tsuboi S, Matsumoto H, Yamazaki A. Phosphorylation of an inhibitory subunit of cGMP phosphodiesterase in Rana catesbeiana rod photoreceptors. II. A possible mechanism for the turnoff of cGMP phosphodiesterase without GTP hydrolysis. J Biol Chem 1994; 269:15024-15029.
59. Xu LX, Tanaka Y, Bonderenko VA et al. Phosphorylation of the γ subunit of the retinal photoreceptor cGMP phosphodiesterase by the cAMP-dependent protein kinase and its effect on the γ subunit interaction with other proteins. Biochemistry 1998; 37:6205-6213.
60. Hu G, Jang G-F, Cowan CW et al. Phosphorylation of RGS9-1 by an endogenous protein kinase in rod outer segments. J Biol Chem 2001; 276:22287-22295.

S-MODULIN

Satoru Kawamura and Shuji Tachibanaki

ABSTRACT

S-Modulin is a Ca^{2+}-binding protein found in frog rod photoreceptors[1,2] and its bovine homologue is known as recoverin[3,4]. In the Ca^{2+}-bound form, S-modulin inhibits rhodopsin phosphorylation[5] through inhibition of rhodopsin kinase.[6-9] Because rhodopsin phosphorylation is the quench mechanism of light-activated rhodopsin (R*),[10,11] the inhibition of the phosphorylation by S-modulin probably contributes to increase the lifetime of R* to result in sustained hydrolysis of cGMP[5]. The Ca^{2+} concentration decreases in the light in vertebrate photoreceptors,[12-14] and this decrease is essential for light-adaptation.[15,16] Thus, S-modulin is expected to regulate the lifetime of R* and thereby regulate the extent and the time course of hydrolysis of cGMP depending on the intensity of background light. With this mechanism, S-modulin is believed to regulate the waveform of a photoresponse and the efficiency of the light in the generation of a photoresponse.

ISOLATION

The S-modulin activity was first detected in an electrophysiological measurement of the cGMP phosphodiesterase (PDE) activity in a truncated preparation of a frog rod outer segment (ROS) which was internally perfused with a bathing solution.[1] This measurement suggested the presence of a factor that regulates PDE activation in a Ca^{2+}-dependent manner. In the measurement, it was also suggested that the factor binds to the disk membrane at high Ca^{2+} concentrations but becomes soluble at low Ca^{2+} concentrations. This characteristic of Ca^{2+}-dependent binding to disk membranes of the factor was utilized in purification of S-modulin.

Department of Biology, Graduate School of Science, Osaka University, Machikane-yama 1-1, Toyonaka, Osaka 560-0043, Japan.

In our first attempt of purification of S-modulin,[1,2] we isolated proteins that bind to disk membranes at high Ca^{2+} concentrations. S-Modulin, a 26 kDa protein, was purified after anion exchange column chromatography as the fraction that increases PDE activity in a Ca^{2+}-dependent manner. Because the maximum PDE activity was constant irrespective of Ca^{2+} concentrations, S-modulin does not seem to increase the V_{max} of PDE but rather regulate the efficiency of PDE activation.

The characteristic of Ca^{2+}-dependent binding of S-modulin to disk membranes suggested that the binding to the membranes is due to exposure of a hydrophobic region(s) to the surface of S-modulin molecule upon Ca^{2+}-binding. Similar kind of Ca^{2+}-dependent structural changes are thought in a family of proteins that posses Ca^{2+}-binding motifs, EF-hands. In these proteins, a Phenyl Sepharose column is a useful tool to purify them. In our subsequent purification, we used this column instead of disk membranes.[17] Similar procedure is used in purification of recoverin.[18]

In the course of purification of S-modulin, we realized that another 26 kDa protein also binds to disk membranes in a Ca^{2+}-dependent manner.[2] Its chromatographical behavior on an anion exchange column is slightly different from that of S-modulin. Our studies showed that this protein, named s26, is a cone homologue of S-modulin.[19] The molar abundance among rhodopsin, S-modulin and s26 in a bullfrog retina is approximately 100: 7: 5.

FUNCTION

The S-modulin activity was found as a Ca^{2+}-dependent regulator of PDE activation.[1] Because PDE is activated through the phototransduction cascade, the reaction that is regulated by S-modulin should be somewhere in the cascade.

When we measured the effect of S-modulin on rhodopsin phosphorylation, S-modulin inhibited the phosphorylation at high Ca^{2+} concentrations[5]. Rhodopsin phosphorylation is a mechanism to quench light-activated rhodopsin (R*),[10,11] and therefore, the inhibition of the phosphorylation will lead to the increase in the lifetime of R.*[5] If this is the case, the hydrolysis of cGMP continues and a photoresponse develops for a relatively long period. Then, the photoresponse time course is rather slow and the response amplitude is large. Under steady illumination, the Ca^{2+} concentration decreases in the photoreceptor outer segment cytoplasm.[12-14] At low Ca^{2+} concentrations, S-modulin does not inhibit the phosphorylation reaction on rhodopsin. Under steady illumination, therefore, the lifetime of R* seems to be relatively short and the photoresponse terminates more rapidly. Then the photoresponse terminates quickly and the response amplitude is small. This expected regulation of a photoresponse by S-modulin is consistent with the behavior of a photoresponse during dark- and light-adaptation, which is the reason why we believe that S-modulin is involved in the control in photoreceptor adaptation.

The above possible regulation mechanism by S-modulin has been tested in many ways, but the results are still controversial. When recoverin (bovine homologue of S-modulin) was introduced into a salamander ROS with a patch pipette,[20] the photoresponse was altered as one would expect. The time-to-peak of a photoresponse was lengthened and fractional amplitude of a photoresponse increased.

The effect of ATP on a photoresponse was examined in a truncated preparation of an ROS.[21] ATP was introduced into the cytoplasm of an ROS through the truncated end of the preparation. ATP reduced the time-to-peak, duration and amplitude of the flash response. These effects can be explained by quench of R* with phosphorylation. These ATP effects were detected within a few seconds after a light flash at high Ca^{2+} concentrations, but the time was shortened to about 0.5 sec at low Ca^{2+} concentrations. The results showed that the ATP-sensitive step is Ca^{2+}-sensitive. If the ATP-sensitive step is R* phosphorylation, which is very plausible, then the result indicated that the phosphorylation is rapid at low Ca^{2+} concentrations. These observations are consistent with the postulated function of S-modulin.

The cytoplasmic Ca^{2+} concentration in an ROS can be clamped by exposing the ROS to a 0 Ca^{2+}/0 Na^{+} solution. When it was clamped at the dark high level, the duration of a saturating photoresponse was prolonged.[22] The result indicated that the cytoplasmic Ca^{2+} concentration decreases in the light and this decrease shortens the duration of a saturating photoresponse. The prolongation effect was dependent on how long the cytoplasmic Ca^{2+} concentration was kept at the dark level after a light flash.[22] The longer the time was, the larger the effect was. However, the effect was observed in a limited time period after a light flash with a time constant of about 0.5 sec. At about 2 sec after the flash, the clamp effect was saturated. The result showed that only a very brief period after a light flash is Ca^{2+}-sensitive. The responsible reaction would most possibly be R* phosphorylation.

In the study of the recoverin knockout rods, the kinetics of the response were found to be faster and the amplitude was found to be smaller than in control rods.[23] When recoverin was internally dialyzed into the truncated ROS, recoverin prolonged the recovery phase of a bright flash response.[24] These observations are consistent with the postulated function of S-modulin, but the effect was not so large. So far, biochemical attempt to detect the S-modulin effect on R* phosphorylation in intact rod photoreceptors has not been successful.[25]

As summarized in the earlier chapters in this book, Ca^{2+} seems to regulate photoreceptor adaptation in many ways. From quantitative studies estimating relative contribution of these regulations, the effect of S-modulin has been suggested to be small at weak background light and becomes large at strong background light.[26] The relative contribution of S-modulin to light-adaptation has yet to be determined.[27]

MECHANISM OF INHIBITION OF RHODOPSIN PHOSPHORYLATION

There are at least two possible mechanisms that account for the inhibition of rhodopsin phosphorylation by S-modulin. One possibility is the binding of Ca^{2+}-bound form of S-modulin (Ca^{2+}/S-modulin) to R* to interfere the access of rhodopsin kinase to R*. The other possibility is the binding of Ca^{2+}/S-modulin to rhodopsin kinase to inhibit the kinase activity.[5]

To identify the target of S-modulin, a reagent was used to link Ca^{2+}/S-modulin and its target molecule.[28] A 60 kDa protein was linked to S-modulin in a Ca^{2+}-dependent manner. In contrast, rhodopsin was not linked at all. The result suggested that Ca^{2+}/S-modulin binds to rhodopsin kinase to inhibit rhodopsin phosphorylation.

All the other studies from other laboratories reached the same conclusion. With affinity chromatography, Ca^{2+}/recoverin has been shown to bind to rhodopsin kinase.[6,7] The inhibition of the phosphorylation by Ca^{2+}/recoverin does not depend on the ratio of R* : recoverin,[8] which excludes the possibility that recoverin binds to R*. In addition, it has been shown that β-adrenergic receptor kinase, which can phosphorylate R*, is not inhibited by recoverin,[9] which indicated that there is a specific interaction between Ca^{2+}/S-modulin with rhodopsin kinase.

SITE OF THE INTERACTION OF S-MODULIN TO RHODOPSIN KINASE

The S-modulin site interacting with rhodopsin kinase has been identified.[29] In this study, firstly the S-modulin peptides that show an S-modulin effect (inhibition of R* phosphorylation) were identified. Because S-modulin family proteins so far examined all inhibited R* phosphorylation, the interaction site of S-modulin should be conserved among the proteins of the family. The conserved amino acids in the effective peptides were determined and they were mutated with one mutation in each of S-modulin mutant proteins. S-Modulin changes its conformation upon Ca^{2+}-binding.[30-32] In the mutants generated, some of them were not able to change the protein conformation upon Ca^{2+}-binding. These amino acids are therefore essential in the Ca^{2+}-dependent conformational change of S-modulin.

With careful determination, several amino acids were identified as the interaction site with rhodopsin kinase (Fig. 1). These amino acids are buried in the inactive, Ca^{2+}-free form of S-modulin molecule, but they are exposed to the surface of the molecule in the active, Ca^{2+}-bond form of S-modulin. These amino acids are located at the surface of a groove.

So far many S-modulin-like proteins were found. All of the family proteins examined inhibited rhodopsin phosphorylation in a Ca^{2+}-dependent manner.[33] The result suggests that all of the family members have similar sites interacting with rhodopsin kinase and that the mechanism of the actions of these proteins is similar to that of the S-modulin-rhodopsin kinase system.

STRUCTURE

Visinin is the protein first described in the S-modulin family,[34] although its function was not known at the time of the finding. S-Modulin, recoverin and visinin contain four potential Ca^{2+}-binding motifs known as EF-hands.[3,4] However, only EF2 and 3 (numbered from the N-terminus) are functional.[35] The N-terminus glycine is modified with a lipid, mainly myristic acid.[36] This myristoylation is essential

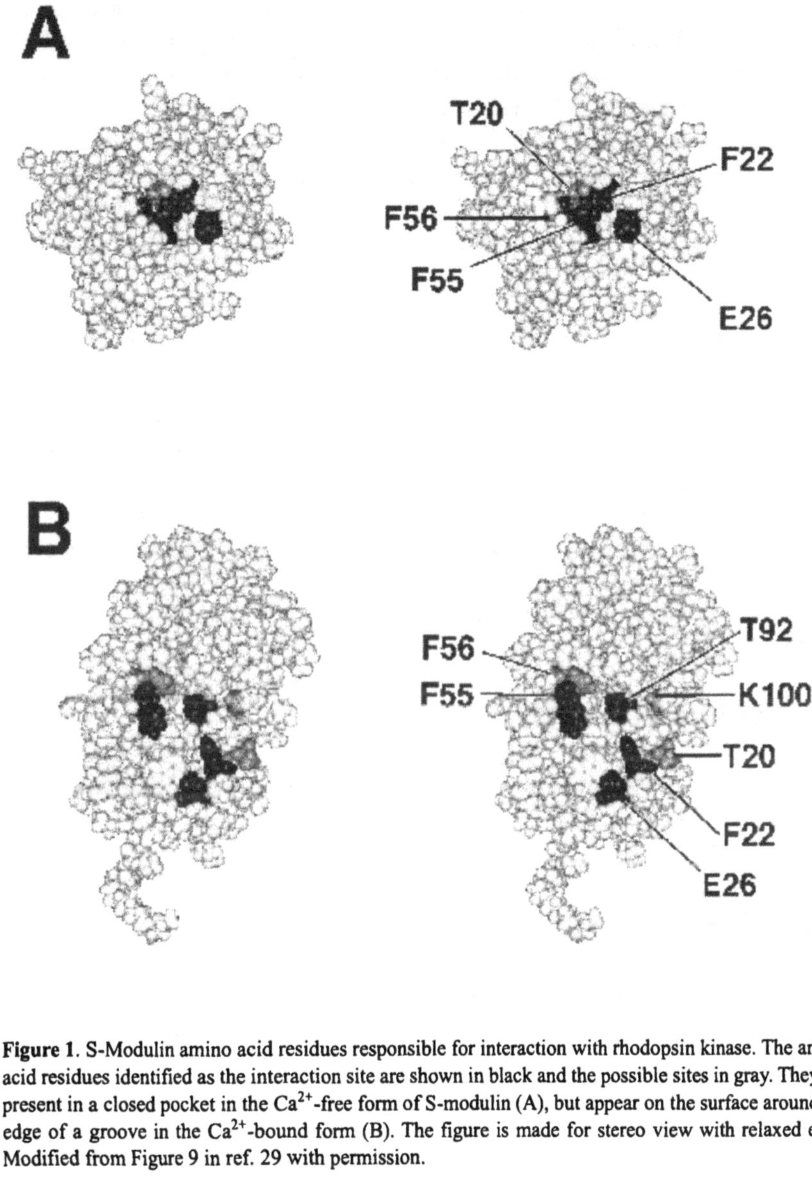

Figure 1. S-Modulin amino acid residues responsible for interaction with rhodopsin kinase. The amino acid residues identified as the interaction site are shown in black and the possible sites in gray. They are present in a closed pocket in the Ca^{2+}-free form of S-modulin (A), but appear on the surface around the edge of a groove in the Ca^{2+}-bound form (B). The figure is made for stereo view with relaxed eyes. Modified from Figure 9 in ref. 29 with permission.

for the Ca^{2+}-dependent binding of recoverin to ROS membranes,[37-39] but not for the function of S-modulin. Non-myristoylated recoverin can also inhibit R* phosphorylation at high Ca^{2+} concentrations.[39,40] The N-terminal myristoylation has been indicated to be essential for cooperative Ca^{2+}-binding to recoverin.[40,41] The first Ca^{2+}-binding to EF3 probably induces the second Ca^{2+}-binding to EF2 in a cooperative manner.[42]

The Ca^{2+}-binding of myristoylated (native) S-modulin to a membrane preparation was observed only in the membranes derived from ROS.[17] Hypotonically washed ROS membranes, proteolysed ROS membranes, phospholipase A2-treated ROS membranes, phospholipase C-treated ROS membranes, and vesicles made from chloroform/methanol-extracted ROS membranes were all good substrates of the binding of Ca^{2+}/S-modulin. However, bovine brain extracts[1,3,5-8,10] or vesicles made from purified phospholipids (PA, PC, PE, PS) or sphyngomyelin did not allow the binding (but see ref. 37 for the case of recoverin). The positive charges near the C-terminus have been suggested to be involved in the binding.[43]

LOCALIZATION

Immunohistochemical studies showed that S-modulin and recoverin are present uniformly throughout the photoreceptors.[3,19] S-Modulin and recoverin are postulated to act on rhodopsin kinase, and therefore it is reasonable that they are present in ROS. The uniform distribution may indicate that S-modulin and recoverin have other roles in other part of a photoreceptor. Because an S-modulin homologue, frequenin,[44] is believed to function in the transmitter release at the neuromuscular junction, S-modulin and recoverin may have similar function at the synaptic terminal of a photoreceptor.

ACKNOWLEDGMENT

This work was supported by Research for the Future Program of the Japan Society for the Promotion of Science under the Project “Cell Signaling” JSPS-RFTF97L00301.

REFERENCES

1. Kawamura S, Murakami M. Calcium-dependent regulation of cyclic GMP phosphodiesterase by a protein from frog retinal rods. Nature 1991; 349:420-423.
2. Kawamura S. Light-sensitivity modulating protein in frog rods. Photochem Photobiol 1992; 56:1173-1180.
3. Dizhoor AM, Ray S, Kumar S et al. Recoverin: a calcium sensitive activator of retinal guanylate cyclase. Science 1991; 251:915-918.
4. Kawamura S, Hisatomi O, Kayada S et al. Recoverin has S-modulin activity in frog rods. J Biol Chem 1993; 268:14579-14582.
5. Kawamura S. Rhodopsin phosphorylation as a mechanism of cGMP phosphodiesterase regulation by S-modulin. Nature 1993; 362:855-857.
6. Gorodovikova EN, Philippov PD. The presence of a calcium-sensitive p26-containing complex in bovine retina rod cells. FEBS Lett 1993; 335:277-279.
7. Chen C-K, Inglese J, Lefkowitz RJ et al. Ca^{2+}-dependent interaction of recoverin with rhodopsin kinase. J Biol Chem 1995; 270:18060-18066.
8. Klenchin VA, Calvert PD, Bownds MD. Inhibition of rhodopsin kinase by recoverin. J Biol Chem 1995; 270:16147-16152.
9. Sanada K, Shimizu F, Kameyama K et al. Calcium-bound recoverin targets rhodopsin kinase to membranes to inhibit rhodopsin phosphorylation. FEBS Lett 1996; 384:227-230.

10. Chen J, Makino CL, Peachey NS et al. Mechanisms of rhodopsin inactivation in vivo as revealed by a COOH-terminal truncation mutant. Science 1995; 267:374-377.
11. Chen CK, Burns ME, Spencer M et al. Abnormal photoresponses and light-induced apoptosis in rods lacking rhodopsin kinase. Proc Natl Acad Sci USA 1999; 96:3718-3722.
12. McCarthy ST, Younger JP, Owen WG. Dynamic, spatially nonuniform calcium regulation in frog rods exposed to light. J Neurophysiol 1996; 76:1991-2004.
13. Sampath AP, Matthews HR, Cornwall MC et al. Bleached pigment produces a maintained decrease in outer segment Ca^{2+} in salamander rods. J Gen Physiol 1998; 111:53-64.
14. Gray-Keller MP, Detwiler PB. The calcium feedback signal in the phototransduction cascade of vertebrate rods. Neuron 1994; 13:849-861.
15. Matthews HR, Murphy RLW, Fain GL et al. Photoreceptor light adaptation is mediated by cytoplasmic calcium concentration. Nature 1988; 334:67-69.
16. Nakatani K, Yau KW. 1988 Calcium and light adaptation in retinal rods and cones. Nature 314:69-71.
17. Kawamura S, Takamatsu K, Kitamura K. Purification and characterization of S-modulin, a calcium-dependent regulator on cGMP phosphodiesterase in frog rod photoreceptors. Biochem Biophys Res Commun 1992; 186:411-417.
18. Polans AS, Buczytko J, Crabb J et al. A photoreceptor calcium binding protein is recognized by autoantibodies obtained from patients with cancer-associated retinopathy. J Cell Biol 1991; 112:981-989.
19. Kawamura S, Kuwata O, Yamada M et al. Photoreceptor protein s26, a cone homologue of S-modulin in frog retina. J Biol Chem 1996; 271:21359-21364.
20. Gray-Keller MP, Polans AS, Palczewski K et al. The effect of recoverin-like calcium-binding proteins on the photoresponse of retinal rods. Neuron 1993; 10:523-531.
21. Sagoo MS, Lagnado L. G-protein deactivation is rate-limiting for shut-off of the phototransduction cascade. Nature 1997; 389:392-394.
22. Matthews HR. Actions of Ca^{2+} on an early stage in phototransduction revealed by the dynamic fall in Ca^{2+} concentration during the bright flash response. J Gen Physiol 1997; 109:141-146.
23. Dodd RL, Makino CL, Chen J et al. Visual transduction in transgenic mouse lacking recoverin. Invest Ophthalmol Vis Sci 1995; 36:S641.
24. Erickson MA, Lagnado L, Zozulya S et al. The effect of recombinant recoverin on the photoresponse of truncated rod photoreceptors. Proc Natl Acad Sci USA 1998; 95:6474-6479.
25. Otto-Bruc AE, Fariss RN, Van Hooser JP et al. Phosphorylation of photolyzed rhodopsin is calcium-insensitive in retina permeabilizesd by α-toxin. Proc Natl Acd Sci USA 1998; 95:15014-15019.
26. Koutalos Y, Yau KW. Regulation of sensitivity in vertebrate rod photoreceptors by calcium. Trends Neurosci 1996; 19:73-81.
27. Mendez A, Burns ME, Sokal et al. Role of guanylate cyclase-activating proteins (GCAPs) in setting the flash sensitivity of rod photoreceptors. Proc Natl Acad Sci USA 2001; 98:9948-9953.
28. Sato N, Kawamura S. Molecular mechanism of S-modulin action: Binding target and effect of ATP. J. Biochem 1997; 122:1139-1145.
29. Tachibanaki S, Nanda K, Sasaki K et al. Amino acid residues of S-modulin responsible for interaction with rhodopsin kinase. J Biol Chem 2000; 275:3313-3319.
30. Tanaka T, Ames JB, Harvey TS et al. Sequestration of the membrane-targeting myristoyl group of recoverin in the calcium-free state. Nature 1995; 376:444-447
31. Ames JB, Ishima R, Tanaka T et al. Molecular mechanics of calcium-myristoyl switches. Nature 1997; 389:198-202.
32. Johnson WC, Palczewski K, Gorczyca WA et al. Calcium binding to recoverin: implications for secondary structure and membrane association. 1997; Biochimica et Biophysica Acta 1997; 1342:164-174.

33. De Castro E, Nef S, Fiumelli H et al. Regulation of rhodopsin phosphorylation by a family of neuronal calcium sensors. Biochem Biophys Res Commun 1995; 216:133-140.
34. Yamagata K, Goto K, Kuo CH et al. Visinin: a novel calcium binding protein expressed in retinal cone cells. Neuron 1990; 2:469-476.
35. Flaherty KM, Zozulya S, Stryer L et al. Three-dimentional structure of recoverin, a calcium sensor in vision. Cell 1993; 75:709-716.
36. Dizhoor MD, Ericsson LH, Johnson RS et al. The NH_2 terminus of retinal recoverin is acylated by a small family of fatty acids. J Biol Chem 1992; 267:16033-16036.
37. Zozulya S, Stryer L. 1992 Calcium-myristoyl switch. Proc Natl Acad Sci USA 89:11569-11573.
38. Dizhoor AM, Chen CK, Olshevskaya E et al. Role of the acylated amino terminus of recoverin in Ca^{2+}-dependent membrane interaction. Science 1993; 259:829-832.
39. Kawamura S, Cox JA, Nef P. Inhibition of rhodopsin phosphorylation by non-myristoylated recombinant recoverin. Biochem Biophys Res Commun 1994; 203:121-127.
40. Calvert PD, Klenchin VA, Bownds MD. Rhodopsin kinase inhibition by recoverin. J Biol Chem 1995; 270:24127-24129.
41. Ames JB, Porumb T, Tanaka T et al. Amino-terminal myristoylation induces cooperative calcium binding to recoverin. J Biol Chem 1995; 270:4526-4533.
42. Matsuda S, Hisatomi O, Ishino T et al. The role of calcium-binding sites in S-modulin function. J Biol Chem 1998; 273:20223-20227.
43. Matsuda S, Hisatomi O, Tokunaga F. Role of carboxyl-terminal charges on S-modulin membrane affinity and inhibition of rhodopsin phosphorylation. Biochemistry 1999; 38:1310-1315.
44. Pongs O, Lindemeier J, Zhu XR, et al. Frequenin-a novel calcium-binding protein that modulate synaptic efficacy in the Drosophila nervous system. Neuron 1993; 11:15-28.

Ca^{2+}-DEPENDENT CONTROL OF RHODOPSIN PHOSPHORYLATION: RECOVERIN AND RHODOPSIN KINASE

Ivan I. Senin,[1] Karl-Wilhelm Koch,[2] Mohammad Akhtar[3] and Pavel P. Philippov[1]

Over many years until the middle of the 1980s, the main problem in vision research had been the mechanism of transducing the visual signal from photobleached rhodopsin to the cationic channels in the plasma membrane of a photoreceptor to trigger the electrophysiological response of the cell. After cGMP was proven to be the secondary messenger, the main intriguing question has become the mechanisms of negative feedback in photoreceptors to modulate their response to varying conditions of illumination. Although the mechanisms of light-adaptation are not completely understood, it is obvious that Ca^{2+} plays a crucial role in these mechanisms and that the effects of Ca^{2+} can be mediated by several Ca^{2+}-binding proteins. One of them is recoverin. The leading candidate for the role of an intracellular target for recoverin is believed to be rhodopsin kinase, a member of a family of G-protein-coupled receptor kinases. The present review considers recoverin, rhodopsin kinase and their interrelationships in the in vitro as well as in vivo contexts.

ABBREVIATIONS

CAR, cancer-associated retinopathy
EF, EF-hand Ca^{2+}-binding motif
G_t, transducin
GRK, G-protein-coupled receptor kinase
NCS, neuronal calcium sensor
Rho-P, phosphorylated non-bleached rhodopsin
Rho*-P, phosphorylated bleached rhodopsin
RK, rhodopsin kinase

[1]Department of Cell Signalling, A.N.Belozersky Institute of Physico-Chemical Biology, Moscow State University, Moscow 119899, Russia; [2]Institut für Biologische Informationsverarbeitung (IBI-1), Forschungszentrum Jülich GmbH, 52425 Jülich, Germany; [3]Division of Biochemistry and Molecular Biology, School of Biological Sciences, University of Southampton, Basset Crescent East, Southampton S 016 7PX, United Kingdom.

Rho, nonbleached rhodopsin
Rho*, bleached rhodopsin
ROS, rod outer segments
$[Ca^{2+}]_f$, free Ca^{2+} concentration

INTRODUCTION

A photoreceptor cell, similar to any neuron, can exist in two basic states: unstimulated and stimulated. The transit of a vertebrate photoreceptor from unstimulated (dark) state to the photoexited (bleached) one is initiated by a light quantum. The recovery of a bleached photoreceptor to the dark state does not demand an additional external stimulus and proceeds due to the intrinsic mechanisms as a consequence of excitation.

Visual transduction is catalyzed by an enzyme cascade that consists of rhodopsin, responsible for a photon capture, G-protein transducin (G_t) and an effector enzyme cGMP-phosphodiesterase. Activation of the cascade causes a fall in the cytoplasmic concentration of the second messenger cGMP. As a consequence, cationic channels in the photoreceptor plasma membrane are closed and the membrane is hyperpolarized, which is the first step on the way from the photoexcited visual cell—via the optic nerve—to the brain.

An additional important consequence of the events described is a fall in the cytoplasmic concentration of free calcium ions ($[Ca^{2+}]_f$), which, as it will be shown below, has a crucial importance for the recovery of the visual cell to the unstimulated state.

In order to return the cell to the dark state, it is necessary that all the components of the visual cascade should be switched off and the dark level of cGMP be restored. The switching off proceeds due to intrinsic mechanisms of the cascade components. Photoexited rhodopsin (Rho*) is inactivated by means of its phosphorylation by rhodopsin kinase (RK) that is followed by binding of arrestin to phosphorylated Rho* to quench it completely. The active form of G_t has GTP bound; it becomes inactive by an intrinsic GTPase activtiy which is facilitated by an GTPase accelerating factor, RGS-9. Activated cGMP-phosphodiesterase is turned off when it reassociates with its inhibitory γ–subunits. In addition, guanylate cyclase catalyses cGMP synthesis from GTP to restore the cGMP dark level and to reopen the plasma membrane cationic channels (for a review, see refs. 1-4).

Photoreceptor cells recovery and light adaptation are controlled by at least nine different steps,[2] of which five to six are Ca^{2+}-sensitive and suggested to be regulated by Ca^{2+}-binding proteins recoverin, GCAPs (guanylate cyclase-activating proteins) and calmodulin. This chapter is focussed on the role of recoverin and its Ca^{2+}-dependent control of RK. The function of other Ca^{2+}-binding proteins in photoreceptor physiology is covered in other chapters.

RECOVERIN

Recoverin was almost simultaneously discovered and purified from bovine rod outer segments (ROS) by several groups in Russia, USA and Germany.

In 1989, A.Dizhoor from P.Philippov's group suggested to use a column with delipidated visual rhodopsin immobilized on Concanavalin A Sepharose in order to purify G_t and some other G-proteins.[5] Chromatography of ROS extract on the column allowed them to obtain G_t subunits which were contaminated with cGMP-phosphodiesterase and an unknown protein with an apparent molecular weight of 26 K, to which a preliminary name «p26» was given[6]. Using rabbit antibodies against purified p26, the protein was demonstrated to be highly specific for retina, in particular it was found in both rod and cone photoreceptors.[6,7] The amino acid sequence of p26 exibited several calcium binding sites of the EF-hand type and its ability to bind Ca^{2+} was confirmed in experiments with calcium-45. In addition, p26 was suggested to be a Ca^{2+}-specific regulator of photoreceptor guanylate cyclase, a key enzyme of photoreceptors recovery, and due to this ability it was renamed as recoverin.[7]

In 1988, Koch and Stryer[8] directly demonstrated that a soluble factor different from calmodulin and capable of activating quanylate cyclase at physiologically relevant $[Ca^{2+}]_f$ was present in bovine ROS. Later Lambrecht and Koch[9] isolated this factor as «a 26 kd protein» and attributed its activity to the protein described by Dizhoor et al.[7]

That recoverin is capable of activating quanylate cyclase was, however, disproved in subsequent works.[10-12] (As is described in "Recoverin as a Ca^{2+}-Sensor of Rhodopsin kinase in vitro" of this chapter, recoverin continues to be considered as a participant of the photoreceptor recovery, but now as a putative Ca^{2+}-sensor of RK). The mistake in the initial assignment on the function of recoverin may be explained by the presence of endogeneous guanylate cyclase activator(s), GCAP1[13] and/or GCAP2[14], in the recoverin preparations used in the preceding works[7,9].

In 1987, Thirkill et al[15] found autoantibodies against an antigen having a molecular weight of 23 K in sera of patients with cancer-associated retinopathy (CAR) and named that antigen as «CAR-antigen». Later CAR-antigen was purified from bovine ROS by Palczewski's group and shown to be identical to recoverin.[16] (The function of recoverin as CAR-antigen is discussed in another chapter of this book).

It should be added that in 1991, Kawamura and Murakami[17] purified a 26 K protein from frog ROS, which modulated cGMP hydrolysis in a Ca^{2+}-dependent manner; later this protein was named as S-modulin and was shown to have a primary structure similar to recoverin.[18,19]

Tissue and Cellular Distribution

Immunochemical analysis demonstrated the presence of recoverin in retina of the following species investigated so far: man,[20-23] monkey,[21] bull,[6,7,24] sheep,[20] mouse,[25] rat,[20,21,24,25] rabbit,[25] pigeon,[20] frog,[20,25] chameleon,[25] lamprey[26] and newt.[27] In the case of chicken retina, contradictory data were obtained: although recoverin-positive immunoreaction was described in one case,[24] it was not found in other works.[20,25] The positive reaction in the former case could be a explained[24] by unspecific interaction of anti-recoverin antibodies with a related protein, visinin, which is known to be present in chicken retina.[28] In addition to retina, recoverin was found in ocular ciliary epitelium,[29] pinealocytes of the pineal organ[20,30] and the rat olfactory epithelium.[31]

Within retina, recoverin immunoreactivity was observed within photoreceptor cells (in their ROS, inner segments and synaptic terminals) as well as in higher order neurons: bipolar and ganglion cells of a number of species,[21,22,24,25] and in amacrine cells of lamprey *Lampreta fluviatilis*.[26]

In photoreceptor cells, the immunopositive recoverin staining was demonstrated for bovine,[6,7,24] human,[20,21,23] monkey,[21] mouse,[25] rat,[25] rabbit,[25] frog[25] and chameleon[25] ROS. No immunostaining was shown for chicken ROS.[25] (In the works cited, antibodies against bovine recoverin were used to visualize recoverin immunochemically). In addition to ROS, recoverin-positive reaction was revealed in inner segments, cell bodies and synaptic pedicles of both rods and cones.[7] In ROS, recoverin was suggested to regulate rhodopsin phosphorylation in a Ca^{2+}-dependent manner (see "Recoverin as a Ca^{2+}-Sensor of Rhodopsin kinase in vitro" in this chapter), but there are yet no experimentally justified hypotheses concerning the recoverin's function(s) in the other parts of photoreceptor cells and in higher order neurons.

Gene, Structure and Molecular Properties

The gene encoding human recoverin was cloned and localized to chromosome 17p13.1; the coding sequence of recoverin consists of three exons, thereby the sequences encoding the Ca^{2+}-binding sites are present in exon 1.[32-35] In mouse, recoverin gene assignes to chromosome 11.[36]

The electrophoretic mobility of recoverin on SDS polyacrylamide gels corresponds to a 26-K protein, and its gel filtration elution volume corresponds to that of a 28-K sphere.[7] The complete amino acid sequence of recoverin deduced from the cDNA contains 202 amino acid residues and has a calculated mass of 23.3 K.[7,37] During posttranslational modification, a consensus sequence for myristoylation present at the N-terminus of recoverin, undergoes cleavage of the N-terminal methionine residue and the resulting N-terminal glycine is acylated by N-myristoyl transferase present in photoreceptors. Finally, N-acylated retinal recoverin contains 201 residues and its M_r is determined to be 23.4 K. (Recombinant recoverin expressed in *E. coli* which lack N-myristoyl transferase has M_r equal to 23.2 K). The N-terminal acylation of recoverin is heterogeneous: the electrospray mass spectrometry demonstrated that it is modified by myristic acid (14:1), and three other fatty acids (14:0, 14:2 and 12:0 acyl residues).[38] These fatty acyl modifications of recoverin are essential for its Ca^{2+}-dependent binding to ROS membranes and phospholipid vesicles (see "Recoverin, *Ca^{2+}-Dependent Binding to Membranes*" in this chapter).

The X-ray crystal structure of recoverin shows that the protein molecule is composed of two domains (N-terminal and C-terminal) separated by a narrow cleft, each containing two EF-hand motifs. These motifs consist of 29 residues that form two perpendicularly placed a-helices and a connecting loop which can be represented as a helix–loop–helix structure. Calcium ions interact with residues that are mostly within the loop region. Calcium is coordinated by side chain oxygen in position 1, 3 and 5 of residues within the loop region, by a backbone carbonyl group in position 7, a water molecule at position 9 and by a carboxylate in position 12. Potential Ca^{2+}-binding sites are distributed relatively uniformly within the recoverin sequence and include amino acids 36 - 48 (EF1), 73 - 85 (EF2), 109 - 121 (EF3) and 159 - 170 (EF4). Of the four potential Ca^{2+}-binding sites, only two (the 2nd and 3rd) EF-hands are capable of binding Ca^{2+}, whereas the remaining two sites (the 1st and 4th) do not possess this ability. EF1 is distorted by Pro40 at residue 4 of position 12. In EF4, (i) residues in positions 2 (Lys161) and 12 (Glu171) form a salt bridge, (ii) important residues in calcium ligand positions 1 (Gly160) and 3 (Lys162) lack oxygen in their side chains and cannot participate in calcium binding, (iii) highly conserved Gly in position 6 is substituted by Asp 165. In the 2nd and 3rd EF-hands of recoverin, six amino acid residues are involved in Ca^{2+} binding; five of them are located in the loop, while the sixth residue, glutamate, is in the 2nd helix of the helix–loop–helix motif.[39]

Recoverin contains three tryptophan residues (Trp 31, 104 and 156) whose fluorescence emission is sensitive to Ca^{2+} binding: the fluorescence spectrum of acylated (retinal and recombinant myristoylated) recoverin shifts to the red, and the fluorescence intensity at the peak decreases when Ca^{2+} binds to the protein. In contrast, addition of Mg^{2+} does not alter the fluorescence properties of the protein.[7,37,40]

Non-myristoylated recombinant recoverin exhibits heterogeneous and uncooperative binding of Ca^{2+} ions with dissociation constants of 0.11 and 6.9 μM, whereas two Ca^{2+} bind cooperatively to myristoylated recombinant recoverin with a Hill coefficient of 1.75 and an apparent dissociation constant of 17 μM. Hence, myristoylation lowers the Ca^{2+} affinity of recoverin and induces cooperativity in Ca^{2+} binding. A concerted allosteric mechanism is proposed to explain how the myristoyl group induces cooperative Ca^{2+} binding to recoverin. The existence of two protein conformations is supported by the fluorescence and NMR data. One of the states places the myristoyl group outside so that it can interact with photoreceptor membranes. The intrinsic dissociation constants of this state for Ca^{2+} are assumed to be 0.11 and 6.9 μM, i.e., the same as in the case of non-myristoylated recoverin.[40]

NMR studies suggest that the binding of Ca^{2+} to EF3 induces EF2 to adopt a conformation favorable for the binding of a second Ca^{2+} to recoverin.[41] This suggestion was experimentally confirmed more recently[42] by measurements of intrinsic fluorescence of recoverin mutants –EF2 and –EF3[43] in which the Ca^{2+} binding EF2 and EF3 structures were destroyed. It was shown that the substitution E121Q, which destroyed the EF3 ability to bind calcium, totally abolished Ca^{2+} binding to the mutant, whereas the mutation E85Q, in EF2, caused only a moderate decrease in the binding. Based on this result, one may conclude that the binding of Ca^{2+} to recoverin is a sequential process with EF3 being filled first. The substitutions G160D, K161E, K162N, D165G and K166Q in EF4 gave the mutant (+EF4) with three active Ca^{2+} binding sites[43] in contrast to retinal recoverin having only two such sites. Measurements of thermal denaturation, intrinsic fluorescence and CD spectra of +EF4 revealed that these substitutions decreased the thermostability of the protein in the absence of Ca^{2+}.[42]

Finally, it is necessary to add that the recoverin/rhodopsin ratio in osmotically intact ROS preparations is approximately 1 : 174.[44]

Ca^{2+}-Dependent Binding to Membranes

The myristoyl group in recoverin serves an important function as an anchor for lipid membranes. In the Ca^{2+}-free state of recoverin, the myristoyl group is sequestered in a hydrophobic cleft in the N-terminal part of the protein. Binding of calcium ions to recoverin leads to the extrusion of the myristoyl group which can then interact with membranes or hydrophobic targets. This Ca^{2+}-triggered exposure of the buried myristoyl group was named as Ca^{2+}-myristoyl switch.[45]

Therefore, recoverin belongs to a group of myristoylated proteins that undergo signal controlled membrane binding. Several homologues of recoverin also display properties typical for a Ca^{2+}-myristoyl switch.[46]

Binding of recoverin to ROS membranes strictly depends on the free Ca^{2+} concentration. Binding was half maximal at 2.1 μM free Ca^{2+} as determined by an equilibrium centrifugation assay.[45] Surface plasmon resonance measurements with immobilized phospholipid liposomes, a constant amount of recoverin and varying free Ca^{2+} concentration resulted in similar EC_{50} of 4 μM; affinity of recoverin for phospholipid liposomes was rather low with an apparent K_D of 140-160 μM.[47] The affinity was higher, when native ROS membranes were immobilized (app. K_D = 27 μM) or when a different immobilization technique for liposomes was used (K_D = 18 μM).[48] In comparison with studies on myristoylated model peptides, it was concluded that the hydrophobic interaction of the myristoyl group in recoverin with lipid membranes is the driving force of membrane association, which means that electrostatic interactions are less important.[47]

Orthologues and Homologues

A number of orthologues and homologues of recoverin containing EF-hands have been cloned and identified and are classified as the neuronal calcium sensor (NCS) protein family.

Recoverin and its chicken orthologue visinin were the first identified members of the family. A search in currently available data bases shows a wide distribution of NCS proteins among species from yeast to man. Most family members share 25-32% of sequence identity with calmodulin, and a phylogenetic tree analysis of the known amino acid sequences of NCS proteins from different species results in five subfamilies named as recoverins, VILIPs, frequenins, GCAPs and KchIPs (Fig. 1). All the proteins harbor 4 EF-hands, but most of these are capable of only binding 2 or 3 Ca^{2+} ions. NCS proteins show a widespread distribution in the nervous system, but the expression pattern of each protein is restricted to a specific set of cells (for a

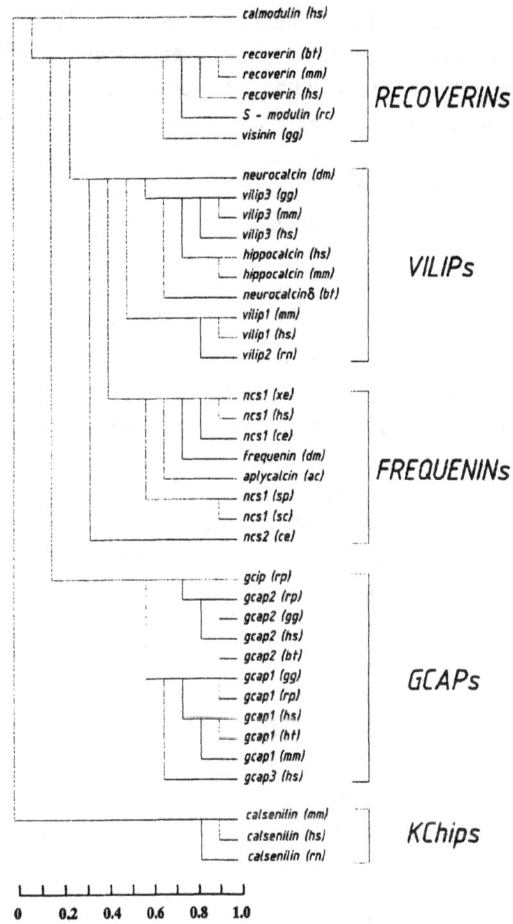

Figure 1. A phylogenetic tree comparing the sequence of recoverin with those of other coat proteins from different species was constructed using the Darwin program. GenBank accession codes are: Bos taurus recoverin P21457[GenBank], Mus musculus recoverin P34057 [GenBank] , Homo sapiens Recoverin P35243 [GenBank], Rana catesbeiana S-modulin P31227 [GenBank], Gallus gallus visinin P22728 [GenBank], Drosophila melanogaster neurocalcin P42325 [GenBank], Gallus gallus vilip3 P42324 [GenBank], Mus musculus vilip3 P35333 [GenBank], Homo sapiens vilip3 P37235 [GenBank], Homo sapiens hippocalcin P41211 [GenBank], Mus musculus hippocalcin P32076 [GenBank], Bos taurus neurocalcin δ P29554 [GenBank], Homo sapiens.vilip1 P 28667 [GenBank], Rattus norvegicus vilip2 P35332 [GenBank], Xenopus laevis ncs-1 Q91614 [GenBank], Homo sapiens ncs-1 P36610 [GenBank], Schizosaccharomyces pombe ncs-1 Q09971 l [GenBank], Caenorhabditis elegans ncs-1 P36608 [GenBank], Saccharomyces cerevisiae ncs-1 Q06389 [GenBank], Drosophila melanogaster frequenin P37236 [GenBank], Aplysia californica aplycalcin Q16982 [GenBank], Caenorhabditis elegans ncs-2 P36609 [GenBank], Rana pipiens GCIP 073763 [GenBank], Rana pipiens GCAP2 073762 [GenBank], Gallus gallus GCAP2 P79881 [GenBank], Gallus gallus GCAP1 P79880 [GenBank], Rana pipiens GCAP1 073761 [GenBank], Homo sapiens GCAP1 P43080 [GenBank], Bos taurus GCAP1 P46065 [GenBank], Mus musculus GCAP1 P43081 [GenBank], Homo sapiens GCAP2 Q9UMX6 [GenBank], Bos taurus GCAP2 P51177 [GenBank], Homo sapiens GCAP3 095843 [GenBank], Homo sapiens calsenilin Q9Y2W7 [GenBank], Mus musculus calsenilin Q9QXT8 [GenBank], Rattus norvegicus calseninlin Q9JM47 [GenBank], Homo sapiens calmodulin . hs - Homo sapiens, bt - Bos taurus, m - Mus musculus, rp - Rana catesbeiana, gg - Gallus gallus, dm - Drosophila melanogaster, G- Gallus , rn - Rattus norvegicus, xe - Xenopus laevis, ce - Caenorhabditis elegans, ac - Aplysia californica, sc - Saccharomyces cerevisiae, rp - Rana pipiens.

review, see refs. 49 and 50). For example, recoverins and GCAPs are specifically expressed in photoreceptor cells in the retina, whereas some members of the VILIP and frequenin subfamilies are also found in nonphotoreceptor cells of the retina. In addition, expression outside the nervous system had been reported in some cases (see "Recoverin, *Tissue and Cellular Distribution*" in this chapter).

A common property of most NCS proteins is their association with membranes. Members of all subfamilies except the KChIPs contain a consensus site for N-terminal acylation and could target to membranes by an extruded fatty acid residue (mainly myristoyl). Members of the KChIP subfamily are tightly associated with A-type voltage-gated potassium channels (Kv4 channels) as they co-localize and co-immunoprecipitate with Kv4 α-subunits. Thus, these proteins are targeted to the membrane by strong protein-protein interactions.[51]

Known targets of NCS proteins are rather diverse in their biological function, and regulation by a member of one of the five subfamilies appears to be very specific. GCAPs regulate the activity of retina specific guanylate cyclases (see other chapters in this book for more details). No other function has been described for GCAPs so far. KChIPs modulate the inactivation kinetics of Kv4 channels. Almost identical with KChIP3 is a protein called downstream regulatory element antagonist modulator or DREAM, which acts as a DNA-binding protein and repressor of transcription.[52] The frequenin and VILIP subfamilies display the most diverse putative target regulation. Yeast frequenin binds to phosphatidylinositol 4-hydroxykinase (PI-4-K) in a Ca^{2+}-independent way and activates this enzyme.[53] Overexpression of the *Drosophila* orthologue in a mutant fly named V7 enhances a frequenin-dependent facilitation of neurotransmitter release;[54] a similar function was reported for *Xenopus* frequenin.[55] Other effects of frequenin orthologues include the regulation of Ca^{2+}-dependent exocytosis in PC12 cells[56] and a modulation of several calmodulin targets.[57] VILIP-1 modulates adenylate cyclase activity in olfactory knobs[58] and transfected C6 glioma cells[59] in a Ca^{2+}-dependent way; at the same time, it can inhibit RK as well as recoverin does.[60]

RHODOPSIN KINASE

Light dependent phosphorylation of rhodopsin by RK was discovered serendipitously in three laboratories in the early 1970s[61-63] and later shown to involve multiphosphorylations on the C-terminal tail of the receptor.[64-68] Progress in the field of seven-helix receptors in the ensuing years then highlighted that RK is a typical member of an expanding family of G protein-coupled receptor kinases (GRKs) and is classified as GRK-1.[69] Like other members of the family, RK phosphorylates only the stimulated form of the receptor (designated as Rho* and considered to be equivalent to metarhodopsin II) that is produced by the action of light on unbleached rhodopsin (Rho). In other cases the receptors become substrates for their cognate GRKs, following agonist occupation.

Gene

Incremental progress in the purification of RK, from bovine retina, eventually led to the isolation of a homogeneous protein of apparent M_r 67 – 70 K.[70] The latter preparation was used to isolate and sequence a cyanogen bromide peptide of 30 amino acids, and the information then exploited to characterize and sequence the cDNA for RK that encoded for a protein of 561 amino acids. The deduced sequence showed that bovine retinal RK has 34% overall similarity with the related β-adrenergic receptor kinase (GRK-2) and the similarity increases to 42% when the comparison is restricted to the putative catalytic domains of the two kinases.[71]

Using a bovine RK cDNA fragment as a probe, the gene for the corresponding human enzyme was found to be located on chromosome 13, at 13q34. The gene is composed of 6 introns and 7 exons and encodes a 564-amino acid polypeptide showing 92% identity in amino acid sequence to bovine RK. The gene in addition to transcribing the predominant RNA species which produces GRK-1, is also processed to give rise to a minor splice variant producing low levels of a protein, designated as GRK-1b, with low catalytic activity.[72]

Post-Translational Modifications

The amino acid sequence of RK contains a CAAX isoprenylation motif (cysteine -aliphatic amino acid - aliphatic amino acid - any amino acid) at the C-terminal, which has been shown to lead to the incorporation of a farnesyl moiety at the C-terminal cysteine residue of the mature protein and also the methylation of its carboxyl group.[71,73,74] During the overall conversion three amino acids from the primary translation product are excised, reducing the length of the protein to that of 559 amino acids. The deduced sequence of RK also shows the presence of a possible myristoylation signal at the N-terminal, MDFGSL.......[71] The latter sequence is somewhat untypical, since in this case the glycine to be modified is preceded by three amino acids, instead of one in the typical motif found in recoverin (see "Recoverin, *Gene, Structure and Molecular Properties*" in this chapter). Indeed, as yet this post-translational modification has not been identified in the mature protein and does not appear to be necessary for its catalytic activity. This is shown by the mutation of Gly 4 to alanine, which gave a fully active protein (M.E.Akhtar, N.E.M.McCarthy, P.Lee-Robichaud, M.Akhtar, unpublished work). On the other hand, the mutation of the cysteine residue (at position 558) to serine in the CAAX motif and the expression of the protein in COS-7 cells, gave a protein which had about 25% of the wild type RK activity.[75] The behavior of this mutant was attributed to its inability to target to the ROS membranes. However, in our hands the expression of the Ser 558 to alanine mutant, in insect cells, gave a protein with less than 5% of the activity of the wild-type enzyme, whether the activity measurement was performed using bleached ROS membranes or detergent solubilized bleached rhodopsin (M.E.Akhtar, N.E.M.McCarthy, P.Lee-Robichaud, M.Akhtar, unpublished work). This finding suggests that the farnesyl modification may be necessary not only for the membrane targeting of RK but also for promoting favorable Rho*-RK interaction. Further support for the latter view is provided by studies using farnesylated peptides corresponding to the C-terminal sequence of RK, which inhibited the activity of the enzyme and were shown to bind to metarhodopsin II retarding its conversion into metarhodopsin III.[76]

The incubation of purified RK with ATP leads to its autophosphorylation[77,78] resulting in the modification of three serine residues in positions 21, 488 and 489.[79] The implication of this finding is that in vivo, since ATP is present, RK exists in an autophosphorylated state and the isolation of the nonphosphorylated form is due to the action of a phosphatase, presumably type-2A protein phosphatase, during the lengthy purification procedure. The biochemical function of the autophosphorylation has been studied using a partial reaction in which membrane-bound Rho* and RK, in the absence of ATP, form a complex. The dissociation constant for the complex formation, between Rho* and RK, is around 0.5 μM.[80] The stability of the complex has been studied using a combination of variously phosphorylated proteins and it was estimated that the affinity decreases in the following order: Rho*-RK > Rho*-phosphorylated RK > phosphorylated Rho*-phosphorylated RK.[81] The trend has been interpreted to hypothesise that the two proteins dissociate when these become phosphorylated, allowing the binding of arrestin to phosphorylated Rho*.[81] Notwithstanding this explanation, it should be born in mind that the dissociation of the autophosphorylated kinase from phosphorylated Rho* is a mandatory requirement of catalytic turnover since these are related to each other, as an enzyme and its product. The mutants of two of the autophosphorylation sites, Ser 488 to Ala or Ser 489 to Ala, had increased activity for the phosphorylation of rhodopsin in the dark thus suggesting that the presence of phosphoryl groups at these positions may play a regulatory role.[82]

The Enzymology and Activation

Enzymological studies on RK have been performed predominantly on the enzyme isolated from bovine retina,[77,80] via ROS, with some interlocking observations made using the corresponding enzyme from ovine retina.[76] ROS from bovine retina contain between 8 to 20 units of RK activity/mg of protein.[80,83] (1 unit represents the incorporation 1 nmol of phosphate, from ATP, into Rho*/min in an assay mixture containing 3 mM ATP, about 10 μM rhodopsin when the incubation is performed under continuous illumination at 37 °C). The most highly purified RK has been reported to have a specific activity of about 500 units/mg of protein,[79] however, for most enzymological studies, when large amounts of the enzyme are required, preparations of modest purity with a specific activity of 100 – 130 units/mg of protein have been routinely used.[84] With the aim of obtaining larger amounts of the enzyme than is isolable from retinas, RK has been expressed in insect[85] and animal cells;[86] from the insect system a yield of up to 0.350 mg of RK, from 250 ml of culture medium, with a specific activity of 96 units/mg of protein was reported.[85]

The fact that RK acts not on Rho but on its light-activated derivative metarhodopsin II, Rho*, was originally explained by a "substrate activation" mechanism which assumed that the C-terminal of the dark adapted form of rhodopsin, is inaccessible to RK and becomes available only following the conformational rearrangement accompanying the formation of Rho*.[87,88] Then using synthetic peptides corresponding to the C-terminal phosphorylation domain of rhodopsin (Fig. 2), it was surprisingly observed that the peptides were extremely poorly handled as substrates by the RK alone, but in presence of Rho and light – generating Rho* – a dramatic enhancement of the phosphorylation reaction, by nearly 50-fold, could be achieved.[89-91] These finding was confirmed and extended by Palczewski et al.[92]

This phenomenon was interpreted by an alternative hypothesis, which envisaged that the kinase normally exists in an inactive resting state and is activated following interaction with Rho*. In principle, there are two pathways through which the activation may be achieved. One of these assumes (equation 1) that the activated form of the enzyme is present only within a Rho*-RK complex, while the second proposes that RK may dissociate and independently exist as an activated entity (equation 2).

$$\text{Rho* + inactive RK} \rightarrow \text{Rho*-inactive RK} \rightarrow \text{Rho*-active RK} \qquad (1)$$

$$\text{Rho* + inactive RK} \rightarrow \text{Rho*-inactive RK} \rightarrow \text{Rho*-active RK} \leftrightarrow \text{Rho* + active RK} \qquad (2)$$

The ability of RK to phosphorylate peptide derived from the C-terminal region of rhodopsin, has been exploited for the study of several facets of the enzyme. The activation and catalytic steps are subsumed into a single event when Rho* is the substrate to be phosphorylated. On the other hand, these two steps are separable in the phosphorylation of peptide substrates, since this process can only occur when RK has already been activated. This feature has enabled us to establish that metarhodopsin II, metarhodopsin III, and their fully phosphorylated derivatives are equally effective in the activation of RK, for peptide phosphorylation.[91] It hardly needs emphasizing that the phosphorylated rhodopsin derivatives are activators only, and not substrates, since their available phosphorylation sites are already full. Opsin, on the other hand, was neither phosphorylated by RK nor could it activate the enzyme for peptide phosphorylation. In other words, the activation of RK is promoted by the interaction of the enzyme with a range of derivatives of bleached rhodopsin, and also by a truncated derivative,[92] which retain the Schiff base linkage between the all-*trans*-retnylidene moiety and opsin. From other studies it can be inferred that metarhodopsin I, II and III not only must activate the kinase but also act as substrates for the enzyme.[91,93]

The Hierarchical Order and Recognition Motif for Phosphorylation

The specificities of several protein kinases are known, and in most cases the target hydroxyl group of serine/threonine/tyrosine is found to be the part of a

Figure 2. The sequence of the 19 C-terminal amino acids of Rho. The peptide sequences used for phosphorylation studies are shown by thick lines and the sites of phosphorylation are italicized.

specific amino acid sequence. RK is an example of a unique type, since in it each potential phosphorylation site is surrounded by a different amino acid sequence; it thus appears that the specificity resides not in a particular primary sequence but in some form of higher order structure. Nonetheless, a possible approach for dissecting the complex recognition sequence into its individual components involved the use of synthetic peptide substrates. All the potential phosphorylation sites of rhodopsin are located within its last 15 amino acids.[66] A range of peptides of 10 to 12 residues in length, containing the target sites, were used as RK substrates and the phosphorylated products subjected to a range of quantitative measurements[84]. The results showed that all the sites known to be phosphorylated in Rho* were, to various extents, also phosphorylated in the peptide substrates. A global comparison of the results of phosphorylation at each site showed that the propensity of modification is in the order: Ser 343 > Ser 338 > Thr 336 > Ser 334, Thr 342 > Thr 335, Thr 340.[84] That Ser 343 and Ser 338 are the most favored site of phosphorylation in the peptide substrates is in agreement with several studies showing that these amino acid residues are also preferentially modified when rhodopsin is fully bleached and exhaustively phosphorylated in vitro by RK,[94,95] or in living mice subjected to flashes of light to bleach 7-15% of the total rhodopsin.[96]

In the hierarchy of preference, the data from peptide phosphorylation place Thr 336 marginally ahead of Ser 334. However, when Rho* is phosphorylated, in vitro or in vivo, Ser 334 seems to be the next site of phosphorylation,[96] although in an earlier study the same group had found Thr 336 to rank after Ser 343/Ser 338.[95] However, not too much should be read in this minor discrepancy since in both the approaches, using peptides or Rho*, the quantitative conclusions are drawn by somewhat over interpretation of the results. In the case of the experiments with Rho*, there is the possibility of differential dephosphorylations at various sites, by type-2A protein phosphatase,[96,97] furthermore the quantitation is performed by mass spectrometric analysis of posphopeptides, isolated following the proteolytic digestion of the protein. These stages are prone to variable losses. In the case of the conclusion drawn from peptide phosphorylation, the results from several experiments, using peptides of different sequences, were pooled and the underlying assumption being that the sequences are faithful surrogates for the native substrate. In the light of these weaknesses, the realistic conclusion is that Ser 343 and Ser 338 are the most favorable sites of phosphorylation then are Ser 336 and Thr 334 and finally the remaining three residues, Thr 342, Thr 340 and Thr 335.

The cumulative results from peptide phosphorylation also suggested that the optimal modification of the target Ser/Thr residue occurs if it has at least one residue on the amino side and five on the acyl side, and also contains a neutral residue, preferably small (Ala, Pro, Ser, Thr) at the fourth acyl position.[84]

How Many Sites in Rho* Need to be Phosphorylated for Rapid Deactivation?

The hyperpolarisation of retinal cells, which follows the absorption of light, needs to be reversed to allow an unambiguous detection of the subsequent photon absorption. The process, known as deactivation or down-regulation, is achieved by two critical events, involving the phosphorylation of Rho*, by RK, followed by arresting binding. The question as to how many sites on Rho* need to be phosphorylated for the reproducible deactivation has been addressed by several investigators[96,98-102] and briefly reviewed in a recent paper by Mendez et al.[102] The latter workers using transgenic mice containing various mutant forms of rhodopsin have concluded that a rapid and reproducible deactivation of Rho* requires multiple phosphorylations. This conclusion is based on the following findings. A single-photon stimulation of rod cells containing the wild type rhodopsin gave a recovery time constant (τ) of about 0.15 seconds. The latter value was increased to about 3 second when rhodopsin contained only two phosphorylation sites, Ser 334 and Ser 338, the other five residues being replaced by alanine. However, the serine triple mutant, in which all the three serine residues (Ser 343, Ser 338 and Ser 334) were replaced by alanine, retaining the four threonine residues, showed a greatly improved recovery time constant of 0.38 seconds.[102] The conclusion from this study that, larger the number of potential phosphorylation sites in rhodopsin shorter is the recovery time is, thus not unreasonable. However, whether this correlation is the consequence of multiphosphorylations in vivo or merely the reflection of the fact that the constructs, like the wild-type rhodopsin, which contain a large number of phosphorylation sites also have a higher propensity for generating a family of monophosphorylated species, remains to be established. It should be mentioned in passing that when the serine triple mutant was bleached in the isolated eye for 10 minutes and the material then analyzed by isoelectrofocusing it showed predominantly the presence of monophosphorylated opsin, with a small amount of di- and a trace of the tri-phosphorylated species.[102] The issue of in vivo phosphorylation sites had previously been investigated by Ohguro et al[96] who following the exposure of mice to light flashes, to bleach 7-15 % of rhodopsin, analyzed the material to determine the number of modified sites, by mass spectrometry, and deduced that the material consisted of a family of monophosphorylated species only.

Since the binding of arrestin to phosphorylated Rho* is an important step in down regulation, this property has been used in several in vitro studies to examine the contribution which the various phosphorylation sites make to the binding interactions. Here, the conclusions range from assigning role to only two residues, Ser 343 and Thr 340,[99] to considering that all the sites are important and make additive contribution.[98,101] In this connection, attention may also be drawn to the fact that the phosphorylation of Rho*, by RK, displays cooperativity, by which once a phosphorylated species is formed, there is a greater possibility that this is used for further phosphorylation than the original unphosphorylated substrate.[65,103] This phenomenon depends on the increase of the affinity of the substrate, for RK, with increasing phosphorylation[104] and should tend to favor the formation of multiphosphorylated species.

In conclusion, since rhodopsin contains seven phosphorylation sites, the view that the phosphorylation of most of these may be physiologically important though teleologically attractive—and may eventually prove to be correct—is not consistently supported when the results from all the approaches are considered together.

RECOVERIN AS A Ca^{2+}-SENSOR OF RHODOPSIN KINASE IN VITRO

By now a great body of in vitro data have been obtained on the ability of recoverin to be a Ca^{2+}-dependent inhibitor of light-dependent rhodopsin phosphorylation catalyzed by RK. In contrast, fewer works have been published to prove or disprove whether these findings are the true reflection of the physiological role of recoverin in vivo. So far there is no general agreement among authors regarding the in vivo function of recoverin. For this reason, we will mainly focus our attention on the results of the in vitro experiments which will be considered separately from the data obtained under physiological conditions.

A Ca^{2+}-Dependent Complex between Recoverin and Rhodopsin Kinase

A Ca^{2+}-dependent interaction between retinal recoverin and RK first suggested by Gorodovikova and Philippov[11] was then directly demonstrated, using the recombinant myristoylated protein.[105] This interaction was rather strong and used to prepare a column with immobilized recoverin for the purification of RK from cellular extracts.[105] Half-maximal binding of RK to recoverin immobilized to a sensor chip was estimated as 0.5 - 1 μM by surface plasmon resonance technique.[106] In addition, using this method, it was demonstrated that ATP, a substrate of RK, substantially inhibited the RK-recoverin interaction whereas 0.1 mM ADP enhanced it. Both effects were specific for adenyl nucleotides while GTP, GDP, CTP, and XTP were ineffective. The effect of ATP was found to be a result of ATP-dependent autophosphorylation of RK. ADP did not affect the autophosphorylation state of RK and did not enhance the binding of autophosphorylated RK to recoverin. The authors explained the stimulatory action of ADP by its binding to RK and thereby changing the RK conformation in favor of that with the higher affinity to recoverin.[106]

No information on the identification of sites in recoverin responsible for its interaction with RK is yet available, but by analogy with S-modulin[107] this site could be in a region near the amino terminus of recoverin. As for RK, its contact site for recoverin seems to be localized in the amino terminus of the enzyme.[108]

Recoverin Inhibits Light-Dependent Rhodopsin Kinase

The first evidence for the ability of recoverin to inhibit the RK activity was obtained by Kawamura et al[109] who demonstrated a Ca^{2+}-dependent inhibitory effect of recoverin upon cGMP hydrolysis in a suspension of frog ROS and suggested this effect to be a result of the recoverin action upon rhodopsin phosphorylation. Simultaneously, a complex of p26 (the initial name of recoverin) with a protein, presumably RK, was found to exist in bovine ROS, in which Ca^{2+} favored formation of the complex whereas EGTA dissociated it.[11] More recently it was directly shown that light-dependent rhodopsin phosphorylation in bovine ROS suspension was inhibited by Ca^{2+} (Fig. 3A) and this effect was reversed by the addition of polyclonal monospecific antibodies against recoverin.[110] Additional evidence for the ability of recoverin to inhibit RK in a Ca^{2+}-dependent manner was obtained with the use of a reconstituted system consisting of rhodopsin (contained in urea-washed bovine photoreceptor membranes) and purified retinal recoverin and RK;[111] in this system, the level of light-dependent rhodopsin phosphorylation decreased with an increase of Ca^{2+} concentration only in the presence of recoverin (Fig. 3B). These data are consistent with subsequent works which used similar reconstituted systems consisting of retinal proteins[44] or of recombinant preparations of RK and recoverin together with bovine photoreceptor membranes.[105]

In order for recoverin to inhibit RK, both its Ca^{2+}-binding sites should be in a native state: the substitutions in the 2nd (E85Q) and 3rd (E121Q) EF-hands, which «spoil» these sites, completely abolish the inhibitory ability of the corresponding constructs. At the same time, the E85Q mutant, of EF2, is capable of binding Ca^{2+} despite its inability to inhibit RK.[43] Thus, there is no strict correlation between Ca^{2+}-binding and inhibitory capacities of recoverin mutants.

Similar to retinal recoverin which is heterogeneously acylated,[38] recombinant nonacylated recoverin is capable of inhibiting light-dependent rhodopsin phosphorylation in a Ca^{2+}-dependent manner.[105,112-114] However, N-terminal acylation of recoverin enhances its inhibitory efficiency with respect to RK. Inhibition of rhodopsin phosphorylation at saturating Ca^{2+} can be described as a hyperbolic function of the recoverin concentration yielding IC_{50} values of 6.5[114] - 8 μM[105] for its non-myristoylated form, 0.8[105] - 0.9 μM[114] for the recombinant myristoylated form and 1 - 6 μM[44,111,114] for retinal recoverin. At saturating recoverin concentrations, IC_{50} values for Ca^{2+} in the reaction of rhodopsin phosphorylation are similar (2 - 3 μM) in the case of the acylated protein (retinal[44,114] and recombinant myristoylated[105,114]) and its nonacylated form[105,114] (the values of 140 - 250 nM[110,111] obtained previously were apparently underestimated as $[Ca^{2+}]_f$ in these works was only calculated without verification by its direct measurements). However, acylated

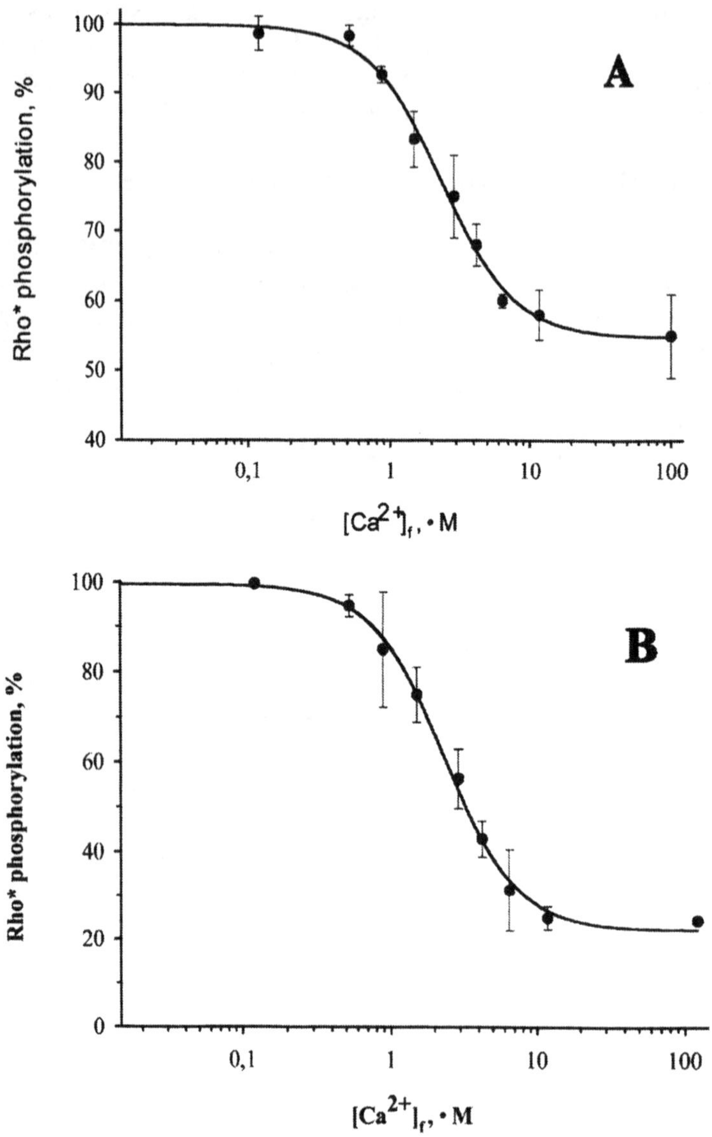

Figure 3. Dependence of the light-stimulated RK activity on $[Ca^{2+}]_f$ concentration in (A) ROS suspension and (B) the reconstituted system consisting of urea-washed ROS membranes, rhodopsin kinase and recoverin. A. Rhodopsin phosphorylation was carried out: A. at 25°C in 75 mM Tris-HCl (pH 7.4) containing 3 mM $MgCl_2$, 0.5 mM GTP, 100 μM [γ-^{32}P]ATP 1 mM EGTA, $CaCl_2$ to create $[Ca^{2+}]_f$ shown, and ROS suspension (10 μM rhodopsin); B. at 25°C in 50 mM Tris-HCl (pH 7.4) containing 100 mM NaCl, 3 mM $MgCl_2$, 1 mM dithiotreitol, 100 μM [γ-^{32}P]ATP, 1 mM EGTA, $CaCl_2$ to create $[Ca^{2+}]_f$ shown, about 1 μg RK, and urea-washed ROS membranes (10 μM rhodopsin). The dark level of ^{32}P incorporation was subtracted in both cases. Free calcium concentrations shown in the figure were directly determined by Ca^{2+}-sensitive electrode to correct the values calculated in the original experiments by Gorodovikova et al (see refs. 104 and 105).

and nonacylated forms of recoverin have different Hill coefficients estimated from the Ca^{2+}-dependence of the recoverin inhibitory effects: 0.7 - 0.9 in the case of the nonacylated form[105,114] and 1.5 - 1.7 in the case of both the retinal[114] and recombinant myristoylated[105,114] forms, i.e., N-acylation confers cooperativity to recoverin with respect to binding of Ca^{2+}

The IC_{50} values for recombinant myristoylated recoverin mentioned above were obtained with preparations which could contain a significant fraction of nonmyristoylated recoverin[115]. As the myristoylated form of recoverin, in contrast to its non-myristoylated counterpart, is capable of binding to photoreceptor membranes in a Ca^{2+}-dependent manner,[45,116] Ca^{2+}-dependent differential adsorption on these membranes was used by us to separate recombinant myristoylated recoverin from the non-myristoylated protein (I.I.Senin, M.Akhtar and P.P.Philippov, unpublished data). Analytical HPLC showed that the preparations of myristoylated recoverin thus obtained were free from non-myristoylated recoverin (Fig. 4). In the RK assay, IC_{50} for purified myristoylated recoverin was found to be 0.4 - 0.5 μM, that is, about 2 times lower than the values 0.8 - 0.9 mM determined earlier.[105,114]

In concluding this subsection, let us touch on specificity of the Ca^{2+}-dependent inhibitory effect of recoverin. The family of G-protein-coupled receptor kinases are composed of six cloned members named GRK1 (this is another designation of RK) to GRK6 (for a review, see ref. 117). Of these GRK1, but not GRK2[105] and GRK5,[117] is sensitive to recoverin (the sensitivity of other GRKs to recoverin has not been determined so far). Similar to recoverin, other members of the family of neuronal calcium sensors, including NCS 1, VILIP 1 and hippocalcin, are able to inhibit RK in a Ca^{2+}-dependent manner.[118]

Mechanism of the Inhibitory Action of Recoverin Upon Light-Dependent Rhodopsin Phosphorylation

The data that have been accumulated so far do not shed light on the mechanism of the inhibitory action of recoverin upon rhodopsin phosphorylation. However, they provide the basis for hypothetic mechanisms which, to a first approximation, can be divided into two following types.

The 1st type is not genuine inhibition in the kinetic sense and can be named as «inhibition by compartmentalization». It is known that in the presence of Ca^{2+}, recoverin: (i) exposes its myristoylated N-terminus which allows the protein to bind to photoreceptor membranes,[45,116] (ii) forms a complex with RK,[105] and (iii) targets RK to the membranes.[119] Although the existence of the binary RK recoverin complex attached to the lipid matrix of photoreceptor membranes has never been directly demonstrated, the preceding considerations may be used to argue that this Ca^{2+}-dependent complex could actually exist and would help recoverin to keep RK away from Rho* which would prevent Rho* phosphorylation.

The 2nd type involves a genuine inhibition of Rho* phosphorylation by direct interference of recoverin with the RK•Rho* complex. In this case, several subtypes of the inhibitory mechanism could be considered, but all of them imply a two-step

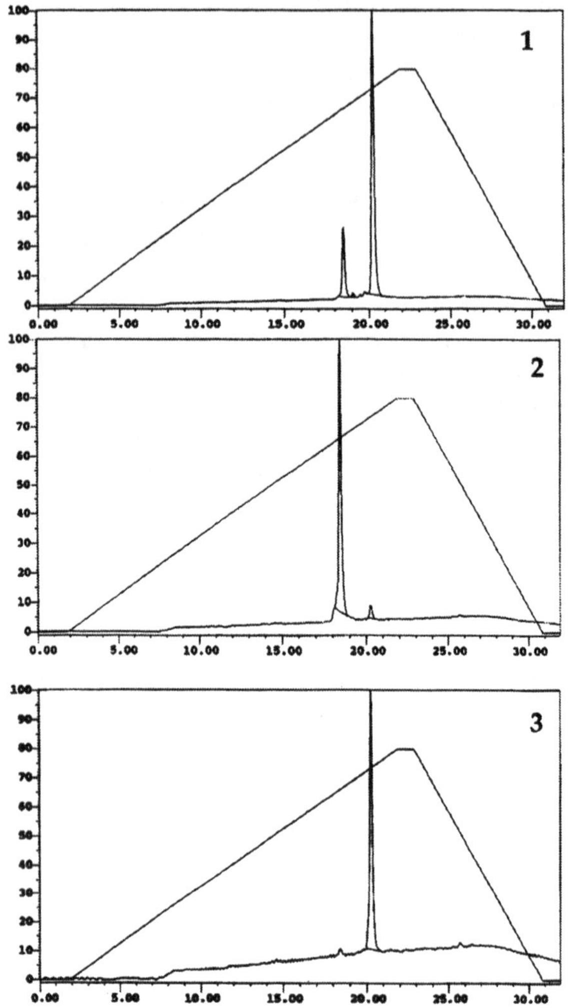

Figure 4. Separation of myristoylated and non-myristoylated forms of recoverin using Ca^{2+}-dependent binding of recoverin to ROS membranes. Preparations of recoverin were analyzed with HPLC chromatography: 1, initial myristoylated recoverin; 2, non-myristoylated recoverin; 3, purified myristoylated recoverin. The procedure of myristoylated recoverin purification from an admixure of non-myristoylated recoverin using urea-washed ROS membranes included: (i) incubation (20 min at 4 °C) of the recombinant myristoylated recoverin preparation (2 mg/ml) and bleached ROS membranes (4 mg rhodopsin per ml) in 20 mM Tris-HCl, pH 7.5, containing 0.2 mM $CaCl_2$, 140 mM NaCl, 1 mM dithiotreitol and 1 mM PMSF; (ii) washing the membranes with the same buffer and their centifugation (250,000 x g, 40 min) to remove non-myristoylated recoverin; (iii) ewashing the membranes with the same buffer, in which $CaCl_2$ was changed for 2 mM EGTA, to extract bound myristoylated recoverin. Before the analysis of the recoverin preparation obtained, the samples were loaded on Hi-trapQ column (AmershamPharmacia Biotech) and recoverin was eluted with a gradient containing 0-300 mM NaCl in 20 mM Tris-HCl, pH 7.5. Analysis of non-myristoylated and myristoylated recoverin preparations was performed with the use HPLC system (Gilson). Briefly, 25 μg of the purified recoverin were applied to analytical C18 reverse-phase column and recoverin was eluted with a gradient containing 0-70% acetonitrile in 0.1% trifluoroacetic acid at a flow rate of 1 ml/min. (I.I.Senin, M.Akhtar and P.P.Philippov, unpublished data).

scheme of Rho* phosphorylation by RK. According to this scheme[89,90,92] (see "Rhodopsin kinase, *The Enzymology and Activation*" in this chapter), before RK phosphorylates the C-terminus of Rho* (the 2nd step) it is activated by interaction with the 3rd cytoplasmic loop of Rho* (the 1st step). For further consideration, it would be useful to subdivide this two-step mechanism into the following events: (i) inactive RK binds to the 3rd cytoplasmic loop of Rho*, and (ii) due to this binding, conformational changes occur leading to the activation of RK; (iii) the C-terminal part of Rho* comes into contact with the substrate-binding site of RK that allows the C-terminus to occupy the proper position in the catalytic site of the enzyme; (iv) RK starts to catalyze the phosphate incorporation into the C-terminus of Rho*. In principle, any one of these events (or their combinations) could be sensitive to a Ca^{2+}-dependent inhibition by recoverin. In addition, recoverin could interfere with the binding of ATP (the second substrate) to RK and/or with the dissociation of ADP (the product) from RK.[106] The choice between options mentioned above is as yet not possible.

Non-Myristoylated Recoverin Detaches Rhodopsin Kinase from ROS Membranes and Myristoylated Recoverin Enhances the Membrane Targeting of the Enzyme

The inhibitory effect of non-myristoylated recoverin upon RK activity has already been described by several groups[105,113,114] (see "Recoverin as a Ca^{2+}-Sensor of Rhodopsin kinase in vitro, *Recoverin Inhibits Light-Dependent Rhodopsin kinase*" in this chapter). It is, however, yet unknown how this species affects the RK binding to ROS membranes.

Figure 5 shows the effect of non-myristoylated and myristoylated forms of recoverin on the membrane localization of RK when the total amount of RK added to the samples is taken as 100% (bar 1). One can see that the percentage of RK bound to bleached ROS membranes is significantly lower (about 30%) in the presence of non-myristoylated recoverin (bar 3) than it is (about 80%) without recoverin (bar 2). In contrast to non-myristoylated recoverin, its myristoylated form (bar 4) increases the amount of RK bound to the bleached membranes to 90%. The results of experiments with unbleached ROS membranes (bars 5-7) are shown for comparison. At 10 μM rhodopsin concentration, only a small portion of RK is bound to the unbleached membranes (bar 5). Myristoylated recoverin (bar 7) drastically increases the binding of RK (up to 75%), whereas non-myristoylated recoverin slightly decreases the binding (bar 6). (I.I.Senin, M.Akhtar and P.P.Philippov, unpublished data).

High-Gain Rhodopsin Phosphorylation and Recoverin

One rhodopsin molecule contains 9 potential sites for phosphorylation, and at a high bleaching level all these sites can be phosphorylated by RK in vitro.[65] However, at a low bleaching level, the phosphate incorporation into rhodopsin, calculated

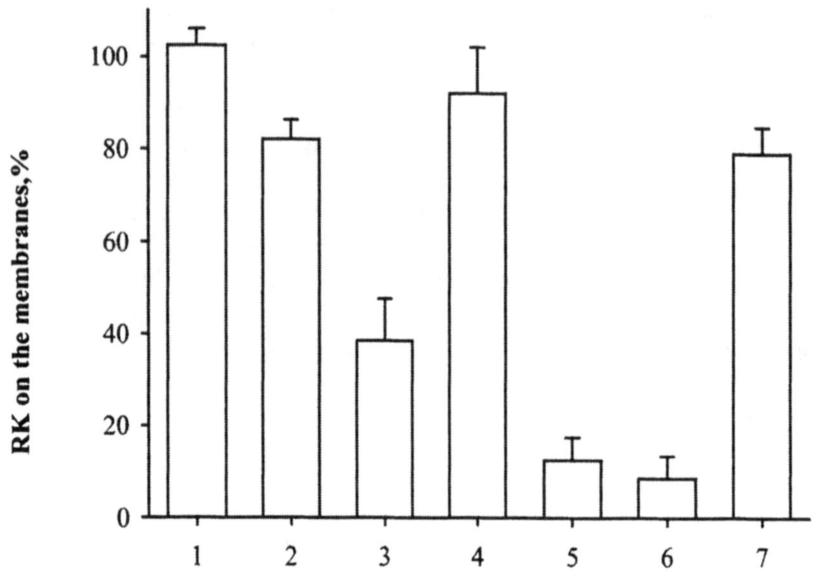

Figure 5. Effect of myristoylated and non-myristoylated recoverin upon interaction of RK with ROS membranes. The total amount of RK added in the samples was taken as 100% (bar 1). RK in the presence of bleached (2-4) or unbleached (5-7) ROS membranes without additions (2,5), with non-myristoylated (3,6) or myristoylated (4,7) recoverin. The mesurements were made in the presence of 5.6 μM Ca^{2+}, 15 μM non-myristoylated or myristoylated recoverin, 300 nM RK and ROS membranes (10 μM rhodopsin). (I.I.Senin, M.Akhtar and P.P.Philippov, unpublished data).

per Rho*, can reach tens in reconstituted systems[105,114,120-122] or even hundreds in suspension of electropermeabilized ROS.[123] This phenomenon dubbed «high-gain rhodopsin phosphorylation»[123] implies that at low bleaching, phosphoryl groups are incorporated not only into Rho*, but also into neighbouring nonbleached rhodopsin molecules, Rho, as well. There are several possible explanations to acount for the phenomenon of high-gain rhodopsin phosphorylation.

1. The presence of a light-independent protein kinase(s), for example protein kinase C,[124] different from RK is responsible for high gain phosphorylation detected in the reconstituted system containing purified retinal[120-122] or recombinant[105] RK. However, an argument against this possibility is the fact that the total amount of phosphorylated Rho increases with the increase of Rho*,[120] i.e., the process of nonbleached rhodopsin phosphorylation seems to be light-dependent.
2. RK once activated by Rho* is also capable of phosphorylating Rho.[120] In other words, this mechanism of high-gain phosphorylation suggests that «trans-phosphorylation»[125] occurs due to collisions of RK, previously activated by a Rho*, with Rho molecules, in the neighbourhood to Rho* molecule that had catalyzed RK activation.

Recent studies[125] have thrown doubt on the fact that the trans-phosphorylation exists at all. Authors of this work suggested an elegant method of distinguishing bleached and nonbleached forms of rhodopsin present in the same reaction mixture. In addition to recombinant wild type rhodopsin (WT-rhodopsin), the method exploited the functionally active split rhodopsin (SR) assembled from two separately expressed fragments of a rhodopsin, one containing transmembrane segments I - IV and the other containing transmembrane segments V - VII. Since only the C-terminal part of rhodopsin contains sites for phosphorylation, incubation of SR in the presence of RK and [γ-^{32}P]ATP resulted in ^{32}P incorporation into the smaller fragment containing transmembrane segments V - VII. In this approach, phosphorylated forms of WT-rhodopsin and SR having different electrophoretic mobility could easily be separated, using SDS - PAGE. In one set of experiments, the authors bleached vesicles containing WT-rhodopsin, mixed them with vesicles containing SR and incubated the mixture in the presence of RK and [γ-^{32}P] ATP. (In the reciprocal experiment, SR-containing vesicles were first bleached and then were mixed with vesicules containing nonbleached WT-rhodopsin). In none of the experiments, trans-phosphorylation has been revealed. However, the negative result obtained could be explained by the inability of RK to diffuse from one compartment, i.e., from WT-rhodopsin-containing vesicles, to the other one, i.e., CR-containing vesicles. To exclude this possibility a constitutively active mutant of rhodopsin, K296G, and SR were co-expressed, purified, reconstituted into the same vesicles and incubated with RK and [γ-^{32}P]ATP in the dark. In this system, whereas the phosphorylation of the K296G mutant occurred, the trans-phosphorylation of SR could not be detected. It should, however, be noted that in this work,[125] the K296G mutant (an analog of Rho*) and SR (an analog of Rho) were incorporated in the vesicles in the ratio of 1:1 that corresponds to 50% of rhodopsin bleaching in photoreceptor membranes. However, it is known that high-gain phosphorylation can only be demonstrated at low rhodopsin bleachings (below 10 %); at high bleachings, Rho* is the main or even the only substrate of phosphorylation.[120,121] If so, one may argue that in the experimental conditions used,[125] trans-phosphorylation was excluded because of too high a concentration of a constitutively active mutant of rhodopsin, K296G, which in this experiment fulfilled the role of Rho*.

Direct evidence of the existence of the high-gain phosphorylation was obtained by direct determination of phosphorylated Rho (Rho-P) and phosphorylated Rho* (Rho*-P) produced in photoreceptor membranes by the action of RK at different levels of bleaching.[120] It was observed that at a 5% bleach level, the amount of Rho-P formed was 15% that of Rho*-P, whereas at 60% bleaching Rho-P was less than 1% of Rho*-P. The phosphorylation of Rho did not occur when Rho* and Rho were dissolved in a detergent solution to exclude their compartmentalization, mixed with [γ-^{32}P]ATP and RK, and incubated in the dark. However, in this experiment, it was not possible to achieve a sufficiently high concentration of RK and, as a result, even a maximal concentration of the enzyme used was too low to detect Rho-P (I.I.Senin, M.Akhtar and P.P.Philippov, unpublished data).

Therefore, the necessary conditions to observe transphosphorylation are low bleaching of photoreceptor membranes and a sufficiently high concentration of RK.

What might be the biological sense of high-gain rhodopsin phosphorylation? Recently we suggested[121,122] that the phenomenon is, merely, a consequence of an unwanted side-reaction, that occurs when a huge excess of Rho over Rho* in ROS membranes exists, i.e., at low levels of illumination in the normal working range of retinal rods. When a rod receives a few photons under very low illumination, RK needs to find and phosphorylate a few molecules of Rho* in the presence of an exceeding number (an order of 10^6) of Rho molecules. Thus the possibility exists that RK, despite its preference for Rho* as a substrate compared to Rho,[120] may phosphorylate Rho. If so, it would be reasonably to expect that in vivo a mechanism could exist to prevent RK from the wasteful phosphorylation of Rho directing the kinase to fulfill its «correct» function of Rho* phosphorylation. Indeed, it was demonstrated in in vitro experiments[105,121,122] that recoverin inhibited Rho phosphorylation more effectively than it did that of Rho*. Obviously, the different efficiency of recoverin to inhibit phosphorylation of Rho in comparison with Rho* is based on in vitro data only and further proof that this mechanism actually functions in vivo is needed.

IS RECOVERIN A Ca^{2+}-SENSOR OF RHODOPSIN KINASE IN VIVO?

As it was shown in "Recoverin as a Ca^{2+}-Sensor of Rhodopsin kinase in vitro" of this chapter, recoverin, in vivo, acts as a Ca^{2+}-sensor of RK in a ROS suspension and reconstituted systems consisting of washed ROS membranes, RK and recoverin, i.e., in the in vitro conditions. Is this in vitro potential of recoverin realized in vivo? A limited amount of works have been performed on in vivo models, such as truncated or detached ROS,[126-128] retinal rod cells,[10,129-131] retina preparations[132] or recoverin knock-out mice[133] (for a review, see ref. 134). These studies do not provide a clear, or consistent, answer to this question. Some of the workers suggest recoverin to participate in Ca^{2+}-dependent control of rhodopsin phosphorylation under physiological conditions, and the others reject this possibility.

Arguments against recoverin as an in vivo Ca^{2+}-sensor of RK, which have recently been summarized in Otto-Bruc et al[132] (see Section "Discussion" in this work[132] and references herein), mainly concern the discrepancy between the Ca^{2+} affinity of recoverin experimentally determined in vitro and the suggested level of $[Ca^{2+}]_f$ in ROS cytoplasm in vivo. The point is that, according to the in vitro data (see "Recoverin, *Gene, Structure and Molecular Properties*" and "Recoverin as a Ca^{2+}-Sensor of Rhodopsin kinase in vitro, *Recoverin Inhibits Light-Dependent Rhodopsin kinase*" in this chapter), the binding of Ca^{2+} to recoverin and the Ca^{2+} dependence of the inhibition of rhodopsin phosphorylation occur at unphysiologically high concentrations of Ca^{2+}: experimentally determined values of K_d and IC_{50} lie in the μM range; whereas under physiological conditions, $[Ca^{2+}]_f$ in ROS cytoplasm changes from several hundreds of nM in the dark to several tens of nM in light. Nevertheless, the following arguments can be advanced to explain this discrepancy.

First, the high-affinity K_d for Ca^{2+} calculated for one of the state of myristoylated recoverin is assumed to be 0.11 μM[40] (see "Recoverin, *Gene, Structure and Molecular Properties*" in this chapter). Second, it has been revealed quite recently that a salamander rod illuminated with a visible laser shows a time dependent substantial increase in $[Ca^{2+}]_f$ within ROS.[135] Obviously, this phenomenon should be taken into account in the construction of a model describing the inactivation of the rod phototransduction cascade. Third, as mentioned earlier (see "Recoverin as a Ca^{2+}-Sensor of Rhodopsin kinase in vitro, *Recoverin Inhibits Light-Dependent Rhodopsin kinase*" in this chapter), recombinant myristoylated recoverin, uncontaminated with the non-myristoylated protein, has IC_{50} of 0.4 - 0.5 μM, that is, about 2 times lower than that (0.8 - 0.9 μM) for the preparations usually used. Fourth, in the intact retinal rod cell, the Ca^{2+} dependence of the recoverin effect is suggested to shift toward the dark concentration of $[Ca^{2+}]_f$.[128] Fifth, the high density of membranes, a plausible existense of a local gradient of $[Ca^{2+}]_f$ and the presence of different compartments within ROS might create unique conditions which cannot be reproduced in biochemical experiments.

The arguments mentioned above and those listed below can also help to explain how recoverin might inhibit RK under physiological conditions. (The authors of the discussed work believe that recoverin can not inhibit RK in vivo because «a weak interaction between RK and Ca^{2+}-recoverin»[132]). First, thc levels of rhodopsin bleaching are extremely low in working rods compared to those usually used in biochemical experiments (in the discussed work[132], the bleaching was about 10%). For example it is known that IC_{50} of the recoverin's inhibitory effect upon rhodopsin phosphorylation in the reconstituted system shifted progressively to lower values with decreasing levels of the bleaching.[121] Second, the recoverin effect is specific for RK (see "Recoverin as a Ca^{2+}-Sensor of Rhodopsin kinase in vitro, *Recoverin Inhibits Light-Dependent Rhodopsin kinase*" in this chapter). Finally, polyclonal monospecific antibodies against recoverin suppressed the Ca^{2+}-dependence of rhodopsin phosphorylation in ROS suspension without addition of exogeneous recoverin.[110] The ability of anti-recoverin antibodies to modulate rhodopsin phosphorylation were quite recently confirmed in experiments fulfiled on ROS homogenates and the antibody-treated rat eyes.[136,137]

One may add that the affinity of EF-hand-containing Ca^{2+}-binding proteins for calcium varies over a rather wide range: from 0.3 to 6 μM in the case of GCAP1[138] and calmodulin.[139] The range of values of the affinity of Ca^{2+}-binding proteins for their targets is even wider. Thus, a large number of calmodulin-dependent effectors are regulated by this protein at concentrations between 10^{-12} and 10^{-6} M.[140] For example, IC_{50} of channel activity in ROS plasma membranes occurs at a calmodulin concentration of 1.8 nM[141] while the IC_{50} of calmodulin inhibitory effect upon GRKs varies from ~50 nM for GRK5 to ~2 μM for GRK2 and GRK3.[142]

As for a diverse pattern of tissue expression of recoverin (see "Recoverin, *Tissue and Cellular Distribution*" in this chapter), which is also considered as an argument against recoverin as a Ca^{2+} sensor of RK,[132] calmodulin, a prototypical EF-hand Ca^{2+} sensor, is expressed in all eukaryotic cells where it participates in signalling pathways that regulate many crucial processes such as growth, proliferation and movement.[140]

The final conclusion of[132] is that recoverin may simply play a role as a Ca^{2+} buffer. However, it is known that Ca^{2+}-binding proteins, which are Ca^{2+} buffers (e.g., parvalbumin and calbindin), do not undergo a significant change in their conformation on binding Ca^{2+}.[140] Whereas recoverin similar to known Ca^{2+} sensors undergoes a Ca^{2+}-induced change in conformation (see "Recoverin, *Gene, Structure and Molecular Properties*" in this chapter). Nevertheless, the investigation of the kinetics and sensitivity of photocurrent responses of salamander rods does not exclude two actions of recoverin, the inhibition of RK and calcium buffering.[130] We will not diccuss other electrophysiological experiments aimed at the elucidation of the in vivo function of recoverin, as this aspect has been covered in a critical review.[1]

CONCLUSION. RECOVERIN: MANY FUNCTIONS OR AN UNKNOWN FUNCTION?

The commulative in vitro experiments discussed in this chapter suggest that in ROS, a light antenna of retinal rod cells, recoverin acts as a Ca^{2+}-sensor of RK in the reactions of phosphorylation of Rho* and Rho. It is not unlikely that in ROS, recoverin also fulfills a function of a Ca^{2+} buffer. In addition, Ca^{2+}-recoverin is capable of activating phosphatase 2A in ROS homogenates.[143] However, the question of whether photoreceptors actually use these properties of recoverin in vivo, remains unclear. In addition to ROS, recoverin is present in other parts of retinal rods; it is also found in higher order neurons of retina, ocular ciliary epitelium and the pineal organ. Are unknown protein kinases related to RK the targets of recoverin in these structures? Or, alternatively, does recoverin interact with a quite different target(s) in them? These questions remain to be solved.

ACKNOWLEDGMENTS

We are indebted to E.V. Bragina and Dr. E. E. Skorikova for helping us in preparation of this article, to M. Smith and I. P. Vorojeikina for their excellent assistance, and to Dr. A. M. Dizhoor who made an invaluable contribution in the discovery of recoverin, the initial name of which was «p26». The work was supported by grants from Wellcome Trust (M.A. and P.P.P.), Ludwig Institute for Cancer Research (P.P.P.), Deutscher Akademischer Austauschdienst (P.P.P.) and Russian Foundation for Basic Research N 00-04-48332 (P.P.P.).

REFERENCES

1. Burns ME, Baylor DA. Activation, deactivation, and adaptation in vertebrate photoreceptor cells. Ann Rev Neurosci 2001; 24:779-805.
2. Fain GL, Matthews HR, Cornwall MC et al. Adaptation in vertebrate photoreceptors. Physiol Rev 2001; 81:117-151.
3. Pugh EN, Nikonov S, Lamb TD. Molecular mechanisms of vertebrate photoreceptor light adaptation. Curr Opin Neurobiol 1999; 9:410-418.
4. Kawamura S. Molecular mechanisms of light-adaptation in retinal photoreceptors. Keio J Med 1994; 43:149-154.
5. Dizhoor AM, Nekrasova ER, Philippov PP. The binding of G proteins to immobilized delipidated rhodopsin. Biochem Biophys Res Commun 1989; 162:544-549.
6. Dizhoor AM, Nekrasova ER, Philippov PP. New 26 kDa protein specific for photoreceptor cells, capable of binding to immobilized delipidated rhodopsin. Biokhimia 1991; 56:225-228.
7. Dizhoor AM, Ray S, Kumar S et al. Recoverin: a calcium sensitive activator of retinal rod guanylate cyclase. Science 1991; 251:915-918.
8. Koch K-W, Stryer L. Highly cooperative feedback control of retinal rod guanylate cyclase by calcium ions. Nature 1998; 334:64-66.
9. Lambrecht H-G, Koch K-W. A 26 kd calcium binding protein from bovine rod outer segments as modulator of photoreceptor guanylate cyclase. EMBO J 1991; 10:793-798.
10. Gray-Keller MP, Polans AS, Palczewski K et al. The effect of recoverin-like calcium-binding protein on the photoresponse of rerinal rods. Neuron 1993; 10:523-531.
11. Gorodovikova EN, Philippov PP. The presence of a calcium-sensitive p26-containing complex in bovine retina rod cells. FEBS Lett 1993; 335:277-279.
12. Hurley JB, Dizhoor AM, Ray S et al. Recoverin's role: conclusion withdrawn. Science 1993; 260:740.
13. Gorczyca WA, Gray-Keller MP, Detwiler PB et al. Purification and physiological evaluation of a guanylate cyclase activating protein from retinal rods. Proc Natl Acad Sci USA 1994; 91:4014-4018.
14. Dizhoor AM, Olshevskaya EV, Henzel WJ et al. Cloning, sequencing, and expression of a 24-kDa Ca^{2+}-binding protein activating photoreceptor guanylyl cyclase. J Biol Chem 1995; 270:25200-25206.
15. Thirkill CE, Roth AM, Keltner JL. Cancer-associated retinopathy. Arch Ophtalmol 1987; 105:372-375.
16. Polans AS, Buczylko J, Crabb J et al. A photoreceptor calcium binding protein is recognized by autoantibodies obtained from patients with canser-associated retinopathy. J Cell Biol 1991; 112:981-989.
17. Kawamura S, Murakami M. Calcium-dependent regulation of cyclic GMP phosphodiesterase by a protein from frog retinal rods. Nature 1991; 349:420-423.
18. Kawamura S, Takamatsu K, Kitamura K. Purification and characterization of S-modulin, a calcium-dependent regulator on cGMP phosphodiesterase in frog photoreceptors. Biochem Biophys Res Commun 1992; 186:411-417.
19. Kawamura S. Light-sensitivity modulating protein in frog rods. Photochem Photobiol 1992; 56:1173-1180.
20. Korf HW, White BH, Schaad NC et al. Recoverin in pineal organs and retinae of various vertebrate species including man. Brain Research 1992; 595:57-66.
21. Milam AH, Dacey DM, Dizhoor A. Recoverin immunoreactivity in mammalian cone bipolar cells. Vis Neurosci 1993; 10:1-12.
22. Wiechmann AF, Hammarback JA. Expression of recoverin mRNA in the human retina: localization by in situ hybridization. Exp Eye Res 1993; 57:763-769.
23. Polans A, Burton MD, Haley TL et al. Recoverin, but not visinin, is an autoantigen in the human etina identified with a cancer-associated retinopathy. Invest Ophtalmol Vis Sci 1993; 34:81-90.

24. De Raad S, Comte M, Nef P et al. Distribuion pattern of three neural calcium-binding proteins (NSC-1, VILIP and recoverin) in chicken, bovine and rat retina. Histochem J 1995; 27:524-535.
25. cGinnis JF, Stepanik PL, Jariangprasert S et al. Functional significance of recoverin localization in multiple retina cell types. J Neurosci Res 1997; 50:487-495.
26. Dalil-Thiney N, Bastianelli E, Pochet R et al. Recoverin and hippocalcin distribution in the lamprey (Lampreta fluviatilis) retina. Neurosci Lett 1998; 247:163-166.
27. Bazhin AV, Poplinskaya VA, Tikhomirova NK et al. Immunochemical localization of alcium-binding protein recoverin in retina of newt Pleurodeles walt. Biol Bull (Moscow) (in press).
28. Hatakenaka S, Kiyama H, Tohyama M et a. Immunohistochemical localization of chick retinal 24 kdalton protein (visinin) in various vertebrate retinae. Brain Res 1985; 331:209-215.
29. Bertazolli-Filho R, Ghosh S, Huang W et al. Molecular evidence that human ocularciliary epithelium expresses components involved in phototransduction. Biochem Biophys Res Commun 2001; 284:317-25.
30. Bastianelli E, Pochet R. Calbindin-D28k, calretinin, and recoverin immunoreactivities in developing chick pineal gland. J Pineal Res 1994; 17:103-111.
31. Bastianelli E, Polans AS, Hidaka H et al. Differential distribution of six calcium-binding proteins in the rat olfactory epithelium during postnatal development and adulthood. J Comp Neurol 1995; 354:395-409.
32. Wiechmann AF, Hammarback JA. Mlecular cloning and nucleotide sequence of a cDNA encoding recoverin from human retina. Exp Eye Res 1993; 56:463-470.
33. Wiechmann AF, Akots G, Hammarback JA et al. Genetic and physical mapping of human recoverin: a gene expressed in retinal photoreceptors. Invest Ophtalmol Vis Sci 1994; 35:325-331.
34. Murakami A, Yajima T, Inana G. Isolation of human retinal genes: recoverin cDNA and gene. Biochem Biophs Res Commun 1992; 181:234-244.
35. Dollfus H, Rozet JM, Musarella MA et al. Dinucleotide repeat polymorphism at the human recoverin RCVI gene lcus on chromosome 17p. Hum Mol Genet 1993; 7:1081.
36. McGinnis JF, Lerious V, Pazik J et al. Chromosomal assignment of the recoverin gene and cancer-associated retnopathy. Mamm Genome 1993; 4:43-45.
37. Ray S, Zozulya S, Niemi GA et al. Proc Natl Acad Sci USA 1992; 89:5705-5709.
38. Dizhoor AM, Ericsson LH, Johnson RS et a. The NH_2 terminus of retinal recoverin is acylated by a small family of fatty acids. J Biol Chem 1992; 267:16033-16036.
39. Flaherty KM, Zozulya S Stryer L et al. Three-dimensional structure of recoverin, a calcim sensor in vision. Cell 1993; 75:709-716.
40. Ames JB, Porumb T, Tanaka T et al. Amino-terminal myristoylation induces cooperative calcium binding to recoverin. JBiol Chem 1995; 270:4526-4533.
41. Ames JB, Tanaka T, Stryer L et al. Secondary structure of myristoylated recoverin determined by three-dimensional heteronuclearNMR: implication for the calcium-myristoyl switch. Biochemistry 1994; 33:10743-10753.
42. Permyakov SE, Cherskaya AM, Senin II et al. Effects of mutations in the calcium-bnding sites of recoverin on its calcium affinity: evidence for successive filling of the calcium binding sites. Protein ng 2000; 13:783-790.
43. Alekseev AM, Shulga-Morskoy SV, Zinchenko et al. Obtaining and characterization of EF-hand mutants of recoverin. FEBS Lett 1998; 440:116-118.
44. Klenchin VA, Calvert PD, Bownds MD. Inhibition of rhodopsin kinase by rcoverin. Further evidence for a negative feedback system in phototransduction. J Biol Chem 1995; 270:16147-16152.
45. Zozulya S, Stryer L. Calcium-myristoyl protein switch. Proc Natl Acad Sci USA 1992; 89:11569-11573.
46. Ames JB, Tnaka T, Stryer L et al. Portrait of a myristoyl switch protein. Curr Opin Struct Biol 1996; 6:432-438.

47. Lange C,Koch K-W. Calcium-dependent binding of recoverin to membranes monitored by surface plasmon resonance spectroscopy in real time. Biochemistry 1997; 36:12019-12026.
48. Senin II, Fisher T, Komolov KE et al. Mechanism of a Ca^{2+}-myristoyl switch in recoverin (in preparation).
49. Braunewell K-H, Gundelfinger ED. Intracellular neuronal calcium sensor proteins: a family of EF-hand calciumbinding proteins in search of a function. Cell Tissue Res 1999; 295:1-12.
50. Burgoyne RD, Weiss JL. The neuronal calcium sensor family of Ca^{2+}-binding proteins. Biochem J 2001; 353:1-12.
51. An WF, Bowlb MR, Betty M et al. Modulation of A-type potassium channels by a family of calcium sensors. Nature 2000; 403:553-556.
52. Carrion AM, Link WA, Ledo F et al. DREAM is a Ca^{2+}-regulated transcriptional repressor. Nature 1999; 398:80-84.
53. Hendricks KB, Wang BQ, Schnieders EA et al. Yeast homologue of neuronal frequenin is a regulator of phophatidylinositol-4-OH kinase. Nat Cell Biol 1999; 1:234-241.
54. Pongs O, Lindemeier J, Zhu XR et al. Frequenin – a novel calcium-binding protein that modulates synaptic efficacy in the Drosophila nervous system. Neuron 1993; 11:15-28.
55. Olafsson P, Wang T, Lu B. Molecular cloning and functional characterization of the Xenopus Ca^{2+}-binding protein frequenin. Proc Natl Acad Sci USA 1995; 92:8001-8005.
56. McFerran BW, Graham ME, Burgoyne RD. Neuronal Ca^{2+} sensor 1, the mammalian homologue of frequenin, is expressed in chromaffin and PC12 cells and regulates neurosecretion from dense-core granules. J Biol Chem 1998; 273:22768-22772.
57. Schaad NC, De Castro E, Nef S et al. Direct modulation of calmodulin targets by the neuronal calcium sensor NCS-1. Proc Natl Acad Sci USA 1996; 93:9253-9258.
58. Boekhoff I, Braunewell KH, Andreini I et al. The calcium-binding protein VILIP in olfactory neurons: requlation of second messenger signaling. Eur J Cell Biol 1997; 72:151-158.
59. Braunewell KH, Spilker C, Behnisch T et al. The neuronal calcium-sensor protein VILIP modulates cyclic AMP accumulation in stably transfected C6 glioma cells: amino-terminal myristoylation determines functional activity. J Neurochem 1997; 68:2129-2139.
60. De Castro E, Nef S, Fiumelli H et al. Regulation of rhodopsin phosphorylation by a family of neuronal calcium sensors. Biochem Biophys Res Commun 1995; 216:133-140.
61. Bownds MD, Dawes J, Miller JL et al. Phosphorylation of frog photoreceptor membranes induced by light. Nature New Biol 1972; 237:125-127.
62. Kuhn H, Dreyer WJ. Light-dependent phosphorylation of rhodopsin by ATP. FEBS Lett 1972; 20:1-6.
63. Frank RN, Cavanagh HD, Kenyon KR. Light-stimulated phosphorylation of bovine visual pigment by adenosine triphosphate. J Biol Chem 1973; 248:596-609.
64. Sale GJ, Towner P, Akhtar M. Topography of the rhodopsin molecule. Identification of the domain phosphorylated. Biochem J 1978; 175:421-430.
65. Wilden U, Kühn H. Light-dependent phosphorylation of rhodopsin: number of phosphorylation sites. Biochemistry 1982; 21:3014-3022.
66. Thompson P, Findlay JB. Phosphorylation of ovine rhodopsin. Identification of the phosphorylated sites. Biochem J 1984; 220:773-780.
67. Shichi H, Somers RL Light-dependent phosphorylation of rhodopsin. Purification and properties of rhodopsin kinase. J Biol Chem 1978; 253:7040-7046.
68. Hargrave PA, Fong SL, McDowell JH et al. The partial primary structure of bovine rhodopsin and its topography in the retinal rod cell disc. Neurochemistry 1980; 1:231-244.
69. Palczewski K, Benovic JL. G-protein-coupled receptor kinases. Trends Biochem Sci 1991; 16:387-391.
70. Palczewski K, McDowell JH, Hargrave PA. Purification and characterization of rhodopsin kinase. J Biol Chem 1988; 263:14067-14073.
71. Lorenz W, Inglese J, Palczewski K et al. The receptor kinase family: primary structure of rhodopsin kinase reveals similarities to the beta-adrenergic receptor kinase. Proc Natl Acad Sci USA 1991; 88:8715-8719.

72. Khani SC, Abitbol M, Yamamoto S et al. Characterization and chromosomal localization of the gene for human rhodopsin kinase. Genomics 1996; 35:571-576.
73. Anant JS, Fung BK. In vivo farnesylation of rat rhodopsin kinase. Biochem Biophys Res Commun 1992; 183:468-473.
74. Inglese J, Glikman JF, Lorenz W et al. Isoprenylation of a protein kinase. Requirement of farnesylation/alpha-carboxyl methylation for full enzymatic activity of rhodopsin kinase. J Biol Chem 1992; 267:1422-1425.
75. Inglese J, Koch WJ, Caron MG et al. Isoprenylation in regulation of signal transduction by G-protein-coupled receptor kinases. Nature 1992; 359:147-150.
76. McCarthy NE, Akhtar M. Function of the farnesyl moiety in visual signalling. Biochem J 2000; 347:163-171.
77. Kelleher DJ, Johnson GL. Characterization of rhodopsin kinase purified from bovine rod outer segments. J Biol Chem 1990; 265:2632-2639.
78. Lee RH, Brown BM, Lolley RN. Autophosphorylation of rhodopsin kinase from retinal rod outer segments. Biochemistry 1982; 21:3303-3307.
79. Palczewski K, Buczylko J, Van Hooser P et al. Identification of the autophosphorylation sites in rhodopsin kinase. J Biol Chem 1992; 267:18991-18998.
80. Pulvermuller A, Palczewski K, Hofmann KP. Interaction between photoactivated rhodopsin and its kinase: stability and kinetics of complex formation. Biochemistry 1993; 32:14082-14088.
81. Buczylko J, Gutmann C, Palczewski K. Regulation of rhodopsin kinase by autophosphorylation. Proc Natl Acad Sci USA 1991; 88:2568-2572.
82. Palczewski K, Ohguro H, Premont RT et al. Rhodopsin kinase autophosphorylation. Characterization of site-specific mutations. J Biol Chem 1995; 270:15294-15298.
83. McCarthy NEM. Mechanistic studies on rhodopsin kinase: a farnesylated protein. Ph D Thesis. Southampton University, Southampton. 1998.
84. Pullen N, Akhtar M. Rhodopsin kinase: studies on the sequence of and the recognition of and the recognition motif for multiphosphorylations. Biochemistry 1994; 33:14536-14542.
85. Cha K, Bruel C, Inglese J et al. Rhodopsin kinase: expression in baculovirus-infected insect cells, and characterization of post-translational modifications. Proc Natl Acad Sci USA 1997; 94:10577-10582.
86. Bruel C, Cha K, Reeves PJ et al. Rhodopsin kinase: expression in mammalian cells and a two-step purification. Proc Natl Acad Sci USA 2000; 97:3004-3009.
87. Frank RN, Buzney SM. Mechanism and specificity of rhodopsin phosphorylation. Biochemistry 1975; 14:5110-5117.
88. McDowell JH, Kuhn H. Light-induced phosphorylation of rhodopsin in cattle photoreceptor membranes: substrate activation and inactivation. Biochemistry 1977; 146:4054-4060.
89. Fowles C, Sharma R, Akhtar M. Mechanistic studies on the phosphorylation of photoexcited rhodopsin. FEBS Lett 1988; 238:56-60.
90. Brown NG, Fowles C, Sharma R et al. Mechanistic studies on rhodopsin kinase. Light-dependent phosphorylation of C-terminal peptides of rhodopsin. Eur J Biochem 1992; 208:659-657.
91. McCarthy NEM, Akhtar M. Concerning the activation of rhodopsin kinase. Biochem J (submitted for publication).
92. Palczewski K, Buczylko J, Kaplan MW et al. Mechanism of rhodopsin kinase activation. J Biol Chem 1991; 266:12949-12955.
93. Paulsen R, Bentrop J. Activation of rhodopsin phosphorylation is triggered by the lumirhodopsin-metarhodopsin I transition. Nature 1983; 302:417-419.
94. McDowell JH, Nawrocki JP, Hargrave PA. Phosphorylation sites in bovine rhodopsin. Biochemistry 1993; 32:4968-4974.
95. Ohguro H, Johnson RS, Ericsson LH et al. Control of rhodopsin multiple phosphorylation. Biochemistry 1994; 33:1023-1028.

96. Ohguro H, Van Hooser JP, Milam AH et al. Rhodopsin phosphorylation and de phosphorylation in vivo. J Biol Chem 1995; 270:14259-14262
97. King AJ, Andjelkovic N, Hemmings BA et al. The phospho-opsin phosphatase from bovine rod outer segments. An insight into the mechanism of stimulation of type-2A protein phosphatase activity by protamine. Eur J Biochem Biochemistry 1994; 225:383-394.
98. Wilden U. Duration and amplitude of the light-induced cGMP hydrolysis in vertebrate photoreceptors are regulated by multiple phosphorylation of rhodopsin and by arrestin binding. Biochemistry 1995; 34:1446-1454.
99. Zhang L, Sports CD, Osawa S et al. Rhodopsin phosphorylation sites and their role in arrestin binding. J Biol Chem 1997; 272:14762-14768.
100. Krupnick JG, Gurevich VV, Benovic JL. Mechanism of quenching of phototransduction. Binding competition between arrestin and transducin for phosphorhodopsin. J Biol Chem 1997; 272:18125-18131.
101. Brannock MT, Weng K, Robinson PR. Rhodopsin's carboxyl-terminal threonines are required for wild-type arrestin-mediated quench of transducin activation in vitro. Biochemistry 1999; 38:3770-3777.
102. Mendez A, Burns ME, Roca A et al. Rapid and reproducible deactivation of rhodopsin requires multiple phosphorylation sites. Neuron 2000; 28:153-164.
103. Aton BR, Litman BJ, Jackson ML. Isolation and identification of the phosphorylated species of rhodopsin. Biochemistry 1984; 23:1737-1741.
104. Pullen N, Brown NG, Sharma RP et al. Cooperativity during multiple phosphorylation catalyzed by rhodopsin kinase: supporting evidence using synthetic phosphopeptides. Biochemistry 1993; 32:3958-3964.
105. Chen C-K, Inglese J, Lefkowitz RJ et al. Ca^{2+}-dependent interaction of recoverin with rhodopsin kinase. J Biol Chem 1995; 270:18060-18066.
106. Satpaev DK, Chen C-K, Scotti A et al. Autophosphorylation and ADP regulate the Ca^{2+}-dependent interaction of recoverin with rhodopsin kinase. Biochemistry 1998; 37:10256-10262.
107. Tachibanaka S, Nanda K, Sasaki K et al. Amino acid residues of S-modulin resposible for interaction with rhodopsin kinase. J Biol Chem 2000; 275:3313-3319.
108. Levay K, Satpaev DK, Pronin AN et al. Localization of the sites for Ca^{2+}-binding proteins on G protein-coupled recepor kinases. Biochemistry 1998; 37:13650-13659.
109. Kawamura S, Hisatomi O, Kayada S et al. Recoverin has S-modulin activity in frog rods. J Biol Chem 1993; 268:14579-14582.
110. Gorodovikova EN, Gimelbrant AA, Senin II et al. Recoverin mediates the calcium effect upon rhodopsin phosphorylation and cGMP hydrolysis in bovine retina rod cells. FEBS Lett 1994; 349:187-190.
111. Gorodovikova EN, Senin II, Philippov PP. Calcium-sensitive control f rhodopsin phosphorylation in the reconstituted system consisting of photoreceptor membranes, rhodopsin kinase and recoverin. FES Lett 1994; 353:171-172.
112. Zargarov AA, Senin II, Alekseev AM et al. Preparation of the myristoylated and nonmyristoylated form of recombinant recoverin in E. coli cells and comparison of their functional activity. Biorg Khim 1996; 22:483-488.
113. Kawamura S, Cox JA, Nef P. Inhibition of rhodopsin phosphorylation by non-myritoylated recombinant recoverin. Biochem Biophys Res Commun 1994; 203:121-127.
114. Senin II, Zargarov AA, Alekseev AM et al. N-myristoylation of recoverin enhaces its efficiency as an inhibitor of rhodopsin kinase. FEBS Lett 1995; 376:87-90.
115. Neubert TA, Walsh KA, Hurley JB et al. Monitoring calcium-induced conformational changes in recoverin by electrospay mass spectrometry. Protein Sci 1997; 6:843-850.
116. Dizhoor AM, Chen C-K, Sinelnikova VV et al. Role of the acylated amio terminus of recoverin in Ca^{2+}-dependent membrane interaction. Science 1993; 259:829-832.
117. Sallese M, Iacovelli L, Cumashi A et al. Regulation of G protein-coupled receptor kinase subtypes by calcium ensor proteins. Biochim Biophys Acta 2000; 1498:112-121.

118. De Castro E, Nef S, Fiumelli H et al. Regulation of rhodopsin phosphorylation by a family of neuronl calcium sensors. Biochem Biophys Res Commun 1995; 216:133-140.
119. Sanada K, Shimizu F, Kameyama K et al. Calcium-bound recoverin targets rhodopsin kinase to membranes to inhibit rhodopsin phosphorylation. FEBS Lett 1996; 384:227-230.
120. Dean KR, Akhtar M. Phosphorylation of solubilized dark-adapted rhodopsin. Insights into the activation of rhodopsin kinase. Eur J Biochem 1993; 21:881-890.
121. Senin II, Dean KR, Zargarov AA et al. Recoverin inhibits the phoshorylation of dark-adapted rhodopsin more than it does that of bleached rhodopsin: a possible mechanism through which rhodopsin kinase is prevented from participation in a side reaction. Biochem J 1997; 321:551-555.
122. Senin II, Zargarov AA, Akhtar M et al. Rhodopsin phosphorylation in bovine rod outer segments is mor sensitive to the inhibitory action of recoverin at the low rhodopsin bleaching than it is at the high bleaching. FEBS Lett 1997; 408:251-254.
123. Binder BM, Biernbaum MS, Bownds MD. Light activation of one rhodopsin molecule causes the phosphorylation of hundreds of others. A reaction observed in electropermeabilized frog rod outer segments exposed to dim illumination. J Biol hem 1990; 265:15333-15340.
124. Newton AC, Williams DS. Rhodopsin is the major in situ substrate of protein kinase C in rod outer segments of photoreceptors. J Biol Chem 1993; 268:18181-18186.
125. Rim J, Faurobert E, Hurley JB et al. In vitro assay for trans-phosphorylation of rhodopsin by rhodopsin kinase. Biochemistry 1997; 36:7064-7070.
126. Koutalos Y, Nakatani K, Yau K-W. The cGMP-phosphodiesterase and its contribution to sensitivity regulation of retinal rods. J Gen Physiol 1995; 106:891-921.
127. Gray-Keller MP, Detwiler PB. Ca2+ dependence of dark- and light-adapted flash responses in rod photoreceptors. Neuron 1996; 17:323-331.
128. Erickson MA, Lagnado L, Zozulya S et al. The effect of recombinant recoverin on the photoresponse of truncated rod photoreceptors. Proc Natl Acad Sci USA 1998; 95:6474-6479.
129. Lubarsky A, Nikonov S, Pugh EN. The kinetics of inactivation of the rod phototransduction cascade with constant $Ca^{2+}{}_i$. J Gen Physiol 1996; 107:19-34.
130. Nikonov S, Lamb TD, Pugh EN. The role of steady phosphodiesterase activity n the kinetics and sensitivity of the light-adapted salamander rod photoresponse. J Gen Physiol 2000; 116:795-824.
131. Matthews HR. Actions of Ca^{2+} on an early stage in phototransduction revealed by the dynamic fall in Ca^{2+} concentration during the bright flash response. J Gen Physiol 1997; 109:141-146.
132. Otto-Bruc AE, Fariss RN, Van Hooser JP et al. Phosphorylation of photolyzed rhodopsin is calcium-insensitive in retina permeabilized by a-toxin. Proc Natl Acad Sci USA 1998; 95:15014-15019.
133. Baylor D. How photon start vision. Proc Natl Acad Sci USA 1996; 93:560-565.
134. Lem J, Makino C. Phototransduction in transgenic mice. Curr Opin Neurobiol 1996; 6:453-458.
135. Matthews HR, Fain GL. A light-dependent increase in free Ca^{2+} concentration in the salamander rod outer segment. J Physiol 2001; 532:305-321.
136. Maeda T, Maeda A, Maruyama I et al. Mechanisms of photoreceptor cell death in cancer-associated retinopathy. Invest Ophtalmol Vis Sci 2001; 42:705-712.
137. Ohguro H, Ogawa K, Maeda T et al. Retinal dysfunction in cancer-associated retinopathy is improved by Ca^{2+} antagonist administration and dark adaptation. Invest Ophtalmol Vis Sci 2001; 42:2589-2595.
138. Gorczyca WA, Polans AS, Surgucheva IG et al. Guanylyl cyclase activating protein. A calcium-sensitive regulator of phototransduction. J Biol Chem 1995; 270:22029-22036.
139. Cox JA, Malnoe A, Stein EA. Regulation of brain cyclic nucleotide phosphodiesterase by calmodulin. A quantitative analysis. J Biol Chem 1981; 256:3218-3222.
140. Chin D, Means AR. Calmodulin: a prototipical calcium sensor. Trends Cell Biol 2000; 10:322-328.

141. Hsu YT, Molday RS. Interaction of calmodulin with the cyclic GMP-gated channel of rod photoreceptor cells. Modulation of activity, affinity purification, and localization. J Biol Chem 1994; 269:29765-29770.
142. Pronin AN, Satpaev DK, Slepak VZ et al. Regulation of G protein-coupled receptor kinases by calmodulin and localization of the calmodulin binding domain. J Biol Chem 1997; 272:18273-18280.
143. Ohguro H, Rudnicka-Nawrot M, Buczylko J et al. Structural and enzymatic aspects of rhodopsin phosphorylation. J Biol Chem 1996; 271:5215-5224.

RECOVERIN AND RHODOPSIN KINASE

Ching-Kang Jason Chen

ABSTRACT

The majority of proteins involved in vertebrate phototransduction are expressed specifically in photoreceptors. Recoverin and rhodopsin kinase are expressed primarily in retinal photoreceptors and they interact with each other in a Ca^{2+}-dependent manner. This Ca^{2+}-dependent interaction has been studied extensively in vitro. Experiments utilizing animal models and electrophysiological approaches have started to provide important insight regarding its in vivo function. Recoverin can be viewed as a negative regulator of rhodopsin kinase in vertebrate phototransduction. This interaction imparts a negative feedback loop at the receptor level and may play an important role in light adaptation and in recovery.

INTRODUCTION

The visual pigments play a catalytic role in rod and cone phototransduction cascades that transduce incident photons into macroscopic neural signals. For effective vision, photoexcited visual pigments must be deactivated in a timely manner.[1] In mouse rod and cone photoreceptors, the deactivation is initiated by G protein-coupled receptor kinase 1 (GRK1 or commonly known as rhodopsin kinase, RK).[2,3] Extensive studies have been reported for rhodopsin deactivation in rod phototransduction where RK phosphorylates the C-terminal serine/threonine residues of rhodopsin in a light-dependent manner. Phosphorylated rhodopsin is then capped by a protein called arrestin that abolishes rhodopsin's ability to catalyze GTP/GDP exchange on the α-subunit of transducin.[4,5] Similar mechanisms are thought to be involved in deactivating cone visual pigments.[1] The Ca^{2+}-dependent interaction and inhibition of RK by recoverin provide a feedback mechanism by the

Department of Ophthalmology and Visual Sciences and Human Genetics, University of Utah, Salt Lake City, UT 84112, U.S.A.

light-induced fall of intracellular Ca^{2+} on receptor activity.[6-9] This may play an important role in photoreceptor light adaptation and in recovery. In this chapter, the evidences for and against the physiological relevance of such an interaction in phototransduction will be discussed. The readers are encouraged to consult other related chapters for a broader coverage of the history of recoverin research and its possible functions in the retina.

RK AND PHOTOTRANSDUCTION

The importance of RK in phototransduction is best demonstrated by the rectangular shape of rod single photon responses with an averaged 3-second response duration in the RK knockout (RK-/-) mice, a response that is in contrast to the typical normal flash responses that last less than 0.5 seconds.[2] The abnormal wave form of RK-/- rod single photon responses is due to the lack of light-dependent rhodopsin phosphorylation, as is demonstrated biochemically and by mass-spectrometry in rhodopsin derived from RK-/- retinas.[2] This rectangular wave form of photoresponses can also be found in transgenic rods where the phosphorylation sites of rhodopsin are removed[10] or mutated.[11] The human Oguchi disease patients with defective RK or arrestin genes suffer from congenital stationary night blindness.[12-14] This is because their rods are constantly saturated due to the inability to effectively quench photoexcited rhodopsin by phosphorylation, and thus must rely on the complete regeneration of rhodopsin via the much slower visual cycle to regain their rod sensitivity.[15]

The recovery of cone phototransduction, as revealed by electroretinographic (ERG) analyses, is greatly delayed in the RK-/- mouse retinas.[3] As RK is present in both rod and cone photoreceptors in retinas of mouse and human,[3, 16-18] it is worthwhile to note that Oguchi patients with defective RK gene reported normal or slightly abnormal photopic vision.[12] This may be explained by the existence of another G protein-coupled receptor kinase, GRK7, in human cones but not in mouse cones.[17, 18] In human, GRK7 may play a compensatory or a complementary role to RK in cone photoreceptors, while in mouse RK is solely responsible for the deactivation of visual pigments in both rod and cone photoreceptors.

RECOVERIN

Recoverin is a Ca^{2+}-binding protein,[19] whose N-terminus is heterogeneously acylated by a group of short chain fatty acids.[20] The N-terminal acyl moiety plays an essential role for recoverin to interact with rod outer segment (ROS) membranes in a Ca^{2+}-dependent manner [21] through a mechanism known as the "calcium-myristoyl switch".[22] The binding of Ca^{2+} ions to recoverin's two functional EF-hand motifs causes the release of the acyl moiety from within a hydrophobic cleft enabling use of the acyl moiety as an anchor to the membranes.[23] Upon binding to Ca^{2+}, the N-terminal region of both non-acylated and myristoylated recombinant recoverin becomes sensitive to limited proteolysis (i.e., has similar conformation changes), but because of the lack of an acyl moiety, non-acylated recoverin has negligible membrane affinity.[21] The acyl moiety may significantly alter recoverin's property in a membranous environment such as the ROS. It has been shown that an N-terminal myristoyl moiety reduces recoverin's affinity for Ca^{2+} but increase the Hill coefficient for its Ca^{2+} binding.[24] A similar gain of cooperativity is also found in the Ca^{2+}-dependent inhibition of rhodopsin phosphorylation by recoverin in a reconstituted system.[6]

Recoverin is primarily expressed in retinal photoreceptors, but it can also be found in a subset of retinal bipolar cells.[25] It was first characterized as a mediator of the negative feedback regulation imposed by Ca^{2+} on the de novo synthesis of cGMP in vertebrate photoreceptors,[19,26] a role now known to be played by the GCAPs (guanylyl cyclase activating proteins).[27-29] The frog counterpart of recoverin, the S-modulin, was identified at roughly the same time (1991) as a Ca^{2+}-dependent activator of cGMP-phosphodiesterase (PDE) in frog ROS.[30] The effect of S-modulin was later shown to be on its Ca^{2+}-dependent inhibition of rhodopsin phosphorylation.[31,32] Since native recoverin was initially purified on immobilized and delipidated rhodopsin columns until the use of Phenyl Sepharose was introduced,[33] the interaction of recoverin and rhodopsin may account for the inhibitory effect of recoverin on rhodopsin phosphorylation. However, as an independent attempt to identify recoverin's function, RK was found to be one of the Ca^{2+}-dependent recoverin binding proteins that can be purified by immobilized recoverin in the presence of Ca^{2+} and released in the presence of Ca^{2+}-chelator, EGTA.[6] Such a specific and Ca^{2+}-dependent interaction makes immobilized recoverin an affinity matrix for rapid isolation of RK.[34] More importantly, because recoverin neither binds to nor inhibits GRK2 (the β-adrenergic receptor kinase, or more commonly known as βARK) when rhodopsin is the substrate, the interaction between recoverin and RK was shown to account for the inhibitory effect of recoverin on rhodopsin phosphorylation.[6] In frog ROS, S-modulin was found crosslinked to a 60 KDa protein (presumably frog rhodopsin kinase) and not to rhodopsin,[8] further supporting the view that recoverin is a Ca^{2+}-dependent negative regulator of RK in vertebrate photoreceptors.

PHYSIOLOGICAL ROLE OF Ca^{2+}-DEPENDENT RECOVERIN/RHODOPSIN KINASE INTERACTION IN PHOTOTRANSDUCTION

Results from several elegant electrophysiological studies aiming to determine the physiological relevance of the interaction are so far consistent with the notion that recoverin confers Ca^{2+}-sensitivity to RK and thus prolongs PDE activation by inhibiting rhodopsin phosphorylation. These experiments are mentioned below. First, when internally dialyzed into functionally intact gecko rods through a patch pipette, purified retinal recoverin prolonged the rising phase of the photoresponses.[35] Second, recombinant acylated and non-acylated recoverin were dialyzed into truncated salamander ROS at different Ca^{2+} concentrations. It was found that at high Ca^{2+}, myristoylated recoverin prolonged the recovery phase of the bright flash response but had less effect on the dim flash response. The prolongation was shown to be mediated by inhibition of rhodopsin deactivation. Acylated recoverin is about an order of magnitude more efficient than non-acylated recoverin.[36] Third, the amplitude of dim flash responses of recoverin knockout rods were found to be smaller than the control rods.[37] Finally, the ERG analyses of recoverin knockout mice showed a reduced flash response amplitude as compared to normal mice under a steady background light,[38] implying that the loss of recoverin affects the transduction gain during light adaptation.

While the above experiments support the notion that recoverin mediates Ca^{2+}-dependent inhibition of rhodopsin phosphorylation by binding and inhibiting RK and thereby controls light-triggered phosphodiesterase activity, there were reports that cast doubts to this conclusion. First, the amount of Ca^{2+} required by recoverin to exert its inhibitory effect on rhodopsin phosphorylation in vitro is about an order of magnitude higher than the resting Ca^{2+} concentration measured for the dark-adapted rod photoreceptors.[6] This is in sharp contrast to the amount of Ca^{2+} required by GCAPs to regulate photoreceptor guanylyl cyclases, which falls within the physiological range of Ca^{2+} changes in the ROS. The controversy may be resolved by taking into account the membranous environment of the ROS, as suggested by Erickson et al[36] and by Ames et al.[24] They suggested that a membranous environment like the ROS may increase recoverin's affinity for Ca^{2+} as acylated recoverin binds to the disc membranes through the Ca^{2+}-myristoyl switch mechanism. Second, in bovine ROS permeabilized by staphylococcal α-toxin treatment, the light-induced phosphorylation of rhodopsin was found similar for Ca^{2+} concentration buffered at 1 μM or at 30 nM.[39] These results caution the use of Ca^{2+}-dependent recoverin/RK interaction to endorse the electrophysiological observations mentioned above.

CONCLUDING REMARKS

The major problem of using recoverin/rhodopsin kinase interaction to account for the results obtained by electrophysiological analyses is the lack of biochemical evidence of such a Ca^{2+}-dependent regulation of rhodopsin phosphorylation in intact retina.[39] However, it is noteworthy to mention that the lack of biochemical demonstration of light-dependent drop in cGMP concentration in intact ROS[40] does not obliterate cGMP as the second messenger in phototransduction. Biochemical assays, although used extensively for the study of signal transduction, are unfortunately limited in providing accurate accounts of local cellular events by many factors like the sensitivities of the instrumentations, necessary immense dilutions, arbitrary loss/addition of components, and destruction of complex biological environments. Therefore the negative results ought to be interpreted cautiously.

With regard to the physiological relevance of recoverin/rhodopsin kinase interaction, an important task now is to develop more sensitive biochemical assays under conditions used for electrophysiological analyses (e.g., low flash intensity and a faster reaction time course).[41] Alternatively, one can try to generate and characterize the flash responses of photoreceptors derived from transgenic or knock-in mice harboring mutation in the RK gene that makes it insensitive to recoverin binding and inhibition. A unique site in the N-terminal region of RK was reported to bind recoverin specifically.[42] Swapping the first 180 amino acids of RK to βARK confers the recoverin binding activity to βARK (Chen et al., unpublished result). Therefore it seems possible to generate recoverin-insensitive RK to further explore the physiological relevance of recoverin/RK interaction.

ACKNOWLEDGMENT

The author is supported in part by an unrestricted fund from the Research to Prevent Blindness to the Department of Ophthalmology and Visual Sciences of the University of Utah.

REFERENCES

1. Burns ME, Baylor DA. Activation, deactivation, and adaptation in vertebrate photoreceptor cells. Annu Rev Neurosci 2001; 24:779-805.
2. Chen CK, Burns ME et al. Abnormal photoresponses and light-induced apoptosis in rods lacking rhodopsin kinase. Proc Natl Acad Sci USA 1999; 96(7):3718-3722.
3. Lyubarsky AL, Chen C, Simon MI et al. Mice lacking G-protein receptor kinase 1 have profoundly slowed recovery of cone-driven retinal responses. J Neurosci 2000; 20(6):2209-2217.
4. Xu J, Dodd RL, Makino CL et al. Prolonged photoresponses in transgenic mouse rods lacking arrestin. Nature 1997; 389(6650):505-509.
5. Wilden U, Hall SW, Kuhn H. Phosphodiesterase activation by photoexcited rhodopsin is quenched when rhodopsin is phosphorylated and binds the intrinsic 48-kDa protein of rod outer segments. Proc Natl Acad Sci USA 1986; 83(5):1174-1178.
6. Chen CK, Inglese J, Lefkowitz RJ et al. Ca(2+)-dependent interaction of recoverin with rhodopsin kinase. J Biol Chem 1995; 270(30):18060-18066.
7. Klenchin VA, Calvert PD, Bownds MD. Inhibition of rhodopsin kinase by recoverin. Further evidence for a negative feedback system in phototransduction. J Biol Chem 1995; 270(27):16147-16152.
8. Sato N, Kawamura S. Molecular mechanism of S-modulin action: binding target and effect of ATP. J Biochem (Tokyo) 1997; 122(6):1139-1145.
9. Gorodovikova EN, Senin Ii, and Philippov PP. Calcium-sensitive control of rhodopsin phosphorylation in the reconstituted system consisting of photoreceptor membranes, rhodopsin kinase and recoverin. FEBS Lett 1994; 353(2):171-172.
10. Chen J, Makino CL, Peachey NS et al. Mechanisms of rhodopsin inactivation in vivo as revealed by a COOH-terminal truncation mutant. Science 1995; 267(5196):374-377.
11. Mendez A., Burns ME, Roca A et al. Rapid and reproducible deactivation of rhodopsin requires multiple phosphorylation sites. Neuron 2000; 28(1):153-164.
12. Cideciyan AV, Zhao X, Nielsen L et al. Null mutation in the rhodopsin kinase gene slows recovery kinetics of rod and cone phototransduction in man. Proc Natl Acad Sci USA 1998; 95(1):328-333.

13. Yamamoto S, Sippel KC, Berson EL et al. Defects in the rhodopsin kinase gene in the Oguchi form of stationary night blindness. Nat Genet 1997; 15(2):175-178.
14. Yamada T, Matsumoto M, Kadoi C et al. 1147 del A mutation in the arrestin gene in Japanese patients with Oguchi disease. Ophthalmic Genet 1999; 20(2):117-120.
15. Dryja TP. Molecular genetics of Oguchi disease, fundus albipunctatus, and other forms of stationary night blindness: LVII Edward Jackson Memorial Lecture. Am J Ophthalmol 2000; 130(5):547-563.
16. Zhao X, Huang J, Khani SC et al. Molecular forms of human rhodopsin kinase (GRK1). J Biol Chem 1998; 273(9):5124-5131.
17. Weiss ER, Ducceschi MH, Horner TJ et al. Species-specific differences in expression of G-protein-coupled receptor kinase (GRK) 7 and GRK1 in mammalian cone photoreceptor cells: implications for cone cell phototransduction. J Neurosci 2001; 21(23):9175-9184.
18. Chen C-K, Zhang K, Church-Kopish J et al. Characterization of human GRK7 as a potential cone opsin kinase. Molecular Vision 2001; 7:305-313.
19. Dizhoor AM, Ray S, Kumar S et al. Recoverin: a calcium sensitive activator of retinal rod guanylate cyclase. Science 1991; 251(4996):915-918.
20. Dizhoor AM, Ericsson LH, Johnson RS et al. The NH2 terminus of retinal recoverin is acylated by a small family of fatty acids. J Biol Chem 1992;. 267(23):16033-16036.
21. Dizhoor AM, Chen CK, Olshevskaya E et al. Role of the acylated amino terminus of recoverin in Ca(2+)-dependent membrane interaction. Science 1993; 259(5096):829-832.
22. Zozulya S, Stryer L. Calcium-myristoyl protein switch. Proc Natl Acad Sci USA 1992; 89(23):11569-11573.
23. Ames JB, Ishima R, Tanaka T et al. Molecular mechanics of calcium-myristoyl switches. Nature 1997; 389(6647):198-202.
24. Ames JB, Porumb T, Tanaka T et al. Amino-terminal myristoylation induces cooperative calcium binding to recoverin. J Biol Chem 1995; 270(9):4526-4533.
25. Milam AH, Dacey DM, Dizhoor AM. Recoverin immunoreactivity in mammalian cone bipolar cells. Vis Neurosci 1993; 10(1):1-12.
26. Hurley JB, Dizhoor AM, Ray S et al. Recoverin's role: conclusion withdrawn. Science 1993; 260(5109):40.
27. Dizhoor AM, Olshevskaya EV, Henzel WJ et al. Cloning, sequencing, and expression of a 24-kDa Ca(2+)-binding protein activating photoreceptor guanylyl cyclase. J Biol Chem 1995; 270(42):25200-25206.
28. Palczewski K, Subbaraya I, Gorczyca WA et al. Molecular cloning and characterization of retinal photoreceptor guanylyl cyclase-activating protein. Neuron 1994; 13(2):395-404.
29. Gorczyca WA, Polans AS, Surgucheva IG et al. Guanylyl cyclase activating protein. A calcium-sensitive regulator of phototransduction. J Biol Chem 1995; 270(37):22029-22036.
30. Kawamura S, Murakami M. Calcium-dependent regulation of cyclic GMP phosphodiesterase by a protein from frog retinal rods. Nature 1991; 349(6308):420-423.
31. Kawamura S. Rhodopsin phosphorylation as a mechanism of cyclic GMP phosphodiesterase regulation by S-modulin. Nature 1993. 362(6423):855-857.
32. Kawamura S, Hisatomi O, Kayada S et al. Recoverin has S-modulin activity in frog rods. J Biol Chem 1993; 268(20):14579-14582.
33. Polans AS, Buczylko J, Crabb J et al. A photoreceptor calcium binding protein is recognized by autoantibodies obtained from patients with cancer-associated retinopathy. J Cell Biol 1991; 112(5):981-989.
34. Chen CK, Hurley JB. Purification of rhodopsin kinase by recoverin affinity chromatography. Methods Enzymol 2000; 315:404-410.
35. Gray-Keller MP, Polans AS, Palczewski K et al. The effect of recoverin-like calcium-binding proteins on the photoresponse of retinal rods. Neuron 1993; 10(3):23-531.
36. Erickson MA, Lagnado L, Zozulya S et al. The effect of recombinant recoverin on the photoresponse of truncated rod photoreceptors. Proc Natl Acad Sci USA 1998; 95(11):6474-6479.

37. Dodd RL, Makino CL, Chen J et al. Visual transduction in transgenic mouse rods lacking recoverin. Invest Ophthalmol Vis Sci 1995; 36:S641.
38. Hurley JB, Chen J. Evaluation of the contributions of recoverin and GCAPs to rod photoreceptor light adaptation and recovery to the dark state. Prog Brain Res 2001; 131:395-405.
39. Otto-Bruc AE, Fariss RN, Van Hooser JP, Palczewski K. Phosphorylation of photolyzed rhodopsin is calcium-insensitive in retina permeabilized by alpha-toxin. Proc Natl Acad Sci USA 1998; 95(25):15014-15019.
40. Kilbride P, Ebrey TG. Light-initiated changes of cyclic guanosine monophosphate levels in the frog retina measured with quick-freezing techniques. J Gen Physiol 1979; 74(3):415-426.
41. Kennedy MJ, Lee KA, Niemi GA et al. Multiple phosphorylation of rhodopsin and the in vivo chemistry underlying rod photoreceptor dark adaptation. Neuron 2001. 31(1):87-101.
42. Levay K, Satpaev DK, Pronin AN et al. Localization of the sites for Ca2+-binding proteins on G protein-coupled receptor kinases. Biochemistry 1998; 37(39):13650-13659.

PATHOLOGICAL ROLES OF RECOVERIN IN CANCER-ASSOCIATED RETINOPATHY

Hiroshi Ohguro and Mitsuru Nakazawa

ABSTRACT

Cancer associated retinopathy (CAR) is an ocular manifestation of paraneoplastic syndrome clinically characterized by progressive visual impairment similar to retinitis pigmentosa. As a possible mechanism causing the retinal degeneration, the presence of serum autoantibodies against recoverin and other retinal antigens are involved. The molecular pathology in CAR by anti-recoverin antibody is considered to occur in the following steps: Firstly, recoverin aberrantly expressed in cancerous tissues is recognized by immunocytes by some unknown mechanisms and then a specific antibody toward recoverin is produced. Secondly, the anti-recoverin antibody reaches the retina via the peripheral circulation and is taken up into photoreceptor cells. Lastly, the antibody blocks recoverin function (inhibition of rhodopsin phosphorylation in a calcium dependent manner), and enhancement of rhodopsin phosphorylation induces retinal apoptosis.

CAR IS AN OCULAR PARANEOPLASTIC MANIFESTATION

It has been reported that the central nervous system could be targets of the remote effect of a malignant tumors, and this may cause a variety of paraneoplastic syndromes, such as the Lambert-Eaton myasthenic syndrome, paraneoplastic cerebellar degeneration and paraneoplastic sensorimotor neuropathy (summarized in Table 1). In terms of the pathogenesis of the paraneoplastic syndrome, it was suggested that an immune reaction toward antigens shared by the tumor cells and the

Departments of Ophthalmology, Hirosaki University School of Medicine, 5 Zaifu cho, Hirosaki 036-8652, Japan.

Table 1. Paraneoplastic syndrome

Paraneoplastic Syndrome	**Associated Malignancy**	**Associated Antibodies**
CAR	SCLC, gastric, gynecologic and other tumors	Anti-recoverin, anti-hsc70, anti-enolase, anti-neurofilament
MAR	Cutaneous malignant melanoma	Anti-retinal bipolar cell
Cortical cerebellar degeneration	SCLC, breast, gynecologic, Hodgkin's	Anti-Yo (anti-Prukinje cell)
Encephalomyelitis/pure sensory neuropathy	SCLC, occasionally Hodgkin's	Anti-Hu (ANNA-1)
Progressive sensorimotor neuropathy	SCLC, breast, and other tumors	
Guillan-Barre	Hodgkin's	
Relapsing-remitting neuropathy	Lung, breast, lymphoma, myeloma	
Subcutaneous motor neuropathy	Lymphoma	
Opsoclonus-myoclonus	Neuroblastoma, lung and breast	Anti-Ri (ANNA-2)
Lumbert-Eaton syndrome	SCLC, breast, gastrointestinal	Anti-voltage-gated calciumchannels
Inflammatory myopathy	Breast, lung ovarian, gastric	

SCLC, small cell carcinoma of the lung

nervous system is involved. Clinically, paraneoplastic retinopathy is known to be associated in patients with malignancies. So far two types of the diseases have been described, cancer-associated retinopathy (CAR), and melanoma-associated retinopathy (MAR). CAR is characterized by sudden and progressive visual loss, ring scotoma, photopsia, impairment of dark adaptation, and abnormalities of the a- and b-waves of the electroretinogram (ERG). Serum autoantibody against retinal antigens including recoverin, a retina specific Ca^{2+} binding regulatory protein, has been identified in CAR patients. Among the various cancers in CAR, small-cell lung carcinoma has reported most frequently. In most cases, CAR is diagnosed before an underlying primary cancer is discovered and such patients show a relatively good prognosis. MAR is associated with cutaneous malignant melanoma and is characterized clinically by night blindness, a sensation of shimmering light, impaired dark adaptation, and an electroretinogram (ERG) resembling that in some forms of congenital stationary night blindness (CSNB) (a preserved a-wave with absent or diminished b-wave). It was found that MAR patients have circulating autoantibodies, presumably generated against a melanoma antigen, that cross react with bipolar cells of the retina. However, in contrast to CAR, little is known in terms of the molecular pathology of MAR (Table 2).

In this Chapter, we review clinical and pathological aspects in CAR and present our data elucidating what mechanisms are involved in the pathogenesis by the serum autoantibody against recoverin, the most common autoantigen found in CAR patients, and examine possible treatments for CAR using an animal model.

CLINICAL ASPECTS OF CAR

Sawyer et al[1] first reported three patients with malignancies (65 year-old male with lung adenocarcinoma, 76 year-old female with lung small cell carcinoma, and 62 year-old female with endometrial stromal sarcoma) who lost visual acuity by idiopathic retinal degeneration. Histopathology examination by using electron microscopy demonstrated destruction of photoreceptor cells and granule cells in the retina in these patients. In this report, they emphasized that these malignant tumors were clinically detected after the visual symptoms appeared. Keltner et al[2] reported similar clinical cases in whom serum autoantibodies reacted with retinal cells as tested by immunocytochemistry study. Therefore, they suggested that serum autoantibodies toward retinal antigens may contribute to the retinal degeneration. Until now, over a hundred CAR cases have been reported, and several retinal proteins including a 23- kDa (recoverin), 46kDa, 58-62kDa, 65kDa, 145kDa, and 205kDa proteins were detected as autoantigens in these cases by western blot analysis. In terms of the criteria of diagnosis of CAR, Jacobson et al[3] proposed a clinical triad of photosensitivity, ring scotomatous visual field loss and attenuated retinal arteriole with the presence of serum autoantigen toward retinal antigens. Importantly, retinal vasculitis was also found to be associated in CAR similar to uveoretinitis. In fact, recoverin was found to be a highly uveitogenic molecule that can cause uveoretinitis by immunization in Lewis rats. Furthermore, CAR-like retinopathy has recently been found in non-cancerous patients as well, and this is called "recoverin-associated retinopathy".

Table 2. Clinical and pathological aspects of CAR and MAR

	CAR	MAR
Legion	photoreceptor	retinal bipolar cell
Night blindness	+	+
Photophobia	+	+
Impairment of dark adaptation	+	+
Visual field	ring-like scotoma	peripheral constriction
ERG abnormality	a and b wave	b wave
Retinal vasculitis	+	-
Associated malignancy	lung, gastric, gynecologic and other tumors	cutaneous malignant melanoma
Associated antibody anti-enolase, anti-neurofilament	anti-recoverin, anti-hsc70,	anti-retinal bipolar cell
Prognosis of tumor	good	good

Retinal Autoantigens Recognized by Sera from CAR Patients

In terms of retinal autoantigens recognized in sera from CAR patients by western blot analysis, several protein bands including 23kDa, 46kDa, 58-62kDa, 65kDa, 145kDa, and 205kDa have been reported. Among these antigens, 23kDa and 65kDa band were most frequently identified. Polans et al[4] identified the 23kDa protein as a retinal calcium binding regulatory protein called recoverin. Recoverin is found to be functionally involved in the adaptation to dark and light by regulating activity of rhodopsin kinase and PrP2A in a calcium dependent manner (See other Chapters related to recoverin). Recently, we analyzed sera from four Japanese patients with CAR associated with different types of cancers (2 patients, small cell lung carcinoma; 1 patient, lung adenocarcinoma; 1 patient, gastric cancer) and found that the sera (1:500 dilution) from the four patients with CAR commonly reacted with both 23 kDa and 65 kDa retinal soluble proteins as above. However such immunoreactivities were not found in any control subjects including 20 cancerous patients at the same serum dilution in western blot analysis (Fig. 1A and B). Immunoreactivity of the 23 kDa protein was identical to that of recoverin (Fig. 1A lanes 1 and lanes 2-5). The in-gel digestion of partially purified 65kDa retinal antigen from fresh bovine retinas with endoproteinase Lys C and the peptide sequence analysis identified as a heat shock cognate protein 70 (Hsc 70) (Fig. 2A and B).[5] Therefore, among the reports presenting the data of western blot analysis, 23 kDa and 65 kDa protein which were most frequently identified as immunoreactive bands, are most likely to be recoverin and hsc 70, respectively. In terms of the other antigens, Adamus et al[6] and Korngruth et al[7] identified the 46 kDa and 58-52kDa, 145 kDa and 205kDa as enolase and neurofilament, respectively.

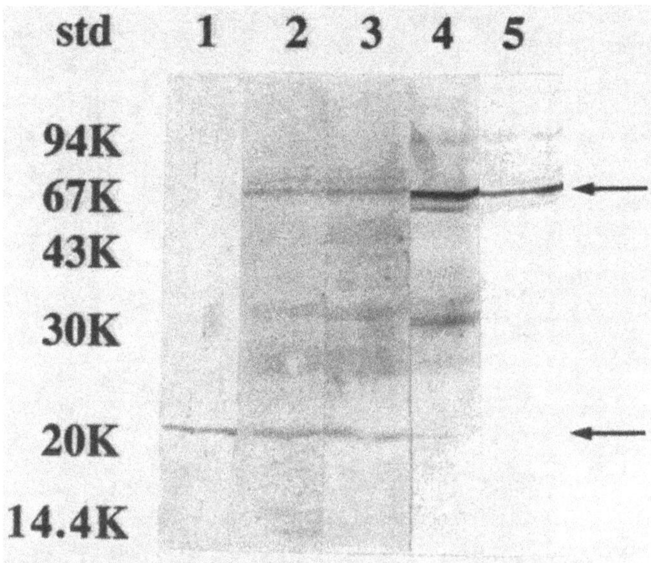

Figure 1A. Western blot analysis of sera from CAR patients. Anti-recoverin serum (lane 1, 1:3000 dilution) and sera from CAR patients (#1-4; lanes 2-5, 1:500 dilution) were tested with bovine retinal soluble extract. The 23kDa and 65kDa proteins are commonly probed by patients' sera (indicated by arrows). (Reprinted with permission from Ohguro H, Ogawa K, Nakagawa T, Invest Ophthalmol Vis Sci 1999;40:82-89. Copyright 1999 Association for Research in Vision and Ophthalmology)

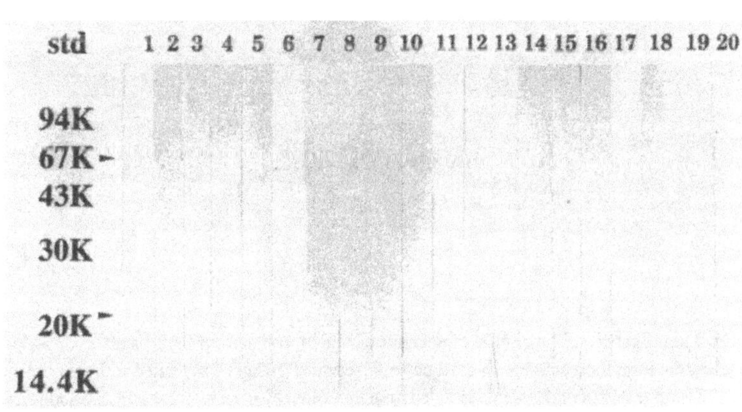

Figure 1B. Western blot analysis of sera from CAR patients. Sera of non-CAR cancerous patients were tested with bovine retinal soluble extract. The type of cancer, age and sex in each lane are shown as follows: small cell lung carcinoma, lanes 1, 66M; 2, 60M; 3, 68M; 4, 58F; 5, 70F; 6, 68F; 7, 62M; 8, 65M; 9, 69F; 10, 63F; lung adenocarcinoma, lanes 11, 70M; 12, 78M; 13, 62M; 14, 65M; 15, 60M; gastric carcinoma, lane 16, 65F; 17, 60F; 18, 68M; 19, 70F; 20, 69F. The positions of the 23-kDa and 65-kDa proteins are indicated by arrows. (Reprinted with permission from Ohguro H, Ogawa K, Nakagawa T, Invest Ophthalmol Vis Sci 1999;40:82-89. Copyright 1999 Association for Research in Vision and Ophthalmology)

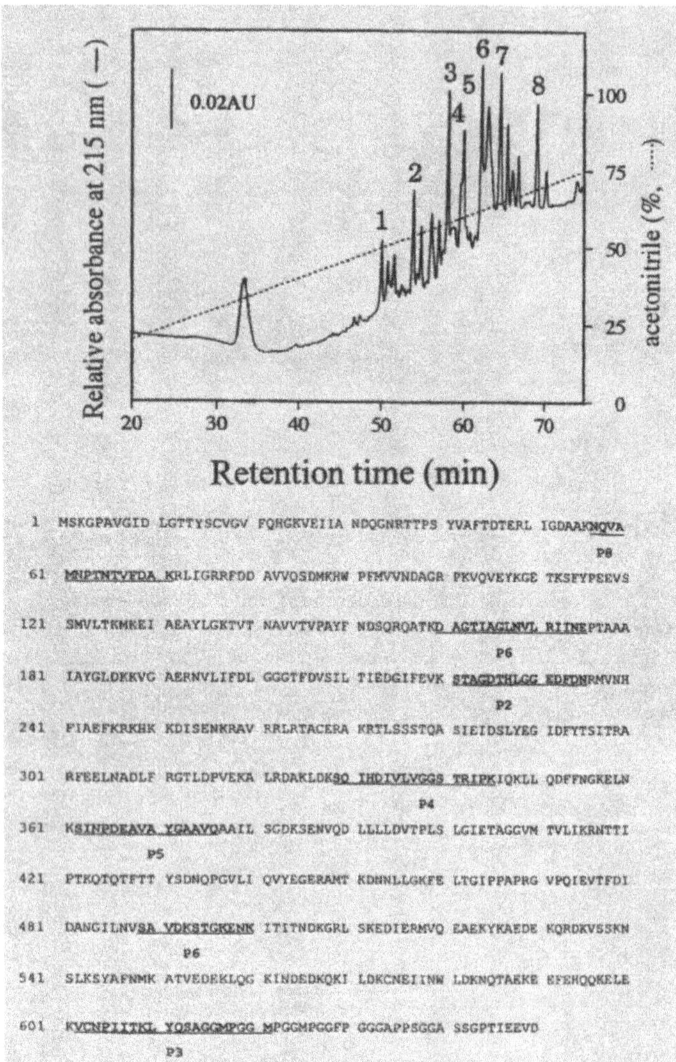

Figure 2. Identification of the 65-kDa protein recognized by serum from CAR patients. **Upper panel**: The 65-kDa protein immunoreactive with CAR patients' serum in 2-D gel was treated with endoproteinase Lys C (1 mg). The digested peptides were separated from each other by reverse-phase HPLC C8 column employing a linear gradient of acetonitrile from 0 to 70 % in 0.05 % trifluoroacetic acid at a flow rate of 0.2 ml/min. Major peaks designated 1 through 8 were subjected to the Edman sequence analysis. **Lower panel**: The amino acid sequences of eight proteolytic peptides from 65-kDa protein are indicated in bold letters with underlines in the bovine heat shock cognate 70-kDa protein sequence. The peptide designations (P1-P8) correspond to those in panel A. (Reprinted with permission from Ohguro H, Ogawa K, Nakagawa T, Invest Ophthalmol Vis Sci 1999;40:82-89. Copyright 1999 Association for Research in Vision and Ophthalmology)

MOLECULAR PATHOLOGY IN CAR

Mechanism of Antibody Generation in CAR; Aberrant Expression of Recoverin

In the initial step of onset of paraneoplastic syndrome, presence of shared antigen between tumor cells and neuronal cells in the central nervous system triggers an autoimmune response. In CAR, retina-specific recoverin was found to be aberrantly expressed in malignant tumors. Preliminary studies have revealed that such aberrant expression of retinal specific recoverin is not identified in cancer cells without retinopathy. These observations suggest that aberrant expression of recoverin in cancer cells is an initial and critical step in the cause of retinopathy. However, in contrast, we found aberrant expression of recoverin in approximately 50% of cancer tissues from non-CAR cancerous patients (Fig. 3).[8-10] Therefore, additional unknown mechanisms for the aberrant expression must be involved in the antibody generation in CAR.

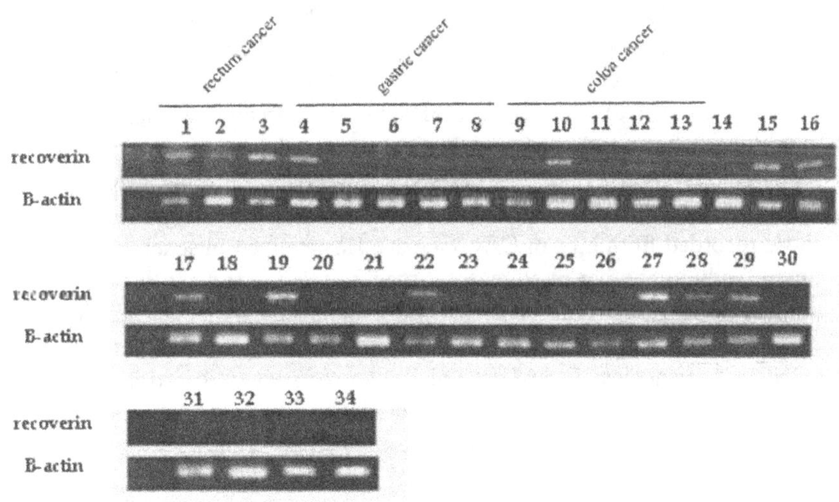

Figure 3. Aberrant expression of recoverin in various tumor tissues by RT-PCR. (Reprinted with permission from Maeda T, Maeda A, Maruyama I et al., Invest Ophthalmol Vis Sci 2001; 42:705-712. Copyright 2001 Association for Research in Vision and Ophthalmology)

At present, the functional roles of aberrantly expressed recoverin in cancer cells are unknown. However, functional roles of recoverin, regulation of rhodopsin phosphorylation in a calcium dependent manner in photoreceptor cells, allowed us to speculate that recoverin may effect calcium-dependent protein phosphorylation in cancer cells. We recently found that protein phosphorylation patterns in A549 cells, which is a cell culture line established from lung adenocarcinoma not expressing recoverin, are modulated by the presence of recoverin in a calcium dependent manner. Furthermore, it was also found that introduction of cDNA encoding recoverin into A549 cell caused significant decrease in the cell growth.[9] Therefore, taken together, we speculated that aberrantly expressed recoverin may be involved in a calcium signaling pathway in cancer cells.

Possible Mechanisms Causing Retinal Degeneration in CAR by Anti-Recoverin Antibody and Anti-hsc 70 Antibodies

The next step of the CAR pathology, penetration of the circulating antibody toward recoverin into retina, causes photoreceptor cell dysfunctions and degeneration. In order to understand the molecular mechanisms involved in this step, we investigated the pathogenic effects of anti-recoverin and anti-hsc 70 antibodies on retinal cells by injecting affinity purified antibodies into the vitreous cavity of Lewis rats and thereafter performing functional and morphological evaluations.[11] As shown in Figure 4, measuring ERG responses in animals at three weeks after the vitreous injections, no changes in ERG were detected from eyes injected with anti-hsc 70 antibody as compared with control (PBS injection). In contrast, significantly lower amplitudes of a- and b-waves in ERG were recorded in eyes injected with anti-recoverin IgG. Furthermore, ERG responses were almost lost in eyes injected with both anti-recoverin and anti-hsc 70 antibodies. These ERG changes closely resemble those recorded from human CAR patients. However, co-injection of anti-arrestin, an antibody to another photoreceptor specific protein, and anti-hsc 70 antibodies showed no significant ERG changes. Morphologically, light microscopic examination of the retinal sections showed significant thinning of the outer nuclear layer (ONL), inner nuclear layer (INL) and outer segment (OS) and accumulation of TUNEL positive cells within the ONL and INL in the antibody treated retina (Fig. 5). Similar to this, Adamus and her colleagues reported anti-recoverin antibody induced apoptotic cell death using a retinal cell culture system and intravitreous administration of monoclonal antibody against recoverin induced apoptosis of ONL detected by TUNEL labeling, DNA fragmentation and electron microscopic features in Lewis rat eyes in vivo.[12,13] Functionally, based upon these observations, anti-recoverin antibody caused retinal dysfunction and anti-hsc 70 antibody enhanced this dysfunction, and morphologically these antibodies induced apoptotic cell death within the outer retinal layers.

Next, in order to elucidate mechanism involved in the anti-recoverin induced retinal dysfunction, the distribution of the antibody after the intravitreous administration was examined. Immunocytochemistry revealed that anti-recoverin antibody

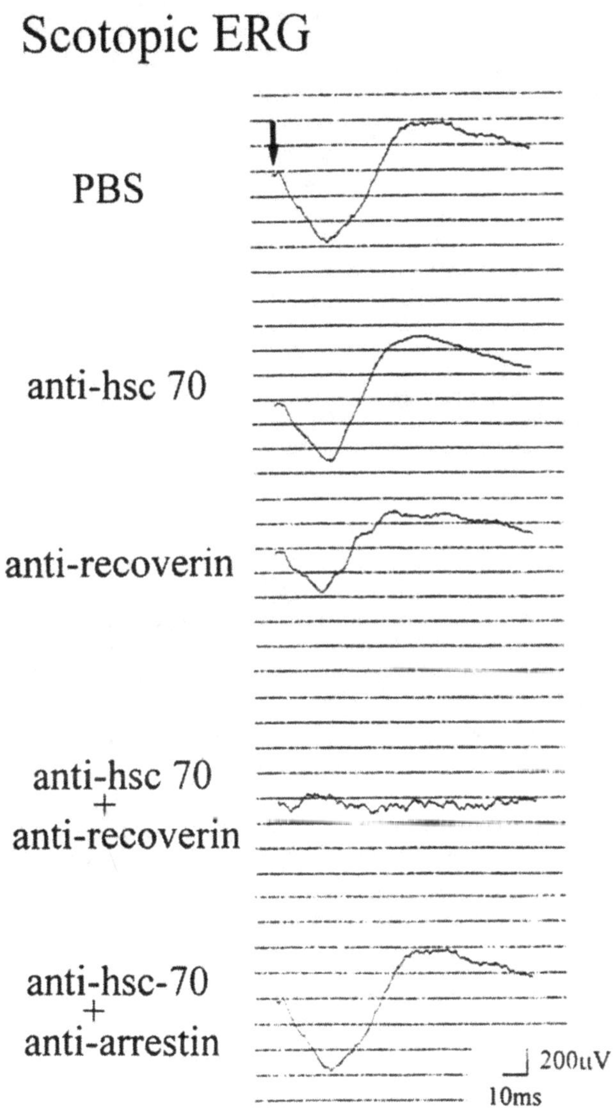

Figure 4. Effects of intravitreal penetration of antibodies against recoverin, hsc 70, and/or arrestin on scotopic ERG.Either PBS, anti-hsc 70 serum (5 μg of IgG), anti-recoverin IgG (5 μg), a mixture of anti-hsc 70 serum (5 μg of IgG) and anti-recoverin IgG (5 μg), or a mixture of anti-hsc 70 serum (5 μg of IgG) and anti-arrestin IgG (5 μg) was injected intravitreously into Lewis rat eyes. Three weeks after the injection, scotopic ERG was recorded. The arrow indicates the timing of the light flash. (Reprinted with permission from Ohguro H, Ogawa K, Maeda T et al., Invest Ophthalmol Vis Sci 1999;40:3160-3167. Copyright 1999 Association for Research in Vision and Ophthalmology)

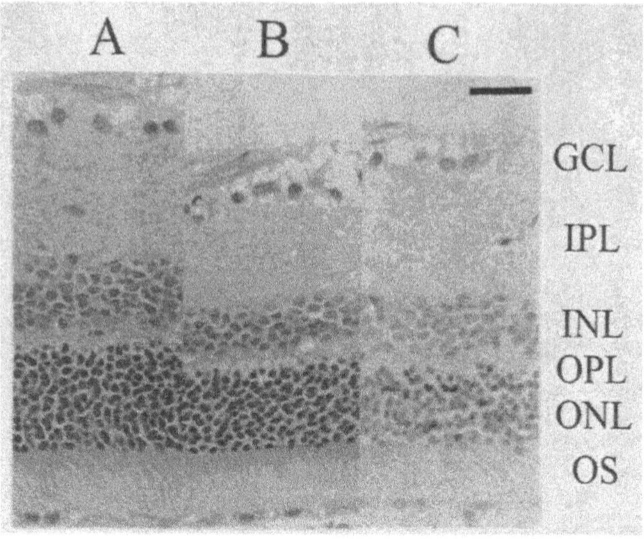

Figure 5. Histopathologic changes in the retina of Lewis rats treated with anti-recoverin and anti-hsc 70 antibodies.Hematoxyline-eosin staining (A and B) or TUNEL staining (C) of retinal sections near the posterior pole from Lewis rat eyes, which were treated with PBS (A) or anti-recoverin and anti-hsc 70 antibodies (B and C), respectively. GCL, ganglion cell layer; IPL, inner plexiform layer; INL, inner nuclear layer; OPL, outer plexiform layer; ONL, outer nuclear layer; OS, outer segment; scale bar, 50 _m. (Reprinted with permission from Ohguro H, Ogawa K, Maeda T et al., Invest Ophthalmol Vis Sci 1999;40:3160-3167. Copyright 1999 Association for Research in Vision and Ophthalmology)

migrated within the inner parts of the retina at 3 hs, and then antibody localization shifted toward the outer parts of the retina during 12 hs. During 12-24 hs, the antibody accumulated within both ONL and photoreceptor layer, and thereafter the antibody diminished slowly from the retina within 6 days. These data clearly show that anti-recoverin antibody is present near the photoreceptor cells for a period of time after the administration. Recoverin is known to function as an inhibitor of rhodopsin kinase in a calcium dependent manner as shown in Figure 6. If the intravitreously injected anti-recoverin antibody penetrates into photoreceptor and binds with recoverin, we speculated that blocking recoverin function by trapping recoverin by the antibody causes activation of rhodopsin kinase resulting in enhancement of rhodopsin phosphorylation levels. In order to elucidate the molecular mechanisms causing apoptosis of retinal photoreceptors by anti-recoverin antibody, we examined the effects of the anti-recoverin antibody on light-dependent rhodopsin phosphorylation in vitro and in vivo.[10] For the in vitro experiment, freshly isolated bovine ROS were pre-incubated with anti-recoverin antibody for 1h in the dark on ice in the presence or absence of Ca^{2+}. Thereafter 1 mM [γ-^{32}P] ATP was incubated with the mixture at 30 °C for 10 min in the dark after a flash. After the incubation, the phosphorylation reaction was terminated by addition of 10% TCA

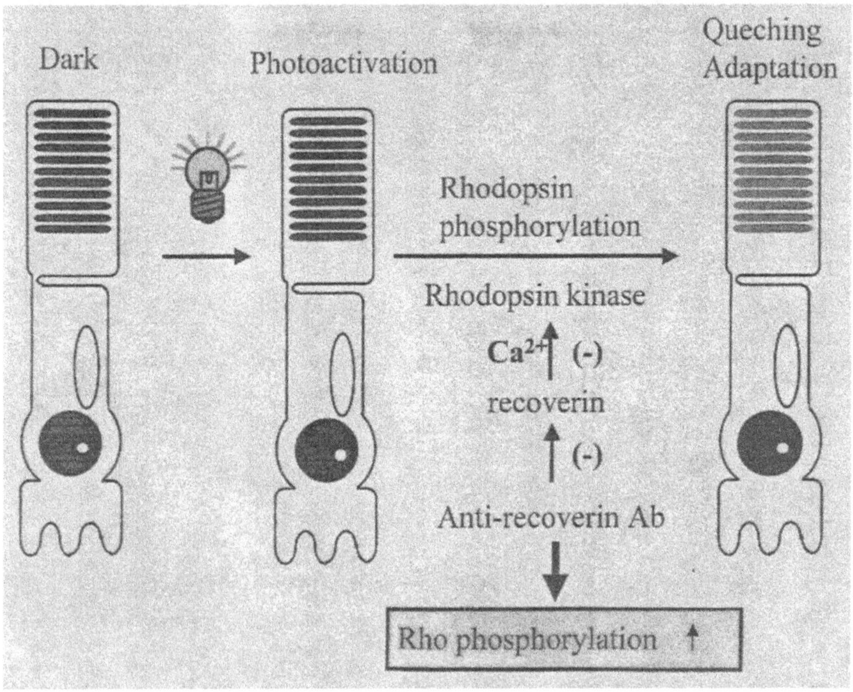

Figure 6. Regulation of phototransduction of rod photoreceptor by recoverin in a calcium dependent manner and possible mechanisms of anti-recoverin antibody on it.

and washed with fresh 10% TCA followed by scintillation counting. As shown in Figure 7A, rhodopsin phosphorylation levels were significantly enhanced by the presence of Ca^{2+} and anti-recoverin antibody. For the *in vivo* experiment, eyes were enucleated 24 hs after administration of either anti-recoverin antibody or preimmune rabbit IgG, and incubated on ice for 1h at dark followed by separation of ROS by sucrose gradient centrifugation method. Then, ROS was incubated with 1mM [γ-^{32}P] ATP in 100 mM sodium phosphate buffer, pH 7.2 containing 5 mM $MgCl_2$ for 10 min at 30 °C under 150-w lamp illumination. Phosphorylated ROS was analyzed by SDS-PAGE in which radioactivity of the corresponding rhodopsin band was counted. As shown in Figure 7B, rhodopsin phosphorylation levels of the anti-recoverin antibody treated retinas were significantly higher than those in control, as expected. In addition, anti-recoverin antibody induced reduction of ERG response and enhancement of rhodopsin phosphorylation were not observed after co-injection of the antibodies with caspase inhibitors, Z-DEVD-FMK (Z-Asp(OMe)-Glu(OMe)-Val-Asp(Ome)-fluoromethyl ketone) (Caspase inhibitor I), and Z-VAD-FMK (Z-Val-Ala-Asp(OMe)-fluoromethyl ketone) (Caspase inhibitor II), with the antibodies (Fig. 8), suggesting that the anti-recoverin antibody induced retinal degeneration is caspase dependent.

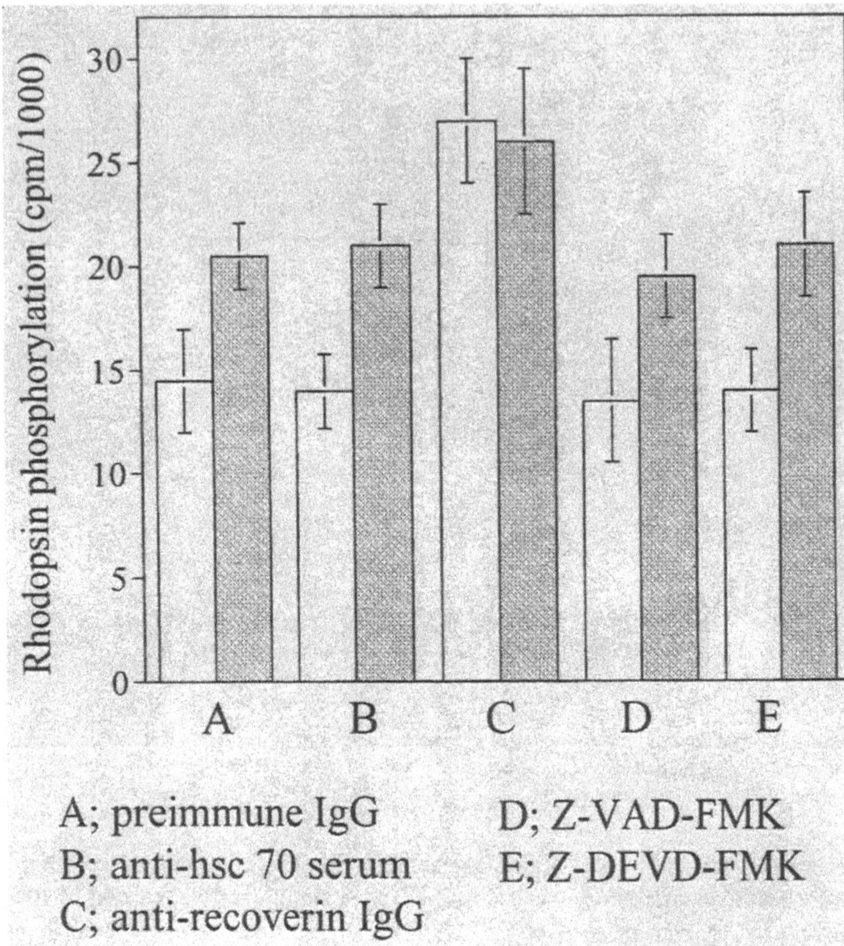

Figure 7. Effects of anti-recoverin and anti-hsc 70 antibodies, or caspase inhibitors on rhodopsin phosphorylation in bovine ROS homogenate. Freshly isolated bovine ROS homogenate was preincubated with antibody, 5 μg preimmune rabbit IgG (A), anti-hsc 70 serum (containing 5 μg IgG) (B) or 5 μg anti-recoverin IgG (C), or 1mM caspase inhibitors, Z-VAD-FMK (D) or Z-DEVD-FMK (E). Then rhodopsin phosphorylation was carried out by addition of 0.5 mM [γ-^{32}P] ATP in the presence of 0.1 mM $CaCl_2$ (open column) or 1 mM EGTA (shaded column) under flash illumination. The reaction was terminated by addition of 10 % TCA and the radioactivity was counted in a scintillation cocktail and the radioactivities were plotted. Experiments were performed in triplicate. (Reprinted with permission from Maeda T, Maeda A, Maruyama I et al., Invest Ophthalmol Vis Sci 2001;42:705-712. Copyright 2001 Association for Research in Vision and Ophthalmology)

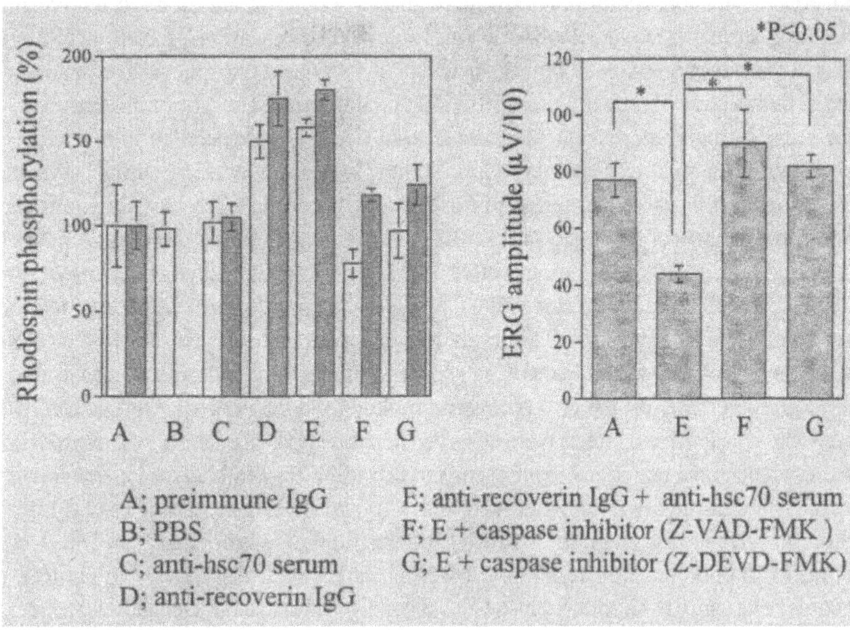

Figure 8. Effects of antibodies and caspase inhibitors on rhodopsin phosphorylation and ERG in rat eyes. Either 5 μg preimmune IgG (A), PBS (B), anti-hsc 70 serum (5 μg IgG) (C), 5 μg anti-recoverin IgG (D), or anti-recoverin IgG and anti-hsc 70 serum (5 μg IgG each) without (E) or with caspase inhibitors (F, G) was injected intravitreously in Lewis rat eyes. After the administration for 36 hours (open column) or 3 weeks (shaded column), ROS was prepared and light dependent phosphorylation by [γ-^{32}P] ATP (left panel) and ERG measurement (right panel) were examined. In condition B, 5 μg anti-recoverin IgG was mixed with fresh dissected retinas followed by the ROS preparation. After the reaction, samples were analyzed by SDS-PAGE followed by autoradiogram. The radioactivity of rhodopsin bands in SDS-PAGE was counted in a scintillation cocktail and the radioactivities were plotted (left). Experiments were performed in triplicate. ERG measurements were performed in 10 eyes in each condition and the amplitudes of b-wave were plotted (right). (Reprinted with permission from Maeda T, Maeda A, Maruyama I et al., Invest Ophthalmol Vis Sci 2001;42:705-712. Copyright 2001 Association for Research in Vision and Ophthalmology)

Possible Therapy for CAR

In terms of therapy for patients with CAR and other paraneoplastic syndromes, such as paraneoplastic cerebellar degeneration and Lambert-Eaton myasthenic syndrome, steroid administration, immunomodulation, and plasmapheresis have been clinically performed in conjunction with anti-neoplastic therapy. For CAR, no definitive therapy has been established, although it has been reported that these treatments may be effective in some patients. During our research related to CAR pathology, we found that anti-recoverin antibody administrated intravitreously internalized into photoreceptors, bound recoverin and blocked the recoverin function that inhibits rhodopsin phosphorylation in a Ca^{2+} dependent manner. Therefore, based upon this observation, we speculate that the effects of inhibition of recoverin function by anti-recoverin antibody resulting in missregulation of the quenching visual transduction cascade by higher levels of rhodopsin phosphorylation including accumulation of intracellular Ca^{2+} within photoreceptor cells may represent critical steps in the photoreceptor degeneration in CAR. If our speculation is correct, decrease of the light-dependent rhodopsin phosphorylation levels by suppression of the increase of intracellular levels of Ca^{2+} by Ca^{2+} antagonist may have a beneficial effect on the retinal dysfunction. In order to test this hypothesis, several drugs including steroid, cyclosporin A and nilvadipine, a Ca^{2+} antagonist which is known to be a highly effective penetrator of the CNS, were intraperitoneally administrated to anti-recoverin and anti-hsc 70 antibody treated Lewis rats everyday for two weeks after the antibodies were injected. As shown in Figure 9, nilvadipine demonstrated remarkable effects on the anti-recoverin antibody induced retinal dysfunction. In contrast, no significant effects were observed in rats treated with steroid or cyclosporin A. At present, the beneficial mechanisms of nilvadipine are unknown. However, we speculate that the lowering of intracellular Ca^{2+} levels by nilvadipine may inhibit the Ca^{2+} dependent apoptotic pathway. Interestingly, Frasson et al[14] recently reported rod photoreceptor rescue by lowering intracellular Ca^{2+} levels in photoreceptor cells using D-cis-diltiazem, a Ca^{2+} channel blocker in a different animal model of RP, rd mouse, in which the gene encoding β-subunit of cGMP phosphodiesterase is affected. However, in contrast, Bush et al[15] reported that a Ca^{2+} antagonist, diltiazem had no effects on the photoreceptor degeneration in the rhodopsin P23H rat. Therefore, it seems likely that Ca^{2+} channel blockers have protective effects on the retinal degeneration in some disease models but these effects may be variable among different models, different species, diseases' stages, and different Ca^{2+} antagonists.

CONCLUSION

Based upon the above observations, we propose the molecular pathology mechanisms of retinal photoreceptor degeneration in CAR as shown in Figure 10 can be summarized as follows:

1. Retina specific recoverin is aberrantly expressed in cancerous tissues.
2. The aberrantly expressed recoverin recognized by immunocytes causes production of specific antibody toward recoverin.
3. Circulating antibody toward recoverin reaches the retina and penetrates into photoreceptor cells.
4. The anti-recoverin antibody specifically binds and blocks recoverin function resulting in significant enhancement of rhodopsin phosphorylation.
5. Misregulation of phototransduction pathway by the elevated levels of rhodopsin phosphorylation induces caspase dependent apoptosis in photoreceptor cells.

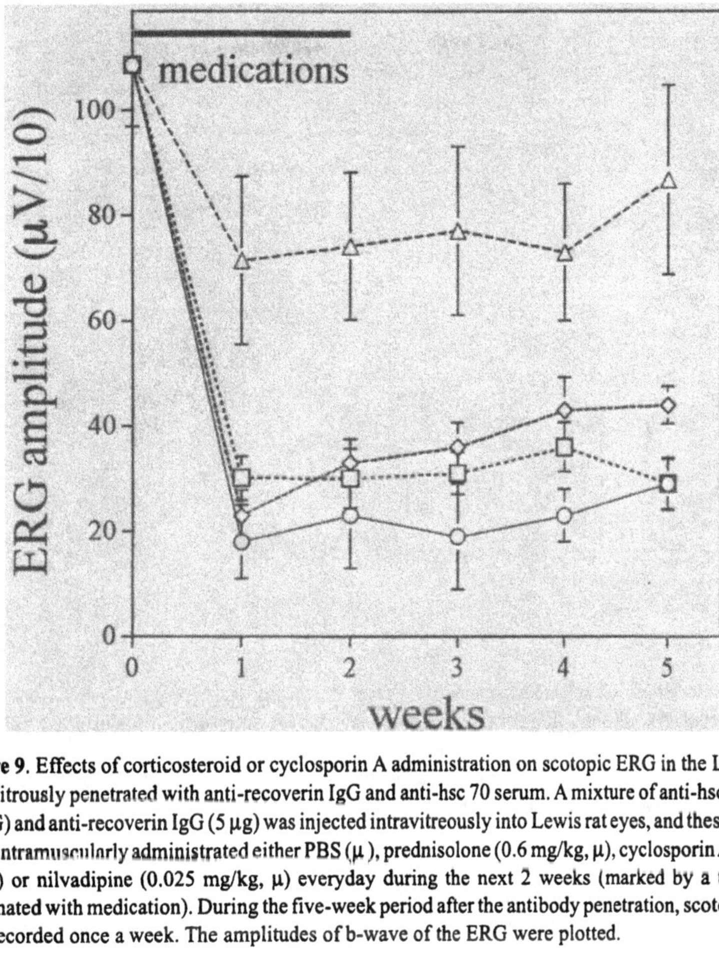

Figure 9. Effects of corticosteroid or cyclosporin A administration on scotopic ERG in the Lewis rats intravitrously penetrated with anti-recoverin IgG and anti-hsc 70 serum. A mixture of anti-hsc 70 (5 μg of IgG) and anti-recoverin IgG (5 μg) was injected intravitreously into Lewis rat eyes, and these animals were intramuscularly administrated either PBS (μ), prednisolone (0.6 mg/kg, μ), cyclosporin A (10 mg/kg, μ) or nilvadipine (0.025 mg/kg, μ) everyday during the next 2 weeks (marked by a thick line designated with medication). During the five-week period after the antibody penetration, scotopic ERG was recorded once a week. The amplitudes of b-wave of the ERG were plotted.

REFERENCES

1. Sawyer RA, Selhorse JB, Zimmerman LE. Blindness caused by photoreceptor degeneration as a remote effect of cancer. Am J Ophthalmol 1976; 90:606-613.
2. Keltner JL, Roth AM, Chang RS. Photoreceptor degeneration: Possible autoimmune disorder. Arch Ophthalmol 1983; 101:564-569.
3. Jacobson DM, Thirkill CE, Tipping SJ. A clinical triad to diagnose paraneoplastic retinopathy. Ann Neurol 1990; 28:162-167.
4. Polans AS, Buczylko J, Crabb J et al. A photoreceptor calcium-binding protein is recognized by autoantibodies obtained from patients with cancer-associated retinopathy. J Cell Biol 1991; 112:981-989.
5. Ohguro H, Ogawa K, Nakagawa T. Both recoverin and hsc 70 are found as autoantigens in patients with cancer-associated retinopathy. Invest Ophthalmol Vis Sci 1999; 40:82-89.
6. Adamus, G, Aptsiauri N, Guy J et al. The occurrence of serum autoantibodies against enolase in cancer-associated retinopathy. Clin Immunol Immunopathol 1996; 78:120-129.

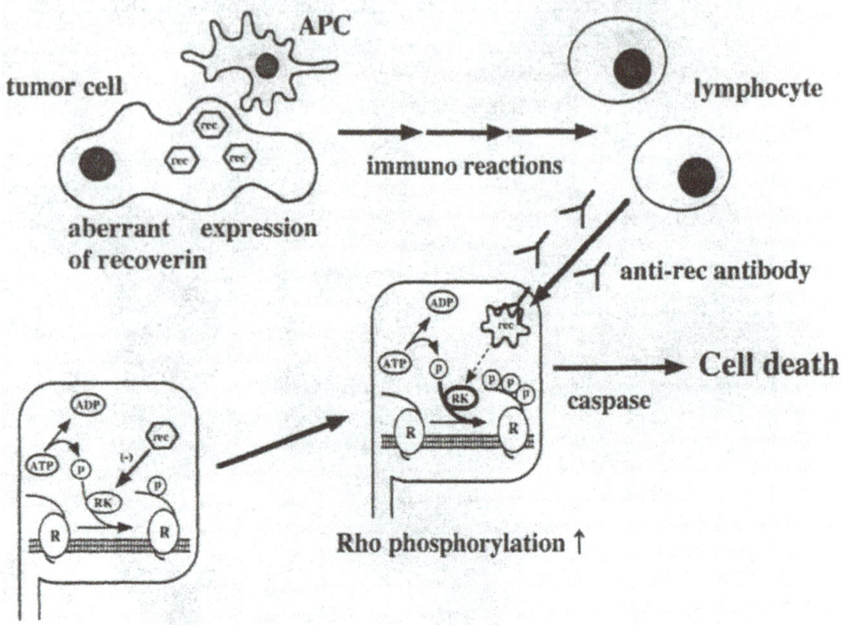

Figure 10. Possible molecular mechanisms of retinal photoreceptor degeneration in CAR. R, rhodopsin; rec, recoverin; RK, rhodopsin kinase; p, phosphorylation; ATP, adenosine 5'-triphosphate; ADP, adenosine 5'-diphosphate; APC, antigen presenting cell. (Reprinted with permission from Maeda T, Maeda A, Maruyama I et al., Invest Ophthalmol Vis Sci 2001;42:705-712. Copyright 2001 Association for Research in Vision and Ophthalmology)

7. Korngruth SE, Kalinke T, Grunwald GB et al. Anti-neurofilament antibodies in the sera of patients with small cell carcinoma of the lung and with visual paraneoplastic syndrome. Cancer Res 1986; 46:2588-2595.
8. Polans AS, Witkowska D, Haley TL et al. Recoverin, a photoreceptor-specific calcium-binding protein, is expressed by the tumor of a patient with cancer-associated retinopathy. Proc Natl Acad Sci USA 1995; 92:9176-9180.
9. Maeda A, Ohguro H, Maeda T et al. Aberrant expression of photoreceptor-specific calcium-binding protein (recoverin) in cancer cell lines. Cancer Res 2000; 60:1914-1920.
10. Maeda T, Maeda A, Maruyama I et al. Mechanisms of photoreceptor cell death in Cancer-associated retinopathy. Invest Ophthalmol Vis Sci 2001; 42:705-712.
11. Ohguro H, Ogawa K, Maeda T et al. Cancer-associated retinopathy induced by both anti-recoverin and anti-hsc70 antibodies in vivo. Invest Ophthalmol Vis Sci 1999; 40:3160-3167.
12. Adamus G, Machnicki M, Seigel GM. Apoptotic retinal cell death induced by antirecoverin autoantibodies of cancer-associated retinopathy. Invest Ophthalmol Vis Sci 1997; 38:283-291.
13. Adamus G, Machnicki M, Elerding H et al. Antibodies to recoverin induce apoptosis of photoreceptor and bipolar cells in vivo. J Autoimmun 1998; 11:523-533.
14. Frasson M, Sahel JA, Fabre M et al. Retinitis pigmentosa: rod photoreceptor rescue by a calcium-channel blocker in the rd mouse. Nature Med 1999; 5:1183-1187.
15. Bush RA, Kononen L, Machida S et al. The effect of calcium channel blocker diltiazem on photoreceptor degeneration in the rhodopsin Pro23His rat. Invest Ophthalmol Vis Sci 2000; 41:2697-2701.

RGS9-1 PHOSPHORYLATION AND Ca^{2+}

Theodore G. Wensel

ABSTRACT

The duration of photoresponses in vertebrate rods and cones is controlled at the level of GTP hydrolysis by a GTPase accelerating protein (GAP) whose catalytic core is provided by RGS9-1. RGS9-1 is in turn regulated by phosphorylation on serine 475, in a reaction that is dependent on Ca^{2+}. In living mice, the level of phosphorylation at this site is reduced by light. Thus RGS9-1 phosphorylation provides a potential mechanism by which light-regulated changes in intracellular [Ca^{2+}] may feed back on phototransduction through effects on the lifetime of activated G protein and cGMP phosphodiesterase.

INTRODUCTION: GTP HYDROLYSIS AND PHOTORESPONSE KINETICS

Because photoexcitation relies on activation of the heterotrimeric G protein, transducin, hydrolysis of GTP bound to this protein is one of the key determinants of recovery kinetics. Experiments with non-hydrolyzable analogues of GTP many years ago[1-3] made it clear that prolonged excitation due to prolonged activation of cGMP phosphodiesterase (PDE6, here referred to simply as PDE) results when R* is formed but GTP hydrolysis is prevented. Thus GTP hydrolysis represents a point in the cascade where Ca^{2+} might potentially exert a regulatory influence. Attempts to demonstrate an effect of Ca^{2+} on PDE inactivation or GTPase kinetics have so far yielded mostly negative results.[4,5] However, recent findings have suggested that there may indeed be a link between light-controlled changes in intracellular [Ca^{2+}] and molecules regulating GTPase kinetics.

Department of Biochemistry and Molecular Biology, Baylor College of Medicine, Houston, TX, U.S.A.

RGS9-1, THE GTPASE ACCELERATING PROTEIN (GAP) FOR PHOTOTRANSDUCTION

The major regulator of GTP hydrolysis in rods and cones is a member of the large RGS (Regulator of G protein signaling)[6,7] family of GTPase accelerating proteins (GAPs),[8] called RGS9-1.[9-11] This protein has been found only in rod and cone photoreceptor cells,[12] and its removal by gene deletion in mice leads to profoundly slowed recovery of light responses in both rods and cones.[13,14]

DOMAIN AND SUBUNIT STRUCTURE OF RGS9-1

The GTPase accelerating function of RGS9-1 is localized to a catalytic core domain, known as the RGS domain or RGS box, whose basic structure is conserved among all RGS proteins. This domain is necessary and sufficient for acceleration of GTP hydrolysis by the photoreceptor G protein, transducin.[9,12,15-17]

What is less certain than the basic catalytic activity of RGS9-1 and the domain responsible, is how RGS9-1 is regulated. In particular, the roles of several other domains of RGS9-1 , and proteins that bind RGS9-1 (Fig. 1), are active areas of current investigations. However, it seems reasonable that because of its ability to control the timing and potentially the sensitivity as well of photoresponses, RGS9-1 activity is likely to be regulated. Likely candidates for stimuli leading to such regulation are light and intracellular calcium, based on the numerous other examples of calcium feedback signals modulating light responses described in this volume.

UNIQUE C-TERMINUS OF RGS9-1

One of the interesting features of RGS9-1 is its unique C-terminal tail. Neither this 18 amino acid peptide nor any closely related sequence is found in any other RGS proteins. It is in this tail region that RGS9-1 differs from a brain-specific protein encoded by the same gene, RGS9-2. The differences between RGS9-2 and RGS9-1 arise because of differences in RNA processing in the striatum and the retina.[10,11,18] The functions of these C-terminal domains are not completely known, but there is evidence suggesting that the RGS9-1 C-terminal domain is close to the membrane attachment site,[19] and it is important for coupling the GAP activity of RGS9-1 to interactions with the effector subunit, cGMP phosphodiesterase γ (PDEγ).[17,20]

PHOSPHORYLATION OF RGS9-1

Treatment of ROS membranes with radiolabeled ATP led to phosphorylation of a protein with electrophoretic mobility identical to that of RGS9-1. Unfortunately, in many gel systems RGS9-1 co-migrates with tubulin, known to be an abundant

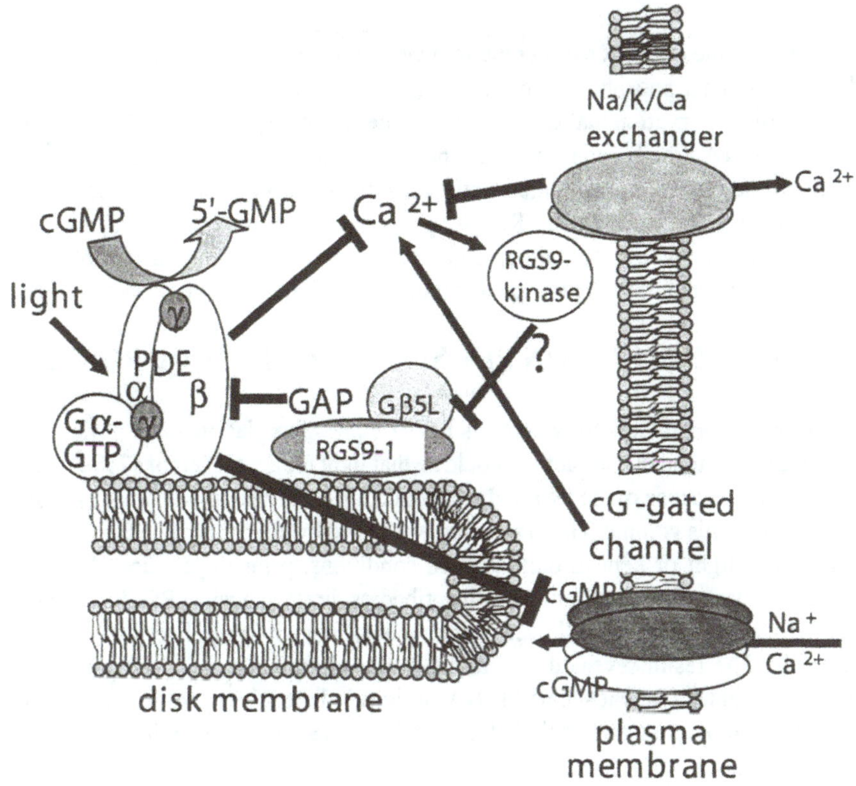

Figure 1. Possible role of RGS9-1 phosphorylation in Ca^{2+} feedback circuitry. When light, through activation of rhodopsin, stimulates transducin (Gα-GTP) and PDE, Ca^{2+} levels fall as cGMP-gated cation channels close and Ca^{2+} continues to be extruded by the exchange protein. The GAP complex, with RGS9-1 as its catalytic core, accelerates inactivation of the Gα-GTP/PDE complex by speeding GTP hydrolysis. Lowering of $[Ca^{2+}]$ leads to inhibition of the kinase responsible for phosphorylating RGS9-1 on serine 475, suggesting a possible link between light-induced changes in intracellular $[Ca^{2+}]$ and the duration of PDE activity stimulated by Gα-GTP.

protein in rods, and a substrate for phosphorylation. Therefore it was necessary to use immunoprecipitation to verify not only that RGS9-1 is indeed phosphorylated by an endogenous protein kinase in ROS, but that it is one of the most strongly phosphorylated proteins in the membranes. Proteolytic digestion of phosphorylated RGS9-1 followed by HPLC and mass spectrometric analysis of tryptic peptides revealed a single site of phosphorylation, 475Ser. Identification of the site allowed to the synthesis of the corresponding peptide, with and without phosphate, and the preparation of phospho-specific monoclonal antibodies.[21] These were then used to monitor RGS9-1 phosphorylation in vivo and in vitro.

EFFECT OF Ca^{2+} ON PHOSPHORYLATION

Tests of the divalent cation requirements for 475Ser phosphorylation revealed that as with other phosphorylation reactions, this one requires Mg^{2+}. More interesting was the observation that Ca^{2+} also stimulates the phosphorylation. The calcium chelator EGTA greatly reduces 475Ser phosphorylation, and restoration of Ca^{2+} at levels in the physiological range of hundreds of nanomolar is sufficient to restore maximal phosphorylation. Thus RGS9-1 phosphorylation, at least in vitro, is stimulated by Ca^{2+} at the levels obtaining under dark-adapted conditions, and is inhibited when Ca^{2+} levels fall to the low concentrations associated with responses to light.

EFFECT OF LIGHT ON PHOSPHORYLATION OF RGS9-1

If intracellular Ca^{2+} truly controls RGS9-1 phosphorylation under physiological conditions, then the prediction would be that light reduces levels of 475Ser phosphorylation, because of the light induced reduction in intracellular $[Ca^{2+}]$ discussed elsewhere in this volume. This prediction was tested by collecting retinas from mice exposed to light or kept in dark-adapted conditions, prior to sacrifice, detergent extraction, immunoprecipitation (with antibodies directed against RGS9-1, but not phosphorylation-specific) followed by immunoblotting with phosphopeptide-specific antibody. The result revealed that RGS9-1 phosphorylation is indeed greatly reduced by light.[21] Thus it seems likely that light-induced decreases in intracellular $[Ca^{2+}]$ in rod outer segments do indeed regulate RGS9-1 phosphorylation in vivo.

CONCLUSION AND REMAINING QUESTIONS

The observation that RGS9-1 phosphorylation is regulated by light and intracellular calcium is tantalizing, because it suggests that changes in $[Ca^{2+}]$ feed back onto the phototransduction cascade at the level of the lifetimes of activated transducin and PDE. However, our current understanding leaves many questions unanswered. Key among these are: what are the functional consequences of 475Ser phosphorylation ? Does phosphorylation increase or reduce GAP activity, or does it affect some other (possibly unknown) function of the protein? What protein kinase is responsible? What is the molecular mechanism for coupling Ca^{2+} concentrations to RGS9-1 phosphorylation? What Ca^{2+}-binding proteins are involved? Research aimed at answering these questions is in progress, and the answers should provide new insights into the link between Ca^{2+} and recovery kinetics in phototransduction.

REFERENCES

1. Sather WA, Detwiler PB. Intracellular biochemical manipulation of phototransduction in detached rod outer segments. Proc Natl Acad Sci USA 1987; 84:9290-4.
2. Lamb TD, Matthews HR. Incorporation of analogues of GTP and GDP into rod photoreceptors isolated from the tiger salamander. J Physiol 1988; 407:463-87.
3. Sagoo MS, Lagnado L. G-protein deactivation is rate-limiting for shut-off of the phototransduction cascade. Nature 1997; 389:392-5.
4. Lyubarsky A, Nikonov S, Pugh EN Jr. The kinetics of inactivation of the rod phototransduction cascade with constant Ca2+i. J Gen Physiol 1996; 107:19-34.
5. Barkdoll AE 3rd, Pugh EN Jr., Sitaramayya A. Calcium dependence of the activation and inactivation kinetics of the light-activated phosphodiesterase of retinal rods. J Gen Physiol 1989; 93:1091-108.
6. Koelle MR, Horvitz HR. EGL-10 regulates G protein signaling in the C. elegans nervous system and shares a conserved domain with many mammalian proteins. Cell 1996; 84:115-25.
7. Ross EM, Wilkie TM. GTPase-activating proteins for heterotrimeric G proteins: regulators of G protein signaling (RGS) and RGS-like proteins. Annu Rev Biochem 2000; 69:795-827.
8. Berman DM, Wilkie TM, Gilman AG. GAIP and RGS4 are GTPase-activating proteins for the Gi subfamily of G protein alpha subunits. Cell 1996; 86:445-52.
9. He W, Cowan CW, Wensel TG. RGS9, a GTPase accelerator for phototransduction. Neuron 1998; 20:95-102.
10. Rahman Z, Gold SJ, Potenza MN et al. Cloning and characterization of RGS9-2: a striatal-enriched alternatively spliced product of the RGS9 gene. J Neurosci 1999; 19:2016-26.
11. Zhang K, Howes KA, He W et al. Structure, alternative splicing, and expression of the human RGS9 gene. Gene 1999; 240:23-34.
12. Cowan CW, Fariss RN, Sokal I et al. High expression levels in cones of RGS9, the predominant GTPase accelerating protein of rods. Proc Natl Acad Sci USA 1998; 95:5351-5356.
13. Chen CK, Burns ME, He W et al. Slowed recovery of rod photoresponse in mice lacking the GTPase accelerating protein RGS9-1. Nature 2000; 403:557-560.
14. Lyubarsky AL, Naarendorp F, Zhang X et al. RGS9-1 is required for normal inactivation of mouse cone phototransduction. Mol Vis 2001; 7:71-78.
15. Sowa ME, He W, Wensel TG et al. A regulator of G protein signaling interaction surface linked to effector specificity. Proc Natl Acad Sci USA 2000; 97:1483-1488.
16. Sowa ME, He W, Slep KC et al. Prediction and confirmation of a site critical for effector regulation of RGS domain activity. Nat Struct Biol 2001; 8:234-237.
17. He W, Lu L, Zhang X et al. Modules in the photoreceptor RGS9-1.Gbeta 5L GTPase-accelerating protein complex control effector coupling, GTPase acceleration, protein folding, and stability. J Biol Chem 2000; 275:37093-37100.
18. Thomas EA, Danielson PE, Sutcliffe JG. RGS9: A regulator of G-protein signalling with specific expression in rat and mouse striatum. J Neurosci Res 1998; 52:118-124.
19. He W, Cowan CW, Melian TJ, Jr. et al. Dependence of RGS9-1 membrane attachment on its C-terminal tail. J Biol Chem 2002; 276:48961-48966.
20. Skiba NP, Martemyanov KA, Elfenbein A et al. RGS9-G beta 5 substrate selectivity in photoreceptors. Opposing effects of constituent domains yield high affinity of RGS interaction with the G protein-effector complex. J Biol Chem 2001; 276:37365-37372.
21. Hu G, Jang GF, Cowan CW et al. Phosphorylation of RGS9-1 by an endogenous protein kinase in rod outer segments. J Biol Chem 2001; 276:22287-22295.

PHOSPHORYLATION BY CYCLIN-DEPENDENT PROTEIN KINASE 5 OF THE REGULATORY SUBUNIT (Pγ) OF RETINAL cGMP PHOSPHODIESTERASE (PDE6): ITS IMPLICATIONS IN PHOTOTRANSDUCTION

Akio Yamazaki,[1,2] Oleg Moskvin[1] and Russell K. Yamazaki[2]

ABSTRACT

Cyclic GMP phosphodiesterase (PDE6) is a key enzyme in vertebrate retinal phototransduction. After GTP/GDP exchange on the α subunit of transducin (Tα) by illuminated rhodopsin, the GTP-bound form Tα (GTP/Tα) interacts with the regulatory subunit (Pγ) of PDE6 to activate cGMP hydrolytic activity. The regulatory mechanism of PDE6 has been believed to be a typical G protein-mediated signal transduction process. We found that cyclin-dependent protein kinase 5 (Cdk5) phosphorylates Pγ complexed with GTP/Tα in vitro and in vivo. Phosphorylated Pγ dissociates from GTP/Tα without GTP hydrolysis and interacts effectively with catalytic subunits of PDE6 to inhibit the enzyme activity. These observations provide new twists to the current model of retinal phototransduction. In this article, in addition to the details of Pγ phosphorylation by Cdk5, we review previous studies implying the Pγ phosphorylation and the turnoff of PDE6 without GTP hydrolysis and indicate the direction for future studies of Pγ phosphorylation, including the possible involvement of Ca^{2+}/Ca^{2+}-binding proteins.

Departments of [1]Ophthalmology and [2]Pharmacology, [1]the Kresge Eye Institute, Wayne State University, School of Medicine, Detroit, Michigan 48201, U.S.A.

INTRODUCTION

In outer segments of retinal photoreceptor, illuminated rhodopsin stimulates GTP/GDP exchange on Tα, the α subunit of transducin, followed by dissociation of GTP/Tα from Tβγ, the βγ subunits of transducin. GTP/Tα activates cGMP phosphodiesterase (PDE6), resulting in a fall of cytoplasmic cGMP concentration, closure of cGMP-gated channels, and hyperpolarization of plasma membranes.[1,2] In rod outer segments (ROS), inhibited PDE is composed of two homologous catalytic subunits, Pα and Pβ[3,4] and two identical regulatory subunits, Pγs.[5,6] Thus, PDE is depicted as Pαβγγ. Each catalytic subunit contains a catalytic and two non-catalytic cGMP binding (GAF) domains [7-9] although whether both of the two catalytic domains and the four GAF domains in Pαβ are individually functional remains as open question.

Pγ inhibits cGMP hydrolysis by Pαβ.[10] For activation of amphibian Pαβγγ, GTP/Tα releases Pγ to remove its inhibitory effect on Pαβ (Fig. 1). The Pγ release has been shown by finding a close relationship between Pγ release and PDE activation[11] and by isolation of Pγ complexed with Tα(GTP- and GDP-bound forms).[12] In mammalian PDE, the Pγ removal by GTP/Tα has been suggested as the mechanism for PDE activation;[5,6,13,14] however, several previous studies also denied the Pγ release.[15,16] In any case, it is obvious that GTP/Tα interacts with Pαβγγ[17] and activates cGMP hydrolytic activity of Pαβ by releasing the inhibitory effect of Pγ. Thus, the PDE activation has been considered to be a typical G protein-mediated signal transduction process and hydrolysis of GTP bound to Tα has been believed as to be the mechanism for the turnoff of GTP/Tα-activated PDE.[18]

Pγ also regulates cGMP binding to GAF domains on Pαβ. In amphibian rod ROS, Pγ stimulates cGMP binding.[19,20] In mammalian ROS, Pγ also appears to regulate cGMP binding to GAF domains. However, the means by which Pγ might regulate cGMP binding are not clear since tight binding of cGMP to mammalian GAF domains has been reported,[21] but no mechanism to cause cGMP release has been proposed. Thus, the role of the cGMP binding to GAF domains in PDE regulation and in phototransduction still remains unclear.

It is clear that Pγ interactions with GTP/Tα and Pαβ are essential for the regulation of PDE. In this review, we summarize the effect of Pγ phosphorylation by cyclin-dependent protein kinase 5 (Cdk5) on Pγ interactions with GTP/Tα and Pαβ. We also review previous studies suggesting PDE regulation by Pγ phosphorylation and turnoff of GTP/Tα-activated PDE without GTP hydrolysis. In addition, as for future studies on Pγ phosphorylation, we describe a potential mechanism for Cdk5 regulation in ROS; possible involvement of Ca^{2+}/Ca^{2+}-binding proteins.

MECHANISM AND FUNCTION OF Pγ PHOSPHORYLATION BY CDK5

The Pγ phosphorylation leads to a turnoff of GTP/Tα-activated PDE without GTP hydrolysis through processes described below. Thus, GTP/Tα-activated PDE

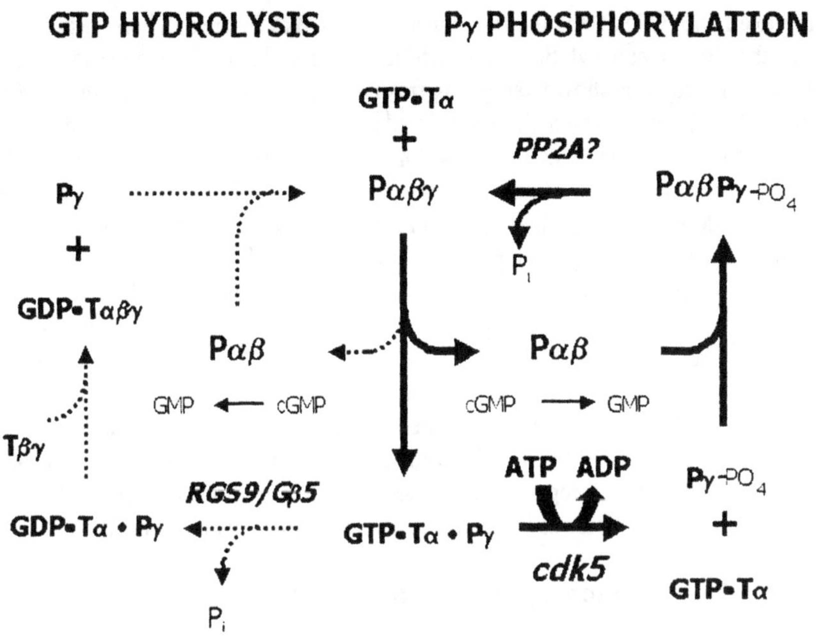

Figure 1. Schematic representation of PDE activation by GTP/Tα and turnoff of GTP/Tα-activated PDE through GTP hydrolysis and Pγ phosphorylation. Although PDE has two Pγs, interactions of one Pγ with other proteins for the regulation of PDE activity are depicted for the simplicity of the figure. Pγ is also involved in the regulation of cGMP binding to GAF domain on Pαβ; however, these Pγ interactions are not shown here. This presentation was based on our studies using frog ROS. However, similar steps are expected in mammalian ROS.

could be deactivated by both GTP hydrolysis by Tα and Pγ phosphorylation by Cdk5 (Fig. 1). When the mechanism of this Pγ phosphorylation was first explored,[22,23] the protein kinase responsible for the Pγ phosphorylation was not identified. Subsequent studies have shown that the protein kinase is Cdk5 complexed with its activator.[24,25]

Pγ Complexed with GTP/Tα is the True Substrate for Cdk5 in vitro and in vivo

Free Pγ is a good substrate for Cdk5 in vitro.[22] However, the lifetime of free Pγ, if present in ROS, is expected to be extremely short since Pγ interacts with GTP/Tα, GDP/Tα[2,26,27] and Pαβ. Therefore, identification of the form of Pγ that serves as the true Cdk5 substrate is crucial to elucidate the role of Pγ phosphorylation in PDE regulation. Comparison of free Pγ and Pγ complexed with GTPγS/Tα or GDP/Tα indicates that the Pγ complex with GTPγS/Tα is the best substrate for Cdk5 (GTPγS is a hydrolysis-resistant GTP analogue).[23] In addition, GTP (or GTPγS) is required for the Pγ phosphorylation in ROS membranes.[23] These results indicate that Pγ complexed with GTP/Tα is the true substrate for Cdk5 in vitro. These results also imply that Pγ phosphorylation by Cdk5 in vivo should be light-dependent. The light-dependency was shown using frog fresh retinas.[25] Under dark conditions, less than 0.5% of the total Pγ was phosphorylated in retinas. A stimulation of phosphorylation was observed when 0.03% of rhodopsin was bleached in 1 s. Moreover, significant amounts of Pγ (~ 4% of the total Pγ) were phosphorylated after less than 0.3% of rhodopsin was bleached in less than 10 s. Furthermore, the Pγ phosphorylation is reversed quickly by changing light-conditions to dark-conditions. These results clearly indicate that Pγ complexed with GTP/Tα is the true substrate in vivo and that Pγ phosphorylation is significant, reversible and rapid enough to be involved in the recovery phase of phototransduction. Recently, Paglia et al[28] re-evaluated Pγ phosphorylation and reported that only small portion (about 10 %) of endogenous Pγ is phosphorylated under light conditions. They incubated a frog ROS homogenate with 100 μM ATP in the presence of GTPγS for 5 hours at room temperature without ATP-regenerating system. These conditions are totally different from ours and may not be suitable to measure endogenous Pγ phosphorylation.

Dissociation of Phosphorylated Pγ from GTP/Tα

Phosphorylation of Pγ decreases its affinity for GTP/Tα. The following five observations[23,25] definitively support this conclusion.

a. Phosphorylated Pγ was physically released from GTPγS/Tα.
b. Pγ complexed with GTPγS/Tα could not inhibit cGMP hydrolytic activity of Pab. However, the Pγ complex could inhibit Pαβ if the Pγ complex was incubated with ATP and Cdk5.
c. GTPγS/Tα could not activate Pαβ complexed with phosphorylated Pγ.
d. In a frog ROS homogenate, Pγ phosphorylated by Cdk5 inhibited light/GTPγS-activated PDE activity as measured using a pH electrode.
e. Under our conditions, Pγ inhibited both GTP hydrolysis by Tα and GTPγS/GDP exchange on Tα,[26] while phosphorylated Pγ did not inhibit either. Paglia et al[28] observed that phosphorylation of Pγ by mitogen-activated protein kinase (MAP kinase) decreased the ability of Pγ to interact with GTP/Tα. While their overall conclusion appears to be similar to ours, several questions regarding their experimental designs and explanations of their data have arisen as described below.

Interaction of Phosphorylated Pγ with Pαβ

The phosphorylated form of Pγ inhibits cGMP hydrolytic activity of Pαβ more effectively than does the unphosphorylated form of Pγ. This conclusion was based on the following experiments.

a. Phosphorylated Pγ had a higher inhibitory activity than unphosphorylated in a reconstituted system containing Pγ-depleted Pαβ/membranes.[23] The comparison was made under the conditions where cGMP hydrolytic activity of Pγ-depleted Pαβ/membranes was essentially linear with the concentration of Pγ-depleted Pab/membranes. The cGMP hydrolytic activity was measured by assaying conversion of [^{3}H]cGMP to [^{3}H]GMP.

b. Lesser amounts of phosphorylated Pγ were required to inhibit cGMP hydrolytic activity of light/GTPγS-activated PDE in a frog ROS homogenate than that of unphosphorylated Pγ.25 The comparison was made under conditions where the light/GTPγS-activated PDE was inhibited by Pγ, but not by Pγ complexed with GTPγS/Tα or GDP/Tα, and the PDE activity was measured using a pH electrode. Paglia et al[28] reported that phosphorylation of Thr22 in Pγ by MAP kinase caused a minor decrease in the ability of Pγ to inhibit cGMP hydrolysis by Pαβ As described below, Thr22 is also phosphorylated by Cdk5 under our conditions. Thus, their conclusion is opposite to ours. They incubated Pγ with MAP kinase obtained from a commercial source for 10-25 hours at room temperature without ATP-regenerating system. In addition, ~ 30% of a Thr22-replaced Pγ mutant was phosphorylated, suggesting that an unknown amino acid(s) other than Thr22 might be phosphorylated under their conditions. They argued that their phosphorylated Pγ was pure since measurement of its molecular weight by matrix-assisted laser-desorption ionization time-of-flight (MALDI-TOF) mass spectrometry showed only one molecular mass. However, MALDI-TOF mass spectrometry is not suitable for establishing protein purity since different forms of a protein may have different volatilities during desorption. Thus, their conclusion may not be supported by suitable experimental data. The possibility of Pγ phosphorylation by MAP kinase is reviewed below.

As summarized above, Pγ phosphorylated by Cdk5 has two special characteristics; phosphorylated Pγ dissociates from GTP/Tα without GTP hydrolysis and phosphorylated Pγ effectively interacts with Pαβ (Fig. 1). These two characteristics provide new twists to the current model of retinal phototransduction, previously thought to be a typical G protein-mediated signal transduction process. Turnoff by GTP hydrolysis of a G protein-regulated effector molecule, the most fundamental characteristic of G protein-mediated signal transduction, may not be necessary in retinal phototransduction, because GTP/Tα-activated PDE can be deactivated without GTP hydrolysis (Fig. 1). This implies that the lifetime of GTP/Tα-activated PDE may also be regulated without GTP hydrolysis. It is also possible that one GTP/Tα, if its lifetime is long, can regulate multiple effector molecules (Pαβγ or Pαβγγ). Thus, regulation of one effector molecule by one active G protein, another fundamental characteristic of G protein-mediated signal transduction, may also not be necessary in retinal phototransduction. This assumption implies that signal amplification in ROS may occur not only in the rhodopsin-transducin interaction[29] but also in the interaction between GTP/Tα and Pαβγγ. This type of system is suited to rapid and sensitive signal transduction mechanisms such as retinal phototransduction.

ATP-Mg^{2+} is Essential for Pγ Phosphorylation

Basic requirements of protein phosphorylation are a protein kinase, ATP-Mg^2 and a substrate protein. In order to explore the complete mechanism of Pγ phosphorylation, we also examined the requirement of ATP-Mg^{2+} for the Pγ phosphorylation (Fig. 2). Pγ phosphorylation was not detected in the absence of Mg^{2+} (i.e., in the presence of excess of EDTA). In the presence of 5 mM Mg^{2+}, Pγ phosphorylation was inhibited competitively by millimolar Ca^{2+} and 10 mM Ca^{2+} inhibited ~ 80% of the Pγ phosphorylation. However, excess of Mg^{2+} (20 mM) overcame the inhibition by 10 mM Ca^{2+}. These observations indicate, as expected, that ATP-Mg^{2+} is an essential requirement for Pγ phosphorylation, and that Pγ phosphorylation by Cdk5 meets all requirements for protein phosphorylation. It should be noted that, in these studies, millimolar Ca^{2+} was used as a competitive inhibitor to Mg^{2+}; therefore, this result does not imply a direct inhibition of Cdk5 by Ca^{2+} in ROS.

PROTEIN KINASES FOR Pγ PHOSPHORYLATION

Cdk5 is the Physiologically Relevant Protein Kinase for the Phosphorylation of Pγ Complexed with GTP/Tα

In the early studies on Pγ phosphorylation,[22, 23] the protein kinase responsible for the Pγ phosphorylation was named "Pγ kinase". Later we identified Pγ kinase as Cdk5, as follows.[24, 25]

Pγ Phosphorylation Requires a Specific Motif for Cdk5

First, a single phosphorylation site in Pγ was indicated by data showing that the phosphorylation caused an increase of only 80 Da in the Pγ molecular weight as measured by laser mass spectrometry. Secondly, the phosphorylation site was identified as Thr22 by analysis of the phosphoamino acid residue in peptides derived from phosphorylated Pγ. The result was also confirmed by using mutants Pγs in which Thr22 is replaced by various amino acids. Extension of the mutant study indicates that the sequence, Pro20-Xaa21-Thr22-Pro23-Arg24, is required for the Pγ phosphorylation. The sequence is conserved in all Pγs. This sequence clearly suggests that Pγ kinase is a proline-directed protein kinase. Proline-directed kinases include the MAP kinases[30] and the cyclin-dependent protein kinases, Cdks,[31] which play critical roles in signal transduction and in regulation of the eukaryotic cell cycle, respectively. As described below, MAP kinases isolated from frog ROS did not phosphorylate Pγ under the conditions that Pγ kinase phosphorylates Pγ. Thus, we focused on Cdk, especially on Cdk5, a member of Cdk found in neuronal cells. Cdk5 phosphorylates serine or threonine residue immediately upstream of a proline residue (at position +1). A proline residue at the –2 position is also involved in the

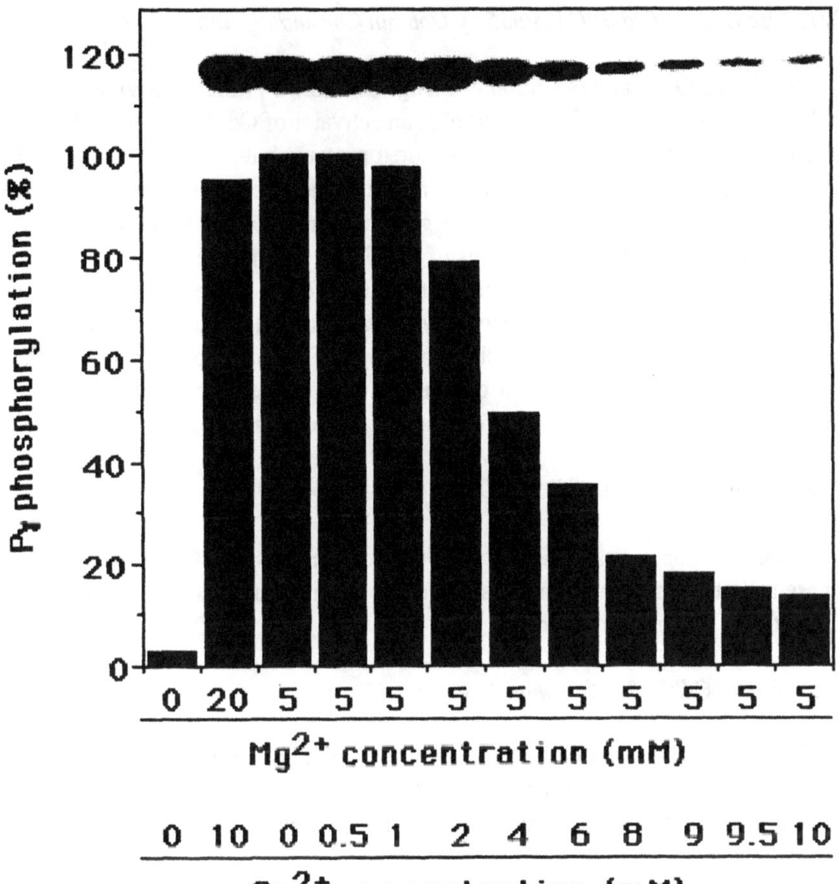

Figure 2. ATP-Mg^{2+} is essential for Pγ phosphorylation by Cdk5. Cdk5 was extracted from frog ROS (8.5 mg protein) by washing with 1.5 ml of Buffer A (10 mM Tris/HCl (pH 7.5), 5 mM dithiothreitol (DTT), 5 mM $MgCl_2$, 0.1 mM phenylmethylsulfonyl fluoride (PMSF), and 0.1 mM $CaCl_2$ (x7). The extract was applied to a Mono Q column (5 x 50 mm) that had been equilibrated with Buffer B (20 mM Tris/HCl (pH 7.5), 1 mM DTT and 0.1 mM PMSF). After washing unbound proteins by Buffer B, proteins were eluted with 300 ml of a NaCl gradient (0–2M) of Buffer B. Flow rate for the protein elution was 1 ml/min and fraction volume was 0.5 ml. Fractions detected Cdk5 activity (fractions 11-15) were combined and used to measure the effect of various concentration of $CaCl_2$ on Pγ phosphorylation in the presence or absence of $MgCl_2$. Pγ phosphorylation activity was assayed using 0.1 μg of frog Pγ complexed with GTPγS/Tα and 33 μl of the Cdk5 preparation (22). After SDS-PAGE and autoradiography, the radioactive protein band (Mr 13,000) was excised from a gel and its radioactivity was quantified. The insert is an autoradiogram of the 13 kDa protein (Pγ).

phosphorylation.[32] The phosphorylation motif is exactly found in the sequence, Pro-Xaa-Thr-Pro-Arg.

Pγ Kinase is Identified as Cdk5/p35 by Column Chromatography

In a Mono Q column (an anion exchange column), Pγ kinase activity coeluted with immunoreactivities of Cdk5 and p35, an activator of Cdk5. The peak fraction of Cdk5 contained less than 3% of the eluted proteins but ~ 30% of the total Pγ kinase activity. When the Cdk5 fraction of the Mono Q column chromatography was reapplied to a TSK-250 column (a gel-filtration column), Pγ kinase activity also coeluted with immunoreactivities of Cdk5 and p35. In the TSK-250 column chromatography, ~ 5% of the eluted proteins was found in the peak fraction although ~ 15% of the total Pγ kinase activity was detected in the peak fraction. In heparin-Sepharose CL-6B column chromatography, Pγ kinase coeluted with Cdk5/p35 and the peak fraction was not overlapped with the peak fraction of proteins. Together, these results strongly suggest that Pγ kinase is Cdk5 complexed with p35.

Pγ Kinase is Identified as Cdk5 by Using Cdk5-Antibodies

When Pγ kinase was incubated with various amounts of a Cdk5 antibody, Pγ kinase activity was proportionally sedimented and immunoreactivity of p35 was also detected in the precipitation.

Recombinant Cdk5/p35 Phosphorylates Pγ

Isolation of clones of both Cdk5 and p35 from a bovine retina cDNA library and sequencing of these clones indicate that these proteins have the same nucleotide sequences as those of bovine brain. Cdk5/p35 expressed in E. coli phosphorylated Pγ and its mutants similarly to Pγ kinase. We note that p35 expressed in E. coli has been digested to the p25 form.

Cdk5 is the Protein Kinase for Pγ Phosphorylation In Vivo

Cdk inhibitors, olomoucine and rescovitine, inhibited Pγ phosphorylation by Pγ kinase isolated from frog ROS and by recombinant Cdk5/p25. The Pγ phosphorylation in vivo was also sensitive to these Cdk inhibitors, indicating that the kinase for Pγ phosphorylation is a Cdk. Two dimensional peptide map analysis showed that the Pγ site phosphorylated in vivo is same as that phosphorylated by Cdk5 in vitro, indicating that Cdk for the Pγ phosphorylation in vivo is Cdk5.

Localization of Cdk5 in Retina

In an immunohistochemical search for Cdk5 in frog retina, strong signals were observed in the inner plexiform layer, in the outer plexiform layer and at the interface between the inner segments and the outer nuclear layer. The signal in the outer segment layer was relatively less intense than that of the layers, but clearly stronger than the background. Similar immunological signals were observed in bovine retina.

Cdk5 has been found to be one of the protein kinases responsible for abnormal phosphorylation of tau, a microtubule-associated protein.[33-35] The abnormal tau phosphorylation appears to cause a loss of tau function and cytoskeletal instability in many neuronal degenerative diseases including Alzheimer's disease, Down's syndrome, progressive supranuclear palsy and Parkinson's disease[36,37] Gene-targeting experiments demonstrate an essential role of Cdk5 in the cyto-architecture of the central nervous system, but Cdk5 is also involved in the regulation of the cytoskeleton, axon guidance, membrane transport, synaptic function and drug addiction.[38,39] In visual tissues other than retina, Cdk5/p35 has been found in lens and cornea.[40,41] Recently, Cdk5 is also reported in non-neuronal tissues.[37]

As described above, Cdk5 is a member of the Cdk family. Like other Cdks, Cdk5 itself shows no kinase activity and requires association with an activator for its enzymatic activity. Cdk5 has two homologous activators, p35 and p39.[42-44] p35 and p39 appear to be redundant in function. Although these activators have little sequence similarity to cyclins, computer modeling and mutagenesis studies have predicted that p35 might adopt a cyclin-like tertiary structure. In retina, p35 is present,[24] but presence of p39 has not been proved. p35 is a very unstable protein and conversion to p25 is induced by various treatments.[45-47] p25 contains all elements necessary for association with and activation of Cdk5 in vitro and in vivo; however, p25, not p35, appears to be involved in the abnormal protein phosphorylation in neuronal cells.[46, 48] Cdk5 prepared from frog and bovine ROS also contains p25,[24] as described above, indicating that ROS also contain a system to degrade p35 to p25. Neither the significance of p25 nor the truncation mechanism of p35 to p25 in ROS is known. In addition, involvement of p25 in retinal degeneration, a very attractive subject, has never been studied.

Possibility for the Involvement of Protein Kinases other than Cdk5 in the Pγ Phosphorylation

Several protein kinases other than Cdk5 have been reported to phosphorylate Pγ by various groups including ours. Are they involved in Pγ phosphorylation in ROS?

Phosphatidylinositol-Stimulated Kinase

Pγ was first reported to be phosphorylated by this enzyme found in intact frog ROS.[49] This Pγ phosphorylation was stimulated by phosphatidylinositol but not by cAMP or cGMP. This Pγ phosphorylation appeared to be inhibited by transducin subunits (Tα and Tβγ) and Pγ-less PDE. Studies using peptides suggest that either Thr35 or Ser40 is phosphorylated by this protein kinase. However, effect of this phosphorylation on Pγ function is unknown. In addition, neither characterization of this protein kinase nor identification of the step for this Pγ phosphorylation in the current model of phototransduction was carried out. Thus, the physiological function of this Pγ phosphorylation in phototransduction is not clear. However, it should be emphasized that this study foreshadowed subsequent findings about change in Pγ functions by Pγ phosphorylation.

Protein Kinase C (PKC)

Udovichenko et al[50] reported that PKC purified from bovine ROS phosphorylated Thr35 in Pγ in vitro. The Pγ phosphorylation increased its ability to inhibit Pαβ catalytic activity with an IC_{50} for phosphorylated Pγ of ~ 26 pM and an IC_{50} of ~ 60 pM for unphosphorylated Pγ. They also suggested that free Pγ was a better substrate for PKC than Pγ complexed with Pαβ, indicating that Pγ complexed with Pαβ may not be the real substrate. This is supported by our observation that no Pγ phosphorylation was detected under dark conditions.[25] Later, they reported that GTPγS/Tα was a less effective activator of Pαβ complexed with phosphorylated Pγ than Pαβ complexed with unphosphorylated Pγ,[51] suggesting that after phosphorylation, Pγ may not interact with GTPγS/Tα. These results suggest that Pγ phosphorylation by PKC may be involved in the regulation of PDE. However, several points should be made in trying to establish a role for Pγ phosphorylation by PKC in the current model of phototransduction. First, the true substrate form of Pγ should be identified for PKC. Pγ complexed with Pαβ may not be a substrate for PKC, as suggested by them.[51] However, Pγ complexed with Tα (GTP- or GDP-bound) will not also be a substrate for PKC because Thr35 in Pγ appears to be masked by Tα in the Pγ complex with Tα.[52,53] In addition, Cdk inhibitors completely inhibited light-dependent Pγ phosphorylation in vivo,[25] indicating that Pγ phosphorylation in illuminated ROS is due to Cdk, and not to PKC or other protein kinases. This also implies that Pγ complexed with GTP/Tγ is not phosphorylated by PKC in vivo. Together, these speculations suggest that only free Pγ is the substrate for PKC. However, presence of free Pγ in ROS has not been proved and is unlikely. Second, activation of PKC by physiological Ca^{2+} concentration should be established for the Pγ phosphorylation. If the α isoform of PKC is involved in the Pγ phosphorylation, as described,[51] activation of PKCα would require a Ca^{2+} concentration higher than that of dark ROS, the maximum physiological concentration of Ca^{2+} in ROS.

Cyclic AMP-Dependent Protein Kinase (PKA)

PKA phosphorylates Thr35 in Pγ in vitro.[53] The Thr35-phosphorylated Pγ showed an increased ability to inhibit catalytic activity of Pαβ. This is in agreement with the result from phosphorylation by PKC of Thr35 in Pγ.[50] The important point is that Pγ phosphorylation by PKA is clearly inhibited by GTPγS/Tα and GDP/Tα,[53] indicating that PKA should not phosphorylate Pγ complexed with Tα during activation of PDE. The mechanism for the inhibition appears to be the masking of Thr35 by Tα. The masking of Thr35 by Tα was also suggested by inhibition of ADP-ribosylation of Arg33 or 36 in Pγ by Tα.[52] However, Pγ complexed with Pαβ is not a substrate for PKA as well because Pγ phosphorylation under the dark conditions was not observed in ROS. Based on these results, we believe that PKA phosphorylates only Thr35 in free Pγ in vitro, and that this Pγ phosphorylation does not function in phototransduction in ROS.

MAP Kinase

MAP kinases are important intermediates in signal transduction pathways that are mediated by many types of cell surface receptors.[30] Signal acquisition by the receptor is a trigger to activate this protein kinase through its phosphorylation. Protein phosphorylation studies by activated MAP kinase indicate that, while many substrates conform to the Pro-Xaa-Ser/Thr-Pro motif, the minimal consensus sequence is Ser/Thr-Pro. In addition to the primary sequence requirement, recognition of protein substrates by MAP kinase also requires structural determinants. Thus, protein phosphorylation by MAP kinase will be established if following criteria are met;

a. target proteins should have the minimal consensus sequence (Ser/Thr-Pro),
b. the substrate form of the target protein should be present in the system where MAP kinase is involved,
c. a mechanism for the activation of MAP kinase should be present in target cells, and
d. protein phosphorylation by MAP kinase should be established in vivo.

Since all known Pγs, as described above, have the consensus sequence Pro-Xaa-Thr-Pro, it is possible that MAP kinase is involved in Pγ phosphorylation. We have studied the possibility of Pγ phosphorylation by MAP kinase using the criteria described above. As shown in Figure 3, Mono Q column chromatography showed that a soluble fraction of illuminated ROS contained a protein kinase for Pγ phosphorylation (fractions 11-13) and the highest activity is observed in fraction 12. This kinase, as summarized above, is Cdk5 complexed with its activator (p35 or p25). Detection using antibodies against MAP kinase indicates that MAP kinase is also present in fractions 15-17 and the largest amount of MAP kinase is in fraction 16. However, these MAP kinase fractions did not phosphorylate Pγ under conditions where Cdk5 phosphorylates Pγ. Since Pγ appears not to have a rigid conformation,[22] Pγ itself may be a MAP kinase substrate (without establishing a specific form by interacting with other proteins). Moreover, these MAP kinase fractions did not phosphorylate Pγ complexed with GTPγS/Tα (data not shown). Thus, these results suggest that Pγ is not a substrate for MAP kinase or MAP kinase in the fraction is not active. Preincubation of the ROS soluble fraction with ATP did not produce the Pγ phosphorylation activity in MAP kinase fractions, suggesting that components required for the MAP kinase activation may not be present in the soluble fraction. Thus, we conclude that Pγ is not a substrate for MAP kinase under the conditions where Cdk5 phosphorylates Pγ. It should be noted that we did not investigate the possibility of the MAP kinase activation system in a homogenate of ROS. Thus, we refrain from judging whether a MAP kinase activation system is present in ROS.

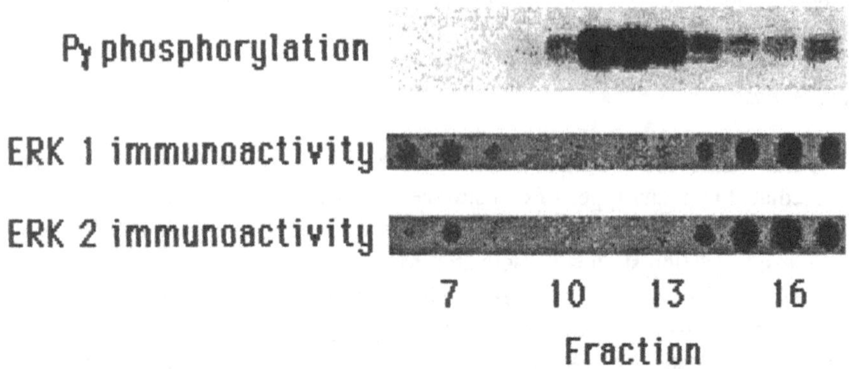

Figure 3. Pγ is not a substrate for MAP kinase. Cdk5 was isolated from frog ROS (8.0 mg protein) using a Mono Q column as described in the legend for Figure 2. Cdk5 activity in each fraction (20 μl, fraction # 6-17) was assayed using 0.1 μg of Pγ purified from frog ROS. After separation of phosphorylated Pγ by SDS-PAGE, autoradiography was carried out. Fractions containing MAP kinase were identified by dot blotting of 50 μl of each fraction (fraction # 6-17) to an Immobilon-P membrane and antibodies (ERK 1 (C-16) and ERK 2 (C-14), Santa Cruz Biotechnology).

Immunohistochemistry using MAP kinase-antibodies did not show a clear signal for MAP kinase in frog and bovine ROS although definite signals for MAP kinase were obtained in other cells of these retinas.[25] At the same time, Harada et al[54] investigated the localization of MAP kinase in retina in their study of involvement of MAP kinase in the circadian clock of retina. Their data indicate that MAP kinase, if present, is very minor in photoreceptors (They did not describe in which segments of photoreceptor MAP kinase is localized). In addition, the active form of MAP kinase was not detected in photoreceptors under the conditions where a definite signal from active MAP kinase was detected in other cells of retina. Their results are in agreement with our results (Y. Fukada, personal communication). Thus, we are skeptical now about the presence of MAP kinase in ROS. It is possible that MAP kinase fractions in Mono Q column chromatography (Fig. 3) may be contaminated from other cells of the retina. Paglia et al[28] reported that Pγ is phosphorylated by MAP kinase. They measured Pγ phosphorylation by MAP kinase by incubation of ROS for 5 to 25 hours. These conditions are totally different from ours and suggest that Pγ phosphorylation by MAP kinase is not kinetically relevant in phototransduction.

Rhodopsin Kinase (RK)

Regulatory mechanisms and roles of RK have been studied extensively because RK is the protein kinase responsible for phosphorylation of the light-activated form of rhodopsin, the major protein in ROS, and RK is a typical protein kinase for a receptor of G protein-mediated signal transduction. We have also investigated the possibility of Pγ phosphorylation by RK (Fig. 4). Cdk5 was isolated from the soluble fraction of frog ROS by Mono Q column chromatography. RK extracted from ROS membranes was mixed with the Cdk5 preparation. The mixture was applied to a heparin-Sepharose column (Cdk5[24] and RK[55] are known to bind to the column) and both kinases were eluted by a NaCl gradient. Activities of RK and Cdk5 in fractions were assayed using a mixture of illuminated ROS membranes (opsin) and Pγ as substrates. We found that these two protein kinase activities were observed in different fractions (peak activities of RK and Cdk5 were in fraction 26 and 40, respectively). We note that the assay mixture used contained an ATP regenerating system, and that even if each protein kinase was separately assayed, the same elution profiles for the two kinases were obtained (data not shown). These results indicate that the lack of Pγ phosphorylation in the RK fraction is not due to the competitive consumption of ATP for the rhodopsin phosphorylation. Together, this column chromatography clearly shows that RK is not a kinase for Pγ phosphorylation.

PREVIOUS STUDIES SUGGESTING PDE REGULATION BY Pγ PHOSPHORYLATION AND TURNOFF OF GTP/Tα-ACTIVATED PDE WITHOUT GTP HYDROLYSIS

Previous Studies Suggesting PDE Regulation by Pγ Phosphorylation

Rhodopsin phosphorylation has been extensively studied because the phosphorylation has been believed to be one of the important mechanisms for regulation of light/GTP-activated PDE in ROS. Liebman and Pugh[56] first suggested that addition of ATP to ROS suspensions leads to highly accelerated deactivation of GTP/Tα-activated PDE (especially at low bleaching of rhodopsin), and that the PDE deactivation is due to rhodopsin phosphorylation. Subsequent studies have supported this role of rhodopsin phosphorylation in PDE deactivation.[57,58] This is especially true when soluble fractions of ROS were added to the quenching reaction[59,60] or ROS preparations contained large amounts of non-rhodopsin proteins were used.[61] We note the similarity in conditions for PDE quenching by rhodopsin phosphorylation and by Pγ phosphorylation since Pγ phosphorylation also deactivates light/GTP-activated PDE, as described above. Components used in the PDE quenching by rhodopsin phosphorylation are basically same as those used in Pγ phosphorylation. It is also possible that soluble fractions of ROS,[59-61] which were believed to contain rhodopsin kinase and to stimulate PDE quenching by enhancing rhodopsin phosphorylation, might also contain Cdk5 and its activator, because Cdk5 complexed with p35 (p25) can be isolated from ROS soluble fractions prepared similarly.[22,24] In addition, it is clear now that rhodopsin kinase does not phosphorylates Pγ (Fig. 4). Thus, it is possible that stimulation of PDE quenching by addition of the soluble fraction may be in part due to the addition of Cdk5/p35. Although we believe that rhodopsin phosphorylation is involved in the PDE quenching in these previous studies, results about quenching of GTP/Tγ-activated PDE by addition of ATP may be explained in part by Pγ phosphorylation by Cdk5.

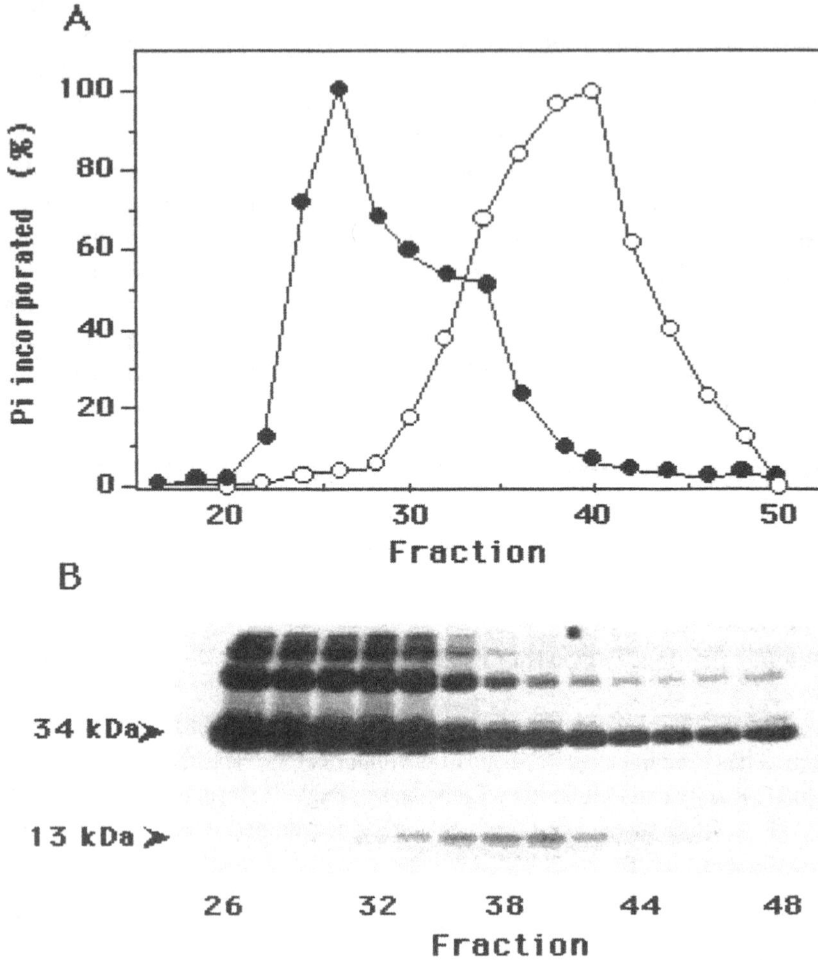

Figure 4. Pγ is not a substrate for rhodopsin kinase (RK). Cdk5 was isolated from frog ROS (18.3 mg protein) using a Mono Q column as described in Figure 2. Fractions containing Cdk5 activity were combined and used as Cdk5 preparation. Following extraction of Cdk5, RK was extracted from ROS membranes with 20 ml of Buffer C (10 mM Tris/HCl (pH 7.5), 5 mM DTT, 1 mM EDTA, 200 mM NaCl, 1 mM benzamidine, 25 μg/ml leupeptin and 0.25 % Tween 80). After dilution of the RK preparation with 60 ml of 10 mM Tris/HCl (pH 7.5) containing 6.7 mM $MgCl_2$ and 1 mM benzamidine, the RK preparation was mixed with the Cdk5 preparation and then the mixture was applied to a heparin-Sepharose column (9 x 150 mm) which had been equilibrated with Buffer D (20 mM Tris/HCl (pH 7.5), 1 mM DTT, 2 mM $MgCl_2$ and 1 mM EGTA). After washing the column with 50 ml of Buffer D and 50 ml of Buffer D containing 0.4 % Tween 80 and 100 mM NaCl, proteins were eluted with a NaCl gradient (0.1-1.0 M/100 ml) of Buffer D containing 0.4 % Tween 80 and 0.2 mM ATP. Phosphorylation of rhodopsin and Pγ in each fraction was measured using opsin (bovine washed ROS membranes, 10 μg), and Pγ purified from frog ROS (0.2 μg) under the conditions for Pγ phosphorylation described previously (22). After SDS-PAGE and antoradiography, the radioactive bands (Mr 34,000 (opsin) and 13,000 (Pγ)) were excised from a gel and their radioactivities were measured. A, radioactivity incorporated into rhodopsin and Pγ. One hundred % indicates that 61.5 and 0.48 pmole of radioactive Pi were incorporated into rhodopsin and Pγ, respectively. (●) rhodopsin; (○) Pγ. Aggregated opsin bands were detected but their radioactivities were not measured. B, phosphorylation profile of rhodopsin and Pγ in fractions (# 26-48).

S-modulin is a frog homologue of recoverin, a Ca^{2+}-binding protein present in vertebrate photoreceptors. Kawamura and Murakami[62] found that S-modulin lengthens the lifetime of active PDE at high Ca^{2+} concentration in frog ROS. Kawamura[63] further suggested that the Ca^{2+}-dependent regulation of active PDE is mediated by regulation of the lifetime of active rhodopsin (Meta II form) through regulation of rhodopsin phosphorylation by S-modulin. Subsequent studies have shown that the regulation of rhodopsin phosphorylation is mediated by Ca^{2+}-dependent interaction of recoverin with rhodopsin kinase in vitro.[64-66] However, it should be emphasized that the lifetime of active rhodopsin does affect the number of active transducin molecules (GTP/Tα), but not the lifetime of active PDE in the current concept of phototransduction.[67] The lifetime of active PDE is regulated by the period for the interaction of Tα with Pγ. As summarized above, Pγ phosphorylation by Cdk5 can decrease the time for the interaction of Pγ with GTP/Tα and subsequently can decrease the lifetime of active PDE. Thus, we thought that Pγ phosphorylation by Cdk5 might be involved in the regulation of the lifetime of active PDE by S-modulin. We found that the reaction mixture for the Pγ phosphorylation contains components similar to those in the reaction mixture used by Kawamura and Murakami to measure the effect of S-modulin on the lifetime of GTP/Tα-activated PDE.[62] It is possible that the regulation of the lifetime of active PDE by Ca^{2+}/S-modulin[62] is in part due to the regulation by Ca^{2+} of Pγ phosphorylation by Cdk5. The possibility that Cdk5 is regulated by Ca^{2+} or Ca^{2+}/Ca^{2+}-binding protein(s) is described below.

It should be emphasized that the lifetime of GTP/Tα-activated PDE is not simply dependent upon the period for the interaction of GTP/Tα with Pγ. In the turnoff mechanism of active PDE by GTP hydrolysis in amphibian ROS (Fig. 1), GTP hydrolysis takes place but is not sufficient for the turnoff, because after GTP hydrolysis, GDP/Tα still binds Pγ.[12, 26] Pγ release from GDP/Tα is essential. We have proposed that Tβγ may stimulate Pγ release from GDP/Tα by competitive interaction with GDP/Tα.[12] Requirement of Tβγ for the turnoff of light signals have also been reported in other species,[27,68] suggesting that a system similar to the frog system may function in other species. However, it should be noted that Tβγ is not required for the turnoff of GTP/Tα-activated PDE by Pγ phosphorylation by Cdk5, because phosphorylated Pγ is directly released from GTP/Tα.

Previous Studies Suggesting Turnoff of GTP/Tα-Activated PDE without GTP Hydrolysis

Introduction of pH measurement for PDE activity assay allows us to measure subtle changes in the initial velocity of cGMP hydrolysis by light/GTP-activated PDE in ROS homogenates. When Liebman and Pugh[56] measured, using the pH assay, the acceleration by ATP of the quenching of light/GTP-activated PDE, their results also suggested that light/GTP-activated PDE was slightly quenched by ATP even in the presence of GMP PNP (GppNHp), a hydrolysis-resistant GTP analogue. They did not comment on the data because the level of PDE quenching in the presence of GppNHp was much lower than that with GTP. Subsequently, Clack et al[69] found that light-dependent hyperpolarization of ROS membranes was terminated even after injection of GppNHp-bound Tα into a toad rod. Then, several studies suggested that light-suppressed dark current was restored even if hydrolysis-resistant GTP analogues were injected into rods, although much slower than in control cells.[70,71] Erickson et al[72] also found that light-stimulated signal was also deactivated even when GTPγS was injected into Limulus ventral photoreceptors. Based on biochemical data, one might have expected that PDE should be permanently active if PDE is activated by a combination of light and hydrolysis-resistant GTP analogues. However, these results clearly suggest that deactivation of visual signal transduction may occur without GTP hydrolysis in photoreceptors. At this point, we cannot say that these results are totally explained by Cdk5 phosphorylation of Pγ. However, it is very inviting for us to speculate that Pγ phosphorylation by Cdk5 may be a mechanism explaining previous results showing the turnoff of light/GTP-activated PDE without GTP hydrolysis. We note that the difference between GTP and GppNHp in PDE quenching, found by several groups,[56,70,71] may be due to the fact that Pγ complexed with GppNHp/Tα is a poor substrate for Cdk5 (AY, unpublished observation). Moreover, it is possible that the slow recovery of the dark current in the presence of hydrolysis-resistant GTP analogues[70,71] might be due to the loss of Cdk5 activity with the whole-cell patch technique.

REGULATION BY Ca^{2+} OR Ca^{2+}-BINDING PROTEIN OF Pγ PHOSPHORYLATION BY CDK5-PRELIMINARY STUDIES

In dark ROS, cytoplasmic cGMP concentration is high enough to bind to and open cGMP-gated channels on plasma membranes. However, under light the cGMP concentration is decreased by PDE activation and the reduction of cGMP concentration leads to closure of cGMP-gated channels. This closure blocks Na^{+} and Ca^{2+} influx, and allows a Na^{+}/Ca^{2+}, K^{+}-exchanger to decrease cytoplasmic Ca^{2+} concentration in ROS.73 This change of Ca^{2+} concentration affects critical enzyme systems to change cytoplasmic cGMP concentration in ROS. For example, Ca^{2+}/recoverin regulates the lifetime of GTP/Tα-activated PDE[62] and Ca^{2+}/GCAPs regulate retinal guanylyl cyclase activity.[74-77]

We also speculate that a Ca^{2+}-binding protein(s) associates with Cdk5/p35 and regulates protein kinase activity of Cdk5/p35 in a Ca^{2+}-dependent manner. As a preliminary experiment, using a molecular sieve column, we investigated the possibility that Cdk5 forms a complex with a Ca^{2+}-binding protein(s) (Fig. 5). Cdk5 was isolated from fresh ROS in the presence of Ca^{2+}. The Cdk5 sample was applied to a molecular sieve column in the presence or absence of Ca^{2+} and a change in the apparent molecular weight of Cdk5 was investigated. The Cdk5 activity was measured by Pγ phosphorylation activity in each fraction. We found that the apparent molecular weight of Cdk5 eluted with a buffer containing Ca^{2+} (~ 80 kDa) was slightly, but consistently larger than that eluted without Ca^{2+}(~ 60 kDa). The molecular weight of Cdk5 isolated from frog ROS in the presence of Ca^{2+} is similar to that previously reported.[22] We note that the molecular weight of Ca^{2+}-binding

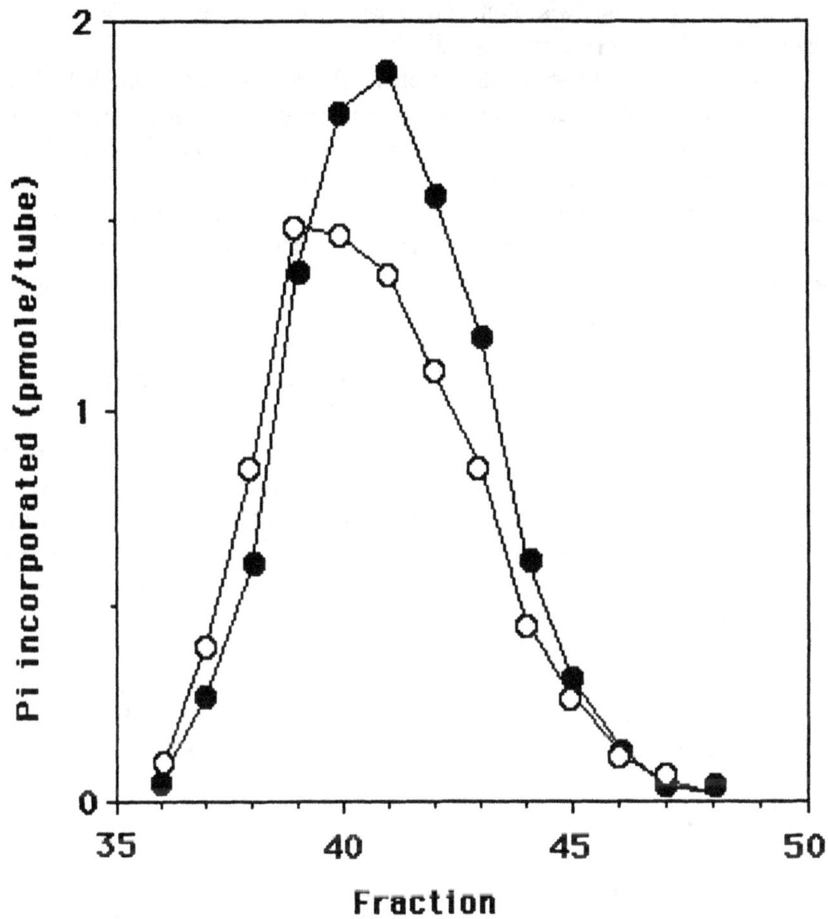

Figure 5. Apparent molecular weight of Cdk5 isolated from frog ROS is changed by Ca^{2+}. Cdk5 was extracted from frog ROS (8.0 mg protein) with 1.5 ml of 10 mM Tris/HCl (pH 7.5) containing 5 mM DTT, 5 mM $MgCl_2$, 0.1 mM PMSF and 1 mM $CaCl_2$ (x7). After concentration with Centricon 3 to 400 μl, the Cdk5 extract was divided into two portions. The buffer of one portion was adjusted to 2 mM EGTA by adding 100 mM EGTA and to another portion the same volume of 1 mM $CaCl_2$ was added. The Ca^{2+}- and EGTA-contained samples were applied to TSK 250 columns that had been equilibrated with 10 mM Tris/HCl (pH 7.5) containing 1 mM DTT, 5 mM $MgCl_2$, 0.1 mM PMSF, 150 mM NaCl and 1 mM $CaCl_2$ or 1 mM EGTA. The flow rate was 0.5 ml/min and the fraction volume was 250 μl. Pγ phosphorylation activity was measured in the presence of 2 mM EGTA using 37.5 ml of each fraction and 0.1 μg of Pγ complexed with GTPγS/Tα. After SDS-PAGE and autoradiography, the radioactive band (Mr 13,000) was excised from a gel and its radioactivity was measured. As molecular weight standards, bovine serum albumin and ovalbumin were used. (○) + Ca^{2+}; (●) – Ca^{2+}.

proteins is usually small and the difference observed may be significant. Thus, these data may suggest that Cdk5 isolated under our conditions forms a complex with a Ca^{2+}-binding protein(s) in the presence of Ca^{2+}. Alternatively, Cdk5/p35 may have a Ca^{2+}-binding site(s) and Ca^{2+} binding to the site changes their conformation.

However, a search of the Swiss-Protein database annotations for Ca^{2+}-binding domains suggests that this possibility is unlikely.

Binding of Ca^{2+} to a Ca^{2+}-binding protein often exposes its hydrophobic domain and the hydrophobic domain exposed stimulates its binding to target proteins. The binding of a Ca^{2+}-binding protein to the target protein changes the profile of target protein in column chromatography. Using a Phenyl-Sepharose column, we investigated the possibility that Cdk5 forms a complex with a Ca^{2+}-binding protein(s) in a Ca^{2+}-dependent manner (Fig. 6). Cdk5 was isolated from fresh ROS in the presence of Ca^{2+}. When Cdk5 was applied to the column and eluted with the buffer containing Ca^{2+}, Pγ phosphorylation activity remained bound to the column after washing with ten column volumes (data not shown). However, Cdk5 bound to the column was released if the washing buffer was replaced by the buffer containing same components except no Ca^{2+} (i.e., in the presence of excess EGTA) (Fig. 6). Moreover, if the sample buffer was replaced by the buffer containing EGTA and the sample was applied to the column equilibrated with the buffer containing EGTA, Pγ phosphorylation activity was found in a flow-through fraction (data not shown). Since Cdk5 and p35 (or p25) appear not to have a Ca^{2+}-binding site, the simplest explanation of these results may be that Cdk5 isolated from ROS in the presence of Ca^{2+} associates with a Ca^{2+}-binding protein.

Although results described here are very preliminary and indirect, these data suggest that Ca^{2+}/Ca^{2+}-binding protein(s) may interact with Cdk5 in ROS. We present these results as initial studies about possible involvement of Ca^{2+}/Ca^{2+}-binding protein in the regulation of Pγ phosphorylation. These results appear to be promising.

CONCLUSIONS

We have firmly shown the significance of the Pγ phosphorylation in the regulation of PDE6 and the protein kinase involved in the Pγ phosphorylation. These studies indicate that Pγ phosphorylation by Cdk5 is a potentially important mechanism in phototransduction. Our observations also imply that GTP/Tα-activated PDE6 can be deactivated without GTP hydrolysis and signal amplification may occur in the interaction between GTP/Tα and Pαβγγ. These implications are quite unexpected in the current model of PDE regulation: a typical G protein-mediated signal transduction process. However, it is also clear that more work is necessary to firmly establish the role of the Pγ phosphorylation in phototransduction. Especially, the regulatory mechanism of the Pγ phosphorylation by Ca^{2+}/Ca^{2+}-binding protein is one of the subjects in need of urgent study. Roles of Cdk5 in neuronal degenerative diseases such as Alzheimer's disease also imply the possibility that Cdk5 may also be involved in some forms of retinal degeneration.

ACKNOWLEDGMENTS

We thank our group members and collaborators who have studied Pγ phosphorylation. We particularly thank Dr. W. H. Miller for his constant support in these

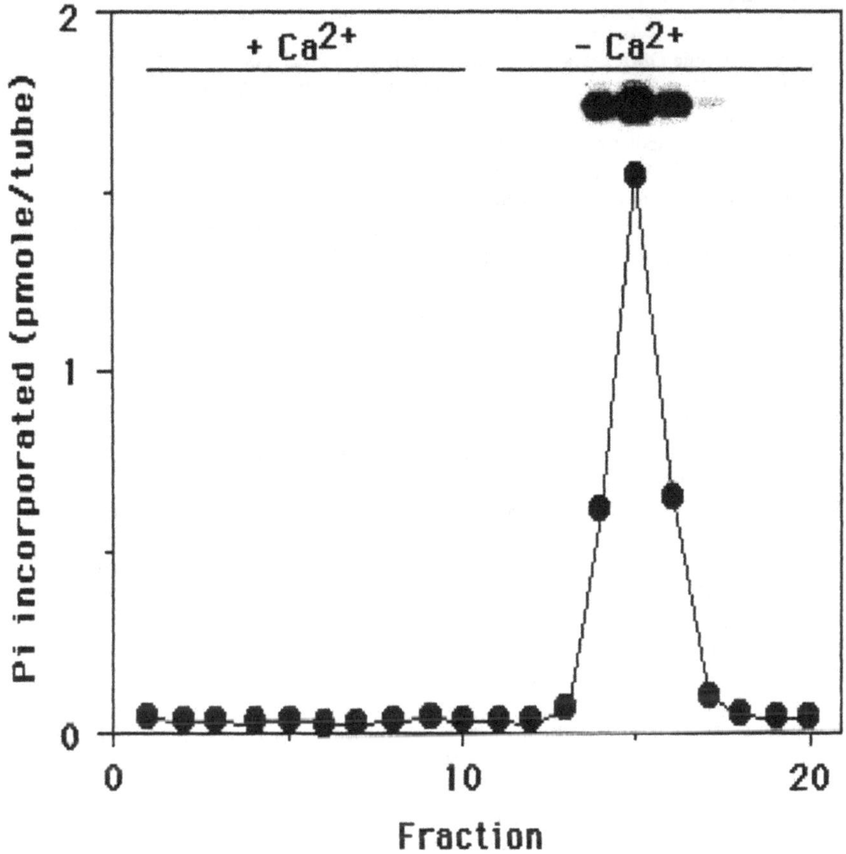

Figure 6. Binding of Cdk5 isolated from frog ROS to Phenyl-Sepharose resin is dependent on Ca^{2+}. Cdk5 was extracted from frog ROS (9 mg protein) as described in the legend for Figure 2. After addition of $CaCl_2$ (final concentration of $CaCl_2$ is 2 mM), the Cdk5 extract was applied to a Mono Q column that had been equilibrated by Buffer E (20 mM Tris/HCl (pH 7.5), 1 mM DTT, 0.1 mM PMSF, and 0.1 mM $CaCl_2$). After washing with Buffer E, Cdk5 was eluted by a NaCl gradient (0-2 M/300 ml) of Buffer E and fractions containing Cdk5 activity were combined and applied to a Phenyl-Sepharose column (2 ml) that had been equilibrated with Buffer F (10 mM Tris/HCl (pH 7.5), 1 mM DTT, 0.1 mM PMSF and 1 mM $CaCl_2$). First, proteins were eluted with 10 ml of Buffer F, and followed with Buffer G (10 mM Tris/HCl (pH 7.5), 1 mM DTT, 0.1 mM PMSF and 5 mM EGTA). Fraction volume was 1 ml. The activity for Pγ phosphorylation was assayed using 40 μl of each fraction and 0.1 μg of Pγ isolated from frog ROS, as described in Figure 2. After SDS-PAGE and autoradiography, the radioactive band (Mr 13,000) was also excised and its radioactivity was quantified. Insert, phosphorylation of 13 kDa protein (Pγ).

studies. Work on Pγ phosphorylation has been supported in part by the National Institute of Health Grants, EY07546 and EY09631, a Jules and Doris Stein Professorship from Research to Prevent Blindness (to AY) and an unrestricted grant from Research to Prevent Blindness to the Kresge Eye institute.

REFERENCES

1. Stryer L. Cyclic GMP cascade of vision. Ann Rev Neurosci 1986; 9:87-119.
2. Miller HW. Dark Mimic. Invest Ophthalmol Vis Sci 1990; 31:1664-1673.
3. Miki N, Baraban JM, Keirns JJ et al. Purification and properties of the light-activated cyclic nucleotide phosphodiesterase of rod outer segments. J Biol Chem 1975; 250:6320-6327.
4. Baehr W, Devlin MJ, Applebury ML. Isolation and characterization of cGMP phosphodiesterase from bovine rod outer segments. J Biol Chem 1979; 254:11669-11677.
5. Deterre P, Bigay J, Forquet F et al. cGMP phosphodiesterase of retinal rods is regulated by two inhibitory subunits. Proc Natl Acad Sci USA 1988; 85:2424-2428.
6. Fung BKK, Young JH, Yamane HK et al. Subunit stoichiometry of retinal rod cGMP phosphodiesterase Biochemistry 1990; 29:2657-2664.
7. Ovchinnikov YA, Gubanov VV, Khramtsov NV et al. Cyclic GMP phosphodiesterase from bovine retina: Amino acid sequence of the α-subunit and nucleotide sequence of the corresponding cDNA. FBE Lett 1987; 223:169-173.
8. Lipkin VM, Khramtsov NV, Vasilevskaya IA et al. β-Subunit of bovine rod photoreceptor cGMP phosphodiesterase: Comparison with the phosphodiesterase family. J Biol Chem 1990; 265:12955-12959.
9. Yamazaki A, Sen I, Bitensky MW et al. Cyclic GMP-specific, high affinity, noncatalytic binding sites on light-activated phosphodiesterase. J Biol Chem 1980; 255:11619-11624.
10. Hurley JB, Stryer L. Purification and characterization of the γ regulatory subunit of the cyclic γGMP phosphodiesterase from retinal rod outer segments. J Biol Chem 1982; 257:11094-11099.
11. Yamazaki A, Stein PJ, Chernoff N et al. Activation mechanism of rod outer segment cyclic GMP phosphodiesterase: Release of inhibitor by GTP/GTP-binding protein. J Biol Chem 1983; 258:8188-8194.
12. Yamazaki A, Hayashi F, Tatsumi M et al. Interaction between the subunits of transducin and cyclic GMP phosphodiesterase in Rana catesbiana rod photoreceptors. J Biol Chem 1990; 265:11539-11548.
13. Wensel TG, Stryer L. Reciprocal control of retinal rod cyclic GMP phosphodiesterase by its γ subunit and transducin. Protein Struct Funct Genet 1986; 1:90-99.
14. Fung BKK, Griswold-Prenner I. G protein-effector coupling: Binding of rod phosphodiesterase inhibitory subunit to transducin. Biochemistry 1989; 28:3133-3137.
15. Clerc A, Bennett N. Activated cGMP phosphodiesterase of retinal rods: A complex with tranducin a subunit. J Biol Chem 1992; 267:6620-6627.
16. Clerc A, Catty P, Bennett N. Interaction between cGMP-phosphodiesterase and transducin α-subunit in retinal rods: A cross-linking study. J Biol Chem 1992; 267:19948-19953.
17. Liu W, Clark WA, Sharma P et al. Mechanism of allosteric regulation of the rod cGMP phosphodiesterase activity by the helical domain of transducin α subunit. J Biol Chem 1998; 273:34284-34292.
18. Fung BKK, Hurley JB, Stryer L. Flow of information in the light-triggered cyclic nucleotide cascade of vision. Proc Natl Acad Sci USA 1981; 78:152-156.
19. Yamazaki A, Bartucca F, Ting A et al. Reciprocal effects of an inhibitory factor on catalytic activity and noncatalytic cGMP binding sites of rod phosphodiesterase. Proc Natl Acad Sci USA 1982; 79:3702-3706.
20. Cote RH, Bownds MD, Arshavsky VY. cGMP binding sites on photoreceptor phosphodiesterase: Role in feedback regulation of visual transduction. Proc Natl Acad Sci USA 1994; 91:4845-4849.
21. Gillespie PG, Beavo JA. cGMP is tightly bound to bovine retinal rod phosphodiesterase. Proc Natl Acad Sci USA 1989; 86:4311-4315.
22. Tsuboi S, Matsumoto H, Jackson KW et al. Phosphorylation of an inhibitory subunit of cGMP phosphodiesterase in Rana catesbiana rod photoreceptors: I. Characterization of the phosphorylation. J Biol Chem 1994; 269:15016-15023.

23. Tsuboi S, Matsumoto H, Yamazaki A. Phosphorylation of an inhibitory subunit of cGMP phosphodiesterase in Rana catesbiana rod photoreceptors: II. A possible mechanism for the turnoff of cGMP phosphodiesterase without GTP hydrolysis. J Biol Chem 1994; 269:15024-15029.
24. Matsuura I, Bondarenko VA, Maeda T et al. Phosphorylation by cyclin-dependent protein kinase 5 of the regulatory subunit of retinal cGMP phosphodiesterase. I. Identification of the kinase and its role in the turnoff of photoreceptor in vitro. J Biol Chem 2000; 275:32950-32957.
25. Hayashi F, Matsuura I, Kachi S et al. Phosphorylation by cyclin-dependent protein kinase 5 of the regulatory subunit of retinal cGMP phosphodiesterase: II. Its role in the turnoff of phosphodiesterase in vivo. J Biol Chem 2000; 275:32958-32965.
26. Yamazaki A, Yamazaki M, Tsuboi S et al. Regulation of G protein function by an effector in GTP-dependent signal transduction: An inhibitory subunit of cGMP phosphodiesterase inhibits GTP hydrolysis by transducin in vertebrate rod photoreceptors. J Biol Chem 1993; 268:8899-8907.
27. Kutuzov M, Pfister C. Activation of the retinal cGMP-specific phosphodiesterase by the GDP-loaded α-subunit of transducin. Eur J Biochem 1994; 220:963-971.
28. Paglia MJ, Mou H, Cote RH. Regulation of photoreceptor phosphodiesterase (PDE6) by phosphorylation of its inhibitory γ subunit-re-evaluated. J Biol Chem 2002; In press
29. Fung BKK, Stryer L. Photolyzed rhodopsin catalyzes the exchange of GTP for bound GDP in retinal rod outer segments. Proc Natl Acad Sci USA 1980; 77:2500-2504.
30. Davis RJ. The mitogen-activated protein kinase signal transduction pathay. J Biol Chem 1993; 268:14553-14556.
31. Pines J. Cyclins and cyclin-dependent kinases: A biochemical view. Biochem J 1995; 308:697-711.
32. Beaudette K, Lew J, Wang JH. Substrate specificity characterization of a CDC2-like protein kinase purified from bovine brain. J Biol Chem 1993; 268:20825-20830.
33. Ishiguro K, Ihara Y, Uchida T et al. A novel tubulin-dependent protein kinase forming a paired helical filament epitope on tau. J Biochem (Tokyo) 1988; 104:319-321.
34. Ishiguro K, Takamatsu M, Tomizawa K et al. Tau protein kinase I converts normal tau protein into A68-like component of paired helical filaments. J Biol Chem 1992; 267:10897-10901.
35. Paudel HK, Lew J, Ali Z et al. Brain proline-directed protein kinase phosphorylates tau on sites that are abnormally phosphorylated in Tau associated with Alzheimer's paired helical filaments. J Biol Chem 1993; 268:23512-23518.
36. Spillantini MG, Goedert M. Tau protein pathology in neurodegenerative diseases. Trends Neurosci 1998; 21:428-433.
37. Dhavan, R., Tsai, L.-H. A decade of Cdk5. Nature Reviews 2001; 2:749-759.
38. Ohshima T, Ward JM, Huh CG et al. Targeted disruption of the cyclin-dependent kinase 5 gene results in abnormal corticogenesis, neuronal pathology and perinatal death. Proc Natl Acad Sci USA 1996; 93:11173-11178.
39. Ahlijanian MK, Barrezueta NX, Williams RD et al. Hyperphosphorylated tau and neurofilament and cytoskeletal disruption in mice overexpressing human p25, an activator of cdk5. Proc Natl Acad Sci USA 2000; 97:2910-2915.
40. Gao CY, Zakeri Z, Zhu Y et al. Expression of Cdk5, p35, and Cdk5-associated kinase activity in the developing rat lens. Dev Genet 1997; 20:267-75.
41. Zelenka PS, Gao CY, Stepp MA. Overexpression of Cdk5 in corneal epithelium of transgenic mice inhibits corneal wound healing. Invest Ophthalmol Vis Sci 2001; 42:S104.
42. Tsai LH, Delalle I, Caviness VS et al. p35 is a neural-specific regulatory subunit of cyclin-dependent kinase 5. Nature 1994; 371:419-423.
43. Lew J, Huang QQ, Qi Z et al. A brain-specific activator of cyclin-dependent kinase 5. Nature 1994; 371:423-426.

44. Humbert S, Dhavan R, Tsai L. p39 activates cdk5 in neurons, and is associated with the actin cytoskeleton. J Cell Sci 2000; 113:975-983.
45. Kusakawa G, Saito T, Onuki R et al. Calpain-dependent proteolytic cleavage of the p35 cyclin-dependent kinase 5 activator to p25. J Biol Chem 2000; 275:17166-17172.
46. Lee MS, Kwon YT, Li M et al. Neurotoxicity induces cleavage of p35 to p25 by calpain. Nature 2000; 405:360-364.
47. Nath R, Davis M, Probert AW et al. Processing of cdk5 activator p35 to its truncated form (p25) by calpain in acutely injured neuronal cells. Biochem Biophys Res Commun 2000; 274:16-21.
48. Patrick GN, Zukerberg L, Nikolic M et al. Conversion of p35 to p25 deregulates Cdk5 activity and promotes neurodegeneration. Nature 1999; 402:615-622.
49. Hayashi F, Lin GY, Matsumoto H et al. Phosphatidylinositol-stimulated phosphorylation of an inhibitory subunit of cGMP phosphodiesterase in vertebrate rod photoreceptors. Proc Natl Acad Sci USA 1991; 88:4333-4337.
50. Udovichenko IP, Cunnick J, Gonzalez K et al. Functional effect of phosphorylation of the photoreceptor phosphodiesterase inhibitory subunit by protein kinase C. J Biol Chem 1994; 269:9850-9856.
51. Udovichenko IP, Cunnick J, Gonzalez K et al. Protein kinase C in rod outer segment: effects of phosphorylation of the phosphodiesterase inhibitory subunit. Biochem J 1996; 317:291-295.
52. Bondarenko VA, Desai M, Dua S et al. Residues within the polycationic region of cGMP phosphodiesterase α subunit crucial for the interaction with transducin γ subunit: Identification by endogenous ADP-ribosylation and site-directed mutagenesis. J Biol Chem 1997; 272:15856-15864.
53. Xu LX, Tanaka Y, Bondarenko VA et al. Phosphorylation of the _ subunit of the retinal photoreceptor cGMP phosphodiesterase by the cAMP-dependent protein kinase and its effect on the γ subunit interaction with other proteins. Biochemistry 1998; 37:6205-6213.
54. Harada Y, Sanada K, Fukada Y. Circadian activation of bullfrog retinal mitogen-activated protein kinase associates with oscillator function. J Biol Chem 2000; 275:37078-37085.
55. Palczewski K. Purification of rhodopsin kinse from bovine rod outer segments. Method Neurosci 1993; 15:217-225.
56. Liebman PA, Pugh Jr EN. ATP mediates rapid reversal of cyclic GMP phosphodiesterase activation in visual receptor membranes. Nature 1980; 287:734-736.
57. Miller JL, Dratz EA. Phosphorylation at sites near rhodopsin's carboxyl-terminus regulates light initiated cGMP hydrolysis. Vision Res 1984; 24:1509-1521.
58. Miller JL, Fox DA, Litman B. Amplification of phosphodiesterase activation is greatly reduced by rhodopsin phosphorylation. Biochemistry 1986; 25:4983-4988.
59. Sitaramayya A, Liebman PA. Mechanism of ATP quench of phosphodiesterase activation in rod disc membranes. J Biol Chem 1983; 258:1205-1209.
60. Sitaramayya A. Rhodopsin kinase prepared from bovine rod disk membranes quenches light activation of cGMP phosphodiesterase in a reconstituted system. Biochemistry 1986; 25:5460-5468.
61. Sitaramayya A, Liebman PA. Phosphorylation of rhodopsin and quenching of cyclic GMP phosphodiesterase activation by ATP at weak bleaches. J Biol Chem 1983; 258:12106-12109.
62. Kawamura S, Murakami M. Calcium-dependent regulation of cyclic GMP phosphodiesterase by a protein from frog retinal rods. Nature 1991; 349:420-423.
63. Kawamura S. Rhodopsin phosphorylation as a mechanism of cyclic phosphodiesterase regulation by S-modulin. Nature 1993; 362:855-857.
64. Chen CK, Inglese J, Lefkowitz RJ et al. Ca^{2+}-dependent interaction of recoverin with rhodopsin kinase. J Biol Chem 1995; 270:18060-18066.
65. Calvert PD, Klenchin VA, Bownds MD. Inhibition of rhodopsin kinase by recoverin: Further evidence for a negative feedback system in phototransduction. J Biol Chem 1995; 270:24127-24129.

66. Sanada K, Shimizu F, Kaneyama K et al. Calcium-bound recoverin targets rhodopsin kinase to membranes to inhibit rhodopsin phosphorylation. FEBS Lett 1996; 384:227-230.
67. Yamazaki A. Is the lifetime of light-stimulated cGMP phosphodiesterase regulated by recoverin through its regulation of rhodopsin phosphorylation? Behavioral Brain Sci 1995; 18:494.
68. Dolph PJ, Man-Son-Hing H, Yarfitz S et al. An eye-specific Gβ subunit essential for termination of the phototransduction cascade. Nature 1994; 370:59-61.
69. Clack JW, Oakley II B, Stein PJ. Injection of GTP-binding protein or cyclic GMP phosphodiesterase hyperpolarizes retinal rods. Nature 1983; 305:50-52.
70. Kondo H, Miller WH. Rod light adaptation may be mediated by acceleration of the phosphodiesterase-guanylate cycalse cycle. Proc Natl Acad Sci USA 1988; 85:1322-1326.
71. Lamb TD, Matthews HR. Incorporation of analogues of GTP and GDP into rod photoreceptors isolated from the tiger salamander. J Physiol (London) 1988; 407:463-487.
72. Erickson MA, Robinson P, Lisman J. Deactivation of visual transduction without guanosine triphosphate hydrolysis by G protein. Science 1992; 257:1255-1258.
73. Yau KW, Baylor DA. Cyclic GMP-activated conductance of retinal photoreceptor cells. Ann Rev Neurisci 1989; 12:289-327.
74. Palczewski K, Subbaraya I, Gorczyca WA et al. Molecular cloning and characterization of retinal photoreceptor guanylyl cyclase-activating protein. Neuron 1994; 13:395-404.
75. Dizhoor AM, Olshevskaya EV, Henzel WJ et al. Cloning, sequencing, and expression of a 24-kDa Ca^{2+}-binding protein activating photoreceptor guanylyl cyclase. J Biol Chem 1995; 270:25200-25206.
76. Gorczyca WA, Polans AS, Surgucheva IG et al. Guanylyl cyclase activating protein: A calcium-sensitive regulator of phototransduction. J Biol Chem 1995; 270:22029-22036.
77. Haeseleer F, Sokal I, Li N et al. Molecular characterization of a third member of the guanylyl cyclase-activating protein subfamily. J Biol Chem 1999; 274:6526-6535.

CENTRINS, A NOVEL GROUP OF Ca^{2+}-BINDING PROTEINS IN VERTEBRATE PHOTORECEPTOR CELLS

Uwe Wolfrum,[1] Andreas Gießl,[1] Alexander Pulvermüller[2]

ABSTRACT

Changes in the intracellular Ca^{2+}-concentration affect the visual signal transduction cascade directly or more often indirectly through Ca^{2+}-binding proteins. Here we review recent findings on centrins in photoreceptor cells of the mammalian retina. Centrins are members of a highly conserved subgroup of the EF-hand superfamily of Ca^{2+}-binding proteins commonly associated with centrosome-related structures. In vertebrate photoreceptor cells, centrins are also prominent components in the connecting cilium linking the light sensitive outer segment with the biosynthetically active inner segment compartment. Recent findings demonstrate that Ca^{2+}-activated centrin forms a complex with the visual G-protein transducin in photoreceptor cells. This Ca^{2+}-dependent assembly of G-proteins with centrin is a novel aspect of the supply of signaling proteins in sensory cells, and a potential link between molecular translocations and signal transduction in general.

INTRODUCTION

Vertebrate rod and cone photoreceptor cells are highly specialized, polarized neurons, which consist of morphologically and functionally distinct cellular compartments (see Fig. 1E, 5D). The light sensitive photoreceptor outer segment is linked with an inner segment via a modified, non-motile cilium, the so-called connecting cilium. The inner segment contains the organelles typical for the metabolism of a

[1]Institut für Zoologie, Johannes Gutenberg-Universität Mainz, 55099 Mainz, Germany. [2]Institut für Medizinische Physik und Biophysik, Humboldt-Universität zu Berlin, Universitätsklinikum Charité, 10098 Berlin, Germany.

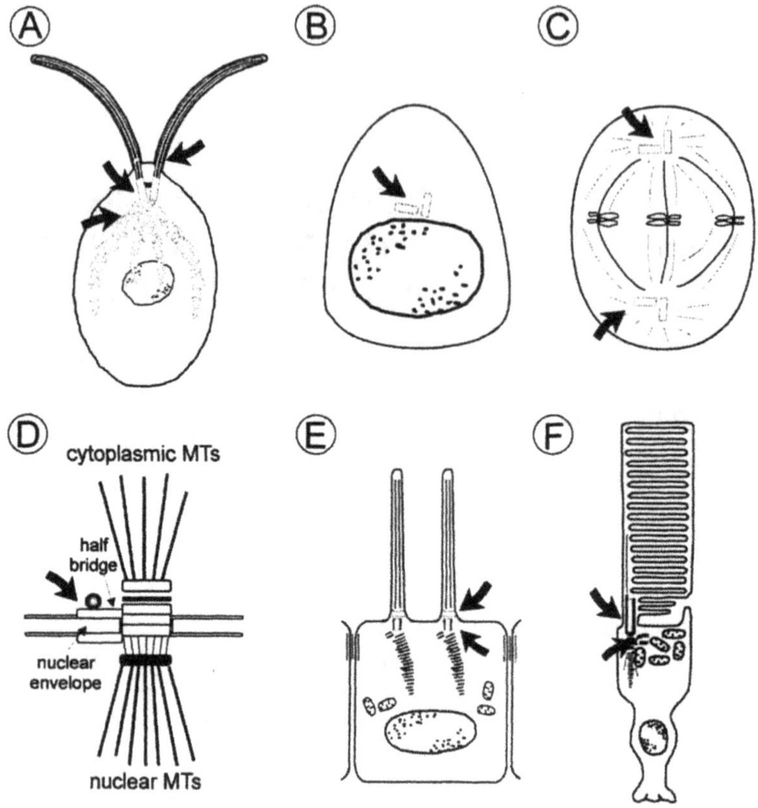

Figure 1. Localization of centrin in diverse cell types. Schematic diagrams of A unicellular green algae (e.g., *Chlamydomonas reinhardtii*); B animal cell in G_1 or G_0 phase (e.g., retinal non-photoreceptor cells, cells of the retinal pigment epithelium) and C in metaphase; D spindle pole body of the yeast *Saccharomyces cerevisiae*, MTs = microtubules; E ciliated epidermal cell; F vertebrate photoreceptor cell. Centrin cellular localization is coloured and indicated by arrows. In the yeast, cdc31p (yeast centrin) is associated with the half bridge of the spindle pole body which accts as the major microtubule organizing centre (MTOC). Centrin is also commonly found at the MTOC, the centrosome of animal cells and at the centrosome-related basal bodies of ciliated cells. In cilia, centrin is also a component of the transition zone which links the basal body region with the axoneme.

eukaryotic cell and continues into the perikaryon and the synaptic region where the electrical signal generated in the photoreceptor cell is transmitted to secondary neurons of the neuronal retina. The outer segment contains all components of the visual transduction cascade (see below) which are arranged disconnected from the plasma membrane bound to hundreds of stacked membrane discs. These membraneous discs are continually renewed throughout lifetime. Newly synthesized membrane is added at the base of the outer segment by the expansion of the plasma membrane[1] or by incorporation of vesicular structures into nascent disc membranes,[2] whereas discs at the distal tip of the outer segment are phagocyted by the cells of the retinal pigment epithelium.[3]

At the outer segment disk membrane, photoexcitation of the visual pigment rhodopsin activates a heterotrimeric G protein cascade leading to cGMP hydrolysis in the cytoplasm and closing of cGMP-gated channels (CNG channels) in the plasma membrane.[4,5] By its rapid lateral diffusion in the membrane, a single molecule of activated rhodopsin (Rho*) can successively activate hundreds of copies of the tissue-specific G-protein (G_t, transducin, composed of an $G_t\alpha$-subunit bearing the guanine nucleotide binding site with GDP attached and an undissociable $\beta\gamma$-complex), thus amplifying the light signal. The activated, GTP-binding α-subunits holds the effector, a cGMP specific phosphodiesterase (PDE), in an enzymatically active form before GTP hydrolysis terminates the interaction and the active state of the PDE. As long as active PDE is present, it decreases the cGMP concentration resulting in closure of the CNG channels and a drop of the cationic current through the channels, which is mainly carried by Na^+ and Ca^{2+}. This hyperpolarizes the cell membrane thus providing the neuronal signal by decreasing transmitter release from the synaptic terminal.

The recovery phase of the phototransduction cascade and the adjustment to background light (light adaptation) of photoreceptor cells rely on changes in the intracellular Ca^{2+}-concentration, $[Ca^{2+}]_i$. It is well established that changes in $[Ca^{2+}]_i$ affects portions of the visual transduction cascade directly or more often indirectly through Ca^{2+}-binding proteins.[6] As a consequence of photoabsorption the efflux of Ca^{2+} (via a light-insensitive plasma membrane Na/Ca-K-exchanger, termed NCKX) exceeds the influx, resulting in $[Ca^{2+}]_i$ decrease, which in turn increases the sensitivity of the cGMP-gated channel to cGMP and accelerates the recovery of the dark current by the release of the Ca^{2+}-binding protein calmodulin (CaM) from the β subunit of the CNG channel (chapter 18 of the present book).[7,8] Lowering of $[Ca^{2+}]_i$ also stimulates the production of cGMP through activation of a photoreceptor-specific particulate guanylate cyclase (GC).[9] The feedback is mediated by one or more Ca^{2+}-binding proteins, termed guanylate cyclase-activating proteins (GCAPs) or GCAP-like proteins (GLPs) described in detail in other chapters of the present book.[6] Besides this well-established role of Ca^{2+} in restoring the dark level of cGMP, yet another mechanism is discussed in the literature, which is thought to act at the level of the activated receptor. It is mediated by another Ca^{2+}-binding protein, recoverin, and affects the phosphorylation of rhodopsin by rhodopsin kinase and thus the quench of light-activated rhodopsin.[10] Furthermore, other Ca^{2+}-binding proteins may also regulate the light insensitive NCKX-exchanger.[6]

The Ca^{2+}-binding proteins involved in the regulation of phototransduction described above are all members of the large EF-hand superfamily of Ca^{2+}-binding proteins which includes besides calmodulin, parvalbumin, troponin C and S100 Ca^{2+}-binding proteins, but also the highly conserved proteins of the centrin subgroup.[11,12] We have recently also identified members of the centrin subgroup as structural proteins in vertebrate retinas.[13-15] The prominent localization of centrin in cytoskeleton of the connecting cilium of vertebrate photoreceptor cells indicated a role in the intracellular transport between the inner segment and the outer segment of the photoreceptor cell. In view of the importance of Ca^{2+}-binding proteins in the regulation of photoreceptor function, centrin's strategic localization and the small knowledge on centrins in the field will present the most recent information on the centrin subgroup of Ca^{2+}-binding proteins. Beside examining the role of centrins in photoreceptor function, we will also provide new insights linkages between the signal transduction cascade with the cytoskeleton.

WHAT ARE CENTRINS?

Centrins, also termed "caltractins", are highly conserved low molecular weight proteins of a subfamily of EF-hand Ca^{2+}-binding proteins.[11,12] The first centrin was discovered as the major component of striated flagellar rootlets associated with the basal bodies of unicellular green algae where it participates in Ca^{2+}-dependent and ATP-independent rootlet contractions.[16] Centrins have since been found to be ubiquitously associated with centrioles of basal bodies and centrosomes, and mitotic spindle poles in cells from diverse organisms, including yeast, ciliates, green algae, higher plants, invertebrates, and vertebrates (Fig. 1).[11,12]

CENTRIN GENES AND MOLECULAR STRUCTURE OF CENTRIN PROTEINS

Cloning efforts in recent years have resulted in the identification of centrin genes in a variety of species from all kingdoms of eukaryotic organisms, protists, fungi, plants, and animals.[17-27] Analyses of amino acid sequences deduced from the cDNA clones demonstrates that centrins are a highly conserved, yet distinct subfamily of the EF-hand superfamily of Ca^{2+}-binding proteins (Fig. 2). Centrins are acid proteins, about 170 amino acids in length, which is in good agreement with their apparent molecular mass of about 20 kDa.[11,12] To date, in lower eukaryotes like the yeast *Saccharomyces cerevisiae* or the unicellular green algae *Chlamydomonas reinhardtii* only one centrin gene (ScCDC31 and CrCEN, respectively) has been identified, whereas in the genome of vertebrates at least three centrin genes (e.g., HsCEN1, HsCEN2, and HsCEN3) are present.[17-19,21,22,26] Clustal analyses of deduced amino acid sequences of centrins from different organisms reveal several phylogenetic groups of centrins (Fig. 3). While some protist centrin species can not be grouped into homogeneous groups, most centrins of higher plants, green algae centrins, and all three known vertebrate centrin isoforms form a phylogenetic group. In vertebrates, Cen1p isofoms and Cen2p isoforms are very close related showing amino acid identities of about 80 percent to 90 percent, whereas sequences of the yeast centrin (ScCdc31p) related vertebrate Cen3p isoforms have only amino acid identities of about 55 percent to both other isoforms. Interestingly, in vertebrate species Cen1p and Cen2p isoforms are closer related to algal centrin (e.g., CrCenp) than to Cen3p isoform of the same species, strongly suggesting two divergent subfamilies.[26]

```
                                                                EF-Hand
                                                         α-Helix      Loop
                                                  : ::     *. :**.*** :     :  :**
RnCen1p --------MASTFRKSNVASTSYKKKVGPKPELTEDQKQEVREAFDLFDSDGSGTIDVKELK   54
MmCen1p --------MASTFRKSNVASTSYKRKVGPKPELTEDQKQEVREAFDLFDSDGSGTIDVKELK   54
HsCen1p --------MASGFKKPSAASTGQKRKVAPKPELTEDQKQEVREAFDLFDVDGSGTIDAKELK   54
HsCen2p --------MASNFKKANMASSSQRKRMSPKPELTEEQKQEIREAFDLFDADGTGTIDVKELK   54
  XlCenp --------MASNYKKPSLGVTTQRKKPVPKTELTEEQKQEIREAFDLFDTDGTGTIDVKELK   54
  CrCenp ----------MSYKAKTVVSARRDQKKGRVG-LTEEQKQEIREAFDLFDTDGSGTIDAKELK   51
  DsCenp ----------MSYR-KTVVSARRDQKKGRVGGLTEEQKQEIREAFDLFDTDGSGTIDAKELK   51
  Sdcenp ----------MSYR--KAASARRDKAKTRSAGLTEEQKQEIREAFDLFDTDGSGTIDAKELK   50
  NgCenp -------MQKYGSKKIGATSATSSNKQKVQIELTDEQRQEIKEAFDLFDMDGSGKIDAKELK   55
MmCen3p --------MSLALR---GELVVDKTKRKKRRELSEEQKQEIKDAFELFDTDKDQAIDYHELK   51
HsCen3p --------MSLALR---SELLVDKTKRKKRRELSEEQKQEIKDAFELFDTDKDEAIDYHELK   51
  GiCenp -----MNRAAIAAGKPSGSISTGKPRRKTRAEVSEEMKHEIREAFDLFDADRSGRIDFHELK   57
ScCdc31p --------MSKNR----SSLQSGPLN----SELLEEQKQEIYEAFSLFDMNNDGFLDYHELK   46
  AnCenp --------------MSSARTVRKDKPRGRHHGLTQQKRQEIKEAFELFDTDGSGTIDAKELN   48
  RnCaMp ---------------------------MADQLTEEQIAEFKEAFSLFDKDGDGTITTKELG   34
Position: 1.......10........20........30........40........50........60..

                                  EF-Hand
                    α-Helix        Loop        α-Helix          α-Helix

RnCen1p VAMRALGFEPRKEEMKKMISEVDKEATGKISFNDFLAVMTQKMAEKDTKEEILKAFRLFDDD   116
MmCen1p VAMRALGFEPRKEEMKKMISEVDKEATGKISFNDFLAVMTQKMAEKDTKEEILKAFRLFDDD   116
HsCen1p VAMRALGFEPRKEEMKKMISEVDREGTGKISFNDFLAVMTQKMSEKDTKEEILKAFRLFDDD   116
HsCen2p VAMRALGFEPKKEEIKKMISEIDKEGTGKMNFGDFLTVMTQKMSEKDTKEEILKAFKLFDDD   116
  XlCenp VAMRALGFEPKKEEIKKMIADIDKEGTGKIAFSDFMCAMTQKMAEKDSKEEIMKAFRLFDDD   116
  CrCenp VAMRALGFEPKKEEIKKMISEIDKDGSGTIDFEEFLTMMTAKMGERDSREEILKAFRLFDDD   113
  DsCenp VAMRALGFEPKKEEIKKMIADIDKAGSGTIDFEEFLQMMTSKMGERDSREEIIKAFKLFDDD   113
  SdCenp VAMRALGFEPKKEEIKKMIADIDKDGSGTIDFEEFLQMMTAKMGERDSREEIMKAFRLFDDD   112
  NgCenp VAMRALGFEPKKEEIKKMISGIDN-GSGKIDFNDFLQLMTAKMSEKDSHAEIMKAFRLFDED   116
MmCen3p VAMRALGFDVKKADVLKILKDYDREATGKITFEDFNEVVTDWILERDPHEEILKAFKLFDDD   113
HsCen3p VAMRALGFDVKKADVLKILKDYDREATGKITFEDFNEVVTDWILERDPHEEILKAFKLFDDD   113
  GiCenp VAMRALGFDVKKEEIQRIMNEYDRDQLGEITFQDFEEVMIEKISNRDPTEEILKAFRLFDDD   119
ScCdc31p VAMKALGFELPKREILDLIDEYDSEGRHLMLYDDFYIVMGEKILKRDPLDEIKRAFQLFDDD   108
  AnCenp VAMRALGFEMTEEQINQMIADVDKDGSGAIDFDEFCHMMTAKIGERDTKEELMKAFRIIDQD   110
  RnCaMp TVMRSLGQNPTEAELQDMINEVDADGNGTIDFPEFLTMMARKMKDTDSEEEIREAFRVFDKD    96
Position: ......70........80........90.......100.......110.......120....

EF-Hand                           EF-Hand
  Loop     α-Helix        α-Helix      Loop         α-Helix
  * *:  :::::  :*.*.:: :::  *   * * . :.::. ::*  :    .
RnCen1p ETGKISFKNLKRVANELGESLTDEELQEMIDEADRDGDGEVNEEEFLKIMKKTNLY---   172
MmCen1p ETGKISFKNLKRVANELGESLTDEELQEMIDEADRDGDGEVNEEEFLKIMKKTNLY---   172
HsCen1p ETGKISFKNLKRVANELGENLTDEELQEMIDEADRDGDGEVNEEEFLRIMKKTSLY---   172
HsCen2p ETGKISFKNLKRVAKELGENLTDEELQEMIDEADRDGDGEVSEQEFLRIMKKTSLY---   172
  XlCenp ETGKISFKNLKRVAKELGENLTDEELQEMIDEADRDGDGEVNEQEFLRIMKKTSLF---   172
  CrCenp NSGTITIKDLRRVAKELGENLTEEELQEMIAEADRNDDNEIDEDEFIRIMKKTSLF---   169
  DsCenp NTGFITLKNLKRVAKELGENLTDEELQEMTDEADRNGDGQIDEDEFYRIMKKTSLF---   169
  SdCenp ETGKISFKNLKRVAKELGENMTDEELQEMIDEADRDGDGEVNEEEFFRIMKKTSLF---   168
  NgCenp DSGFITFANLKRVAKDLGENMTDEELREMIEEADRSNQGQISKEDFLRIMKKTNLF---   172
MmCen3p DSGKISLRNLRRVARELGENMSDEELRAMIEEFDKDGDGEINQEEFIAIM--TGDI---   167
HsCen3p DSGKISLRNLRRVARELGENMSDEELRAMIEEFDKDGDGEINQEEFIAIM--TGDI---   167
  GiCenp ATGRISLKNLRRVAKELSENISDEELLAMIQEFDRDGDGEIDEEDFIAILRSTSAFS--   176
ScCdc31p HIGKISIKNLRRVAKELGETLTDEELRAMIEEFDLDGDGEINENEFIAIC--TDS----   161
  AnCenp NNGKISPEDIQRIAKELGENFTVKDIQDMIEEADRDRDGEVNVEEFLRMMKRTSYAY--   167
  RnCaMp GNGYISAAELRHVMTNLGEKLTDEEVDEMIREADIDGDGQVNYEEFVQMMTAK------   149
Position: ...130.......140.......150.......160.......170.......180...
```

Figure 2. Protein alignment of centrin isoforms and calmodulin from diverse species. ClustalX alignment of 15 different amino acid sequences of centrin species and rat calmodulin. (RnCaMp = rat calmodulin Accession Number (AN): CAA32120; NgCenp = *Naegleria gruberi* centrin AN: AAA75032; DsCenp = *Dunaliella salina* centrin AN: AAB67855; HsCen1p, 2p, 3p = human centrins 1, 2, 3 AN: AAC27343, AAH13873, AAH05383; MmCen1p, 3p = mouse centrins 1, 3 AN: AAD46390, AAH02162; RnCen1p, = rat centrins AN AAK20385, AnCenp = *Atriplex nummularia* centrin AN: P41210; CrCenp = *Chlamydomonas reinhardtii* centrin AN: CAA41039; SdCenp = *Scherffelia dubia* centrin AN: CAA49153; ScCdc31p = *Saccharomyces cerevisiae* ("yeast centrin") AN: P06704; XlCenp = *Xenopus laevis* centrin AN: AAA79194; GiCenp = *Giardia intestinalis* centrin AN: AAB05594). EF-hand domains are indicated as a block above the sequence alignment. EF-hands are composed of an a-helix and a loop. Note, that the EF-hands 2 and 3 of most centrins appear most probably non-functional.

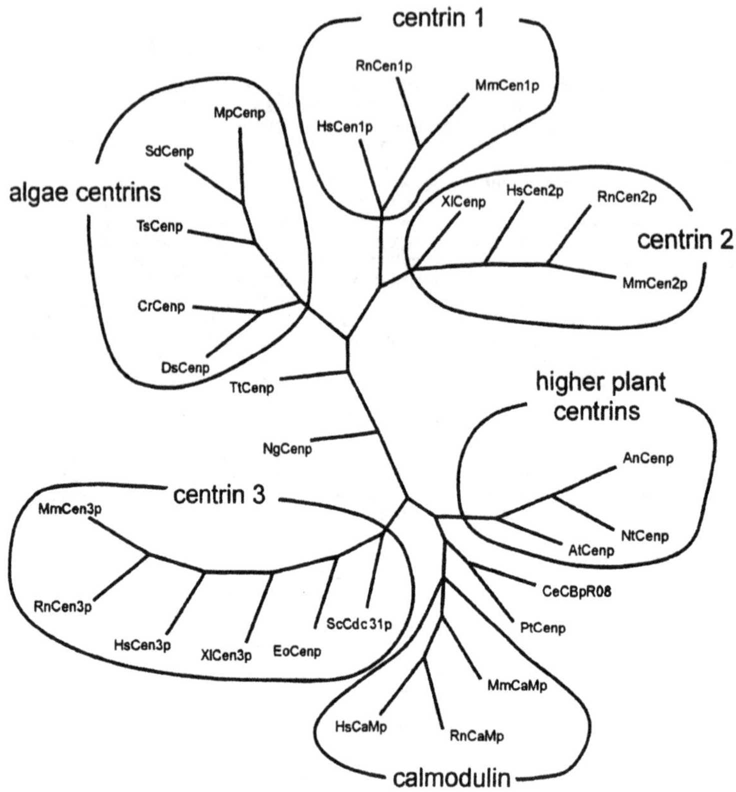

Figure 3. Comparison of centrin isoforms of diverse species. Comparison (using programs: Omiga 2.0, Genedoc 2.5.006 and phylip) of 28 different amino acid sequences of centrins and calmodulins. The phylogram shows a consensus tree which shows the highest frequency of each node of 1000 repetitions. Phylip divids the centrins into subgroups of centrin isoforms 1, 2, 3 , algae centrins, higher plant centrins and a group of calmodulin (RnCaMp = rat calmodulin Accsession Number (AN): CAA32120; MmCaMp = mouse calmodulin AN: NP_033920; HsCaMp = human calmodulin AN: BAA08302; NgCenp = *Naegleria gruberi* centrin AN: AAA75032; XlCenp = *Xenopus laevis* centrin AN: AAA79194; XlCenp3 = *Xenopus laevis* centrin 3 AN AAG30507; PtCenp = *Paramecium tetrauelia* centrin AN: AAB188752; DsCenp = *Dunaliella salina* centrin AN: AAB67855; HsCen1p, 2p, 3p = human centrins 1, 2, 3 AN: AAC27343, AAH13873, AAH05383; MmCen1p, 2p, 3p = mouse centrins 1, 2, 3 AN: AAD46390, AAD46391, AAH02162; RnCen1p, 2p, 3p = rat centrins AN AAK20385, AAK20386, AAK83217; AtCenp = *Arabidopsis thaliana* centrin AN: CAB16762, AnCenp = *Atriplex nummularia* centrin AN: P41210; NtCenp = *Nicotiana tabacum* centrin AN AAF07221; CrCenp = *Chlamydomonas reinhardtii* centrin AN CAA41039; SdCenp = *Scherffelia dubia* centrin AN CAA49153; MpCenp = *Micromonas pusilla* centrin AN CAA58718; EoCenp = *Euplotes octocarinatus* centrin AN CAB40791; TsCenp = *Tetraselmis striata* centrin AN P43646; ScCdc31p = *Saccharomyces cerevisiae* AN P06704; CeCBpR08 = *Caenorhabditis elegans* AN P30644; TtCenp = *Tetrahymena thermophila* AN AAF66602. Tree is not complete.

As members of the parvalbumin superfamily of Ca^{2+}-binding proteins, centrins contain four helix-loop-helix EF-hands consensus domains which may each bind one Ca^{2+}.[28-31] Protein sequence comparisons between different centrin species reveal that the EF-hand consensus motifs are the most highly conserved domains (Fig. 2). Further phylogenetic analyses indicate that the EF-hand domains arose from two-fold duplication of an ancestral EF-hand motif.[32] However, during molecular phylogenesis EF-hand motifs in centrins lost their ability to bind Ca^{2+}. Binding studies indicate that in the green algae *Chlamydomonas* CrCenp and *Tetraselmis* TsCenp all four EF-hands bind a Ca^{2+}, two EF-hands bind Ca^{2+} with high affinity and two EF-hands bind Ca^{2+} with low affinity,[33,34] whereas other green algae possess two or three functional EF-hands.[27] Sequence analysis of vertebrate centrin isoforms suggests that Cen1p and Cen2p molecules bind two Ca^{2+} with their first and the fourth EF-hand and in Cen3p the fourth is the last remaining functional EF-hand motif as it is the case in the yeast centrin ScCdc31p.[11,26,27,35] There are several lines of evidence that Ca^{2+}-binding to centrins induces drastic conformation changes in centrin molecules [11,12,36,37] as previously demonstrated for the related EF-hand protein calmodulin.[38-40] In contrast to calmodulin, centrin molecules become more compact upon Ca^{2+}-binding and Ca^{2+}-activated centrins form dimers and oligomers.[36,37] In polymerization assays, Ca^{2+}-binding induces even centrin polymers, not only with green algae centrins but also with mammalian centrin 1,[36] which may be the structural basis for contractile centrin-fiber systems (see above). Furthermore, Ca^{2+}-binding to centrins increases the affinity of centrin-binding proteins to centrins [36,37,41] which we recently also demonstrated in mammalian photoreceptor cells (Gießl et al, in preparation).[35] To understand Ca^{2+}-induced conformation changes in centrins and binding characteristics of target proteins of centrins, data from high resolution structural analysis are required.

The amino-terminal subdomain of centrins is unique for small Ca^{2+}-binding proteins, unlike those found in, e.g., calmodulin or GCAPs. It is also the most distinctive and variable region of centrins and it has been suggested to be responsible for some functional diversity among centrin species.[11,32,36] Studies on the polymerization properties of centrins indicate that the Ca^{2+}-induced polymerization of centrins, e.g., the formation of contractile centrin-fibers in green algae, is mainly dependent on the amino-terminal domain.[36] In the green algae, it has been demonstrated that centrin phosphorylation correlates with centrin-fiber elongation (relaxation).[16,42] Although conserved potential sites for phosphorylation by protein kinase A (PKA) and $p34^{cdc2}$ kinase are located in the amino-terminal of centrins,[11] direct evidence for in vivo phosphorylation at the amino-terminus of centrins is missing. However, aberrant centrin phosphorylation has been shown under pathogenic conditions in human breast cancer cells that have amplified centrosomes containing supernumerary centrioles.[43] Furthermore, recent studies by Lutz and coworkers[44] indicate that vertebrate centrins are phosphorylated by PKA at conserved PKA consensus sequences present in the carboxy-terminal of centrin molecules. These results suggest that centrin phosphorylation in centrioles signals the separation of centrosomes during the prophase of the cell cycle.

CENTRIN'S CELLULAR LOCALIZATION AND FUNCTION

Centrin was first described as the major component of the massive striated flagellar rootlets of the Prasinophocean unicellular green alga *Tetraselmis striata*.[16] Centrin containing striated rootlets are commonly found in unicellular green algae.[45] They originate near the centrioles of the the basal body apparatus, project into the cytoplasm of the cell body and extend to the plasma membrane, the nucleus or other organelles. In the algal model system *Chlamydomonas*, descending centrin-based fibers connect the basal body apparatus with the nucleus (Fig. 1A).[46,47] In addition to these descending fibers, in *Chlamydomonas* at least two other fiber systems contain centrin: the distal fibers which connect both adjacent basal bodies to one another[48] and the stellate fibers in the transition zone in the plane between the basal body an the axoneme of the flagella.[49] In green algae, all of these centrin fibers have in common that they contract in response on an increase of the intracellular Ca^{2+}-concentration, $[Ca^{2+}]_i$. Most interestingly Ca^{2+}-triggered contraction of centrin fibers of the transition zone may induce microtubule severing and thereby the excision of the flagellum.[49,50] Present microtubule severing mediated by Ca^{2+}-activated centrin may be a more wide spread phenomenon proceeding the massive reorganization of the microtubule cytoskeleton during cell migration[51] or contributing to the microtubule released from the centrosome, the major microtubule organizing center (MTOC) of higher eukaryotic cells.[52]

Major contributions to evaluate the function of centrins in the cell cycle are provided by intensive studies on yeast centrin.[12,36,41,53,54] In baker's yeast *S. cerevesiae*, centrin encoded by the CDC31 gene functions in the duplication of the spindle pole body, the structural equivalent of the centrosome in higher eukaryotic cells. During the first steps of the yeast spindle pole body duplication the binding of Cdc31p to Kar1p is required. Furthermore, Cdc31p specifically interacts with other yeast proteins including an essential kinase (Kic1p) which activity probably regulates the spindle pole body duplication.[53,55]

In vertebrates, centrin proteins are ubiquitously expressed commonly associated with centrosome-related structures such as spindle poles of dividing cells or centrioles in centrosomes and basal bodies.[11,12] As discussed above, in mammals at least three centrin genes are expressed which may cluster to two divergent subfamilies.[26] As a consequence of the isoform diversity the three mammalian centrin isoforms may also exhibit differences in their subcellular localization as well as in their cellular function. Unfortunately, little is known about the specific subcellular localization of the different centrin isoforms in diverse cell types and tissues. Most studies on the localization of the centrin in mammalian cells and tissues have been performed with polyclonal and monoclonal antibodies raised against green algae centrins which do not discriminate between the centrin isforms. Using these antibodies, centrins were detected in the centrioles of centrosomes and in the pericentriolar matrix.[56-58] Further immunological experiments show that antibodies to yeast Ccd31p or mammalian Cen3p react exclusively with Cen3p[26,59] whereas, to our knowledge, to date all of the antibodies raised against the close related mammalian

Cen1p or Cen2p isoforms react with both isoforms[59] (Gießl et al in preparation). Nevertheless, recent studies by immunoelectron microscopy demonstrate that Cen1p/Cen2p and Cen3p are localized in the central lumen of the centrioles of centrosomes and basal bodies.[59,60] In human ciliated tracheal cells, immunoelectron microscopy reveals that the isoform Cen3p is exclusively a core component of the basal body centriole, antibodies to Cen1p/Cen2p additionally decorate epitopes in transition zone of motile cilia.[59] Furthermore, comparative RT-PCR experiments (combined reverse transcriptase reaction and polymerase chain reaction) using isoform specific primers demonstrate that CEN2 is ubiquitously expressed, whereas CEN1 expression is restricted to ciliated cells.[15,59] Thus, it is likely that Cen1p functions as a centrin isoform in compartments of cilia and flagella. Functional analyses indicate that ciliary centrins are involved in the beating of cilia which is controlled by the intraciliary Ca^{2+}-concentration.[59]

The prominent localization of centrins at the centrosomes and basal bodies gave the rise for several hypothesis of the function of centrins. In interphase cells or in arrested cells of differentiated tissue, the centrosome functions as the major microtublule organizing center determining the number and polarity of cytoplasmic microtubules. Polymerization of novel microtubules at the centrosome is preceded by the de novo nuclation of microtubules in the pericentriolar matrix that surrounds and connects the centriole pair of an individual centrosome. It has been suggested that centrins are involved in the microtubule severing which should occur to release de novo synthesized microtubules from the pericentriolar origin.[52] However, more liable evidence was gathered that centrins may play important, but probably distinct roles at the centrosome during the cell cycle. Once in the cell cycle, the centrosome is duplicated to give rise to two spindle poles that organize the microtubules array of the mitotic spindle. While Cen3p, as its yeast relative Cdc31p, participates in centrosome reproduction and duplication,[61] Cen1p/Cen2p may play a role in centriole separation preceding centrosome duplication.[44]

CENTRINS IN THE VERTEBRATE RETINA

RT-PCR studies with centrin isoform specific primers reveal that all three centrin isoforms are expressed in the mammalian retina, which has been confirmed by Western blot analysis using antibodies specific for Cen3p and Cen1p/Cen2p, respectively[15] (A. Gießl, A. Schmitt, and U. Wolfrum, unpublished results). Further studies showed that centrins are expressed in the retina of species distributed throughout the subphylum of vertebrates (Fig. 4). Thus, centrins are probably ancient cytoskeletal proteins in the vertebrate retina indicating this conserved basic function in retinal cells.

Immunocytochemical studies demonstrate that centrins are concentrated in the cells of the vertebrate retina in two basically distinct structural domains (Fig. 5). As in other cell types of animals, centrins are components of the centrioles of centrosomes and basal bodies in the retinal neurons contributing to centrosome function discussed above. However, in all of our studies on numerous different vertebrate

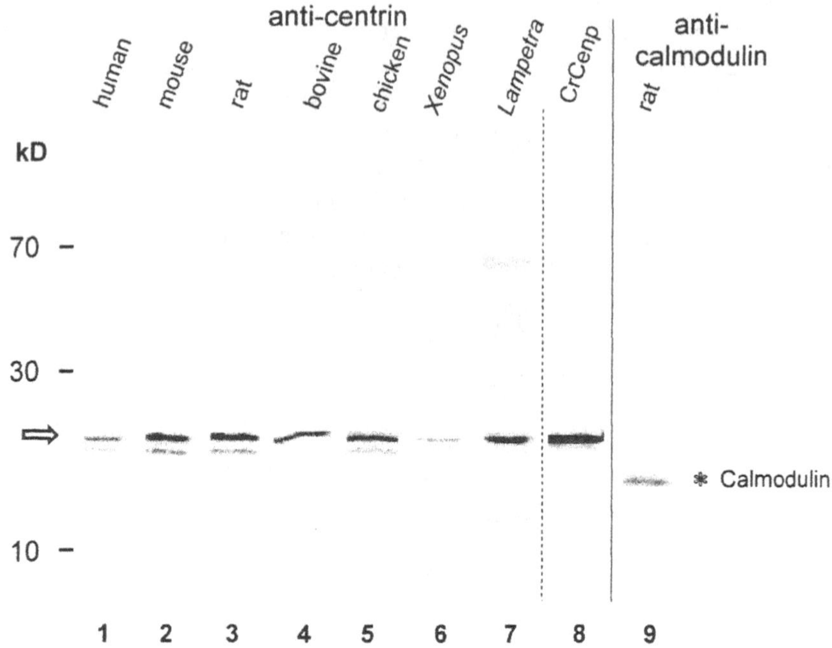

Figure 4. Western blot analysis reveals centrin expression in retina of various vertebrate species. Lane 1—Lane 8: Anti-centrin (mAb clone 20H5) Western blots. Lane 9: Anti- calmodulin Western blot of rat retina. Lane 1: human retina. Lane 2: mouse retina. Lane 3: rat retina. lane 4: bovine retina. Lane 5: chicken retina. Lane 6: *Xenopus* retina. Lane 7: *Lampetra* retina. Lane 8: Bacterially expressed *Chlamydomonas* centrin. Anti-centrin antibodies detect bands at about the predicted molecular weight of 20 kDa (arrow) and do not crossreact with the calmodulin migrating at 17 kDa. Note: in some lanes (e.g., lane 5) several bands around 20 kDa are anti-centrin positive. These bands do neither represent different centrin isoforms nor different Ca^{2+}-binding status of centrin. The higher bands most probably resemble phosphorylated centrn[43,44] and some lower bands may result from proteolytic digestion.

species, indirect anti-centrin immunofluorescene was most prominent in photoreceptor cell layer (Fig. 5).[13-15]

CENTRIN FUNCTIONS AS A CYTOSKELETAL COMPONENT THE CONNECTING CILIUM OF THE PHOTORECEPTOR CELL

Higher magnification of anti-Cen1p/Cen2p stained cryosections through vertebrate retinas shows that centrins are localized at the photoreceptor layer at the joint between the photoreceptor inner segment and outer segment (Fig. 5). Analysis of immunolabled isolated photoreceptor fragments reveals that centrins are not only present in the basal body, but also localized along the entire longitudinal extension

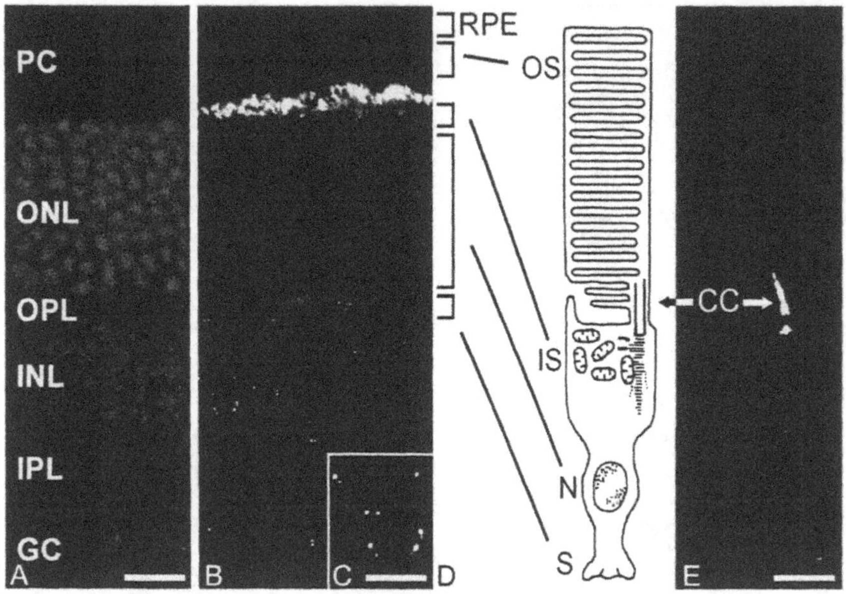

Figure 5. Localization of centrin in the mammalian retina and in photoreceptor fragments by indirect immunofluorescence. A DAPI-staining of a longitudinal cryosection through the rat retina. Staining of nuclei DNA demonstrates the retinal layers: PC = layer of outer and inner segments of photoreceptor cells; ONL = outer nuclear layer where nuclei of photoreceptors are localized; OPL = outer plexiform layer; INL = inner nuclear layer; IPL = Inner plexiform layer; GC = ganglian cell layer. B Indirect anti-centrin immunofluorescence in the cryosection through rat retina. Anti-centrin antibodies predominantly react within the photoreceptor cell layer at the joint between the inner and out segment of the photoreceptors. In addition, indirect immunofluorescence is present in dot pairs in the inner nuclear layer and ganglion cell layer. C Higher magnification of immunofluorescent staining with antibodies against centrin in the inner nuclear layer of the section shown in figure B. Centrin is present in the centrioles of the centrosomes present in the perikaryon of retinal neurons. Note that as a rule one centriole of a single centrosome shows brighter anti-centrin immunofluorescence. D Schematic representation of a mammalian rod photoreceptor cell. The light sensitive outer segment (OS) is linked via the non-motile connecting cilium (CC) with the inner segment (IS) where the protein synthesis machinery is localized. N = nuclear region; S = synaptic region. Centrin localization is indicated by the green color of centrin in the PRC. E Indirect anti-centrin immunofluorescence of a photoreceptor fragment of the rat retina analysed by confocal laser scanning microscopy. RPE = rod pigment epithelium. The figure shows the a labeling of the connecting cilium and the basal body.
Bars in B = A: 20 μm; C: 7 μm, E: 2 μm.

of the connecting cilium (Fig. 5E).[14,15] Precise subcellular localization by immunoelectron microscopy and the quantification of silver-enhanced immunogold labeling show that centrin is localized in the subciliary domain of the inner face of the microtubule doublets of the connecting cilium of rod and cone photoreceptor cells (Fig. 6).[35] As in other ciliated cells, in photoreceptor cells the centrin decorated by immunolabeling in the connecting cilium most likely resembles the centrin 1 isoform.

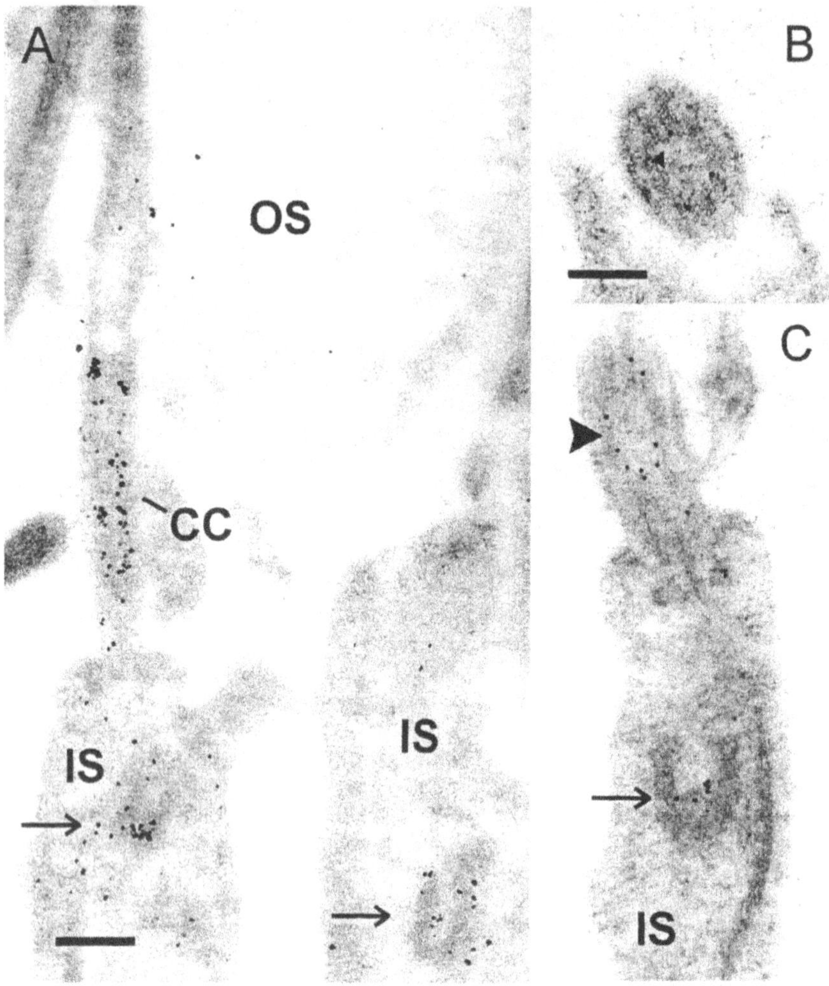

Figure 6. Immunoelectron microscopic localization of centrin in the connecting cilium of rod photoreceptor cells. A Silver-enhanced immunogold labeling of centrin in a longitudinal section of parts of rat rod photoreceptor cell. Centrin labeling is exclusively localized in the connecting cilium (CC) and the basal body complex (arrow) in the inner segment (IS) of photoreceptors. B Transversal section through the connecting cilium reveals that centrin is localized in the sub-ciliary domain of the ciliary lumen encircled by axonemal microtubule doublets. C Slightly tangential section through the apical part of rat rod photoreceptor cell inner segment. Centrin antibodies react in the connecting cilium at the inner surface of the axonemal microtubule doublets (arrowhead). The arrow indicates basal body labeling.
Bars: A: 265 nm, B, C: 175 nm

The modified connecting cilium of vertebrate photoreceptor cells is the structurally equivalent of an extended transition zone present at the base of a common motile cilium.[62] Therefore, the presence of centrin (most probably centrin 1) along the entire extension of the connecting cilium of photoreceptor cells is in agreement with the centrin localization in the transition zone of motile cilia or the sensory cilia of mammalian olfactory cells.[59] In photoreceptor cells, the connecting cilium links the morphological and functional distinct cellular compartments of the light sensitive outer segment with the biosynthetically active inner segment. The connecting cilium serves as an active barrier for membrane components and soluble proteins regulating free diffusion between the inner and the outer segment of photoreceptor cells.[62,63] Since it is also the only intracellular bridge between both segments, intracellular exchanges between the inner segment and the outer segment are forced to occur through the slender connecting cilium.[62] Recently, we and others have shown that the visual pigment opsin is translocated to its final destination at the base of the photoreceptor outer segment along the membrane of the connecting cilium.[64-66] Actin filament-based and microtubule-associated transport processes seem to be involved in the unidirectional ciliary transport of opsin: The membrane associated molecular motor protein myosin VIIa has been shown to participate in ciliary transport of rhodopsin.[64-66] Marszalek and co-workers[67] gathered indications by a genetic approach that the microtubule-based heterotrimeric kinesin II-motor might be additionally involved in ciliary transport of rhodopsin but also of arrestin. However, cytoskeletal molecules associated with other proteins of the visual transduction cascade and which, therefore, are probably involved in the ciliary translocation of these proteins, have not yet been identified. The prominent localization of centrin in the connecting cilium of photoreceptor cells obviously indicates a specific role of centrin in the function of the photoreceptor cilium. Besides its possible role in ciliary transport, an involvement of centrin in retinomotor movement and in the photoreceptor outer segment alignment or orientation has been discussed.[14] If any of these processes are based on the centrin system of the cilium they should be dependent on and regulated by changes of the intracellular Ca^{2+}-concentration. Our recent results, as discussed below provide striking evidence for Ca^{2+}-dependent interaction between centrin 1 and the visual G-protein transducin on its pathway through the inner lumen of connecting cilium of mammalian photoreceptor cells.[35]

CENTRIN-INTERACTING PROTEINS IN MAMMALIAN PHOTORECEPTOR CELLS

In the context of the cell, protein function and its regulation is determined by the binding proteins to the target protein. Unfortunately, little is know about centrin-binding proteins in mammalian photoreceptor cells. To evaluate centrin functions in vertebrates, in other experimental systems, different strategies for the identification of centrin-associated or centrin-interacting proteins were applied. Analysis of proteins in co-immunoprecipitations performed with antibodies against algae centrin has revealed centrin interaction with the heat shock proteins HSP70 and HSP90 in cytoplasm of arrested *Xenopus* oocytes.[68] The centrin/HSP-complex may sequestrate centrin in a non-active form until Ca^{2+}-activation of the oocyte causes the dissociation of the complex making centrin available for subsequent centrosome assembly. In yeast 2-hybrid screens the laminin-binding protein (LBP) of the basal lamina and the cytoplasmic receptor protein tyrosine kinase k have been identified as proteins interacting with HsCen2p, the ubiquitously expressed centrin isoform.[69] Although, there is no specific experimental evidence, none of the proteins identified as centrin-binding proteins has an obvious function in the connecting cilium of the photoreceptor cells. Nevertheless, Western blot overlay assays of retinal proteins with recombinant expressed MmCen1p indicate the presence of several centrin 1-binding proteins in the mammalian retina (Fig. 7). However, only centrin 1 in its Ca^{2+}-activated form interacts with the several polypeptides. A Ca^{2+}-dependent increase of the affinity of centrin to target proteins is known from binding of the yeast centrin Cdc31p to Kar1p[12] which has been confirmed in in vitro binding studies of diverse recombinant expressed centrin species to the yeast target protein.[36,41] Further analysis of the proteins identified by the MmCen1p overlay assay are currently performed. However, the centrin 1-binding protein p37 has been already identified as the β-subunit of the visual G-protein transducin (G_t) (Fig. 8,9B).

CENTRIN/TRANSDUCIN COMPLEX

Recently, evidence was provided that MmCen1p interacts with the visual G-protein transducin with high affinity, and thereby form functional protein-protein complexes in photoreceptor cells in a Ca^{2+}-dependent manner.[35] Transducin is the tissue-specific G-protein of the visual signal transduction cascade of the photoreceptor cells in the vertebrate retina (see also introduction). Upon light-activation, rhodopsin (Rho*) activates hundreds of G-protein molecules and the light signal is amplified. This receptor-G-protein interaction requires the intact G_t holoprotein, composed of an α-subunit bearing the guanine nucleotide binding site with GDP bound and an undissociable βγ-complex, and initiates the intermolecular transduction of the light signal by catalyzing the exchange of GDP for GTP in the α-subunit of the G-protein. Activated, GTP-binding α-subunits are free to couple to the effector, a cGMP specific phosphodiesterase (PDE).

In vertebrate photoreceptor cells the subcellular localization of transducin is modulated by light: in the dark G_t is highly concentrated in outer segments while in light, the majority of G_t is translocated and abundantly localized in the inner segment and the cell body of photoreceptor cells (Fig. 8).[35,70,71] Light-induced exchanges and movements of the cytoplasmic components between the photoreceptor segments have to occur through the connecting cilium, since the slender cilium serves as the only intracellular linkage between both photoreceptor compartments. As described above, centrin 1 is a prominent component of the cytoskeleton of the non-motile motified cilium and immunufluorscence double labelings of tranducin and centrin indicate that transducin and centrin 1 co-localize in the connecting cilium (Fig. 8C). Further immunoelectron microscopical analysis and the quantification of silver enhanced immunogold decorations reveal that centrin and transducin are not only exist parallel in the cilium, but share also the same subciliary domain, the inner ciliary lumen of the connecting cilium.[35] Their spatial co-distribution indicate that both

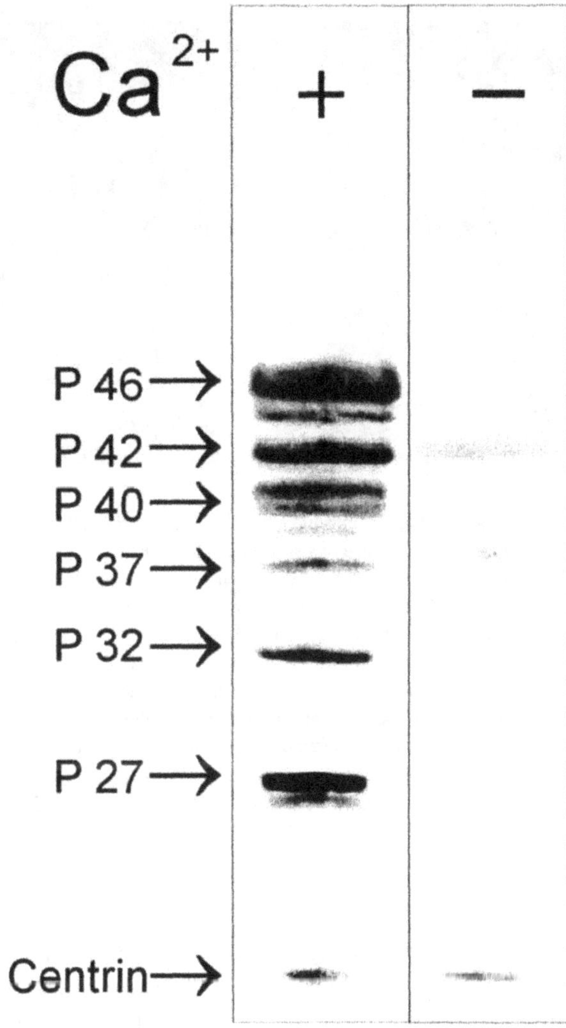

Figure 7. Analysis of centrin blot overlays with bovine retinal proteins. Western blot of retinal proteins overlay with recombinant expressed MmCen1p (67 µg/ml). Bound centrin was detected in a 2nd step by immunolabeling with anti-centrin antibodies (mAb clone 20H5). Lane 1: Centrin overlay assay in the presence of Ca^{2+} (1 µM $CaCl_2$). Lane 2: Centrin overlay assay in the absence of Ca^{2+} (6 mM EGTA). MmCen1p interaction with retinal proteins is dependent on the presence of Ca^{2+}. In the absence of Ca^{2+} MmCen1p-binding was dramatically reduced. Centrin interacting proteins are named according to their molecular weight (P 27, P 32, P 37, P 40, P 42, P46).

proteins may physically interact during the exchange of transducin between the photoreceptor segments through the cilium.

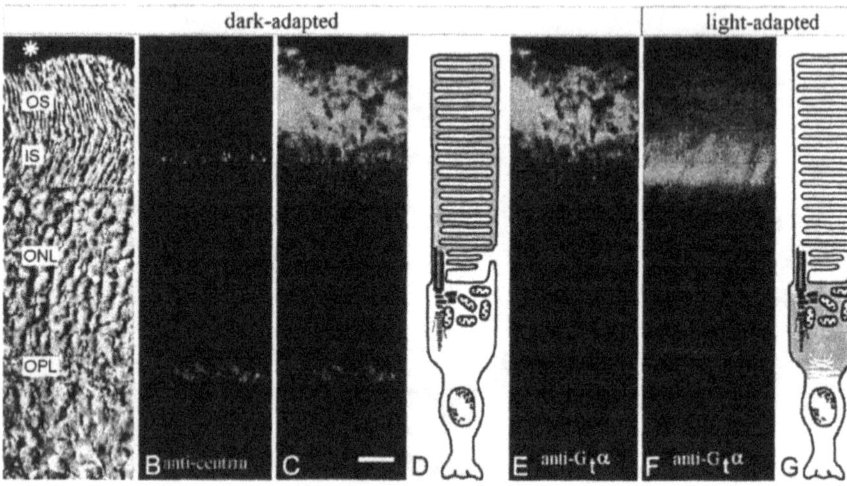

Figure 8. Light-dependent translocations of transducin in the mammalian retina. A-E dark-adapted mouse retina. F, G light adapted retina. A DIC-image of a cryosection through mouse retina. Asterisk indicates retinal pigment epithelium, OS: photoreceptor outer segment; IS: photoreceptor inner segment, ONL: outer nuclear layer, OPL: outer plexiform layer. B anti-centrin immunofluorescence (Alexa,546) is concentrated in the connecting cilium between IS and OS of photoreceptor cells. C Merged images of B and E suggest partial co-localization of $G_t\alpha$ and centrin in the joint between both photoreceptor segments. D Schematic representation of a dark- adapted rod photoreceptor cell. Green colour indicates $G_t\alpha$ distribution. E Indirect anti-$G_t\alpha$ immunofluorescence in the double labeled cryosection through the dark-adapted mouse retina shown in A-C. F Indirect anti-$G_t\alpha$ immunofluorescence in the section through the light-adapted mouse retina. G Schematic representation of a light-adapted rod photoreceptor cell. Green colour indicates $G_t\alpha$ distribution. In dark adapted photoreceptor cells, $G_t\alpha$ is predominantly localized in the OS where as in the light-adapted condition $G_t\alpha$ is most prominent stained in the IS of photoreceptor cells.
Bar: 10 μm

Recently, we have gathered striking evidence that centrin 1 and transducin indeed interact with high affinity.[35] In vitro assays including co-immunoprecipitation, overlay and co-sedimentation assays as well as size exclusion chromatography and kinetic light scattering experiments independently demonstrate that centrin 1 binds with high affinity to transducin (Fig. 9).[35] Our studies also show that the protein-protein interaction centrin 1 and transducin is highly specific: centrin 1 specifically interacts with transducin and does not bind to other components of the visual signal transduction cascade (e.g., arrestin, rhodopsin, rhodopsin kinase, PDE). The centrin relatives recoverin and calmodulin do not show significant affinities to transducin. The analyses of MmCen1p overlay assays with antibodies specific to transducin subunites and size exclusion chromatographies further demonstrate that assembly of centrin 1/G-protein complex is mediated by the βγ-complex (see also Fig. 9B, D-G). Our data also reveal that the assembly of the centrin 1/G-protein protein complex is strictly dependent on the Ca^{2+}-concentration and that at least two Ca^{2+}-ions are required for the activation of centrin 1 necessary for the formation of centrin 1/G-protein complex. Moreover, further analysis indicates that activated centrin 1 binds as a homooligomer to the βγ-complex of transducin.[35]

What is the role of the centrin 1/G-protein complex in the photorecptor cell? Current working hypotheses of the centrin 1/G-protein complex function in the photorecptor cell are summarized in the cartoon in Figure 10. The spatial co-localization of centrin 1 and transducin in the lumen of the connecting cilium emphasizes that in photoreceptor cells, the formation of centrin 1/G-protein complex should occur in this ciliary compartment. An increase of the intracellular Ca^{2+}-concentration in the photoreceptor cell should cause the activation of centrin 1 in the connecting cilium and in turn induce the binding of centrin 1 oligomers to transducin passing through the ciliary. As a consequence of the assembly of centrin/transducin complexes the movement of transducin should be effected. In photoreceptors, light modulated changes of free Ca^{2+} in the outer segment which include the well-studied the Ca^{2+}-drop within the operating (single quantum detective) range of the rod[10] and recent observations of Ca^{2+} increase in bright light (rod saturated conditions)[72] should also effect the free Ca^{2+} in the connecting cilium. In the cilium the assembly of centrin 1/G-protein complexes may contribute to a Ca^{2+}-induced barrier for further exchange of transducin between the photoreceptor inner and outer segment (barrier hypothesis Fig. 10B). A drop of Ca^{2+} should induce the disassembly of the complex, thus providing a necessary condition for the light-modulated exchange of transducin between the inner and the outer segment of photoreceptor cells described above.[35,70,71] However, Ca^{2+} triggered sequential binding of transducin to centrin 1 may although contribute to the transport of transducin though the photoreceptor connecting cilium (Ca^{2+}-gradient hypothesis Fig. 10C). The Ca^{2+}-dependent assembly of a G-protein with centrin is a novel aspect of the supply of signaling proteins in sensory cells, and a potential link between molecular translocations and signal transduction in general.

CONCLUSION

Centrins are members of a conserved subfamily of EF-hand Ca^{2+}-binding proteins. During the past years, 3 centrin isoforms have been found to be ubiquitously associated with the centrioles of centrosomes or centrosome related structures in diverse vertebrate cells. Our work on centrins in photoreceptor cells has revealed that centrins are prominent components of the ciliary apparatus of photoreceptor cells. Although several lines of evidence indicate defined spatial distributions of the known centrin isoforms, the differential localization of centrin isoforms by using isoform specific antibodies or by the transfection of the retina with tagged-centrin constructs should provide more liable information on the specific localization and function of the centrin isoforms in photoreceptor cells. Our recent findings reveal that the centrin isoform centrin 1 binds with high affinity to transducin in a strict Ca^{2+}-dependent manner. Additional experimental efforts are necessary to resolve the question whether transducin binding is restricted to centrin 1. If so, what are the functions of the other centrin isoforms in photoreceptor cells? The results of the

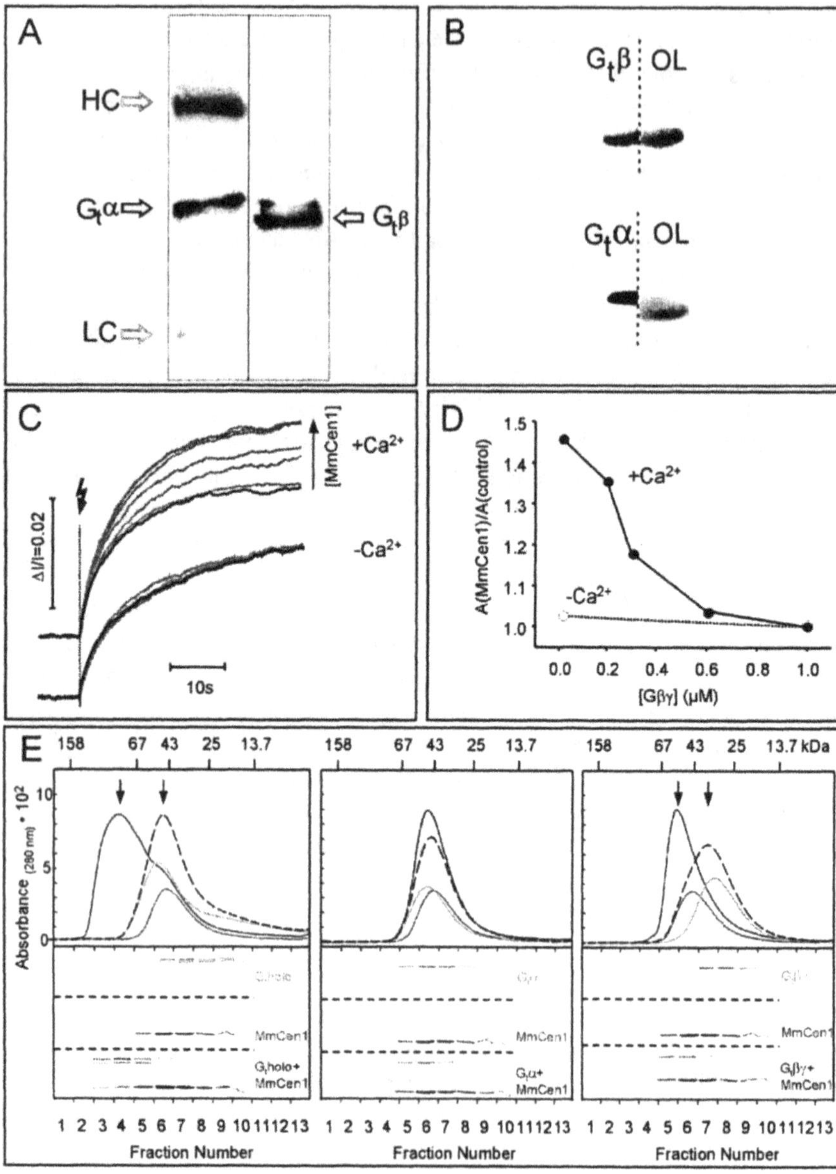

Figure 9. Calcium dependent assembly of a centrin1/G-protein complex. A) Co-Immunoprecipitation of transducin with centrin from lysed retinal photoreceptor cell fragments.

Lanes 1: Western blot analysis with mAb anti-G_ta of an immunoprecipitation with mAb anti-centrin (clone 20H5) from photoreceptor cell fragments of bovine retina. (Upper and lower bands in lane 1 correspond to the heavy (HC) and light chains (LC) of mouse antibodies.) Lane 2: Western blot analysis with polyclonal anti-$G_t\beta$ of anti-centrin of an immunoprecipitation with mAb anti-centrin (clone 20H5) from photoreceptor cell fragments of bovine retina. $G_t\alpha$ and $G_t\beta$ co-immunoprecipitate with centrin. The upper and lower bands in lane 1 correspond to the heavy (HC) and light chains (LC) of the mouse antibodies.

continued on next page

continued from previous page

B)Combined Western blot-overlay analysis identifies retinal centrin-interacting protein P37 as $G_t\beta$ subunit of transducin.

For specific determination of the centrin binding protein Western blotted lanes were cut in half and parallel processed for immunolabeling with subunit specific antibodies against Gtβ (upper lane 1), and Gtα transducin (lower lane 1) and for overlays with recombinantly expressed MmCen1p (67 μg/ml) (OL). The 37 kDa centrin-binding protein (P37 in Fig. 7) is identified by centrin overlays had the exact mobility as the $G_t\beta$ subunit.

C) Calcium-dependent enhancement of kinetic light-scattering (KLS) G_t-binding signal in the presence of MmCen1p. Upper panel represent KLS binding signals (3 μM rhodopsin, 0.5 μM G_t) in the presence of calcium, and 0 (control, black curve), 0.6, 1.2, 2.5, 3.6, 5, 7.3, and 10 μM MmCen1p (gray curves), respectively. Lower panel represent KLS binding signals under the same conditions as in the upper panel, but with EGTA instead of calcium. Experimental conditions were 50 mM BTP, pH 7.5 containing 80 mM NaCl, 5 mM $MgCl_2$ and either 100 μM $CaCl_2$ or 1 mM EGTA at 20°C, sample volume of 300 μL, and cuvette path length of 1 cm; 32% rhodopsin was photolyzed per flash (500±20 nm).

D) Competition between $G_t\beta\gamma$-subunit and G_t for binding to MmCen1p. Calcium dependent inhibition of the MmCen1p enhanced amplitude of flash-induced KLS G_t-binding signals by the βγ-subunit of G_t. The KLS assay was carried out as described in (A). Experiments were performed with the bg-subunit of G_t. Data points represent the normalized amplitude of the MmCen1 dependent enhancement of the G_t-binding signal (AMmCen1) divided to the control G_t-binding signal without added MmCen1 (control). Filled and empty circles indicate the results obtained from experiments with calcium and with EGTA, respectively.

E) Calcium-dependent interaction of MmCen1p with G_t and its subunits analyzed by size-exclusion chromotography and SDS-PAGE.

Upper panels represent elution profiles of MmCen1p alone (L), G_t or its subunits alone (—) and the mixture of MmCen1p with G_t or its subunits (æ) in the presence of calcium. The gray dotted lines are the calculated superpositions of the respective single component profiles (MmCen1p plus G_t or its subunits) yielding the predicted profiles for the mixture of the two non-interacting components. Arrows indicate the shift of the formed complexes. In the lower panel the SDS-PAGE analysis of the fractions of the size-exclusion chromatography is shown. Interaction of MmCen1p with the transducin holoprotein is shown in 1st panel with the $G_t\alpha$-subunit in 2nd panel and with the $G_t\beta\gamma$-subunit in 3rd in the presence of calcium.

Experimental conditions: 10 μg of MmCen1p and 10 μg of $G_t$holo (or G_t subunits) were incubated in 50 mM BTP, pH 7.0 containing 80 mM NaCl, 1 mM $MgCl_2$ and 100 μM $CaCl_2$ for 5 min at room temperature, loaded on a Superose TM 12 column (using the Smart System, Pharmacia Biotech. Inc., flow rate, 40 μL/min) equilibrated with the same buffer, eluted by monitoring the absorbance at 280 nm and subsequent analyzed by SDS-PAGE. Note: G_t holoprotein elutes at an apparently lower MW, as compared to its subunits.[73]

current analysis of putative centrin-associated proteins (other than transducin) in the mammalian retina will most probably also provide further insights in the role of centrins in photoreceptor cell function. In the future, we will also address the question whether Ca^{2+}-activation of centrins is the only post-translational modification regulating the function(s) of centrin in photoreceptor cells. And finally, the clarification of the structure of centrin isoforms will also enlighten the molecular mechanisms of the diverse functions of centrins in photoreceptors.

ACKNOWLEDGMENTS

This work was supported by grants of the Deutsche Forschungsgemeinschaft (DFG) to U.W. (Wo548/1), the DFG-SPP 1025 "Molecular Sensory Physiology" to

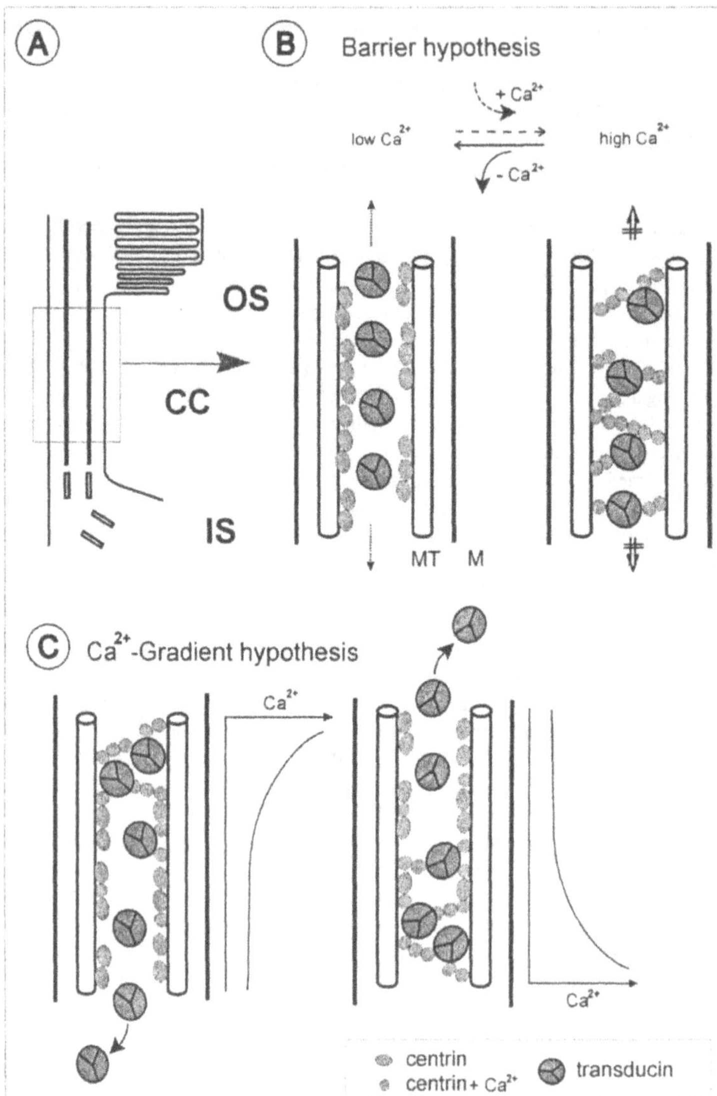

Figure 10. Models for Ca^{2+}-dependent centrin-transducin assembly in the connecting cilium of vertebrate photoreceptor cell. A) Schematic representation of a part of a rod photoreceptor cell shows the linkage between the outer segment (OS) and inner segment (IS) by the connecting cilium (CC). (B and C) Enlargement of CC indicated in Figure A. B) Barrier hypothesis: under low free Ca^{2+} centrin is not activated and transducin floats through the inner lumen of the connecting cilium. If free Ca^{2+} increases in the outer segment and, in the cilium, centrin is activated by Ca^{2+} which induces Ca^{2+}-centrin-transducin complex assembly and centrin fiber contraction. Thus transducin is trapped in the connecting cilium and a barrier between inner and outer segment raises. C) Ca^{2+}-gradient hypothesis: transducin may bind to centrin 1 dependent on free Ca^{2+} concentration actually present in the ciliary domains. A putative Ca^{2+}-gradient along the ciliary lumen may cause sequential assembly of the centrin 1-transducin complex and the release of transducin from the complex.

A.P. (Ho832/6) and U.W. (Wo548/3)), and the FAUN-Stiftung, Nürnberg, Germany to U.W. The authors thank Dr. K. P. Hofmann, Humboldt-Universität, Universitätsklinikum Charité, Berlin, Germany, for helpful comments and critical discussions of the manuscript and Brenda K. Huntley, Mayo Clinic Foundation, Rochester, MN, USA, for attentive linguistic corrections. We also thank Dr. H. E. Hamm, Northwestern University Institute of Neuroscience, University of Texas at Austin, USA, for kindly supplying the monoclonal antibody to frog a-transducin and to Dr. J. L. Salisbury, Mayo Clinic Foundation, Rochester, MN, USA, for providing us mouse centrin 1 cDNA and antibodies raised against algae centrin.

REFERENCES

1. Steinberg RH, Fisher SK, Anderson DH. Disc morphogenesis in vertebrate photoreceptors. J Comp Neurol 1980; 190:501-18.
2. Usukura J, Obata S. Morphogenesis of photoreceptor outer segments in retinal development. Prog Retin Eye Res 1995; 15:113-25.
3. Young RW. Visual cells and the concept of renewal. Invest Ophthalmol 1976; 15:700-25.
4. Heck M, Hofmann KP. G-protein-effector coupling: a real-time light-scattering assay for transducin-phosphodiesterase interaction. Biochemistry 1993; 32:8220-7.
5. Okada T, Ernst OP, Palczewski K et al. Activation of rhodopsin: new insights from structural and biochemical studies. Trends Biochem Sci 2001; 26:318-24.
6. Palczewski K, Polans AS, Baehr W et al. Ca2+-binding proteins in the retina: structure, function, and the etiology of human visual diseases. Bioessays 2000; 22:337-50.
7. Hsu YT, Molday RS. Modulation of the cGMP-gated channel of rod photoreceptor cells by calmodulin. Nature 1993; 361:76-9.
8. Weitz D, Zoche M, Muller F et al Calmodulin controls the rod photoreceptor CNG channel through an unconventional binding site in the N-terminus of the beta-subunit. EMBO J 1998; 17:2273-84.
9. Koch KW, Stryer L. Highly cooperative feedback control of retinal rod guanylate cyclase by calcium ions. Nature 1988; 334:64-6.
10. Molday RS, Kaupp UB. Ion channels of vertebrate photoreceptors. In: Stavenga DG, Degrip WJ, Pugh ENJr, eds. Molecular mechanism in visual transduction. Amsterdam: Elsevier Science Publishers, 2000:143-82.
11. Salisbury JL. Centrin, centrosomes, and mitotic spindle poles. Curr Opinion Cell Biol 1995; 7:39-45.
12. Schiebel E, Bornens M. In search of a function for centrins. Trends Cell Biol 1995; 5:197-201.
13. Wolfrum U. Cytoskeletal elements in arthropod sensilla and mammalian photoreceptors. Biol Cell 1992; 76:373-81.
14. Wolfrum U. Centrin in the photoreceptor cells of mammalian retinae. Cell Motil Cytoskeleton 1995; 32:55-64.
15. Wolfrum U, Salisbury JL. Expression of centrin isoforms in the mammalian retina. Exp Cell Res 1998; 242:10-7.
16. Salisbury JL, Baron A, Surek B et al. Striated flagellar roots: isolation and characterization of a calcium-modulated contractile organelle. J Cell Biol 1984; 99:962-70.
17. Huang B, Mengerson A, Lee VD. Molecular cloning of cDNA for caltractin, a basal body-associated Ca2+-binding protein: homology in its protein sequence with calmodulin and the yeast CDC31 gene product. J Cell Biol 1988; 107:133-40.
18. Baum P, Furlong C, Byers BE. Yeast gene required for spindle pole body duplication: homology of its product with Ca2+-binding proteins. Proc Natl Acad Sci USA 1986; 83:5512-6.

19. Baum P, Yip C, Goetsch L et al. A yeast gene essential for regulation of spindle pole duplication. Mol Cell Biol 1988; 8:5386-97.
20. Zhu JA, Bloom SE, Lazarides E et al. Identification of a novel Ca2+-regulated protein that is associated with the marginal band and centrosomes of chicken erythrocytes. J Cell Sci 1995; 108:685-98.
21. Lee VD, Huang B. Molecular cloning and centrosomal localization of human caltractin. Proc Natl Acad Sci USA 1993; 90:11039-43.
22. Errabolu R, Sanders MA, Salisbury JL. Cloning of a cDNA encoding human centrin, an EF-hand protein of centrosomes and mitotic spindle poles. J Cell Sci 1994; 107:9-16.
23. Meng TC, Aley SB, Svard SG et al. Immunolocalization and sequence of caltractin/centrin from the early branching eukaryote Giardia lamblia. Mol Biochem Parasitol 1996; 79:103-8.
24. Madeddu L, Klotz C, Lecaer JP et al. Characterization of centrin genes in Paramecium. Eur J Biochem 1996; 238:121-8.
25. Levy YY, Lai EY, Remillard SP et al. Centrin is a conserved protein that forms diverse associations with centrioles and MTOCs in Naegleria and other organisms. Cell Motil Cytoskeleton 1996; 33:298-323.
26. Middendorp S, Paoletti A, Schiebel E et al. Identification of a new mammalian centrin gene, more closely related to Saccharomyces cerevisiae CDC31 gene. Proc Natl Acad Sci USA 1997; 94:9141-6.
27. Wottrich R. Klonierung und computergestützte Strukturanalyse von Centrinisoformen der Ratte (Rattus norvegicus).—Cloning and computer based structural analysis of centrin isoforms in the rat (Rattus norvegicus). Diploma thesis 1998; University of Karlsruhe, Germany.
28. Kretsinger RH. Evolution and function of calcium-binding proteins. Int Rev Cytol 1976; 46:323-93.
29. Kretsinger RH. Calcium-binding proteins. Annu Rev Biochem 1976; 45:239-66.
30. Moncrief ND, Kretsinger R, Goldman M. Evolution of EF-hand calcium-modulated proteins. I. Relationships based on amino acid sequences. J Mol Evol 1990; 30:522-62.
31. Nakayama S, Moncrief ND, Kretsinger RH. Evolution of EF-hand calcium-modulated proteins. II. Domains of several subfamilies have diverse evolutionary histories. J Mol Evol 1992; 34(5):416-48.
32. Bhattacharya D, Steinkötter J, Melkonian M. Molecular cloning and evolutionary analysis of the calcium-modulated contractile protein, centrin, in green algae and land plants. Plant Mol Biol 1993; 23(6):1243-54.
33. Coling DE, Salisbury JL. Characterization of the calcium-binding contractile protein centrin from Tetraselmis striata (Pleurastrophyceae). J Protozool 1992; 39:385-91.
34. Weber C, Lee VD, Chazin WJ, et al. High level expression in Escherichia coli and characterization of the EF-hand calcium-binding protein caltractin. J Biol Chem 1994; 269:15795-802.
35. Pulvermüller A, Gießl A, Heck M et al. Calcium dependent assembly of centrin/G-protein complex in photoreceptor cells. Mol Cell Biol 2002; in press.
36. Wiech H, Geier BM, Paschke T et al. Characterization of green alga, yeast, and human centrins. J Biol Chem 1996; 271:22453-61.
37. Durussel I, Blouquit Y, Middendorp S et al. Cation- and peptide-binding properties of human centrin 2. FEBS Lett 2000; 472:208-12.
38. Barbato G, Ikura M, Kay LE et al. Backbone dynamics of calmodulin studied by 15N relaxation using inverse detected two-dimensional NMR spectroscopy: the central helix is flexible. Biochemistry 1992; 31(23):5269-78.
39. Meador WE, Means AR, Quiocho FA. Modulation of calmodulin plasticity in molecular recognition on the basis of x-ray structures. Science 1993; 262(5140):1718-21.
40. Crivici A, Ikura M. Molecular and structural basis of target recognition by calmodulin. Annu Rev Biophys Biomol Struc 1995; 24:85-116.

41. Geier BM, Wiech H, Schiebel E. Binding of centrin and yeast calmodulin to synthetic peptides corresponding to binding sites in the spindle pole body components Karlp and Spc110p. J Biol Chem 1996; 271:28366-74.
42. Martindale VE, Salisbury JL. Phosphorylation of algal centrin is rapidly responsive to changes in the external milieu. J Cell Sci 1990; 96:395-402.
43. Lingle WL, Lutz WH, Ingle JN et al. Centrosome hypertrophy in human breast tumors: implications for genomic stability and cell polarity. Proc Natl Acad Sci USA 1998; 95(6):2950-5.
44. Lutz W, Lingle WL, McCormick D et al. Phosphorylation of centrin during the cell cycle and its role in centriole separation preceding centrosome duplication. J Biol Chem 2001; 276(23):20774-80.
45. Salisbury JL. Centrin and the algal flagellar apparatus. J Phycol 1989; 25:201-6.
46. Salisbury JL, Sanders MA, Harpst L. Flagellar root contraction and nuclear movement during flagellar regeneration in Chlamydomonas reinhardtii. J Cell Biol 1987; 105:1799-805.
47. Schulze D, Robenek H, McFadden GI, et al. Immunolocalization of a Ca2+-modulated contractile protein in the flagellar apparatus of green algae: the nucleus-basal body connector. Eur J Cell Biol 1987; 45:51-61.
48. McFadden GI, Schulze D, Surek B et al. Basal body reorientation mediated by Ca2+-modulated contractile protein. J Cell Biol 1987; 105:903-12.
49. Sanders MA, Salisbury JL. Centrin-mediated microtubule serving during flagellar excision in Chlamydomonas reinhardtii. J Cell Biol 1989; 108:1751-60.
50. Sanders MA, Salisbury JL. Centrin plays an essential role in microtubuli severing during flagellar excision in Chlamydomonas reinhardtii. J Cell Biol 1994; 124:795-805.
51. Salisbury JL. Algal centrin: Calcium sensitive contractile organelles. Algae as experimental systems. Alan R. Liss Inc., 1989:19-37.
52. Schatten G. The centrosome and its mode of inheritance: the reduction of the centrosome during gametogenesis and its restoration during fertilization. Dev Biol 1994; 165:299-335.
53. Khalfan W, Ivanovska I, Rose MD. Functional interaction between the PKC1 pathway and CDC31 network of SPB duplication genes. Genetics 2000; 155:1543-59.
54. Ivanovska I, Rose MD. Fine structure analysis of the yeast centrin, Cdc31p, identifies residues specific for cell morphology and spindle pole body duplication. Genetics 2001; 157:503-18.
55. Sullivan DS, Biggins S, Rose MD The yeast centrin, Cdc31p, and interacting protein kinase, Kic1p, are required for cell integrity. J Cell Biol 1998; 143:751-65.
56. Salisbury JL, Baron AT, Sanders MA The centrin-based cytoskeleton of Chlamydomonas reinhardtii: distribution in interphase and mitotic cells. J Cell Biol 1988; 107:635-41.
57. Baron AT, Salisbury JL. The centrin-related pericentriolar lattice of metazoan centrosomes. Comparative Spermatology 20 Years After 1991; 75:285-9.
58. Baron AT, Greenwood TM, Bazinet CW, et al. Centrin is a component of the pericentriolar lattice. Biol Cell 1992; 76:383-8.
59. Laoukili J, Perret E, Middendorp S et al. Differential expression and cellular distribution of centrin isoforms during human ciliated cell differentiation in vitro. J Cell Sci 2000; 113(8):1355-64.
60. Paoletti A, Moudjou M, Paintrand M et al. Most of centrin in animal cells is not centrosome-associated and centrosomal centrin is confined to the distal lumen of centrioles. J Cell Sci 1996; 109:3089-102.
61. Middendorp S., Kuntziger T., Abraham Y. et al. A role for centrin 3 in centrosome reproduction. J Cell Biol 2000; 148(3):405-15.
62. Besharse JC, Horst CJ. The photoreceptor connecting cilium—a model for the transition tone. In: Bloodgood RA, ed. Ciliary and flagellar membranes. New York: Plenum, 1990:389-417.
63. Spencer M, Detwiler PB, Bunt-Milam AH. Distribution of membrane proteins in mechanical dissociated retinal rods. Invest Ophthalmol Visual Sci 1988; 29:1012-20.

64. Wolfrum U, Schmitt A. Evidence for myosin VIIa driven rhodopsin transport in the plasma membrane of the photoreceptor connecting cilium. In: Hollyfield JG, Andersson RE, LaVail M, eds. Retinal degeneration diseases and experimental therapy. New York: Plenum Press, 1999:3-14.
65. Wolfrum U, Schmitt A. Rhodopsin transport in the membrane of the connecting cilium of mammalian photoreceptor cells. Cell Motil Cytoskeleton 2000; 46:95-107.
66. Liu X, Udovichenko IP, Brown SD et al. Myosin VIIa participates in opsin transport through the photoreceptor cilium. J Neurosci 1999; 19(15):6267-74.
67. Marszalek JR, Liu X, Roberts EA et al. Genetic evidence for selective transport of opsin and arrestin by kinesin-II in mammalian photoreceptors. Cell 2000; 102:175-87.
68. Uzawa M, Grams J, Madden B et al. Identification of a complex between centrin and heat shock proteins in CSF-arrested Xenopus oocytes and dissociation of the complex following oocyte activation. Dev Biol 1995; 171:51-9.
69. Paschke T. Untersuchungen zur Familie der Ca^{2+}-bindenden Centrine: biochemische Charakterisierung und Identifikation von interagierenden Proteinen. Analysis of the family of Ca^{2+}-binding centrins: biochemical characterization and identification of interacting proteins Dissertation 1997; University of Cologne, Germany
70. Philp NJ, Chang W, Long K. Light-stimulated protein movement in rod photoreceptor cells of the rat retina. FEBS Lett 1987; 225:127-32.
71. Whelan JP, McGinnis JF. Light-dependent subcellular movement of photoreceptor proteins. J Neurosci Res 1988; 20:263-70.
72. Matthews HR, Fain GL. A light-dependent increase in free Ca2+ concentration in the salamander rod outer segment. J Physiol 2001; 532:305-21.
73. Bigay J, Faurobert E, Franco M et al. Roles of lipid modifications of transducin subunits in their GDP-dependent association and membrane binding. Biochemistry 1994; 33:14081-90.

TUNING OUTER SEGMENT Ca^{2+} HOMEOSTASIS TO PHOTOTRANSDUCTION IN RODS AND CONES

Juan I. Korenbrot and Tatiana I. Rebrik

ABSTRACT

Cone photoreceptors respond to light with less sensitivity, faster kinetics and adapt over a much wider range of intensities than do rods. These differences can be explained, in part, by the quantitative differences in the molecular processes that regulate the cytoplasmic free Ca^{2+} concentration in the outer segment of both receptor types. Ca^{2+} concentration is regulated through the kinetic balance between the ions' influx and efflux and the action of intracellular buffers. Influx is passive and mediated by the cyclic-GMP gated ion channels. In cones, Ca^{2+} ions carry about 35% of the ionic current flowing through the channels in darkness. In rods, in contrast, this fraction is about 20%. We present a kinetic rate model of the ion channels that helps explain the differences in their Ca^{2+} fractional flux. In cones, but not in rods, the cGMP-sensitivity of the cyclic GMP-gated ion channels changes with Ca^{2+} at the concentrations expected in dark-adapted photoreceptors. Ca^{2+} efflux is active and mediated by a Na^+ and K^+-dependent exchanger. The rate of Ca^{2+} clearance mediated by the exchanger in cones, regardless of the absolute size of their outer segment is of the order of tens of milliseconds. In rod outer segments, and again independently of their size, Ca^{2+} clearance rate is of the order of hundreds of milliseconds to seconds. We investigate the functional consequences of these differences in Ca^{2+} homeostasis using computational models of the phototransduction signal in rods and cones. Consistent with experimental observation, differences in Ca^{2+} homeostasis can make the cone's flash response faster and less sensitive to light than that of rods. In the simulations, however, changing Ca^{2+} homeostasis is not

Department of Physiology, School of Medicine, University of California at San Francisco, San Francisco, CA 94143, U.S.A.

sufficient to recreate authentic cone responses. Accelerating the rate of inactivation (but NOT activation) of the enzymes of the transduction cascade, in addition, to changes in Ca^{2+} homeostasis are needed to explain the differences between rod and cone photosignals.

The large gain and precise kinetic control of the electrical photoresponse of rod and cone retinal receptors suggested a long time back that phototransduction is mediated by cytoplasmic second messengers that, in turn, control membrane ionic conductance.[1] The unquestionable identification of cyclic GMP as the phototransduction messenger, however, did not come until the mid 1980's with the discovery that the light-regulated membrane conductance in both rods and cones is gated by this nucleotide[2-4] and is, in fact, an ion channel.[5-7] The cyclic nucleotide gated (CNG) channels, now we know, are not just the compliant targets of light-dependent change in cytoplasmic cGMP, but actively participate in the regulation transduction through Ca^{2+} feedback signals.

The precise magnitude and time course of the concentration changes of cGMP and Ca^{2+} in either rods or cones remains controversial. It is clear, however, that whereas cGMP directly controls the opening and closing of the plasma membrane channels, Ca^{2+} controls the light-sensitivity and kinetics of the transduction signal.[8,9] The modulatory role of Ca^{2+} is particularly apparent in the process of light adaptation: in light-adapted rods or cones, the transduction signal generated by a given flash is lower in sensitivity and faster in time course than in dark-adapted cells. Light adaptation is compromised if Ca^{2+} concentration changes are attenuated by cytoplasmic Ca^{2+} buffers[8,10,11] and does not occur if Ca^{2+} concentration changes are prevented by manipulation of the solution bathing the cells.[12-14] Several Ca^{2+}-dependent biochemical reactions have been identified in photoreceptors, among them:

1. ATP-dependent deactivation.[15,16]
2. Rhodopsin phosphorylation, through the action of recoverin (S-modulin).[17-19]
3. Catalytic activity of guanylyl cyclase,[20-22] through the action of GCAP proteins.[23,24,25]
4. cGMP-sensitivity of the CNG channels.[26-29,30]

A challenge in contemporary phototransduction research is to understand the details of these reactions and their role in the control of the phototransduction signal.

Transduction signals in cone photoreceptors are faster, lower in light sensitivity, and more robust in their adaptation features than those in rods (for review see refs. 31;32). A detailed molecular explanation for these differences is not at hand. However, biochemical and electrophysiological[33] studies indicate that the elements in the light-activated pathway that hydrolyzes cGMP are quantitatively similar in their function in rods and cones and unlikely to account for the functional differences. Also, within the limited exploration completed todate, the Ca^{2+} -dependence of guanylyl cyclase[34] and visual pigment phosphorylation[19] do not differ in rods and cones. On the other hand, data accumulated over the past few years indicate that cytoplasmic Ca^{2+} homeostasis, while controlled through essentially identical mechanisms it is quantitatively very different in its features in the two photoreceptor types. Both Ca^{2+} influx through CNG channels and the rate of Ca^{2+} clearance from the outer segment differ between the two receptor cells. Also, the Ca^{2+} -dependent modulation of cGMP sensitivity is larger in extent in cones than in rods. Most significantly, the concentration range of this Ca^{2+} dependence overlaps the physiological range of light-dependent changes in cytoplasmic Ca^{2+} level in cones, but not in rods. We briefly review some of the evidence that supports these assertions and we then provide a quantitative analysis of the possible significance of these known differences. We conclude that while differences in Ca^{2+} homeostasis contribute importantly to explaining the differences between the two receptor types, they are alone not sufficient to explain the differences in the photoreceptor's response. It is likely that Ca^{2+} -independent inactivation of the transduction cascade enzymes is more rapid in cones than in rods.

THE FRACTION OF THE DARK-CURRENT CARRIED BY Ca^{2+} THROUGH cGMP-GATED ION CHANNELS IS HIGHER IN CONES THAN IN RODS

CNG channels allow the flux of cations over anions with almost ideal selectivity, but they select poorly among monovalent and divalent cations. In intact cells and under physiological solutions, therefore, the current through the channels in darkness and at the resting membrane potential is a mixture of the inward fluxes of Na^+, Ca^{2+} and Mg^{2+} and the outward flux of K^+.[35-39] The fraction of the dark-current carried by any given ion depends on the ion selectivity of the CNG channels and the concentration and driving force of that ion. Electrical studies carried out in detached membrane patches have demonstrated that CNG channels in both rods and cones are more permeable to Ca^{2+} than to Na^+, but the relative selectivity between Ca^{2+} and Na^+ (P_{Ca}/P_{Na}) is higher in cone than in rod CNG channels. Under saturating concentrations of cGMP, when the probability of channel opening is about 0.9, PCa/PNa in cones is about 21.7 but in rods it is about 6.5.[37,40,41] The difference in PCa/PNa between CNG channels of rods and cones is also observed in recombinant channels formed from alpha subunits alone.[42] The value of P_{Ca}/P_{Na} changes with channel gating: at the cGMP concentrations expected in intact cells, when probability of channel opening is about 0.03, P_{Ca}/P_{Na} in cones is about 7.4 times larger than that in rods.[41]

Electrical measurements of P_{Ca}/P_{Na}, however, do not provide direct information on the physiologically relevant parameter: the fraction of the cGMP-gated current specifically carried by Ca^{2+}. In rods, this fraction was estimated through indirect electrical methods to be between 10 and 20% of the net current.[43-46] We have recently developed a more direct method to assess this fraction in intact rods and cones. The method depends on the simultaneous measurement of membrane current and cytoplasmic Ca^{2+} following synchronous activation of the CNG channels.[47-49] In brief, under a defined set of experimental conditions, the membrane current and cytoplasmic Ca^{2+} concentration are simultaneously measured in intact photoreceptors loaded with Fura-2 (2 mM) and caged 8-Br-cGMP (50 μM). The caged compound rapidly (<1 msec) releases free nucleotide upon exposure to bright flashes of near-UV light. CNG channels are activated by flash photolysis of the caged compound. If Ca^{2+} ions are the only carriers of the current flowing through the activated channels, then the total change in cytoplasmic Ca^{2+} is the integral of the current change. If Ca^{2+} ions are only a fraction of the charge carriers, then the change in total Ca^{2+} is less than the current integral and the exact value of the fraction can be computed. To assure that the total change in Ca^{2+} is measured, not confounded by intracellular buffering or sequestration or by active transport the fractional flux is measured in the presence of high Fura-2 concentrations. Because of this high concentration, the initial changes in Fura-2 fluorescence reflect the equilibrium conversion of Ca^{2+}-free to Ca^{2+}-bound states of the dye, while the effective free Ca^{2+} concentration remains nearly zero ($<10^{-10}$ M).

Typical experimental measurements of the fraction of cGMP-gated current carried by Ca^{2+} in detached rod outer segments (tiger salamander) or intact single cones (striped bass fish) are shown in Figure 1. Free 8Br-cGMP generated by flash photolysis of the caged compound activated the CNG channels in the outer segment, causing an inward current. The current rapidly reached a peak and then declined with an exponential time course (Fig. 1, panels A and D). The exponential decline in current reveals that the free cyclic nucleotide is rapidly sequestered. Direct experimental tests have shown that this decline is due to binding of the nucleotide by intracellular cGMP binding sites, likely noncatalytic binding sites in PDE.[39] The cGMP-gated current is carried in part by Ca^{2+}, and the entering Ca^{2+} binds to cytoplasmic Fura-2 causing the decrease in cell fluorescence at 380 nm simultaneous with the current change (Fig. 1, panels B and E). The integral of the current change is a measure of the total charge entering the cone outer segment, while the fluorescence change specifically measures the total amount of Ca^{2+} entering the outer segment, provided all the entering Ca^{2+} is bound by Fura-2 and none is lost either to competing intracellular buffers or to sequestration by other compartments. In photoreceptors this necessary condition holds for 5 to 10 seconds following the uncaging flash.[39] Over this limited time period the integral of the current and the change in fluorescence have the same time course (Fig. 1, panels C and F). From analysis of repeated measurements similar to those of Figure 1, we determined that at –35 mV under normal physiological solutions and for membrane currents <200 pA, the Ca^{2+} fractional flux in single cones of striped bass is 33 ± 2% and in rod outer segments of tiger salamander it is 21 ± 1%. The value in rods is similar to that previously known from other measurements and in other species, but the value in cones is significantly higher. To verify that the difference between tiger salamander and striped bass is not simply due to species differences, we carried out studies in catfish rods and cones. In this species, the fractional flux is 34 ± 6% in cones and 14 ± 1% in rods. Finally, in gecko rods we found the Ca^{2+} fractional flux to be about 18%. Thus, in all species studied to date and with two independent methods it is clear that the fractional Ca^{2+} flux in cone CNG channels is far larger than that in rod CNG channels.

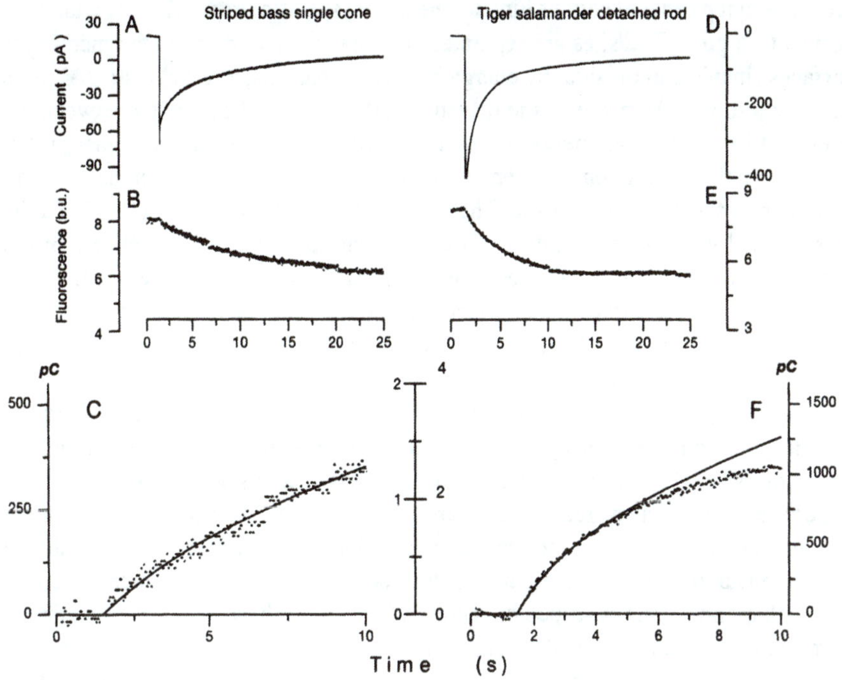

Figure 1. Determination of the fractional Ca^{2+} flux, P_f, in cyclic GMP-gated channels of intact single cones (from striped bass) and detached rod outer segments (from tiger salamander) in normal Ringer's solution. Shown are membrane currents and single cell fluorescence measured simultaneously in rods (A and D) or cones (B and E) loaded with 2 mM Fura-2 and 50 μM caged-8BrcGMP. Uncaging flashes (350-400 nm, 170 msec half bandwidth) activated transient inward currents and a decrease in cell fluorescence. To analyze the data, currents were integrated and the computed net charge (in picoCoulombs, pC) is shown as a continuous line in the lower most panels (C and F). The time course of the current integral matches that of the fluorescence change and the ratio of their amplitudes can be used to compute the Ca^{2+} fractional flux (details in ref. 49). For the rod shown, the holding current was –4 pA, the peak current was 408 pA and P_f was 0.21. For the cone the holding current was 22 pA, the peak current was 75 pA and P^f was 0.39.

The CNG channels of both rods and cones have been cloned (review in refs. 50;51), but it is not yet possible to define the channel's structural features that explain their function, though this is an area of intense investigation (review in refs. 50;52). It is possible, nonetheless, to define a thermodynamic model of ion transport to help understand the functional differences in the CNG channels of rods and cones. Such insight is possible, using the theory of rate processes,[53-55] if one postulates a specific mechanism to describe ion fluxes through open CNG channels. A successful rate theory model should simulate I-V curves that match experimental results.

Using such rate theory, we have completed a theoretical analysis of ion permeation in CNG channels of rods and cones.[39] Our model is similar to permeation models previously developed for rod channels.[56,57] In our theoretical analysis we presume that permeant cations bind to one of two specific sites within the channel's permeation pore. The sites are separated from each other and from the membrane surfaces (inside and outside) by energy barriers (3 barriers, 2 wells) (Fig. 2A). Ions flow by jumping from site to site driven by differences in free energy between the sites and biased by the energy of applied membrane voltage and the energy that arises from the interaction between the jumping ions and other ion that may already be bound to their permeation site. The sites are located at positions defined in units of electrical membrane thickness. This electrical thickness is the thickness of the membrane over which the applied voltage drops; its value is 0 at the extracellular surface and 1 at the intracellular surface. The task is to define the rates at which ions jump between the sites and from the rates to predict current-voltage (I-V) curves that match the experimental data.

Omitting here mathematical details (see details in ref. 39), we calculated the kinetic rates in the model by computer-assisted, optimized fit between simulated and experimental data. In Figure 2B we illustrate typical fits between simulated and experimental data measured in membrane patches detached from the outer segment of tiger salamander rods or striped bass cones. The experimental I-V curves shown were measured in membrane patches detached from the outer segment of either tiger salamander rods or striped bass cones. In each patch, we measured membrane currents under saturating cGMP concentrations in the presence of symmetric Na^+ solutions with varying amounts of Ca^{2+} added to the cytoplasmic membrane surface (details in ref. 40). The experimental data (continuous lines) are well fit by the results of the model computations (filled symbols) indicating that the permeation model is reasonable.

The features of the rate model can be diagrammatically presented as an energy profile[55] that defines the location of the wells and barriers and their specific size for each permeant ion. Figure 2A illustrates the kinetic model of CNG channels in both rods and cones inferred from our data. By definition the profile for Na^+ is the same in both channels. In both rods and cones the Ca^{2+} energy profiles have deeper wells than the Na^+ profile, indicating the interaction between Ca^{2+} and the channels is favored over that between Na^+ and the channels. This, of course, is manifested by the higher selectivity for Ca^{2+} over Na^+ in both channel types. The Ca^{2+} energy profiles differ in rods and cones allowing a mechanistic view of the Ca^{2+} permeation differences between the channel types. The height of the outermost barriers and the depth of both wells are rather similar. This indicates that in rods and cones, Ca^{2+} interacts with the internal and external binding sites with nearly the same characteristics. This is consistent with the fact that in both rods and cones current block by intracellular Ca^{2+} is described by a single Langmuir absorption isotherm with the same value for the concentration that half-blocks the open channel conductance, about 1 mM.[58] On the other hand, the middle barrier (G23) is much smaller in cones than in rods, indicating that Ca^{2+} can jump between internal binding sites more readily in cones than in rods. This feature explains why the Ca^{2+} fractional flux of cone CNG channels is higher than that of rod channels.

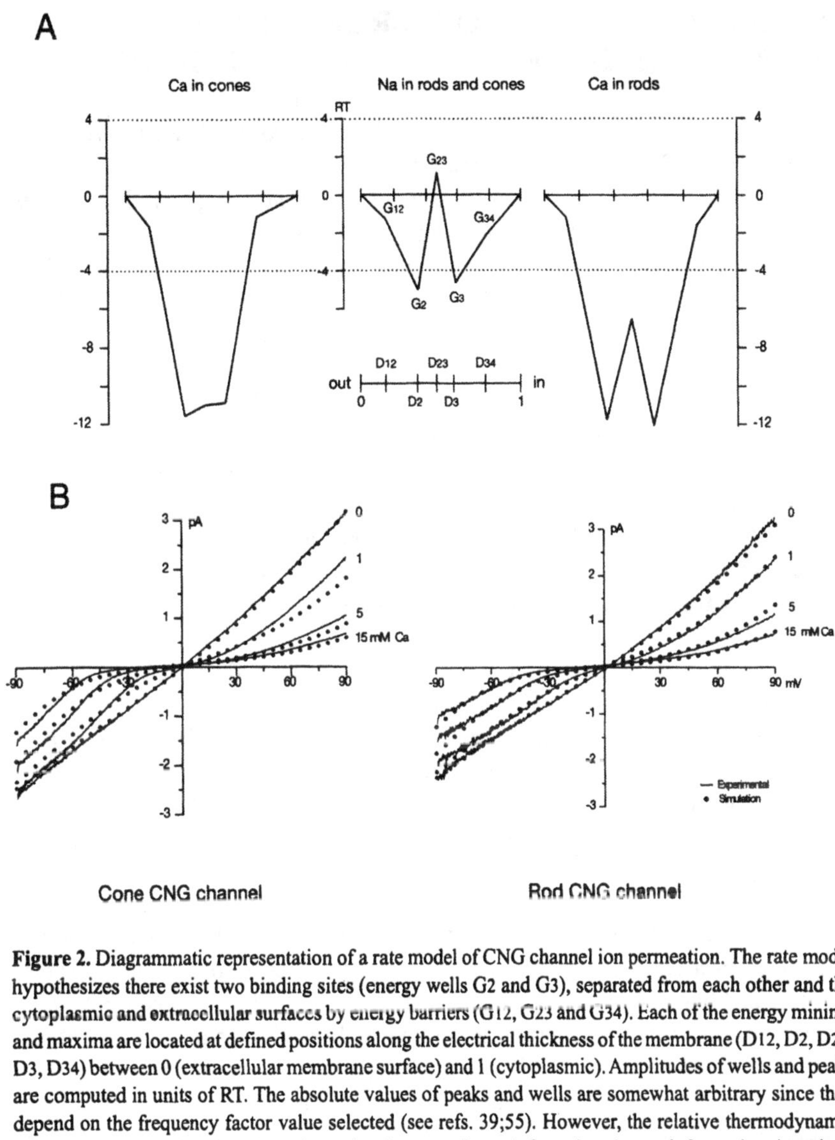

Figure 2. Diagrammatic representation of a rate model of CNG channel ion permeation. The rate model hypothesizes there exist two binding sites (energy wells G2 and G3), separated from each other and the cytoplasmic and extracellular surfaces by energy barriers (G12, G23 and G34). Each of the energy minima and maxima are located at defined positions along the electrical thickness of the membrane (D12, D2, D23, D3, D34) between 0 (extracellular membrane surface) and 1 (cytoplasmic). Amplitudes of wells and peaks are computed in units of RT. The absolute values of peaks and wells are somewhat arbitrary since they depend on the frequency factor value selected (see refs. 39;55). However, the relative thermodynamic differences between ions in a given channel, or between channels for a given ion are informative since these differences will be the same regardless of the frequency factor value. The accuracy of the model is tested by its ability to simulate I-V curves that fit experimentally recorded data. The lower panels illustrate I-V curves measured in membrane patches detached from either tiger salamander rods or striped bass cones. In both experiments shown currents were measured in the presence of 1 mM cGMP while bathing the membrane with symmetric Na^+ solutions. In successive trials in the same membrane, the concentration of Ca^{2+} in the cytoplasmic surface was increased in steps up to 15 mM, as labeled. The experimental data (continuous lines) is well fit by the results of model simulations (filled symbols) computed from the thermodynamic model depicted in the top panel.

THE MAXIMUM RATE OF OUTER SEGMENT Ca^{2+} CLEARANCE IS HIGHER IN CONES THAN IN RODS

Outer segment free Ca^{2+} is controlled by the continuous balance between the ion's influx and efflux across the plasma membrane, buffered by the action of endogenous buffering sites. Ca^{2+} flows passively into the outer segment through the CNG channels and flows actively out through Na^+-dependent exchangers.[59-61] Interestingly, the two Ca^{2+} transporting molecules appear to be structurally linked in the membrane.[62] In darkness, Ca^{2+} influx and efflux are in kinetic balance and the free Ca^{2+} concentration in rod and cone outer segments is between 500 and 700 nM.[45,63-65] Brief, dim illumination causes a decrease in Ca^{2+} concentration because, as channels close, influx declines more rapidly than does efflux. In time, efflux also decreases and, again, comes to be in equilibrium with influx. Bright light rapidly closes all open channels and saturates the amplitude of the photocurrent. Under saturating light, Ca^{2+} concentration decreases with an exponential time course.[45,61,66,65] The rate of this exponential decline, the time it takes for Ca^{2+} to be cleared from the outer segment, is determined by the molecular transport rate of the $Na^+/Ca^{2+},K^+$ exchanger, the number of molecules and the volume of the outer segment.

The maximum rate of outer segment Ca^{2+} clearance in rod and cone outer segments has been quantified through three principal methods.

1. Net Ca^{2+} efflux observed when all CNG channels are suddenly closed by a bright light is measured as a membrane current that arises from the electrogenic transport through the $Na^+/Ca^{2+},K^+$ exchanger.[33,44,59,67,68]
2. Light-dependent changes in extracellular Ca^{2+}.[60]
3. Light-dependent changes in intracellular Ca^{2+} [45,61,66,65]

Figure 3 illustrates the maximum rate of Ca^{2+} efflux measured through the electrogenic $Na^+/Ca^{2+},K^+$ transport current recorded in response to a bright flash. Shown are currents measured in striped bass single cones and Gecko rods under the same recording conditions. In both cells the current was measured at the holding voltage as a component of the photocurrent generated by brief flashes of bright light, as described before by others.[67,69,44,68] The exchanger current amplitude decays from an initial large value towards zero. The time course of the exchanger current is fit by a single exponential component in cones, but two components are needed to fit the data in rods. The exchanger current in striped bass single cones is fit by an exponential with a time constant of 25.1 msec ± 7.4 (n=11, range 17.5 to 38.8). The exchanger current in Gecko rods is much slower. On average the fast exponential component is 276 msec (range 243 to 308) and the slow one 1.614 sec (range 1.448 to 1.778). Our data in striped bass single cones are similar to those in tiger salamander cones.[33,70] In Gecko rods, Gray-Keller and Detwiler[45] have determined the Ca^{2+} clearance rate by simultaneously measuring the exchanger current and the cytoplasmic Ca^{2+} in the presence of 150 μM Indo, a fluorescent Ca^{2+}-indicator. They also observed two exponential components, but the time constants of these compo-

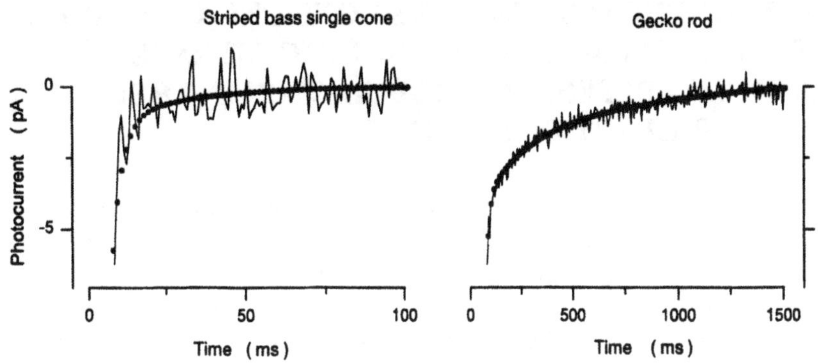

Figure 3. Maximum rate of Ca^{2+} efflux measured by the electrogenic transport through the $Na^+/Ca^{2+},K^+$ exchanger. Shown are currents measured under voltage-clamp in either a striped bass cone or a Gecko rod, as labeled. CNG channel activity was suppressed nearly instantaneously by 10 ms bright flashes presented at time 0 (1.5×10^4 Rh* for rods and 3×10^4 Rh* for cones). The saturated photocurrent amplitude defines the zero current value. The time course of the declining electrogenic current (details in ref. 43) was fit by the sum of independent exponentials with a least squares minimization algorithm. The functions that best fit the experimental data (filled symbols) are superimposed on the current tracings. The time course of the $Na^+/Ca^{2+},K^+$ exchanger current was best fit in cones by a single exponential of 24.3 ms time constant and in rods by the sum of two exponentials, of 0.243 s and 1.7 s time constants.

nents were 2- to 3-fold slower than those we report here. This difference, of course, is not surprising since intracellular Indo, an effective Ca^{2+} buffer, will slow down the time course and attenuate the amplitude of changes in cytoplasmic Ca^{2+}. Similar slow, exponential Ca^{2+} clearance rates in other non-mammalian rods have also been measured by others.[60,66] Sampath et al,[65] measuring cytoplasmic free Ca^{2+} with fluorescent dyes, report that Ca^{2+} clearance rates in non-mammalian cones are well described by the sum of two exponentials of approximately 0.045 s and 0.65 s time constants. While the specific values are not in full agreement when comparing among the various techniques, the evidence is inescapable: Ca^{2+} clearance rate in cones are much faster than in rods, about 5 fold.

Differences in Ca^{2+} efflux rates between rods and cones are not simply a consequence of differences in the surface to volume ratio of their outer segments. In Gecko rod outer segments the surface to volume ratio is about 0.44 whereas in striped bass single cones it is about 65. The Ca^{2+} efflux rate in rods, however, is only about 4 to 5 fold slower than in cones. The membrane packing density and/or the intrinsic transport rates of the $Na^+/Ca^{2+},K^+$ transporters, therefore, must be different in rods and cones.

IN INTACT PHOTORECEPTORS CONE CNG CHANNELS, BUT NOT ROD CNG CHANNELS, DEMONSTRATE A LARGE Ca^{2+} -DEPENDENT MODULATION IN THEIR CYCLIC NUCLEOTIDE SENSITIVITY

Ca^{2+} modulates the ligand sensitivity of CNG channels: the value of $K_{1/2}$, the ligand concentration necessary to obtain half maximal channel activation, increases as Ca^{2+} rises.[26-29] The features of this modulation, however, are significantly different in intact rods and cones, as illustrated in Figure 4. Using suction electrodes, we measured the activity of CNG channels in the outer segments of tiger salamander rods or striped bass cones (see details in ref. 30). We then gained access to the cytoplasmic space by electropermeabilizing the inner segment membrane with a brief electrical pulse. Cytoplasmic composition was controlled by superfusing the electropermeabilized photoreceptors with solutions designed to mimic the intracellular milieu and containing varying concentrations of cGMP. To generate an electromotive force to drive cGMP-gated currents at 0 mV holding potential, measurements were conducted with Na^+, a permeable cation, in the extracellular medium and choline$^+$, an impermeant cation, in the intracellular medium. Under these conditions, activation of CNG channels by cytoplasmic cGMP generated an inward current (Fig. 4). The amplitude of this current increases with cGMP with a dependence well described by the Hill equation:

$$\frac{I(cGMP)}{I_{\max}} = \frac{[cGMP]^n}{[cGMP]^n + K_{1/2}^n} \tag{1}$$

where $I(cGMP)$ is the cGMP-dependent current amplitude, $I_{\max}$ is the maximum current measured, $K_{1/2}$ is the cGMP concentration at which I has halve its maximum value and n is an adjustable parameter that denotes cooperative interaction. The same function applied to data measured in the presence of all Ca^{2+} concentrations tested, but the values of $K_{1/2}$ (and not n) changed with Ca^{2+}. In intact rods, the Ca^{2+}-dependent change in $K_{1/2}$ is about 1.5-fold between it extreme values.[30,71,72] In intact single cones, in contrast, the total change in $K_{1/2}$ is about 4-fold.[30]

More dramatic than the rod-cone difference in the extent of channel modulation is their difference in the Ca^{2+} -dependence of this modulation At a fixed cGMP concentration, the Ca^{2+} dependence of ligand sensitivity has a midpoint at about 50 nM in rods,[71,72] but at 290 nM in cones.[30] Hence, in rods the extent of change in $K_{1/2}$ is small and it occurs at Ca^{2+} concentrations expected only under bright illumination. Therefore, Ca^{2+} -dependent channel modulation is expected to play a very limited role in the control of gain and kinetics of the transduction signal in rods. To better asses the potential role of channel modulation in intact cones, we investigated in detail its Ca^{2+}-dependence and kinetics. Our findings are highlighted in Figure 5A (see details in ref. 73). In electropermeabilized striped bass cones we measured the cGMP-dependent currents in the presence of various fixed Ca^{2+} concentrations.

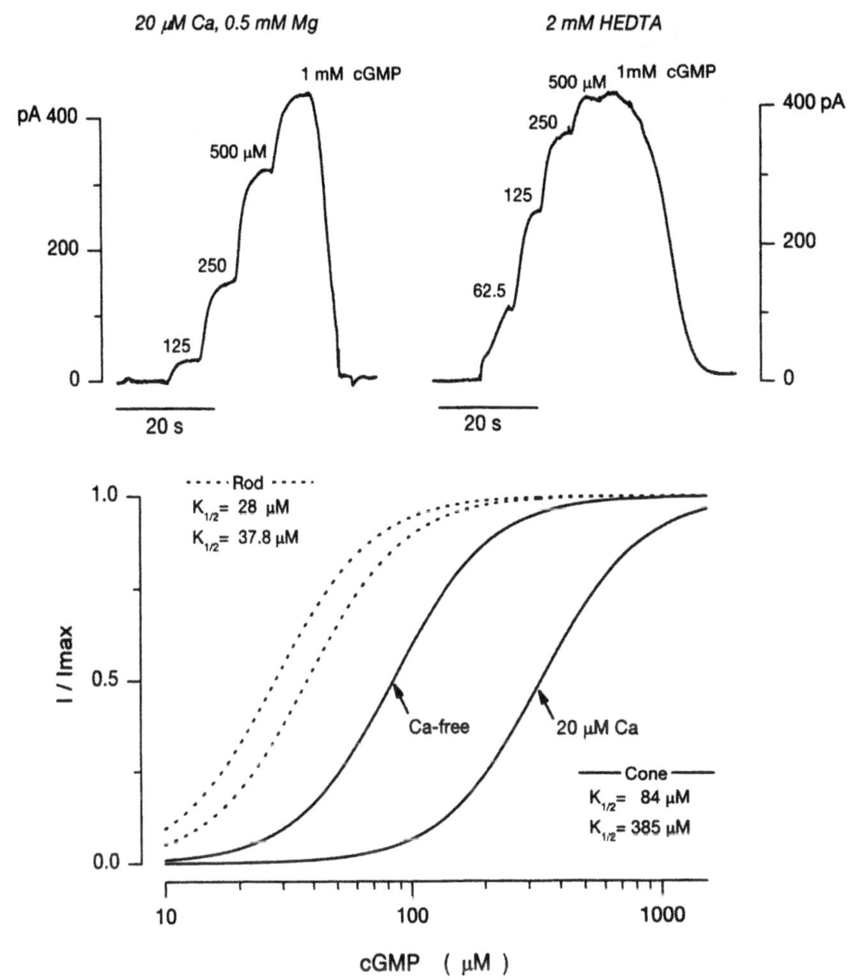

Figure 4. Activation of cGMP-dependent currents in the outer segment of intact, electropermeabilized rods and cones. The **top, left panel** illustrates inward cone outer segment currents activated by increasing concentrations of cGMP in the presence of 20 μM Ca^{2+} and 0.5 mM Mg^{++}, while the **top, right panel** illustrates currents measured in response to cGMP, but in the absence of divalent cations (with 2 mM HEDTA). The **bottom panel** illustrates the dependence of normalized current amplitude on cGMP concentration in electropermeabilized rods (dotted line) and cones (continuous line). The curves illustrate the Hill equation (text Eq 1) optimally fit to mean experimental data. The optimal $K_{1/2}$ values for each cell type in the presence and absence of divalent cations are shown in the Figure.

The Ca^{2+} dependence of $K_{1/2}$ is well described by a modified Michaelis-Menten equation:

$$K_{1/2}(Ca) = K_{1/2}^{low} + \left(K_{1/2}^{hi} - K_{1/2}^{low}\right)\frac{[Ca]}{[Ca] + {}^{Ca}K_m} \qquad (2)$$

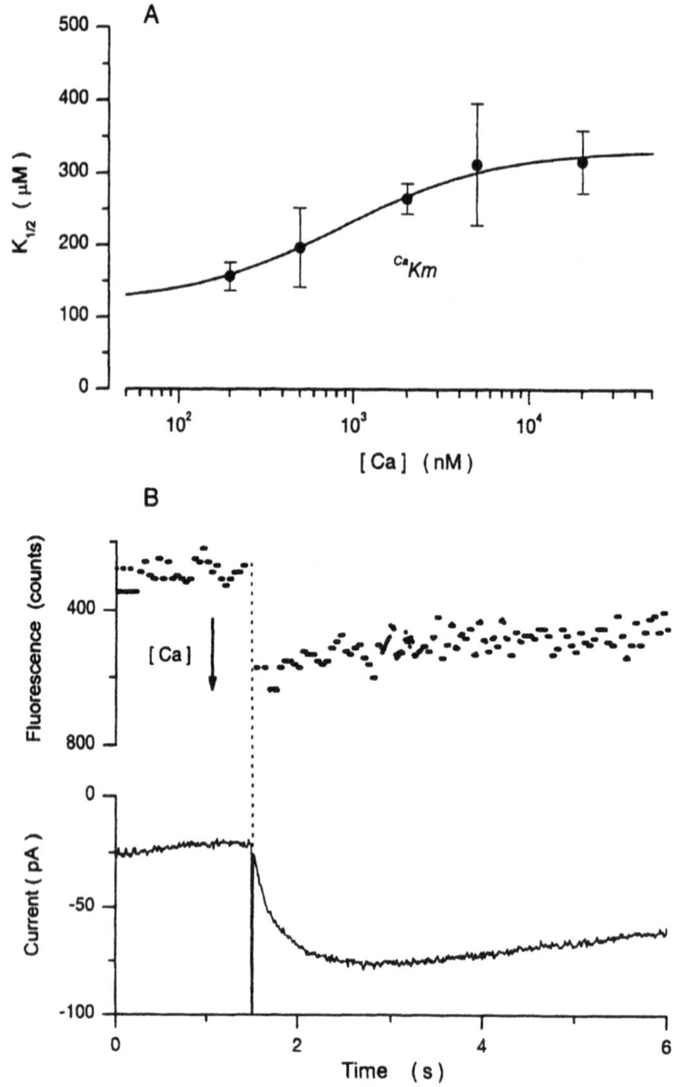

Figure 5. Top Panel: Ca^{2+}-dependence of CNG channel agonist sensitivity. The panel illustrates the dependence on Ca^{2+} of the $K_{1/2}$ value of cGMP in the Hill equation (text Eq. 1). The continuous line is a modified Michaelis-Menten function (text eqn.2) that best fits the experimental data. The value of $^{Ca}K_m$, the concentration at the midpoint of the curve, is given in the figure. **Lower Panels:** Time course of Ca^{2+} -dependent CNG current modulation in an intact single cone. Using tight-seal electrodes, the cells was loaded with standard electrode-filling solution containing 15 μM 8Br-cGMP, 1 mM Diazo-2, 0.1 mM bis-Fura-2 with 600 nM free Ca^{2+} and 1 mM free Mg^{++}. At the instant marked by the vertical dotted line, a Xenon flash uncaged diazo-2, causing a nearly instantaneous decrease in free cytoplasmic Ca^{2+} and a much slower increase in inward current. Current reached a peak in about 1.4 sec and membrane conductance was enhanced 2.26-fold. Ca^{2+} concentration decrease was detected as a change in bis-Fura-2 fluorescence excited at 380 nm and emitted in the 410-600 nm range. Fluorescence intensity was measured with photon counting instrumentation in counts per 50 msec bin.

where $K_{1/2}(Ca)$ is the Ca^{2+}-dependent value of $K_{1/2}$, $K^{hi}_{1/2}$ and $K^{low}_{1/2}$ are the extreme values of $K_{1/2}$, at 20 and 0 μM Ca^{2+}, respectively. ${}^{Ca}Km$ is the Ca^{2+} concentration at the midpoint between $K^{hi}_{1/2}$ and $K^{low}_{1/2}$. Over a large number of cells, the mean values in single cones were: $K^{low}_{1/2} = 120.6 \pm 9.3$ μM, $K^{hi}_{1/2} = 316.4 \pm 43.9$ μM and ${}^{Ca}Km = 857 \pm 68$ nM.

We also analyze the kinetics of channel modulation in intact cones. We measured changes in channel activity caused by a sudden decrease in free cytoplasmic Ca^{2+} achieved with the use of a "caged" Ca^{2+} buffer, diazo-2 (see details in ref. 73). In darkness, this compound binds Ca^{2+} with relatively low affinity because the buffer molecule, BAPTA, is linked by a photo-labile ester to a protecting group ("cage"). Upon absorption of ultraviolet light, the ester is cleaved; BAPTA is freed and able to sequester available free Ca^{2+}. We simultaneously measured the free Ca^{2+} concentration by following the fluorescent signal of the Ca^{2+}-indicator dye bis-Fura-2 also loaded into the cytoplasm of the intact cell. Figure 5B illustrates a typical result. In the presence of 15 μM cytoplasmic 8Br-cGMP, an uncaging flash caused a slow increase in current that reached a minimum in about a 1 second then slowly returned towards its starting value. In contrast to the slow changes in current, the uncaging flash caused a near instantaneous (<50 msec) decrease in cytoplasmic free Ca^{2+}, which then very slowly (> 20 sec) recovered to its starting value. Detailed analysis of the time course of current and Ca^{2+} concentration changes[73] reveals that the Ca^{2+} -dependent channel modulation is well described as a pseudo first order reaction

$$g_1 + Ca(t) \underset{k_{21}}{\overset{k_{12}}{\rightleftarrows}} g_2 \tag{3}$$

where channels change from conductance state g_1 to conductance state g_2 upon interacting with Ca^{2+}. The time constant of this reaction has a mean value of $\tau g = 0.40 \pm 0.14$ sec. That is, the speed of channel modulation is not limited by the rate at which Ca^{2+} concentration changes, but follows the concentration change with a first order rate of about 2.5 sec^{-1}.

We are currently investigating the molecular mechanism(s) of Ca^{2+} -dependent modulation in cone CNG channels. Modulation almost certainly depends on a diffusible factor that mediates the interaction between Ca^{2+} and channel gating. This diffusible factor is slowly and inevitably lost from electropermeabilized cones by simply waiting.[30] The rate of loss is significantly slower in the presence of elevated divalent cation concentrations (Mg^{2+} at 1 mM and Ca^{2+} at > 20 μM) than in their absence. The diffusible factor is entirely lost from membrane patches detached from cone outer segments.[29,74] In rods, the known channel modulation can be mimicked by exogenous calmodulin and this molecule is the likely modulator in the intact cells. In cones, the modulator is unlikely to be calmodulin because

1. exogenously added calmodulin does not restore modulation to channels in detached membrane patches[29] and
2. drugs that are known to interfere with calmodulin action do not affect channel modulation in electropermeabilized cones.[73]

FUNCTIONAL CONSEQUENCE OF ROD-CONE DIFFERENCES IN Ca^{2+} HOMEOSTASIS

The differences in CNG channel permeation and modulation and Ca^{2+} clearance rate between rods and cones exist, but are they of consequence to the phototransduction function of the receptor cells? It is already clear that while important, differences in Ca^{2+} homeostasis cannot alone explain the different transduction signals in the two receptor types: Blocking light-dependent changes in cytoplasmic Ca^{2+} slows down the photoresponse and changes its sensitivity in both rods and cones, but does not make the signals identical to each other. Direct experimental assessment of the effects of molecular changes in the CNG channels, their modulator and the Na^+-dependent exchanger will require future experimental work, possibly through the use of "transgenic" photoreceptors. At this time, however, possible physiological consequences of the known molecular differences between rods and cones can be assessed through computational simulations. These simulations, it must be remembered, are not better than the limits of reality imposed by the mathematical simplifications necessary in the simulations. Nonetheless, simulations offer instructional insight.

Reasonably successful computational simulations of the enzymatic cascade of phototransduction and the regulation of membrane current were first developed for rods by Forti et al[75,76] and for cones by Sneyd and Tranchina.[77,78] These models have served as the basis for further elaboration[79,80] and Hamer[81] has recently considered in great details the faults and merits of the various available models. Based on these previous models, we have implemented a model to simulate the flash response of dark-adapted rods in non-mammalian vertebrates. The activation phase of the phototransduction cascade can be adequately represented by two sequential differential equations:[79,81]

$$\frac{dRh^*}{dt} = \phi - \frac{Rh^*(t)}{\tau_{Rh^*}} \tag{4}$$

The first equation (Eq. 4) posits that the rate of production of photoexcited rhodopsin (Rh*) is a first order process of time constant τ_{Rh^*} limited by the delivery of ϕ photons/s. The second equation (Eq. 5) describes the subsequent activation of phoshodiesterase (PDE) (this term stands for the rate limiting event in the true activation of transducin and/or phosphodiesterase) . It proposes that a photoexcited PDE* is produced at a rate limited by the conversion of Rh* into PDE* (with stoichiomety $vPDE$) and it is also a first order process of time constant τ_{PDE^*}.

$$\frac{dPDE^*}{dt} = vPDE \bullet Rh^*(t) - \frac{PDE^*(t)}{\tau_{PDE^*}} \tag{5}$$

The model proposes that the rate of production of cytoplasmic cGMP arises from the kinetic balance of its synthesis by guanylyl cyclase and its hydrolysis by PDE. The velocity of each of the two enzymes is independently defined. The velocity of guanylyl cyclase (*VGC(Ca)*) depends on cytoplasmic free Ca^{2+} and is given by:

$$VGC(Ca) = VGC\max \frac{1}{1+\left(\frac{Ca}{KGC}\right)^n} \tag{6}$$

where *VGCmax* is the maximum enzymatic velocity, *Ca* is the free cytoplasmic Ca^{2+}, *KGC* is the Ca^{2+} concentration at which the velocity is half its maximal value and n denotes cooperative interaction in enzyme regulation by Ca^{2+}. This equation is based on results from direct biochemical assays.[22,82] The velocity of PDE, *VPDE* depends on cytoplasmic free cGMP and is given by:

$$VPDE(t) = cGMP(t)(\beta dark + \beta sub \bullet PDE^*(t)) \tag{7}$$

where *βdark* is the PDE rate in darkness, *βsub* is the rate of PDE*(t). Assuming there are no intracellular cGMP buffers, it follows that:

$$\frac{dcGMP}{dt} = VGC(Ca) - VPDE(t) \tag{8}$$

Finally, differential equations to describe the time-dependent changes in cytoplasmic Ca^{2+} concentration are needed. All simulation models agree on the fundamental concept that cytoplasmic Ca^{2+} is controlled by the kinetic balance of its influx through the CNG channels and its efflux through the $Na^+/Ca^{2+}, K^+$ exchanger.[60] Ca^{2+} influx, ${}^{Ca}Jin(t)$ is given by the fraction of the outer segment current carried by Ca^{2+}. Thus,:

$${}^{Ca}Jin(t) = \frac{Pf\,\mathrm{Im}(t)}{zFVosBuff} \tag{9}$$

where *Pf* is the fractional Ca^{2+} flux through the CNG channels, Im(t) is the outer segment current, *z* and *F* are valence and Faraday's constant, *Vos* is the outer segment volume and *Buff* is the Ca^{2+} buffering capacity in the outer segment. The definition of Ca^{2+} efflux is more ambiguous. The problem is that while all models agree the efflux rate is proportional to free cytoplasmic Ca^{2+}, the form of this dependence and the role of intracellular buffers is controversial. This controversy reflects a significant gap in our understanding of Ca^{2+} homeostasis: while there is general agreement on the fact that Ca^{2+} decreases with light, the exact time course and extent of this change are not known with certainty. More than that, a direct measurement of light-dependent changes in cytoplasmic Ca^{2+} in rods[45] proposes that, contrary to all common wisdom understanding, the light-dependent change in free Ca^{2+} far outlasts the photocurrent. This controversy needs experimental resolution. For now, it is clear that the kinetics of Ca^{2+} concentration change in the cytoplasm are poorly understood and that the role of intracellular buffers or intracellular compartmentalization need to be brought into a fully correct model.[66,81] In this exercise, we will assume Ca^{2+} efflux, ${}^{Ca}Jeff(t)$ is directly proportional to free cytoplasmic Ca^{2+} and will use the formalism developed before.[34,77] Thus,

$$ {}^{Ca}Jeff(t) = \frac{kexch \bullet Ca(t)}{VosBuff} \tag{10} $$

where *kexch* is the rate of Ca^{2+} clearance from the outer segment. This value lumps both the transport rate and the rate of buffer binding. It is a free parameter empirically adjusted to optimize the fit between simulated and experimental data. It follows that,

$$ \frac{dCa}{dt} = {}^{Ca}Jin(t) - {}^{Ca}Jeff(t) \tag{11} $$

We solved the differential equations listed above (Eqs. 4,5, 8 and 11) using simulation algorithms (20Sim Software, Twente University, The Netherlands). The simulations yield a time continuous track of free cytoplasmic cGMP concentration, which we converted into membrane current using Eq. 1.

The difficult task in simulations, of course, is the assignment of specific values to the many independent variables called for in the model. Many of these can be assigned from direct biochemical and biophysical experiments available in the literature.[34,79] Others, however, have to be "best guessed", for example, by optimizing the fit between experimental and simulated data.[81] Using a collection of reasonable values derived from the literature or from optimization[34,73,81] (and listed in Table 1), we computed simulated photocurrents of a dark-adapted tiger salamander rod. The results of computations are illustrated in Figure 6, alongside a set of experimental photocurrents measured in these cells (data from ref. 34). The computations generate photocurrents reasonably similar to, but not identical with experimental data.

In lieu of yet unavailable transgenic cones, the ability to reasonably simulate rod photocurrents provides us with a tool to investigate possible functional consequences of the rod-cone differences in Ca^{2+} homeostasis. Figure 7A illustrates simulations of flash photocurrents at a fixed light intensity of a "pseudo-cone" computed by leaving the values of all variables in the rod model unchanged, except those we can define as rod-cone differences: the fractional Ca^{2+} flux in the CNG channels

Table 1. Values of adjustable variables used in photocurrent simulations

		Rod Tiger Salamander	Pseudo-Cone	Pseudo-Cone Enhanced PDE* Inactivation
CNG ion channels				
Dark current	pA	72.6	same	same
Maximal cGMP-gated current	pA	2500	same	same
[cGMP] at $K_{1/2}$	μM	35	same	same
[cGMP] in darkness	μM	7.1	same	same
Cooperativity coefficient, n		2.2	same	same
Ca^{2+} homeostasis				
$[Ca^{2+}]$ in darkness	μM	0.5	same	same
Fractional Ca^{2+} flux, P_f		0.2	0.35	0.35
OS volume, Vos	plt	1	same	same
Ca^{2+} buffering, $Buff$	Total/free	5	same	same
$kexch$ Ca^{2+} clearance rate,	1/s	0.14	0.52	0.52
Guanylyl cyclase				
Maximum Velocity, $Vmax$	μM/s	32.5	same	same
$[Ca^{2+}]$ at midpoint, K_{GC}	μM	0.1	same	same
Cooperativity coefficient, n		2	same	same
Phosphodiesterase				
Rate constant in darkness, $\beta dark$	1/s	0.176	same	same
Rate constant of activated PDE*, βsub	1/s	1.68E-05	same	same
Production of PDE* per Rh*, $vPDE$		1920	same	same
Time constant of PDE* inactivation, τPDE^*	s	1.051	same	0.105
Rhodopsin				
Time constant of Rh* inactivation τ_{Rh}*	s	0.35	same	same

and the Ca^{2+} clearance rate (Table 1). Although the $K_{1/2}$ for cGMP of the CNG channels (Eq. 1) differs in rods and cones[73] this parameter need not be changed in the simulations because in both photoreceptor types the dark current is about 2% of the maximal cGMP-gated current. Each of the changes is important. In Figure 7A we illustrate the response of a "pseudo-cone" when only P_f is changed (from 0.2 to

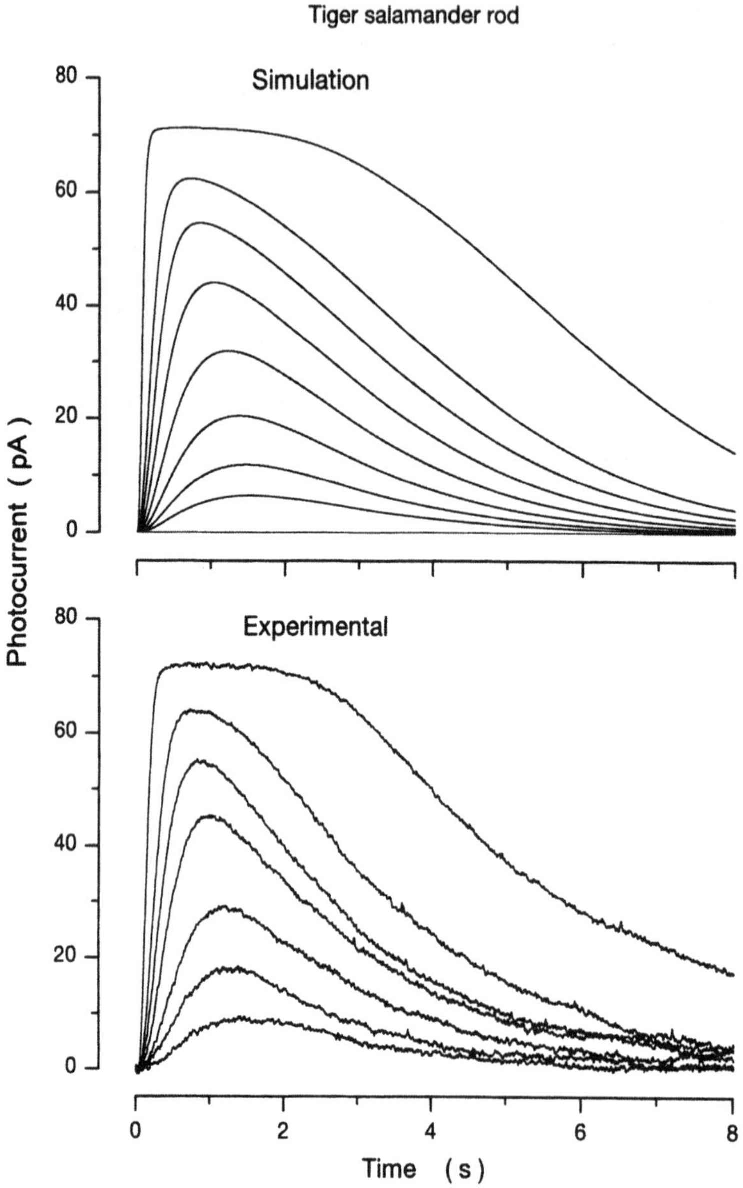

Figure 6. Photocurrents measured in dark-adapted single rods under voltage clamp at room temperature. The upper panel displays photocurrents simulated by the model detailed in the text. Simulations were generated in response to 20 msec duration flashes that increased in intensity by a factor of 2 for each of the 8 dimmest responses. Saturated photocurrent (71.6 pA) was selected to match the amplitude of the experimental data set. The lower panel shows experimental photocurrents measured in normal Ringer's solution with a tight-seal electrode using the perforated-patch technique (details in ref. 34). The responses shown were generated with 20 msec duration flashes that delivered 0.59, 1.21, 2.38, 6.53, 13.72, 27.42 and 156.67 photons μm^{-2} at 520 nm.

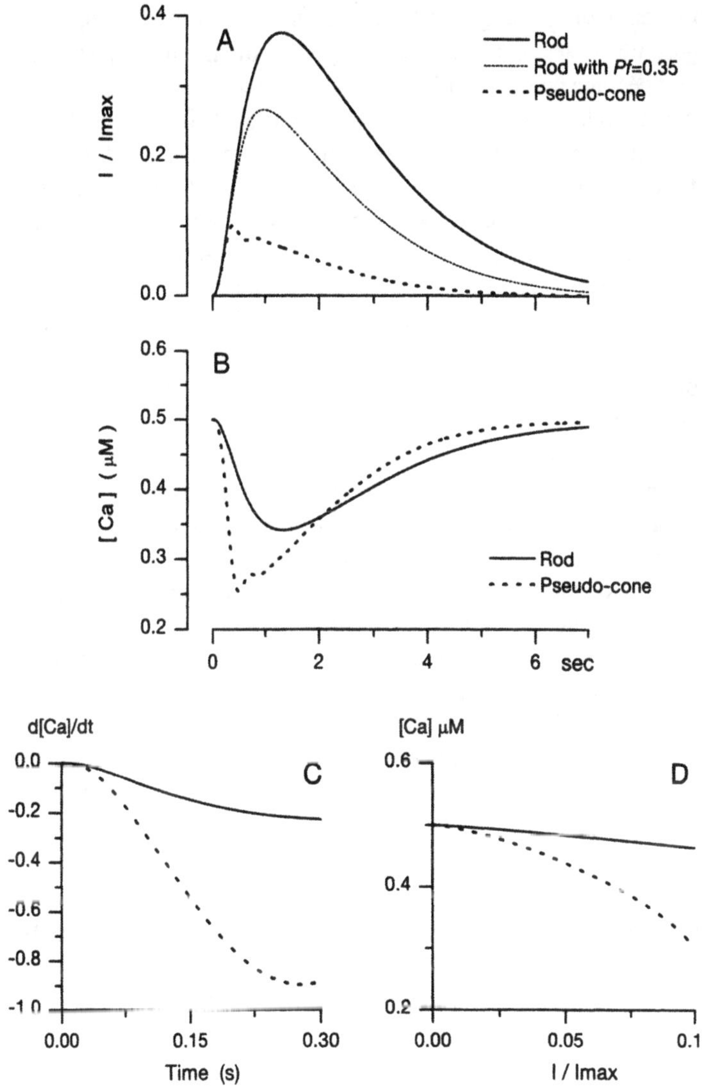

Figure 7. Simulation of the photocurrent in a model rod and a "pseudo-cone" created by leaving all the rod model variables unchanged, except those that reflect the differences in Ca^{2+} homeostasis between rods and cones: the Ca^{2+} fractional flux in CNG channels and the rate of Ca^{2+} clearance. **A** compares the normalized photocurrent of the rod with photocurrents simulated in response to identical light flashes in a pseudo-cone with only P_f changed and one in which both P_f and Ca^{2+} clearance rate are changed. **B** compares the simulated light-dependent changes in cytoplasmic Ca^{2+} in a rod and a pseudo-cone with both P_f and Ca^{2+} clearance rates changed. Both models start at a dark Ca^{2+} concentration of 0.5 μM (a definition in the model). In the pseudo-cone the decline in Ca^{2+} is, initially, faster and larger than that in the rod. These features are analyzed in panels **C** and **D**. **C** compares the rate of decrease in cytoplasmic Ca^{2+} ($d[Ca^{2+}]/dt$) for the first 0.3 sec of the photoresponse. The pseudo-cone initially losses cytoplasmic Ca^{2+} concentration nearly ten times faster than the rod. **D** plots the change in Ca^{2+} concentration as a function of the normalized photocurrent, for small current changes (up to about 0.1). For a given current, the decline in Ca^{2+} concentration is larger in the pseudo-cone than the rod.

0.35) and when both P_f changes and the clearance rate is accelerated about 4-fold. As the figure illustrates, these simple changes are of significant consequence: the photocurrent become much faster in time course, it reaches to a peak earlier and it is less sensitive to light (a smaller pA/photon response). This is simply because for a given change in current, the Ca^{2+} concentration decrease in the "pseudo-cone" is much faster and much larger than in the comparable rod (Fig. 7B). These features can be graphically demonstrated by comparing for rods and pseudo-cones: (1) the rate of loss of Ca^{2+} ($d[Ca^{2+}]/dt$) for the early part of the simulated response (Fig. 7C); and (2) the rate of change of Ca^{2+} concentration as a function of the current change for small currents (Fig. 7D). In words, the faster and larger decrease in cytoplasmic Ca^{2+} causes a larger and faster activation of guanylyl cyclase in the pseudo-cone than in the rod, defeating the activated PDE* and terminating the photocurrent at an earlier time.

The response of the "pseudo-cone" is still far from that of an authentic cone. This is illustrated in Figure 8 in which photocurrents measured under voltage clamp in a tiger salamander single cone are shown next to the simulated tiger salamander "pseudo-cone" data. The "pseudo-cone" photocurrent, though it reaches to a peak rapidly, is much longer in duration than that of the authentic cone. We investigated whether the Ca^{2+} -dependent modulation of sensitivity plays a role in the flash photocurrent by introducing this modulation into the simulated model. To this end, Eq. 2 was introduced into the computational simulations, and allowed to follow the free cytoplasmic Ca^{2+} low band-passed through a single-pole filter of 0.4 sec time constant (to account for the known kinetics of modulation). The channel modulation is of little consequence to the flash response of the "pseudo-cone" (data not shown). This is not surprising since the modulation is slow in comparison with the time course of the photocurrent. Channel modulation, on the other hand, strongly affects the response to light steps. Again, this is as expected, channel modulation is likely important in explaining the large difference between rods and cones in their range of light-adaptation.[73] Finally, the "pseudo-cone" flash response can be made to more nearly approximate the authentic cone response by accelerating the rate of decay of the photoexcited PDE*, decreasing by 10-fold the value of τ_{PDE^*} in Table 1 (Eq. 5) (Fig. 8). That is, in addition to changes in Ca^{2+} homeostasis cones are likely to exhibit more rapid inactivation of the enzymes of the transduction cascade than do rods.

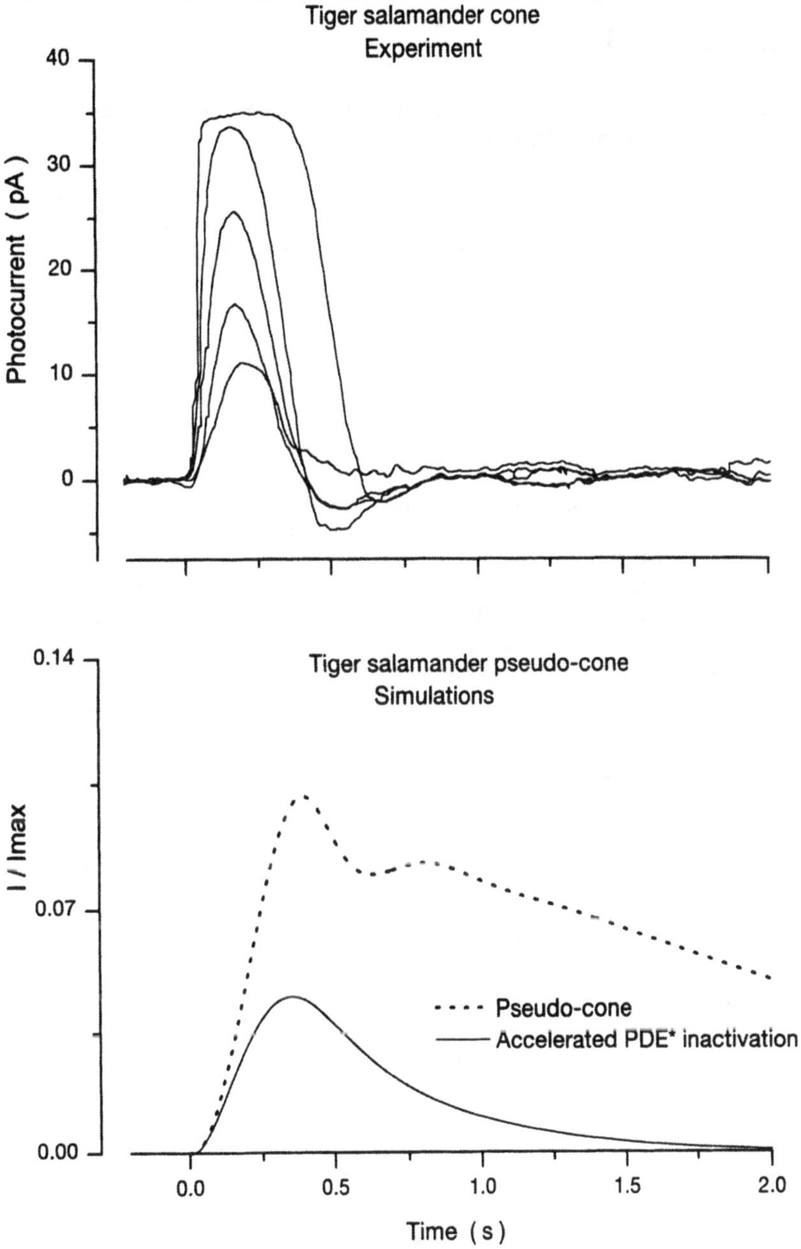

Figure 8. Top panel: Photocurrents measured in dark-adapted red-sensitive single cones from tiger salamander under voltage clamp at room temperature. Currents were measured with a tight-seal electrode at –30 mV holding voltage in normal Ringer's solution (details in ref. 33). The cone was stimulated with 10 msec flashes at 620 nm of intensity (in photons μm^{-2}) 118.5, 354.3, 758.1, 1659 and 17770. **Bottom panel:** Model simulated photocurrent. The dotted tracing is the same "pseudo-cone" signal displayed in Figure 7. While it reaches to a peak rapidly, its duration is far longer than that of the photocurrent in the authentic cone. The simulated pseudo-cone signal can be made to more closely approximate the experimental data by simply enhancing 10-fold the inactivation rate of PDE* (text Eq. 5).

REFERENCES

1. Hagins WA. The visual process: Excitatory mechanisms in the primary receptor cells. Annu Rev Biophys Bioeng 1972; 1:131-58.
2. Fesenko EE, Kolesnikov SS, Lyubarski AL. Induction by cyclic GMP of cationic conductance in plasma membrane of retinal rod outer segments. Nature 1985; 313:310-13.
3. Koch KW, Kaupp UB. Cyclic GMP directly regulates a cation conductance in membranes of bovine rods by a cooperative mechanism. J Biol Chem 1985; 260:6788-800.
4. Matthews G. Comparison of the light-sensitive and cyclic GMP-sensitive conductances of the rod photoreceptor: noise characteristics. J Neurosci 1986; 6:2521-6.
5. Zimmerman AL, Baylor DA. Cyclic GMP-sensitive conductance of retinal rods consists of aqueous pores. Nature 1986; 321:70-2.
6. Bodoia RD, Detwiler PB. Patch-clamp recordings of the light-sensitive dark noise in retinal rods from the lizard and frog. J Physiol 1985; 367:183-216.
7. Haynes LW, Kay AR, Yau KW. Single cyclic GMP-activated channel activity in excised patches of rod outer segment membrane. Nature 1986; 321:66-70.
8. Korenbrot JI, Miller DL. Calcium ions act as modulators of intracellular information flow in retinal rod phototransduction. Neurosci Res Suppl 1986; 4:S11-34.
9. Torre V, Matthews HR, Lamb TD. Role of calcium in regulating the cyclic GMP cascade of phototransduction in retinal rods. Proc Natl Acad Sci USA 1986; 83:7109-13.
10. Lamb TD, Matthews HR, Torre V. Incorporation of calcium buffers into salamander retinal rods: a rejection of the calcium hypothesis of phototransduction. J Physiol 1986; 372:315-49.
11. Matthews HR. Incorporation of chelator into guinea-pig rods shows that calcium mediates mammalian photoreceptor light adaptation. J Physiol 1991; 436:93-105.
12. Nakatani K, Yau KW. Calcium and light adaptation in retinal rods and cones. Nature 1988; 334:69-71.
13. Fain GL, Lamb TD, Matthews HR et al. Cytoplasmic calcium as the messenger for light adaptation in salamander rods. J Physiol 1989; 416:215-43.
14. Matthews HR, Fain GL, Murphy RL et al. Light adaptation in cone photoreceptors of the salamander: a role for cytoplasmic calcium. J Physiol (Lond) 1990; 420:447-69.
15. Lagnado L, Baylor DA. Calcium controls light-triggered formation of catalytically active rhodopsin. Nature 1994; 367:273-7.
16. Sagoo MS, Lagnado L. G-protein deactivation is rate-limiting for shut-off of the phototransduction cascade. Nature 1997; 389:392-5.
17. Kawamura S. Rhodopsin phosphorylation as a mechanism of cyclic GMP phosphodiesterase regulation by S-modulin. Nature 1993; 362:855-7.
18. Chen CK, Inglese J, Lefkowitz RJ et al. Ca^{2+}-dependent interaction of recoverin with rhodopsin kinase. J Biol Chem 1995; 270:18060-6.
19. Kawamura S, Kuwata O, Yamada M et al. Photoreceptor protein s26, a cone homologue of S-modulin in frog retina. J Biol Chem 1996; 271:21359-64.
20. Lolley RN, Racz E. Calcium modulation of cyclic GMP synthesis in rat visual cells. Vision Res 1982; 22:1481-6.
21. Pepe IM, Boero A, Vergani L et al. Effect of light and calcium on cyclic GMP synthesis in rod outer segments of toad retina. Biochim Biophys Acta 1986; 889:271-6.
22. Koch KW, Stryer L. Highly cooperative feedback control of retinal rod guanylate cyclase by calcium ions. Nature 1988; 334:64-6.
23. Dizhoor AM, Olshevskaya EV, Henzel WJ et al. Cloning, sequencing, and expression of a 24-kDa Ca^{2+}-binding protein activating photoreceptor guanylyl cyclase. J Biol Chem 1995; 270:25200-6
24. Palczewski K, Subbaraya I, Gorczyca WA et al. Molecular cloning and characterization of retinal photoreceptor guanylyl cyclase-activating protein. Neuron 1994; 13:395-404.
25. Gorczyca WA, Polans AS, Surgucheva I et al. Guanylyl cyclase activating protein (GCAP): a calcium sensitive regulator of phototransduction. J Biol Chem 1995; 270:22029-22036.

26. Hsu YT, Molday RS. Interaction of calmodulin with the cyclic GMP-gated channel of rod photoreceptor cells. Modulation of activity, affinity purification, and localization. J Biol Chem 1994; 269:29765-70.
27. Bauer PJ. Cyclic GMP-gated channels of bovine rod photoreceptors: affinity, density and stoichiometry of Ca^{2+}-calmodulin binding sites. J Physiol (Lond) 1996; 494:675-85.
28. Gordon SE, Downing-Park J, Zimmerman AL. Modulation of the cGMP-gated ion channel in frog rods by calmodulin and an endogenous inhibitory factor. J Physiol (Lond) 1995; 486 :533-46.
29. Hackos DH, Korenbrot JI. Calcium modulation of ligand affinity in the cyclic GMP-gated ion channels of cone photoreceptors. J Gen Physiol 1997; 110:515-28.
30. Rebrik TI, Korenbrot JI. In intact cone photoreceptors, a Ca^{2+}-dependent, diffusible factor modulates the cGMP-gated ion channels differently than in rods. J Gen Physiol 1998; 112:537-48.
31. McNaughton PA. Light response of vertebrate photoreceptors. Physiol Rev 1990; 70:847-83.
32. Burns ME, Baylor DA. Activation, deactivation, and adaptation in vertebrate photoreceptor cells. Annu Rev Neurosci 2001; 24:779-805.
33. Hestrin S, Korenbrot JI. Activation kinetics of retinal cones and rods: response to intense flashes of light. J Neurosci 1990; 10:1967-73.
34. Miller JL, Korenbrot JI. Differences in calcium homeostasis between retinal rod and cone photoreceptors revealed by the effects of voltage on the cgmp-gated conductance in intact cells. J Gen Physiol 1994; 104:909-40.
35. Hodgkin AL, McNaughton PA, Nunn BJ. The ionic selectivity and calcium dependence of the light-sensitive pathway in toad rods. J Physiol (Lond) 1985; 358:447-68.
36. Cervetto L, Menini A, Rispoli G et al. The modulation of the ionic selectivity of the light-sensitive current in isolated rods of the tiger salamander. J Physiol (Lond) 1988; 406:181-98.
37. Wells GB, Tanaka JC. Ion selectivity predictions from a two-site permeation model for the cyclic nucleotide-gated channel of retinal rod cells. Biophys J 1997; 72:127-40.
38. Zimmerman AL, Baylor DA. Cation interactions within the cyclic GMP-activated channel of retinal rods from the tiger salamander. J Physiol (Lond) 1992; 449:759-83.
39. Ohyama T, Picones A, Korenbrot JI. Voltage-dependence of Ion Permeation in Cyclic GMP-gated ion channels is optimized for cell function in rod and cone photoreceptors. J Gen Physiol 2002; 119:341-54.
40. Picones A, Korenbrot JI. Permeability and interaction of Ca2+ with cGMP-gated ion channels differ in retinal rod and cone photoreceptors. Biophys J 1995; 69:120-7.
41. Hackos DH, Korenbrot JI. Divalent cation selectivity is a function of gating in native and recombinant cyclic nucleotide-gated ion channels from retinal photoreceptors [see comments]. J Gen Physiol 1999; 113:799-818.
42. Frings S, Seifert R, Godde M et al. Profoundly different calcium permeation and blockage determine the specific function of distinct cyclic nucleotide-gated channels. Neuron 1995; 15:169-79.
43. Nakatani K, Yau KW. Calcium and magnesium fluxes across the plasma membrane of the toad rod outer segment. J Physiol (Lond) 1988; 395:695-729.
44. Lagnado L, Cervetto L, McNaughton PA. Calcium homeostasis in the outer segments of retinal rods from the tiger salamander. J Physiol (Lond) 1992; 455:111-42.
45. Gray-Keller MP, Detwiler PB. The calcium feedback signal in the phototransduction cascade of vertebrate rods. Neuron 1994; 13:849-61.
46. Younger JP, McCarthy ST, Owen WG. Light-dependent control of calcium in intact rods of the bullfrog Rana catesbeiana. J Neurophysiol 1996; 75:354-66.
47. Dzeja C, Hagen V, Kaupp UB et al. Ca2+ permeation in cyclic nucleotide-gated channels. EMBO J 1999; 18:131-44.
48. Frings S, Hackos DH, Dzeja C et al. Determination of fractional Calcium ion current in cyclic nucleotide-gated channels. Methods Enzymol 2000; 315:797-817.

49. Ohyama T, Hackos DH, Frings S et al. Fraction of the dark current carried by Ca(2+) through cGMP-gated ion channels of intact rod and cone photoreceptors. J Gen Physiol 2000; 116:735-54.
50. Zagotta WN, Siegelbaum SA. Structure and function of cyclic nucleotide-gated channels. Annu Rev Neurosci 1996; 19:235-63.
51. Yau, K-W and Chen, T-Y. Cyclic nucleotide-gated channels. In R. Alan North Ligand- and voltage-gated channels, 307-337. Boca Raton: CRC Press, 1995.
52. Seifert R, Eismann E, Ludwig J et al. Molecular determinants of a Ca2+-binding site in the pore of cyclic nucleotide-gated channels: S5/S6 segments control affinity of intrapore glutamates. EMBO J 1999; 18:119-30.
53. Zwolinski BJ, Eyring H, Reese CE. Diffusion and Membrane Permeability. J. Phys Colloid Chem 1949; 53:1426-53.
54. Hille B. Ionic selectivity, saturation, and block in sodium channels. A four- barrier model. J Gen Physiol 1975; 66:535-60.
55. Andersen OS. Perspectives on ion permeation. J Gen Physiol 1999; 113:763-4.
56. Zimmerman AL, Baylor DA. Cation interactions within the cyclic GMP-activated channel of retinal rods from the tiger salamander. J Physiol (Lond) 1992; 449:759-83.
57. Wells GB, Tanaka JC. Ion selectivity predictions from a two-site permeation model for the cyclic nucleotide-gated channel of retinal rod cells. Biophys J 1997; 72:127-40.
58. Picones A, Korenbrot JI. Permeability and interaction of Ca2+ with cGMP-gated ion channels differ in retinal rod and cone photoreceptors. Biophys J 1995; 69:120-7.
59. Yau KW, Nakatani K. Light-induced reduction of cytoplasmic free calcium in retinal rod outer segment. Nature 1985; 313:579-82.
60. Miller DL, Korenbrot JI. Kinetics of light-dependent Ca fluxes across the plasma membrane of rod outer segments. A dynamic model of the regulation of the cytoplasmic Ca concentration. J Gen Physiol 1987; 90:397-425.
61. McNaughton PA, Cervetto L, Nunn BJ. Measurement of the intracellular free calcium concentration in salamander rods. Nature 1986; 322:261-3.
62. Bauer PJ, Drechsler M. Association of cyclic GMP-gated channels and Na(+)-Ca(2+)-K+ exchangers in bovine retinal rod outer segment plasma membranes. J Physiol 1992; 451:109-31.
63. Korenbrot JI, Miller DL. Cytoplasmic free calcium concentration in dark-adapted retinal rod outer segments. Vision Res 1989; 29:939-48.
64. McCarthy ST, Younger JP, Owen WG. Free calcium concentrations in bullfrog rods determined in the presence of multiple forms of Fura-2. Biophys J 1994; 67:2076-89.
65. Sampath AP, Matthews HR, Cornwall MC et al. Light-dependent changes in outer segment free-Ca2+ concentration in salamander cone photoreceptors. J Gen Physiol 1999; 113:267-77.
66. McCarthy ST, Younger JP, Owen WG. Dynamic, spatially nonuniform calcium regulation in frog rods exposed to light. J Neurophysiol 1996; 76:1991-2004.
67. Cobbs WH, Pugh EN Jr. Kinetics and components of the flash photocurrent of isolated retinal rods of the larval salamander, Ambystoma tigrinum. J Physiol (Lond) 1987; 394:529-72.
68. Nakatani K, Yau KW. Sodium-dependent calcium extrusion and sensitivity regulation in retinal cones of the salamander. J Physiol (Lond) 1989; 409:525-48.
69. Yau KW, Nakatani K. Light-suppressible, cyclic GMP-sensitive conductance in the plasma membrane of a truncated rod outer segment. Nature 1985; 317:252-5.
70. Perry RJ, McNaughton PA. Response properties of cones from the retina of the tiger salamander JT . J Physiol (Lond) 1991; 433:561-87.
71. Nakatani K, Koutalos Y, Yau KW. Ca2+ modulation of the cGMP-gated channel of bullfrog retinal rod photoreceptors. J Physiol (Lond) 1995; 484 (Pt 1):69-76.
72. Sagoo MS, Lagnado L. The action of cytoplasmic calcium on the cGMP-activated channel in salamander rod photoreceptors. J Physiol (Lond) 1996; 497:309-19.

73. Rebrik TI, Kotelnikova EA, Korenbrot JI. Time course and Ca(2+) dependence of sensitivity modulation in cyclic GMP-gated currents of intact cone photoreceptors. J Gen Physiol 2000; 116:521-34.
74. Haynes LW, Stotz SC. Modulation of rod, but not cone, cGMP-gated photoreceptor channel by calcium-calmodulin. Vis Neurosci 1997; 14:233-9.
75. Forti S, Menini A, Rispoli G et al. Kinetics of phototransduction in retinal rods of the newt Triturus cristatus. J Physiol (Lond) 1989; 419:265-95.
76. Torre V, Forti S, Menini A et al. Model of phototransduction in retinal rods. Cold Spring Harb Symp Quant Biol 1990; 55:563-73.
77. Sneyd J, Tranchina D. Phototransduction in cones: an inverse problem in enzyme kinetics. Bull Math Biol 1989; 51:749-84.
78. Tranchina D, Sneyd J, Cadenas ID. Light adaptation in turtle cones. Testing and analysis of a model for phototransduction. Biophys J 1991; 60:217-37.
79. Nikonov S, Engheta N, Pugh EN Jr. Kinetics of recovery of the dark-adapted salamander rod photoresponse. J Gen Physiol 1998; 111:7-37.
80. Tamura T, Nakatani K, Yau KW. Calcium feedback and sensitivity regulation in primate rods. J Gen Physiol 1991; 98:95-130.
81. Hamer RD. Computational analysis of vertebrate phototransduction: combined quantitative and qualitative modeling of dark- and light-adapted responses in amphibian rods. Vis Neurosci 2000; 17:679-99.
82. Duda T, Goraczniak R, Surgucheva I et al. Calcium modulation of bovine photoreceptor guanylate cyclase. Biochemistry 1996; 35:8478-82.

REGULATION OF THE ROD PHOTORECEPTOR CYCLIC NUCLEOTIDE-GATED CHANNEL

René Warren and Robert S. Molday

ABSTRACT

The cGMP-gated channel of rod photoreceptors plays a key role in phototransduction by controlling the flow of cations into the outer segment in response to light-induced changes in cGMP. The channel is a heterotetramer composed of α-subunits required for channel activity and β-subunits that are important in modulating the activity of channel. Earlier studies have shown that exogenous calmodulin binds to the β-subunit of the channel and modulates the sensitivity of the channel for cGMP in a calcium dependent manner. In addition unidentified Ca^{2+}-dependent endogenous proteins have been reported to modulate the activity of the frog rod channel. In this paper, we investigated whether endogenous calmodulin and other Ca^{2+} binding proteins interact with and modulate the cGMP-gated channel in bovine rod outer segments. Using immunoaffinity techniques in conjunction with ion flux assays, we show that endogenous calmodulin, but not other Ca^{2+} dependent proteins, binds and modulates the rod cGMP-gated channel in bovine rod outer segments. We also show that the β-subunit of the channel is phosphorylated by endogenous and exogenous casein kinase 2. This posttranslational modification, however, does not alter the sensitivity of the channel for cGMP.

INTRODUCTION

The cyclic nucleotide-gated (CNG) channel of the rod photoreceptor cell plays a key role in phototransduction by controlling the flow of Na^+ and Ca^{2+} into the rod outer segment (ROS) in response to light-induced changes in intracellular cGMP

Departments of Biochemistry and Molecular Biology, University of British Columbia, Vancouver, B.C. Canada V6T1Z3 concentration. In the dark, a significant number of CNG channels are maintained in their open state by the reversible binding of cGMP to the channel[1,2] (Fig. 1). This allows Na^+ and Ca^{2+} to enter the outer segment establishing a dark current between the outer and inner segment. Under these conditions, the rod photoreceptor is in its depolarized state and there is a continuous release of glutamate transmitter at the rod synapse. Low intracellular Na^+ concentration is maintained by the active transport of Na^+ from the cell through a Na/K ATPase in the plasma membrane of the rod inner segment. The Ca^{2+} concentration in the ROS is kept at 400-500 nM by the balanced extrusion of Ca^{2+} from the ROS by the Na/Ca-K exchanger.

Photobleaching of rhodopsin in the disk membrane leads to the activation of the G-protein mediated visual cascade that culminates in a decrease in cGMP concentration and the closure of the CNG channel to the influx of Na^+ and Ca^{2+} (Fig. 1).[3-5] Suppression of the dark current and the corresponding hyperpolarization of the rod cell inhibits the release of glutamate at the synapse. Closure of the CNG channel also decreases the intracellular Ca^{2+} concentration to below 100 nM as the Na/Ca-K exchanger continues to extrude Ca^{2+} from the ROS.[6,7]

The rod photoreceptor cell returns to its dark state following photoexcitation through a series of biochemical reactions.[1,8] Photoexcited rhodopsin is inactivated by rhodopsin kinase catalyzed phosphorylation followed by the binding of arrestin. The G-protein coupled cascade is inactivated by RGS9 protein mediated hydrolysis of GTP to GDP bound to transducin.[9] This results in the return of phosphodiesterase (PDE) to its prebleach, basal activity and restoration of transducin to its inactive state (Fig. 1). Finally, the cGMP concentration is restored to its dark level through the Ca^{2+}-dependent activation of guanylate cyclase (10). The rise in cGMP concentration reopens the CNG channels and returns the rod cell to its depolarized state.

The decrease in intracellular Ca^{2+} following illumination initiates multiple negative feedback mechanisms that contribute to light adaptation and photorecovery.[1,8,11] These mechanisms typically involve Ca^{2+}-binding proteins that sense and respond to small changes in free intracellular Ca^{2+} levels.[12,13] Modulation of guanylate cyclase by guanylate cyclase activating proteins known as GCAPs is the principal Ca^{2+}-dependent mechanism[14-16] Regulation of the lifetime of PDE presumably through the modulation of rhodopsin kinase by recoverin, also known as S-modulin, has also been suggested to play a role in light adaptation under certain conditions.[1,17] Finally, the sensitivity of the CNG channel to cGMP is altered by the Ca^{2+} dependent binding of calmodulin (CaM) to the channel.[18] Other, as yet unidentified, Ca^{2+}-dependent factors, have also been implicated in the control of rod channel sensitivity.[19,20]

In this chapter, the structural features of the rod CNG channel and its interaction with ROS proteins is discussed. Biochemical experiments designed to determine if endogeneous, as well as exogeneous, CaM binds to and modulates the rod CNG channel in ROS are reported. Finally, studies to assess the regulation of the channel by other Ca-binding factors and phosphorylation are described.

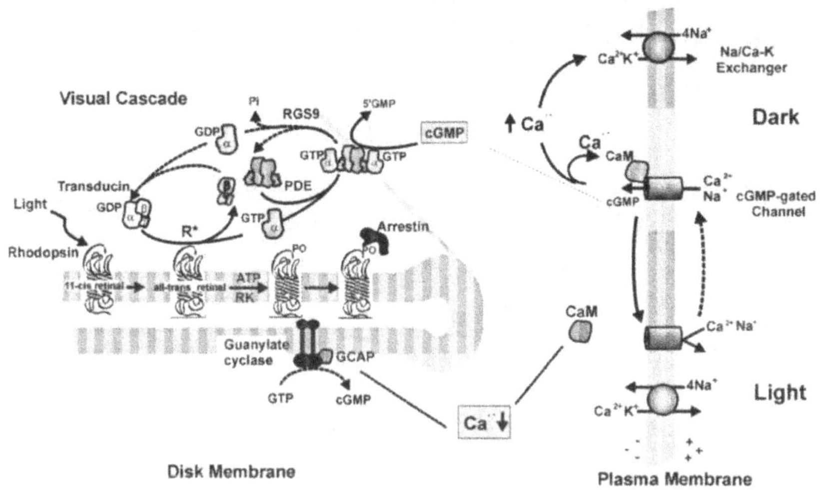

Figure 1. Basic biochemical reactions responsible for phototransduction in rod outer segments. Photoexcitation (solid lines) is initiated when light converts 11-cis retinal to its all-trans isomer in rhodopsin. Activated rhodopsin, known as Meta II rhodopsin or R*, catalyzes the exchange of GDP bound to the α-subunit of transducin for GTP. The α-subunit dissociates from the βγ subunit of transducin and associates with the γ-subunit of phosphodiesterase (PDE). Activated PDE catalyzes the hydrolysis of cGMP to 5'GMP resulting in a decrease in intracellular cGMP concentration, a closure of the rod CNG channel to the influx of Na^+ and Ca^{2+}, and a hyperpolarization of the rod cell. The intracellular concentration of Ca^{2+} decreases as the Na/Ca-K exchanger continues to extrude Ca^{2+} from the outer segment. Recovery (dashed lines) of the ROS to its dark state occurs by the shutoff of the visual cascade and resynthesis of cGMP. Rhodopsin is inactivated through rhodopsin kinase (RK) catalyzed phosphorylation and the binding of arrestin. Transducin and PDE are returned to their inactive state by RGS9 mediated hydrolysis of GTP bound to transducin. Low Ca^{2+} resulting from photoexcitation leads to the activation of guanylate cyclase via GCAP proteins. It also causes the dissociation of calmodulin (CaM) from the channel increasing the sensitivity of the channel of cGMP. The increase in cGMP concentration leads to the reopening of the channel and an increase in intracellular Ca^{2+} as part of the negative feedback mechanism.

ROD CNG CHANNEL—STRUCTURAL FEATURES

The structural and functional properties of CNG channels have been the subject of a number of recent reviews.[21-23,24,25] Briefly, the rod channel is a heterotetrameric complex consisting of two types of subunits designated as α or CNGA1 and β or CNGB1 subunits.[26-29] They are 30% identical in sequence and contain similar structural features including six transmembrane segments (S1-S6), a pore region of approximately 25 amino acids situated between S5 and S6 segments, and a cyclic nucleotide binding domain near the C-terminus (Fig. 2A).

The α subunit is essential for channel activity since it self-assembles into a functional cGMP-gated channel when expressed in Xenopus oocytes or HEK 293 cells.[26,30] In contrast, the β-subunit itself fails to form a functional channel when

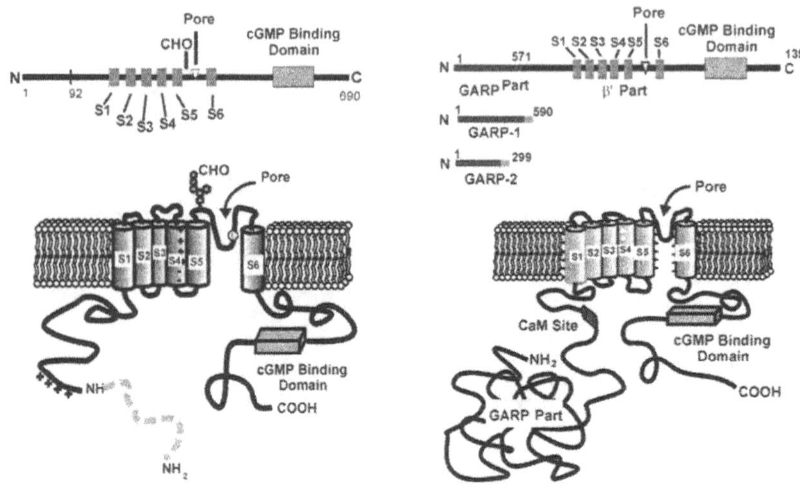

Figure 2A. Schematic diagrams of the rod CNG channel. A. Linear and topological models for the α and β subunits of the bovine rod CNG channel. Basic structural features include 6 transmembrane segments (S1-S6), a pore region, and a cGMP binding domain. In addition, the β-subunit contains the calmodulin (CaM) binding site and the large GARP-part on the N-terminus. The α-subunit contains an oligosaccharide (CHO) chain on the extracellular side of the membrane and a glutamate (E) residue in the pore region that are not present on the β-subunit. GARP, in addition to being part of the β-subunit, is also present in ROS as two spliced variants referred to as GARP-1 and GARP-2. B. Schematic model of the rod CNG channel showing a tetrameric arrangement of two α and two β subunits and a central cavity. The stoichiometry of the native channel, however, remains to be determined.

individually expressed in heterologous cells.[27,28] Co-expression of the α and β subunits, however, produces a channel that has essentially identical physiological and pharmacological properties (cyclic nucleotide dependence, rapid opening-closing or flickering behavior, selective ion permeability, divalent ion blockage, l-cis diltiazem inhibition and Ca-calmodulin modulation) of the native rod channel indicating that the β-subunit is important in fine tuning and modulating the cGMP-dependent ion conducting activity of the channel.[27,28,31]

The α-subunit has several structural features not found in the β-subunit. An oligosaccharide chain is linked to an asparagine residue located between the S5 transmembrane segment and the pore region.[32] The pore region harbors a negatively charged glutamate residue that binds extracellular Ca^{2+} and Mg^{2+}.[33,34] This reduces the current through the channel, a property that contributes to low background noise in rod cells. The glutamate residue within the pore region of the α subunit is replaced with a neutral glycine residue in the β subunit. Finally, the rod α-subunit undergoes a photoreceptor-specific, posttranslational cleavage that removes the N-terminal 92 amino acids.[35] The role of this protein modification is not known at the present time.

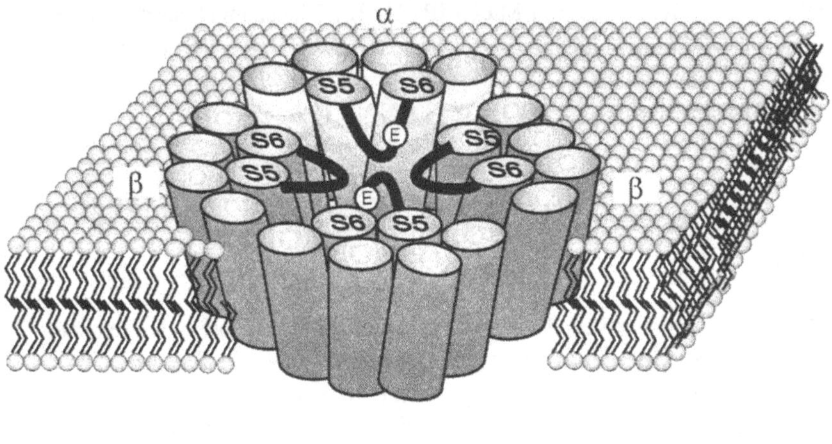

Figure 2B. See figure legend previous page

The bovine rod β-subunit is unique in that it contains a large cytoplasmic N-terminal region (592 amino acids in the bovine β subunit) known as the GARP part[28](Fig. 2A). GARP, in addition to being part of the β subunit, is also expressed in ROS as two alternatively spliced variants referred to as GARP-1 (also known as f-GARP) and GARP-2 (also known as t-GARP).[36-38] GARP-1 is identical in sequence to the GARP part of the channel except for an additional 19 amino acid extension at the C-terminus. GARP-2, the more abundant variant, is a truncated form of GARP-1 consisting of the N-terminal 291 amino acids and an 8 amino acid C-terminal extension. The GARP variants are characterized by a high content of glutamate and proline residues that affect their migration on SDS polyacrylamide gels. Recently, four proline rich repeat regions have been identified within the N-terminal half of the GARP proteins[38] The GARP part of the β subunit does not influence the rod channel activity.[28] Instead, several recent studies indicate that this region plays a role in protein-protein interactions within ROS.[38,39] The β subunit of the CNG channel also contains a binding site for CaM as discussed in detail below.

The rod CNG channel is known to exist in membranes as tetramer. The stoichiometry and arrangement of subunits of the native channel of ROS remains to be determined. However, recent heterologous expression studies suggest that the channel consists of two α and two β subunits with identical subunits arranged diagonal to each other (Fig. 2B).[40] By analogy with the structure of the K^+ channel of Streptomyces lividans,[41] the four subunits are thought to form a central cavity through which cations flow. This cavity is most likely lined by the S5 and S6 transmembrane segments of the α and β subunits (Fig. 2). The pore region of each subunit extends into the central cavity where together they function as the gate and selectivity filter for the flow of ions through the membrane.

The rod channel is cooperatively activated by cGMP with a Hill coefficient greater than 3 and a $K_{1/2}$ of 20-50 μM cGMP (21, 22). The channel has a relatively high probability of being in its open state when 3-4 cGMP molecules are bound.[42]

REGULATION OF THE ROD CHANNEL

Exogenous Calmodulin (CaM) Regulates the Rod Channel Sensitivity for cGMP

CaM is a highly conserved, 148 amino acid Ca^{2+} binding protein that regulates a wide variety of enzymes, channels, transporters, and cytoskeletal proteins in eukaryotic cells. It has a dumbbell-like shape with two globular lobes connected by a seven-turn α helix. Each lobe has two high affinity Ca^{2+} binding sites that are formed by helix-loop-helix motifs known as EF hands. The binding of Ca^{2+} to these sites induces a protein conformational change enabling CaM to interact and regulate target proteins. The site of interaction on the target protein typically consists of an amphipathic helix.

Modulation of the rod CNG channel by exogeneous CaM was first demonstrated using ion flux assays in ROS vesicle preparations.[18] The sensitivity of the channel for cGMP was decreased when CaM was added in the presence of excess Ca^{2+}. This was reflected in an increase in the apparent $K_{1/2}$ for cGMP from 19 μM without CaM to 33 μM with CaM. The cooperativity of the channel for cGMP was not affected by the presence of CaM. Modulation of the sensitivity of the rod channel for cGMP by exogeneous CaM has been reproduced in several different experimental systems including purified and reconstituted channel preparations, excised patches from ROS, and the heteromeric rod CNG channel expressed in HEK 293 cells.[19,28,31,43,44,45] CaM regulation of the channel is Ca^{2+} dependent occurring over the physiological relevant Ca^{2+} concentration range of 50 to 300 nM Ca^{2+}.[43,44] The olfactory channel is also modulated by CaM, but to a significantly larger extent. CaM decreases the cyclic nucleotide sensitivity of the olfactory channel by 10-20 fold.[46] This mechanism plays a dominant role in olfactory adaptation.

The binding site for CaM resides on the β subunit of the rod channel. This was initially determined using gel overlay methods and functional analysis of homomeric and heteromeric channels.[18,31] More recently, Weitz et al[47] and Grunwald et al[48] mapped the location of two CaM binding sites within the β subunit. One site was localized to a 15 amino acid segment just before the first membrane spanning segment (S1) of the β-subunit. Deletion of this site abolished binding and modulation of the channel by CaM. A second site was found downstream of the cyclic nucleotide binding domain. Deletion of this site, however, had little effect on CaM modulation. The sites are unusual since they do not harbor an amphipathic helical binding motif characteristic of most CaM binding proteins. The binding affinity of CaM for these individual target sites is 2-3 orders of magnitude lower than the affinity of CaM for the intact channel, estimated to be approximately 1-5 nM.[43,44] This suggests that the conformation of the protein is important in establishing a high affinity CaM binding site.

Endogeneous CaM is the Major Ca^{2+} Regulator of the Rod Channel

In previous studies, exogenous CaM was added to samples to demonstrate binding and modulation of the rod channel. In order to further evaluate the potential role of this regulatory mechanism in rod photoreceptor function, we have used immunoprecipitation techniques to examine whether CaM in ROS interacts with the rod channel in a Ca^{2+}-dependent manner. In these studies, ROS were washed in hypotonic buffer (10 mM HEPES, pH 7.4, 1 mM DTT) containing either 1 μM $CaCl_2$ or 1 μM EDTA and the membranes were solubilized in CHAPS buffer (10 mM HEPES, pH 7.4, 18 mM CHAPS, 150 mM KCl, 1μM $CaCl_2$). The rod CNG channel was then immunoprecipitated with an anti-channel (α-subunit) monoclonal antibody (PMc 1D1) immobilized on Sepharose beads. Co-precipitation of CaM with the channel was analyzed by Western blotting using a CaM monoclonal antibody. As shown in Figure 3, CaM co-precipitates with the rod channel from ROS washed in the presence of Ca^{2+}, but not in the absence of Ca^{2+} i.e EDTA–treated ROS. This indicates that endogeneous CaM associates with the rod channel in the presence of Ca^{2+} and dissociates when Ca^{2+} is removed.

In a previous study, Gordon et al[19] reported that frog ROS contain an unidentified, endogenous Ca^{2+} binding protein that regulates the channel in a similar way as CaM, but over a different range of Ca^{2+} concentrations. From these studies, they concluded that this endogenous Ca^{2+}-binding protein associates with the channel at the same site as CaM.

To determine if bovine ROS contain a related endogeneous Ca^{2+} binding protein, we investigated the interaction of the channel with CaM-Sepharose matrix in the presence and absence of endogeneous CaM. ROS were washed in hypotonic buffer containing 2 mM EDTA to remove endogenous CaM bound to the channel. The ROS membranes were then solubilized in CHAPS buffer and incubated with HEPES (buffer control), supernatant from hypotonically lysed ROS containing endogenous CaM, and supernatant from hypotonically lysed ROS in which CaM was removed by immunoprecipitation with an anti-CaM antibody (Fig. 4A). As shown in Figure 4B, the channel derived from solubilized ROS treated with CaM-depleted soluble extract quantitatively bound to the CaM-Sepharose matrix. In contrast the channel from solubilized ROS treated with extract containing endogenous CaM did not bind to the CaM-Sepharose matrix. These data indicate that bovine ROS extracts do not contain an endogenous protein that effectively competes with CaM for the same binding site on the channel.

The ability of endogenous CaM or another factor to modulate the sensitivity of the channel for cGMP was investigated using ion flux assays. In these studies, the channel from ROS was reconstituted into Ca^{2+}—containing lipid vesicles as previously described.[49] Cyclic GMP-dependent efflux of Ca^{2+} was measured in the presence and absence of soluble extracts from hypotonically lysed ROS. Figure 5A shows

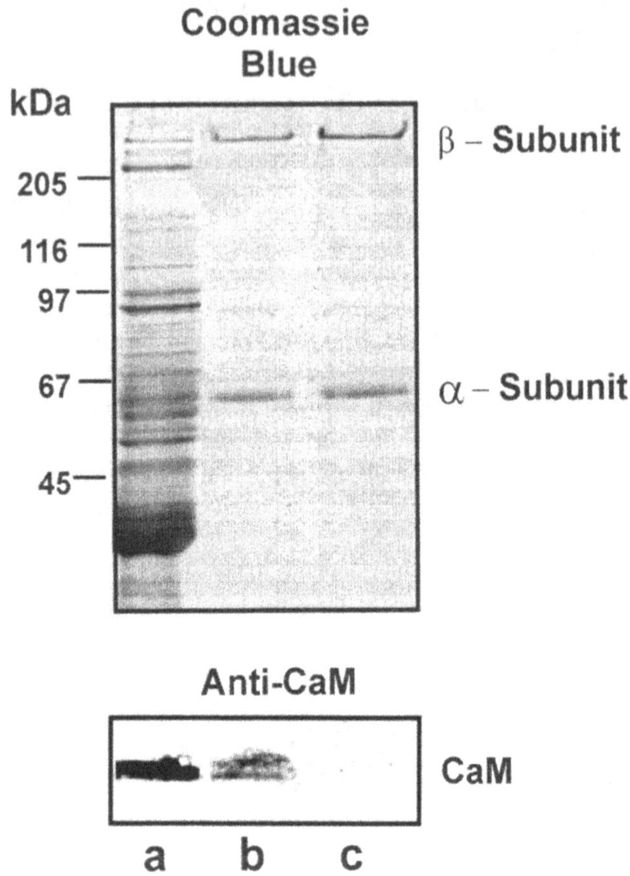

Figure 3. Co-purification of CaM and the rod CNG channel from bovine ROS. ROS membranes were washed in HEPES buffer containing either 1 μM $CaCl_2$ or 2 mM EDTA. After solubilization in CHAPS detergent, the rod CNG channel was purified on a PMc 1D1 anti-channel-Sepharose matrix. The samples were subjected to SDS gel electrophoresis and the proteins were stained with Coomassie Blue (top) and Western blots were labeled with an anti-CaM monoclonal antibody (bottom). Lane a: ROS; lane b: purified channel (α and β subunits) isolated from ROS pre-washed in the presence of $CaCl_2$; lane c: purified channel (α and β subunits) isolated from ROS pre-washed in the presence of EDTA. Either 40 μg of ROS or 1.5 μg of purified channel were applied to the respective lanes. CaM coprecipitated with the channel from ROS washed in the presence of $CaCl_2$, but not in the presnce of EDTA.

cGMP-dependent efflux of Ca^{2+} as a function of time and Figure 5B shows the normalized rate of efflux as a function of cGMP concentration. A decrease in sensitivity of the channel for cGMP is observed only for vesicles treated with soluble extract containing endogenous CaM. Vesicles treated with soluble extract depleted of CaM exhibited the same sensitivity to cGMP as the buffer control. These results indicate that endogenous CaM modulates the rod channel to the same degree as exogenously added CaM. Furthermore, soluble extracts from bovine ROS do not contain other factors that modulate the sensitivity of the channel for cGMP.

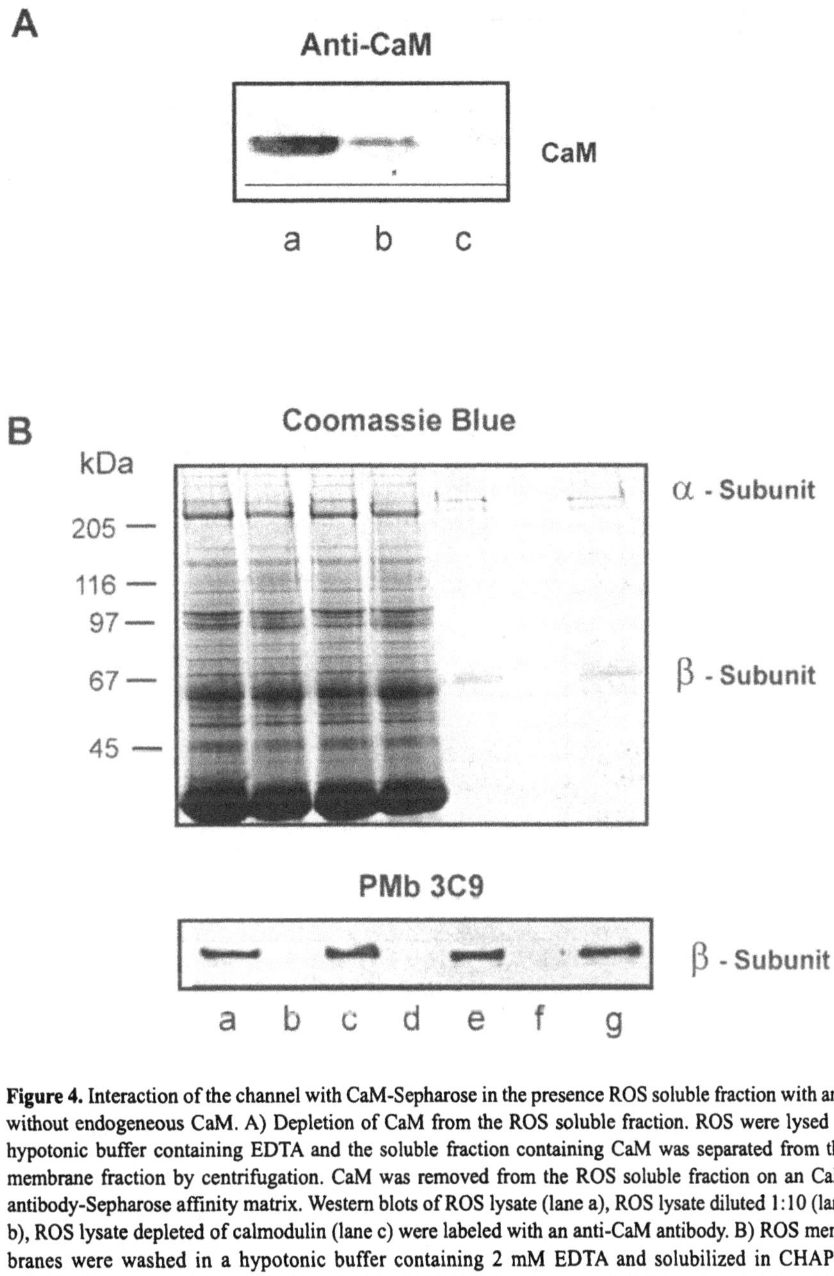

Figure 4. Interaction of the channel with CaM-Sepharose in the presence ROS soluble fraction with and without endogeneous CaM. A) Depletion of CaM from the ROS soluble fraction. ROS were lysed in hypotonic buffer containing EDTA and the soluble fraction containing CaM was separated from the membrane fraction by centrifugation. CaM was removed from the ROS soluble fraction on an CaM antibody-Sepharose affinity matrix. Western blots of ROS lysate (lane a), ROS lysate diluted 1:10 (lane b), ROS lysate depleted of calmodulin (lane c) were labeled with an anti-CaM antibody. B) ROS membranes were washed in a hypotonic buffer containing 2 mM EDTA and solubilized in CHAPS. CaM-Sepharose was incubated with solubilized ROS in the presence of $CaCl_2$ and containing either HEPES buffer (HEPES control); ROS soluble fraction containing endogenous CaM (+CaM); or ROS soluble fraction depleted of CaM (-CaM). The unbound fraction from the CaM-Sepharose was separated from the bound, EDTA eluted fraction. SDS gels were stained with Coomassie Blue and Western blots were labeled for the β-subunit with the PMb 3C9 monoclonal antibody. Lane a: bovine ROS; lane b: unbound fraction (HEPES control); lane c: unbound fraction (+CaM); lane d: unbound fraction (-CaM); lane e: bound fraction (HEPES control); lane f: bound fraction (+CaM); lane g : bound fraction (-CaM). Endogenous CaM inhibited the binding of the channel to CaM-Sepharose.

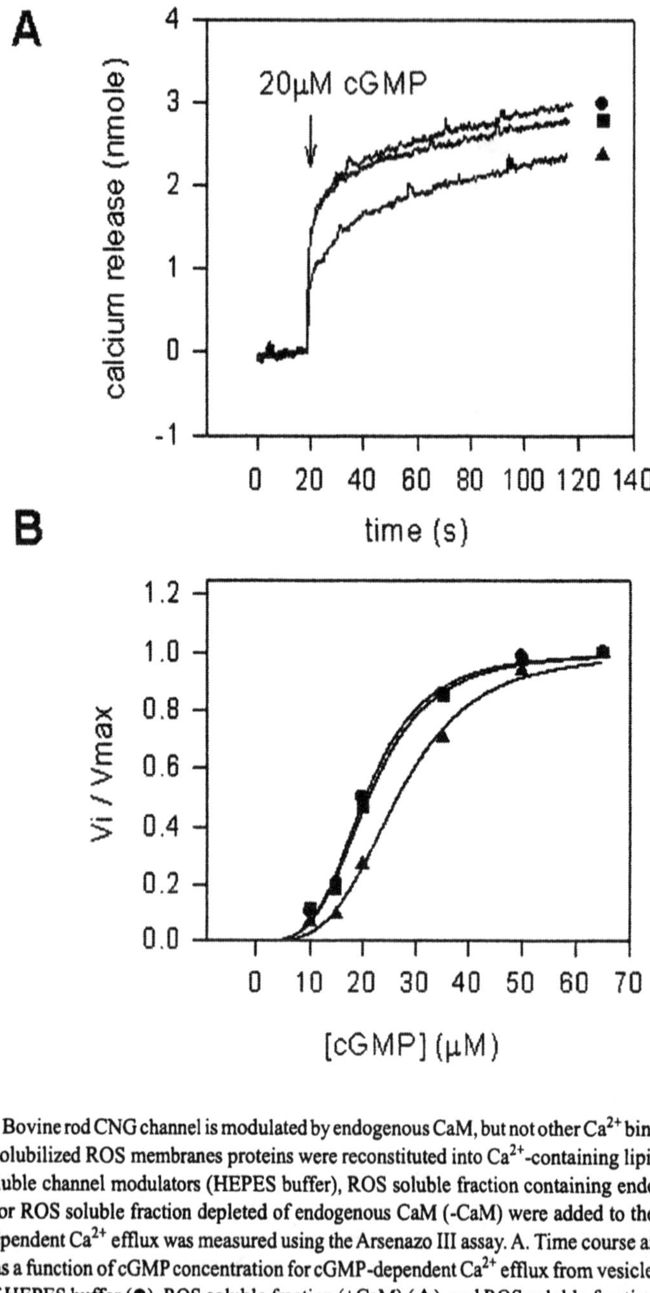

Figure 5. Bovine rod CNG channel is modulated by endogenous CaM, but not other Ca^{2+} binding proteins. CHAPS solubilized ROS membranes proteins were reconstituted into Ca^{2+}-containing lipid vesicle. Potential soluble channel modulators (HEPES buffer), ROS soluble fraction containing endogenous CaM (+CaM), or ROS soluble fraction depleted of endogenous CaM (-CaM) were added to the vesicles and cGMP-dependent Ca^{2+} efflux was measured using the Arsenazo III assay. A. Time course and B. normalized rate as a function of cGMP concentration for cGMP-dependent Ca^{2+} efflux from vesicles treated with 100 ml of HEPES buffer (●), ROS soluble fraction (+CaM) (▲), and ROS soluble fraction (-CaM) (■). The data in B was analyzed with a Sigma Plot curve fit program of using the Hill equation $V/Vmax = [cGMP]^n/([cGMP]^n + Km^n)$ where Km and n are the Michaelis constant and the Hill coefficient, respectively. The cGMP-gated channel incubated with HEPES (●) had a Km = 20.5 mM and n = 3.7, the channel incubated with a ROS lysate containing calmodulin (▲) had a Km = 26.5 mM and n = 3.7 and the channel incubated with a ROS lysate depleted of calmodulin (■) had a Km = 21.1 mM and n = 3.6. These results indicate that endogenous CaM, but not other soluble proteins modulate the activity of the channel.

Phosphorylation of the Channel

A number of ROS proteins that function in phototransduction undergo kinase catalyzed phosphorylation reactions. These include rhodopsin, guanylate cyclase, rhodopsin kinase, RGS9, transducin among others. In some instances, phosphorylation directly regulates the activity of these proteins such as in the case of rhodopsin-kinase catalyzed phosphorylation of rhodopsin.[50]

Several reports have suggested that the activity of the rod CNG channel is also affected by phosphorylation. Using patch clamp techniques, Gordon et al[51] showed that serine/threonine inhibitors or exogenous phosphatase alter the sensitivity of the channel for cGMP in frog ROS. The native and expressed rod channel α-subunit is also modulated by tyrosine kinase and phosphatase inhibitors.[52,53] Regulation of the channel by tyrosine phosphorylation appears to occur through a signal transduction pathway initiated by the binding of insulin-like growth factors to surface rod cells.[54] Although these studies suggest that the channel can be regulated by phosphorylation reactions, direct phosphorylation of the channel was not determined in these studies.

We have carried out a number of studies to determine if the rod CNG channel can be phosphorylated by an endogenous protein kinase in ROS. In one series of experiments, ROS were hypotonically lysed in 0.01M HEPES buffer and the ROS membranes were separated from the soluble extract by centrifugation. The ROS membranes were resuspended in HEPES buffer or ROS soluble extract and the phosphorylation reaction was initiated by the addition of [γ^{32}P]ATP. After 15 min at 30°C the membranes were washed in HEPES buffer and the channel was isolated on a PMc 1D1 anti-channel-Sepharose affinity matrix. As shown in Figure 6, the β-subunit, but not the α-subunit of the rod channel was phosphorylated by an endogenous kinase. The extent of of phosphorylation was significantly greater when the reaction was carried out in the presence of ROS soluble extract. This suggests that the protein kinase that catalyzed the phosphorylation of the β-subunit exists primarily as a soluble protein in ROS.

We examined the effect of a number of factors, known to be important in phototransduction, on channel phosphorylation. The extent of phosphorylation was independent of the light/dark state of ROS or the presence of up to 2 mM Ca^{2+} or EDTA, 100 μM phorbol ester with and without Ca^{2+}, 200 μM cGMP or Br-cGMP, 100 μM cAMP or CaM.

To further identify the protein kinase responsible for phosphorylation of the channel β-subunit, we carried out a number of kinase inhibitor studies. Only inhibitors known to affect casein kinase 2 (CK2) inhibited phosphorylation of the rod channel β-subunit (data not shown).

The expression of casein kinase 2 in ROS was investigated using specific anti-CK2 antibodies. Western blotting studies confirmed the presence of an immunoreactive CK2 variant in the soluble fraction of ROS (Fig. 7). This variant migrated with an apparent molecular mass slightly higher than that of human recombinant CK2. Immunofluorescence labeling studies further confirmed the presence of CK2 in ROS (data not shown).

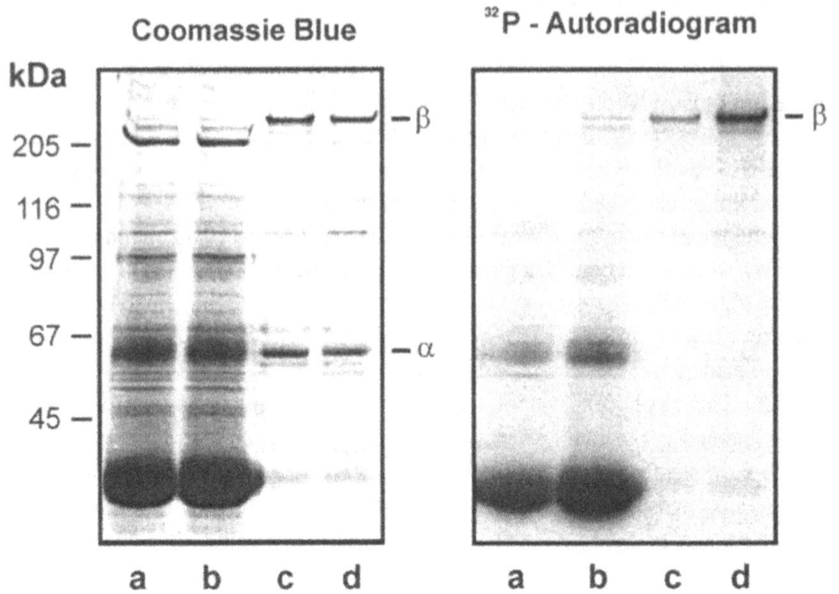

Figure 6. Phosphorylation of the rod cGMP-gated channel β-subunit by protein kinase(s) in the ROS soluble fraction. ROS membranes were hypotonically lysed, washed by centrifugation and resuspended in either HEPES buffer or ROS lysate containing endogenous ROS soluble proteins. The phosphorylation reaction was initiated by addition of [γ-^{32}P]ATP and the samples were incubated for 15 min at 30°C. The phosphorylated membranes were washed in HEPES and solubilized in CHAPS detergent. The cGMP-gated channel complex that bound to the PMc 1D1-Sepharose matrix was separated from the unbound fraction. Proteins were resolved on a 9% SDS-polyacrylamide gel and stained with Coomassie Blue and subjected to autoradiography for analysis of phosphorylated proteins. lane a—unbound fraction for ROS membranes phosphorylated in HEPES; lane b—unbound fraction for ROS phosphorylated in the presence of ROS soluble fraction; lane c—bound fraction containing the channel (α and β subunits) for ROS phosphorylated in HEPES; lane d–bound fraction containing the channel (α and β subunits) for ROS phosphorylated in the presence of ROS soluble fraction. The β-subunit is most intensely labeled when phosphorylation is carried out in the presence of ROS soluble fraction.

The ability of the soluble extract from bovine ROS to phosphorylate bovine casein was examined. As shown in Figure 8A, ROS soluble fraction catalyzed the phosphorylation of bovine casein.

The ability of recombinant CK2 to phosphorylate the β-subunit was also examined. Figure 8B shows that the β-subunit and to a lesser extent the α-subunit of the rod CNG channel is phosphorylated by exogenous CK2. Together, these results suggest that the β-subunit in rod CNG channel can be phosphorylated by a CK2 variant present in ROS.

The effect of phosphorylation on the rod channel activity was determined using ion flux assays. Endogenous CK2 kinase or exogenous recombinant CK2 kinase catalyzed phosphorylation of the channel had no effect on its $K_{1/2}$ or cooperativity for cGMP.

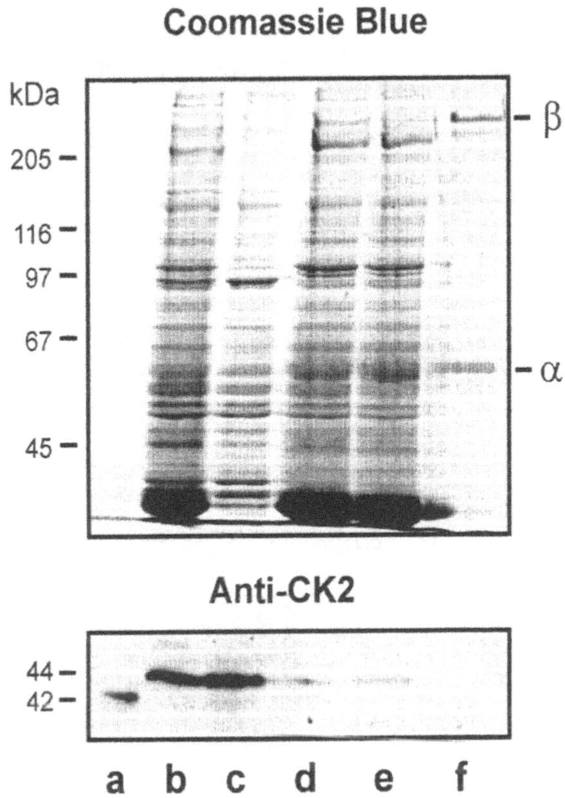

Figure 7. Soluble fraction of ROS contain a CK2 α-subunit variant. Human recombinant casein kinase II (lane a), bovine ROS (lane b), ROS cytosol (lane c), solubilized ROS membranes (lane d), unbound (lane e) and bound (lane f) fraction from PMc 1D1-Sepharose were subjected to SDS gel electrophoresis and either stained with Coomassie Blue or transferred to Immobilon-P membranes and labeled with an anti-CK2α monoclonal antibody. Lanes contained either 40 μg of ROS or 3 μg of purified channel. CK2 α – subunit is present in the soluble fraction from hypotonically lysed ROS and migrates with a slightly higher apparent molecular weight relative to recombinant CK2 α.

These studies indicate that the β-subunit of the rod channel can be phosphorylated by an endogenous CK2, but this post-translational modification does not affect the sensitivity of the channel for cGMP. It is possible that phosphorylation may influence channel properties not measured by these ion flux assays or alter another function of the channel such as its interaction with other ROS proteins. More studies are required to fully assess the possible role of channel phosphorylation in ROS structure and function.

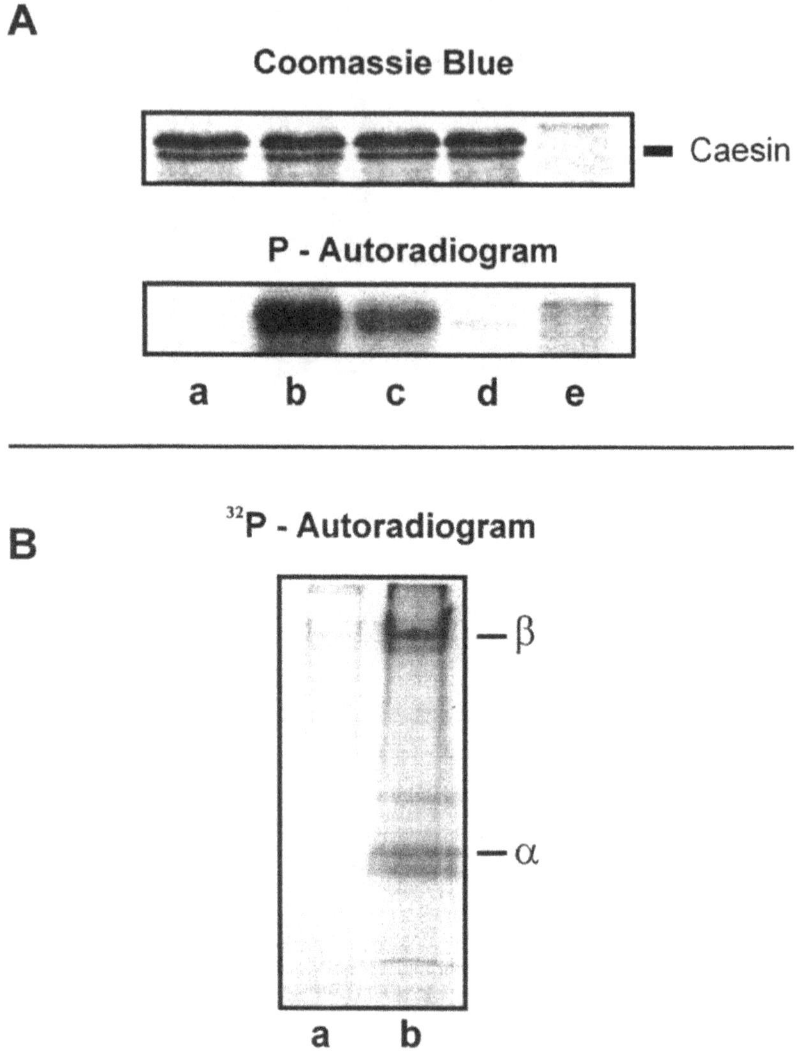

Figure 8. Phosphorylation of caesin by ROS soluble fraction and the cGMP-gated channel by a human recombinant CK2 A. The ability of a soluble fraction from ROS to phosphorylate bovine milk casein was examined by autoradiography. lane a: casein alone; lane b: casein phosphorylated using a hypotonic soluble fraction from ROS; lane c: casein phosphorylated using a diluted (1:30) ROS soluble fraction; lane d: casein phosphorylated using ROS lysate depleted of CK2 by immunoprecipitation; lane e: phosphorylated ROS lysate. B. Phosphorylation of the rod channel by recombinant CK2. Washed ROS membranes were solubilized in CHAPS and the cGMP-gated channel bound to PMc 1D1-Sepharose. The bound channel complex was phosphorylated in the presence of CHAPS detergent by the addition of exogenous CK2 (2.3 U), eluted from the column and subjected to SDS gel electrophoresis. Phosphorylation of the channel was detected by autoradiography. lane a: purified channel incubated with [γ-^{32}P]ATP; lane b: purified channel incubated with exogenous CK2 and [γ-^{32}P]ATP.

INTERACTION OF THE ROD CHANNEL WITH OTHER ROS MEMBRANE PROTEINS

In addition to CaM, the rod channel also interacts with other ROS proteins. Crosslinking and immunoprecipitation studies have revealed that the channel is associated with the Na/Ca-K exchanger in the plasma membrane.[39,55,56] These studies are in agreement with an earlier report suggesting an association of the channel and exchanger based analysis of ion flux measurements in vesicle preparations.[57] The exchanger, which exists as a dimer in the ROS plasma membrane,[58] interacts with the α-subunit of the channel.[56] The stoichiometry of the complex is not known although it has been suggested that one channel interacts with two exchangers on the basis of activity measurements.[56]

More recently, several studies have indicated that the channel-exchanger complex also interacts with disc membrane proteins.[38,39] Using chemical crosslinking and immunoprecipitation approaches, we have shown that the channel associates with peripherin-2 homomeric and peripherin-2:rom-1 heteromeric complexes localized along the rim region of the disc membrane.[39] This interaction appears to be mediated through the GARP part of the β-subunit of the channel since GARP-2 also interacts with peripherin-2 containing oligomeric complexes. In contrast, the channel and soluble GARP proteins do not appear to interact with ABCR or guanylate cyclase based on immunoprecipitation and crosslinking studies.[39]

These studies lead to a model in which the channel associates with the exchanger in the ROS plasma membrane.[39] This complex interacts with peripherin-2:rom-1 complexes to form a large multimeric complex that spans the narrow gap between the disk and plasma membrane (Fig. 9). The interaction between the channel and peripherin-2 may be responsible, at least in part, for maintaining the defined spacing between the disks and plasma membrane and for anchoring the channel:exchanger complex in the ROS plasma membrane.

CONCLUSIONS

When the rod CNG channel was first purified, cloned and expressed in heterologous cells, it was generally thought to be a relatively simple oligomeric complex consisting of a single type of subunit. Since these pioneering studies, it has become clear that the rod channel is exceedingly complex in structure, function and regulation. It consists of two subunits that assemble into a heterotetrameric complex. Each subunit confers different properties to the channel. The α-subunit is critical for cGMP-gated channel activity and for interaction with the Na/Ca-K exchanger. The β-subunit, on the other hand, contributes to the ion permeation properties of the channel, CaM-dependent modulation of the sensitivity of the channel for cGMP, and interactions with peripherin-2 complexes in the disk membrane. Thus, a picture is emerging where the channel is part of a large multi-protein complex that not only controls the flow of cations into and out of the ROS as part of the phototransduction

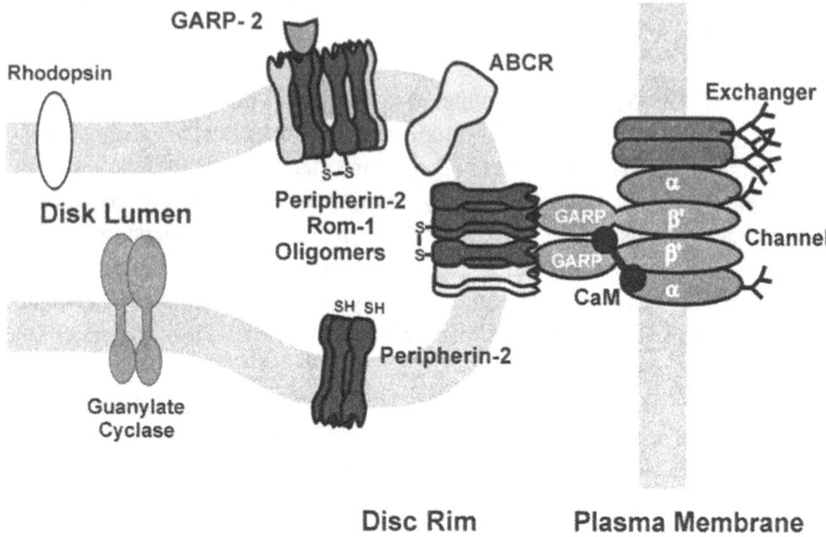

Figure 9. Diagram depicting the interaction of the rod channel with other ROS proteins as revealed by immunoprecipitation and crosslinking studies. In the plasma membrane, the heterotetrameric cGMP-gated channel (depicted as consisting of two α subunits and two β subunits) interacts with Na/Ca-K exchanger[39,55,56] previously shown to exist as a dimer in the membrane.[58] The β-subunit of the channel binds endogenous CaM in a Ca^{2+} dependent manner as shown in this study. The channel:exchanger complex also interacts with peripherin-2 containing oligomers in the rim region of disc membranes. This interaction is mediated by the GARP part of channel β-subunit. GARP-2 and less abundant GARP-1 (not shown) are confined to the rim region of disc membranes through their interactions with peripherin-2 homomeric and peripherin-2:rom-1 hetormeric disulfide-linked oligomeric and core complexes.[38,39] Disc proteins ABCR and GC-1 (guanylate cyclase) do not form stable associations with the cyclic GMP-gated channel or soluble GARP proteins.

process and Ca^{2+} homeostasis, but also stabilizes the unique structure of the ROS via GARP mediated interactions with peripherin-2 complexes in the disk membranes.

Although remarkable progress has been made on characterizing the structure, function and regulation of the channel complex and its interaction with other ROS proteins, one cannot help but think that we have merely scratched the surface. The challenge for the future will be to define the structural, functional and regulatory properties of the rod channel in order to more detail to more fully define its role in both phototransduction and rod photoreceptor cell function and structure.

ACKNOWLEDGMENTS

The work was supported by a grant from the National Eye Institute (EY02422) and Canadian Institutes for Health Research (MT 5822).

REFERENCES

1. Burns ME, Baylor DA. Activation, deactivation, and adaptation in vertebrate photoreceptor cells. Annu Rev Neurosci 2001; 24:779-805.
2. Yau KW, Baylor DA. Cyclic GMP-activated conductance of retinal photoreceptor cells. Annu Rev Neurosci 1989; 12:289-327.
3. Stryer L. Cyclic GMP cascade of vision. Annu Rev Neurosci 1986; 9:87-119.
4. Pugh EN, Jr., Lamb TD. Cyclic GMP and calcium: the internal messengers of excitation and adaptation in vertebrate photoreceptors. Vision Res 1990; 30:1923-48.
5. Yau KW. Phototransduction mechanism in retinal rods and cones. The Friedenwald Lecture. Invest Ophthalmol Vis Sci 1994; 35:9-32.
6. Yau KW, Nakatani K. Light-induced reduction of cytoplasmic free calcium in retinal rod outer segment. Nature 1985; 313:579-82.
7. Schnetkamp PP, Basu DK, Szerencsei RT. Na^+-Ca^{2+} exchange in bovine rod outer segments requires and transports K^+. Am J Physiol 1989; 257:C153-7.
8. Pugh EN Jr, Nikonov S, Lamb TD. Molecular mechanisms of vertebrate photoreceptor light adaptation. Curr Opin Neurobiol 1999; 9:410-8.
9. He W, Cowan CW, Wensel TG. RGS9, a GTPase accelerator for phototransduction. Neuron 1998; 20:95-102.
10. Koch KW, Stryer L. Highly cooperative feedback control of retinal rod guanylate cyclase by calcium ions. Nature 1988; 334:64-6.
11. Koutalos Y,Yau KW. Regulation of sensitivity in vertebrate rod photoreceptors by calcium. Trends Neurosci 1996; 19:73-81.
12. Polans A, Baehr W, Palczewski K. Turned on by Ca^{2+}! The physiology and pathology of $Ca(^{2+})$-binding proteins in the retina. Trends Neurosci 1996; 19:547-54.
13. Palczewski K, Polans AS, Baehr W et al. $Ca(^{2+})$-binding proteins in the retina: structure, function, and the etiology of human visual diseases. Bioessays 2000; 22:337-50.
14. Gorczyca WA, Polans AS, Surgucheva IG et al. Guanylyl cyclase activating protein. A calcium-sensitive regulator of phototransduction. J Biol Chem 1995; 270:22029-36.
15. Dizhoor AM, Olshevskaya EV, Henzel WJ et al. Cloning, sequencing, and expression of a 24-kDa $Ca(^{2+})$-binding protein activating photoreceptor guanylyl cyclase. J Biol Chem 1995; 270:25200-6.
16. Mendez A, Burns ME, Sokal I et al. Role of guanylate cyclase-activating proteins (GCAPs) in setting the flash sensitivity of rod photoreceptors. Proc Natl Acad Sci USA 2001; 98:9948-53.
17. Kawamura S. Rhodopsin phosphorylation as a mechanism of cyclic GMP phosphodiesterase regulation by S-modulin. Nature 1993; 362:855-7.
18. Hsu YT, Molday RS. Modulation of the cGMP-gated channel of rod photoreceptor cells by calmodulin [published erratum appears in Nature 1993 Sep 16;365(6443):279] [see comments]. Nature 1993; 361:76-9.
19. Gordon SE, Downing-Park J, Zimmerman AL. Modulation of the cGMP-gated ion channel in frog rods by calmodulin and an endogenous inhibitory factor. J Physiol 1995; 486:533-46.
20. Sagoo MS, Lagnado L. The action of cytoplasmic calcium on the cGMP-activated channel in salamander rod photoreceptors. J Physiol 1996; 497:309-19.
21. Kaupp UB. Family of cyclic nucleotide gated ion channels. Curr Opin Neurobiol 1995; 5:434-42.
22. Zimmerman AL. Cyclic nucleotide gated channels. Curr Opin Neurobiol 1995; 5:296-303.
23. Zagotta WN, Siegelbaum SA. Structure and function of cyclic nucleotide-gated channels. Annu Rev Neurosci 1996; 19:235-63.
24. Biel M, Zong X, Hofmann F. Cyclic nucleotide gated channels. Adv Second Messenger Phosphoprotein Res 1999; 33:231-50.
25. Molday RS, Kaupp UB. Ion Channels of Vertebrate Photoreceptors. In: Handbook of Biological Physics, Vol. 3. Amsterdam: Elsevier, 2000.

26. Kaupp UB, Niidome T, Tanabe T et al. Primary structure and functional expression from complementary DNA of the rod photoreceptor cyclic GMP-gated channel. Nature 1989; 342:762-6.
27. Chen TY, Peng YW, Dhallan RS et al. A new subunit of the cyclic nucleotide-gated cation channel in retinal rods. Nature 1993; 362:764-7.
28. Korschen HG, Illing M, Seifert R et al. A 240 kDa protein represents the complete beta subunit of the cyclic nucleotide-gated channel from rod photoreceptor. Neuron 1995; 15:627-36.
29. Bradley J, Frings S, Yau KW et al. Nomenclature for ion channel subunits. Science 2001; 294:2095-6.
30. Dhallan RS, Macke JP, Eddy RL et al. Human rod photoreceptor cGMP-gated channel: amino acid sequence, gene structure, and functional expression. J Neurosci 1992; 12:3248-56.
31. Chen TY, Illing M, Molday LL et al. Subunit 2 (or beta) of retinal rod cGMP-gated cation channel is a component of the 240-kDa channel-associated protein and mediates $Ca(^{2+})$-calmodulin modulation. Proc Natl Acad Sci USA 1994; 91:11757-61.
32. Wohlfart P, Haase W, Molday RS et al. Antibodies against synthetic peptides used to determine the topology and site of glycosylation of the cGMP-gated channel from bovine rod photoreceptors. J Biol Chem 1992; 267:644-8.
33. Root MJ, MacKinnon R. Identification of an external divalent cation-binding site in the pore of a cGMP-activated channel. Neuron 1993; 11:459-66.
34. Eismann E, Muller F, Heinemann SH et al. A single negative charge within the pore region of a cGMP-gated channel controls rectification, Ca^{2+} blockage, and ionic selectivity. Proc Natl Acad Sci USA 1994; 91:1109-13.
35. Molday RS, Molday LL, Dose A et al. The cGMP-gated channel of the rod photoreceptor cell characterization and orientation of the amino terminus. J Biol Chem 1991; 266:21917-22.
36. Colville CA, Molday RS. Primary structure and expression of the human beta-subunit and related proteins of the rod photoreceptor cGMP-gated channel. J Biol Chem 1996; 271:32968-74.
37. Sugimoto Y, Yatsunami K, Tsujimoto M et al. The amino acid sequence of a glutamic acid-rich protein from bovine retina as deduced from the cDNA sequence. Proc Natl Acad Sci USA 1991; 88:3116-9.
38. Korschen HG, Beyermann M, Muller F et al. Interaction of glutamic-acid-rich proteins with the cGMP signalling pathway in rod photoreceptors. Nature 1999; 400:761-6.
39. Poetsch A, Molday LL, Molday RS. The cGMP-gated channel and related glutamic acid-rich proteins interact with peripherin-2 at the rim region of rod photoreceptor disc membranes. J Biol Chem 2001; 276:48009-16.
40. He Y, Ruiz M, Karpen JW. Constraining the subunit order of rod cyclic nucleotide-gated channels reveals a diagonal arrangement of like subunits. Proc Natl Acad Sci USA 2000; 97:895-900.
41. Doyle DA, Morais Cabral J, Pfuetzner RA et al. The structure of the potassium channel: molecular basis of K^+ conduction and selectivity. Science 1998; 280:69-77.
42. Ruiz ML, Karpen JW. Single cyclic nucleotide-gated channels locked in different ligand-bound states. Nature 1997; 389:389-92.
43. Hsu YT, Molday RS. Interaction of calmodulin with the cyclic GMP-gated channel of rod photoreceptor cells. Modulation of activity, affinity purification, and localization. J Biol Chem 1994; 269:29765-70.
44. Bauer PJ. Cyclic GMP-gated channels of bovine rod photoreceptors: affinity, density and stoichiometry of $Ca(^{2+})$-calmodulin binding sites. J Physiol 1996; 494:675-85.
45. Molday RS. Calmodulin regulation of cyclic-nucleotide-gated channels. Curr Opin Neurobiol 1996; 6:445-52.
46. Liu M, Chen TY, Ahamed B et al. Calcium-calmodulin modulation of the olfactory cyclic nucleotide-gated cation channel. Science 1994; 266:1348-51.

47. Weitz D, Zoche M, Muller F et al. Calmodulin controls the rod photoreceptor CNG channel through an unconventional binding site in the N-terminus of the beta-subunit. Embo J 1998; 17:2273-84.
48. Grunwald ME, Yu WP, Yu HH et al. Identification of a domain on the beta-subunit of the rod cGMP-gated cation channel that mediates inhibition by calcium-calmodulin. JJ Biol Chem 1998; 273:9148-57.
49. Cook NJ, Hanke W et al. Identification, purification, and functional reconstitution of the cyclic GMP-dependent channel from rod photoreceptors. Proc Natl Acad Sci USA 1987; 84:585-9.
50. Kuhn H, Hall SW et al. Light-induced binding of 48-kDa protein to photoreceptor membranes is highly enhanced by phosphorylation of rhodopsin. FEBS Lett 1984; 176:473-8.
51. Gordon SE, Brautigan DL, Zimmerman AL. Protein phosphatases modulate the apparent agonist affinity of the light-regulated ion channel in retinal rods. Neuron 1992; 9:739-48.
52. Molokanova E, Maddox F, Luetje CW et al. Activity-dependent modulation of rod photoreceptor cyclic nucleotide- gated channels mediated by phosphorylation of a specific tyrosine residue. J Neurosci 1999; 19:4786-95.
53. Molokanova E, Trivedi B, Savchenko A et al. Modulation of rod photoreceptor cyclic nucleotide-gated channels by tyrosine phosphorylation. J Neurosci 1997; 17:9068-76.
54. Savchenko A, Kraft TW, Molokanova E et al. Growth factors regulate phototransduction in retinal rods by modulating cyclic nucleotide-gated channels through dephosphorylation of a specific tyrosine residue. Proc Natl Acad Sci USA 2001; 98:5880-5.
55. Molday RS, Molday LL. Molecular properties of the cGMP-gated channel of rod photoreceptors. Vision Res 1998; 38:1315-23.
56. Schwarzer A, Schauf H, Bauer PJ. Binding of the cGMP-gated channel to the Na/Ca-K exchanger in rod photoreceptors. J Biol Chem 2000; 275:13448-54.
57. Bauer PJ, Drechsler M. Association of cyclic GMP-gated channels and $Na(^{+})$-$Ca(^{2+})$-K^{+} exchangers in bovine retinal rod outer segment plasma membranes. J Physiol 1992; 451:109-31.
58. Schwarzer A, Kim TSY, Hagen V et al. The Na/Ca-K exchanger of rod photoreceptor exists as dimer in the plasma membrane. Biochemistry 1997; 36:13667-76.

Ca^{2+}-CHANNELS IN THE RPE

Rita Rosenthal and Olaf Strauß

SUMMARY

The retinal pigment epithelium closely interacts with photoreceptors and helps to maintain the activity of photoreceptors. Investigations using patch-clamp techniques on cultured or freshly isolated retinal pigment epithelial cells from various species demonstrated the expression of voltage-dependent Ca^{2+} channels with characteristics of L-type channels. Since retinal pigment epithelial cells rarely display changes of the membrane potential which lead to the activation of these Ca^{2+} channels, their function seemed to be unclear. Recent findings shed light onto the possible role of these Ca^{2+} channels. First of all, the subtype of these ion channels could be identified as neuroendocrine subtype of L-type channels. Recent studies demonstrated that the neuroendocrine subtype of L-type channels is regulated by serine/threonine kinases and protein tyrosine kinases. These phosphorylation-dependent regulatory mechanisms lead to Ca^{2+} fluxes into the cell which are independent of changes in the membrane potential and induced by a shift in the voltage-dependence of these ion channels. The regulation modality implied that L-type Ca^{2+} channels play an important role in signal transduction pathways which are important for a communication between retinal pigment epithelium and photoreceptors. L-type Ca^{2+} channels in the retinal pigment epithelium seem to be involved in the regulation of secretion of various factors, in growth factor-dependent intracellular signalling and in the regulation of the phagocytosis of photoreceptor outer membranes. Thus, voltage-dependent Ca^{2+} channels in the retinal pigment epithelium are of importance for the function of photoreceptors.

Institut für Klinische Physiologie, Universitätsklinikum Benjamin Franklin, Freie Universität Berlin, 12200 Berlin, Germany.

THE RETINAL PIGMENT EPITHELIUM

The retinal pigment epithelium is a monolayer of pigmented cells which faces with its apical membrane the photoreceptor outer segments. With its basolateral cell membrane, together with Bruch's membrane, it is connected to the blood supply by the choriocapillaris.[1,2] The retinal pigment epithelium is part of the retina/blood barrier. Further it has multiple functions contributing to the maintenance of excitability of photoreceptors. As a layer of pigmented cells it helps to prevent photooxidation and eliminates adventitious light in the inner bulbus. The retinal pigment epithelium transports nutrients and ions between the blood site and the subretinal space. This nutrient transport is essential for the supply of glucose and vitamin A to the photoreceptors. The ion transport helps to control the subretinal ion homeostasis and to eliminate end products of the photoreceptors metabolism such as lactic acid. In the subretinal space, the retinal pigment epithelium can compensate fast changes in the ion concentration which are induced by the activity of the photoreceptors.[1] This glia-comparable function can be monitored in the signals of the electroretinogram which originate in the electric activity of the retinal pigment epithelium. Another important function of the retinal pigment epithelium is the phagocytosis of photoreceptor outer segments. This process, controlled by the circadian clock, serves for the renewal of the photoreceptor outer segments after bleaching of photopigments.[3-6] In addition to these functions, the retinal pigment epithelium displays secretory activity.[7] In the healthy eye it secretes a variety of growth factors which help to maintain the structural integrity of photoreceptors and to protect them against damaging influences such as light.[7-10] It could be demonstrated that the failure of one of these functions leads to retinal degeneration in consequence of an impaired interaction between retinal pigment epithelium and photoreceptors.[11-21]

Ca^{2+} CHANNELS IN THE RPE

Like many other tissues, the retinal pigment epithelium was investigated using patch-clamp techniques. Under intra- and extracellular K^+-free conditions, using high extracellular concentrations of Ba^{2+} as charge carrier, several groups could measure voltage-dependent whole-cell inward currents in freshly isolated and cultured retinal pigment epithelial cells.[22-25] These currents displayed identical properties. They activate at potentials more positive than -30 mV, reached maximal current amplitudes at +10 mV, and showed a voltage-dependent slow inactivation (Fig. 1B, C). The current amplitudes were higher and the inactivation was much slower when using Ba^{2+} as charge carrier compared to currents with Ca^{2+} as charge carrier. Furthermore, these currents were sensitive to dihydropyridine compounds such as nifedipine, Bay K8644 or nimodipine (Fig.1 D). These are typical characteristics of L-type Ca^{2+} channels. Other types of Ca^{2+} channels could not be confirmed in freshly isolated cells and were only found in cultured cells in long-time cultures.

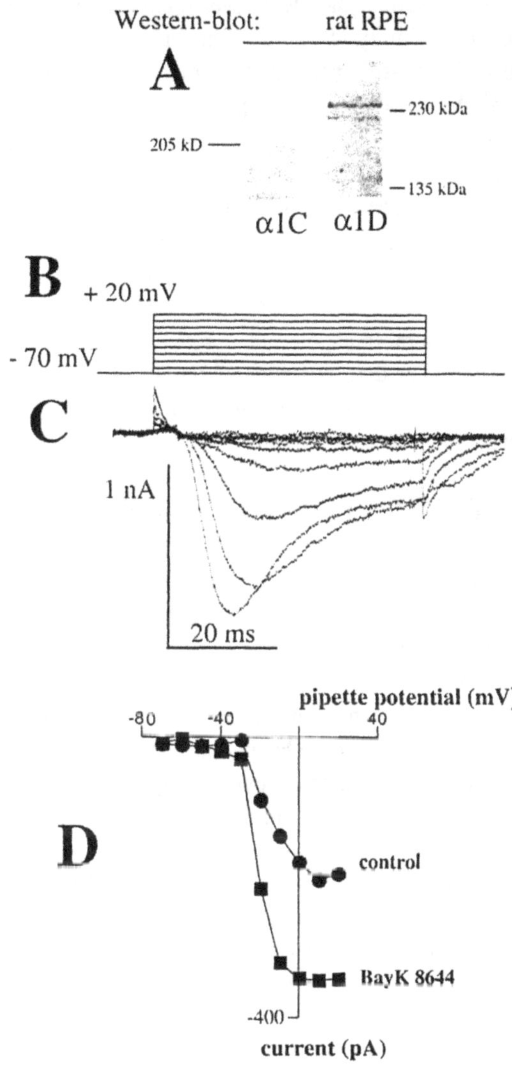

Figure 1. Identification of L-type channels in rat RPE cells. A: Western Blot of membrane proteins of freshly isolated RPE, staining with anti-α1D-subunit and anti-a1C-subunit antibodies.[29] Only with anti-α1D antibodies the 240 kD band characteristic for L-type channels could be detected. B,C,D: Identification of Ba^{2+} currents through L-type channels.[40] B: Pattern of electrical stimulation to induce inward currents. The membrane potential was clamped to -70 mV and the cell was stimulated by 9 voltage steps of 50 ms duration and +10 mV increasing amplitude. C: Inward currents induced by the electrical stimulation that appear in the presence of extracellular Ba^{2+} (10^{-2} M) under extra- and intracellular K^+-free conditions. D: Ba^{2+} currents are dihydropyridine sensitive: Current-voltage plot under control conditions and in the presence of 5 x 10^{-6} M BayK8644. The maximal current amplitudes were plotted against the test potentials of the stimulation protocol shown in panel B.

Ion channels which reflect the electrophysiological characteristics of L-type channels belong to a gene family of four different subtypes which are distinguished by their α1-subunits. The α1-subunit builds the ion channel pore and determines the electrophysiological characteristics of L-type channels.[26,27] Until now, four different subtypes of L-type channels have been described: the skeletal subtype (α1s or $Ca_V1.1$), the cardiac subtype (α1C or $Ca_V1.2$), the neuroendocrine subtype (α1D or $Ca_V1.3$), and the retinal subtype (α1F or $Ca_V1.4$).[28] Using Western blot analysis, it could be shown that retinal pigment epithelial cells exclusively express the neuroendocrine subtype (α1D or $Ca_V1.3$) of L-type channels (Fig. 1A).[29] However, the precise role of these Ca^{2+} channels for the function of the retinal pigment epithelium seems unclear, since the membrane potential of retinal pigment epithelial cells rarely reaches values more positive than -30 mV.[30-33] On the other hand, this ion channel subtype is known to be involved in the regulation of hormone/growth factor secretion and gene expression.[34-39]

THE REGULATION OF L-TYPE CHANNELS

In order to gain information about the role of L-type channels for the function of retinal pigment epithelial cells, the regulation of these channels was investigated. Several studies showed the phosphorylation-dependent regulation of L-type channels which implies their involvement in Ca^{2+} second messenger systems.[24] In resting cells, the L-type channels are independently activated by protein kinase C and by tyrosine kinase of the src-subtype. The regulation by tyrosine kinase was demonstrated by intracellular application of the cytosolic tyrosine kinase $pp60^{c\text{-}src}$ and by immunoprecipitation studies.[29] In immunoprecipitation studies the coprecipitation of [29]1D-subunits and $pp60^{c\text{-}src}$ point to a close interaction of tyrosine kinase and these Ca^{2+} channels. The regulation of the L-type channel seems to be under the control of the protein kinase C. After stimulation of protein kinase C the tyrosine kinase $pp60^{c\text{-}src}$ appears as an inhibitor of the L-type channel. Thus, when both L-type channel activating kinases are stimulated at the same time, an influx of Ca^{2+} into the cell was limited by the inhibitory effect of the src-like tyrosine kinase. Other regulators of L-type channels are receptor tyrosine kinases. In a recent study it could be shown that the growth factor receptor FGFR2 activates L-type Ca^{2+} channels by a close interaction of the receptor with the α1-subunit of the Ca^{2+} channel.[40] This is a regulation mechanism which is independent of the activation of $pp60^{c\text{-}src}$. Thus, the L-type Ca^{2+} channels in the retinal pigment epithelium are able to build signalling complexes with different target proteins.

INVOLVEMENT OF L-TYPE CHANNELS IN INTRACELLULAR SIGNALLING SYSTEMS

With these features of ion channel regulation, the L-type Ca^{2+} channels are involved in the activation of intracellular second messenger pathways by providing an influx pathway for extracellular Ca^{2+} into the cell. Activation of the inositolphosphate/Ca^{2+} second messenger system led to an increase in intracellular free Ca^{2+} by an influx of Ca^{2+} through L-type channels. In this signal cascade the activation of L-type channels was dependent on the depletion of cytosolic Ca^{2+} stores and on the activation of tyrosine kinase. Application of bFGF to retinal pigment epithelial cells also led to an increase in intracellular free Ca^{2+} by an influx of Ca^{2+} through L-type channels (Fig. 2).[25] This process was also dependent on the activation of tyrosine kinase. However, these changes in intracellular free Ca^{2+} mediated by an influx of Ca^{2+} through L-type channels were induced without changing the membrane potential. For example, Ca^{2+} influx through L-type channels was achieved by stimulation of the inositolphosphate/Ca^{2+} second messenger system while the membrane potential was clamped to the resting potential of retinal pigment epithelial cells. An explanation for the fact that L-type channels can mediate an influx of Ca^{2+} into cells without changing the membrane potential was found when examining the voltage-dependence of the L-type channels during this process. Both stimulation of the bFGF receptor FGFR2 or increase in intracellular inositol-1,4,5-trisphosphate shifted the activation curve towards more negative values. In addition, in some cases also the inactivation curve was shifted towards more positive values. The result of these shifts in the voltage-dependence is a higher number of spontaneous open L-type channels at the resting potential of retinal pigment epithelial cells. The subsequent enhancement in the membrane conductance for Ca^{2+} is able to generate Ca^{2+} influx into the cell which increases the concentration of free cytosolic Ca^{2+} as second messenger. Similar effects of L-type Ca^{2+} channel activation have been observed in other cell types.

L-TYPE CHANNELS AND RETINAL DEGENERATION

The importance of L-type channels for the function of the retinal pigment epithelium becomes obvious in an animal model for human autosomal recessive retinitis pigmentosa, the Royal College of Surgeons rat (RCS rat). The retinal degeneration in the RCS rat is caused by a mutation in a gene expressed in the retinal pigment epithelium.[41] The consequence of the gene defect is the absence of a receptor tyrosine kinase (MerTK). This leads to an impaired interaction between retinal pigment epithelium and the photoreceptors because the retinal pigment epithelium is unable to phagocytose shed photoreceptor outer segments.[42,43] Investigation of the membrane conductance of retinal pigment epithelial cells from RCS rats revealed that these cells express an increased activity of L-type channels.[25,44] The current density of L-type channels was approximately two times higher in retinal pigment epithelial cells from RCS rats compared to cells from control rats. The increased current is due to a changed regulation of L-type channels by tyrosine kinase. In retinal pigment epithelial cells from RCS rats the cytosolic tyrosine kinase $pp60^{c\text{-}src}$ acts as an inhibitor of the L-type Ca^{2+} channel. Furthermore, the cells are not able to recruit L-type channels to generate an influx of Ca^{2+} into the cell as a second messenger in response to stimulation by bFGF. The connection between the mutation on

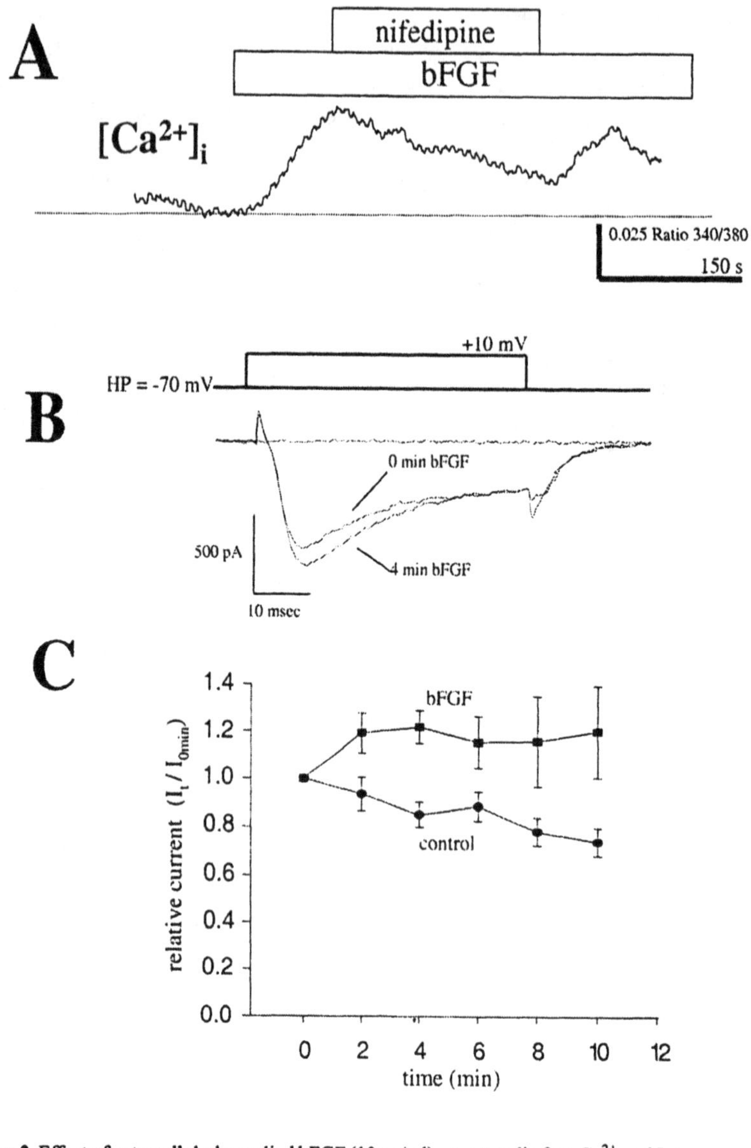

Figure 2. Effect of extracellularly applied bFGF (10 ng/ml) on cytosolic free Ca^{2+} and L-type currents in rat RPE cells. A: Effect of nifedipine (10^{-6} M) on the bFGF induced rise in cytosolic free Ca^{2+}. Changes of cytosolic free Ca^{2+} are depicted as the fluorescence ratio of the Ca^{2+} indicator fura-2 induced by the excitation wavelengths 340 and 380 nm.[25] Application of bFGF led to an increase in cytosolic free Ca^{2+} which could be reduced by block of L-type channels. B: Effect of bFGF on L-type currents. Currents (induced by voltage-steps from -70 mV to +10 mV) measured directly after establishing the perforated-patch configuration (0 min) and 4 min later (4 min) in the presence of bFGF. Cells were maintained serum-free for 24 h prior to patch-clamp recordings. C: Summary of experiments investigating the effect of bFGF. Maximal peak currents (voltage-step from -70 mV to +10 mV) were measured every two minutes and plotted against the experiment time. Currents were normalized to the value which was measured directly before application of bFGF. The upper trace was determined in the presence of bFGF (n = 4) and the lower in the absence of bFGF (n = 5).[40] bFGF leads to activation of L-type Ca^{2+} channels in RPE cells.

the MerTK gene and the impaired regulation of L-type channels in the retinal pigment epithelium of the RCS rats is not known.

THE PHYSIOLOGICAL ROLE OF L-TYPE CHANNELS IN THE RETINAL PIGMENT EPITHELIUM

In the previous Chapters the following conclusions were made. The retinal pigment epithelium expresses the neuroendocrine subtype of L-type Ca^{2+} channels which is activated by different types of protein kinases. The regulation of these Ca^{2+} channels enables their participation in second-messenger systems by providing an influx pathway for Ca^{2+} into the cell. On the basis of these conclusions several roles of this channels for the function of retinal pigment epithelial cells can be suggested.

Some aspects of the physiological role of L-type channels for the retinal pigment epithelium can be deduced from the functional role of neuroendocrine L-type channels in other tissues. This type of L-type channels has been described for the first time in the insulin secreting cells of the pancreas.[45,46] Here, the Ca^{2+} channel regulates the insulin secretion. Activation of this channel by depolarization leads to an increase in the cytosolic free Ca^{2+} which represents the trigger to release insulin from the cell. Since the retinal pigment epithelium secretes a variety of growth factors,[9,10,47-51] one possible role of the L-type channels is the regulation of this secretory activity. The secretion by the retinal pigment epithelium is often stimulated by other growth factors known to act via activation of tyrosine kinases.[52,53] Therefore, the tyrosine kinase activated L-type channels represent an ideal target for the stimulation of secretory activity by growth factors.

In neurons, the neuroendocrine type of Ca^{2+} channel serves a different function. These ion channels are located on the dendrites and on the cell body of neurons. Here the channels are involved in a neuronal activity-dependent control of gene expression.[34,35,37-39] A high frequency of action potentials increases the open probability of L-type channels followed by a stimulation of Ca^{2+} influx into the cell. This activates a signal cascade resulting in an increased expression of immediate early genes such as c-fos. The effect is mediated by Ca^{2+} influx-dependent phosphorylation of cAMP-response-element binding protein (CREB).

In addition, L-type channels are integrated into the inositolphosphate/Ca^{2+} second messenger system. In this signalling pathway an increase in intracellular inositol-1,4,5-trisphosphate results in an influx of extracellular Ca^{2+} into the cell via tyrosine kinase-dependently activated L-type channels. Thus, the L-type channels are also activated in response to growth factors. In the case of bFGF, the L-type channels are directly activated by the growth factor receptor. For these cases of L-type channel activation the result might be a change in the regulation of gene expression which is necessary to induce growth factor-dependent differentiation processes. For example, the tyrosine kinase-dependent activation of L-type channels in response to stimulation by the insulin-like growth factor 1 has been shown in neurons.[54] The change in the expression of transcription factors in consequence to the stimulation of L-type Ca^{2+} channels has been demonstrated as a function of L-type channels in neurons. Thus, L-type channels in the retinal pigment epithelium might transduce growth factor-dependent influences on differentiation and cell behaviour. The changed regulation of L-type channels in the retinal pigment epithelial cells from RCS rats which are unable to phagocytose shed sections of photoreceptor outer segments also indicates their involvement in the regulation of phagocytosis.

In summary, L-type channels in retinal pigment epithelial cells are involved in second messenger systems stimulated in response to growth factors. They possibly regulate the secretory activity of retinal pigment epithelial cells. Furthermore, they may participate in processes concerning gene expression and phagocytosis. Thus, the L-type channels are involved in functions of the retinal pigment epithelium which are of importance for the maintenance of retinal function. In the healthy eye, a basic secretion of growth factors by the retinal pigment epithelium protects retinal cells against light damage. This secretion is controlled by factors which are secreted by the neuronal retina. Thus, L-type channels mediate the growth factor-dependent communication between the neuronal retina and the retinal pigment epithelium. These Ca^{2+} channels are involved in the control of the differentiation and secretory activity of the retinal pigment epithelium influenced by the neuronal retina.

Therefore, the L-type Ca^{2+} channels might also play a role in the etiology of retinal degenerations based on an impaired interaction between the retinal pigment epithelium and the neuronal retina. One example is the RCS rat where the retinal pigment epithelium is unable to phagocytose shed photoreceptor outer segments. Other examples are retinal diseases accompanied by cell proliferation such as age-related macular degeneration or proliferative vitreoretinopathy. In these diseases, the proliferation is due, in part, to a changed growth factor secretion by the retinal pigment epithelium.

REFERENCES

1. Steinberg RH. Interactions between the retinal pigment epithelium and the neural retina. Doc Ophthalmol 1985; 60(4):327-46.
2. Bok D. Retinal photoreceptor-pigment epithelium interactions. Invest Ophthalmol Vis Sci 1985; 26(12):1659-94.
3. Young RW, Bok D. Participation of the retinal pigment epithelium in the rod outer segment renewal process. J Cell Biochem 1969; 42:392-403.
4. LaVail MM. Circadian nature of rod outer segment disc shedding in the rat. Invest Ophthalmol Vis Sci 1980; 19:407-411.
5. Timmers AM, Van Groningen-Luyben DA, de Grip WJ. Uptake and isomerization of all-trans retinal by isolated bovine retinal pigment epithelial cells: further clues to the visual cycle. Exp Eye Res 1991; 52:129-138.
6. Gordon WC, Rodriguez de Turco EB, Bazan NG. Retinal pigment epithelial cells play a central role in the conservation of docosahexanoic acid by photoreceptor cells after shedding and phagocytosis. Curr Eye Res 1992; 11:73-83.
7. Tanihara H, Inatani M, Honda Y. Growth factors and their receptors in the retina and pigment epithelium. Prog Ret Eye Res 1997; 16:271-301.
8. Tanihara H, Yoshida M, Matsumoto M et al. Identification of transforming growth factor-beta expressed in cultured human retinal pigment epithelial cells. Invest Ophthalmol Vis Sci 1993; 34(2):413-419.

9. Bost LM, Aotaki-Keen AE, Hjelmeland LM. Cellular adhesion regulates bFGF gene expression in human retinal pigment epithelial cells. Exp Eye Res 1994; 58:545-552.
10. Wen R, Song Y, Cheng T, et al. Injury-induced upregulation of bFGF and CNTF mRNAs in the rat retina. J Neurosci 1995; 15:7377-7385.
11. Edwards RB, Szamier RB. Defective phagocytosis of isolated rod outer segments by RCS rat retinal pigment epithelium in culture. Science 1977; 197(4307):1001-3.
12. Goldman AI, PJ OB. Phagocytosis in the retinal pigment epithelium of the RCS rat. Science 1978; 201(4360):1023-5.
13. Young RW. Pathophysiology of age-related macular degeneration. Surv Ophthalmol 1987; 31(5):291-306.
14. Amin R, Puklin JE, Frank RN. Growth factor localization in choroidal neovascular membranes of age-related macular degeneration. Invest Ophthalmol Vis Sci 1994; 35(8):3178-88.
15. Reddy VM, Zamora RL, Kaplan HJ. Distribution of growth factors in subfoveal neovascular membranes in age-related macular degeneration and presumed ocular histoplasmosis syndrome. Am J Ophthalmol 1995; 120(3):291-301.
16. Frank RN. Growth factors in age-related macular degeneration: pathogenic and therapeutic implications. Ophthalmic Res 1997; 29(5):341-53.
17. Kliffen M, Sharma HS, Mooy CM et al. Increased expression of angiogenic growth factors in age-related maculopathy. Br J Ophthalmol 1997; 81(2):154-62.
18. Redmond TM, Yu S, Lee E et al. RPEG5 is necessary for production of 11-cis-vitamin A in the retinal visual cycle. Nat Genet 1998; 20(4):344-51.
19. Redmond TM, Hamel CP. Genetic analysis of RPE65: from human disease to mouse model. Methods Enzymol 2000; 316:705-24.
20. Bosl MR, Stein V, Hubner C et al. Male germ cells and photoreceptors, both dependent on close cell-cell interactions, degenerate upon ClC-2 Cl(-) channel disruption. Embo J 2001; 20(6):1289-1299.
21. Seeliger MW, Grimm C, Stahlberg F et al. New views on RPE65 deficiency: the rod system is the source of vision in a mouse model of Leber congenital amaurosis. Nat Genet 2001; 29(1):70-4.
22. Ueda Y, Steinberg RH. Voltage-operated calcium channels in fresh and cultured rat retinal pigment epithelial cells. Invest Ophthalmol Vis Sci 1993; 34(12):3408-18.
23. Strauß O, Wienrich M. Ca^{2+}-conductances in cultured rat retinal pigment epithelial cells. J Cell Physiol 1994; 160:89-96.
24. Strauß O, Mergler S, Wiederholt M. Regulation of L-type calcium channels by tyrosine kinase and protein kinase C in cultured rat and human retinal pigment epithelial cells. FASEB J 1997; 11:859-867.
25. Mergler S, Steinhausen K, Wiederholt M, Strauß O. Altered regulation of L-type channels by protein kinase C and protein tyrosine kinases as a pathophysiologic effect in retinal degeneration. FASEB J. 1998; 12:1125-1134.
26. Perez-Reyes E, Schneider T. Molecular biology of calcium channels. Kidney International 1995; 48:1111-1124.
27. Catterall WA. Structure and function of neuronal Ca^{2+} channels and their role in neurotransmitter release. Cell Calcium 1998; 24(5-6):307-23.
28. Catterall WA. Structure and regulation of voltage-gated Ca^{2+} channels. Annu Rev Cell Dev Biol 2000; 16:521-55.
29. Strauss O, Buss F, Rosenthal R et al. Activation of neuroendocrine L-type channels (alpha1D subunits) in retinal pigment epithelial cells and brain neurons by $pp60^{c-src}$. Biochem Biophys Res Commun. 2000; 270(3):806-10.
30. Steinberg RH, Linsenmeier RA, Griff ER. Retinal pigment epithelial cell contributions to the electroretinogram and electrooculogram. In: Osborne N, Chader GJ, eds. Progress in Retinal and Eye Research. Vol 4. Oxford: Pergamon Press, 1985:33-66.

31. LaCour MD, Lund-Andersen H, Zeuthen T. Potassium transport of the frog retinal pigment epithelium: autoregulation of potassium activity in the subretinal space. J Physiol 1986; 375:461-479.
32. Gallemore RP, Steinberg RH. Light-evoked modulation of basolateral membrane Cl- conductance in chick retinal pigment epithelium: the light peak and fast oscillation. J Neurophysiol 1993; 70(4):1669-80.
33. LaCour MD. Ion transport in the retinal pigment epithelium. A study with double barrelled ion-selective microelectrodes. Acta Ophthalmol Suppl 1993, 209:1-32.
34. Murphy TH, Worley PF, Baraban JM. L-type voltage-sensitive calcium channels mediate synaptic activation of immediate early genes. Neuron 1991; 7(4):625-35.
35. Bading H, Ginty DD, Greenberg ME. Regulation of gene expression in hippocampal neurons by distinct calcium signaling pathways. Science 1993; 260(5105):181-6.
36. Ashcroft FM, Proks P, Smith PA et al. Stimulus-secretion coupling in pancreatic beta cells. J Cell Biochem 1994; 55:54-65.
37. Rosen LB, Ginty DD, Greenberg ME. Calcium regulation of gene expression. Adv in Second Messenger Phosphoprotein Res 1995; 30:225-53.
38. Bito H, Deisseroth K, Tsien RW. CREB phosphorylation and dephosphorylation: a Ca^{2+}- and stimulus duration-dependent switch for hippocampal gene expression. Cell 1996; 87(7):1203-14.
39. Deisseroth K, Heist EK, Tsien RW. Translocation of calmodulin to the nucleus supports CREB phosphorylation in hippocampal neurons. Nature 1998; 392(6672):198-202.
40. Rosenthal R, Thieme H, Strauss O. Fibroblast growth factor receptor 2 (FGFR2) in brain neurons and retinal pigment epithelial cells act via stimulation of neuroendocrine L-type channels (Ca(v)1.3). FASEB J 2001; 15(6):970-7.
41. D' Cruz PM, Yasumura D, Weir J et al. Mutation of the receptor tyrosine kinase gene Mertk in the retinal dystrophic RCS rat. Mol Genet 2000; 9(4):645-51.
42. Herron WL, Riegel BW, Myers OE et al. Retinal dystrophy in the rat—a pigment epithelial disease. Invest Ophthal/mol Vis Sci 1969; 8:595-604.
43. Bok D, Hall MO. The role of the pigment epithelium in the etiology of inherited retinal dystrophy in the rat. J Cell Biol 1971; 49:664-682.
44. Strauss O, Wienrich M. Cultured retinal pigment epithelial cells from RCS rats express an increased calcium conductance compared with cells from non-dystrophic rats. Pfl""gers Arch 1993; 425(1-2):68-76.
45. Seino S, Chen L, Seino M et al. Cloning of the alpha 1 subunit of a voltage-dependent calcium channel expressed in pancreatic beta cells. Proc Natl Acad Sci USA 1992; 89(2):584-588.
46. Yaney GC, Wheeler MB, Wei X et al. Cloning of a novel alpha 1-subunit of the voltage-dependent calcium channel from the beta-cell. Mol Endocrinol 1992; 6(12):2143-52.
47. Bost LM, Aotaki-Keen AE, Hjelmeland LM. Coexpression of FGF-5 and bFGF by the retinal pigment epithelium in vitro. Exp Eye Res 1992; 55:727-734.
48. Yoshida M, Tanihara H, Yoshimura N. Platelet-derived growth factor gene expression in cultured retinal pigment epithelial cells. Biochem Biophys Res Commun 1992; 189:66-71.
49. Yoshimura N, Kuriyama S, Iwaki M et al. Growth factor dependent phosphorylation of membrane proteins in cultured human retinal pigment epithelial cells. Curr Eye Res 1992; 11:997-1004.
50. Steele FR, Chader GJ, Johnson LV et al. Pigment epithelium-derived factor: Neurotrophic activity and identification as a member of the serine protease inhibitor gene family. Proc Natl Acad Sci USA 1993; 90(4):1526-1530.
51. Shima DT, Adamis AP, Ferrara N et al. Hypoxic induction of endothelial cell growth factors in retinal cells: identification and characterization of vascular endothelial growth factor (VEGF) as the mitogen. Mol Med 1995; 1:182-193.
52. Ullrich A, Schlessinger J. Signal transduction by receptors with tyrosine kinase activity. Cell 1990; 61(2):203-12.

53. Fantl WJ, Johnson DE, Williams LT. Signalling by receptor tyrosine kinases. Annu Rev Biochem 1993; 62:453-81.
54. Bence-Hanulec KK, Marshall J, Blair LA. Potentiation of neuronal L calcium channels by IGF-1 requires phosphorylation of the alpha1 subunit on a specific tyrosine residue. Neuron 2000 Jul; 27(1):121-31.

THE RETINAL ROD AND CONE Na^+/Ca^{2+}-K^+ EXCHANGERS

Clemens F.M. Prinsen,[1] Conan B. Cooper,[1]
Robert T. Szerencsei,[1] Sabita K. Murthy,[2]
Douglas J. Demetrick[2] and Paul P.M. Schnetkamp[1]

ABSTRACT

The past few years has seen significant progress in our understanding of the retinal rod and cone Na^+/Ca^{2+}-K^+ exchanger (NCKX) genes. The human rod and cone NCKX genes were localized to chromosomes 15q22 and 9p22, respectively. In situ hybridization localized the rod and cone NCKX transcripts in both human and chicken retinas: rod NCKX transcripts were found only in the inner segments of rods, whereas cone NCKX transcripts were found in a subset of retinal ganglion cells as well as in the inner segments of cones. We identified two sets of putative transmembrane spanning segments (TM's) as the only sequence elements strongly conserved between the rod and cone NCKX cDNAs, as well as between mammalian NCKX cDNAs and NCKX cDNAs cloned from lower organisms (*C. elegans* and *Drosophila*). The two sets of TM's make up less than onethird of the rod NCKX sequence and less than half of the cone NCKX sequence. Basic cation binding properties as inferred from an analysis of ^{45}Ca transport rates and NCKX currents were very similar for all our NCKX clones, implying that conserved residues within the two sets of TM's contain all the residues involved in cation binding and cation transport.

Departments of [1]Biophysics & Physiology and [2]Pathology, Faculty of Medicine, University of Calgary, Calgary, Canada.

INTRODUCTION

Ca^{2+}-ions play an important role as intracellular messengers in most cells in our body. As has been described in other chapters of this book, Ca^{2+}-ions also play significant roles in visual transduction in the outer segments of retinal rod and cone photoreceptors. In darkness, a significant Ca^{2+} influx occurs via the cGMP-gated channels. It has been shown in amphibian rod photoreceptors that regulation of cytosolic Ca^{2+} in the outer segment is compartmentalised from that in the rest of the rod cell;[1] this necessitates the presence of an effective Ca^{2+} extrusion mechanism in the outer segment plasma membrane to extrude Ca^{2+} entering the outer segment via the cGMP-gated channels. Ca^{2+} extrusion across the plasma membranes of eukaryotic cells is carried out either by ATP-driven Ca^{2+} pumps or by Na^+-dependent Ca^{2+} exchangers. It has been long known that the rod outer segment plasma membrane contains a Na^+/Ca^{2+} exchanger,[2,3] and it is now thought that the Na^+/Ca^{2+} exchanger and cGMP-gated channels are the only two ion transport proteins present in the plasma membrane of both mammalian and amphibian rod outer segments.[4,5] More recently, the Na^+/Ca^{2+} exchanger in rod outer segments was shown to differ from those found in cardiomyocytes, kidney and other tissues, since it uses both the inward Na^+ gradient as well as the outward K^+ gradient to drive Ca^{2+} extrusion at a stoichiometry of $4Na^+:(1Ca^{2+} + 1K^+)$.[6,7] Molecular cloning showed that the heart Na^+/Ca^{2+} exchanger (NCX) and the bovine rod Na^+/Ca^{2+}-K^+ exchanger (NCKX) represent completely distinct gene families with sequence similarity found only in two short stretches of about 20 amino acids each.[8,9] The bovine rod NCKX cDNA does not result in functional activity when expressed in cell lines,[10,11] and it has been suggested that co-expression of another yet to be identified protein may be required to mediate K^+ dependence and K^+ transport.[12] Subsequent cloning of several other mammalian rod NCKX cDNAs has resulted in at least one cDNA, dolphin rod NCKX, that yields robust Na^+/Ca^{2+}-K^+ exchange transport function after heterologous expression.[10] We recently identified by molecular cloning the distinct NCKX gene product expressed in the cone photoreceptors of both the human and chicken retina;[13] heterologous expression of the retinal cone NCKX cDNAs resulted in strong expression of Na^+/Ca^{2+}-K^+ exchange activity operating with a $4Na^+:(1Ca^{2+} + 1K^+)$ stoichiometry,[14] the same stoichiometry we reported before for the in situ bovine rod NCKX .[7] It is the objective of this chapter to review recent work carried out in our laboratory on the retinal rod and cone NCKX gene products.

FUNCTIONAL CHARACTERISTICS OF IN SITU ROD NCKX

The in situ rod Na^+/Ca^{2+}-K^+ exchanger has been characterized in significant detail in bovine and amphibian rods by a variety of methods including ^{45}Ca fluxes, absorption spectrophotometry with Arsenazo III, use of fluorometric Ca^{2+}-indicating dyes, and electrophysiological techniques, as has been reviewed previously.[15-17] In contrast, very little has been reported on the in situ cone Na^+/Ca^{2+}-K^+ exchanger. Here, we will briefly summarize the salient features that have emerged from our studies on the in situ rod Na^+/Ca^{2+}-K^+ exchanger:

- NCKX is a bidirectional transporter, it carries out both $Na_{outside}$-dependent Ca^{2+} efflux (forward exchange) and Na_{inside}-dependent Ca^{2+} influx (reverse exchange); the direction of Ca^{2+} transport depends on the direction of the transmembrane gradients of the respective ions and can be calculated from thermodynamic considerations of a $4Na^+$:($1Ca^{2+}$ + $1K^+$) stoichiometry.[18] Perhaps the most significant physiological consequence of bidirectional Na^+/Ca^{2+}-K^+ exchange transport is that removing extracellular Na^+ will lead to a rapid influx of Ca^{2+} into rod and cone photoreceptors via reverse exchange.
- NCKX displays an absolute ion selectivity for Na^+ ions, i.e., no other cations such as Li^+ or K^+ can substitute for Na^+ in counter transport with Ca^{2+}; the rate of Na^+-dependent Ca^{2+} extrusion shows a distinct sigmoidal relationship with the Na^+ concentration in the medium with a dissociation constant of 30-40 mM. In this respect, NCX and NCKX are very similar.
- K^+ ions can bind to the Na^+/Ca^{2+}-K^+ exchanger and are co-transported with Ca^{2+}; Rb^+ and NH_4^+ ions can substitute for K^+, while Sr^{2+} can substitute for Ca^{2+}.
- Li^+ ions can affect the Na^+ and K^+ dissociation constants, but are not transported by the Na^+/Ca^{2+}-K^+ exchanger.
- Rod NCKX is thought to be inactivated in a time- and/or Ca^{2+}-dependent manner, which is thought to prevent lowering of free rod cytosolic Ca^{2+} concentration to levels below 1 nM when rod photoreceptors are saturated for prolonged periods of time in bright daylight illumination.

Two distinct features of the rod Na^+/Ca^{2+}-K^+ exchanger are that it is electrogenic and carries net current (a feature shared with the Na^+/Ca^{2+} exchanger), and that Ca^{2+} transport is not only coupled to the transmembrane Na^+ gradient but also to the transmembrane K^+ gradient (unique to Na^+/Ca^{2+}-K^+ exchange). Here, we illustrate those features by examining the effect of changes in K^+ gradient and changes in membrane potential on Ca^{2+} fluxes (measured with ^{45}Ca) in purified intact bovine rod outer segments (Fig. 1). The experiment was carried out in the light to close ion fluxes via cGMP-gated channels; under these conditions, the Na^+/Ca^{2+}-K^+ exchanger is the only cation transporter in the rod outer segment plasma membrane and all ^{45}Ca fluxes and changes in ^{45}Ca content are mediated only by the Na^+/Ca^{2+}-K^+ exchanger. The membrane potential in a suspension of cells or cellular organelles such as rod outer segments can be controlled by the application of electrogenic ionophores such as the electrogenic protonophore FCCP or valinomycin, an electrogenic carrier selective for K^+ ions.[7,15,19] The electrogenic protonophore FCCP or the electrogenic K^+-selective ionophore valinomycin were used to clamp the membrane potential to the proton or K^+ Nernst potentials, respectively. Next, we tested the addition of several test cations that would bring about changes in the K^+ and proton gradients, respectively. Tetramethylammoniun ion was used as a negative control. In the presence of FCCP, addition of 20 mM KCl caused a tenfold change in the

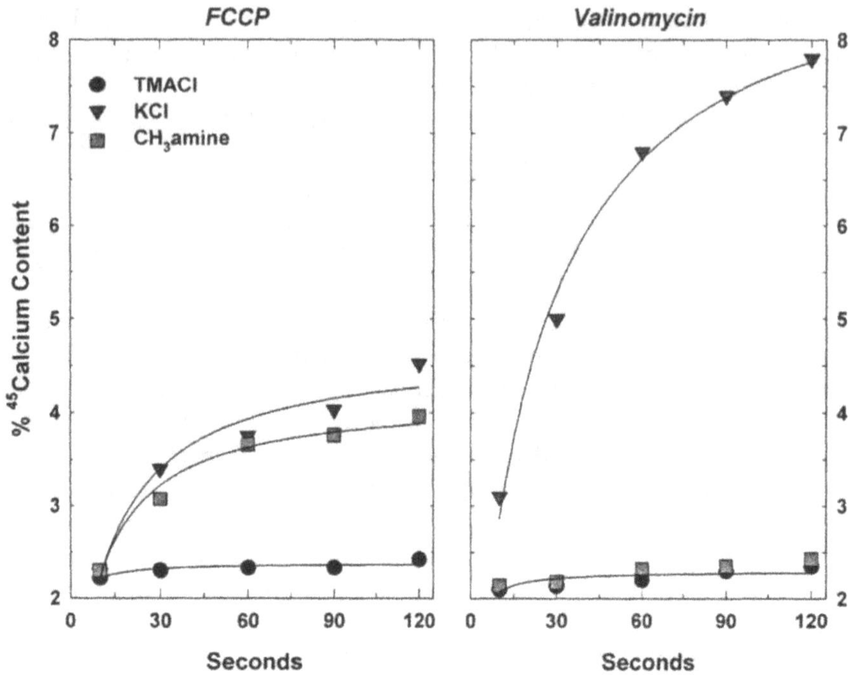

Figure 1. Effect of changes in membrane potential and transmembrane K^+ gradient on Ca^{2+} fluxes in rod outer segments. Intact bovine rod outer segments were purified on mixed sucrose/Ficoll 400 gradients as described.[33] A rod outer segment suspension containing 40 :M rhodopsin was equilibrated for 15 minutes in a solution containing 600 mM sucrose, 20 mM Hepes (adjusted to pH 7.4 with arginine) and low concentrations of external NaCl (10 mM), K^+ chloride (2 mM). FCCP or valinomycin was present at 0.5 µM as indicated. Free Ca^{2+} concentration was 1.5 µM (0.2 mM HEDTA and 0.2 mM Ca^{2+}HEDTA) and 1µCi ^{45}Ca was added to monitor net Ca^{2+} fluxes. ^{45}Ca content of rod outer segments was determined after rapid filtration technique over borosilicate glass fibre filters and liquid scintillation counting.[5] The experiment was started at time zero by addition of 20 mM tetramethyl ammonium (TMA) chloride, 20 mM methylamine hydrochloride, or 20 mM K^+ chloride. Temperature 25°C.

transmembrane K^+ gradient (without a concomitant change in membrane potential) and resulted in a significant influx of Ca^{2+} (as monitored with ^{45}Ca) consistent with co-transport of Ca^{2+} and K^+ by the Na^+/Ca^{2+}-K^+ exchanger. In the presence of valinomycin, addition of 20 mM KCl caused both a tenfold change in the transmembrane K^+ gradient and an accompanying membrane depolarization by electrogenic K^+ influx via valinomycin. K^+-induced ^{45}Ca influx was about twice as large when valinomycin controlled membrane potential indicating that both the change in K^+ gradient and membrane depolarization contributed to Ca^{2+} influx via the Na^+/Ca^{2+}-K^+ exchanger. Addition of methylamine hydrochloride is expected to cause an alkalinization of the cytosol due to the rapid permeation of the neutral methylamine and reprotonation in the cytosol. In the presence of valinomycin this should not affect membrane potential and no methylammonium-induced ^{45}Ca flux was observed consistent with no change in membrane potential and consistent with the fact that the Na^+/Ca^{2+}-K^+ exchanger does not transport protons.[20] In contrast, addition of 20 mM methyl ammonium chloride in the presence of FCCP is expected to cause membrane depolarization by the alkalinization of the cytosol and electrogenic proton influx via FCCP and this was accompanied by a significant ^{45}Ca influx, consistent with electrogenic Na^+/Ca^{2+}-K^+ exchange transport. In all cases, addition of 20 mM tetramethylammonium chloride was without effect. In summary, the experiment illustrated in Fig. 1 demonstrates that the direction of Ca^{2+} flux via the Na^+/Ca^{2+}-K^+ exchanger is affected by changes in K^+ gradient and by changes in membrane potential, consistent with a $4Na^+:(1Ca^{2+} + 1K^+)$ stoichiometry.

INTRARETINAL LOCALIZATION OF THE HUMAN RETINAL ROD AND CONE Na^+/Ca^{2+}-K^+ EXCHANGERS

The first NCKX cDNA was cloned from a bovine retinal cDNA library and identified as the rod Na^+/Ca^{2+}-K^+ exchanger by Western blotting.[9] In our laboratory we cloned the cDNAs of both the human and chicken retinal rod and cone Na^+/Ca^{2+}-K^+ exchangers.[13,21] No antibodies are available to determine the intraretinal localization of either the human rod or cone NCKX. Therefore, intraretinal localization of human rod and cone NCKX transcripts was determined by in situ hybridization (Fig. 2). The full-length human cone NCKX antisense riboprobe strongly labeled the base of the inner segments of cone photoreceptors (Fig. 2A and C). Also, a clear but weaker staining was observed in some of the cells in the retinal ganglion cell layer (GCL) as observed previously in the chicken retina.[13] Some weak staining was observed as well in the inner nuclear layer (INL) of human tissue in contrast to the chicken retina where no staining was observed in the INL. We used partial primate opsin antisense riboprobes to label cone inner segments containing transcripts of red/green opsin (Fig. 2E) or blue opsin (Fig. 2G). From a comparison of the number of inner segments that are positive with our cone NCKX probe to the three primate opsin probes, we estimate that the cone NCKX antisense probe stains the inner segments of all three types of cones. Sense probes were used as a control and never resulted in any specific staining as illustrated here for the red/green opsin sense probe (Fig. 2H).

The rod NCKX antisense riboprobe showed an even distribution of positive cells in the outer nuclear layer, consistent with the localization of the rod photoreceptor cell (Fig. 2B and D). Staining observed with the rod NCKX antisense riboprobe showed a very similar pattern compared to staining with the rhodopsin antisense riboprobe (compare panels B and F).

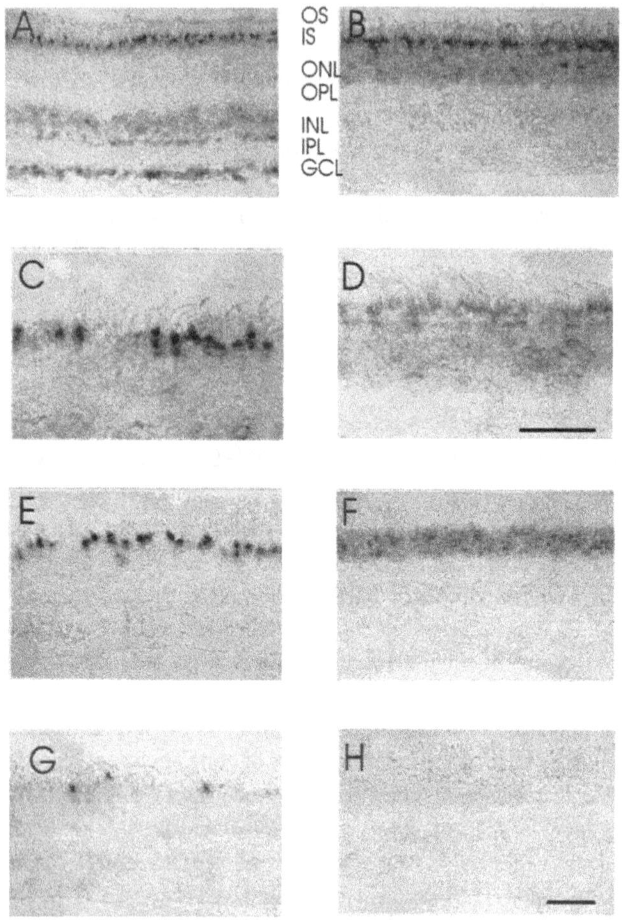

Figure 2. Intraretinal localization of the human retinal rod and cone $Na^+/Ca^{2+}-K^+$ exchangers. Eyeballs (obtained from the Lions Eye Bank Alberta at the Rockyview General Hospital, Calgary) of a 46-years old Caucasian male were enucleated and the gel vitreous was removed. Eyeballs were fixed for 12 hr at 4°C in buffered neutral formalin (10%)(BDH, Toronto, ON). Fixed samples were washed three times in phosphate-buffered saline (PBS; 0.05 M phosphate buffer and 195 mM NaCl, pH 7.4), dehydrated in a graded ethanol series and embedded in paraffin. Vertical sections (5 μm) were mounted on Micro Slides Superfrost Plus slides (VWR Canlab, Mississauga, ON), dried at room temperature for 18 h and stored at -20°C. cRNA probes: a partial human rod NCKX1 cDNA (nt 2122-2795 of Acc. No. AF062922) and a full length human cone NCKX cDNA in Bluescript were linearized using BamHI and HindIII (rod NCKX1)(New England Biolabs, Mississauga, ON) and BamHI and Acc65I (cone NCKX) for in vitro transcription using T7 and T3 RNA polymerase, respectively. Human blue opsin cDNA (224 nt) and cyno monkey red/green opsin cDNA (197 nt) cloned in pBK-CMV were used to make antisense probes using T3 and T7 RNA polymerase, respectively, and sense probes using T7 and T3 RNA polymerase, respectively. A 1 kb human rhodopsin cDNA cloned into pCRII-TOPO was used to make antisense and sense rhodopsin probes using T7 RNA polymerase and SP6 RNA polymerase, respectively. Transcription reactions were carried out according to the manufacturer's instructions using either T7 or T3 RNA polymerase in the presence of 11-digoxygenin UTP (DIG RNA labeling kit (SP6/T7), Boehringer). In situ hybridization: The RNA in situ hybridization was performed as described previously.[34] The retinal pigment epithelium cell layer was usually lost during the in situ procedure. Results were documented with Elite Chrome 160 Thungsten (Eastman Kodak, New Haven, CT) at ASA 160. See text for explanation of panels

CHROMOSOMAL LOCALIZATION OF THE HUMAN RETINAL ROD AND CONE Na^+/Ca^{2+}-K^+ EXCHANGERS

We previously used fluorescence in situ hybridization (FISH) to localize the rod NCKX gene to chromosome 15q22.[22] Here, FISH was used to localize the cone NCKX gene using the methods previously used for the rod NCKX gene.[22] The cDNA of cone NCKX was isolated and characterized from human retina by RT-PCR (GenBank Accession No. AF097366) as described before.[13] The two probes used in FISH analysis were two non-overlapping lambda FIX II phages HG51 and HG91, containing the 3' and 5' ends of the genomic cone NCKX region, respectively. These probes were obtained as follows. A human placental genomic library in λ FIX II vector (9-22 kb insert length, Stratagene) was screened with the human partial cDNA sequence (1442 nt), which was obtained by PCR (primers 402, 5'-GAATGATTTACATGTTCATAGCCTTAGC and 280, 5'-AGTGGCTGTCAGCAGCAATGGC) from retinal cDNA. Approximately $6.6x10^5$ genomic clones were screened using a partial (1442 nt) human cone NCKX cDNA as probe under medium stringency (100 mM Na^+ phosphate, pH 7.2/0.1% SDS at 65°C). Eleven positive clones were isolated, their DNA purified (Qiagen lambda mini kit) and analyzed by PCR and direct sequencing (Thermosequenase, Amersham) to define exon-intron boundaries. Positive phages were isolated by the plate lysate method and subsequent polyethylene glycol (MW=8,000/ NaCl precipitation); phage DNA was subsequently isolated by phenol/chloroform extraction and ethanol precipitation. Purified DNAs of two of the eleven positive genomic clones (e.g., HG51 and HG91) were used in fluorescence in situ hybridization, as described in the figure legend. Clone HG91 contains the 5' part of the gene whereas clone HG51 contains the 3' part of the gene (non-overlapping and approximately 20 kb apart). The cone NCKX gene was localized to chromosome 9p22 using either of the two clones as probe in FISH (Fig. 3). This localization is consistent with that obtained from the human genome sequencing project. No hereditary retinal diseases have yet been mapped to the positions of the rod and cone NCKX genes. A first screen of patients with retinal disease for mutations in the rod and cone NCKX genes revealed a number of sequence changes, especially for the rod NCKX gene, but it remains to be established whether any of these sequence changes are pathogenic.[23]

ANALYSIS OF THE ROD AND CONE Na^+/Ca^{2+}-K^+ EXCHANGER SEQUENCES

The bovine rod Na^+/Ca^{2+}-K^+ exchanger was the first NCKX cDNA cloned and its sequence did not match anything in the data base at the time except for two short stretches found in the dog heart NCX1 Na^+/Ca^{2+} exchanger comprising in total only 40 out of 1210 residues.[9] In the past few years we have cloned rod and cone NCKX cDNAs from several species[10,13,21,24] as well as two NCKX cDNAs found in lower

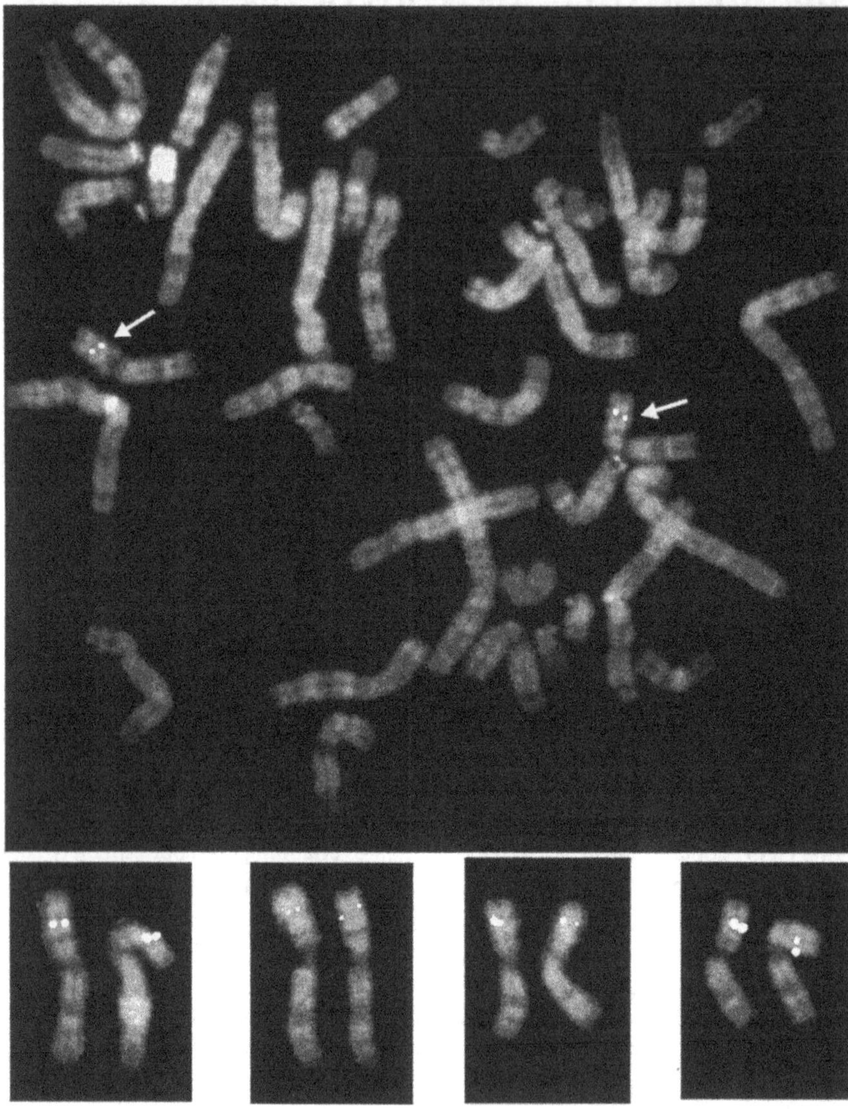

Figure 3. Chromosomal localization of the human retinal cone NCKX gene. Fluorescence in situ hybridization (FISH) was performed on methotrexate/thymidine synchronized, phytohemagglutinin stimulated, normal peripheral blood lymphocytes as described before.[35] Suppression for 30 minutes with a mixture of sonicated human DNA (Sigma) and cot1 DNA (Gibco/BRL) was required to reduce the background. The stained slides were counterstained with DAPI and actinomycin D (for a DA-DAPI banding pattern) and were mounted in antifade medium and visualized utilizing a Zeiss Axioplan 2 microscope. Approximately 30 metaphase spreads were examined for probe localization. Images of representative mitoses were captured using a cooled CCD camera (Photometrics PXL1400). Digital alignment of the images from each fluor was done after registration calibration through a triple bandpass filter (FITC/Texas Red/DAPI) to minimize registration error, utilizing commercial software (Electronic Photography v1.3, Biological Detection Inc., Pittsburgh PA).

organisms, *Caenorhabditis elegans*[11] and *Drosophila*.[25] Figure 4 shows a diagram representing the rod and cone NCKX sequences. Two sets of putative transmembrane spanning segments (TM1 and TM2) represent the only sequence elements that show conservation in all our NCKX cDNAs, and are the only domains that show sequence similarity between retinal NCKX sequences and those of *C. elegans* and *Drosophila*, although the overall topology of all NCKX sequences appears very similar. The TM's are thought to consist of membrane spanning alpha helices with short (<25 residues) connecting loops. The TM's contain two sequence elements with internal similarity, the alpha repeats, which are thought to have arisen from an ancient gene duplication event.[26] In addition to the TM's, NCKX sequences contain two large hydrophilic loops, one at the N-terminus and one between TM1 and TM2, and located in the cytosol. The N-terminal hydrophilic loop is thought to contain a cleavable signal peptide and is quite variable in length, ranging from 450 residues in the mammalian rod NCKX, to 250 residues in *Drosophila* NCKX, to 135 residues in chicken rod NCKX and in cone NCKX, to only 102 residues in *C. elegans* NCKX. The large extracellular loop is relatively poorly conserved (~50 percent identity) among mammalian rod NCKX, and does not contain sequences conserved between mammalian rod NCKX and chicken rod NCKX, or between mammalian rod and cone NCKX. The N-terminal loop is poorly conserved among the cone NCKX sequences as well. The large cytosolic loops are not conserved between rod and cone NCKX sequences, respectively, with the exception of the short sequence: LQRGxGSSxSLHNSxMRxS/TIFQLMIHTLPDA, located in the N-terminal part of the large cytosolic loop and found in the retinal NCKX sequences, but not in the NCKX sequences from *C. elegans* and *Drosophila*. As is observed for the N-terminal hydrophilic loop, the sequences of large cytosolic loop of the mammalian rod NCKX stand apart from those of the chicken rod and the various cone NCKX sequences: a large and highly acidic C-terminal portion of this cytosolic loop is found in the mammalian rod NCKX sequences, but not in any of the other NCKX sequences obtained so far. This acidic C-terminal part of the large cytosolic loop together with the large N-terminal hydrophilic loop accounts for the greatly increased size of the mammalian rod NCKX sequences compared with the other NCKX sequences. The acidic portion of mammalian rod NCKX sequence often, but not always (e.g., human rod NCKX), contains several repeats of an acidic motif of variable length and composition. These repeats appear to have arisen quite recently as the buffalo rod NCKX sequence contains the bovine motif but with two fewer repeats.[21]

In addition to the variability of different NCKX sequences, additional complexity can be obtained through alternatively spliced transcripts. When the genomic organization of the human NCKX rod gene was analyzed, it became clear that the three "full-length" rod NCKX clones we were using in our laboratory represented three different splice variants; the bovine rod NCKX is the full-length clone, the human NCKX clone lacks exon III (coding for 18 amino acids), and the dolphin NCKX clone lacks exons III, IV, V, and VI (coding for 114 amino acids).[22] RT-PCR of human retinal RNA showed six discrete splice products of the rod NCKX, and sequencing showed that all six represented alternate splice combinations of the

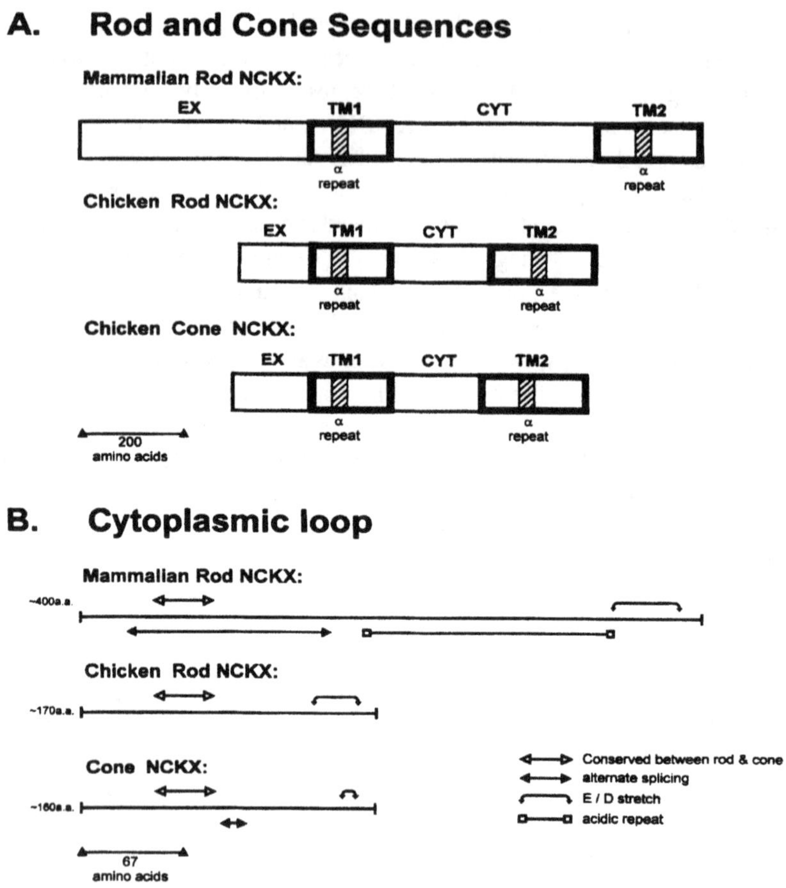

Figure 4. Sequence analysis of retinal rod and cone NCKX proteins. Diagram of rod and cone sequences drawn to reflect the relative length of each domain. EX denotes the N-terminal loop located in the extracellular space; TM1 and TM2 represent two clusters of hydrophobic transmembrane spanning segments; CYT indicates a large hydrophilic loop located in the cytoplasm; the hatched segments within the TM's represent the alpha repeats (see text).

exons located in the large cytosolic loop (Tucker, J.E. and Schnetkamp, P.P.M., unpublished observations). Similarly, RT-PCR of rat retinal RNA resulted in five different splice variants for the rat rod NCKX.[24] For the much shorter chicken rod NCKX we did not find any evidence for alternate splicing, while two splice variants were found for both the chicken and human cone NCKX, the shorter splice variant lacked exon 5, coding for 17 amino acids and located in the middle of the large cytosolic loop.[13,23]

In summary, sequence comparison has revealed a remarkable variability in rod NCKX sequences, whereas the much shorter cone NCKX sequences were less variable. This lack of conservation is quite unusual when compared with other

Na^+-coupled transporters such as the Na^+-K^+ pump, the Na^+/Ca^{2+} exchanger or the Na^+/H^+ exchanger, in which cases sequence identity is >90 percent over the entire sequence when comparing mammalian sequences from the same isoform. Both rod and cone NCKX transcripts can undergo alternate splicing which results in deletion of short stretches of variable length in the rod NCKX or of a stretch of 17 residues in the cone NCKX sequence, in both cases located in the N-terminal part of the large cytosolic loop. The physiological relevance of alternate splicing is currently unclear nor is it clear which splice variants are normally expressed as rod or cone NCKX protein.

FUNCTIONAL CHARACTERIZATION OF HETEROLOGOUSLY EXPRESSED ROD AND CONE NCKX

The first NCKX cDNA cloned, containing the full-length bovine rod NCKX, did not result in $Na^+/Ca^{2+}-K^+$ exchange function when expressed in heterologous systems despite the fact that protein expression was readily detected by Western blotting.[10,11] This created some considerable initial confusion and led to the suggestion in one study that K^+ dependence might require an additional protein component.[12] However, more recently we have shown that upon deletion of the two large hydrophilic loops from the bovine rod NCKX cDNA, expression of the resultant bovine rod NCKX deletion mutant in cultured insect cells yields both K^+-dependent Na^+/Ca^{2+} exchange transport and Ca^{2+}-dependent $^{86}Rb^+$ (used as a K^+ congener) transport; moreover, the *C. elegans* NCKX cDNA we cloned resulted in both K^+-dependent Na^+/Ca^{2+} exchange transport and Ca^{2+}-dependent ^{86}Rb transport.[11,14] The two sets of TM's are the only sequence elements in common between the bovine rod NCKX and the *C. elegans* NCKX, and are the only the sequence elements left in the rod deletion mutant. Therefore, we believe these results clearly demonstrate that a single polypeptide encoded for by NCKX genes is sufficient for $Na^+/Ca^{2+}-K^+$ exchange transport. Of all our NCKX clones, the human and chicken cone NCKX cDNAs, especially the short splice variants, have consistently given the best functional expression.[13] In contrast, the human and bovine rod NCKX cDNAs did not result in $Na^+/Ca^{2+}-K^+$ exchange transport function when expressed in either insect or HEK293 cells, whereas the dolphin and chicken rod NCKX clones did result in $Na^+/Ca^{2+}-K^+$ exchange transport function when expressed in either insect or HEK293 cells,[10,13] Interestingly, the three different mammalian rod clones studied in our laboratory represent three different splice variants as is discussed above, the functional dolphin rod NCKX lacking exons III through VI. It is tempting to speculate that sequences in exons III through VI prevent proper targeting of NCKX cDNA in heterologous systems and might cause the lack of functional expression observed for our human and bovine NCKX clones. Consistent with this, deletion of the cytosolic loop from the bovine NCKX or deleting the residues encoded for by exons III through VI restored NCKX function when these mutated bovine NCKX cDNAs were expressed in either insect cells or HEK cells.[10,11] On the other hand, functional expression of various splice variants of the rat rod NCKX did not follow this pattern.[24]

We characterized the K^+, Na^+, and Ca^{2+} dissociation constants of several different rod, cone and *C. elegans* NCKX clones when expressed in the insect cell system we developed. The values for the external Na^+ dissociation constant (30-50 mM), the external K^+ dissociation constant (see below) and both the internal (~1 μM) and external (~2 μM) Ca^{2+} dissociation constants were very similar for all the NCKX clones tested, and similar to those previously observed for the in situ bovine rod NCKX.[11,13,14,27] Moreover, the K^+ dissociation constants of all NCKX clones showed a feature quite characteristic for the in situ bovine rod NCKX: dissociation constants measured in choline medium (~2 mM) were 5- to 10-fold lower compared with the case when the dissociation constants were measured in Li^+ medium (~15 mM).[11,13]

Here, the properties of the K^+ binding site on the rod Na^+/Ca^{2+}-K^+ exchanger in situ in bovine rod outer segments is compared with properties of the heterologously expressed dolphin rod Na^+/Ca^{2+}-K^+ exchanger. Na_{inside}-dependent Ca^{2+} influx (reverse Na^+/Ca^{2+}-K^+ exchange) was measured as a function of the extracellular K^+ concentration by measuring the rise in cytosolic free Ca^{2+} concentration with the fluorescent Ca^{2+}-indicating dye fluo 3 following protocols developed previously.[10,28,29] Figure 5 shows that the K^+ dependence of reverse Na^+/Ca^{2+}-K^+ exchange in bovine rod outer segments was very similar to that observed in HEK293 cells transfected with the dolphin rod Na^+/Ca^{2+}-K^+ exchanger; the K^+ dissociation constant observed under these experimental conditions was 1.5 mM for the in situ bovine rod Na^+/Ca^{2+}-K^+ exchanger and 2.7 mM for the dolphin rod Na^+/Ca^{2+}-K^+ exchanger expressed in HEK293 cells. Figure 6 illustrates the ion selectivity of the K^+ binding site as measured with the same experimental protocol as illustrated in Figure 5. With both the in situ bovine rod Na^+/Ca^{2+}-K^+ exchanger and the heterologously expressed dolphin rod Na^+/Ca^{2+}-K^+ exchanger, Rb^+ and NH_4^+ ions could replace K^+, whereas Cs^+, Li^+ and choline could not.

The Na^+/Ca^{2+}-K^+ exchange transport activity obtained with the human and chicken cone NCKX clones expressed in insect cells was sufficiently high to determine a transport stoichiometry of $4Na^+:(1Ca^{2+} + 1K^+)$ by direct measurements of the $K^+:Ca^{2+}$ and $Na^+:Ca^{2+}$ coupling ratios,[14] the same stoichiometry we previously determined for the in situ bovine rod NCKX.[7] The NCKX clones used in these studies include the double deletion bovine rod NCKX construct in which both large hydrophilic loops were deleted. These results demonstrate that the two sets of TM's are sufficient for Na^+/Ca^{2+}-K^+ exchange transport and that residues contributing to cation binding and transport are confined two the TM's.

In summary, our work on the functional characterization of the retinal rod and cone NCKX proteins, and a comparison with NCKX proteins found in *C. elegans* and *Drosophila*, has clearly shown that two sets of transmembrane spanning segments contain all the residues that encode for cation binding and cation transport functions associated with Na^+/Ca^{2+}-K^+ exchange. Although the maximal transport

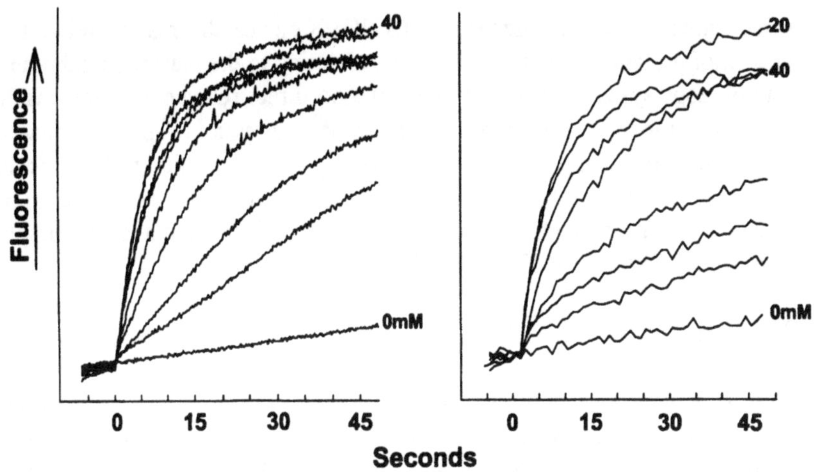

Figure 5. K^+ dependence of in situ rod NCKX and rod NCKX expressed in HEK293 cells. K^+-induced Ca^{2+} influx was measured as a rise in cytosolic free Ca^{2+}, indicated as by the rise in fluorescence with the fluorescent Ca^{2+} indicating dye fluo 3, in bovine rod outer segments (left panel) or HEK293 cells transfected with the dolphin rod NCKX (right panel). Protocols and instrumental details have been described elsewhere.[10,28,29] Incubation media contained 50 mM NaCl, 400 mM sucrose (rod outer segments) or 100 mM LiCl (HEK cells), 20 mM Hepes (adjusted to pH 7.4 with Tris), and 100 µM EDTA. Gramicidin was added to the cell suspension at a concentration of 100 nM (rod outer segments) or 2.5 µM (HEK cells) and extracellular and intracellular Na^+ was allowed to equilibrate for 2 minutes. Ca^{2+} was added (150 µM for rod outer segments, 1 mM for HEK cells), and K^+-dependent reverse Na^+/Ca^{2+}-K^+ exchange was initiated at time zero by addition of KCl. Rates of Ca^{2+} influx increased as the K^+ concentration was raised to: 0, 0.1, 0.2, 0.5, 1, 2, 5, 10, 20, 40 mM (rod outer segments); 0, 0.2, 0.5, 1, 2, 5, 10, 20, 40 mM (HEK cells). Temperature 25°C.

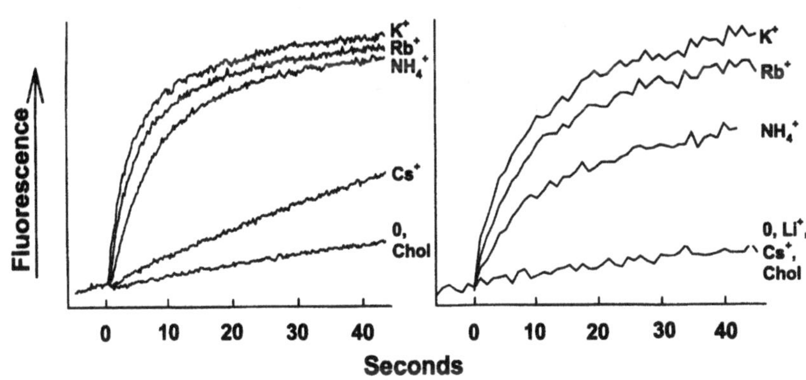

Figure 6. Ion selectivity of the rod NCKX K^+ binding site. Experimental procedures were as described in the legend of Figure 5. Cation-dependent Ca^{2+} influx was initiated at time zero by addition of 20 mM of the indicated chloride salts. Temperature 25°C.

rate can be quite different among the various NCKX clones, this may merely reflect the amount of protein expression and correct targeting to the plasma membrane in the heterologous systems used, and may have no direct relevance to the case of expression and function in retina rod and cones. It is tempting to speculate that the large cytosolic loops found in all NCKX sequences impart regulatory features that have yet to be explored. It has been shown that the Na^+/Ca^{2+}-K^+ exchanger in bovine rod outer segment membranes occurs as a dimer and is part of a larger complex with the two subunits of the cGMP-gated channel.[30-32]

ACKNOWLEDGMENTS

This work was supported through operating grants from the Canadian Institutes for Health Research (PPMS) and the Canadian Genetic Disease Network Center of Excellence (DJD). PPMS is a Senior Scientist of the Alberta Heritage Foundation for Medical Research, CFMP was a recipient of a PDF from the Alberta Heritage Foundation for Medical Research. DJD was an MRC/CIHR Clinician Scientist. We thank Dr. Lorraine Aha for carefully reading the manuscript.

REFERENCES

1. Krizaj D,.Copenhagen DR. Compartmentalization of calcium extrusion mechanisms in the outer and inner segments of photoreceptors. Neuron 1998; 21:249-56.
2. Schnetkamp PPM. Ion selectivity of the cation transport system of isolated cattle rod outer segments: evidence of a direct communication between the rod plasma membrane and the rod disk membranes. Biochim Biophys Acta 1980; 598:66-90.
3. Yau K-W, Nakatani K. Cation selectivity of light-sensitive conductance in retinal rods. Nature 1984; 309:352-354.
4. Lagnado L, Cervetto L, McNaughton PA. Ion transport by the Na:Ca exchange in isolated rod outer segments. Proc Natl Acad Sci 1988; 85:4548-52.
5. Schnetkamp PPM, Szerencsei RT, Basu DK. Unidirectional Na^+, Ca^{2+} and K^+ fluxes through the bovine rod outer segment Na-Ca-K exchanger. J.Biol.Chem. 1991; 266:198-206.
6. Cervetto L, Lagnado L, Perry RJ et al. Extrusion of calcium from rod outer segments is driven by both sodium and potassium gradients. Nature 1989; 337:740-3.
7. Schnetkamp PPM, Basu DK, Szerencsei RT. Na-Ca exchange in the outer segments of bovine rod photoreceptors requires and transports potassium. Am.J.Physiol.(Cell Physiol.) 1989; 257:C153-C157.
8. Nicoll DA, Longoni S, Philipson KD. Molecular cloning and functional expression of the cardiac sarcolemmal Na^+-Ca^{2+} exchanger. Science 1990; 250:562-5.
9. Reilander H, Achilles A, Friedel U et al. Primary structure and functional expression of the Na/Ca,K-exchanger from bovine rod photoreceptors. EMBO J. 1992; 11:1689-95.
10. Cooper CB, Winkfein RJ, Szerencsei RT, Schnetkamp PPM. cDNA-Cloning and Functional Expression of the Dolphin Retinal Rod Na-Ca+K ExchangerNCKX1: Comparison with the Functionally Silent Bovine NCKX1. Biochemistry 1999; 38:6276-83.
11. Szerencsei RT, Tucker JE, Cooper CB et al. Minimal Domain Requirement for Cation Transport by the Potassium-dependent Na/Ca-K Exchanger: Comparison with an NCKX Paralog from *Caenorhabditis elegans*. J Biol Chem 2000; 275:669-76.
12. Navangione A, Rispoli G, Gabellini Ne et al. Electrophysiological characterization of ionic transport by the retinal exchanger expressed in Human embryonic kidney cells. Biophys J 1997; 73:45-51.

13. Prinsen CFM, Szerencsei RT, Schnetkamp PPM. Molecular cloning and functional expression the potassium-dependent sodium-calcium exchanger from human and chicken retinal cone photoreceptors. J.Neurosci. 2000; 20:1424-34.
14. Szerencsei RT, Prinsen CFM, Schnetkamp PPM. The Stoichiometry of the Retinal Cone Na/Ca-K Exchanger Heterologously Expressed in Insect Cells: Comparison with the Bovine Heart Na/Ca Exchanger. Biochemistry 2001; 40:6009-15.
15. Schnetkamp PPM. Na-Ca or Na-Ca-K exchange in the outer segments of vertebrate rod photoreceptors. Prog.Biophys Mol Biol 1989; 54:1-29.
16. Schnetkamp PPM. Calcium homeostasis in vertebrate retinal rod outer segments. Cell Calcium 1995; 18:322-30.
17. Schnetkamp PPM. How does the retinal rod Na-Ca + K exchanger regulate free cytosolic Ca^{2+}? J Biol Chem 1995; 270:13231-9.
18. Schnetkamp PPM, Basu DK, Szerencsei RT. The stoichiometry of Na-Ca+K exchange in rod outer segments isolated from bovine retinas. Ann NY Acad Sci 1991; 639:10-21.
19. Schnetkamp PPM. Optical measurements of Na-Ca-K exchange currents in intact outer segments isolated from bovine retinal rods. J Gen Physiol 1991; 98:555-73.
20. Schnetkamp PPM. Chelating properties of the Ca^{2+} transport site of the retinal rod Na-Ca+K exchanger: Evidence for a common Ca^{2+} and Na^+ binding site. Biochemistry 1995; 34:7282-7.
21. Tucker JE, Winkfein RJ, Cooper CB, Schnetkamp PPM. cDNA Cloning of the Human Retinal Rod Na-Ca+K Exchanger: Comparison with a Revised Bovine Sequence. IOVS 1998; 39:435-40.
22. Tucker JE, Winkfein RJ, Murthy SK et al. Chromosomal localization and genomic organization of the human retinal rod Na—Ca+K exchanger. Human Genetics 1998; 103:411-4.
23. Sharon D, Yamamoto H, McGee et al. Mutated alleles of the rod and cone Na/Ca+K exchanger genes in patients with retinal diseases. IOVS 2002; 43:1971-1979.
24. Poon S, Leach S, Li XF et al. Alternatively spliced isoforms of the rat eye sodium/calcium+potassium exchanger NCKX1. Am J Physiol Cell Physiol 2000; 278:C651-C660.
25. Haug-Collet K, Pearson B, Park S et al. Cloning and Characterization of a Potasium-Dependent Sodium/Calcium Exchanger In *Drosophila*. J Cell Biol 1999; 147:659-69.
26. Schwarz EM,.Benzer S. Calx, a Na-Ca exchanger gene of *Drosophila melanogaster*. Proc. Natl. Acad.Sci. USA 1997;94:10249-54.
27. Sheng J-Z, Prinsen CFM, Clark RB, Giles WR, Schnetkamp PPM. Na^+-Ca^{2+}-K^+ Currents Measured in Insect Cells Transfected with the Retinal Cone or Rod Na^+-Ca^{2+}-K^+ Exchanger cDNA. Biophys J 2000;79:1945-53.
28. Schnetkamp PPM, Tucker JE, Szerencsei RT. Ca^{2+} influx into bovine retinal rod outer segments mediated by Na-Ca+K exchange. Am J Physiol (Cell Physiol) 1995; 269:c1153-c1159.
29. Cooper CB, Szerencsei RT, Schnetkamp PPM. Spectrofluorometric detection of Na-Ca+K exchange. Methods Enzymol. 2000; 315:847-64.
30. Schwarzer A, Kim TSY, Hagen V, Molday RS, Bauer PJ. The Na/Ca-K exchanger of rod photoreceptor exists as dimer in the plasma membrane. Biochemistry 1997; 36:13667-76.
31. Schwarzer A, Schauf H, Bauer PJ. Binding of the cGMP-gated channel to the Na/Ca-K exchanger in rod photoreceptors. J Biol Chem 2000; 275:13448-54.
32. Poetsch A, Molday LL, Molday RS. The cGMP-gated channel and related glutamic acid-rich proteins interact with peripherin-2 at the rim region of rod photorceptor disc membranes. J Biol Chem 2001; 276:48009-16.
33. Schnetkamp PPM. Sodium-Calcium exchange in the outer segments of bovine rod photoreceptors. J Physiol 1986; 373:25-45.
34. Breitschopf H, Suchanek G. Nonradioactive in situ hybridization application manual. Mannheim, Germany: Boehringer Mannheim GmbH, 1996:136-40.
35. Demetrick DJ. Fluorescence in situ hybridization and human cell cycle genes. In: Pagano M, ed. The cell cycle-Materials and methods. Berlin/Heidelberg/New York: Springer Verlag, 1995:29-45.

THE COMPLEX OF cGMP-GATED CHANNEL AND Na^+/ Ca^{2+},K^+ EXCHANGER IN ROD PHOTORECEPTORS

Paul J. Bauer

ABSTRACT

Ca^{2+} is an important signal ion in photoreceptors for recovery after excitation and light adaptation. It enters the outer segment as a minor fraction of the dark current through cGMP-gated channels and is extruded in the same cell compartment by Na^+/Ca^{2+}, K^+ exchange. Channel and exchanger are located exclusively in the plasma membrane, but not in the cytoplasmic membrane stack, the discs, which contain the visual pigment rhodopsin. The channel consists presumably of two α-subunits and two β-subunits, whereas the exchanger is a monomeric protein. Recently, considerable evidence has been accumulated indicating that both proteins form a complex which is bound to peripherin/rds, an integral protein of the disc rim. This review focuses on the complex of cGMP-gated channel and Na^+/Ca^{2+}, K^+ exchanger. The possibility of direct functional interaction between channel and exchanger is discussed. Furthermore, the consequences of different subunit arrangements of the channel for the channel-exchanger complex are considered. Finally, a Ca^{2+} diffusion model is presented which examines the possibility that Ca^{2+} currents are locally restricted to the close vicinity of the channel.

INTRODUCTION

Vertebrate photoreceptors constitute an impressive paradigm of ultimate functional performance: the sensitivity of these cells reaches the physical limit of single photon detection.[1,2] A key feature to understand this amazing performance appears

Institute for Biological Information Processing (IBI-1), Research Center, D-52425 Juelich, Germany.

to be the high degree of structural organization: outer segment, inner segment, nerve axon, and synaptic terminal are distinctly separated parts of these cells (Fig. 1). The primary task of photon detection and conversion into a voltage signal is exclusively located in the outer segment, whereas the rest of the cell is basically a normal nerve cell.

The vertebrate retina comprises two different types of photoreceptors, rods and cones. There is general agreement that the basic molecular mechanism of excitation of both classes of photoreceptors is similar. We concentrate on rod photoreceptors, the prevalent class of photoreceptors which was most thoroughly investigated. The rod outer segment consists of a stack of several hundred flattened sacks, the so-called discs, surrounded by the cell membrane (Fig. 1). The discs are physically separated from the cell membrane and contain the light-absorbing visual pigment, rhodopsin.

The ion conductance of the cell membrane of the rod outer segment is conspicuously low. The only ion channels are cGMP-gated channels which mediate a standing cation influx in the non-excited photoreceptor. This so-called dark-current consists mainly of Na^+ ions, and to a lesser extent of Ca^{2+} ions. Na^+ ions diffuse through the rod outer segment before they are exported by the Na^+/K^+ pump in the inner segment. In contrast, Ca^{2+} ions are exported in the rod outer segment by Na^+/Ca^{2+}, K^+ exchange. Therefore, the Ca^{2+} flux controlling the intra-cellular Ca^{2+} of the rod outer segment is restricted to this cell compartment.

Light stimulation of the photoreceptor leads to hydrolysis of cGMP resulting in deactivation of cGMP-gated channels and abatement of the dark current.[3-5] Because Ca^{2+} extrusion by Na^+/Ca^{2+}, K^+ exchange continues, the cytoplasmic Ca^{2+} decreases. This light-induced drop of intracellular Ca^{2+} constitutes a major signal for recovery and adaptation following light exposure.[6] At maximal light excitation, all channels are closed and the cytoplasmic Ca^{2+} concentration decreases to about 10-50 nM.[7-11]

Several Ca^{2+} dependent responses of rod photoreceptors have been described, however, the underlying molecular mechanism has not been identified for all of them.[6,12] Most Ca^{2+} signaling mechanisms in photoreceptors are indirect, mediated by the Ca^{2+} binding proteins calmodulin, recoverin, and guanylate cyclase activating proteins (GCAPs).[13-16] While only a small influence of calmodulin has been observed in the electrophysiological response of rod photoreceptors, recoverin and GCAPs play important roles in Ca^{2+} signaling.[17-21] Both proteins are involved in the recovery of the resting state after light excitation: GCAPs regulate guanylate cyclases and recoverin appears to control the activity of the rhodopsin kinase. However, the physiological relevance involving recoverin is still controversial.[22] Calmodulin, recoverin and GCAPs are members of a large family of Ca^{2+} binding proteins containing one or more EF-hand Ca^{2+} binding sites.[23]

Generally, Ca^{2+} dependent pathways in the outer segment of the rod photoreceptor are considered to be controlled by the overall intracellular Ca^{2+} concentration. This view is supported by the fact that there is a steady Ca^{2+} ion influx in the dark which is reduced upon light excitation, being fully shut off upon maximal excitation. For this reason, there is sufficient time for thermal equilibration to take place. This does not necessarily mean that the cytoplasmic Ca^{2+} concentration is

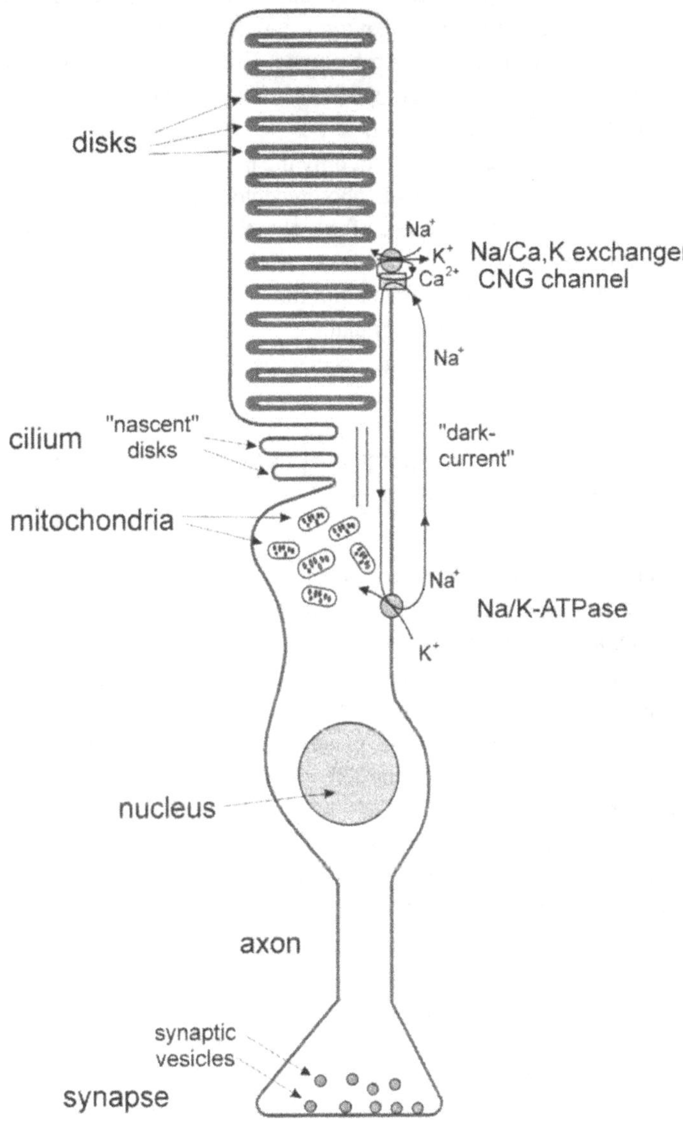

Figure 1. Rod photoreceptor, schematic representation.

spatially constant. In the presence of several concurrent Ca^{2+} regulatory steps, these regulatory mechanisms would be coupled if there would be only a constant intracellular Ca^{2+} concentration.

However, different Ca^{2+} signaling processes are often uncoupled in cells by local changes of the Ca^{2+} concentration. This may also be the case for photoreceptors. Ca^{2+} ions entering the rod outer segment in the resting phase through the cGMP-gated channel may create a local elevation of the Ca^{2+} concentration. Similarly, Ca^{2+} dependent processes may experience a different Ca^{2+} concentration at the plasma membrane than at the disc membrane. Even the Ca^{2+} concentration at the plasma membrane may not be homogeneous, changing with the distance to the channel and the Na^+/Ca^{2+}, K^+ exchanger.

These considerations indicate that not only the molecular components must be identified to understand Ca^{2+} signaling in photoreceptors, but the ultrastructure must be understood, as well. It is the purpose of this chapter to address this issue, and to review the characteristics of disc and plasma membrane. In particular, ultrastructural aspects of the cGMP-gated channel and the Na^+/Ca^{2+}, K^+ exchanger, which constitute source and sink of the intracellular Ca^{2+} in the rod outer segment, will be considered. Finally, the relevance of the occurrence of local Ca^{2+} phenomena will be discussed.

CHANNEL AND Na^+/Ca^{2+}, K^+ EXCHANGER RESIDE ONLY IN THE PLASMA MEMBRANE

In rod photoreceptors, discs are continually renewed at the base (the cilium) and are shed at the tip where they become phagocytosed in the pigment epithelium.[24] The discs are initially contiguous with the plasma membrane (nascent discs) and during maturation become eventually separated from the plasma membrane.[25] The major integral protein of both membranes is rhodopsin which makes up 90 percent of the disc and 60 percent of the plasma membrane proteins. Rhodopsin is structurally and functionally identical in both membranes.[26]

In spite of this common feature, there are distinct differences between these two membranes. Disc membranes consist of a flat lamellar region of two opposed membranes circumscribed by a more bulgy rim region. Rhodopsin is highly concentrated in the lamellar region of the disc but not in the rim region.[27] This special region of the discs contains rim-specific proteins including peripherin/rds and Rom-1 which serve structural purposes,[28-30] and the 220 kDa glycoprotein ABCR, a member of the ATP binding cassette transporters.[31-34] Deep infoldings of the disc rim, incisures, constitute a characteristic feature which is particularly distinct in the large photoreceptors of amphibians. Immunohistochemically, these incisures can be labeled with antibodies against the glutamic acid rich protein (GARP) and with antibodies against tubulin.[35,36] Moreover, the lipid composition of the two membranes is markedly different: disc membranes contain 35 percent of a lipid with highly unsaturated fatty acid side chains, whereas the plasma membrane contains only 5 percent of this lipid.[37] The presence of highly unsaturated lipids in the disc membranes results in high lateral and rotational diffusion constants of rhodopsin, and presumably also in specific hydrophobic interactions of membrane associated proteins.[38-40] Recently, it has been found in transgenic mice that the high rhodopsin concentration in disc membranes actually leads to a reduction of the rhodopsin diffusion due to protein collisions.[41]

The plasma membrane also contains several non-rhodopsin molecules which are not present in the disc membranes.[42] Most remarkably, the ion permeability of disc and plasma membrane is completely different. Initially, this became clear from Ca^{2+} flux studies of bovine ROS membrane vesicles which indicated that only about 6 percent of the vesicle contained cGMP-gated channels and Na^+/Ca^{2+}, K^+ exchangers whereas the remaining vesicle fraction didn't contain either of these proteins.[43] One estimates that in bovine rod outer segments, about 6 percent of the total membrane area is plasma membrane (based on the fraction $2 \cdot d/D$ where d denotes the inter-disc spacing and D the diameter of the bovine ROS with $D = 1\ \mu m$, $d = 0.03\ \mu m$). Since the plasma membrane contains cGMP-gated channels and Na^+/Ca^{2+}, K^+ exchanger, this finding indicates that these proteins are virtually absent in disc membranes.[43]

This conclusion was corroborated further by immunohistochemical identification of cGMP-gated channel in the plasma membranes but not in disc membranes of lysed ROS membranes.[44] Moreover, cGMP-induced Ca^{2+} fluxes were only observed with purified plasma but not with disc membrane vesicles. Similarly, the Na^+/Ca^{2+}, K^+ exchanger was identified as the main ricin-binding protein of the plasma membrane which permitted the biochemical separation of the plasma membrane from the disc membrane.[45] Fluorescein conjugated cGMP also indicated a high density of cGMP binding sites at the cell membrane but not in the disc space of toad rod photoreceptors.[46]

Together, these results indicate that there is a clear functional organization in the rod outer segment. Cation fluxes occur only through the cGMP-gated channels located in the plasma membrane, but there is no evidence for ion fluxes through the disc membrane. Disc membranes contain most of the light absorbing pigment rhodopsin and provide the appropriate membrane surface for association of the proteins of the visual transduction cascade. Regarding the steady renewal of disc and plasma membranes at the base of the ROS, experimental evidence has been reported that the disc rim protein peripherin/rds undergoes light dependent phosphorylation and may be involved in membrane fusion.[47] However, the efficient sorting process which leads to the generation of these two completely different membranes is still largely unknown and remains to be clarified further.

THE DENSITY OF cGMP-GATED CHANNELS

Several attempts have been undertaken to estimate the density of cGMP-gated channels localized in the plasma membrane of rod outer segments. Initially, noise analysis of the dark current of amphibian rod photoreceptors was extensively carried out.[48-51] The dark current of the outer segment exhibits considerable high frequency noise which decreases upon light absorption. Assuming that this noise is generated only by one class of channels which fluctuate independently of each other between an open and a closed state it was estimated that 5000 to 10,000 channels are simultaneously open in the dark. Because only about 5 percent or less of the channels are activated under these conditions, a channel density of at least 100–200 channels per μm^2 was estimated. These studies were carried out in the presence of Ca^{2+} ions which partially block the cGMP-gated channel by binding to the pore region.[52] Therefore, the unitary conductance of the channel was in the range of about 100 fS, far too small to be detectable as single channel conductance.

In the absence of divalent ions, the single channel conductance increases to 25 pS, i.e., by more than two orders of magnitude and can readily be determined in patch clamp experiments.[53] The number of channels in a patch was calculated directly as the ratio of the maximal ion current and the single channel current. The channel density of excised patches of rod outer segment membrane was determined to be 200–600 channels/μm^2, in reasonable agreement with the noise analysis data.[53-55] Remarkably, a wide variation of channel densities was observed in these studies which was inconsistent with a random distribution of channels over the plasma membrane.[55] No evidence for a distinct channel clustering was detected, however, some physical stress during membrane excision could not be excluded if the channels interacted with the cytoskeleton or directly with the disc membrane. In fact, recent evidence clearly indicates that the GARP domain of the cyclic GMP-gated channel binds to peripherin/rds, an integral membrane protein of the disc rim.[56]

When compared to amphibians, warm-blooded vertebrates have very tiny photoreceptors which limit the possibility of electrophysiological studies considerably. Fortunately, the channel density of bovine rod photoreceptors could be estimated biochemically.[57] Starting with purified rod outer segments, the membrane proteins were solubilized in the presence of Ca^{2+} and functionally reconstituted into artificial lipid vesicles after the detergent was removed by dialysis. The Ca^{2+} flux assay indicated that only a small fraction of lipid vesicles contained cGMP-gated channels suggesting that this fraction contained only one channel per vesicle. From the volume fraction of lipid vesicles and the rhodopsin concentration, a channel density of 600 channels/μm^2 was calculated based on the assumption that 3 percent of the membrane area was plasma membrane. If this percentage was assumed to be about 6 percent,[43] then the channel density would be estimated to be about 300 channelsμm^2, assuming that all channels were still functionally active after solubilization and reconstitution.

The channel density of bovine ROS plasma membranes was also estimated without solubilization of the membranes. Ca^{2+} flux experiments indicated that only a small fraction of ROS membrane vesicles contained cGMP-gated channels.[43] However, if these membranes were subjected to a single freeze-thaw cycle before preparation of membrane vesicles, the channel-containing fraction increased almost ten-fold.[58] Because freeze-thawing often leads to membrane fusion, this observation suggested that both disc and plasma membranes became fused and the cGMP-gated channels, which were initially only in the plasma membrane, became distributed over a much larger membrane area. If small vesicles were prepared by prolonged sonication of the fused ROS membranes, there were more vesicles than channel molecules. Assuming that the channels were randomly distributed over the vesicles (i.e., not clustered), a mean density of 0.37 channels per vesicle was obtained for vesicles with a mean diameter of 0.12 nm.[58] These data yielded a channel density of about 16 channels per μm^2 for fused membranes, corresponding to about 270 channels per μm^2 in the plasma membrane, in excellent agreement with the channel density estimated by Cook et al.[57]

Together, these very different experimental approaches led to conspicuously similar values of the channel density between 200 and 600 channels per μm^2 of plasma membrane, irrespective of the animal species. This is rather remarkable, considering that the volumes of the amphibian and the bovine ROS differ by more than two orders of magnitude.

THE DENSITY OF Na^+/Ca^{2+}, K^+ EXCHANGERS: AN INTRIGUING PROBLEM

Similar to the cardiac Na^+/Ca^{2+} exchanger, the retinal Na^+/Ca^{2+}, K^+ exchanger is electrogenic, and therefore, can be investigated electrophysiologically.[59-62] In bright light, all cGMP-dependent channels of the photoreceptor are closed, and the exchange current can be recorded in the absence of any other interfering ion current. Using this approach, salamander rod photoreceptors were loaded in the dark with various amounts of Ca^{2+}, and the kinetics of Ca^{2+} export studied after closing the light sensitive channels.[60-62] The maximal exchange current was proportional to the total number of exchangers present. A rigorous quantitative analysis of the Ca^{2+} transport kinetics, based on a Michaelis type Ca^{2+} binding isotherm in the presence of a low affinity Ca^{2+} buffer[60], yielded a maximal turnover rate of 10 Ca^{2+}/s/exchanger and a very high exchanger density of 35,000 μm^{-2}. However, the latter value which is even greater than the rhodopsin density, seems unlikely. Moreover, the maximal turnover rate seems too low. Nevertheless, this analysis is instructive because it suggests that a simple kinetic model based on overall Ca^{2+} concentrations does not appropriately describe the molecular transport mechanism.

Purification of the exchanger protein yielded a single polypeptide with an apparent molecular mass of 220 kDa which made up about 0.1percent of the total protein mass of ROS membranes.[63,64] Assuming that the exchanger was only localized in the plasma membrane, an exchanger density of 600–1000 μm^{-2} was estimated from these data. This value is somewhat greater than the channel density and in agreement with Ca^{2+} flux data from proteoliposomes.[63] Moreover, we found that upon reconstitution of the ROS membrane proteins in the presence of increasing amounts of lipid, the ratio of the maximal Na^+ induced and the cGMP induced Ca^{2+} release increased in parallel up to about four (see Fig. 10, ref. 58). Assuming random orientation, only half of the channels could be activated with cGMP whereas the exchanger would mediate Ca^{2+} flux regardless of orientation. We concluded from these findings that the exchanger outnumbered the channel by a factor of two.

Surprisingly, channel and exchanger appeared to be associated after reconstitution of ROS membrane proteins into lipid vesicles.[58] This was suggested by the outcome of a competition experiment, where in the same sample cGMP dependent and Na^+ dependent Ca^{2+} releases were induced consecutively. If channel and exchanger would be on different vesicle fractions, then the amplitudes of the Ca^{2+}

release should not depend on the order of addition of cGMP and Na^+. However, this was not observed experimentally (Fig. 2). When cGMP was added first and then Na^+, both agents elicited a distinct Ca^{2+} release, but when cGMP was added *after* a saturating Na^+ induced Ca^{2+} release, little or no cGMP-induced Ca^{2+} release was observed. This finding indicated that vesicles containing a channel also contained at least one exchanger molecule. Because channel and exchanger constitute only minor protein components in the ROS membranes, this finding suggested that both proteins form a complex in the plasma membrane. This suggestion will be further examined in the following sections.

INTERACTION BETWEEN SOLUBILIZED CHANNEL AND EXCHANGER

The cGMP-gated channels of rod photoreceptors bind Ca^{2+}-calmodulin with very high affinity.[13,14] Based on this characteristic property, the cGMP-gated channel could be purified in a single step using calmodulin affinity chromatography[13]. The non-bound fraction contained rhodopsin and other membrane proteins, but not the channel. The bound fraction, which was eluted at zero free Ca^{2+}, contained the channel, but some exchanger co-eluted with the channel, as well.[65] Even a second calmodulin affinity chromatographic run did not fully remove the exchanger suggesting that the exchanger was specifically bound to the channel. In agreement with this finding, it was recently reported that purification of the cGMP-gated channel by immuno-affinity chromatography co-purified the exchanger under some conditions.[66]

In contrast to the cardiac-type Na^+/Ca^{2+} exchanger, the retinal Na^+/Ca^{2+}, K^+ exchanger does not contain a putative calmodulin binding site and, therefore, cannot bind to a calmodulin affinity column.[67-69] In fact, we did not observe any binding of the purified retinal Na^+/Ca^{2+}, K^+ exchanger to a calmodulin affinity column. The finding that purification of the channel either by calmodulin affinity or immuno-affinity chromatography co-purified part of the Na^+/Ca^{2+}, K^+ exchanger indicated that the channel has a specific binding affinity for the exchanger. This idea was examined more directly in a Western blot of the purified channel (Fig. 3). After incubation of this blot with purified exchanger protein, monoclonal antibodies against the exchanger identified distinctly the band of the α-subunit, but not of the β-subunit of the channel. To enhance the detection sensitivity, the exchanger was biotinylated to enable its detection also by streptavidin binding. Horseradish peroxidase conjugated streptavidin detected also a distinct band at the location of the channel α-subunit, indicating that the exchanger specifically bound to the α-subunit, but not to the β-subunit of the channel.

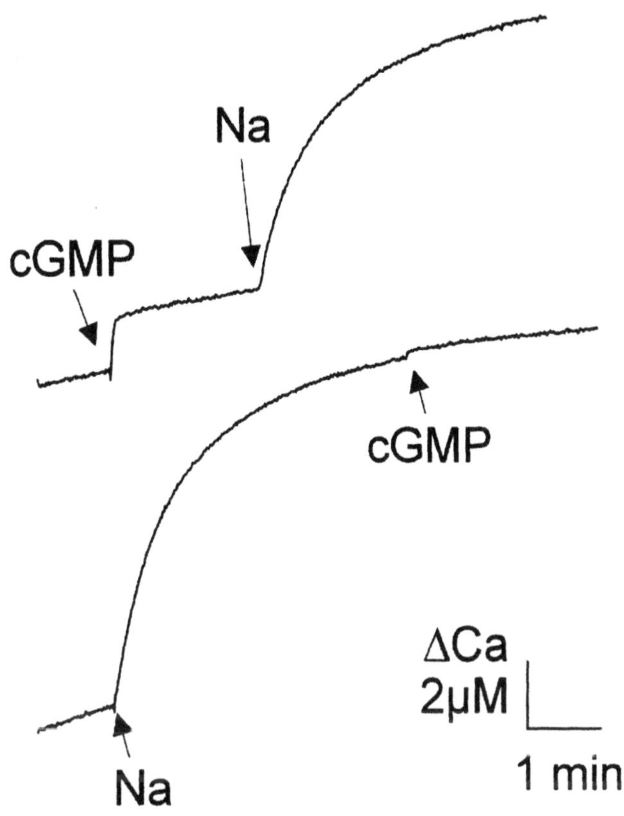

Figure 2. Ca^{2+} flux from ROS proteoliposomes. Na^+ (100 mM) and cGMP (100 μM) was added at the indicated times. Ca^{2+} was monitored photometrically at 652 nm with the Ca^{2+} indicator arsenazo III.

ASSOCIATION OF CHANNEL AND EXCHANGER IN THE PLASMA MEMBRANE

The preceding section indicated that the solubilized exchanger binds to the α-subunit of the cGMP-gated channel. However, this does not necessarily mean that this is also the case in vivo. In fact, many membrane proteins, including rhodopsin and the cGMP-gated channel, tend to aggregate when solubilized. Therefore, we addressed the questions: does the exchanger bind to the channel in the plasma membrane? If so, how many exchangers bind to one channel?

A suitable technique to address these questions is covalent cross-linking of adjacent proteins. Thiol-specific cross-linking appeared to be particularly suited because both channel and exchanger contain only a few cysteines. Therefore, the probability of accidental cross-links with non-associated proteins, or intramolecular cross-links was small.

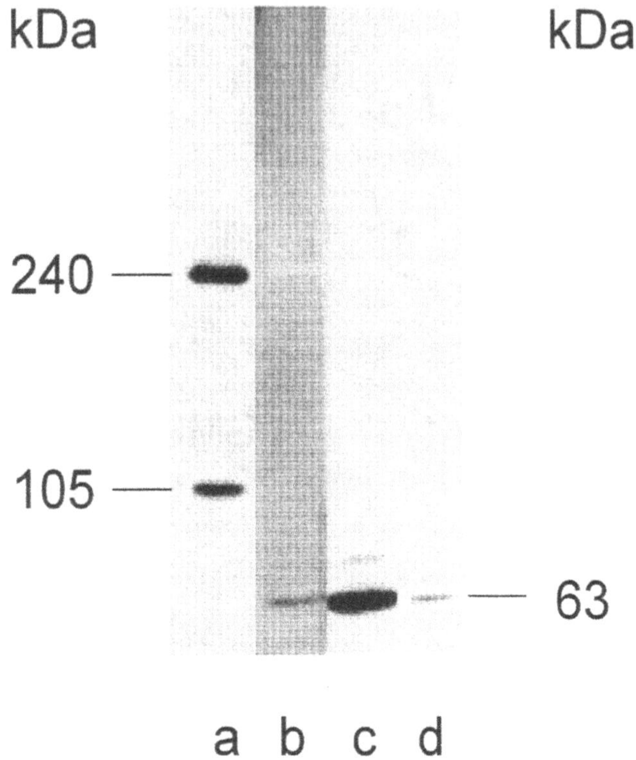

Figure 3. Western blot of a purified cGMP-gated channel labeled with biotinylated purified exchanger. The same blot was re-probed with a polyclonal antibody (PPc 32K) against the channel β-subunit (lane a), with streptavidin horseradish peroxidase (lane b), with a monoclonal antibody (PMc 2G11) against the channel α-subunit (lane c), and with a monoclonal antibody (PMe 1B3) against the exchanger (lane d). Reproduced from Schwarzer et al[65] with permission of the American Society for Biochemistry and Molecular Biology.

Covalent modification of ROS membranes with thiol-specific bivalent reagents yielded cross-links of the exchanger as well as cross-links of the β-subunit, but no cross-links of the α-subunit of the channel. We investigated thoroughly the cross-links of the exchanger.[70] The main cross-link band of the exchanger had an apparent molecular mass of 490 kDa, which was about twice the molecular mass of the monomeric exchanger. Under certain conditions, the exchanger could be almost completely cross-linked to the 490 kDa band. Purification of the cross-linked exchanger and subsequent cleavage of the cross-link indicated that the exchanger existed as a homo-dimer. The purified cross-linked exchanger indicated no accidental cross-linking between exchanger and rhodopsin, the latter being by far the prevalent protein in the plasma membrane. These findings led us to conclude that in the plasma membrane the exchanger molecules exist as dimers.[70]

In the absence of cGMP, thiol-specific reagents did not yield any cross-link between channel and exchanger. Realizing that both α- and β-subunit of the channel contain three cysteines in or near the cGMP-binding domains, we carried out thiol-specific cross-linking in the presence of cGMP. Figure 4 shows that cross-linking with dimaleimides in the presence of cGMP yielded indeed a strong dimer band of the α-subunit at about 130 kDa, whereas in the absence of cGMP only the monomer band at 63 kDa was detected.[65]

Cross-linking in the presence of cGMP also yielded additional bands of the exchanger.[65] These additional bands were observed at about 60 kDa greater molecular masses than the monomer and the dimer band of the exchanger, suggesting that these bands were due to cross-links between the α-subunit of the channel and the exchanger. In agreement with this suggestion purification of the cross-linked channel led to an enrichment of these cross-link bands of the exchanger whereas the other cross-link bands and the monomer band disappeared. Furthermore, chemical cleavage of the cross-links in the polyacrylamide gel and analysis by electrophoresis in a second polyacrylamide gel identified these bands as cross-links between the α-subunit and the exchanger.[65]

The β-subunit of the cGMP-gated channel was almost completely cross-linked with thiol-specific reagents to a host of adduct bands with apparent molecular masses up to about 700 kDa. However, we detected no cross-links between β-subunit and exchanger. No obvious difference was discernible between the cross-link band patterns of the β-subunit observed in the absence and in the presence of cGMP.[65] Presumably, most of these bands originate from cross-links of the β-subunit with peripherin/rds and/or cytoskeletal proteins.[56]

DIFFERENT SUBUNIT ARRANGEMENTS OF THE CHANNEL ENTAIL DIFFERENT QUATERNARY STRUCTURES

Summarizing this cross-linking study, we found that exchanger and channel are associated in the plasma membrane. Moreover, both the α-subunits of the channel and the exchanger can be cross-linked to homo-dimers. The simplest picture compatible with these findings is depicted in Figure 5A: an exchanger dimer is associated with a dimer of α-subunits of the channel. The cGMP-gated channel is assumed to consist of a dimer of α-subunits and a dimer of β-subunits, as suggested recently.[71,72]

It should be stressed that there is direct evidence of (at least) two α-subunits in the natural cGMP-gated channel of the rod photoreceptor, however, the evidence of two β-subunits is indirect. The β-subunit binds Ca^{2+}-calmodulin with a Hill coefficient of 1.36 suggesting binding of two molecules of Ca^{2+}-calmodulin per channel.[14] However, this finding does not necessarily indicate the presence of two β-subunits in the channel, because two Ca^{2+}-calmodulin binding sites have been identified on

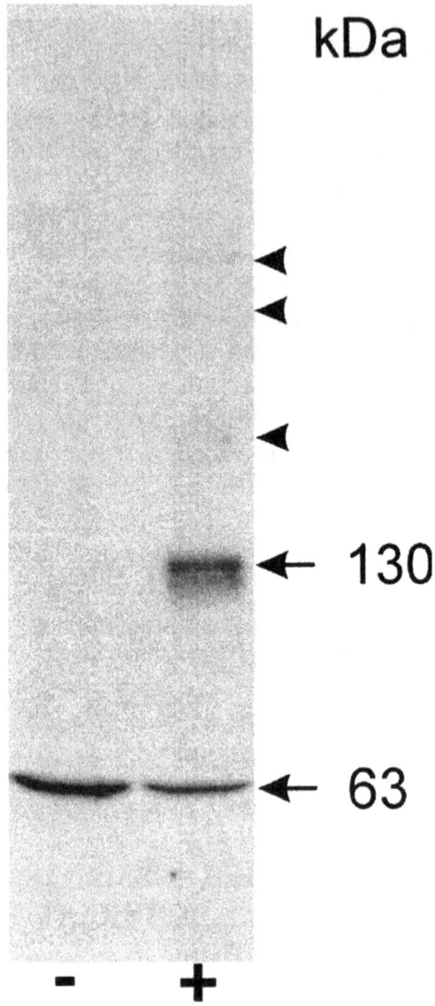

Figure 4. Western blot of ROS membrane proteins, labeled with a monoclonal antibody against the channel α-subunit (PMc 2G11). The ROS membranes were cross-linked with a thiol-specific reagent (BMDB) in the absence (lane labeled: -) and in the presence of cGMP (lane labeled: +). Several weak adduct bands of the α-subunit at higher molecular masses were also observed after prolonged exposure to the cross-linking reagent in the presence of cGMP (arrowheads).

the β-subunit.[73] A more direct indication of a β-subunit dimer derives from the recent finding that the GARP domain of the β-subunit interacts with peripherin/rds of the disc membrane.[56] Peripherin/rds has a strong tendency to dimerize and to associate to higher oligomers.[74] Therefore, it seems likely that the cGMP-gated channel of rod photoreceptors contains two β-subunits which are dimerized.

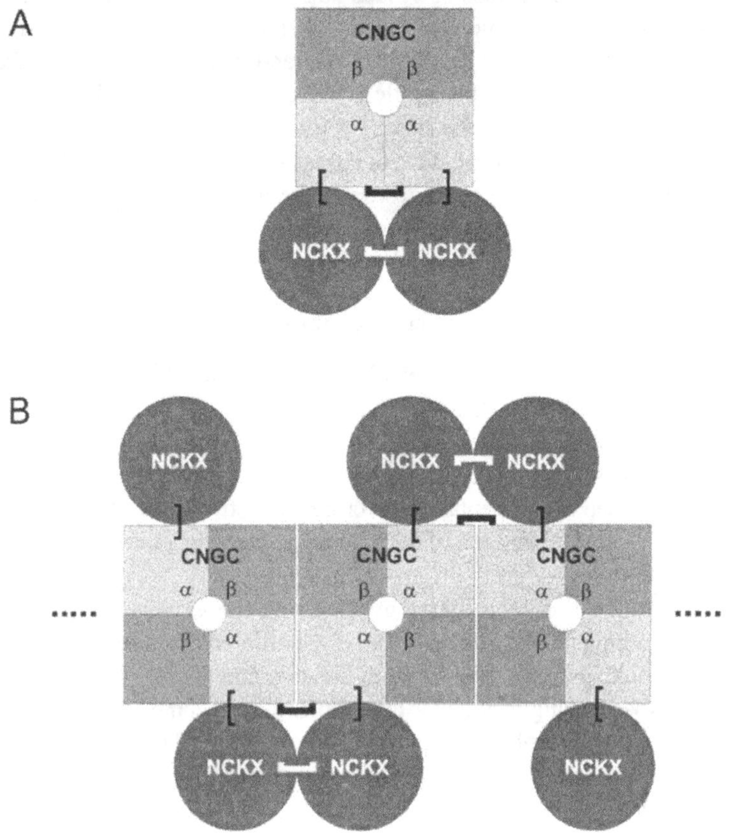

Figure 5. The complex of cGMP-gated channel (CNGC) and Na^+/Ca^{2+}, K^+ exchanger (NCKX) in rod photoreceptors. Brackets indicate thiol-specific cross-links. Bold brackets denote high cross-linking efficiency, thin brackets denote low cross-linking efficiency. Black brackets signify cross-links obtained only in the presence of cGMP, whereas the white brackets signify cross-links obtained independent of cGMP. A) The subunit arrangement of the cGMP-gated channel is assumed to be $\alpha_2\beta_2$ as supported by several reports.[65,71,72] B) A linear cluster of channels and exchangers has to be assumed in order to explain dimers of the α-subunits and the exchangers if the subunit arrangement of the cGMP-gated channels is αβαβ as suggested by He et al.[75] For a discussion of both models, see text.

A different subunit arrangement of the cGMP-gated channel has recently been suggested by He et al.[75] These authors conclude that like subunits of the channel are diagonally opposed. Assuming that only adjacent proteins can be cross-linked, our result would only be compatible with an αβαβ subunit arrangement if the channels were linearly assembled, as depicted in Figure 5B. Although channel clustering is a common phenomenon, clustering of cGMP-gated channels in photoreceptors has not been detected electrophysiologically in "loose-patch" recordings.[55] Clustering of cGMP-gated channels was also not observed when the channel/exchanger complex was allowed to diffuse in a much larger membrane area by fusion of disc and plasma membrane:[58] as pointed out above (see: "The density of cGMP-gated channels"), the results of this study indicated a random distribution of the channels in the membrane. Together, there is little evidence of clustering of the cGMP-gated channels in the plasma membrane of the photoreceptor.

What is the physiological meaning of the association of channel and exchanger? Two mutually non-exclusive hypotheses will be addressed in the following sections: (1), channel and exchanger influence each other functionally, and/or (2), local Ca^{2+} signaling occurs near the channel-exchanger complex.

SELF-INHIBITION OF THE EXCHANGER—AN ALLOSTERIC REGULATORY MECHANISM?

The exchanger can only be studied functionally when the channel is closed. The question whether activation of the channel influences the exchanger activity cannot be examined experimentally. Nevertheless, activation of the channel might entail a small change in distance of the associated exchanger dimer influencing the exchanger activity.

We examined this hypothesis by studying the exchanger activity after cross-linking with thiol-specific reagents which differed in the length of the spacer arm. If cross-linking was carried out with dimaleimides there was only little variation in the cross-linking efficiency between different compounds. Surprisingly, cross-linking generally enhanced the exchanger activity.[76] The enhancement of the activity correlated with the length of the spacer arm of the cross-linking reagent. Considering that only about 50 percent of the exchangers were cross-linked, the maximal enhancement of the exchange rate was estimated to be almost a factor of two. Only cross-linking by disulfide formation led to a reduction of the exchanger activity. These findings suggest that the contact sites of the exchanger dimer are inhibitory.

LOCAL Ca^{2+} SIGNALING IN PHOTORECEPTORS

Generally, microdomains of elevated Ca^{2+} are known to be generated for a short period of time in the close vicinity of Ca^{2+} channels during activation.[77] However, the cGMP-gated channels of photoreceptors are continuously activated in the dark. The close apposition of Ca^{2+} source (the channel) and Ca^{2+} sink (the exchanger) hints to a novel mechanism which restricts part of the Ca^{2+} spatially to the close vicinity of the channel. In the following, the possibility of local Ca^{2+} currents in photoreceptors will be considered on the basis of a simple Ca^{2+} flux model.

To restrict a continuous Ca^{2+} current to a microdomain, two necessary conditions must be met: (1) the Ca^{2+} efflux must balance the Ca^{2+} influx (mass conservation), and (2), the diffusion kinetics of the fluxes be consistent with such a mechanism.

The following calculation shows that the first condition is met in amphibian rod photoreceptors. The plasma membrane area is about 1200 μm^2 and the dark current is 25 pA, or $1.3 \cdot 10^5$ positive charges/μm^2/s.[78] Using the mean channel density of 650 channels/μm^2 reported for the rod photoreceptor of salamander,[55] this value corresponds to an ion flux of 200 positive charges/channel/s. Considering that only 15 percent of the dark current is carried by Ca^{2+} ions,[79] this means that only 30 positive charges/channel/s, i.e., 15 Ca^{2+} ions/channel/s enter the rod outer segment. This value is considerably smaller than the maximal Ca^{2+} export rate of the exchanger of amphibians.[60] For mammalian photoreceptors, the Ca^{2+} influx through the cGMP-gated channel is likely to be similar to that of amphibian photoreceptors. Therefore, the Ca^{2+} export by the two exchangers associated with the channel occurs far from saturation and can readily keep up with the Ca^{2+} influx through the channel.

Are the kinetics of intracellular Ca^{2+} diffusion compatible with the idea of a local Ca^{2+} current? To address this question, two Ca^{2+} concentration profiles will be considered: first, the stationary Ca^{2+} concentration near a point source alone, and second, near a point source coupled with a point sink (Fig. 6). The total Ca^{2+} content of a hemisphere centered at the point source, J_1, and at a point source plus a point sink, J_2, can be calculated.[80] The difference of the two values divided by J_1 yields the fraction of Ca^{2+} which does not leave the close vicinity of the channel.

It should be noted that the combined effect of source and sink can be obtained by linear superposition of the solutions of the source and the sink alone because the differential equations of Ca^{2+} diffusion are linear.[81] The rate of Ca^{2+} extrusion by the exchanger is given by a Michaelis relation with a Ca^{2+} binding constant of 1.6 μM.[82]

The most important parameter in this consideration is the intracellular Ca^{2+} diffusion constant. Therefore, the fraction of local Ca^{2+} was calculated for different diffusion constants (Fig. 7). In the cytoplasm, the Ca^{2+} diffusion constant is about 220 μm^2/s at high Ca^{2+} concentrations (i.e., 1 mM); however, it strongly depends on the Ca^{2+} concentration, and drops under physiological conditions to about 13 μm^2/s due to buffered diffusion.[83] Figure 7, curve a, shows the result if a distance of 5 nm between the channel pore and the Ca^{2+} binding sites of the exchangers is assumed, and the maximal exchange rate of the two exchangers is taken to be 120 Ca^{2+}/s.[84] If, as discussed above, the exchange rate would increase by a factor of two upon channel activation, the result would be that depicted in curve b of Figure 7. The gray line marks the diffusion constant of 13 μm^2/s determined by Allbritton et al for cytoplasmic Ca^{2+} diffusion under physiological conditions.[83]

The main result of this estimation is that the association of channel and exchanger presumably results in a considerable fraction of localized Ca^{2+} of at least 30%, assuming a Ca^{2+} diffusion constant of 13 μm^2/s (Fig. 7, red line). This Ca^{2+} diffusion constant can also vary locally. For the following reasons it is likely that in the close vicinity of channel-exchanger-complex this constant is in fact smaller than 13 μm^2/s. The β-subunit of the channel has a very acidic stretch on the carboxyl terminus, the GARP domain. The amino terminus of the β-subunit is very acidic, as well.

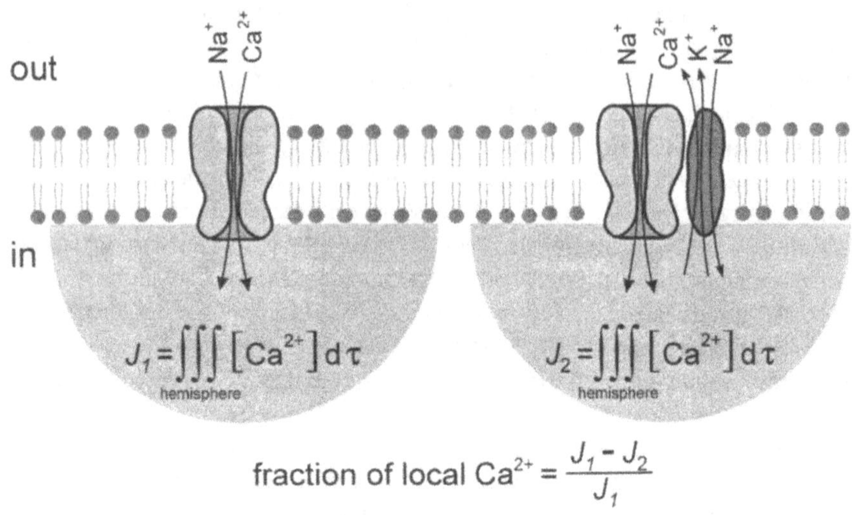

Figure 6. Schematic representation of ion fluxes through the cGMP-gated channel alone (left), and when associated with the exchanger (right) to estimate the fraction of local Ca^{2+} flux.

Furthermore, the intracellular loop of the Na^+/Ca^{2+}, K^+ exchanger is highly acidic. Interestingly, the carboxyl terminus of the α-subunit is highly basic, so that possible electrostatic interaction between positive and negative domains may occur. Nevertheless, the acidic amino acids prevail by far, so that the local Ca^{2+} diffusion constant is likely to be even below 13 $\mu m^2/s$. Consequently, the fraction of local Ca^{2+} should be even higher than the values calculated with the diffusion constant 13 $\mu m^2/s$ (see Fig. 7).

Interestingly, the fraction of local Ca^{2+} current turns out to be almost independent of the value assumed for the Ca^{2+} influx as long as the Ca^{2+} influx is much smaller than the maximal exchange rate. The reason for this result is that, e.g., an increase in the Ca^{2+} influx entails a corresponding increase of the rate of Na^+/Ca^{2+}, K^+ exchange, resulting in an almost identical fraction of local Ca^{2+}.

In summary, the necessary conditions for a local Ca^{2+} current are met. At least 30 percent of the Ca^{2+} entering the outer segment of the rod photoreceptor are presumably confined to the microdomain of the channel exchanger complex.

CONCLUSION AND OUTLOOK

In the plasma membrane of photoreceptors, cGMP-gated channel and Na^+/Ca^{2+},K^+ exchanger form a complex of two exchangers binding to one channel. Binding occurs to the α-subunit, but not to the β-subunit of the channel. Both, α-subunit and exchanger form dimers. Assuming that the channel has a tetrameric structure, the channel exchanger complex consists presumably of a stoichiometry of $x_2\ \alpha_2\beta_2$

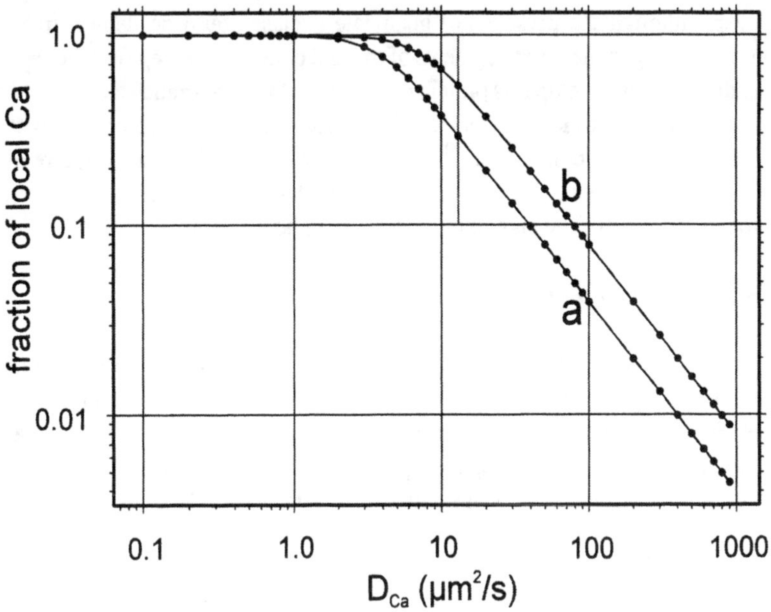

Figure 7. Estimation of the fraction of localized Ca^{2+} current. For the calculation, a distance of 5 nm between the channel pore and the cytoplasmic Ca^{2+} binding site of the exchanger was assumed. The maximal exchange rate was taken to be 120 Ca^{2+}/s (curve a) and 240 Ca^{2+}/s (curve b). The gray line indicates the Ca^{2+} diffusion constant of 13 μm^2/s determined for the cytoplasm of *Xenopus* oocytes at physiological Ca^{2+} concentration by Allbritton et al.[83]

where "x" denotes the exchanger, "α" the α-subunit, and "β" the β-subunit of the channel (Fig. 5).

The association of channel and exchanger hints to a novel mechanism to create a local Ca^{2+} microdomain which is realized in rod photoreceptors. As opposed to the "classical" mechanism, where Ca^{2+} microdomains at Ca^{2+} channels are shaped by mobile Ca^{2+} buffers,[85] restriction of the Ca^{2+} current to a microdomain is achieved by local Ca^{2+} extrusion.[86] Immobile Ca^{2+} buffering domains increase the fraction of localized Ca^{2+} flux. At least part of the Ca^{2+} signaling in rod photoreceptors appears to be based on such a mechanism of localization of the Ca^{2+} flux. The characteristics of local Ca^{2+} domains is fast signaling and the avoidance of cross-coupling.[77] Furthermore, the complex of channel and exchanger enables an allosteric regulation by direct interaction of these proteins.[76]

$Na^+/Ca^{2+},K^+$ exchanger and cGMP-gated channels are only part of a greater protein assembly. Recent evidence strongly indicates that the GARP domains of the cGMP-gated channel interact with peripherin/rds, a major disc rim protein.[56] Therefore, the channel-exchanger complex is apparently tethered to the disc rim. It seems reasonable to assume that the guanylate cyclase, a major effector protein for Ca^{2+} signaling in rod photoreceptors, is close to the channel-exchanger complex.

However, although the presence of guanylate cyclase could be demonstrated by immuno-labeling in the outer segments of rod and cone photoreceptors, the reported results do not allow to map this protein definitely in the membrane stack.[87-89] Moreover, a protein tyrosine kinase has recently been suggested to be associated with the α-subunit of the channel, as well.[90] Further knowledge of the ultrastructure of the proteins assembled with the cyclic GMP-gated channel is required to understand the molecular basis of the signaling processes in photoreceptors.

ACKNOWLEDGMENT

I thank Dr. W. Baehr, Dr. K. Palczewski and Dr. D. Schulze for comments on the manuscript and Dr. R.S. Molday for generously providing the monoclonal antibodies PMc 2G11 and PMe 1B3 against the α-subunit of the cGMP-gated channel and the Na^+/Ca^{2+}, K^+ exchanger, respectively. This work was supported by a grant of the Deutsche Forschungsgemeinschaft (Ba 721/1-3).

REFERENCES

1. Hecht S, Shlaer S, Pirenne MH. Energy, quanta, and vision. J Gen Physiol 1942; 25:819-40.
2. Rieke F, Baylor DA. Single photon detection by rod cells of the retina. Reviews of Modern Physics 1998; 70:1027-36.
3. Burns ME, Baylor DA. Activation, deactivation, and adaptation in vertebrate photoreceptor cells. Annual Review of Neuroscience 2001; 24:779-805.
4. Pugh ENJr, Nikonov S, Lamb TD. Molecular mechanisms of vertebrate photoreceptor light adaptation. Current Opinion in Neurobiology 1999; 9:410-8.
5. Yau K-W. Phototransduction mechanism in retinal rods and cones. Invest Ophthal Vis Sci 1994; 35:9-32.
6. Detwiler PB, Gray-Keller MP. The mechanisms of vertebrate light adaptation: speeded recovery versus slowed activation. Curr Opin Neurobiol 1996; 6:440-4.
7. Ratto GM, Payne R, Owen WG et al. The concentration of cytosolic free calcium in vertebrate rod outer segments measured with Fura-2. J Neurosci 1988; 8:3240-6.
8. Gray-Keller MP, Detwiler PB. The calcium feedback signal in the phototransduction cascade of vertebrate rods. Neuron 1994; 13:849-61.
9. McCarthy ST, Younger JP, Owen WG. Free calcium concentrations on bullfrog rods determined in the presence of multiple forms of Fura-2. Biophys J 1994; 67:2076-89.
10. Sampath AP, Matthews HR, Cornwall MC et al. Bleached pigment produces a maintained decrease in outer segment Ca^{2+} in salamander rods. J Gen Physiol 1998; 111:53-64.
11. Detwiler PB, Gray-Keller MP. Measurement of light-evoked changes in cytoplasmic calcium in functionally intact isolated rod outer segments. Meth Enzymol 2000; 316:133-46.
12. Koutalos Y, Yau K-W. Regulation of sensitivity in vertebrate rod photoreceptors by calcium. *TINS* 1996; 19:73-81.
13. Hsu Y-T, Molday RS. Modulation of the cGMP gated channel of rod photoreceptor cells by calmodulin. Nature 1993; 361:76-9.
14. Bauer PJ. Cyclic GMP-gated channels of bovine rod photoreceptors: affinity, density and stoichiometry of Ca^{2+}-calmodulin binding sites. J Physiol 1996; 494:675-85.
15. Ray S, Zozulya S, Niemi GA et al. Cloning, expression, and crystallization of recoverin, a calcium sensor in vision. Proc Natl Acad Sci USA 1992; 89:5705-9.

16. Gorczyca WA, Gray-Keller MP, Detwiler PB et al. Purification and physiological evaluation of a guanylate cyclase activating protein from retinal rods. Proc Natl Acad Sci USA 1994; 91:4014-8.
17. Kawamura S. Rhodopsin phosphorylation as a mechanism of cyclic GMP phosphodiesterase regulation by S-modulin. Nature 1993; 362:855-7.
18. Koutalos Y, Nakatani K, Tamura T et al. Characterization of guanylate cyclase activity in single retinal rod outer segments. J Gen Physiol 1995; 106:863-90.
19. Koutalos Y, Nakatani K, Yau K-W. The cGMP-phosphodiesterase and its contribution to sensitivity regulation in retinal rods. J Gen Physiol 1995; 106:891-921.
20. Koch K-W. Control of photoreceptor proteins by Ca^{2+}. Cell Calcium 1995; 18:314-21.
21. Senin II, Zargarov AA, Alekseev AM et al. N-myristoylation of recoverin enhances its efficiency as an inhibitor of rhodopsin kinase. FEBS Letters 1995; 376:87-90.
22. Otto-Bruc AE, Fariss RN, Van Hooser JP et al. Phosphorylation of photolyzed rhodopsin is calcium-insensitive in retina permeabilized by α-toxin. Proc Natl Acad Sci USA 1998; 95:15014-9.
23. Ikura M. Calcium binding and conformational response in EF-hand proteins. TIBS 1996; 21:14-7.
24. Young RW. The renewal of photoreceptor cell outer segments. J Cell Biol 1967; 33:61-72.
25. Steinberg RH, Fisher SK, Anderson DH. Disc morphogenesis in vertebrate photoreceptors. J Comp Neurol 1980; 190:501-18.
26. Hsu Y-T, Wong SYC, Connell GJ, Molday RS. Structural and functional properties of rhodopsin from rod outer segment disk and plasma membrane. Biochim Biophys Acta 1993; 1145:85-92.
27. Molday RS, Hicks D, Molday LL. Peripherin. A rim-specific membrane protein of rod outer segment discs. Invest Ophthal Vis Sci 1987; 28:50-61.
28. Arikawa K, Molday LL, Molday RS et al. Localization of peripherin/rds in the disk membranes of cone and rod photoreceptors:relationship to disk membrane morphogenesis and retinal degeneration. J Cell Biol 1992; 116:659-67.
29. Moritz OL, Molday RS. Molecular cloning, membrane topology, and localization of bovine Rom-1 in rod and cone photoreceptor cells. Invest Ophthal Vis Sci 1996; 37:352-62.
30. Goldberg AFX, Molday RS. Subunit composition of the peripherin/rds-rom-1 disk rim complex from rod photoreceptors: Hydrodynamic evidence for a tetrameric quaternary structure. Biochemistry 1996; 35:6144-9.
31. Papermaster DS, Schneider BG, Zorn MA et al. Immunocytochemical localization of a large intrinsic membrane protein to the incisures and margins of frog rod outer segment disks. J Cell Biol 1978; 78:415-25.
32. Illing M, Molday LL, Molday RS. The 220-kDa rim protein of retinal rod outer segments is a member of the ABC transporter superfamily. J Biol Chem 1997; 272:10303-10.
33. Sun H, Molday RS, Nathans J. Retinal stimulates ATP hydrolysis by purified and reconstituted ABCR, the photoreceptor-specific ATP-binding cassette transporter responsible for Stargardt disease. J Biol Chem 1999; 274:8269-81.
34. Bungert S, Molday LL, Molday RS. Membrane topology of the ATP binding cassette transporter ABCR and its relationship to ABC1 and related ABCA transporters. J Biol Chem 2001; 276:23539-46.
35. Körschen HG, Beyermann M, Müller F et al. Interaction of glutamic-acid-rich proteins with the cGMP signalling pathway in rod photoreceptors. Nature 1999; 400:761-6.
36. Eckmiller MS. Microtubules in a rod-specific cytoskeleton associated with outer segment incisures. Visual Neuroscience 2000; 17:711-22.
37. Boesze-Battaglia K, Albert AD. Fatty acid composition of bovine rod outer segment plasma membrane. Exp Eye Res. 1989; 49:699-701.
38. Cone RA. Rotational diffusion of rhodopsin in the visual receptor membrane. Nature New Biology 1972; 236:39-43.

39. Liebman PA, Entine G. Lateral diffusion of visual pigment in photoreceptor disk membranes. Science 1974; 185:457-9.
40. Poo M, Cone RA. Lateral diffusion of rhodopsin in the photoreceptor membrane. Nature 1974; 247:438-41.
41. Calvert PD, Govardovskii VI, Krasnoperova N et al. Membrane protein diffusion sets the speed of rod phototransduction. Nature 2001; 411:90-4.
42. Molday RS, Molday LL. Differences in the protein composition of bovine retinal rod outer segment disk and plasma membranes isolated by ricin-gold-dextran density perturbation method. J Cell Biol 1987; 105:2589-601.
43. Bauer PJ. Evidence for two functionally different membrane fractions in bovine retinal rod outer segments. J Physiol 1988; 401:309-27.
44. Cook NJ, Molday LL, Reid D et al. The cGMP-gated channel of bovine rod photoreceptors is localized exclusively in the plasma membrane. J Biol Chem 1989; 264:6996-9.
45. Reid DM, Friedel U, Molday RS et al. Identification of the sodium-calcium exchanger as the major ricin-binding glycoprotein of bovine rod outer segments and its localization to the plasma membrane. Biochemistry 1990; 29:1601-7.
46. Caretta A, Saibil HR. Visualization of cyclic nucleotide binding sites in the vertebrate retina by fluorescence microscopy. J Cell Biol 1989; 108:1517-22.
47. Boesze-Battaglia K, Kong F, Lamba OP et al. Purification and light-dependent phosphorylation of a candidate fusion protein, the photoreceptor cell peripherin/rds. Biochemistry 1997; 36:6835-46.
48. Fesenko EE, Kolesnikov SS, Lyubarsky AL. Induction by cyclic GMP of cationic conductance in plasma membrane of retinal rod outer segment. Nature 1985; 313:310-3.
49. Bodoia RD, Detwiler PB. Patch-clamp recordings of the light-sensitive dark noise in retinal rods from the lizard and frog. J Physiol 1985; 367:183-216.
50. Gray P, Attwell D. Kinetics of light-sensitive channels in vertebrate photoreceptors. Proc R Soc Lond B 1985; 223:379-88.
51. Zimmerman AL, Baylor DA. Electrical properties of the light-sensitive conductance of salamander retinal rods. Biophys J 1985; 47:357a.
52. Eismann E, Müller F, Heinemann SH et al. A single negative charge within the pore region of a cGMP-gated channel controls rectification, Ca^{2+} blockage, and ionic selectivity. Proc Natl Acad Sci USA 1994; 91:1109-13.
53. Haynes LW, Kay AR, Yau K-W. Single cyclic GMP-activated channel activity in excised patches of rod outer segment membrane. Nature 1986; 321:66-70.
54. Zimmerman AL, Baylor DA. Cyclic GMP-sensitive conductance of retinal rods consists of aqueous pores. Nature 1986; 321:70-2.
55. Karpen JW, Loney DA, Baylor DA. Cyclic GMP-activated channels of salamander retinal rods: spatial distribution and variation of responsiveness. J Physiol 1992; 448:257-74.
56. Poetsch A, Molday LL, Molday RS. The cGMP-gated channel and related glutamic acid rich proteins interact with peripherin-2 at the rim region of rod photoreceptor disc membranes. J Biol Chem 2001; 276:48009-16.
57. Cook NJ, Zeilinger C, Koch K-W et al. Solubilization and functional reconstitution of the cGMP-dependent cation channel from bovine rod outer segments. J Biol Chem 1986; 261:17033-9.
58. Bauer PJ, Drechsler M. Association of cyclic GMP-gated channels and Na^{+}-Ca^{2+} -K^{+} exchangers in bovine retinal rod outer segment plasma membranes. J Physiol 1992; 451:109-31.
59. Yau K-W, Nakatani K. Electrogenic Na-Ca exchange in retinal rod outer segment. Nature 1984; 311:661-3.
60. Hodgkin AL, McNaughton PA, Nunn BJ. Measurement of sodium-calcium exchange in salamander rods. J Physiol 1987; 391:347-70.
61. Hodgkin AL, Nunn BJ. The effect of ions on sodium-calcium exchange in salamander rods. J Physiol 1987; 391:371-98.

62. Lagnado L, Cervetto L, McNaughton PA. Ion transport by the Na-Ca exchange in isolated rod outer segments. Proc Natl Acad Sci USA 1988; 85:4548-52.
63. Cook NJ, Kaupp UB. Solubilization, purification, and reconstitution of the sodium-calcium exchanger from bovine retinal rod outer segments. J Biol Chem 1988; 263:11382-8.
64. Nicoll DA, Applebury ML. Purification of the bovine rod outer segment Na^{+}/Ca^{2+} exchanger. J Biol Chem 1989; 264:16207-13.
65. Schwarzer A, Schauf H, Bauer PJ. Binding of the cGMP-gated channel to the Na/Ca-K exchanger in rod photoreceptors. J Biol Chem 2000; 275:13448-54.
66. Molday RS, Molday LL. Molecular properties of the cGMP-gated channel of rod photoreceptors. Vision Res 1998; 38:1315-23.
67. Li Z, Nicoll DA, Collins A et al. Identification of a peptide inhibitor of the cardiac sarcolemmal Na^{+}-Ca^{2+} exchanger. J Biol Chem 1991; 266:1014-20.
68. He Z, Petesch N, Voges K-Pet al. Identification of important amino acid residues of the Na^{+}-Ca^{2+} exchanger inhibitory peptide, XIP. J Membr Biol 1997; 156:149-56.
69. Hale CC, Bliler S, Quinn TP et al. Localization of an exchange inhibitory peptide (XIP) binding site on the cardiac soidum-calcium exchanger. Biochem Biophys Res Commun 1997; 236:113-7.
70. Schwarzer A, Kim TSY, Hagen V et al. The Na/Ca-K exchanger of rod photoreceptor exists as dimer in the plasma membrane. Biochemistry 1997; 36:13667-76.
71. Shammat IM, Gordon SE. Stoichiometry and arrangement of subunits in rod cyclic nucleotide-gated channels. Neuron 1999; 23:809-19.
72. Liu DT, Tibbs GR, Paoletti P et al. Constraining ligand-binding site stoichiometry suggests that a cyclic nucleotide-gated channel is composed of two functional dimers. Neuron 1998; 21:235-48.
73. Weitz D, Zoche M, Müller F et al. Calmodulin controls the rod photoreceptor CNG channel through an unconventional binding site in the N-terminus of the β-subunit. EMBO J 1998; 17:2273-84.
74. Loewen CJR, Molday RS. Disulfide-mediated ologomerization of peripherin/rds and Rom-1 in photoreceptor disk membranes. J Biol Chem 2000; 275:5370-8.
75. He Y, Ruiz ML, Karpen JW. Constraining the subunit order of rod cyclic nucleotide-gated channels reveals a diagonal arrangement of like subunits. Proc Natl Acad Sci USA 2000; 97:895-900.
76. Bauer PJ, Schauf H. Mutual inhibition of the dimerized Na/Ca-K exchanger in rod photoreceptors. Biochim Biophys Acta 2002; 1559:121-34.
77. Neher E. Vesicle pools and Ca^{2+} microdomains: new tools for understanding their roles in neurotransmitter release. Neuron 1998; 20:389-99.
78. Stryer L. Cyclic GMP cascade of vision. Annu Rev Neurosci 1986; 9:87-119.
79. Nakatani K, Yau K-W. Calcium and magnesium fluxes across the plasma membrane of the toad rod outer segment. J Physiol 1988; 395:695-729.
80. Bauer PJ. Binding of the retinal rod Na^{+}/Ca^{2+}-K^{+} exchanger to the cGMP-gated channel indicates local Ca^{2+}-signaling in vertebrate photoreceptors. Ann NY Acad Sci 2002; in press.
81. Neher E. Usefulness and limitations of linear approximations to the understanding of Ca^{++} signals. Cell Calcium 1998; 24:345-57.
82. Lagnado L, Cervetto L, McNaughton PA. Calcium homeostasis in the outer segments of retinal rods from the tiger salamander. J Physiol 1992; 455:111-42.
83. Allbritton NL, Meyer T, Stryer L. Range of messenger action of calcium ion and inositol 1,4,5-trisphophate. Science 1992; 258:1812-5.
84. Huppertz B, Bauer PJ. Na^{+}-Ca^{2+},K^{+} exchange in bovine retinal rod outer segments: quantitative characterization of normal and reversed mode. Biochim Biophys Acta 1994; 1189:119-26.
85. Naraghi M, Neher E. Linearized buffered Ca^{2+} diffusion in microdomains and its implication for calculation of $[Ca^{2+}]$ at the mouth of a calcium channel. J Neurosci 1997; 17:6961-73.
86. Bauer PJ. The local Ca concentration profile in the vicinity of a Ca channel. Cell Biochem Biophys 2001; 35:49-61.

87. Liu X, Seno K, Nishizawa Y et al. Ultrastructural localization of retinal guanylate cyclase in human and monkey retinas. Exp Eye Res 1994; 59:761-8.
88. Cooper N, Liu L, Yoshida A et al. The bovine rod outer segment guanylate cyclase, ROS-GC, is present in both outer segment and synaptic layers of the retina. J Mol Neurosci 1996; 6:211-22.
89. Yang R-B, Garbers DL. Two eye guanylyl cyclases are expressed in the same photoreceptor cells and form homomers in preference to heteromers. J Biol Chem 1997; 272:13738-42.
90. Molokanova E, Savchenko A, Kramer RH. Interaction of cyclic nucleotide-gated channel subnits and protein tyrosine kinase probed with genistein. J Gen Physiol 2000; 115:685-96.

REGULATION OF VOLTAGE-SENSITIVE Ca^{2+} CHANNELS IN BIPOLAR CELLS BY DIVALENT CATIONS AND POLYAMINES

Eric M. Lasater[1] and Eduardo Solessio[2]

ABSTRACT

Ca^{2+} plays a key role in intracellular signal transduction in neurons but in excess it can lead to cell death. Thus its entry into cells is highly regulated by both extrinsic and intrinsic mechanisms. Little is known of the regulation of Ca^{2+} entry into retinal neurons. Here we describe the role of divalent cations and polyamines as intrinsic modulators of Ca^{2+} entry into retinal bipolar cells. Cone-dominant (small) bipolar cells of the white bass retina were studied using whole cell patch clamp techniques. With biophysical and pharmacological tools it was determined that these cells expressed a Ca^{2+} current similar to an L-type current. This current was very susceptible to blockage by divalent cations including Ca^{2+}. In addition, when tested with the polyamines, spermine, spermidine and putrescine, only spermine effectively inhibited the current. When the dose response curve was fit with the Hill function we found an EC_{50} of 28 μM and a Hill-coefficient of about 2. Our results indicate that divalent cations and the polyamine, spermine, are effective modulators of calcium entry into cone-dominated bipolar cells. The in vivo regulation of the concentrations of these molecules provides an exquisitely sensitive mechanism for regulating Ca^{2+} entry into bipolar cells under different conditions.

[1]Department of Ophthalmology and Visual Sciences, John Moran Eye Center, University of Utah, Salt Lake City, UT 84132, U.S.A.; [2]Center for Vision Research SUNY, Upstate Medical University, Syracuse, NY, U.S.A.

INTRODUCTION

Transient changes in cytosolic Ca^{2+} play a pivotal role in intracellular signaling in neurons. Ca^{2+} modulates pathways such as dendritic signal integration, neurotransmitter release, synaptic plasticity, and ligand-gated ion channel function. The primary mechanism for increasing intracellular Ca^{2+} levels is by Ca^{2+} entry through voltage-activated Ca^{2+} channels.[1] Neuronal Ca^{2+} channel activity is regulated as a means of controlling the amount of Ca^{2+} entering the cell. For example, neurotransmitters and neuromodulators as well as intrinsic agents such as divalent cations and polyamines have been shown to modulate the flow of Ca^{2+} through various types of Ca^{2+} channels.[2,3] At present, our understanding is incomplete of the mechanism(s) of Ca^{2+} entry into retinal neurons and, in particular, the regulation of this entry.

In the retina, bipolar cells are second order neurons that transfer information from the distal to the proximal retina. In the distal retina they make synaptic contact with photoreceptor cells and horizontal cells and in the inner retina they are in contact with amacrine cells and ganglion cells.[4] In some species they are also in contact with interplexiform cells.[5] As with all neurons, Ca^{2+} entry into bipolar cells plays an important role in intracellular signal transduction processes including neurotransmitter release. This entry is regulated by a number of processes including the actions of neurotransmitters and neuromodulators. As mentioned above, in other neuronal systems, mechanisms intrinsic to the cell also have been shown to modulate Ca^{2+} channel activity[6] but little is known about this in retinal bipolar cells. In fact, such intrinsic mechanisms have not been studied in detail in retinal neurons other than in photoreceptor cells.

In an effort to better understand intrinsic mechanisms that might regulate Ca^{2+} entry into bipolar cells, we studied Ca^{2+} currents found in isolated teleost bipolar cells. Here we report on divalent cations and the polyamine spermine as a mechanism for the bipolar cell to regulate its Ca^{2+} current and thus, the amount of Ca^{2+} entry into these cells.

METHODS

Preparation

This study was conducted ON bipolar cells isolated from the retina of the white bass (*Morone chrysops*). We have described the isolation procedure in detail in the past.[7] In brief, eyes were removed from the fish following euthanasia in accordance with the guidelines set out by the Association for Research in Vision and Ophthalmology for the Use of Animal in Research. Retinas were removed from the eyes and placed in an enzyme (papain) solution. Following incubation, the retinas were then triturated and plated out into cell culture dishes. The cells were then used after between one to seven days in culture. Two types of bipolar cells result from the trituration procedure. Bipolar cells were identified by their characteristic morphology[7] (see below).

Solutions

At the beginning of each experiment the tissue culture media in a dish was replaced with white bass Ringer's containing the following (in mM):130 NaCl, 2.5 $CaCl_2$, 2.8 KCl,1 $MgCl_2$,10 glucose and 8.4 HEPES. The pH was adjusted to 7.4 using NaOH. The recording pipette solution contained (in mM):120 cesium chloride, 4NaCl, 11EGTA, 1$MgCl_2$, 8.4HEPES and 20 tetraethylammonium. For the divalent free solution, both magnesium chloride ($MgCl_2$) and calcium chloride ($CaCl_2$) were omitted and EDTA and EGTA were added at 5mM each. In the Na^+ free solution, NaCl was replaced with an equimolar amount of choline chloride.

Solutions were applied from a micro-perfusion system (DAD-12, ALA Scientific Instruments, Westbury, NY). Drugs could be rapidly applied and the solutions completely bathed the cell under study. Unless otherwise indicated, cells were perfused with the divalent free solution prior to test agent application. The test agents were dissolved in the divalent free media.

Recording Procedures

Whole cell recordings were made with micro-pipettes that had been pulled on a two-stage puller (Narishige Instruments, Tokyo, Japan) and were used unpolished. The electrode tip resistance when measured in the bath solution was from 5 to 7 MΩ. Electrode capacitance and series resistance were compensated for electronically. Electrical potentials were recorded using an Axopatch 200A amplifier (Axon Instruments, Union City, California). A personal computer interfaced to a Digidata 1200 data acquisition system driven by the pClamp suite of programs (Axon Instruments) controlled data collection and analysis. The recordings shown are representative of three or more samples.

RESULTS

In isolation, bipolar cells were identified by their characteristic shape. There are two types of bipolar cells found in the white bass retina following isolation.[7] One has been termed a large bipolar and the other a small bipolar. The large bipolar corresponds to the mixed rod-cone bipolar found in the teleost retina[8] which is characterized by a relatively large cell body with several apical processes opposite of which is a long thin axon that ends in a relatively large oval terminal. Small bipolar cells have a small oval or tear shaped soma and on the apical end, contain several small fine dendritic processes. A long thin axon emanates from opposite the apical end to terminate in a small axonal terminal. These likely correspond to cone-dominated teleost bipolar cells. Our studies here were confined to the small bipolar cell type.

The complete membrane properties of bipolar cells in the bass retina have been previously described[7] so we won't go into the details here. In Figure 1, in the upper left corner, is illustrated a control voltage clamp recording from a small bipolar cell. The current response to voltage commands is characterized by a slowly activating inward current that is characteristic of an inwardly rectifying H-type K^+ current.[9] Depolarizing responses are characterized by a slowly activating outward K^+ current (IK_V). Also present in these cells, but not visible in the control recordings because of the large K^+ currents, is a small inward Ca^{2+} current which is elicited at voltage levels more positive than the holding potential.[7]

Since it was our desire to study the Ca^{2+} current, we tried to bring out the current by taking advantage of the fact that in a zero divalent cation solution, Ca^{2+} channels are highly permeable to monovalent ions and will conduct Na^+ ions readily, generating a measurable inward current.[10] It also took Ca^{2+} and Mg^{2+} away as possible modulators of the channel allowing us to study their action in isolation. So, voltage clamp recordings in a divalent free extra-cellular solution were made and revealed small voltage-dependent changes in the bipolar cell currents. This is illustrated in the upper right panel of Figure 1. In the lower half of the figure are shown current-voltage relationships measured 5 milliseconds after the onset of the voltage command (left) and 50 milliseconds following the onset of the voltage command (right). What was seen was a change in the outward currents evoked by the voltage steps. They increased in magnitude over control values at more depolarized levels. However, no inward current appeared in the voltage range typical of Ca^{2+} currents. It seemed likely that this was due to masking by K^+ currents which were still relatively large.

We next blocked the inward and outward K^+ currents with 20mM TEA in addition to cesium replacing K^+ ions in the intracellular recording solution. This revealed a reduction in the K^+ currents observed. The effect is illustrated in Figure 2. In the upper left corner is shown a control recording; the Ca^{2+} current was still not very prominent. We went on and tried the intracellular TEA in the presence of zero divalent Ringer's. The result of this manipulation is shown the upper right corner of the figure. In the zero divalent solution, both the inward and outward currents have increased in magnitude. The difference between the control currents (TEA and cesium in the pipette) and the zero Ca^{2+}, zero Mg^{2+}/control currents revealed the true inward current. In the lower right corner of Figure 2 is a current-voltage relationship illustrating the three recording conditions. One can see that the inward current was activated at around –50 mVs and reached a peak at about –20 mVs. Hence forth, all future recordings were made with TEA in addition to cesium in the recording pipette. Also, the I-V curves illustrate that the magnitude of the current was relatively small. We found this to be the case in all small bipolar cells from which we recorded.

The I-V relationship and sustained time course of the currents that we recorded was similar to that of a high voltage-activated, L-type Ca^{2+} current, although the peak current occurred at a more hyperpolarized level than is characteristic for most L-type currents. However, we recorded this current in a zero Ca^{2+} medium. Thus,

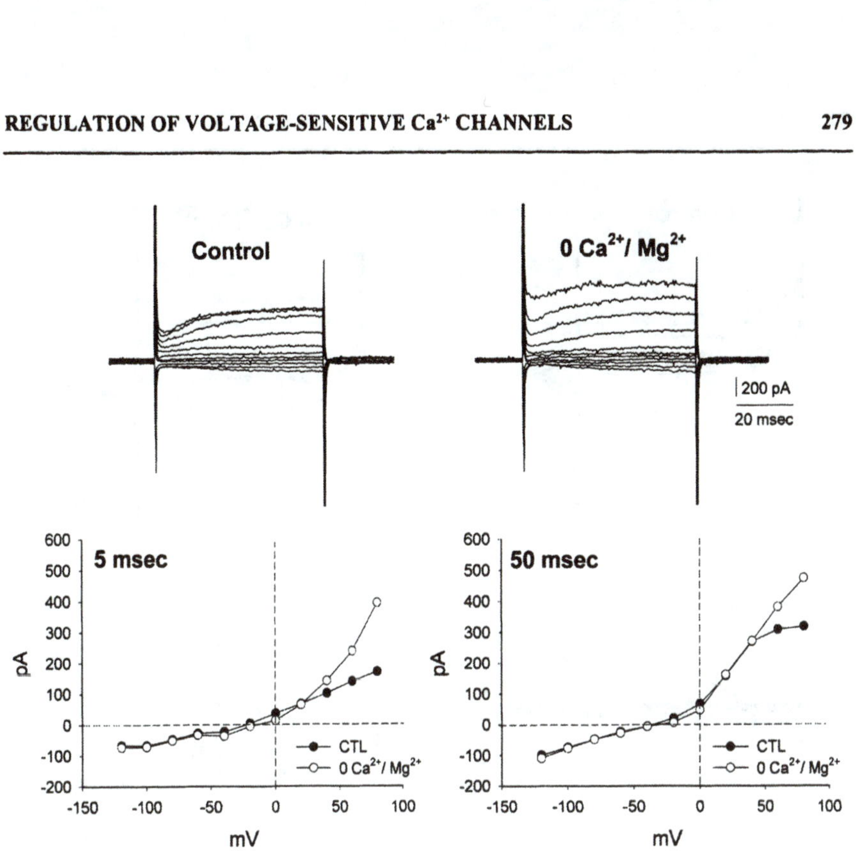

Figure 1. Typical voltage clamp responses of a cone-dominated bipolar cell. Upper left. The cell was held at -70mV and stepped in 20mV increments above and below the holding potential, ranging from -120mV to +80mV. The response is characterized by a slowly activating inward current, which corresponds to an H-type K^+ current. The outward current is a slowly activating outward K^+ current. Upper right. Voltage clamp response of the bipolar cell when superfused with a zero Ca^{2+}/Mg^{2+} solution. Here, the inward current has changed little, however, the outward current has increased in magnitude. Lower left plot. Plotted is an I-V relationship measured 5 ms after the onset of the command potential. The filled circles show the recordings in control solution (CTL in this and all following figures) and the open circles plot the responses in zero Ca^{2+}/Mg^{2+} solution. Lower right. Plotted is the response measured 50 ms following the onset of the command potential. Filled circles represent the control responses and open circles represent the responses recorded in zero Ca^{2+}/Mg^{2+}. The outward current increased in magnitude in the zero divalent solution.

this current may be carried by Na^+ ions flowing through Ca^{2+} channels.[10] We next determined that in fact, it was the case that Na^+ was flowing into these cells through voltage-activated Ca^{2+} channels and we further characterized the channel. Figure 3 (left) illustrates the effects of the Na^+ channel blocker TTX on the current. Recordings were done in control conditions, zero Ca^{2+}/Mg^{2+} solution and in the presence of TTX (10 μM). The I-V relationship shows the current in zero Ca^{2+}/Mg^{2+} solution (open triangles) and the results of the application of TTX (solid squares). The TTX had no effect on the current measured, indicating that the current was not flowing through voltage-dependent Na^+ channels. We then subjected the cells to a solution containing zero Na^+ (see methods). Under these conditions, no current was observed to flow into the cell over the activation range for the Ca^{2+} channels (open diamonds).

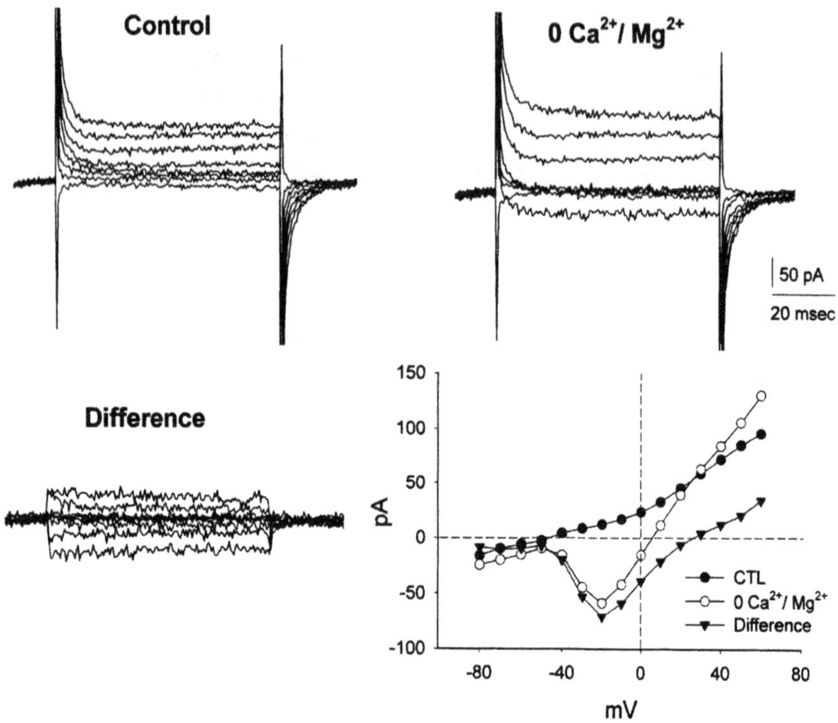

Figure 2. Typical voltage clamp recordings from a small bipolar cell with 10 mM TEA in the intracellular pipette solution. Upper left is the control recording. The cell was held at -70mV and stepped from -80mV to +60mV in 10mV increments. Only every other trace is shown. The I-V relationship for this control recording is shown in the lower right-hand plot with the filled circles. Upper right-hand. Illustrated is the response in the presence of zero Ca^{2+}/Mg^{2+} solution. Under these conditions, a prominent inward current was observed. Lower left-hand corner. The traces in the upper right-hand figure were subtracted from those in the upper left-hand figure and the result is shown as the difference currents. The difference current was used for the analysis in Figures 3 and 4 as well. The I-V relationship in zero Ca^{2+}/Mg^{2+} solution and the difference currents are plotted in the lower right-hand figure as the open circles and the filled triangles respectively. Under conditions where K^+ currents are blocked with intracellular TEA, in a zero divalent solution, an inward current is present as seen in the I-V relationship.

In a second set of experiments, we found that the Ca^{2+} channel blocker, nifedipine (10 μM), substantially inhibited the current. This is shown in Figure 3 to the right. The application of nifedipine resulted in an I-V curve in which the current was reduced by about 80% (solid triangles). Washing the cell with zero Ca^{2+}/Mg^{2+} media resulted in a return to control as illustrated by the curve plotted with the open triangles. The L-type Ca^{2+} channel blocker diltiazam had a similar effect (not shown). The results from Figure 3 illustrate that indeed we are observing Na^+ ions flowing through the bipolar cell Ca^{2+} channels in a zero Ca^{2+}/Mg^{2+} solution and that these channels are likely to be similar to L-type channels. Consequently, we are able to

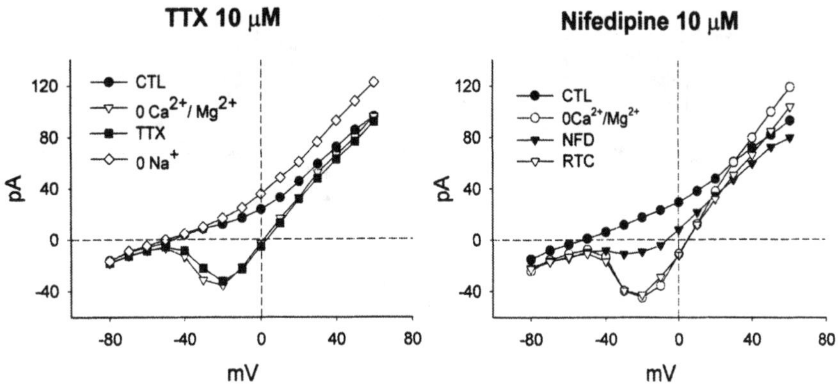

Figure 3. Effect of TTX and nifedipine on the inward current recorded from a small bipolar cell. Left figure shows the effects of TTX. Plotted with the closed circles is the control response and plotted with the open triangles is the response in the presence of zero Ca^{2+}/Mg^{2+} solution. Application of TTX (filled squares) had no effect on the inward current. However, removing Na^+ from the bathing media (open diamonds) abolished the inward current, indicating that the current was being carried by Na^+ ions. Leftside I-V relationship illustrates the effects of the Ca^{2+} channel blocker, nifedipine, on the inward current. Plotted with the filled circles is the response of the I-V relationship under control condition. Plotted with the open circles is the relationship while the cell was subject to zero divalent solution. The application of nifedipine (filled triangles) reduced the current by about 80%. Washing the nifedipine from around the cell resulted in a return to control (RTC), plotted with the open triangles. The results from these experiments indicate that under zero Ca^{2+}/Mg^{2+} conditions, Na^+ ions flow through the Ca^{2+} channel into the bipolar cell.

measure the bipolar cell Ca^{2+} current in an indirect fashion and examine the modulatory actions of Ca^{2+} and Mg^{2+} in isolation.

We next investigated the susceptibility of the current to block by divalent cations. Surprisingly, we found that as little as 0.1mM Ca^{2+} in the extracellular media would substantially reduce the current. This is shown in Figure 4 (left). In this I-V plot, the current carried by Na^+ through the Ca^{2+} channels is illustrated by the trace with the open circles. The trace with the solid triangles illustrates the current flowing through these channels when 0.1mM Ca^{2+} is present in the bathing media. The current has been reduced by about 90% (the degree of block remained unchanged in 2.5 mM Ca^{2+}). In a similar fashion, the divalent cation Ba^{2+} is also able to inhibit the current, but not to the same degree as Ca^{2+}. This is illustrated in Figure 4, right, by the trace with the open triangles. Interestingly, increases in the concentration of Ba^{2+} did not improve the block. Rather, there was a depolarizing shift in the activation voltage and an increase in the peak current observed. Increasing the Ba^{2+} concentration from 0.1mM to 1.0 mM produced a reduction in the size of the observed current. However, further increasing the concentration of Ba^{2+} to 10 mM caused a subsequent increase in the size of the current as well as a depolarizing shift in the activation range.

The results of Figure 4 (right), can be explained in terms of an anomalous mole-fraction effect between Na^+ and Ba^{2+} ions[9] in a multi-ion pore: in a divalent

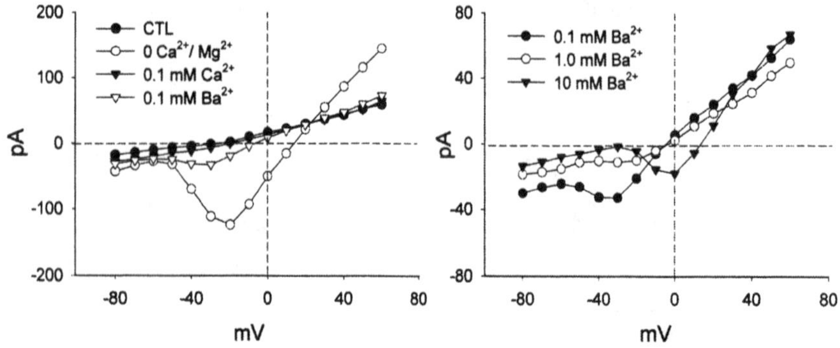

Figure 4. Block of the inward current by divalent cations. Left panel shows the I-V relationship for small bipolar cell responses in zero Ca^{2+}/Mg^{2+} solution when being blocked by low Ca^{2+} (filled triangles) or Ba^{2+} (open triangles). Ca^{2+}, 0.1 mM, substantially blocked the current into these cells. A similar concentration of Ba^{2+} reduced the current by about 90%. Right panel illustrates the I-V curve for three different concentrations of Ba^{2+}. (See text for details.)

free solution, Na^+ ions flow freely through the pore. At low concentration, Ba^{2+} is entering the Ca^{2+} channel but likely binding, thus producing a small block of the Na^+ currents. At higher concentrations the repulsion between Ba^{2+} ions drives the current across the channels, which is now carried by Ba^{2+} ions. Anomalous molar fraction effects have been described in L- and N- type Ca^{2+} channels[11,12] and in fact accounts for why Na^+ can flow through the channel in the absence of divalents.[13] The shift in the activation range has been previously observed and is probably related to surface charge effects.

Intracellular polyamines at endogenous concentrations have been shown in past studies to block or modulate K^+ channels,[14-16] glutamate activated channels[17] and nicotinic acetylcholine activated ion channels[18] in a voltage dependent fashion. Polyamines applied extracellularly can also block K^+ currents[19] and L-type Ca^{2+} currents.[20] So, we next investigated the effects of polyamine application. Figure 5A shows the effect of the polyamine spermine in zero Ca^{2+}/Mg^{2+} media. The closed circles represent the control Na^+ current. Application of spermine reduced the current by about 80%. We went on and tested other polyamines. Extracellular application of 200 μM putrescine had no effect on the divalent free currents (not shown). The polyamine spermidine (200 μM) did not show a blocking action although it tended to broaden the I-V relationship somewhat.

We found the block by spermine to be concentration dependent. This is illustrated in Figure 5B. Here the control I-V relationship of the Na^+ current flowing through Ca^{2+} channels is compared to the currents measured with the application of 15, 30 and 120 μM of spermine. As can be seen from the I-V traces, as the concentration of the applied spermine increased, the peak of the I-V curve systematically became smaller. This dose response relationship is plotted in Figure 5C. The response curve was then fit with the Hill function. From this we obtained an EC_{50} of 28 μM and a Hill-coefficient (n) of ~2.0.

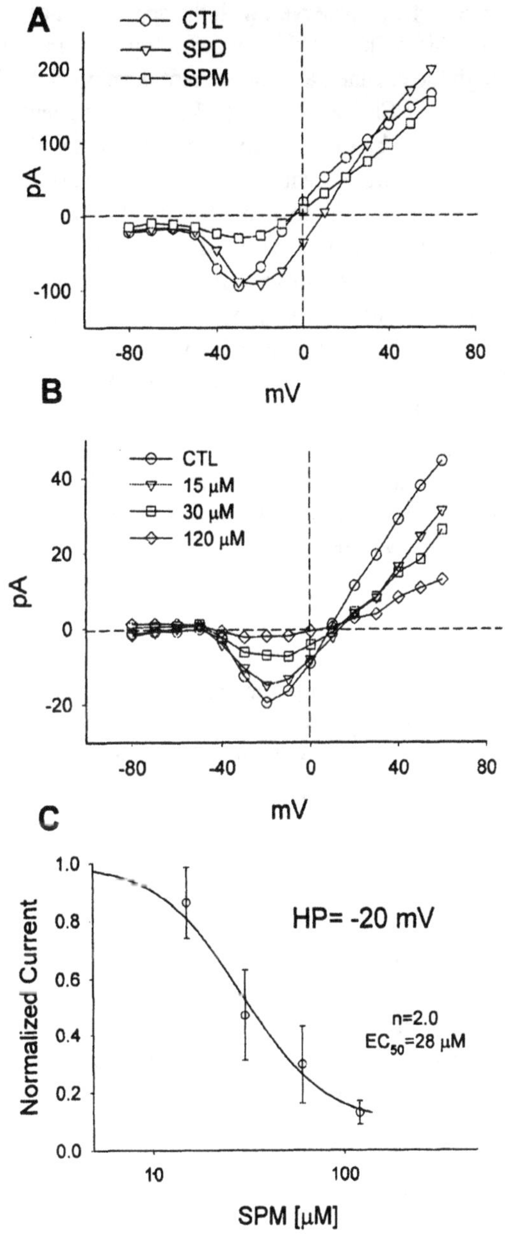

Figure 5. Effects of polyamines on the inward current. The I-V relationship plotted with the open triangles is that which was obtained when the cell was superfused with 200 μM of the polyamine spermidine (SPD). Spermidine had little effect on the current flow into the cell. However, 200 μM spermine (SPM) block the current by about 80% (open squares). In panel B is shown the block by varying concentrations of spermine. As the concentration of spermine superfusing the cell is raised, a greater degree of blockage of the inward current is produced. Panel C illustrates this dose response curve. (See text for details.) Here control (CTL) refers to the difference current.

An unexpected effect was observed with the application of very high concentrations of spermine. This is illustrated in Figure 6. Rather than increasing the block of the current through the channel, a linear current developed. In the presence of a very low concentration of Ca^{2+} and Mg^{2+} (0.5mM) the current through the Ca^{2+} channel carried by Na^{+} was blocked. In this same solution, with the application of 2mM spermine, a large inward current was seen. When 2mM of spermine was applied in the presence of a normal concentration of Ca^{2+} and Mg^{2+} (2.5mM) the inward current was reduced by more than two-fold. The high concentration of spermine also eliminated the voltage-sensitive characteristics of the channel for unknown reasons. This may be due to surface charge effects or mass action simply driving the spermine through the channel. When the cells were returned to the control Ca^{2+}/Mg^{2+} solution in the absence of spermine, the inward current was once again blocked. This indicates that the cell remains viable and this current was not due to some kind of toxic effect by the spermine. Results from this experiment suggest that at very high concentrations, spermine is able to flow through the Ca^{2+} channel without binding and producing a block. This is analogous to the block of rod photoreceptor cGMP-gated channels by spermine.[21]

CONCLUSION

Intrinsic mechanisms play an important role in altering an ion channel's ability to conduct current and thus regulate ion entry into a cell. In this study we have shown that divalent cations and the polyamine spermine play a central role in the modulation of Ca^{2+} channel conductance in teleost, cone-dominated bipolar cells (small bipolars).

Bipolar Cell Ca^{2+} Channels

Previous studies have demonstrated several types of Ca^{2+} channels in retinal bipolar cells. In rodents, both low voltage activated (LVA) and high voltage activated (HVA) currents are present.[22,23] In lower vertebrates only HVA currents have been reported in bipolar cells of goldfish[24,25] salamander[26] and in 'large' (mixed rod-cone) bipolar cells of white bass,[7] HVA and LVA Ca^{2+} currents are closely associated with neurotransmitter release in bipolar cell synaptic terminals.[24,27,28]

Consistent with this, HVA Ca^{2+} channels locate at higher densities in the synaptic terminal.[24,25] Althought we did not record specifically from axon terminals, a presynaptic localization of Ca^{2+} channels is also likely in the case of the small bipolar cells in bass as can be inferred from our results in figures 1 and 2. As noted, the cells were clamped at the soma and not the synaptic terminal which here is very small in size (<3-4 μm). This is unlike the mixed rod-cone, Ab, bipolars in goldfish and carp that can have a terminal of ~10 μm in diameter. In our studies, the currents through the Ca^{2+} channels were only evident (of sufficient magnitude) after blocking the K^{+} channels, suggesting that the channel blockage improved the capacity to adequately clamp the terminal at the end of the long and thin axonal process. While

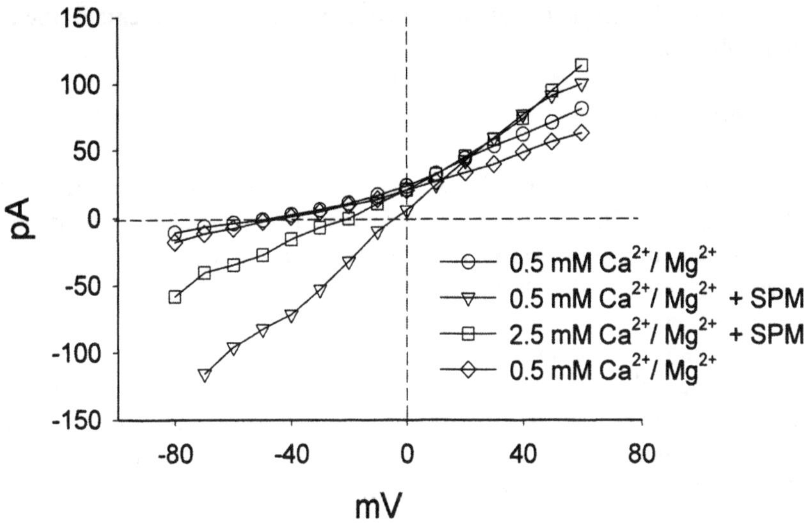

Figure 6. Permeation of the bipolar cell Ca^{2+} channel by spermine. At high concentrations the spermine (SPM) is able to permeate the channel rather than produce a block (open triangles) in low divalent extra-cellular solution. The current produced by a high concentration of spermine is inhibited but not completely blocked in a normal Ca^{2+}/Mg^{2+} containing Ringer's solution (open squares). (See text for details.)

there is a high concentration of Ca^{2+} channels in the axon terminal, this is not to suggest that they solely contributed to the currents we recorded. It is likely that some of the Ca^{2+} channels underlying the Ca^{2+} currents, were located in other regions of the cell.

Despite the improved clamping conditions, we had to go to a divalent free media to enhance the magnitude of the currents flowing through the Ca^{2+} channels.[10] Under these conditions the currents are carried by monovalent ions, predominantly Na^+. The currents are sustained in nature, and triggered at relatively depolarized levels. They are sensitive to nifedipine and not to TTX, consistent with the idea that although the currents are not carried by divalent ions, they flow through Ca^{2+} channels exhibiting anomalous mole fraction effects and increased permeability to Ba^{2+} ions. The I-V curve and the pharmacological sensitivity of the current suggest that it is flowing through channels that are, or closely resemble, L-type Ca^{2+} channels. This is consistent with results from other lower vertebrates.[24,25]

It is intriguing however that under normal conditions the magnitude of the current is negligible and cannot readily be measured. One possibility is that the small size of the synaptic terminal in these particular cells, if this is where the channels are predominantly located, results in a scaled down magnitude of the current. A second possibility is that conditions in culture may alter expression of the channels, or it may be possible that these are 'silent' currents that a regulatory agent , may modulate in amplitude[25,29] and are only active under certain conditions. Lastly, they may simply be blocked in culture by the concentrations of divalents in the recording solution.

Ca^{2+} Channel Block

In cone-dominated bipolar cells, we found that the extracellular divalent cations Ca^{2+}, Mg^{2+} and Ba^{2+}, blocked the Ca^{2+} channels at physiological concentrations. In the absence of divalents, Na^{+} permeates the channel. This is due to the permeation properties of the channel. Energy profiles suggest that two Ca^{2+} ions are involved in the permeation process, whereas for Na^{+}, three ions are involved.[13] Because of the energy profiles in the channel, it is easy for a Ca^{2+} ion to enter and drive out one or more Na^{+} ions, but once a Ca^{2+} ion is in the channel Na^{+} ions cannot do the same.[13] A similar situation exists for Mg^{2+} and Ba^{2+}, however, as illustrated by the 'anomalous mole-fraction effect', Ba^{2+} is less effective as a blocker.

We found that extracellular application of the polyamine spermine can reduce the magnitude of these currents at relatively low concentrations (EC_{50} = 28 μM). Polyamines are a group of amine containing compounds that are synthesized as an intermediate as part of the catabolism of arginine.[30] Normally the intracellular concentration is highly regulated.[31] Levels can rise under various conditions including in a circadian cycle [32]and interact with cellular processes in a variety of ways.[3,31] As we showed here, in retinal bipolar cells they seem to interact with Ca^{2+} channels and alter the magnitude of the Ca^{2+} currents. Along these lines, in the case of certain disease states[33] and cell lysis, local extracellular levels of spermine could rise to micromolar levels blocking entry of Ca^{2+}, and perhaps exert a neuroprotective effect.

The mechanism by which polyamines act as Ca^{2+} channel blockers is related to their binding within the pore. We found spermine to be the most effective polyamine blocker. Spermine has four positive charges whereas spermidine has three [3], so spermidine may not bind as effectively. As mentioned, two Ca^{2+} ions seem to be involved in the permeation process,[13] suggesting that there are two binding sites for Ca^{2+} within the channel. Spermine more effectively interacts with these binding sites to effect a block. In fact, we found a Hill coefficient of ~ 2 for spermine block suggesting that a spermine molecule is binding at each site. This mechanism appears to be the way by which the polyamine toxin FTX acts.[2,34,35] Or, it may be that spermidine binds, but because of its smaller size is not a sterically effective blocker.[36] It is interesting that at low concentrations spermine blocks the channel but at high concentrations permeates it. This suggests that the block is probably only weakly to moderately voltage-dependent[36] although we did not test this specifically. Permeation of the channel also indicates that the block may not be difficult to overcome under certain circumstances.

Functional Implications

Our results here suggest that divalent cations and polyamines are effective agents for modulating Ca^{2+} entry into bipolar cells. In vivo, at rest it is likely little or no

Ca^{2+} would enter the cell. Upon depolarization, only small amounts would enter, assuming normal levels of extracellular cations and physiological levels of polyamines.[31] Should these levels be decreased, more Ca^{2+} would flow into the cell. In photoreceptors, it has been shown that polyamines block photoreceptor cyclic nucleotide-gated channels from both sides of the membrane.[21] From the outside it acts as a permeant blocker. Lu and Ding [21] suggest that the polyamine block may play a role in suppressing noise in the signal transduction cascade in that it would prevent indiscriminant entry of Ca^{2+} which has a major role in the phototransduction cascade. Polyamines may be acting similarly in preventing Ca^{2+} entry into bipolar cells under particular conditions or at particular membrane potential levels. This then might prevent 'noise' in synaptic transmission under certain circumstances such as low light levels.

Obviously the concentration of extracellular cations and polyamines will determine the level of block that occurs. Along these lines, it is interesting that the concentrations of polyamines at which we observed a block, is roughly 30-300 fold lower than that which was needed for block in NMDA-activated channels.[2,37-39] This suggests that extracellular polyamine concentrations are highly regulated in the retina and may exist at lower levels than found in the CNS, otherwise these channels might be permanently blocked. Unfortunately we know little of how or under what circumstances retinal levels of divalents and polyamines might be regulated. In other systems a variety of mechanisms exist for up- and down-regulating cation levels (transport, intracellular release, sequestering, etc.)[1,40-42] and polyamine levels (ie. synthesis, transport, degradation, etc).[3,30,31] These mechanisms may be active in the retina or it may be possible that function is regulated by activity.[17] In this instance, increased levels of activity could lead to a decrease of synaptic levels of divalents and polyamines, resulting in greater influx of Ca^{2+}. It is possible that Müller cells provide such a mechanism for clearing divalents and polyamines from the synaptic cleft.[43-45] Whatever the case, it is clear that divalent cations and polyamines are powerful intrinsic modulators of bipolar cell Ca^{2+} channel activity.

ACKNOWLEDGEMENTS

This work was supported by an NIH grant, EY05972 (E.M.L.) from the National Eye Institute and an unrestricted grant to the Department of Ophthalmology and Visual Sciences, University of Utah from Research to Prevent Blindness, Inc. (RPB). E.M.L. is an RPB Lew R. Wasserman Merit awardee.

REFERENCES

1. Bertolino M, Llinas RR. The central role of voltage-activated and receptor-operated calcium channels in neuronal cells. Ann Rev Pharmacol Toxicol 1992; 32:399-421.
2. Scott RH, Sutton KG, Dolphin AC. Interactions of polyamines with neuronal ion channels. Trends Neurosci 1993; 16:153-60.
3. Johnson TD. Modulation of channel function by polyamines. Trends Pharmacol Sci 1996; 17:22-7.
4. Dowling JE. The Retina: An approachable part of the Brain. Cambridge: The Belknap Press of Harvard University Press, 1987.
5. Dowling JE, Ehinger B. Synaptic Organization of the Amine-Containing Interplexiform Cells of the Goldfish and Cebus Monkey Retinas. Science 1975; 188:270-3.
6. Scott RH, Pearson HA, Dolphin AC. Aspects of vertebrate neuronal voltage-activated calcium currents and their regulation. Prog Neurobiol 1991; 36:485-520.
7. Lasater EM. Membrane currents of retinal bipolar cells in culture. J Neurophysiol 1988; 60:1460-80.
8. Ishida AT, Stell WK, Lightfoot DO. Rod and cone inputs to bipolar cells in goldfish retina. J Comp Neurol 1980; 191:315-335.
9. Hille B. Ionic Channels of Excitable Membranes. Second Ed. Sunderland: Sinauer Associates Inc, 1992.
10. Almers W, McCleskey EW. Non-selective conductance in calcium channels of frog muscle: calcium selectivity in a single-file pore. J Physiol 1984; 353:585-608.
11. Friel DD, Tsien RW. Voltage-gated calcium channels: Direct observation of the anomalous mole fraction effect at the single-channel level. Proc Natl Acad Sci USA 1989; 86:5207-11.
12. Wakamori M, Strobeck M, Niidome T et al. Functional characterization of ion permeation pathway in the N-type Ca^{2+} channel. J Neurophysiol 1998; 79:622-34.
13. Corry B, Allen TW, Kuyucak S et al. A model of calcium channels. Biochim Biophys Acta 2000; 1509:1-6.
14. Wible BA, Taglialatela M, Ficker E et al. Gating of inwardly rectifying K^+ channels localized to a single negatively charged residue. Nature 1994; 371:246-249.
15. Fakler B, Brandle U, Glowatzki E et al. Strong voltage-dependent inward rectification of inward rectifier K^+ channels is caused by intracellular spermine. Cell 1995; 80:149-54.
16. Nichols CG, Lopatin AN. Inward rectifier potassium channels. Annu Rev Physiol 1997; 59:171-91.
17. Bowie D, Lange GD, Mayer ML. Activity-Dependent Modulation of Glutamate Receptors by Polyamines. J Neurosci 1998; 18:8175-85.
18. Haghighi AP, Cooper E. Neuronal nicotinic acetylcholine receptors are blocked by intracellular spermine in a voltage-dependent manner. J Neurosci 1998; 18:4050-62.
19. Solessio E, Rapp K, Perlman I et al. Spermine mediates inward rectification in potassium channels of turtle retinal Muller cells. J Neurophysiol 2001; 85:1357-67.
20. Herman MD, Reuveny E, Narahashi T. The effect of polyamines on voltage-activated calcium channels in mouse neuroblastoma cells. J Physiol 1993; 462:645-60.
21. Lu Z, Ding L. Blockade of a retinal cGMP-gated channel by polyamines. J Gen Physiol 1999; 113:35-43.
22. Pan ZH. Differential expression of high- and two types of low-voltage-activated calcium currents in rod and cone bipolar cells of the rat retina. J Neurophysiol 2000; 83:513-27.
23. de l V, Vaquero CF, Kaneko A. Two types of calcium currents of the mouse bipolar cells recorded in the retinal slice preparation. Eur J Neurosci 1998; 10:317-23.
24. Tachibana M, Okada T, Arimura T et al. Dihydropyridine-sensitive calcium current mediates neurotransmitter release from retinal bipolar cells. Ann NY Acad Sci 1993; 707:359-61.
25. Heidelberger R, Matthews G. Calcium influx and calcium current in single synaptic terminals of goldfish retinal bipolar neurons. J Physiol (Lond) 1992; 447:235-56.

26. Maguire G, Maple B, Lukasiewicz P et al. Gamma-aminobutyrate type B receptor modulation of L-type calcium channel current at bipolar cell terminals in the retina of the tiger salamader. Proc Natl Acad Sci USA 1989; 86:10144-7.
27. von Gersdorff H, Matthews G. Dynamics of synaptic vesicle fusion and membrane retrieval in synaptic terminals. Nature 1994; 367:735-9.
28. Pan ZH. Voltage-activated Ca^{2+} channels and ionotropic GABA receptors localized at axon terminals of mammalian retinal bipolar cells. Vis Neurosci 2001; 18:279-88.
29. Maguire G, Maple B, Lukasiewicz P et al. y-Aminobutyrate type B receptor modulation of L-type calcium channel current at bipolar cell termimnals in the retina of the tiger salamander. Proc Natl Acad Sci USA 1989; 86:10144-7.
30. Tabor CW, Tabor H. Polyamines. Annu Rev Biochem 1984; 53:749-90.
31. Morgan DM. Polyamines. An overview. Mol Biotechnol 1999; 11:229-50.
32. Macaione S, Cangemi F, Fabiano C et al. Putrescine, polyamines, and N1-acetylpolyamine levels in retina, visual cortex and cerebellum of free-running mice kept under continuous light or darkness. Ital J Biochem 1993;42:151-64.
33. Haghighi AP, Cooper E. Neuronal nicotinic acetylcholine receptors are blocked by intracellular spermine in a voltage-dependent manner. J Neurosci 1998;18:4050-62.
34. Llinas R, Sugimori M, Lin J-W et al. Blocking and isolation of a calcium channel from neurons in mammals and cephalopods utilizing a toxin fraction (FTX) from funnel-web spider poison. Proc Natl Acad Sci USA 1989; 86:1689-93.
35. Williams K. Interactions of polyamines with ion channels [published erratum appears in Biochem J 1997 Sep 15;326(Pt 3):943]. Biochem J 1997; 325:289-97.
36. Oliver D, Baukrowitz T, Fakler B. Polyamines as gating molecules of inward-rectifier K+ channels. Eur J Biochem 2000; 267:5824-9.
37. Rock DM, Macdonald RL. Polyamine regulation of n-methyl-d-aspartate receptor channels. Annu Rev Pharmacol Toxicol 1995; 35:463-82.
38. Williams K. Modulation and block of ion channels: a new biology of polyamines. Cell Signal 1997; 9:1-13.
39. Bahring R, Bowie D, Benveniste M et al. Permeation and block of rat GluR6 glutamate receptor channels by internal and external polyamines. J Physiol (Lond) 1997; 502:575-89.
40. Llinas R, Moreno H. Local Ca^{2+} signaling in neurons. Cell Calcium 1998; 24:359-66.
41. Dubinsky JM. Intracellular calcium levels during the period of delayed excitotoxicity. J Neurosci 1993; 13:623-31.
42. Bindokas VP, Yoshikawa M, Ishida AT. Na^+-Ca^{2+} exchanger-like immunoreactivity and regulation of intracellular Ca^{2+} levels in fish retinal ganglion cells. J Neurophysiol 1994; 72:47-55.
43. Newman E, Reichenbach A. The Muller cell: a functional element of the retina. Trends Neurosci 1996; 19:307-12.
44. Newman EA, Zahs KR. Modulation of neuronal activity by glial cells in the retina. J Neurosci 1998; 18:4022-8.
45. Biedermann B, Skatchkov SN, Brunk I et al. Spermine/spermidine is expressed by retinal glial (Muller) cells and controls distinct K^+ channels of their membrane. Glia 1998; 23:209-20.

SITE-DIRECTED AND NATURAL MUTATIONS IN STUDYING FUNCTIONAL DOMAINS IN GUANYLYL CYCLASE ACTIVATING PROTEINS (GCAPs)

Alexander Dizhoor

ABSTRACT

Guanylyl cyclase activating proteins (GCAPs) are Ca^{2+}-binding proteins of the EF-hand superfamily, through which the intracellular calcium regulates cGMP synthesis in vertebrate photoreceptors. GCAPs play an essential role in the calcium feedback mechanism that controls recovery and light adaptation of rods and cones. Moreover, mutations in at least one of the GCAPs have already been linked to two forms of congenital human retinal diseases. The GCAPs represent a separate small subfamily among the EF-hand proteins that are structurally similar to recoverin, but demonstrate a number of unique regulatory properties. When in the Ca^{2+}-free conformation (as in light-adapted photoreceptors), GCAPs stimulate photoreceptor membrane guanylyl cyclase (retGC), but when the intracellular free Ca^{2+} concentrations ($[Ca^{2+}]_{free}$) rise (as in dark-adapted photoreceptors), GCAPs turn into retGC inhibitors. In GCAPs, site-directed mutagenesis has been successfully used to identify a number of structural elements that contribute to their specific function as guanylyl cyclase regulators. These elements include EF-hand Ca^{2+}-binding loops and various other regions in the GCAP primary structure involved in multiple protein-protein interactions within the retGC/GCAP complex.

Departments of Ophthalmology/Kresge Eye Institute, Pharmacology and Anatomy and Cell Biology, Wayne State University School of Medicine, Detroit, Michigan 48201, U.S.A.

INTRODUCTION

Photoisomerized rhodopsin triggers hydrolysis of cGMP through activation of a G protein, transducin, that stimulates cGMP phosphodiesterase and thus causes cGMP-gated channels to close (see refs. 1-2 for review). Rods and cones rapidly recover to their resting potential after excitation induced by a brief non-saturating exposure to light. In addition to their ability to recover after a single test flash, photoreceptors are also capable of light adaptation to a constant background illumination. Compared to a dark-adapted retina, light-adapted photoreceptors decrease the amplitude and accelerate the kinetics of their electrical responses to a standard test flash, and the degree of such change depends on the intensity of the background light (see ref. 4 for review). Reopening of the cGMP-gated channels during recovery and light adaptation occurs as a result of multiple mechanisms that include resynthesis of cGMP by guanylyl cyclase. The major factor that controls the activity of guanylyl cyclase in photoreceptors is fluctuation in the level of $[Ca^{2+}]_{free}$ in their light-adapted versus dark-adapted states. In the dark, Ca^{2+} extruded from the photoreceptor outer segment by a Na^{+}/K^{+}, Ca^{2+} -exchanger can re-enter the outer segment with the cation current through the open cGMP-gated channels. Therefore, the free intracellular Ca^{2+} concentrations lie between 500 and 600 nM.[5] In the light, when cGMP is hydrolyzed and the channels are closed, the exchanger continues to extrude Ca^{2+} ions from the photoreceptor outer segment, therefore, the intracellular Ca^{2+} concentrations in illuminated rods and cones decreases over 10-fold,[5, 6] and this causes acceleration of cGMP synthesis by guanylyl cyclase[7] (see refs. 1, 8-10).

Two isozymes of membrane guanylyl cyclase, retGC-1 and retGC-2, have been identified in photoreceptors.[11,12] Unlike many other members of the membrane guanylyl cyclase family, both retGC isozymes are regulated by Ca^{2+}-binding proteins termed GCAPs (guanylyl cyclase activating proteins), that interact with the cytoplasmic portion of the cyclase molecule (see refs. 9 and 10).

Two homologous 24-kDa proteins, GCAP-1 and GCAP-2, that impart Ca^{2+}-sensitivity to retGC isozymes were first isolated from a bovine retina in 1994.[13,14] Amino acid sequences of GCAP-1 and GCAP-2 are over 40% identical.[15,16] More recently, cDNA clone for the third homologue, GCAP-3, has been found in a human retinal cDNA library.[17]

GCAPs form a separate subfamily within a broader family of recoverin-like proteins, a distinct group within the EF-hand superfamily.[10] Unlike ubiquitous members of the remotely related to them calmodulin family, recoverin-like proteins are predominantly localized to neurons (see ref. 18), they include three distinct EF-hand Ca^{2+} -binding domains, and their N-terminal Gly residue is acylated with either saturated or unsaturated fatty acyl groups.[19,20]

Although all recoverin-like proteins have four EF-hand motifs (named EF-1 through EF-4), only three (or sometimes just two) of them create true Ca^{2+}-binding domains. The N-terminal EF-hand-related motif, EF-1, deviates from the consensus sequence lacking several key side chain residues required for coordinating Ca^{2+} ions,[10,18] and therefore cannot bind Ca^{2+}.

The solution structure of the Ca^{2+}-bound GCAP-2 established by Ames et al[21] is also drastically different from calmodulin and closely resembles recoverin and neurocalcin. The GCAP-2 molecule consists of two globular halves connected by a flexible "hinge" region between the exiting helix of the EF-2 and the entering helix of the EF-3. The NMR structure is currently available only for the Ca^{2+}-bound, inhibitory, form of GCAP-2, therefore, the exact conformational changes caused by the release of Ca^{2+} ions upon its transition into the activator conformation remain unclear, as well as the structure of GCAP-2 bound to its target cyclase. Structural-functional analysis of GCAP-1 is even more complicated by the lack of direct three-dimensional model based on X-ray crystallography, or NMR spectroscopy.

Therefore, site-directed mutagenesis remained the main tool for the functional analysis of GCAPs over the last five years. This approach has been fairly successful in revealing the presence of various functionally important domains in both GCAP-1 and GCAP-2.

MUTATIONS IN Ca^{2+}-BINDING DOMAINS IN GCAP-1 AND GCAP-2

Based on the functional study in vitro, GCAP-2 has three functional EF-hands as revealed by the directed mutagenesis. Inactivation of the EF-hands can be efficiently achieved by substituting the first invariant Asp or the last invariant Glu in the 12-amino acid consensus motif of the Ca^{2+}-binding loop of the EF-hand by their amides, Asn or Gln, respectively.[22,23] When such substitutions are introduced in any of the three "conventional" EF-hands (i.e., EF-2, -3, or -4), it decreases Ca^{2+}-sensitivity of retGC activation by GCAP-2[24] (Fig. 1). Inactivation of the EF-3 alone requires that 2-3- fold higher concentrations of Ca^{2+} be added to the in vitro assay in order to suppress the activity of retGC. Yet, the inhibitory effect of Ca^{2+} largely remains within the physiologically relevant range of the freeCa^{2+} concentrations. Inactivation of the EF-2 and EF-4 has stronger effect and the freeCa^{2+} concentrations required for the half-maximal inactivation of retGC exceed the submicromolar range estimated for the dark-adapted photoreceptors in vivo. Inactivation of all three active EF-hands in GCAP-2 renders GCAP-2 in a constitutively active form that efficiently stimulates RetGC both at low and high Ca^{2+} concentrations.[24] A similar strategy was applied to the study of EF-hands in GCAP-1,[25] when substitutions Glu->Asp in positions 75, 111, and 155 were used for inactivation of the EF-hands (Fig. 1). Based on that study, EF-3 and EF-4 are both essential for the Ca^{2+}-sensitivity of GCAP-1, while the EF-2 makes little or no contribution to the overall Ca^{2+}-sensitivity of the retGC regulation.[25]

Binding of Ca^{2+} makes GCAPs undergo an "activator-to-inhibitor" transition, so that fully Ca^{2+}- loaded GCAPs inhibit basal activity of RetGC in photoreceptor membranes.[24,25] However, at the $[Ca^{2+}]_{free}$ expected in the dark-adapted photoreceptors, the ratio between the Ca^{2+}-free (activator) and the Ca^{2+} bound (inhibitor) forms of GCAP-2 is such that the activity of retGC is relatively close to its basal activity observed in the absence of GCAPs.[24]

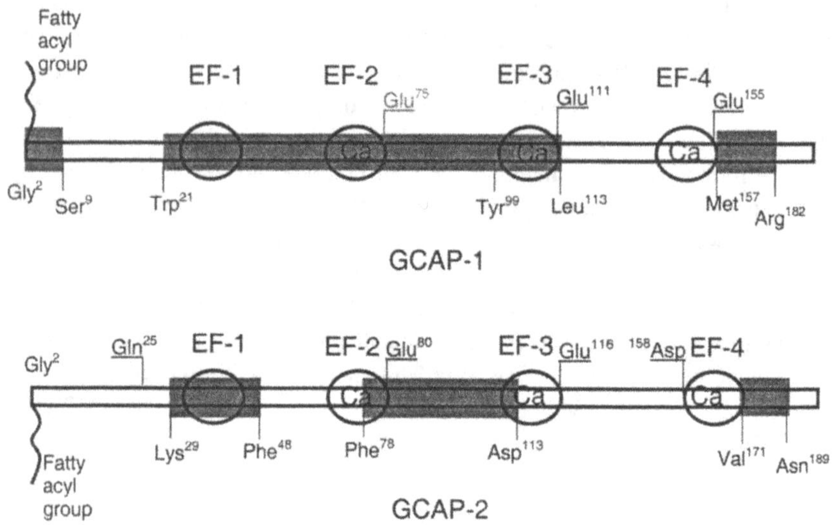

Figure. 1. Mutations at the amino acid residues positions in GCAPs that affect their activity (the shaded areas in both GCAP-1 and GCAP-2 mark the regions that are most important for the cyclase regulation based on the chimera protein analyses[29,30] as explained in the text).

MUTATIONS THAT AFFECT THE N-FATTY ACYLATION

Unlike calmodulin or many other EF-hand Ca^{2+}-binding proteins, GCAPs, as a part of the recoverin family, have their N-terminal Gly in a consensus motif that is recognized by the N-myristoyl transferase (NMT). Myristoyl (C14:0) and similar fatty acyl groups (C12:0; C14:1 and C14: 2) can therefore be covalently attached to the N-terminus of GCAPs both in vivo and in vitro [19,20] (for simplicity, this modification is often referred to as just "myristoylation"). To prevent GCAPs from myristoylation, the recombinant GCAPs can either be expressed in bacteria that do not express their own NMT, or expressed in eukaryotic cells as a mutant in which the N-terminal Gly is substituted by Ala (the latter is not a substrate for the NMT). This approach allowed to demonstrate that the N-fatty acyl group significantly contributes to the activity of GCAP-1,[26] but makes relatively minor contribution to the activity or Ca^{2+}-sensitivity of GCAP-2.[20]

FUNCTIONAL REGIONS IN GCAPs THAT CONTRIBUTE TO THE RETGC ACTIVATION OR INHIBITION

Structural and functional properties of GCAPs as Ca^{2+}-sensor proteins for retGC are not limited to the presence of the Ca^{2+}-binding EF-hand motifs. While it is relatively easy to evaluate the contribution of the EF-hands to the overall regulatory properties of GCAPs, identification of other structural elements that are directly involved in regulatory interactions with retGC is a more complex task. In order to avoid, or at least to minimize the potential threats from artifacts that may result from a general misfolding of the protein in response to relatively large deletions, localization of their essential functional regions often requires chimera mutants, where parts of the GCAP molecule are substituted by the corresponding regions from a closely related recoverin-like protein, such as recoverin or neurocalcin (neither of which regulates retGC within the physiologically relevant range of $[Ca^{2+}]_{free}$). This approach is based on the close similarity between the different members of the recoverin family that can be clearly seen not only from the comparison of their primary structures,[15] but also their overall 3-dimensional backbone folding.[21,27-28] Results combined from the study of deletion mutants and chimera proteins strongly indicate that in addition to the Ca^{2+}-binding loops of the EF-hands the following multiple regions in GCAPs are most important for the regulation of retGC.[29,30] In GCAP-1, the 9-amino acid fragment at the N-terminus of the molecule is most likely involved in the interactions between the Ca^{2+}-loaded (inhibitory) form of GCAP-1 and the cyclase, because a substitution of this fragment in GCAP-1 with the corresponding fragment from recoverin does not prevent the chimera protein from activating retGC in the absence of Ca^{2+}, but substantially diminishes retGC basal activity in the presence of saturating Ca^{2+} concentrations.[29] The very N-terminal segment of GCAP-2 is also sensitive to substitution with the corresponding neurocalcin fragment—such a chimera strongly inhibits retGC activation. However, it is more likely that the N-terminal region in the neurocalcin/GCAP-2 chimera interferes with the cyclase activation by adding an extra length to the N-terminus, rather than by the removal of an essential sequence required for retGC activation.[30] In both GCAP-1 and GCAP-2, the region adjacent to the last EF-hand (EF-4) was required for the cyclase activation by Ca^{2+}-free GCAP, but it did not significantly contribute to the inhibitory interactions between retGC and Ca^{2+}-loaded GCAPs.[29,30] The central portion of the molecule in both GCAP-1 and GCAP-2 also appears to be essential for the cyclase regulation. In GCAP-2, a region that includes the exiting helix of the EF-2 and the entering helix of the EF-3 (that also forms a flexible "hinge" between two globular parts of the molecule [30]), contributes not only to the activation of retGC, but also to the "direction" of the functional Ca^{2+} switch (Fig. 1). Activation of retGC by the chimera mutant in which this part of the molecule is substituted with the corresponding segment from neurocalcin is completely suppressed at low free Ca^{2+} concentrations. However, this chimera can slightly stimulate retGC in the presence of Ca^{2+}. Although neurocalcin has no effect on retGC activity at physiologically relevant submicromolar concentrations of Ca^{2+},[30] this unusual property of the chimera may reflect the reported ability of neurocalcin to slightly increase retGC activity in vitro in the presence of saturating Ca^{2+} concentrations.[28] Substitution of the central part of the molecule with recoverin sequence inhibits activation of retGC by GCAP-1, although the weak stimulation of retGC by the Ca^{2+}-loaded chimera was not observed in that case.

As the most intriguing observation, substitution of the N-terminal EF-hand-related region, EF-1, in both GCAP-1 and GCAP-2 completely suppressed cyclase activation.[29,30] Unlike other three EF-hands, the first EF-hand-related motif lacks several key side chain residues required for high-affinity binding of Ca^{2+}. In order to efficiently coordinate Ca^{2+} ion in the EF-hand loop,[29,30] it is essential that the first Ca^{2+}-coordinating position (X) of the 12-amino acid consensus motif to be occupied by Asp, while the positions 3 (Y) and 9 (-X) must include oxygen-containing side chain residues.[22,23,31] The absence of the first invariant Asp[33] of the consensus motif and replacement of the obligatory oxygen-containing side chain residues by Cys35 and Phe41[19,15] prohibit Ca^{2+} from binding within the loop structure of the EF-1 domain,[22,23,31] Although the EF-1 region is unable to bind Ca^{2+}, its amino acid sequence remains fairly homologous between recoverin-like proteins and, except for several variable residues that are specific for each particular member of that family, it includes several highly conserved amino acid residues.[10,18,32] Also, regardless of its deviation from the EF-hand Ca^{2+} binding consensus motif, EF-1 retains an overall structure of the helix-loop-helix domain and fairly closely resembles other EF-hands.[21] More recent analysis of this domain in GCAP-2 using a series of point mutations clearly indicates that the EF-1 domain in GCAP-2 is directly involved in interaction with retGC.[33] There is an independent observation that truncation of the N-terminal portion of the molecule that included a part of the EF-1 entering helix could also abolish the cyclase activation by GCAP-1.[19] This may indicate that EF-1 domain in GCAP-1 is also directly involved in target recognition, although this hypothesis remains to be further verified.

Although there is an obvious need in further refining the GCAP-2 functional map, it is quite encouraging that the initial analysis of their functional domains using a combination of chimera proteins[29,30,34] and point mutations already provided significant clues on where such functional domains are located.

MUTATIONS IN GCAP-2 THAT AFFECT ITS DIMERIZATION

Recent observations also strongly indicate that the retGC complex, at least in case of GCAP-2, likely consists of a dimerized cyclase and dimerized GCAP-2:

1) Both RetGC-1 and RetGC-2 have a putative "dimerization domain" typical for peptide receptor guanylyl cyclases (see also ref. 32). Co-expression of recombinant RetGC-1 and RetGC-2 in transfected cells can result in a complex that contains both isozymes, but they are likely to form only homodimers in vivo.[35]
2) The efficiency of the covalent cross-linking of retGC into a dimer increases in the presence of the Ca^{2+}-free recombinant GCAP-1 and GCAP-2, while Ca^{2+}-loaded GCAPs inhibit it.[36]
3) Catalytic center of retGC-1 is formed by at least two retGC-1 subunits.[37]

Therefore, the functional role of GCAPs is likely to stimulate closer interactions between the catalytic domains within the RetGC homodimer as a function of Ca^{2+} concentrations. The molecular details of this process remain obscure and may be quite different between GCAP-1 and GCAP-2. However, unlike GCAP-1, GCAP-2 can form a relatively stable dimer whose stability is greatly affected by Ca^{2+} binding, so that the Ca^{2+}-loaded form of GCAP-2 completely dissociates into monomers during gel-chromatography.[38] Moreover, the ability of GCAP-2 and some of its chimerical mutants to activate RetGC at low $[Ca^{2+}]_{free}$ apparently correlates with their ability to dimerize in the absence of Ca^{2+}:[38]

1) Constitutively active mutant of GCAP-2 undergoes dimerization regardless of freeCa^{2+} concentrations.
2) A chimera that has the entire N-terminal fragment (Gly1-Gln25) substituted with the homologous fragment from neurocalcin both lack the ability to activate retGC and to demonstrate stable dimerization in the absence of Ca. Same is true for a chimera that contains a C-terminal fragment, Glu159-Phe204, derived from neurocalcin. The lack of activity in these chimeras cannot be attributed to a general non-specific misfolding of the proteins, because both chimeras were able to inhibit the cyclase basal activity in the presence of Ca^{2+}. Similarly to that, the substitution of the central part of the molecule, Ala 57-Leu127, with the neurocalcin fragment inactivated both retGC stimulation and GCAP-2 dimer formation at low freeCa^{2+} concentrations.

It is likely that substitutions of the GCAP-2-specific fragments in either case resulted in the absence of amino acid sequence(s) that were essential for dimerization of Ca^{2+}-free GCAP-2. Alternatively, introducing the neurocalcin fragments in those regions, may interfere with dimerization of GCAP-2 by affecting the neighboring region (s) in its molecule. Therefore, although it is obvious that dimerization of GCAP-2 can be affected via multiple regions on the same molecule, the exact side chain residues in GCAP-2 that participate in GCAP/GCAP interaction remain to be identified

These observations can be described as a model, according to which GCAP-2 monomers bound to each molecule of the cyclase would stimulate the interaction between cyclase subunits as a result of a Ca^{2+}-sensitive intramolecular change that both affects conformation of the cyclase and promotes GCAP:GCAP interaction. Forces that "pull" together two GCAP molecules bound to the cyclase subunits can add the energy for the interactions between the retGC subunits (referred to as a dimer-adapter" hypothesis in ref.38)

NATURALLY OCCURRING MUTATIONS IN GCAP-1 ASSOCIATED WITH CONE DEGENERATION

Two cases of congenital dominant cone degeneration have been linked to mutations in GCAP-1. The first of these mutations, Tyr99Cys (Fig. 1), was found within the entering helix of the third EF-hand structure in GCAP-1.[39] This mutation causes gradually progressing degeneration of cones and results in a complete loss of color perception and daylight vision, yet rods are unaffected and their responses to dim light remain virtually normal. Although the Y99C GCAP-1 activates retGC in vitro to the same extent as the wild type GCAP-1, in the presence of the Y99C mutant the cyclase remains active even at the $[Ca^{2+}]_{free}$ above 1 μM).[40, 41] Unlike wild type GCAP-1, the Y99C mutant promotes dimerization of the cyclase not only in the absence but also in the presence of micromolar Ca^{2+}.[36] The Y99C GCAP-1 efficiently competes with the Ca^{2+} -loaded wild type GCAP-1 and GCAP-2 in vitro, and that may account for the dominant nature of this mutation in humans.[40]

The second mutation, Glu155Gly (Fig. 1), has been described more recently also in association with congenital dominant cone dystrophy.[42] This mutation directly inactivates Ca^{2+}-binding loop in the EF-hand 4. Very much like the Tyr99Cys, this mutation causes abnormally low Ca^{2+}-sensitivity of GCAP-1 and makes it constitutively active within the physiologically relevant range of the free Ca^{2+} concentrations.

There are three major questions that remain unresolved with regard to the GCAP-1 mutations linked to the cone dystrophy:

1) Why are only cones affected?
2) Why does the presence of the mutant GCAP-1 in both cases trigger the apoptosis?
3) Why does it take substantial period of time for cone degeneration to develop?

First, despite the variability among the published immunolocalization data,[43-45] there is number of indications that GCAP-1 can be expressed in rods at lower levels than in cones, while GCAP-2, on the contrary, is primarily present in outer segments of rods. On the other hand, Ca^{2+}-loaded inhibitory GCAPs may compete with the constitutively active mutants of GCAP-1, and hence to mitigate to some extent the effect of the mutant GCAP-1 on retGC activity. Although even at the equimolar ratio between mutant GCAP-1 and wild type GCAPs the Ca^{2+}-sensitivity of cyclase remains affected,[40] one can speculate that the ratio between GCAP-2 and GCAP-1 in rods may be more favorable for mitigating the activity of the mutant GCAP-1.

As a possible scenario for the intracellular events that lead to the cell death, high intracellular Ca^{2+} concentrations may alter the permeability of rod mitochondria and promote apoptosis in photoreceptors.[46] Since activation of retGC in vitro by the Tyr99Cys or Glu155Gly GCAP-1 is not fully suppressed even at >>1 μM $[Ca^{2+}]_{free}$, it is very tempting to suggest that in live photoreceptors it also remains active when the $[Ca^{2+}]_{free}$, reach their normal "dark" level. In normal photoreceptors, the "dark" concentrations of the circulating Ca^{2+} are at ~600 nM level because the rate of cGMP synthesis is lowered by Ca/GCAP. This maintains the free cGMP concentrations near 2-4 μM level,[47] so that no more than 3% percent of cGMP-gated channels are open to allow the influx of Na^+ and Ca^{2+}. The increased rate of cGMP synthesis by the Ca^{2+}-insensitive GCAP-1 may lead to opening of the excessive number of cGMP-gated channels in cones and therefore cause much higher than normal influx of Na^+/Ca^{2+}. It may take the intracellular Ca^{2+} concentrations to reach ~10 μM before the Y99C GCAP-1 would finally stop activating the cyclase, but such concentrations of Ca^{2+} may already be sufficient for triggering the apoptosis in cones.

If such were the case, it would be relatively easy to explain the onset of the cone degeneration process. However, it would still be difficult to explain why it should take a relatively long period of human life for the cone death to occur.

The answer to these questions would require further study using transgenic animal models that can mimic photoreceptor degeneration caused by the GCAP-1 mutants.

ACKNOWLEDGMENT

The financial support has been provided in part by the National Institutes of Health EY11522 from NEI and Research to Prevent Blindness.

REFERENCES

1. Pugh EN Jr, Lamb, TD. Amplification and kinetics of the activation steps in phototransduction. Biochim Biophys Acta 1993; 1141:111-149.
2. Baylor D. How photons start vision. Proc Natl Acad Sci USA 1996; 93:560-565.
4. Pugh EN Jr, Nikonov S, LambTD. Molecular mechanisms of vertebrate photoreceptor light adaptation. Curr Opin Neurobiol 1999; 9:410-418.
5. Gray-Keller MP, Detwiler PB. The calcium feedback signal in the phototransduction cascade of vertebrate rods. Neuron 1994; 13:849-861
6. Sampath AP, Matthews HR, Cornwall MC et al. Light-dependent changes in outer segment free-Ca2+ concentration in salamander cone photoreceptors. J Gen Physiol 1999; 113:267-77.
7. Koch KW, Stryer L. Highly cooperative feedback control of retinal rod guanylate cyclase by calcium ions. Nature 1988; 334:64-66
8. Pugh EN Jr, Duda T, Sitaramayya A et al. Photoreceptor guanylate cyclases: a review. Biosci Rep 1997; 17:429-473.
9. Dizhoor AM. Regulation of cGMP synthesis in photoreceptors: role in signal transduction and congenital diseases of the retina. Cell Signal 2000; 12:711-9.
10. Palczewski K, Polans AS, Baehr W et al. Ca(2+)-binding proteins in the retina: structure, function, and the etiology of human visual diseases. Bioessays 2000; 22:337-50.
11. Yang RB, Foster DC, Garbers DL et al. Two membrane forms of guanylyl cyclase found in the eye. Proc Natl Acad Sci USA 1995; 92:602-606.
12. Lowe DG, Dizhoor AM, Liu K et al. Cloning and expression of a second photoreceptor-specific membrane retina guanylyl cyclase (RetGC), RetGC-2. Proc Natl Acad Sci USA 1995; 92:5535-5539.
13. Gorczyca WA, Gray-Keller MP, Detwiler PB et al. Purification and physiological evaluation of a guanylate cyclase activating protein from retinal rods. Proc Natl Acad Sci USA 1994; 91:4014-4018.
14. Dizhoor AM, Lowe DG, Olshevskaya EV et al. The human photoreceptor membrane guanylyl cyclase, RetGC, is present in outer segments and is regulated by calcium and a soluble activator. Neuron 1994; 12:1345-1352.
15. Dizhoor AM, Olshevskaya EV, Henzel WJ et al. Cloning, sequencing, and expression of a 24-kDa Ca(2+)-binding protein activating photoreceptor guanylyl cyclase. J Biol Chem 1995; 270:25200-25206.
16. Gorczyca WA, Polans AS, Surgucheva IG et al, Guanylyl cyclase activating protein. A calcium-sensitive regulator of phototransduction. J Biol Chem 1995; 270:22029-22036.
17. Haeseleer F, Sokal I, Li N et al. Molecular characterization of a third member of the guanylyl cyclase- activating protein subfamily. J Biol Chem 1999; 274:6526-6535.

18. Nef P. Neuron-specific calcium-sensors. In: Celio MR, Pauls TL, Shwaller B, eds. Guidebook to the Calcium-Binding Proteins. New York: Oxford University Press, 1996:15-20.
19. Palczewski K, Subbaraya I, Gorczyca WA et al. Molecular cloning and characterization of retinal photoreceptor guanylyl cyclase-activating protein. Neuron 1994; 13:395-404.
20. Olshevskaya EV, Hughes RE, Hurley JB et al. Calcium binding, but not a calcium-myristoyl switch, controls the ability of guanylyl cyclase-activating protein GCAP-2 to regulate photoreceptor guanylyl cyclase. J Biol Chem 1997; 272:14327-14333.
21. Ames JB, Dizhoor AM, Ikura M et al. Three-dimensional structure of guanylyl cyclase activating protein-2, a calcium-sensitive modulator of photoreceptor guanylyl cyclases. J Biol Chem 1999; 274:19329-19337.
22. Strynadka NC, James MN. Crystal structures of the helix-loop-helix calcium-binding proteins. Annu Rev Biochem 1989; 58:951-998.
23. Babu A, Su H, Ryu Y et al. Determination of residue specificity in the EF-hand of troponin C for Ca2+ coordination, by genetic engineering. J Biol Chem 1992; 267:15469-15474.
24. Dizhoor AM, Hurley JB. Inactivation of EF-hands makes GCAP-2 (p24) a constitutive activator of photoreceptor guanylyl cyclase by preventing a Ca2+-induced "activator- to-inhibitor" transition. J Biol Chem 1996; 271:19346-50.
25. Rudnicka-Nawrot M, Surgucheva I, Hulmes J et al, Changes in biological activity and folding of guanylate cyclase- activating protein 1 as a function of calcium. Biochemistry. 1998; 37:248-257.
26. Otto-Bruc A, Buczylko J, Surgucheva I et al. Functional reconstitution of photoreceptor guanylate cyclase with native and mutant forms of guanylate cyclase-activating protein 1. Biochemistry 1997; 36:4295-4302.
27. Flaherty KM, Zozulya S, Stryer L et al. Three-dimensional structure of recoverin, a calcium sensor in vision. Cell 1993; 75:709-716.
28. Vijay-Kumar S, Kumar VD Crystal structure of recombinant bovine neurocalcin. Nat Struct Biol 1999; 6:80-8.
29. Krylov DM, Niemi GA, Dizhoor AM et al. Mapping sites in guanylyl cyclase activating protein-1 required for regulation of photoreceptor membrane guanylyl cyclases. J Biol Chem 1999; 274:10833-9.
30. Olshevskaya EV, Boikov S, Ermilov A et al. Mapping functional domains of the guanylate cyclase regulator protein, GCAP-2. J Biol Chem 1999; 274:10823-10832.
31. Kawasaki H, Kretsinger RH. Calcium-binding proteins. 1: EF-hands. Protein Profile 1994; 1:343-517.
32. Dizhoor AM, Hurley JB. Regulation of photoreceptor membrane guanylyl cyclases by guanylyl cyclase activator proteins. Methods 1999; 19:521-31.
33. Ermilov AN, Olshevskaya EV, Dizhoor AM. Instead of Binding Calcium, One of the EF-hand Structures in Guanylyl Cyclase Activating Protein-2 Is Required for Targeting Photoreceptor Guanylyl Cyclase. J Biol Chem 2001; 276:48143-48148.
34. Li N, Sokal I, Bronson JD et al. Identification of functional regions of guanylate cyclase-activating protein 1 (GCAP1) using GCAP1/GCIP chimeras. Biol Chem 2001; 382:1179-1188.
35. Yang RB, Garbers DL. Two eye guanylyl cyclases are expressed in the same photoreceptor cells and form homomers in preference to heteromers. J Biol Chem 1997; 272:13738-42.
36. Yu H, Olshevskaya E, Duda T et al. Activation of retinal guanylyl cyclase-1 by Ca2+-binding proteins involves its dimerization. J Biol Chem 1999; 274:15547-15555.
37. Ramamurthy V, Tucker C, Wilkie SE et al. Interactions within the Coiled-coil Domain of RetGC-1 Guanylyl Cyclase Are Optimized for Regulation Rather than for High Affinity. J Biol Chem 2001; 276:26218-26229.
38. Olshevskaya EV, Ermilov AN, Dizhoor AM et al. Dimerization of guanylyl cyclase-activating protein and a mechanism of photoreceptor guanylyl cyclase activation. J Biol Chem 1999; 274:25583-7.

39. Payne AM, Downes SM, Bessant DA et al. A mutation in guanylate cyclase activator 1A (GUCA1A) in an autosomal dominant cone dystrophy pedigree mapping to a new locus on chromosome 6p21.1. Hum Mol Genet 1998; 7:273-277.
40. Dizhoor AM, Boikov SG, Olshevskaya EV. Constitutive activation of photoreceptor guanylate cyclase by Y99C mutant of GCAP-1. Possible role in causing human autosomal dominant cone degeneration. J Biol Chem 1998; 273:17311-4.
41. Sokal I, Li N, Surgucheva I et al, GCAP1 (Y99C) mutant is constitutively active in autosomal dominant cone dystrophy. Mol Cell 1998; 2:129-133.
42. Wilkie SE, Newbold RJ, Deery E et al, Functional characterization of missense mutations at codon 838 in retinal guanylate cyclase correlates with disease severity in patients with autosomal dominant cone-rod dystrophy. Hum Mol Genet 2000; 9:3065-3073.
43. Kachi S, Nishizawa Y, Olshevskaya E. Detailed localization of photoreceptor guanylate cyclase activating protein-1 and -2 in mammalian retinas using light and electron microscopy. Exp Eye Res 1999; 68:465-473.
44. Otto-Bruc A, Fariss RN, Haeseleer F et al, Localization of guanylate cyclase-activating protein 2 in mammalian retinas. Proc Natl Acad Sci USA 1997; 94:4727-4732.
45. Howes K., Bronson JD, Dang YL et al. Gene array and expression of mouse retina guanylate cyclase activating proteins 1 and 2. Invest Ophthalmol Vis Sci 1998; 39:867-875.
46. He L, Poblenz AT, Medrano CJ et al. Lead and calcium produce rod photoreceptor cell apoptosis by opening the mitochondrial permeability transition pore. J Biol Chem 2000; 275:12175-84.
47. Pugh EN Jr, Lamb TD. Cyclic GMP and calcium: the internal messengers of excitation and adaptation in vertebrate photoreceptors. Vision Res 1990; 30:1923-48.

CALMODULIN AND Ca^{2+}-BINDING PROTEINS (CaBPs): VARIATIONS ON A THEME

Francoise Haeseleer[1] and Krzysztof Palczewski[1,2,3]

ABSTRACT

Ca^{2+} is a ubiquitous second messenger that frequently exerts its effects through Ca^{2+}-binding proteins. In response to changes in the intracellular [Ca^{2+}], Ca^{2+}-binding proteins modulate the cellular activities of enzymes, channels and structural proteins. Multiple Ca^{2+}-binding proteins are expressed in the retina and, in most cases, in a unique cellular and sub-cellular manner. CaBPs are retinal Ca^{2+}-binding proteins displaying a high similarity to calmodulin (CaM). CaBPs are able to mimic some of the interactions of CaM with effector enzymes, although their physiological role has not yet been resolved. CaBPs could be cell-type specific proteins that play a key role in the Ca^{2+} signaling of specialized retinal neurons.

INTRODUCTION

Ca^{2+} is a ubiquitous second messenger whose concentration changes as a result of cell stimulation. Within the cell, Ca^{2+} signals originate from changes in Ca^{2+} influx through Ca^{2+} channels, Ca^{2+} release from internal stores, as well as Ca^{2+} removal through monovalent-dependent Ca^{2+} exchangers and re-absorption by the internal stores.[1] Ca^{2+}-binding proteins are important elements of the Ca^{2+} homeostasis. A subset of Ca^{2+}-binding proteins may bind Ca^{2+} without activating any target protein and augment the buffering capacity of the cell. These Ca^{2+} "buffer" proteins play a central role in controlling the amplitude and duration of Ca^{2+} signals. Changes in the intracellular [Ca^{2+}] are also signals "sensed" by Ca^{2+}-binding

Departments of [1]Ophthalmology, [2]Chemistry, [3]Pharmacology, University of Washington, Seattle, Washington 98195, U.S.A.

proteins, which in turn regulate the cellular activities of enzymes, channels and structural proteins. The Ca^{2+} sensors take advantage of transient changes in [Ca^{2+}] within the cell to mediate the Ca^{2+} signal to their target.[2-4]

Over the years hundreds of Ca^{2+}-binding proteins have been identified, reflecting the importance of this cation in the normal function of the cell. While many Ca^{2+}-binding proteins have been identified, functions can be ascribed only for a few of them. Many Ca^{2+}-binding proteins are expressed in a specific cell-type and therefore could be an important element in the physiology of more complex tissues.

Ca^{2+}-binding proteins are very common to all cell-types where Ca^{2+} must be temporally and spatially regulated, including neuronal cells. Neurons from the retina contain multiple forms of Ca^{2+}-binding proteins. The main focus of this chapter will be on the novel subfamily of Ca^{2+}-binding proteins displaying high similarity to CaM, termed CaBPs, that were earlier identified in our laboratory.[5]

Ca^{2+} SIGNALING IN THE RETINA

Visual perception requires the conversion of a light signal, received by the retina, into an electrical signal that is ultimately sent to the brain. Two major types of photoreceptors, rods and cones, capture and convert light into neuronal signals. The absorption of photons by the photoreceptor pigments initiates a cascade of biochemical reactions called phototransduction. The ultimate consequence of light absorption by photoreceptor cells is a decrease of the cytoplasmic [Ca^{2+}], because the Na^+/Ca^{2+}-K^+ exchanger continues to extrude Ca^{2+} from the cell, while the influx of Ca^{2+} through the cGMP-gated channel is reduced. Consequently, in both types of photoreceptor cells, Ca^{2+} plays an important role in adaptation and modulates the physiological activity of several of the biochemical events underlying phototransduction.[6-8]

Ca^{2+} also plays a role in the biochemical processes at synapses, where it triggers the release of the neurotransmitter glutamate. In the dark, the rods and cones are depolarized and the voltage-gated Ca^{2+} channels in their synaptic terminals are opened, allowing Ca^{2+} ions to enter the terminal space and trigger the release of glutamate. When illuminated, photoreceptors become hyperpolarized and the Ca^{2+} influx is reduced through the voltage-gated Ca^{2+} channels, subsequently resulting in a reduced rate of glutamate release. This change in the rate of neurotransmitter released by photoreceptors is decoded by the secondary neurons, termed bipolar cells. Bipolar cells along with amacrine cells and horizontal cells constitute, in part, the inner nuclear layer and are further subdivided into a number of functional and/or morphologically distinct subtypes. This network of retinal interneurons conveys the signal perceived by photoreceptors to ganglion cells. Ganglion cells then further transfer the signal from the retina to the brain.[9]

EF-HAND MOTIF-CONTAINING Ca^{2+}-BINDING PROTEINS EXPRESSED IN THE RETINA

Ca^{2+}-Binding Proteins

The Ca^{2+}-binding domain of Ca^{2+}-binding proteins is composed of a 29-residue helix-loop-helix structure termed the EF-hand motif, present in 2 to 12 copies in the sequence of a number of proteins[10]. Oxygen atoms of the amino acid side-chains or peptide bonds in position 1, 3, 5, 7, 9 and 12 of the EF-hand loops are directly involved in Ca^{2+} coordination; consequently, these motifs are well conserved among Ca^{2+}-binding proteins.[11-13] The most studied EF-hand motif-containing Ca^{2+}-binding proteins is the ubiquitous CaM.[14,15] Frequently, EF-hand motifs occur in pairs and are responsible, in addition to Ca^{2+} coordination, for stabilization of the protein conformational changes upon Ca^{2+} binding/dissociation. An intramolecular interaction can occur between two EF-hand motifs on the same protein, or an intermolecular interaction between EF-hands of two proteins within a dimeric complex can also stabilize the protein conformation.[16] Ca^{2+} binding to Ca^{2+}sensors induces a variety of conformational changes in Ca^{2+}-binding proteins, leading to exposure of hydrophobic residues necessary for interaction with the target proteins. For Ca^{2+} buffer proteins, these conformational changes are generally small.[16,17]

Numerous Ca^{2+}-binding proteins have been identified in the retina, many being specifically expressed in this tissue. However, even widely distributed Ca^{2+}-binding proteins may have cell-restricted expression in the retina.[18] Calbindin D28K, parvalbumin and calretinin are cytosolic proteins that are commonly used as a marker to identify specific cell-types in the retina.[19] These three Ca^{2+}-binding proteins are found in morphologically distinct classes of retinal neurons. Calbindin contains six EF-hand motifs; four are functional and two contain sequence substitutions that disable them from Ca^{2+} coordination. Calbindin is present within the retina in bipolar cells, amacrine cells, cone photoreceptors and the majority of horizontal cells.[19,20] The role of calbindin is most likely restricted to the Ca^{2+} buffer capacity. Calretinin also contains six EF-hand motifs and displays high homology to calbindin. It is expressed in most of bipolar, horizontal and a subset of amacrine cells. Parvalbumin contains three EF-hand motifs two of which are functional, and it is expressed in horizontal, amacrine and ganglion cells. Ca^{2+} buffer proteins, like parvalbumin and calbindin, are important in neurons because they ensure that Ca^{2+} signals are confined to synapses, they allow grading Ca^{2+} signaling, and they affect the kinetics of changes in $[Ca^{2+}]$.[21]

Neuron-Specific Ca^{2+}-Binding Proteins

Neuron-specific Ca^{2+}-binding proteins (NCBPs) are a subset of the EF-hand motif-containing proteins that are found predominantly in neurons.[6,13]

Guanylate cyclase-activating proteins, GCAPs, are the best-characterized retinal NCBPs.[6,13] Three isoforms of mammalian GCAPs, GCAP1,[22] GCAP2[23,24] and GCAP3,[25] have been characterized to date. GCAPs have four EF-hand motifs where EF-hand 1 is disabled by naturally occurring mutations and all proteins are heterogeneously acylated, mostly myristoylated,* at the N-terminal Gly residue.[6,13,26] These proteins stimulate guanylate cyclases (GCs) during phototransduction, leading to the restoration of the cGMP level that has been depleted as a consequence of rhodopsin/G-protein mediated activation of cGMP phosphodiesterase.[6] A significant feature of GCAPs is that they activate their target in a Ca^{2+} free form. GCIP is a Ca^{2+}-binding protein closely related to GCAPs and cloned from frog retina.[27] However, GCIP (Guanylate Cyclase-Inhibitory Protein) inhibits GC in high [Ca^{2+}], compared to GCAPs that stimulate GC in low [Ca^{2+}]. GCAP1,[24] GCAP3 (Imanishi and Palczewski, unpublished) and GCIP[27] are expressed exclusively in photoreceptor cells. GCAP2, in addition to expression in photoreceptor cells, has also been detected in bipolar cells.[28] Mutations in GCAP1 have been found to be associated with autosomal dominant cone dystrophy.[29-32]

Recoverin is a Ca^{2+}-binding protein expressed in rod, cone photoreceptor cells and in cone bipolar cells.[33,34] In general, it has similar topology, post-translational modification and structure to GCAPs (reviewed in ref. 13), however, in addition to the EF-hand 1, the EF-hand 4 motif is also nonfunctional. Originally, recoverin was thought to be involved in the regulation of GC[35] and in the modulation of the activity of rhodopsin kinase.[36] Since then, the GC regulation was proven to be incorrect[37] and the rhodopsin kinase regulation is still controversial.[28,38] The Ca^{2+}-bound form of recoverin is a hydrophobic "sticky" protein and some biochemical observations could be irrelevant for the in vivo function of this protein. This property is in part due to the hydrophobic modification of recoverin. Myristoylation of recoverin enables this protein to reversibly interact with the cell membrane and can be essential for its function.[39] The binding of Ca^{2+} to recoverin leads to a conformational change, a Ca^{2+}-myristoyl switch, resulting in surface exposure of hydrophobic parts of the polypeptide and the myristoyl residue, thereby making these structures available for interaction with membrane and/or target proteins.[39] Recoverin is also an auto-antigen in human autoimmune conditions, known as Cancer-associated retinopathy (CAR).[40] Cellular and humoral responses against recoverin have been shown to be involved in CAR, where recoverin is also expressed in tumors developed by patients afflicted with this disease.[41,42]

Neuronal Ca^{2+} sensor 1 and VILIP (Visinin-Like-Protein) are also found in the inner part of the retina. These proteins have three putative Ca^{2+}-binding domains. Neuronal Ca^{2+} sensor 1 is expressed in ganglion cells, amacrine cells and the inner plexiform layer.[43] The role of these proteins in the retina has not been defined.

CaBPs

Homology Among CaBPs

CaBPs are the members of a subfamily of Ca^{2+}-binding proteins that show similarities to CaM (Figure 1). Initially, we cloned 5 members of this subfamily, with CaBP3 a likely pseudo-gene because despite a high mRNA level, the protein has not been detected.[5] Recently, we have also identified a novel CaBP, termed CaBP8, which also belongs to this family. A protein identical to CaBP8 has been recently cloned and called calneuron.[44] CaBP8 has not been fully characterized. Another protein, termed CaBP7, was also recently cloned in our laboratory but we will only briefly describe its properties, as it is not expressed in the retina. CaBP6 and CaBP9 are Ca^{2+}-binding proteins from an unrelated subfamily and are not presented in this chapter.

CaBP1, 2, 4, and 5 share ~70 % similarity with each other in the conserved domain and about 50 % similarity with CaM and CaBP8. CaBP8 shares ~70% homology with CaBP7 and ~50% homology with CaM. The N-terminal domain encoded by exon 1, absent in CaM, is variable in amino acid sequence and in size. This N-terminal domain of CaBP1, 2, and 4, is also less conserved between species than the C-terminal part of the protein. In contrast, human CaBP8 and CaBP7 are 100% similar to their mouse orthologs. Furthermore, an alternatively spliced exon of about 60 amino acids exists between exon 1 and exon 2 of CaBP1 and CaBP2 (Figure 1). Another feature shared by CaBP1 and CaBP2 is the presence of a putative myristoylation site on the glycine adjacent to the N-terminal Met.

EF-Hand Motifs of CaBPs

Retinal CaBPs, as well as their putative ancestor CaM, contain four EF-hand motifs, although some are not capable of binding Ca^{2+}. CaBPs contain a different combination of functional and nonfunctional EF-hand motifs. CaBP1, 2, 4 and 5 have three putative functional EF-hand motifs, the second one being disabled by deletion or substitution of amino acids critical for Ca^{2+} coordination. In contrast, CaBP8 has two putative functional EF-hand motifs in the N-terminal half of the protein, followed by two disabled EF-hand motifs at the C-terminus.

Phylogenetic Analysis of CaBPs

As expected from the sequence similarity analysis, all CaBPs are grouped together with CaM (Figure 2). CaBP1, 2, 4 and 5 form one subgroup, while CaBP7 and CaBP8 form another branch.

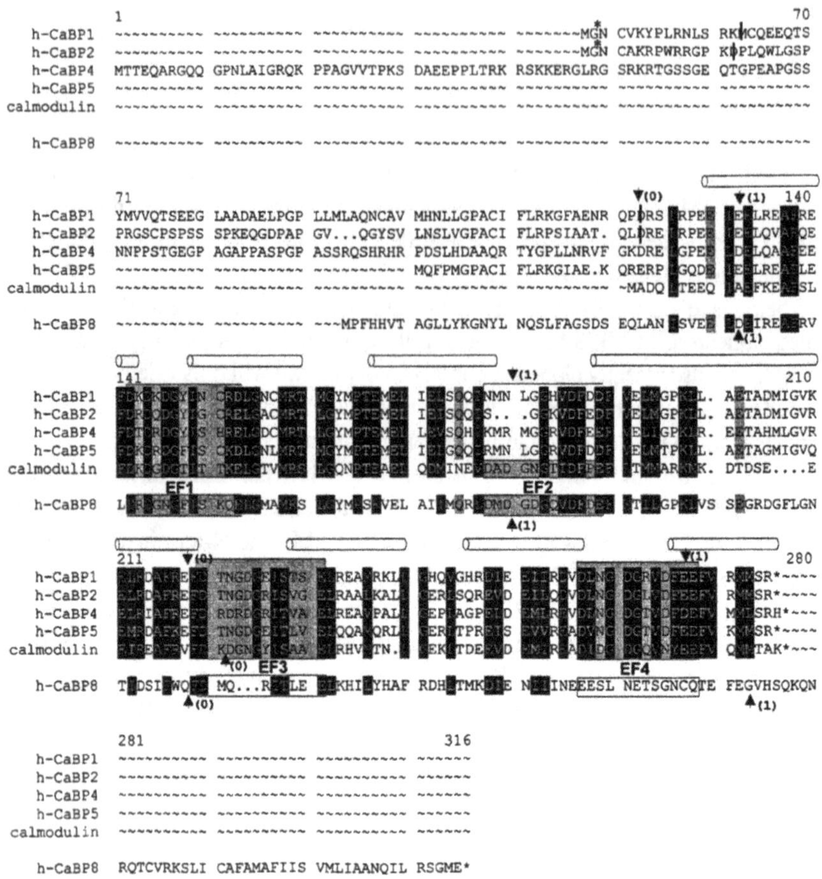

Figure 1. Sequence alignment of the deduced amino acid sequences of human (h-) CaBPs and Calmodulin (CaM). The identical residues in all sequences are shown in white letters on black background. The conservative substitutions are shown in white letters on dark gray background (E=D; S=T; V=M=I=L; K=R). The residues identical among CaBPs are shaded in light gray. Putative functional EF-hand motifs are shown in shaded boxes and open boxes indicate the nonfunctional EF-hand motifs. An asterisk represents the myristoylation site, and arrows indicate positions of introns for the h-CaBP1, 2, 4, 5, 8 and CaM genes. The numbers between brackets adjacent to the arrow indicate the phase of the intron. The symbol ˉ indicate the intron position in alternative splice forms of CaBP1 and CaBP2. α-Helical region of the proteins is indicated as a cylinder above the sequence.

Conformational Changes of CaBPs

CaM, the closest homolog of CaBPs, displays large conformational changes of the EF-hand motifs upon Ca^{2+} binding. In addition, the pair of EF-hands 1 and 2 are connected to EF-hands 3 and 4 through a long α-helix and form two domains that give the dumbbell shape. This central helix is bent in the Ca^{2+} loaded forms. It

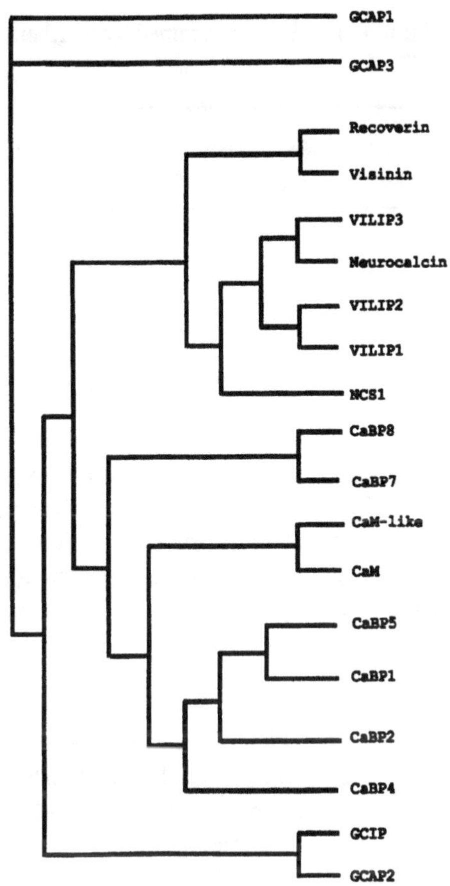

Figure 2. Phylogenetic tree of selected retinal Ca^{2+}-binding proteins. The tree was built with a bootstrap analysis of neighbor-joining distance using the PAUPSearch program from the Genetics Computer Group package (GCG from Accelrys). The sequences included are: h-CaM (A31920); h-CaM-like protein (P27482); human recoverin (S62028); GCAP1: human guanylate cyclase-activating protein (L36859); GCAP2: human guanylate cyclase-activating protein 2 (Surguchov et al, 1997); GCAP3: human guanylate cyclase-activating protein 3 (AF110002); GCIP: frog guanylate cyclase-inhibitory protein (AAC15878); chicken visinin (P22728); neurocalcin: bovine neurocalcin (JH0616); VILIP1: human visinin-like protein 1 (U14747); VILIP2: rat visinin-like protein 2 (P35332); VILIP3: human visinin-like protein 3 (P37235); NCS1: rat neuronal Ca^{2+} sensor 1 (P36610); CaBP1: human CaBP1(AF169148); CaBP2: human CaBP2 (AF169154); CaBP4: human CaBP4: human CaBP4 (AY039217); and CaBP5: human CaBP5 (AF169159). CaBP7 and CaBP8 are novel sequences reported in this chapter: CaBP7: mouse CaBP7 (AF419324); CaBP8: human CaBP8 (AY007302).

consists of 25 amino acid residues in CaM, forming a seven-turn α-helix, compared to troponin C, which has an extended central helix by one turn. CaBPs, like troponin C, has a longer linker than CaM and consequently, forms an 8 turn-long helix. Upon Ca^{2+} binding by CaM and troponin C, this linker bends and assembles both domains, enfolding the target peptide.[45-47] An extra glycine is present in the extra turn

Table 1. Tissue distribution of CaBPs determined by Northern blot and/or RT-PCR

	Retina	Brain	Nasal Epithelium	Liver	Kidney	Heart Muscle	Skeletal	Placenta	Pancreas	Lung
CaBP1	+	+	-	-	-	-	-	-	-	-
CaBP2	+	-	-	-	-	-	-	-	-	-
CaBP4	+	-	-	-	-	-	-	-	-	-
CaBP5	+	-	-	-	-	-	-	-	-	-
CaBP8	+	+	-	-	-	-	-	-	-	-

of CaBPs that might give more flexibility to wrap this region around the target. The amino acid residue is highly conserved in the structurally conserved region of the CaBPs. Further structural studies will determine if a similar structural change of conformation occurs in CaBPs. The conserved EF-hand motifs of CaBPs, four successive helix-loop-helix motifs, are encoded by the highly conserved exons 2 to 5.

Tissue Distribution of CaBPs

CaBPs are neuron-specific, with CaBP2, 4 and 5 being retina-specific (Table 1), and CaBP1 and 8 both expressed in retina and brain. With the exception of CaBP8, the size of CaBPs mRNA detected by Northern blot is ~1.7 kb. CaBP8 mRNAs detected in brain are ~9 kb. This is surprisingly high compared to the other CaBPs and to the size of the coding region. CaBP8 is likely expressed at low levels in the retina because it could only be detected in the retina by RT-PCR, and not by Northern blot.

CaBP5 has been localized primarly to rod bipolar cells, although it can also be detected in cone bipolar cells. CaBP1 is expressed in cone bipolar cells and in amacrine cells. CaBP8 and CaBP2 are also expressed in cells of the inner nuclear layer, including some bipolar cells (Imanishi and Palczewski, personal communication). The precise localization of CaBP4 in the retina has not yet been determined.

The Gene Structures of CaBPs

The exon-intron pattern is largely conserved among CaBPs (Figure 3). However, CaBP8 and 7 differ with exon 1 and 2 being fused together. CaBP7 is not expressed in the retina, but will be included in this gene structure analysis because of its high homology to CaBP8. The CaBP7 human ortholog was also deposited in GenBank database by others (AAF03511).

CaBPs have a gene structure that is very similar, but not identical to C M. CaBP1, 2, 4, 5, 7 and 8 have their 4th intron located 9 bases upstream of the 4th intron of CaM (Figure 1). CaBP8 and 7 also have their 5th intron inserted a few

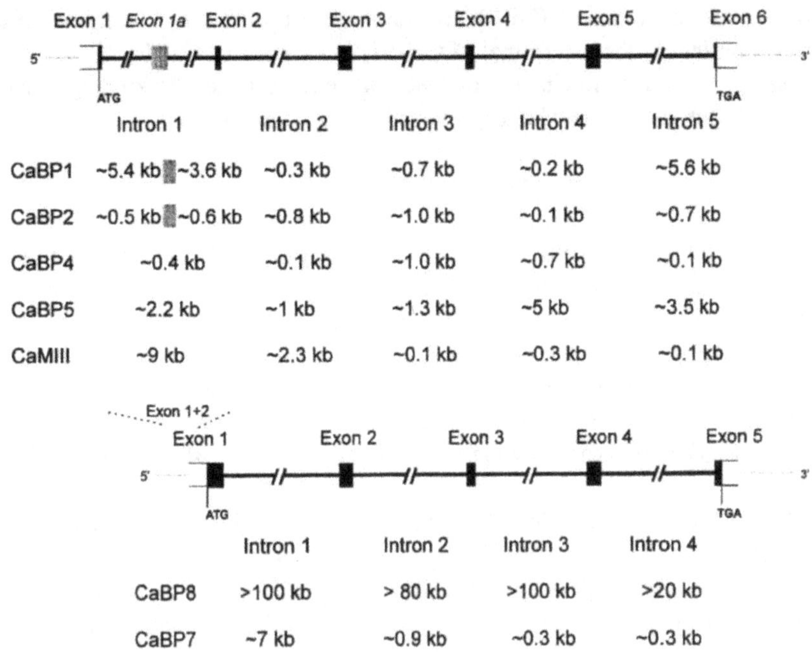

Figure 3. Organization of human CaBP genes. The coding regions are shown in black boxes and the untranslated regions are shown in white boxes. The gray box represents an alternatively spliced exon present in the CaBP1 and CaBP2 genes. The numbers represent the size of the introns. CaBP8 is the largest gene and encompasses more than 300 kb, whereas other CaBP genes cover <12 kb. The CaBP7 and CabP8 genes lack an intron that corresponds to intron 1 in the CaM gene and other CaBP genes, resulting in a gene consisting of 5 exons rather than 6 exons. All these genes, except CaBP7, encode Ca^{2+}-binding proteins expressed in the retina.

bases downstream of the 5th intron of CaBP1-5 and CaM. However, despite the shift of 1 or 2 exon/intron junctions compared to CaM, the phase is conserved between all CaBPs and CaM. CaBP1, 2, 4, 5 and CaM have a 0, 1, 1, 0, 1 phase for introns 1 to 5, respectively, while CaBP8 and 7, encoded by 4 exons, display a 1, 1, 0, 1 phase for introns 1 to 4, respectively (Figure 1). Such conserved gene structure suggests that CaBPs genes are closely related to CaM and are most likely the product of gene duplication events. This is corroborated by the proximity of some of these genes on the chromosome, as described below. It should be noted that the only EF-hand motif, EF-hand 1, functionally conserved among all CaBPs and CaM, is not interrupted by an intron as compared to EF-hands 2 to 4 (Figure 1).

Alternative splicing of the genes encoding CaBP1 and 2 gives two isoforms of these proteins, with or without a 60 amino acid-long additional sequence. This alternatively spliced exon is inserted at the junction between exon 1 and exon 2, and keeps the 0 phase of the surrounding introns. Caldendrin is another spliced variant of CaBP1 that is expressed in the brain.[48] In this variant, exon 1 differs, but exons 2 through 5 are identical with CaBP1. Calbrain, a short protein of 70 amino acids that is identical to the C-terminal of CaBP1, is a truncated CaBP1.[49]

The primary transcript in most of the CaBP genes is in the 4-16 kb range, however, CaBP8 is an unusually large gene of more than 300 kb.

Chromosomal Localization of CaBPs

The chromosomal localization of genes encoding CaBPs have been determined by FISH (CaBP2 and 5) or inferred from the characterization of genomic clones found in GenEMBL databases (CaBP1, 4 and 8).

Most CaBP genes are located on different chromosomes, with the exception of CaBP2 and 4. The CaBP2 gene is located 200 kb distant from the CaBP4 gene in the 3'-CaBP2-5' 5'-CaBP4-3' orientation (Figure 4). The CaBP5 gene is 1,500 kb away from the CaM gene encoded by the CALM3 gene in a 5'-CaBP5-3' 3'-CALM3-5' orientation.

Some CaBP genes co-localize with loci for retinal diseases. Details and references about cloned and/or mapped genes causing retinal diseases can be found at "http://www.sph.uth.tmc.edu/Retnet/disease.htm". A few retinal diseases have been mapped to the same locus as CaBP2 and CaBP4, on chromosome 11 at band q13.1. Forty percent of the patients affected with Bardet-Biedl syndrome show linkage to 11q13. Retinitis pigmentosa has also been mapped to this locus, but is associated with mutations in ROM-1 and peripherin/rds genes. Mutations in Best vitelliform macular dystrophy have also been indicated at the 11q13 locus. Mutations in the bestrophin gene have been found in patients with Bests disease as well as Age-related Macular Degeneration.

Cone-rod dystrophy has been mapped to the same locus as CaBP5, on chromosome 19 at band q13.3. However, mutations in the pineal/retina-specific transcription factor gene have been associated with multiple cases of cone-rod dystrophy. No retinal diseases have been mapped to the CaBP1 (chromosome 12q24.2) or CaBP8 (chromosome 7q11.21-11.23) genetic locus.

The Function of CaBPs

The physiological role of these CaBPs has not yet been deciphered, although they have been shown to modulate various targets. CaBPs are closely related to CaM, and their activity was analyzed on targets known to be modulated by CaM (Table 2). None of the CaBPs significantly inhibited the activity of G protein-coupled receptor kinase 2 (GRK2).[5] However, CaBPs inhibited GRK5-mediated phosphorylation of rhodopsin. CaBP1 has the strongest effect on GRK5, compared to CaBP2 and 5. Out of all CaBPs, CaBP2 was the most effective in the CaM kinase II stimulation. CaBP2 is also able to bind to a peptide derived from the CaM-binding domain of CaM kinase II, whereas CaBP1 and CaBP5 do not bind significantly[5]. CaBP2 and CaBP5, although less effective than CaM, have also been shown to stimulate calcineurin activity.[31]

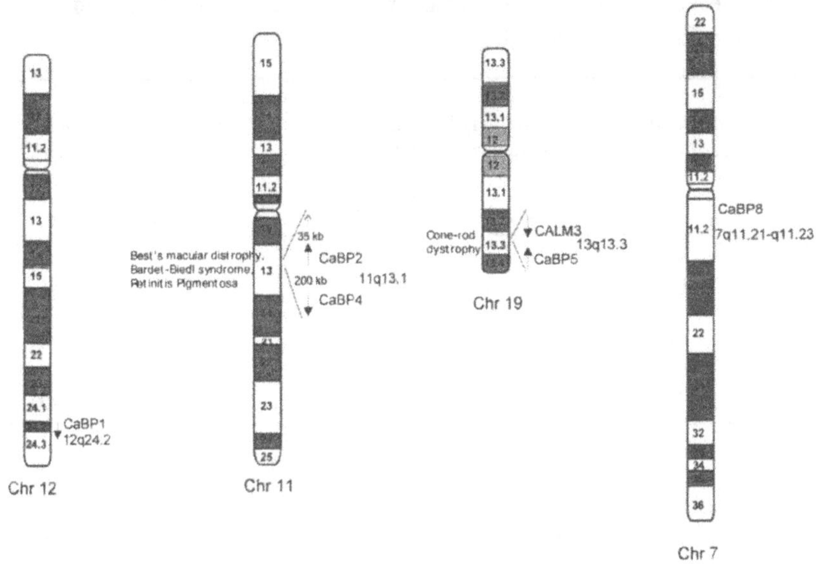

Figure 4. Identification of CaBP genes on the chromosome map. The sketches are based on the chromosome maps from the National Center for Biotechnology Information (http://www.ncbi.nlm.nih/cgi-bin/SCIENCE96/chr (No.)). The CaBP genes and eye disease loci that colocalize with CaBP genes are indicated.

It is possible that CaBPs could also modulate targets like CaM kinase II and/or calcineurin in vivo, in response to a change in Ca^{2+}. In response to a decrease in the concentration of glutamate released by photoreceptors, metabotropic receptors (mGluR6) present in the dendritic terminals of on-bipolar cells activate a transduction cascade ultimately resulting in the closure of a cGMP-activated channel. There is evidence that Ca^{2+} dependent phosphorylation or dephosporylation of a target site on the cGMP-activated channel, resulting from activation of CaM kinase II or calcineurin, can be upregulated by Ca^{2+} concentration. The Ca^{2+} influx through the channel would modify its phosphorylation state and its sensitivity to cGMP, ultimately resulting in the closure of the channel.[50-52] Another putative target for CaBPs would be the cGMP-gated channel itself, by analogy with the cGMP-gated channel known to be modulated by CaM in photoreceptor cells. Further studies of the co-localization of CaBPs with their target proteins will be necessary to determine whether CaBPs modulate these targets in vivo.

Many other neuronal targets modulated by [Ca^{2+}] changes are candidates for a regulation by CaBPs. Ca^{2+} channels are found in all nerve cells and many types exist with specific current properties and functions in cell physiology. The localization of Ca^{2+} channels at active zones provides a high, confined rise in [Ca^{2+}] at the site of the transmitter release, which is perceived by multiple cellular targets. The L-type channel and P/Q type Ca^{2+} channel are modulated by CaM in response to changes in [Ca^{2+}].[53,54] However, CaM fails to completely convey the effects of Ca^{2+}

Table 2. Effects on the target enzymes and binding of CaBPs to CaM-binding domain of CaM kinase II

	Stimulation of CaM Kinase II	Stimulation of Calcineurin	Binding to CaM-Binding Domain of CaM Kinase II	Inhibition of GRK5	Inhibition of GRK2
CaBP1	+	+	-	+++	-
CaBP2	++	++	+	++	-
CaBP4	ND	ND	ND	ND	ND
CaBP5	-	+	-	+	-
CaBP8	ND	ND	ND	ND	ND

on the expressed P/Q type channel as compared to those isolated from native membranes, suggesting that there might be additional modulation by Ca^{2+}, independent of CaM.[55] Preliminary data showed that CaBP1 significantly enhanced inactivation of $Ca_v2.1$ Ca^{2+} currents, caused a depolarizing shift in the voltage-dependence of activation, and prevented Ca^{2+}-dependent facilitation of $Ca_v2.1$.[56] The modulation of voltage-gated Ca^{2+} channel might be particularly important in bipolar synaptic neurotransmission because the influx of Ca^{2+} is believed to be directly correlated to neurotransmitter release. Ca^{2+} entering into bipolar cells through Ca^{2+} channels stimulate the supply of neurotransmitter for the rapidly releasable pool of vesicles.[57] The mechanism that regulates the trafficking and release of transmitter vesicles has not yet been deciphered. Inositol 1,4,5-triphosphate receptors (IP_3R), Ca^{2+} channels that release Ca^{2+} from intracellular stores, are also modulated by CaM.[1,58] There is a controversy, however, about the Ca^{2+} dependency of the CaM effect on IP_3-induced Ca^{2+} release.[59-61] It has been suggested that a protein that shares structural homology with CaM modulates IP_3R.

As a consequence of their similarities to CaM, CaBPs might be the physiological modulator of targets known to be affected by CaM in vitro. Moreover, it is possible that both CaM and CaBPs modulate the same effector enzymes in vitro and in vivo, but by different mechanisms. For example, CaM has a complex effect on the same target (e.g., on the P/Q-type Ca^{2+} channels).[55,62] It is possible that CaBPs might be the physiological modulators that have opposing effects to CaM, acting like antagonists. Further studies of the activity of CaBPs, co-localization with the targets, analysis of the phenotype of CaBP knockout mice, identification and investigation of pathological mutations in humans will help in determining the function of those CaBPs in vivo.

ACKNOWLEDGMENTS

We would like to thank members of our laboratories for comments on the manuscript, in particular Samantha Dovey. This research was supported by NIH grant EY08061; a grant from Research to Prevent Blindness, Inc. (RPB) to the Department of Ophthalmology at the University of Washington; the Alcon Research Institute award, and the E.K. Bishop Foundation.

REFERENCES

1. Berridge MJ, Lipp P, Bootman MD. The versatility and universality of calcium signalling. Nat Rev Mol Cell Biol 2000; 1(1):11-21.
2. Thomas AP, Bird GSJ, Hajnoczky G et al. Spatial and temporal aspects of cellular calcium signaling. FASEB J 1996; 10(13):1505-1517.
3. Berridge MJ. Elementary and global aspects of calcium signalling. J Physiol (Lond) 1997; 499(2):291-306.
4. Allbritton NL, Meyer T, Stryer L. Range of Messenger Action of Calcium-Ion and Inositol 1,4,5- Trisphosphate. Science 1992; 258(5089):1812-1815.
5. Haeseleer F, Sokal I, Verlinde C et al. Five members of a novel Ca2+-binding protein (CABP) subfamily with similarity to calmodulin. J Biol Chem 2000; 275(2):1247-1260.
6. Polans A, Baehr W, Palczewski K. Turned on by Ca2+! The physiology and pathology of Ca2+-binding proteins in the retina. Trends Neurosci 1996; 19(12):547-554.
7. Schnetkamp PPM. Calcium Homeostasis in Vertebrate Retinal Rod Outer Segments. Cell Calcium 1995; 18(4):322-330.
8. Fain GL, Matthews HR, Cornwall MC. Dark adaptation in vertebrate photoreceptors. Trends Neurosci 1996; 19(11):502-507.
9. Rodieck RW. Retinas. In: The First Steps in Seeing. Sunderland: Sinauer Associates, Inc., 1998:36-55.
10. Kretsinger RH. Calcium Coordination and the Calmodulin Fold—Divergent Versus Convergent Evolution. Cold Spring Harbor Symposia On Quantitative Biology 1987; 52:499-510.
11. Heizmann CW, Hunziker W. Intracellular Calcium-Binding Proteins—More Sites Than Insights. Trends Biochem Sci 1991; 16(3):98-103.
12. Kawasaki H, Nakayama S, Kretsinger RH. Classification and evolution of EF-hand proteins. Biometals 1998; 11(4):277-295.
13. Palczewski K, Polans AS, Baehr W et al. Ca2+-binding proteins in the retina: structure, function, and the etiology of human visual diseases. Bioessays 2000; 22(4):337-350.
14. Vogel HJ. The Merck-Frosst-Award-Lecture 1994—Calmodulin—a Versatile Calcium Mediator Protein. Biochem Cell Biol 1994; 72(9-10):357-376.
15. Means AR, Dedman JR. Calmodulin—an Intracellular Calcium Receptor. Nature 1980; 285(5760):73-77.
16. Yap KL, Ames JB, Swindells MB et al. Diversity of conformational states and changes within the EF- hand protein superfamily. Proteins-Structure Function and Genetics 1999; 37(3):499-507.
17. Skelton NJ, Kordel J, Akke M et al. Signal-Transduction Versus Buffering Activity in Ca2+-binding proteins. Nat Struct Biol 1994; 1(4):239-245.
18. Celio MR, Heizmann CW. Calcium-Binding Protein Parvalbumin As a Neuronal Marker. Nature 1981; 293(5830):300-302.
19. Haverkamp S, Wassle H. Immunocytochemical analysis of the mouse retina. JComp Neurol 2000; 424(1):1-23.
20. Deng P, Cuenca N, Doerr T. Localization of neurotransmitters and calcium binding proteins to neurons of salamander and mudpuppy retinas. Vision Research 2001; 41(14):1771-1783.

21. Roberts WM. Localization of Calcium Signals By a Mobile Calcium Buffer in Frog Saccular Hair-Cells. J Neurosci 1994; 14(5):3246-3262.
22. Palczewski K, Subbaraya I, Gorczyca WA et al. Molecular-Cloning and Characterization of Retinal Photoreceptor Guanylyl Cyclase-Activating Protein. Neuron 1994; 13(2):395-404.
23. Dizhoor AM, Olshevskaya EV, Henzel WJ et al. Cloning, Sequencing, and Expression of a 24-Kda Ca2+-binding protein Activating Photoreceptor Guanylyl Cyclase. J Biol Chem 1995; 270(42):25200-25206.
24. Gorczyca WA, Polans AS, Surgucheva IG et al. Guanylyl Cyclase-Activating Protein—a Calcium-Sensitive Regulator of Phototransduction. J Biol Chem 1995; 270(37):22029-22036.
25. Haeseleer F, Sokal I, Li N et al. Molecular characterization of a third member of the guanylyl cyclase-activating protein subfamily. J Biol Chem 1999; 274(10):6526-6535.
26. Boutin JA. Myristoylation. Cell Signal 1997; 9(1):15-35.
27. Li N, Fariss RN, Zhang K et al. Guanylate-cyclase-inhibitory protein is a frog retinal Ca2+-binding protein related to mammalian guanylate-cyclase- activating proteins. Eur J Biochem 1998; 252(3):591-599.
28. Otto-Bruc A, Fariss RN, Haeseleer F et al. Localization of guanylate cyclase-activating protein 2 in mammalian retinas. Proceedings of the National Academy of Sciences of the United States of America 1997; 94(9):4727-4732.
29. Payne AM, Downes SM, Bessant DAR et al. A mutation in guanylate cyclase activator 1A (GUCA1A) in an autosomal dominant cone dystrophy pedigree mapping to a new locus on chromosome 6p21.1. Hum Mol Genet 1998; 7(2):273-277.
30. Sokal I, Li N, Surgucheva I et al. GCAP1(Y99C) mutant is constitutively active in autosomal dominant cone dystrophy. Mol Cell 1998; 2(1):129-133.
31. Sokal I, Li N, Verlinde C et al. Ca2+-binding proteins in the retina: from discovery to etiology of human disease. Biochimica Et Biophysica Acta-Molecular Cell Research 2000; 1498(2-3):233-251.
32. Downes SM, Holder GE, Fitzke FW et al. Autosomal dominant cone and cone-rod dystrophy with mutations in the guanylate cyclase activator 1A gene-encoding guanylate cyclase activating protein-1. Arch Ophthalmol 2001; 119(1):96-105.
33. Milam AH, Dacey DM, Dizhoor AM. Recoverin Immunoreactivity in Mammalian Cone Bipolar Cells. Vis Neurosci 1993; 10(1):1-12.
34. Massey SC, Mills SL. Calbindin-immunoreactive cone bipolar cell type in the rabbit retina. J Comp Neurol 1996; 366(1):15-33.
35. Dizhoor AM, Ray S, Kumar S et al. Recoverin—a Calcium Sensitive Activator of Retinal Rod Guanylate-Cyclase. Science 1991; 251(4996):915-918.
36. Chen CK, Inglese J, Lefkowitz RJ et al. Ca2+-Dependent Interaction of Recoverin With Rhodopsin kinase. J Biol Chem 1995; 270(30):18060-18066.
37. Graykeller MP, Polans AS, Palczewski K et al. The Effect of Recoverin-Like Calcium-Binding Proteins On the Photoresponse of Retinal Rods. Neuron 1993; 10(3):523-531.
38. Ohguro H, Rudnicka-Nawrot M, Buczylko J et al. Structural and enzymatic aspects of rhodopsin phosphorylation. J Biol Chem 1996; 271(9):5215-5224.
39. Ames JB, Ishima R, Tanaka T et al. Molecular mechanics of calcium-myristoyl switches. Nature 1997; 389(6647):198-202.
40. Polans AS, Witkowska D, Haley TL et al. Recoverin, a Photoreceptor-Specific Calcium-Binding Protein, Is Expressed By the Tumor of a Patient With Cancer-associated retinopathy. Proceedings of the National Academy of Sciences of the United States of America 1995; 92(20):9176-9180.
41. Maeda A, Ohguro H, Nabeta Y et al. Identification of human antitumor cytotoxic T lymphocytes epitopes of recoverin, a cancer-associated retinopathy antigen, possibly related with a better prognosis in a paraneoplastic syndrome. Eur J Immunol 2001; 31(2):563-572.
42. Adamus G, Machnicki M, Elerding H et al. Antibodies to recoverin induce apoptosis of photoreceptor and bipolar cells in vivo. J Autoimmun 1998; 11(5):523-533.

43. Reynolds AJ, Bartlett SE, Morgans C. The distribution of neuronal calcium sensor-I protein in the developing and adult rat retina. Neuroreport 2001; 12(4):725-728.
44. Wu YQ, Lin X, Liu CM, Jamrich M, Shaffer LG. Identification of a human brain-specific gene, calneuron 1, a new member of the calmodulin superfamily. Mol Genet Metab 2001; 72(4):343-350.
45. Meador WE, Means AR, Quiocho FA. Target Enzyme Recognition By Calmodulin—2.4-Angstrom Structure of a Calmodulin-Peptide Complex. Science 1992; 257(5074):1251-1255.
46. Vassylyev DG, Takeda S, Wakatsuki S et al. Crystal structure of troponin C in complex with troponin I fragment at 2.3-angstrom resolution. Proceedings of the National Academy of Sciences of the United States of America 1998; 95(9):4847-4852.
47. Finn BE, Forsen S. The Evolving Model of Calmodulin Structure, Function and Activation. Structure 1995; 3(1):7-11.
48. Seidenbecher CI, Langnaese K, Sanmarti-Vila L et al. Caldendrin, a novel neuronal calcium-binding protein confined to the somato-dendritic compartment. J Biol Chem 1998; 273(33):21324-21331.
49. Yamaguchi K, Yamaguchi F, Miyamoto O et al. Calbrain, a novel two EF-hand calcium-binding protein that suppresses Ca2+/calmodulin-dependent protein kinase II activity in the brain. J Biol Chem 1999; 274(6):3610-3616.
50. Nawy S. Regulation of the on bipolar cell mGluR6 pathway by Ca2+. J Neurosci 2000; 20(12):4471-4479.
51. Shiells RA, Falk G. Activation of Ca2+-calmodulin kinase II induces desensitization by background light in dogfish retinal 'on' bipolar cells. J Physiol (Lond) 2000; 528(2):327-338.
52. Walters RJ, Kramer RH, Nawy S. Regulation of cGMP-dependent current in ON bipolar cells by calcium/calmodulin-dependent kinase. Vis Neurosci 1998; 15(2):257-261.
53. Peterson BZ, DeMaria CD, Yue DT. Calmodulin is the Ca2+ sensor for Ca2+-dependent inactivation of 1-type calcium channels. Neuron 1999; 22(3):549-558.
54. Zuhlke RD, Pitt GS, Deisseroth K et al. Calmodulin supports both inactivation and facilitation of L- type calcium channels. Nature 1999; 399(6732):159-162.
55. Lee A, Scheuer T, Catterall WA. Ca2+/calmodulin-dependent facilitation and inactivation of P/Q- type Ca2+ channels. J Neurosci 2000; 20(18):6830-6838.
56. Lee A, Westenbroek RE, Haeseleer F et al. Differential modulation of $Ca_v2.1$ channels by calmodulin and Ca^{2+}-binding protein 1. Nat Neurosci (in press).
57. Gomis A, Burrone J, Lagnado L. Two actions of calcium regulate the supply of releasable vesicles at the ribbon synapse of retinal bipolar cells. J Neurosci 1999; 19(15):6309-6317.
58. Bootman MD, Lipp P, Berridge MJ. The organisation and functions of local Ca2+ signals. J Cell Sci 2001; 114(12):2213-2222.
59. Missiaen L, Parys JB, Weidema AF et al. The bell-shaped Ca2+ dependence of the inositol 1,4,5- trisphosphate-induced Ca2+ release is modulated by Ca2+/calmodulin. J Biol Chem 1999; 274(20):13748-13751.
60. Patel S, Morris SA, Adkins CE et al. Ca2+-independent inhibition of inositol trisphosphate receptors by calmodulin: Redistribution of calmodulin as a possible means of regulating Ca2+ mobilization. Proceedings of the National Academy of Sciences of the United States of America 1997; 94(21):11627-11632.
61. Adkins CE, Morris SA, De Smedt H et al. Ca2+-calmodulin inhibits Ca2+ release mediated by type-1,-2 and-3 inositol trisphosphate receptors. Biochem J 2000; 345:357-363.
62. DeMaria CD, Soong TW, Alseikhan BA, Alvania RS, Yue DT. Calmodulin bifurcates the local Ca2+ signal that modulates P/Q- type Ca2+ channels. Nature 2001; 411(6836):484-489.

GCAPs: Ca^{2+}-SENSITIVE REGULATORS OF retGC

Wojciech A. Gorczyca[1,2] and Izabela Sokal[3]

ABSTRACT

Lowered concentration of Ca^{2+} ions, resulting from illumination of the photoreceptor cell, is the signal for resynthesis of cGMP by retina-specific guanylyl cyclase (retGC). This Ca^{2+} -dependent activation of retGC is mediated by Ca^{2+} -binding proteins named GCAPs (guanylyl cyclase-activating proteins) and contributes to the recovery of photoreceptor cell to the dark state. Three different GCAPs (GCAP1, GCAP2 and GCAP3) are identified in vertebrate retina to date. In this chapter we describe their discovery, methods of purification, properties, and possible modes of action.

INTRODUCTION

Cyclic GMP (cGMP) is the central messenger molecule in phototransduction. It interacts directly with nonspecific cationic channels located in the plasma membrane of the outer segments of photoreceptor cell and keeps the channels open. The cations (mainly Na^+ and Ca^{2+}) entering the cell through opened channels constitute so called dark current and depolarize the cell membrane. Light triggers the cascade of biochemical events (recently reviewed in refs. 1-3), which leads to hydrolysis of cGMP by photoreceptor-specific phosphodiesterase (PDE6) and closure of the channels. In consequence the inward cationic current is interrupted causing hyperpolarization of plasma membrane and neurochemical response at synaptic terminals. At

---

[1]Laboratory of Signaling Proteins, L. Hirszfeld Institute of Immunology and Experimental Therapy, Polish Academy of Sciences, Wroclaw, Poland. [2]Department of Biogenic Amines, Polish Academy of Sciences, Lodz, Poland. [3]Department of Ophthalmology, School of Medicine, University of Washington, Seattle, Washington, USA.

the same time Na^+/K^+-Ca^{2+} exchangers, also present in the outer segment membrane, continuously extrude Ca^{2+} ions outside the cell and cytoplasmic concentration of Ca^{2+} ($[Ca^{2+}]_{free}$), which is high (~600 nM) in the darkness, drops to low level below 50 nM. Lowered concentration of Ca^{2+} ions is the signal for resynthesis of cGMP by guanylyl cyclase (GC) beginning recovery of the photoreceptor cell to the dark state. This interdependence of Ca^{2+} and cGMP concentrations is essential in phototransduction and also in adaptation of the sensitivity of photoreceptor cells to the strength of light signal. Thus, the mechanisms that regulate the GC activity appear to be key elements of both processes.

In vertebrates, guanylyl cyclases are present in many types of cells and are found in two forms: cytosolic (soluble GC) and membrane-bound (particulate GC). Soluble cyclases exist as heterodimers and are activated by nitric oxide (NO). Particulate cyclases exist as homodimers of polypeptide chains containing large extracellular, transmembrane, regulatory kinase-like, dimerization, and catalytic domains (Fig. 5). The mechanism of their activation is complex and still not completely understood. At least three of eight known particulate cyclases are activated through interaction their extracellular domains with peptide ligands.[4,5]

The photoreceptor-specific guanylyl cyclase, termed also retinal GC (retGC), is regulated in a different way. It has been known for a long time that synthesis of cGMP in outer segments of rods (ROS) is negatively regulated by the $[Ca^{2+}]$. Guanylyl cyclase activity is low at high Ca^{2+} concentration and significantly (5-20 times) increases at low Ca^{2+} concentration.[6,7] It has also been known that synthesis of cGMP occurs in the membrane fraction of rod outer segments indicating that retGC is a particulate form of the enzyme. Its structure, however, remained unknown until 1992, when cDNA for a new particulate cyclase was cloned from the human retina library.[8] Analysis of the amino acid sequence revealed that this retina-specific guanylyl cyclase belongs to the family of particulate guanylyl cyclases sharing their overall structure (Fig. 5). The new GC was insensitive to peptide ligands that activated other particulate cyclases and although its Ca^{2+}-sensitivity was not shown, it was the main candidate to be the enzyme responsible for cGMP synthesis in phototransduction. The proof that new GC is regulated by $[Ca^{2+}]$ came together with identification and isolation of Ca^{2+}-binding proteins of the EF-hand family referred to as GCAPs (Guanylyl Cyclase-Activating Proteins).[9-13]

So far two isoforms of retinal GC, retGC1 and retGC2 (in human named RetGC1 and RetGC2, in bovine ROS-GC1 and ROS-GC2, in rat GC-E and GC-F), have been identified in mammalian retina.[8,14-17] Both enzymes were shown to be regulated by GCAPs in a Ca^{2+}-dependent manner.[11,12,15]

DISCOVERY OF GCAPs

The Ca^{2+}-sensitivity of retGC is lost when the ROS membranes are washed out with buffer of low ionic strength but may be restored when the extract is added back to the membranes. Based on this observation, in 1988 Koch and Stryer postulated in reference 7 participation of a soluble factor, possibly protein, in the Ca^{2+}-dependent regulation of the retinal GC activity. A few years later the search of this putative regulator of retGC resulted in identification in the bovine retina of two different Ca^{2+}-binding proteins. Both proteins activated the enzyme in washed ROS membranes at $[Ca^{2+}]_{free}$ below 100 nM. The first of them was isolated from the outer segments of the photoreceptors. It was shown to restore the Ca^{2+} sensitivity of the GC in reconstituted system and to decrease the sensitivity, time-to-peak, and recovery time of the light response in intact rod outer segments.[9] The protein was cloned and named GCAP.[10] The second activator of retinal GC was purified from the whole retinas and originally named p^{24} because of its apparent electrophoretic mobility.[11] Its sequence was found to be 42% identical with GCAP (now termed GCAP1) and protein was named GCAP2.[12,13] Recently, the third member of the GCAPs subfamily, named GCAP3, has been cloned from the human and zebrafish retinas.[18,19]

STRUCTURE AND PROPERTIES OF GCAPs

Bovine GCAP1 and GCAP2 were predicted to consist of 205 amino acids with calculated molecular mass of 23.5 kDa[10,20] and 204 amino acids (calculated molecular mass 23.7 kDa)[12,13], respectively. Human GCAP3 consists of 209 amino acids with calculated molecular mass of 23.8 kDa.[18] GCAP1 and GCAP2 are highly conserved proteins showing at amino acid level about 70 percent identity among vertebrates and even more than 90 percent identity among mammals (Fig. 1). Homology between different GCAPs is lower. Human GCAP1, GCAP2, and GCAP3 share about 40 percent identity with the highest homology (57%) between GCAP1 and GCAP3 (Fig. 2). All GCAPs posses a consensus sequence for N-terminal acylation. In accordance with postulated physiological role, GCAPs contain in their structures three motifs EF-hand that are potential sites of Ca^{2+} binding. The presence of fourth, residual EF-hand motif (located at N-terminus), indicates that GCAPs constitute the subfamily of Ca^{2+}-binding proteins, which originates from common calmodulin-like ancestor[21] and together with recoverins, visinin-like proteins (VILIPs), and frequenins belong to the larger family of structurally related neuronal proteins, named Neuronal Calcium Sensors (NCS) (recently reviewed in ref. 22).

The genes encoding GCAPs reveal identical intron/exon organization and contain 4 exons and 3 introns.[1] In human, genes encoding GCAP1 and GCAP2 are located on the short arm of chromosome 6 (p21.1) in tail-to-tail arrangement.[23,24] The gene for GCAP3 is located on chromosome 3 (q13.1)[18] indicating that different GCAPs genes are the result of ancestral gene duplication and translocation.

The first properties of GCAP1 and GCAP2 were determined using native proteins purified from the bovine retina. More intensive studies were undertaken when functionally active GCAP1 and GCAP2 were successfully expressed in bacteria[20,25,26]and in eukaryotic cells.[12,25] All GCAPs are soluble, highly acidic (pKa 4.1-4.7) and hydrophobic proteins.[1] Their electrophoretic mobility depends on Ca^{2+} ions. In the presence of Ca^{2+} apparent molecular mass of GCAP1 is 20 kDa and in the presence of EGTA shifts to ~25 kDa.[12] GCAP2 appears as a band of 18-19 kDa in the presence of Ca^{2+} or 24 kDa when Ca^{2+} is chelated.[12] Ca^{2+} binding to GCAPs

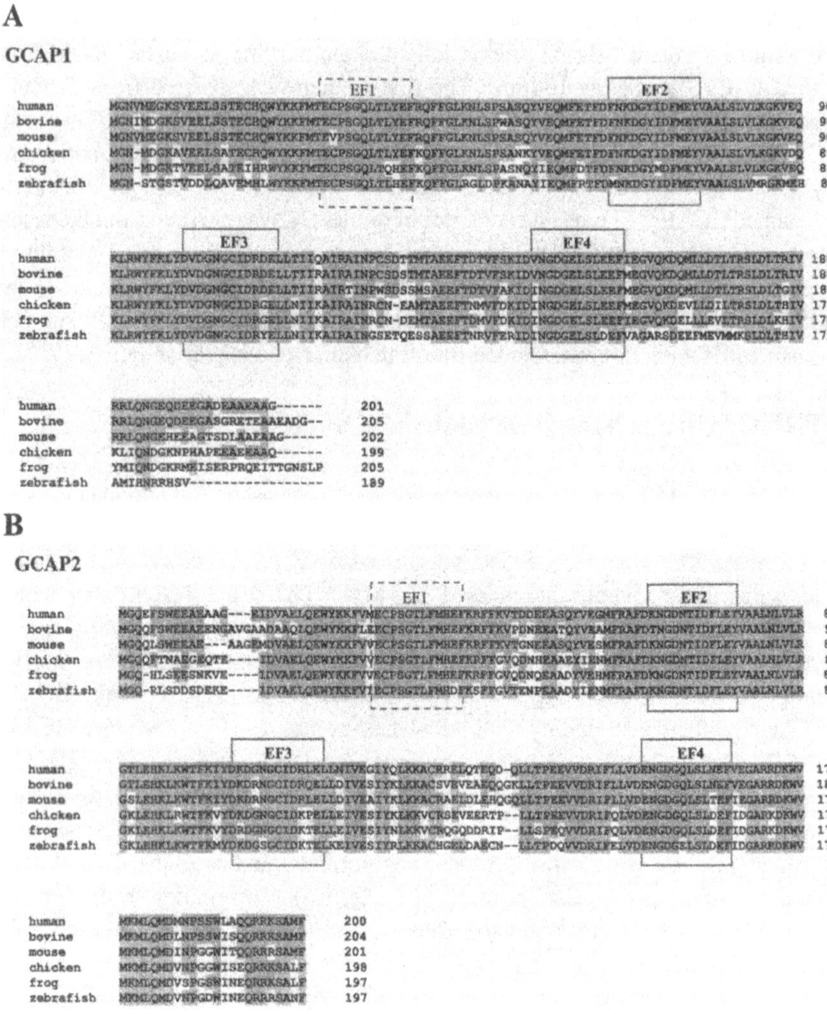

Figure 1. Alignments of GCAP1 A) and GCAP2 B) amino acid sequences from various species. Residues identical with human GCAPs sequences are printed in black on the gray background. The EF-hand motifs are shown in the boxes (dashed for the nonfunctional EF1 motifs). Sequence of mouse GCAP2 was obtained from reference 59. All other sequences were obtained from GenBank (accession numbers: XP004153 for human GCAP1; P46065 for bovine GCAP1; P43081 for mouse GCAP1, P79880 for chicken GCAP1; O73761 for frog GCAP1; AAL05602 for zebrafish GCAP1; Q9UMX6 for human GCAP2; P51177 for bovine GCAP2; P79881 for chicken GCAP2; O73762 for frog GCAP2; AAL05603 for zebra fish GCAP2).

is a cooperative process with Hill coefficients equal to 2-2.5 and 1.7-2.1 for GCAP1 and GCAP2, respectively.[9,11,13,20,27] It induces conformational changes in both proteins as studied by the fluorescence spectroscopy,[10,28-30] NMR,[31] circular dichroism,[31]

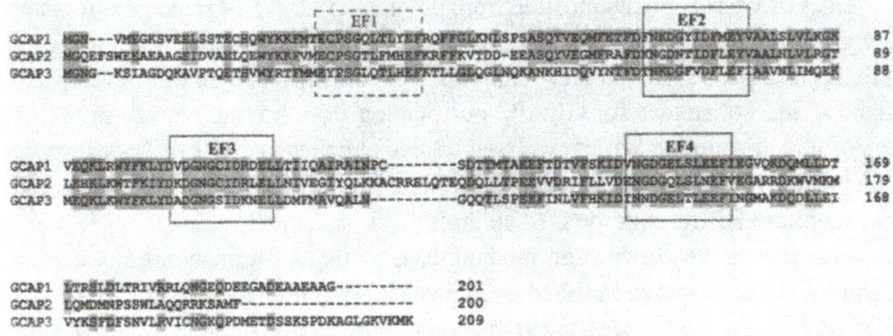

Figure 2. Sequence alignment of human GCAP1, 2 and 3. Residues identical with the GCAP1 sequence are printed in black on the gray background. The four EF-hand motifs are boxed (dashed for the nonfunctional EF1 motif). Sequences of human GCAP1 and GCAP2 are the same as in Figure 1. The GenBank accession number for GCAP3 is NP_005450

chemical modifications,[30] site directed mutagenesis,[29,32,33] and limited proteolysis.[31,33] Ca^{2+}-binding properties of GCAPs are critical in regulation of target cyclase(s). It was shown in experiments that involved mutagenesis in Ca^{2+}-binding regions of GCAP1 and GCAP2.[28,32-35]

GCAP1 and GCAP2, like recoverin, are acylated at N-terminal glycine by fatty acids.[10,25] Also, GCAP3 was found to be myristoylated when expressed in insect cells.[18] The role of myristoylation is not clear yet but it was shown to change Ca^{2+}-binding characteristics of bacteria expressed GCAP1 and thus to influence the ability of protein to activate retGC.[20] Moreover, expressed in insect High Five cells GCAP1 mutants that lacked myristoylation, markedly weaker activated retGC in comparison with the wild-type protein.[28] N-terminal acylation by myristic acid apparently does not influence the functional properties of GCAP2.[25]

PURIFICATION OF GCAPs FROM THE RETINA

Two GCAPs (GCAP1 and GCAP2) have been purified so far from vertebrate retina. It was very critical to employ a simple, fast, and reliable assay to measure the GC activity in ROS homogenates during purification. The major difficulty in the determination of retGC activity was the rapid hydrolysis of cGMP by the photoreceptor phosphodiesterase, which was also present in the homogenates. Therefore, we developed an assay which takes advantage of stereospecific metabolism of the α-thionucleotides.[36,37] In this assay the substrate GTPαS is converted to a stereoisomer of cGMP-S, which is not further degraded by the PDE. The assay proved to be very useful for monitoring GC activity at different stages of GCAP purification. The assay was used in parallel with SDS-PAGE. Thus, the ability of purified fractions to activate the retGC was correlated with the existence in these fractions specific protein band(s), visualized by staining the gels after electrophoretic separation.

GCAP1 was originally purified from bovine ROS by ion-exchange chromatography (DEAE-Sepharose), followed by HPLC on hydroxylapatite column, and finally reverse-phase HPLC on C4 column.[9,36] Another procedure was developed by Dizhoor and colleagues for GCAP2 purification from bovine retinas. In the first step of this method the extract was heat treated and then GCAP2 (p^{24}) was isolated using Phenyl Sepharose chromatography, preparative electrophoresis, and reverse-phase HPLC over the C18 column.[11]

The alternative purification method used by us was immunoaffinity chromatography, which employed antibodies against GCAPs coupled to Sepharose 4B. This relatively simple, fast, and efficient method of purification of proteins have enabled isolation of both, GCAP1 and GCAP2, from the same bovine retinal extract.[12,37] The extract was obtained by homogenization bovine retinas in water containing inhibitors of proteases. The homogenate was then clarified by sequential centrifugation at 30,000g and at 100,000g. The resulting supernatant (the retinal extract) has been used as a source of GCAPs. Separation of GCAP1 from GCAP2 employed two immunoaffinity columns connected in tandem: first binding only GCAP1 and second binding GCAP2. Using this method we showed for the first time that both GCAPs were present in the same retinal extract and were able to stimulate retGC activity independently when separated.[12] In another approach GCAP2 was separated first and then GCAP1 was purified from unbound fraction (Fig. 3). These methods allow to obtain in one step highly purified proteins with yield much higher than using conventional chromatography methods. The following specific antibodies (monoclonal or polyclonal) generated against recombinant GCAPs[10,37] or their peptide fragments[12] have been used by us during purification procedures:

- polyclonal antibody UW14 (pAbUW14) raised against bacterially expressed truncated bovine GCAP1 (GCAP180, Met_{26}-Pro_{205})[12]
- pAbGS31, rabbit antibody raised against bovine C-terminal peptide (Asp_{189}-Asp_{205}-Lys) of GCAP1[12]
- pAbUW50 and pAb850, rabbit antibodies raised against recombinant bovine GCAP2 expressed in E. coli , in which six N-terminal residues are replaced by His_6 tag[26,37]
- monoclonal antibody G2 (mAbG2) raised against truncated bovine GCAP1 (GCAP180) in BALB/c mice.[10] This antibody shows weak cross-reactivity with bovine GCAP2[12] and was used either for isolation GCAP2 or for isolation GCAP1. It can also be used for isolation of both GCAPs at the same time from retinal extract.

These antibodies have also been found useful for immunodetection of GCAPs in retinal extracts and in tissue sections.

LOCALIZATION OF GCAPs

By the Northern blotting and in situ hybridization techniques the photoreceptor cells of the retina were demonstrated to be the main site of GCAPs genes expression.[10,18,23,27] Also by the immunohistochemical techniques GCAP1 and GCAP2

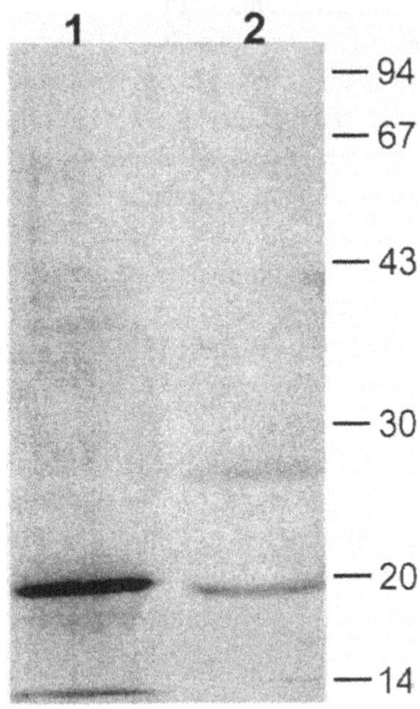

Figure 3. Bovine GCAP1 and GCAP2 purified from retinal extract. GCAP1 and GCAP2 were isolated from bovine retinal extract by immunoaffinity chromatography as described in the text using antibodies G2 and UW50 raised against recombinant GCAP1 and GCAP2, respectively. Fractions containing purified proteins were analyzed in SDS-PAGE. (1)—GCAP1; (2)—GCAP2

were localized predominantly in the photoreceptor layer of retina.[12,13,20,26,38,39] These results are consistent with suggested function of GCAPs in phototransduction.

However, there are still some controversies concerning sites of GCAP1 and GCAP2 expression. Only GCAP1 was purified from the bovine ROS,[9,12] while GCAP2 was purified from the whole retina,[11,12] indicating that GCAP2 might be located mainly in the inner segments of photoreceptor cells or other cells of the retina. Using different antibodies generated against GCAP1 or GCAP2, attempts were undertaken by different groups to establish precisely where in fact the proteins are expressed. Antibodies UW14, specific to GCAP1 (see preceding paragraph), in most analyzed species labeled more intensively outer and inner segments of cones than of rods.[38] Moreover, in the inner segments of cones staining was more intense than in outer segments. In the case of bovine and mouse retinas, however, a similar pattern of staining was observed in rods and cones.[12,38] Interestingly, in another report[39] when other antibodies against GCAP1 were used, more intense staining of GCAP1 was observed in outer segments than in inner segments of cones. Data from the same report showed that antibodies against GCAP2 labeled entire regions of cone and rod photoreceptors of bovine, monkey, and human retinas. In monkey and human labeling of rods and cones was similar, while in bovine retina rods were labeled much stronger than cones.[39] On the other hand, Otto-Bruc et al showed in the reference 26 that GCAP2 was recognized mainly in the cone inner segments, somata, and synaptic terminals of human and monkey retinas. The weaker labeling was detected in rod inner segments and inner retinal regions. In bovine retina the most intense labeling was observed by these authors in rod inner segments.[26] Similarly, in studies described in reference 38, antibodies raised against GCAP2 stained intensively cones of human and monkey retina but weakly cones of bovine, mouse, and rat retinas. In primates GCAP2 was not detected in rod outer segments but was intensively labeled in rod inner segments. In all other species tested (bovine, mouse, rat, cat), GCAP2 staining was observed in rod inner and outer segments. GCAP2 was also detected in other cells of vertebrate retina.[38]

Although the retina is the primary site where GCAPs are expressed, GCAP1 was also detected in the bovine[40] and rat (A. Dejda and W.A. Gorczyca, manuscript in preparation) pineal glands (PG). The expression of genes encoding GCAP1 and GCAP2 was also reported for the chicken pineal gland.[41] The presence of retGC1 in pineal glands of these species indicates that synthesis of cGMP is regulated there also in a Ca^{2+}-dependent manner but the physiological significance of this mechanism is yet unknown. It is only hypothesized that since pinealocytes contain most proteins typical for the phototransduction in photoreceptor cells of the retina and pineal glands of lower vertebrates (e.g., birds) and newborn mammals directly respond to the light, this Ca^{2+}-dependent regulation of GC in PG might be also the part of a phototransduction process.[40,41]

Ca^{2+}-DEPENDENT REGULATION OF retGC ACTIVITY BY GCAPs

Both isoforms of retinal GCs are regulated in a Ca^{2+}-dependent manner by GCAPs,[11,12,15] however, only GCAP2 and GCAP3 were shown to activate human RetGC1 and RetGC2, while GCAP1 activated only RetGC1.[18] It is not known yet, if this observation can be extended for retGCs in other species. All GCAPs at low $[Ca^{2+}]_{free}$ stimulate activity of target cyclase(s) and inhibit it at high $[Ca^{2+}]_{free}$ (it is illustrated for recombinant bovine GCAP1 and GCAP2 in Fig. 4 and reviewed in refs.1 and 3).

The precise mechanism of regulation of retGCs activity by GCAPs is not known. However, it is known that this regulation involves direct interactions between the intracellular region of the cyclase, including kinase-homology domain[42-45] and the catalytic site,[46] and various regions of GCAPs,[10,28,47-51] most notably the N-terminal region in the case of GCAP1.[10,28,49,51]

By means of different approaches (competition of synthetic peptides, deletions, substitutions) the sites that apparently are involved in regulation of effector cyclase were recognized in GCAP1 and GCAP2.[10,28,47-50] In experiments that used synthetic peptides encompassing entire GCAP1 sequence, the peptides corresponding to sequences myrGly$_2$-Glu$_{28}$, Glu$_{57}$-Gly$_{86}$, Asp$_{137}$-Asp$_{163}$ inhibited more than 60% activation of retGC by GCAP1.[10,28] Recently Schrem et al, also using synthetic peptides, showed in reference 48 that the region located between Phe$_{73}$-Lys$_{87}$ in GCAP1 sequence is important for interaction of GCAP1 with target cyclase. In experiments that used chimeric proteins or deletion mutants of GCAP1, regions located between EF1 and EF3, mainly sequences between Trp$_{21}$-Thr$_{27}$ and

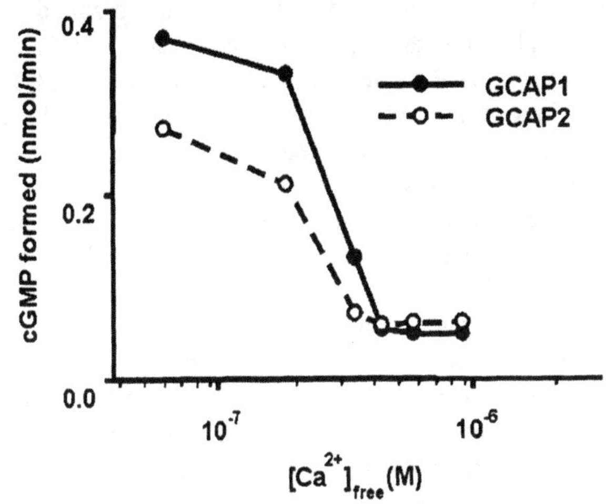

Figure 4. Retinal GC stimulation by GCAP1 and GCAP2 as a function of free Ca^{2+}. GC activity in washed bovine ROS was stimulated in the presence of 3 μM recombinant GCAP1 or recombinant GCAP2 at $[Ca^{2+}]_{free}$ concentrations in the range of 50 nM-2.5 μM.

Met_{157}-Arg_{182}, were shown to be important for activation of GC by GCAP1, while nine N-terminal amino acids were involved in inhibition of the enzyme.[49] Using the same approach, three different regions of GCAP2 were recognized as important for regulation of retGC activity: Phe_{78}-Asp_{113}, Val_{171}-Asn_{189} and Lys_{29}-Phe_{48}.[47] The first two were proposed to be involved in activation of the enzyme at low Ca^{2+} concentration, and the third in activation of retGC at low and inhibition at high Ca^{2+} concentrations. The mutations in the EF1 domain have additionally proved importance of this region in interaction of GCAP2 with retGC.[50]

GCAPs regulate retGCs interacting with the cytoplasmic domains of the enzymes. Deletion of the cyclase extracellular domain and transmembrane domain did not change its sensitivity to activation by GCAPs.[42,43,52] GCAP1 apparently forms a complex with target enzyme independently of Ca^{2+} concentration.[12,42,48] Also GCAP2 was reported to interact constitutively with retinal GC in the presence of Ca^{2+} and in the presence of EGTA.[44] The observation that stimulatory activity of GCAP1 and GCAP2 is not additive at saturation has suggested existence of common or overlapping sites for both GCAPs on retGC.[12]

Once the complex GCAP-retGC is formed, the rate of cGMP production is modulated by the mechanisms that could utilize dimerization of guanylyl cyclase catalytic subunits as a consequence of Ca^{2+}-dependent GCAPs dimerization,[53,54] or by stabilization of guanylyl cyclase-GTP transition state by Ca^{2+}-free GCAP.[46]

The interaction between retinal guanylyl cyclase and GCAP1 within the catalytic region of the enzyme is of particular interest. This region corresponds to the sequence in adenylyl cyclase (AC) that is responsible for the interaction with $G_s\alpha$, which stabilizes the transition state of ATP-cAMP transformation.[55] In retGC1, this fragment is likely to form a loop between α-helix 3 and β-strand 4. When this region in retGC1 is replaced by the corresponding sequence of GCAP-insensitive GC type A (GC-A), GCAP1 does not stimulate the retGC1 mutant.[46] Therefore, the activities of both AC and retGC, possibly are modulated through their respective regions within catalytic domains, through the interactions with activating proteins such as $G_s\alpha$ and GCAP, respectively.

Based on the modeling studies, the structure of GCAP1,[30] which belongs to the NCS family,[22] appears to be similar to that of other members of this family such as recoverin, neurocalcin (VILIP) or frequenin. Two pairs of EF-hands of GCAP1 are organized into N- and C-terminal domains. Under conditions dependent on Ca^{2+} concentration, GCAP1 may undergo a reorientation of helices at the interface of its two domains.[29,30,51] This mechanism is also consistent with earlier studies employing limited proteolysis[33] and fluorescence methods.[29] During GCAP1 transition from inactive to an active conformation, major rotation occurs around helix 6 and causes exposure of hydrophobic residues in this area (EF3 hand).[30] Therefore, in GCAP1 mutation Y99C, which causes autosomal dominant cone dystrophy,[56] reduces sensitivity of GCAP1 for Ca^{2+}, because smaller Cys residue is not able to stabilize Ca^{2+}-bound form of GCAP1 in the inactive conformation.[34]

In summary, in our postulated model of regulation retGC by GCAPs, both proteins form the multi-site complex in which GCAP undergoes Ca^{2+}-dependent rearrangement, that causes reorientation of two structural domains, each containing pair of EF-hand motifs, that leads to changes in the catalytic site of retGC (Fig. 5).

CLOSING REMARKS

Although structurally closely related, GCAPs reveal substantial differences with recoverin and other members of the NCS family. GCAPs are more acidic and more hydrophobic proteins. They activate the effector enzyme (retGC) at low and inhibit it at high $[Ca^{2+}]_{free}$. They apparently form complex with retGC independently of Ca^{2+} concentration. Moreover, GCAP2 more easily dissociates from the ROS membranes at high Ca^{2+},[25] indicating that its interaction with membranes is weaker at high than at low $[Ca^{2+}]_{free}$ what is quite opposite to behavior of recoverin and other recoverin-like proteins.

A still unsolved issue is the functional significance of the existence of multiple GCAPs in photoreceptor cells. In rods of transgenic mice with disrupted expression of GCAP1 and GCAP2 ($GCAP^{-/-}$), Ca^{2+}-dependent regulation of retGC was completely abolished and generation of light response was slower than in wild-type rods. Expression of GCAP2 in $GCAP^{-/-}$ rods restored light-induced retGC activity but did not restore normal kinetics of response to light.[57] On the other hand, expression either GCAP1 or GCAP2 in $GCAP^{-/-}$ mice showed that only GCAP1 restored wild-type responses suggesting different function of both proteins.[58]

These observations together with earlier findings that mutations in human GCAP1 lead to retinal degeneration[56] indicate that GCAP1 is within the NCS family still the only protein, whose physiological function is clearly shown.

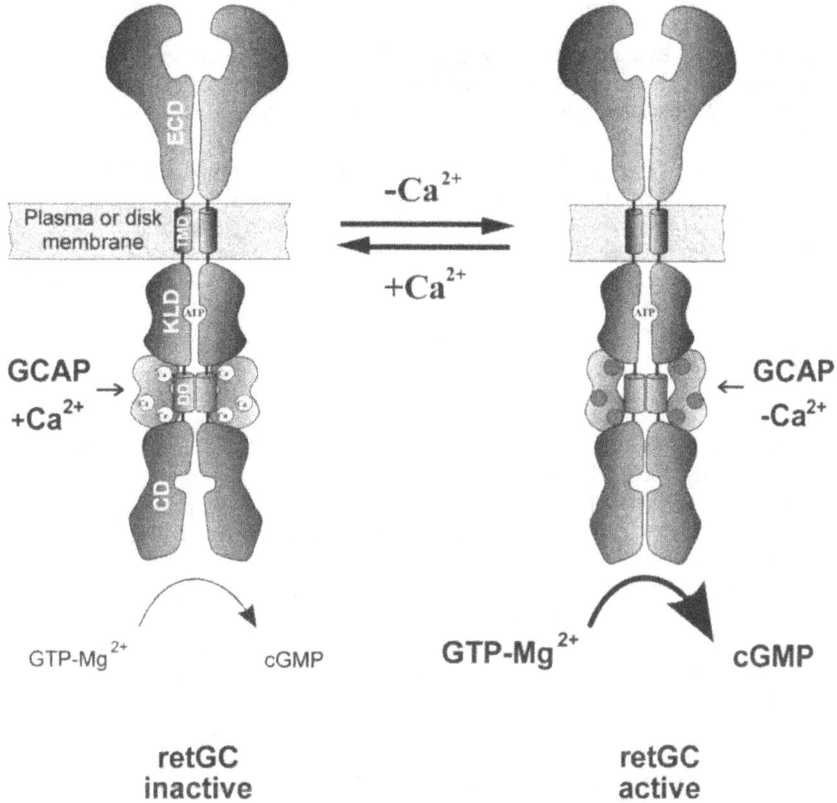

Figure 5. Model of retinal GC activation by GCAPs. Like in other particulate cyclases, the single polypeptide chain of retGC consists of: extracellular domain (ECD), single transmembrane domain (TMD), intracellular kinase-like domain (KLD), and single catalytic domain (CD). Particulate cyclases function as homodimers, whose formation is dependent on the existence of dimerization domain (DD). GCAP interacts with cytoplasmic domain(s) (KLD and/or CD) of retGC independently of $[Ca^{2+}]_{free}$. Thus, the conformational changes induced by Ca^{2+} in GCAP structure are transmitted to the adjacent cyclase and act as a switch between inhibition and activation. At high $[Ca^{2+}]_{free}$ GCAP in its Ca^{2+}-loaded form suppresses activity of the retGC and activates the enzyme in the apo-form at low $[Ca^{2+}]_{free}$.

ACKNOWLEDGMENTS

We would like to thank Anna Kazimierczak for reading the manuscript and Marcin Kobialka for help in preparation of figures.

REFERENCES

1. Palczewski K, Polans AS, Baehr W et al. Ca^{2+}-binding proteins in the retina: structure, function, and the etiology of human visual diseases. BioEssays 2000; 22:337-350.
2. Gorczyca WA. Role of calcium ions in vertebrate phototransduction. Pol J Pharmacol 1999; 51:167-72
3. Dizhoor AM. Regulation of cGMP synthesis in photoreceptors: role in signal transduction and congenital diseases of the retina. Cell Signal 2000; 12: 711-719.
4. Kobialka M, Gorczyca WA. Particulate guanylyl cyclases: multiple mechanisms of activation. Acta Biochim Pol 2000; 47:517-528.
5. Lucas KA, Pitari GM, Kazerounian S et al. Guanylyl cyclases and signaling by cyclic GMP. Pharmacol Rev 2000; 52:375-413.
6. Lolley RN, Racz E Calcium modulation of cyclic GMP synthesis in rat visual cells. Vision Res 1982; 22:1481-1486.
7. Koch K.-W, Stryer L Highly cooperative feedback control of retinal rod guanylate cyclase by calcium ions. Nature 1988; 334:64-66.
8. Shyjan AW, de Sauvage FJ, Gillett NA, et al. Molecular cloning of a retina-specific membrane guanylyl cyclase. Neuron 1992; 9:727-737
9. Gorczyca WA, Gray-Keller MP, Detwiler PB, et al. Purification and physiological evaluation of a guanylate cyclase activating protein from retinal rods. Proc Natl Acad Sci USA 1994; 91:4014-8.
10. Palczewski K, Subbaraya I, Gorczyca WA et al. Molecular cloning and characterization of retinal photoreceptor guanylyl cyclase-activating protein. Neuron 1994; 13:395-404.
11. Dizhoor AM, Lowe DG, Olshevskaya EV et al. The human photoreceptor membrane guanylyl cyclase, RetGC, is present in outer segments and is regulated by calcium and a soluble activator. Neuron 1994; 12:1345-52.
12. Gorczyca WA, Polans AS, Surgucheva IG et al. Guanylyl cyclase activating protein. A calcium-sensitive regulator of phototransduction. J Biol Chem 1995; 270:22029-36.
13. Dizhoor AM, Olshevskaya EV, Henzel WJ et al. Cloning, sequencing, and expression of a 24-kDa Ca(2+)-binding protein activating photoreceptor guanylyl cyclase. J Biol Chem 1995; 270:25200-6.
14. Goraczniak RM, Duda T, Sitaramayya A et al. Structural and functional characterization of the rod outer segment membrane guanylate cyclase. Biochem J 1994; 302:455-461.
15. Lowe DG, Dizhoor AM, Liu K et al. Cloning and expression of a second photoreceptor-specific membrane retina guanylyl cyclase (RetGC), RetGC-2. Proc Natl Acad Sci U S A 1995; 92:5535-9
16. Yang R-B, Foster DC, Garbers DL et al. Two membrane forms of guanylyl cyclase found in the eye. Proc natl Acad Sci USA 1995; 92:602-606.
17. Goraczniak RM, Duda T, Sharma RK Structural and functional characterization of a second subfamily member of the calcium-modulated bovine rod outer segment membrane guanylate cyclase, ROS-GC2. Biochem Biophys Res Commun 1997; 234:666-670.
18. Haeseleer F, Sokal I, Li N et al. Molecular characterization of a third member of the guanylyl cyclase-activating protein subfamily. J Biol Chem 1999; 274:6526-35.
19. Imanishi Y, Li N, Sokal I, et al. Characterization of retinal guanylyl-cyclase activating protein 3(GCAP3) from zebrafish to man. Eur J Neurosci 2002; 15:63-78.
20. Frins S, Bönigk W, Müller F et al. Functional characterization of a guanylyl cyclase-activating protein from vertebrate rods. Cloning, heterologous expression, and localization. J Biol Chem 1996; 271:8022-7.
21. Polans A, Baehr W, Palczewski K. Turned on by Ca2+! The physiology and pathology of Ca(2+)-binding proteins in the retina. Trends Neurosci 1996; 19:547-54.
22. Burgoyne RD, Weiss JL. The neuronal calcium sensor family of Ca^{2+}-binding proteins. Biochem J 2001; 353:1-12.

23. Subbaraya I, Ruiz CC, Helekar BS et al. Molecular characterization of human and mouse photoreceptor guanylate cyclase-activating protein (GCAP) and chromosomal localization of the human gene. J Biol Chem 1994; 269:31080-9.
24. Surguchov A, Bronson JD, Banerjee P et al. The human GCAP1 and GCAP2 genes are arranged in a tail-to-tail array on the short arm of chromosome 6 (p21.1). Genomics 1997; 39:312-22.
25. Olshevskaya EV, Hughes RE, Hurley JB et al. Calcium binding, but not a calcium-myristoyl switch, controls the ability of guanylyl cyclase-activating protein GCAP-2 to regulate photoreceptor guanylyl cyclase. J Biol Chem 1997; 272:14327-33.
26. Otto-Bruc A, Fariss RN, Haeseleer F, et al. Localization of guanylate cyclase-activating protein 2 in mammalian retinas. Proc Natl Acad Sci USA 1997; 94:4727-32.
27. Ames JB, Dizhoor AM, Ikura M et al. Three-dimensional structure of guanylyl cyclase activating protein-2, a calcium-sensitive modulator of photoreceptor guanylyl cyclases. J Biol Chem 1999; 274:19329-37.
28. Otto-Bruc A, Buczylko J, Surgucheva I et al. Functional reconstitution of photoreceptor guanylate cyclase with native and mutant forms of guanylate cyclase-activating protein 1. Biochemistry 1997; 36:4295-302.
29. Sokal I, Otto-Bruc AE, Surgucheva I et al. Conformational changes in guanylyl cyclase-activating protein 1 (GCAP1) and its tryptophan mutants as a function of calcium concentration. J Biol Chem 1999; 274:19829-37.
30. Sokal I, Li N, Klug CS et al. Calcium-sensitive regions of GCAP1 as observed by chemical modifications, flourescence and EPR spectroscopies. J Biol Chem 2001; 46:43361-43373.
31. Hughes RE, Brzovic PS, Dizhoor AM et al. Ca2+-dependent conformational changes in bovine GCAP-2. Protein Sci 1998; 7:2675-80.
32. Dizhoor AM, Hurley JB. Inactivation of EF-hands makes GCAP-2 (p24) a constitutive activator of photoreceptor guanylyl cyclase by preventing a Ca^{2+}-induced, "activator-to-inhibitor" transition. J Biol Chem 1996; 271:19346-50
33. Rudnicka-Nawrot M, Surgucheva I, Hulmes JD et al. Changes in biological activity and folding of guanylate cyclase-activating protein 1 as a function of calcium. Biochemistry 1998; 37:248-57.
34. Sokal I, Li N, Surgucheva I, Warren MJ et al. GCAP1 (Y99C) mutant is constitutively active in autosomal dominant cone dystrophy. Mol Cell 1998; 2:129-33.
35. Dizhoor AM, Boikov SG, Olshevskaya EV. Constitutive activation of photoreceptor guanylate cyclase by Y99C mutant of GCAP-1. Possible role in causing human autosomal dominant cone degeneration. J Biol Chem 1998; 273:17311-4.
36. Gorczyca WA, Van Hooser JP, Palczewski K. Nucleotide inhibitors and activators of retinal guanylyl cyclase. Biochemistry 1994; 33:3217-22.
37. Gorczyca WA. Use of nucleoside alpha-phosphorothioates in studies of photoreceptor guanylyl cyclase: purification of guanylyl cyclase activating proteins. Methods Enzymol 2000; 315:689-707.
38. Cuenca N, Lopez S, Howes K, et al. The localization of guanylyl cyclase-activating proteins in the mammalian retina. Invest Ophthalmol Vis Sci 1998; 39:1243-50.
39. Kachi S, Nishizawa Y, Olshevskaya E, et al. Detailed localization of photoreceptor guanylate cyclase activating protein-1 and -2 in mammalian retinas using light and electron microscopy. Exp Eye Res 1999; 68:465-73.
40. Venkataraman V, Nagele R, Duda T, et al. Rod outer segment membrane guanylate cyclase type 1-linked stimulatory and inhibitory calcium signaling systems in the pineal gland: biochemical, molecular, and immunohistochemical evidence. Biochemistry 2000; 39:6042-52.
41. Semple-Rowland SL, Larkin P, Bronson JD, et al. Characterization of the chicken GCAP gene array and analyses of GCAP1, GCAP2, and GC1 gene expression in normal and rd chicken pineal. Mol Vis 1999; 5:14.
42. Duda T, Goraczniak R, Surgucheva I, et al. Calcium modulation of bovine photoreceptor guanylate cyclase. Biochemistry 1996; 35:8478-82.

43. Laura RP, Dizhoor AM, Hurley JB. The membrane guanylyl cyclase, retinal guanylyl cyclase-1, is activated through its intracellular domain. J Biol Chem 1996; 271:11646-51.
44. Tucker CL, Laura RP, Hurley JB. Domain-specific stabilization of photoreceptor membrane guanylyl cyclase by adenine nucleotides and guanylyl cyclase activating proteins (GCAPs). Biochemistry 1997; 36:11995-2000.
45. Laura RP, Hurley JB. The kinase homology domain of retinal guanylyl cyclases 1 and 2 specifies the affinity and cooperativity of interaction with guanylyl cyclase activating protein-2. Biochemistry 1998; 37:11264-71.
46. Sokal I, Haeseleer F, Arendt A, et al. Identification of a guanylyl cyclase-activating protein-binding site within the catalytic domain of retinal guanylyl cyclase 1. Biochemistry 1999; 38:1387-93.
47. Olshevskaya EV, Boikov S, Ermilov A, et al. Mapping functional domains of the guanylate cyclase regulator protein, GCAP-2. J Biol Chem 1999; 274:10823-32.
48. Schrem A, Lange C, Beyermann M, et al. Identification of a domain in guanylyl cyclase-activating protein 1 that interacts with a complex of guanylyl cyclase and tubulin in photoreceptors. J Biol Chem 1999: 274:6244-9.
49. Krylov DM, Niemi GA, Dizhoor AM, et al. Mapping sites in guanylyl cyclase activating protein-1 required for regulation of photoreceptor membrane guanylyl cyclases. J Biol Chem 1999; 274:10833-9.
50. Ermilov A.N., Olshevskaya E.V., Dizhoor A.M. Instead of binding calcium, one of the EF-hand structures in guanylyl cyclase activating protein-2 is required for targeting photoreceptor guanylyl cyclase. J Biol Chem 2001; 276:48143-48.
51. Li N, Sokal I, Bronson JD, et al. Identification of functional regions of guanylate cyclase-activating protein 1 (GCAP1) using GCAP1/GCIP chimeras. Biol Chem 2001; 382:1179-1188.
52. Sokal I, Alekseev A, Baehr W, et al. Soluble fusion proteins between single transmembrane photoreceptor guanylyl cyclases and their activators. Biochemistry 2002; 41:251-257.
53. Yu H, Olshevskaya E, Duda T, et al. Activation of retinal guanylyl cyclase-1 by Ca2+-binding proteins involves its dimerization. J Biol Chem 1999; 274:15547-55.
54. Olshevskaya EV, Ermilov AN, Dizhoor AM. Dimerization of guanylyl cyclase-activating protein and a mechanism of photoreceptor guanylyl cyclase activation. J Biol Chem 1999; 274:25583-7
55. Tesmer JJ, Sunahara RK, Gilman AG, et al. Crystal structure of the catalytic domains of adenylyl cyclase in a complex with Gsa:GTPgS. Science 1997; 278: 1907-1916.
56. Payne AM, Downes SM, Bessant DA, et al. A mutation in guanylate cyclase activator 1A (GUCA1A) in an autosomal dominant cone dystrophy pedigree mapping to a new locus on chromosome 6p21.1. Hum Mol Genet 1998; 7:273-277.
57. Mendez A, Burns ME, Sokal I, et al. Role of guanylate cyclase-activating proteins (GCAPs) in setting the flash sensitivity of rod photoreceptors. Proc Natl Acad Sci USA 2001; 98:9948-9953.
58. Howes KA, Pennesi ME, Sokal I, et al. GCAP1 rescues rod photoreceptor response in GCAP1/GCAP2 knockout mice. EMBO J, 2002; 21:1545-54.
59. Howes K, Bronson JD, Dang YL, et al. Gene array and expression of mouse retina guanylate cyclase activating proteins 1 and 2. Invest Ophthalmol Vis Sci 1998; 39:867-75.

STRUCTURE AND MEMBRANE-TARGETING MECHANISM OF RETINAL Ca^{2+}-BINDING PROTEINS, RECOVERIN AND GCAP-2

James B. Ames[1] and Mitsuhiko Ikura[2]

ABSTRACT

Recoverin and the guanylate cyclase activating proteins (GCAPs) are calcium-sensing proteins in retinal rod and cone cells that belong to the EF-hand superfamily and serve as important calcium sensors in vision. Recoverin and the GCAP proteins are myristoylated at their amino-terminus and are targeted to retinal disc membranes by a myristoyl switch. Here, we present the three-dimensional, atomic-resolution structures of recombinant myristoylated recoverin containing 0, 1 and 2 calcium ions (Ca^{2+}) bound and unmyristoylated GCAP-2 with 3 Ca^{2+} bound as determined by nuclear magnetic resonance. The Ca^{2+}-induced structural changes in these proteins are important for elucidating their membrane-targeting mechanisms and for understanding the molecular mechanism of Ca^{2+}-sensitive regulation of phototransduction.

INTRODUCTION

An important class of calcium-binding proteins present in retinal photoreceptor cells include recoverin from mammalian rods,[1] guanylate cyclase activating proteins (GCAP-1, GCAP-2, GCAP-3) from mammalian rods and cones,[2-5] and guanylate cyclase inhibitory protein (GCIP) from frog rods.[6] Their amino acid sequences indicate that they are members of the EF-hand superfamily[7] that includes

[1]Center for Advanced Research in Biotechnology, University of Maryland Biotechnology Institute, Rockville, MD 20850; [2]Division of Molecular and Structural Biology, Ontario Cancer Institute and Department of Medical Biophysics, Universiy of Toronto, 610 University Avenue, Toronto, Ontario, Canada M5G 2M9.

calmodulin and troponin C (Fig. 1). The recoverin branch of the EF-hand superfamily includes neuronal Ca^{2+} sensors such as neurocalcin, frequenin, and hippocalcin (reviewed by ref. 8). Indeed, there is a homolog in yeast,[9] indicating that these calcium sensors arose early in the evolution of eukaryotes. The common features of these proteins are (1) an amino-terminal myristoylation consensus sequence, (2) an approximately 200-residue chain containing four EF-hand motifs, and (3) the sequence CPXG in the first EF-hand that markedly impairs its capacity to bind Ca^{2+}.

Mass spectrometric analysis of retinal recoverin and the GCAP proteins revealed that they are myristoylated at the amino terminus.[2,10,11] Recoverin contains an N-terminal myristoyl (14:0) or related fatty acyl group (12:0, 14:1, 14:2). Retinal recoverin and myristoylated recombinant recoverin, but not unmyristoylated recoverin, bind to membranes in a Ca^{2+}-dependent manner.[12,13] Likewise, bovine neurocalcin and hippocalcin contain an N-terminal myristoyl group and both exhibit Ca^{2+}-induced membrane binding.[14,15] These findings led to the proposal that recoverin and the neural homologs possess a Ca^{2+}-myristoyl switch (Fig. 2). The covalently attached fatty acid is highly sequestered in recoverin in the calcium-free state.[16] The binding of calcium to recoverin leads to the extrusion of the fatty acid,[17,18] making it available to interact with lipid bilayer membranes or other hydrophobic sites.

GCAP-2 exhibits very different membrane-targeting properties[11] from those of recoverin. The myristoylated and unmyristoylated forms of GCAP-2 bind to membranes at low Ca^{2+} concentrations (<100 nM). The GCAP-2 membrane interaction is highly sensitive to ionic strength and is much stronger at higher salt concentrations (>50 mM NaCl). Interestingly, the myristoylated form of GCAP-2 appears cytosolic at high Ca^{2+} and low ionic strength (<50 mM NaCl), although the Ca^{2+}-bound GCAP-2 does bind to membranes at physiological ionic strength. Recoverin and the GCAPs regulate different proteins. The Ca^{2+}-bound form of recoverin blocks the phosphorylation of photoexcited rhodopsin in vitro,[19,20] thereby prolonging the photoresponse.[21,22] Myristoylated recoverin is much more effective than unmyristoylated protein in inhibiting the deactivation of photoexcited rhodopsin.[21] By contrast, the GCAP proteins activate photoreceptor guanylate cyclases at low Ca^{2+} (<100 nM).[2,3] GCAP-2 in addition inhibits cylase at high Ca^{2+}.[11] Furthermore, the unmyristoylated form of GCAP-2 is nearly as effective as the myristoylated protein in activating the cyclase in vitro.[11]

In this chapter we present a detailed structural analysis of recoverin and GCAP-2. The Ca^{2+}-induced structural changes in these proteins are important for elucidating their membrane-targeting mechanisms and for understanding the molecular mechanism of Ca^{2+}-sensitive regulation of phototransduction.

THREE-DIMENSIONAL STRUCTURES OF RECOVERIN

Atomic-resolution structures of the various forms of recoverin were determined to understand the detailed molecular mechanism of the Ca^{2+}-myristoyl switch. The x-ray crystal structure of recombinant unmyristoylated recoverin first showed it to contain a compact array of EF-hand motifs,[23] in contrast to the dumbbell shape of

```
              10          20          30          40          50          60          70
GCAP2  GQQFSWEEA EEN----GAVGAAD AAQLQEWYKK FLEECPSGTL FMHEFKRFFK VPDNEE-ATQY VEAMFRAFDT
GCAP1  GNIMDGKSV EE----------LS STECHQWYKK FMTECPSGQL TLYEFRQFFG LKNLSPWASQY VEQMFETFDF
GCAP3  GNGKSIAGD QKA---------VP TQETHVWYRT FMMECPSGLQ TLHEFKTLLG LQGLNQKANKH IDQVYNTFDT
GCIP   GQVASMPHR CG----------TY VLELHEWYRK FVEECPSGLI TLHEFRQFFS DVTVGENSSEY AEQIFRALDN
REC    GNSKSGALS KEILEELQLNTKFT EEELSSWYQS FLKECPSGRI TRQEFQTIYS KFFPEADPKAY AQHVFRSFDA

        80          90         100        110        120        130        140
GCAP2 NGDNTIDFLE YVAALNLVLR GTLEHKLKWT FKIYDKDRNG CIDRQELLDI VESIYKLKKA CSVEVEAEQQ
GCAP1 NKDGYIDFME YVAALSLVLK GKVEQKLRWY FKLYDVDGNG CIDRDELLTI IRAIRAIN-- --------CS
GCAP3 NKDGFIDFLE FIAAVNLIMQ EKMEQKLKWY FKLYDADGNG SIDKNELLDM FMAVQALNG- ----------
GCIP  NGDGIVDFRE YVTAISMLAH GTPEDKLKWS FKLYDKDGDG AITRSEMLEI MRAVYKMSVV ASL-----TK
REC   NSDGTLDFKE YVIALHMTSA GKTNQKLEWA FSLYDVDGNG TISKNEVLEI VTAIFKMI-- -SPEDTKHLP

            150        160        170        180        190        200
GCAP2 GKLLTPEEVV DRIFLLVDEN GDGQLSLNEF VEGARRDKWV MKMLQMDLNP SSWISQQRRK SAMF  204
GCAP1 DSTMTAEEFT DTVFSKIDVN GDGELSLEEF MEGVQKDQML LDTLTRSLDL TRIVRRLQNG EQDEEGASGR
GCAP3 QQTLSPEEFI NLVFHKIDIN NDGELTLEEF INGMAKDQDL LEIVYKSFDF SNVLRVICNG KQPDMETDSS
GCIP  VNPMTAEECT NRIFVRLDKD QNAIISLQEF VDGSLGDEWV RQMLECDLST VEIQKMTKHS HLPARSSRER
REC   EDENTPEKRA EKIWGFFGKK DDDKLTEKEF IEGTLANKEI LRLIQFE--P QKVKEKLKEK KL  202

GCAP1 ETEAAEADG  205
GCAP3 KSPDKAGLGKVKMK  209
GCIP  LFHANT  202
```

Figure 1. Amino-acid sequence alignment of bovine GCAP-2 (accession no. U32856) with bovine GCAP-1 (accession no. P46065), human GCAP-3 (accession no. AF110002), frog GCIP (accession no. AF047884), and bovine recoverin (accession no. P21457). The 29-residue EF-hand motifs are shaded grey: EF-1, EF-2, EF-3, EF-4. Regions of regular secondary structure (α-helices and β-stands) are indicated schematically.

calmodulin[24] and troponin C.[25] The four EF-hands are organized into two domains: The first EF-hand, EF-1 (residues 27-56), interacts with EF-2 (residues 63-92) to form the N-terminal domain, and EF-3 (residues 101-130) and EF-4 (residues 148-177) form the C-terminal domain. The linker between the two domains is U-shaped rather than a-helical. Ca^{2+} is bound to EF-3 and Sm^{3+} (used to derive phases) is bound to EF-2. The other two EF hands possess novel features that prevented ion binding. EF-1 is disabled by a Cys-Pro sequence in the binding loop. EF-4 contains an internal salt bridge in the binding loop that competes with Ca^{2+} binding. Myristoylated recoverin, the physiologically active form, has thus far eluded crystallization.

The structures of myristoylated recoverin in the Ca^{2+}-free and Ca^{2+}-bound states were determined by nuclear magnetic resonance (NMR) spectroscopy (Fig. 3). In the Ca^{2+}-free state, the myristoyl group is sequestered in a deep hydrophobic cavity in the N-terminal domain.[16] The cavity is formed by five α-helices. The two helices of EF-1 (residues 26-36 and 46-56), the exiting helix of EF-2 (residues 83-93), and entering helix of EF-3 (residues 100-109) lie perpendicular to the fatty acyl chain and form a box-like arrangement that surrounds the myristoyl group laterally. A long, amphipathic α-helix near the N-terminus (residues 4-16) packs closely against and runs antiparallel to the fatty acyl group, and serves as a lid on top of the four-helix box. The N-terminal residues Gly 2 and Asn 3 form a tight hairpin turn that connects

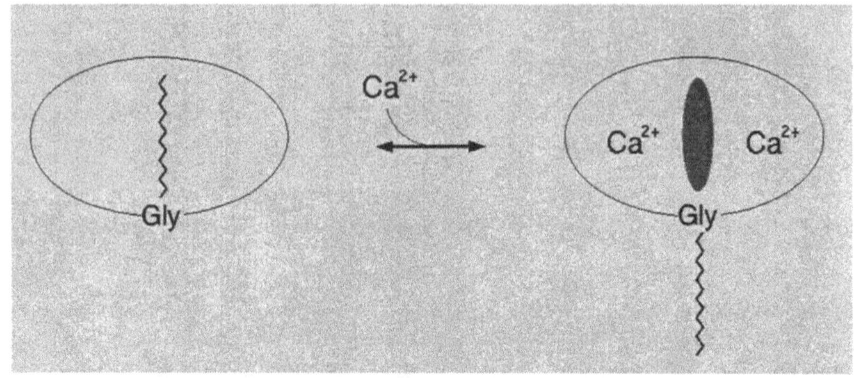

Figure 2. Schematic diagram of calcium-myristoyl switch in recoverin. The binding of two Ca^{2+} ions promotes the extrusion of the myristoyl group and exposure of other hydrophobic residues (marked by the blackened oval).

the myristoyl group to the N-terminal helix. This turn positions the myristoyl group inside the hydrophobic cavity and gives the impression of a cocked trigger. The bond angle strain stored in the tight hairpin turn may help eject the myristoyl group from the pocket once Ca^{2+} binds to the protein.

The structure of myristoylated recoverin with two Ca^{2+} bound shows the amino-terminal myristoyl group to be extruded (Fig. 3).[18] The N-terminal eight residues are solvent exposed and highly flexible and thus serve as a mobile arm to position the myristoyl group outside the protein when Ca^{2+} is bound. The flexible arm is followed by a short α-helix (residues 9-17) that precedes the four EF-hand motifs, arranged in a tandem array as was seen in the x-ray structure. Calcium ions are bound to EF-2 and EF-3. EF-3 has the classic "open conformation" similar to the Ca^{2+} occupied EF-hands in calmodulin[24] and troponin C.[25] EF-2 is somewhat unusual. The root mean squared (RMS) deviation of the main chain atoms of EF-2 is 3.7 Å in comparing recoverin to calmodulin. Interestingly, the helix packing angle of Ca^{2+}-bound EF-2 (120°) in recoverin more closely resembles that of the Ca^{2+}-free EF-hands (in the "closed conformation") found in calmodulin and troponin C. The overall topology of Ca^{2+}-bound myristoylated recoverin is similar to the x-ray structure of unmyristoylated recoverin with one Ca^{2+} bound.[23] The RMS deviation of the main chain atoms in the EF-hand motifs is 1.5 Å in comparing Ca^{2+}-bound myristoylated recoverin to unmyristoylated recoverin.

STRUCTURE OF A RECOVERIN MUTANT (E85Q) IN AN ALLOSTERIC INTERMEDIATE STATE

To gain insight into the kinetic mechanism of the Ca^{2+}-myristoyl switch, we recently determined the structure of an intermediate species of myristoylated recoverin containing only one Ca^{2+} bound.[26] The highly cooperative binding of two

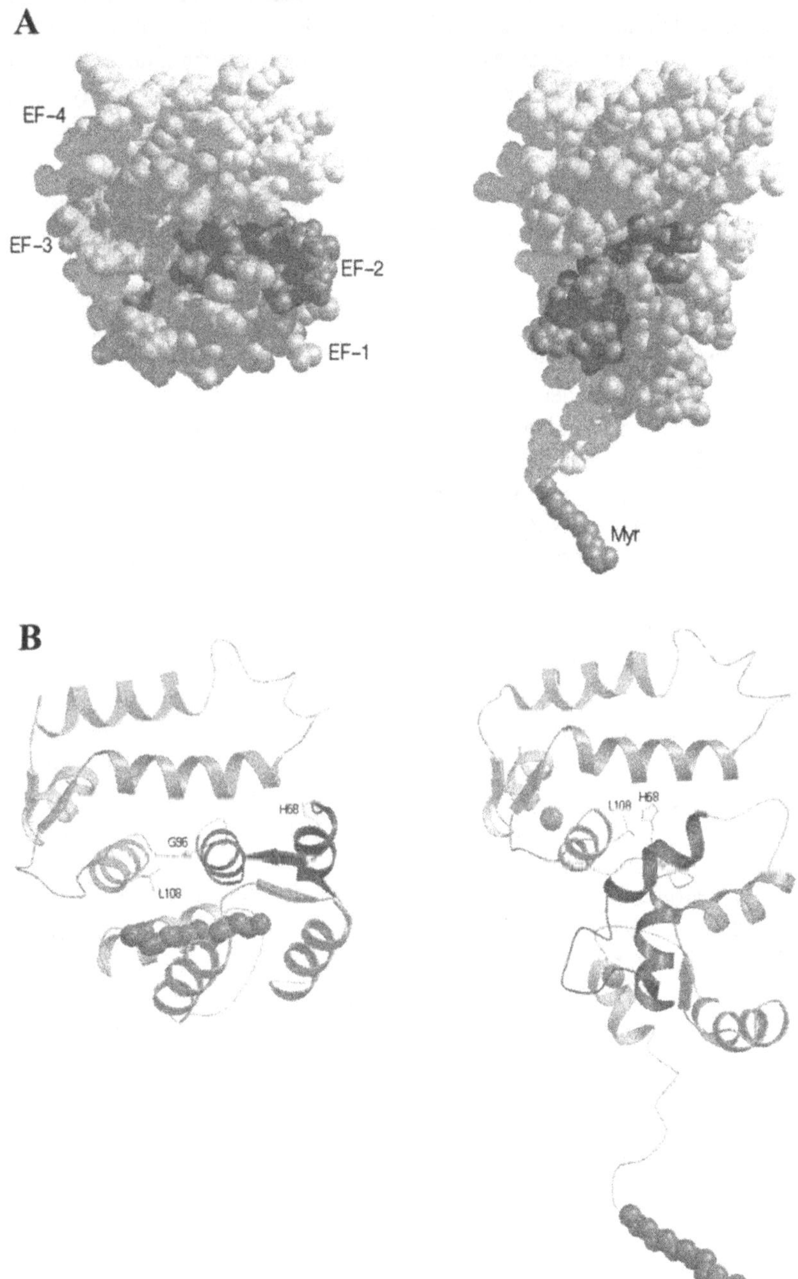

Figure 3. Space-filling model (a) and ribbon diagram (b) of Ca^{2+}-free (left, 1iku.pdb) and Ca^{2+}-bound (right, 1jsa.pdb) myristoylated recoverin. The C-terminal domains of the two forms are aligned to show the Ca^{2+}-induced 45-deg rotation of the N-terminal domain. This figure and other illustrations were generated using Midas[42], Molscript [43] and Raster3d.[44]

Ca^{2+} to myristoylated recoverin prevents the accumulation of allosteric intermediate species containing only one bound Ca^{2+} at equilibrium.[27] To isolate intermediate species and study structural changes in myristoylated recoverin induced solely by the binding of Ca^{2+} to EF-3, we constructed a mutant that lacks the ability to bind Ca^{2+} to EF-2. The substitution of residue, E85 with the neutral residue, glutamine (E85Q) eliminates an important negatively charged chelating group in the Ca^{2+} binding loop of EF-2. The E85Q mutation has been shown previously to completely abolish the binding of calcium to EF-2 while Ca^{2+} binding to EF-3 remains intact.[26,28-30]

The average main chain structure of myristoylated recoverin mutant, E85Q (myr-E85Q) containing Ca^{2+} bound solely to EF-3 was solved recently by NMR[26] and is depicted here as a ribbon diagram (Fig. 4B). The structure of the entire polypeptide main chain has been defined except for disordered regions in the loops of EF-2 (residues 74-78) and EF-4 (residues 160-164), and at the carboxyl terminus (residues 190-202). The secondary structure of Ca^{2+}-bound myr-E85Q is very similar to that seen in previous structures of recoverin.[16,18,23] However, the overall tertiary fold of Ca^{2+}-bound myr-E85Q is very different from the overall folds of recoverin with 0 and 2 Ca^{2+}-bound (Fig. 4). The RMS deviations of all main chain atoms are 9.3 and 7.0 Å in comparing Ca^{2+}-bound myr-E85Q with Ca^{2+}-free and Ca^{2+}-bound wildtype, respectively. Ca^{2+}-bound myr-E85Q, like Ca^{2+}-free and Ca^{2+}-bound wildtype, contains two tightly packed and interacting N-terminal and C-terminal domains.

The structure of the N-terminal domain (residues 2-92, shown green and red in Fig. 4) of Ca^{2+}-bound myr-E85Q partially resembles that of Ca^{2+}-free wildtype (Fig. 4A) and is very different from that of Ca^{2+}-bound wildtype (Fig. 4C). The RMS deviations of the main chain atoms in the N-terminal domain are 1.6 and 7.7 Å in comparing Ca^{2+}-bound myr-E85Q with Ca^{2+}-free and Ca^{2+}-bound wildtype, respectively. Most striking in the N-terminal domain of Ca^{2+}-bound myr-E85Q is that the myristoyl group is flanked by a long N-terminal helix (residues 5-17) and is sequestered in a hydrophobic cavity containing many aromatic residues from EF-1 and EF-2 (F23, W31, Y53, F56, F83, and Y86). An important difference in the structures of Ca^{2+}-bound myr-E85Q and Ca^{2+}-free wildtype is that the carbonyl end of the fatty acyl group in Ca^{2+}-bound E85Q does not interact with any hydrophobic residues from EF-3 (W104 and L108, Figs. 4A-B) and is somewhat solvent exposed.

The structure of the C-terminal domain (residues 101-202, cyan and yellow in Fig. 4) of Ca^{2+}-bound myr-E85Q (Fig. 4B) is almost identical to that of Ca^{2+}-bound wildtype (Fig. 4C). The RMS deviation of the main chain atoms in the C-terminal domain is 1.25 Å in comparing Ca^{2+}-bound myr-E85Q with Ca^{2+}-bound wildtype. In Ca^{2+}-bound myr-E85Q, the entering helix of EF-2 (residues 63-73) packs against the helices of EF-3 (residues 101-109 and 119-130) at the interface between the two domains, similar to that of Ca^{2+}-bound wildtype (see residues H68 and L108 in Figs. 4B and 4C). The linker between the two domains is U-shaped, which positions the four EF-hands in a tandem array like that of Ca^{2+}-bound wildtype, and unlike

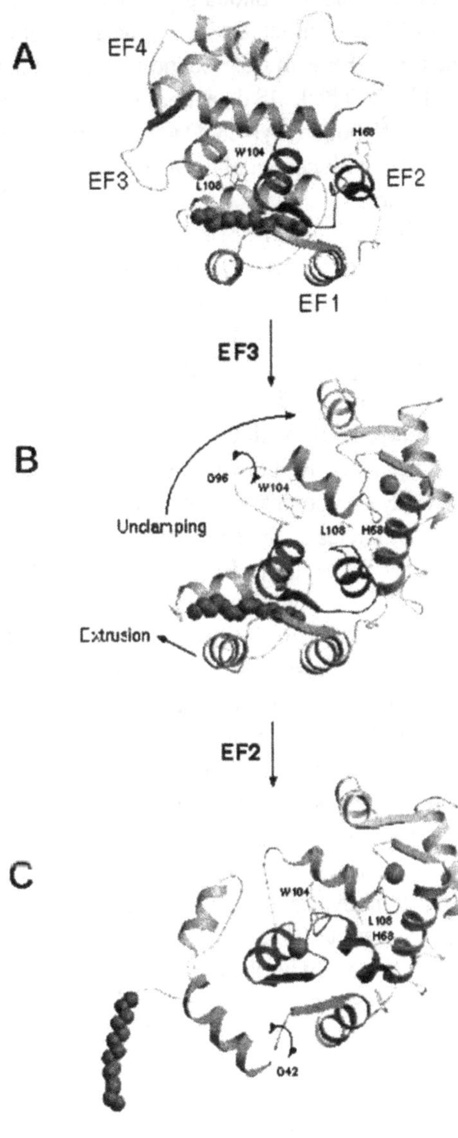

Figure 4. Three-dimensional structures of myristoylated recoverin with 0 Ca^{2+} bound[16] (A), 1 Ca^{2+} bound[26] (B) and 2 Ca^{2+} bound[18] (C). The first step of the mechanism involves the binding of Ca^{2+} to EF-3 that causes minor structural changes within the EF-hand that sterically promote a 45° swiveling of the two domains, resulting in a partial unclamping of the myristoyl group and a dramatic rearrangement at the domain interface. The resulting altered interaction between EF-2 and EF-3 potentiates the binding of a second Ca^{2+} to the protein at EF-2 in the second step, which causes structural changes within the N-terminal domain that directly lead to the ejection of the fatty acyl group.

the dumbbell arrangement found in calmodulin[24] and troponin C.[25] Ca^{2+} is bound to EF-3 and not bound to EF-2 as anticipated from the analysis of the Ca^{2+}-binding data.[26,28-30] The structure of EF-3 in Ca^{2+}-bound myr-E85Q is nearly identical to that of Ca^{2+}-bound wildtype. The RMS deviations of the main chain atoms of EF-3 are 0.6 Å in comparing Ca^{2+}-bound myr-E85Q with wildtype and 0.85 Å in comparing Ca^{2+}-bound myr-E85Q with calmodulin. Likewise, the coordination of Ca^{2+} is virtually identical in all three.

The structures of EF-1 (residues 27-56) and EF-2 (residues 63-92) in Ca^{2+}-bound E85Q are somewhat similar to those of Ca^{2+}-free wildtype and are very different from those of Ca^{2+}-bound wildtype. Conversely, the structures of EF-3 (residues 101-130) and EF-4 (residues 148-177) in Ca^{2+}-bound myr-E85Q are more similar to those of Ca^{2+}-bound wildtype. The RMS deviations of the main chain atoms of the EF-hands are 0.53 Å (EF-1), 1.5 Å (EF-2), 2.5 Å (EF-3) and 2.2 Å (EF-4) in comparing Ca^{2+}-bound myr-E85Q with Ca^{2+}-free wildtype. Conversely, the RMS deviations of the main chain atoms of the EF-hands are 3.9 Å (EF-1), 3.1 Å (EF-2), 0.6 Å (EF-3) and 1.4 Å (EF-4) in comparing Ca^{2+}-bound myr-E85Q with Ca^{2+}-bound wildtype. The interhelical angles of the EF-hands as defined by Zhang [31] follow a similar trend. The interhelical angles of EF-1 and EF-2 of Ca^{2+}-bound myr-E85Q are similar to those of Ca^{2+}-free wildtype and differ from those of Ca^{2+}-bound wildtype. Conversely, the interhelical angles of EF-3 and EF-4 of Ca^{2+}-bound myr-E85Q are more similar to those of Ca^{2+}-bound wildtype.

A detailed comparison of the structures of Ca^{2+}-free and Ca^{2+}-bound EF-3 reveals how a small and localized Ca^{2+}-induced structural change within EF-3 leads to a large relative displacement of EF-2 at the domain interface (Fig. 5). The overall Ca^{2+}-induced structural change within EF-3 is relatively small. The RMS deviation is only 2.5 Å in comparing the main chain atoms of Ca^{2+}-free and Ca^{2+}-bound EF-3 (See Ca^{2+}-free and Ca^{2+}-bound EF-3 represented dark- and light-blue, respectively, in Fig. 5). Closer inspection reveals that the interhelical angle of EF-3 changes noticeably from 118° (Ca^{2+}-free) to 91° (Ca^{2+}-bound). The Ca^{2+}-induced decrease in interhelical angle resembles somewhat the familiar Ca^{2+}-induced "closed" to "open" transition of EF-hands first observed for troponin C[25] and later for calmodulin[31,32] as described by the Herzberg-James Model.[25,33] The Ca^{2+}-induced opening of EF-3 causes the N-terminal E-helix of EF-3 to pull upward (Fig. 5), resulting in a significant displacement of the connected domain linker and EF-2. In addition, the Ca^{2+}-induced opening of EF-3 appears to cause a steric clash of residues at the domain interface (see residues I88 and L108 in Fig. 5) that further promotes the displacement and rearrangement of EF-2 and EF-3. The Ca^{2+}-induced structural changes depicted in Figure 5 illustrate how the local structural change within EF-3 results in the rearrangement of EF-2 and EF-3 at the domain interface that overall can be represented as a swiveling of the two domains.

In summary, the overall three-dimensional structure of Ca^{2+}-bound myr-E85Q is unique and represents a hybrid of the structures of recoverin with 0 and 2 Ca^{2+}-bound. A distinctive feature of Ca^{2+}-bound myr-E85Q is that the carbonyl end of the fatty acyl group is more solvent exposed than it is in Ca^{2+}-free wildtype. In

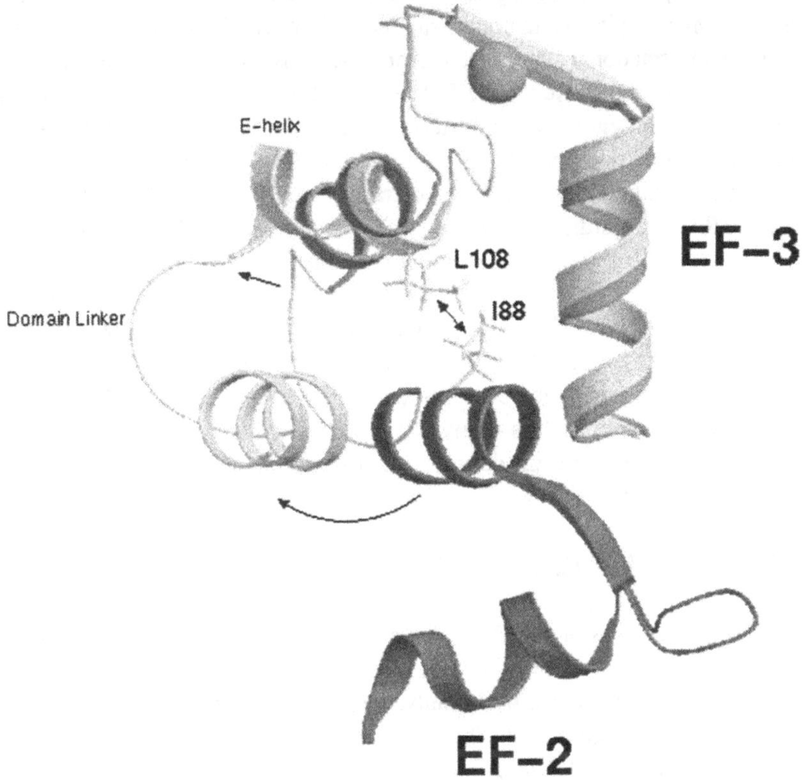

Figure 5. Overlay of the main chain structure of Ca^{2+}-free wildtype[16] compared with the structure of Ca^{2+}-bound myr-E85Q. The structures of Ca^{2+}-free (dark-grey) and Ca^{2+}-bound (light-grey) EF-3 are superimposed (Overall RMS deviation of main chain atoms is 2.5 Å). The N-terminal helix and binding loop of EF-2 from the structure of Ca^{2+}-bound myr-E85Q is not shown for clarity. The side-chain atoms of I88 and L108 are indicated and illustrate possible Ca^{2+}-induced steric interactions that may help promote the rearrangement of EF-2 and EF-3 at the domain interface.

contrast, the C_{14}-methyl group of the myristate is buried inside the protein like that of Ca^{2+}-free wildtype. The binding of Ca^{2+} to EF-3 causes structural changes within the EF-hand itself that sterically promotes a 45° swiveling of the two domains about the central linker. The swiveling of domains is functionally important because it displaces the C-terminal domain away from the amino-terminus (see residues W104 and L108 in Figs. 4A-B), causing a partial unclamping of the myristoyl group.

The structure of Ca^{2+}-bound myr-E85Q (that contains Ca^{2+} bound solely to EF-3) is different from the x-ray structure of unmyristoylated recoverin containing a single Ca^{2+} bound to EF-3.[23] The apparent differences in these structures can be explained by the allostery introduced by the attached myristoyl group.[27] The allosteric effects of the myristoyl group have been described previously by a two-state concerted model,[27] which also explains the dramatic differences in the Ca^{2+} binding properties of myristoylated and unmyristoylated recoverin. In the allosteric model, two stable conformational states of the protein are postulated ("T" and "R"). The myristoyl group is sequestered in the T-state and extruded in the R-state. Therefore, the myristoyl group serves as an allosteric effector that stabilizes the T form. The myristoylated protein occupies the T-state in the absence of Ca^{2+} and switches to the R form upon binding two Ca^{2+}. In contrast, the unmyristoylated protein is always in the R state regardless of the Ca^{2+} level because the T-state is unstable without the fatty acid. This explains the similarity of the structures of unmyristoylated recoverin with one Ca^{2+} bound[23] and two Ca^{2+} bound.[8] In contrast, the structure of myristoylated recoverin with one Ca^{2+} bound (myr-E85Q) is intermediate between the T and R forms. The structures of various intermediate species of hemoglobin were also observed recently to be trapped halfway between the T and R-states.[34,35] To our knowledge, the structure of myristoylated recoverin containing one bound Ca^{2+} is the most detailed structure determination of an intermediate species involved in an allosteric transition.

MECHANISM OF THE Ca^{2+}-MYRISTOYL SWITCH

We propose a plausible kinetic pathway of the Ca^{2+}-induced conformational changes of recoverin described above to illustrate the atomic-level mechanism of the Ca^{2+}-myristoyl switch (Fig. 4). Initially, the protein is in the Ca^{2+}-free state with the myristoyl group sequestered. The first step is the binding of Ca^{2+} to EF-3 as demonstrated in the structures of intermediate species with one Ca^{2+} bound.[23,26] Also, our Ca^{2+} binding studies reveal that EF-3 binds Ca^{2+} with much higher intrinsic affinity than EF-2. The initial binding of Ca^{2+} to EF-3 induces conformational changes in this EF-hand (Fig. 5) that resemble the Ca^{2+}-induced opening of EF-hands observed previously for calmodulin and troponin C.[25,33] The resulting Ca^{2+}-induced change in the interhelical angle of EF-3 causes the E-helix to pull upward on the domain linker (Fig. 5), leading to a dramatic rearrangement of EF-2 and EF-3 at the domain interface and resulting in a 45° swiveling of the two domains centered near Gly 96 (Fig. 4B). An important structural consequence of the domain swiveling is that it leads to a partial unclamping and exposure of the carbonyl end of the myristoyl group. Note that the C-terminal residues W104 and L108 are spatially close to the carbonyl end of the myristoyl group in the Ca^{2+}-free protein (Fig. 4A) and are displaced far away from the fatty acid in the intermediate state (Fig. 4B). In addition, the domain swiveling alters the interaction between EF-2 and EF-3 at the domain interface and may explain, in part, the observed cooperativity of Ca^{2+} binding in the myristoylated wildtype protein.[27] As a consequence of this altered interaction between EF-2 and EF-3 in the intermediate state (see H68 and L108 in Figs. 4A-B), the binding affinity of EF-2 in the intermediate state would be expected to increase and thereby promote the binding of a second Ca^{2+} to the protein at EF-2 (Figs. 4B-C). The binding of a second Ca^{2+} then leads to conformational changes in EF-2 that indirectly result in the swiveling about Gly 42 in EF-1. The swiveling about Gly

42 directly causes the myristoyl group to be lifted out of the binding pocket and places the fatty acid toward the exterior where it can interact with membrane targets.

THREE-DIMENSIONAL STRUCTURE OF GCAP-2

The structures of recoverin whetted our appetite for atomic resolution views of the GCAP proteins. None of these proteins has yet been crystallized. NMR spectroscopic studies of GCAP-2 were performed instead. Ideally, one would like to solve the structure of the Ca^{2+}-free and Ca^{2+}-bound forms of the myristoylated protein, the physiologic species, but this is not yet feasible because of their low solubility. We decided to begin the structural analysis of this important group of retinal calcium sensors by focusing on Ca^{2+}-bound unmyristoylated GCAP-2, which is soluble and gives clearly resolved NMR spectra. Moreover, the structure of the unmyristoylated form of GCAP-2 is likely to be biologically pertinent. Unmyristoylated GCAP-2 is nearly as effective as myristoylated GCAP-2 in activating guanylyl cyclase at low Ca^{2+} and inhibiting it at high Ca^{2+}.[11] Hence, structural studies of unmyristoylated GCAP-2 should reveal the Ca^{2+}-induced conformational changes underlying its regulation of cyclase.

The structure of Ca^{2+}-bound, unmyristoylated GCAP-2 is presented in Figure 6A. The entire polypeptide chain has been traced except for the disordered region at the carboxyl terminus (residues 191-204). The structure near the amino terminus (residues 2-18) and the region between EF-3 and EF-4 (residues 132-144) are rather poorly defined due to dynamical disordering. GCAP-2 is a compact protein (radius of gyration = 17 Å) made of two domains separated by a flexible linker. Each domain contains a pair of EF-hands defined from the amino-terminus: EF-1 (Ala 22-Val 51), EF-2 (Thr 58-Leu 87), EF-3 (Leu 96-Lys 126) and EF-4 (Glu 147-Arg 176). EF-1 and EF-2 interact intimately to form the N-terminal domain, and EF-3 and EF-4 to form the C-terminal domain. The entering helix of EF-2 (residues 58-68) packs against the helices of EF-3 (residues 96-104 and 114-127) at the interface between the two domains, forming a cleft. The linker between the domains is U-shaped, which positions the four EF-hands in a compact tandem array like that found in Ca^{2+}-bound recoverin. Indeed, the overall folding of Ca^{2+}-bound unmyristoylated GCAP-2 closely resembles that of Ca^{2+}-bound unmyristoylated recoverin[23] and Ca^{2+}-bound myristoylated recoverin.[18] The RMS deviation of the main chain atoms (in the EF-hand motifs) is 2.2 Å in comparing GCAP-2 to recoverin.

Three Ca^{2+} ions are bound to GCAP-2, as anticipated on the basis of its amino acid sequence and site-directed mutagenesis.[36] Ca^{2+} is not bound to EF-1 because the binding loop is distorted from a favorable Ca^{2+}-binding geometry by Pro 36 at the fourth position of the 12-residue loop. Also, the third residue in the loop (Cys 35) is not suitable for ligating Ca^{2+}. EF-2 binds Ca^{2+} and adopts a favorable structure despite the tight turn centered at Asn 74 (position 6 of the loop). The structure of EF-3 is strikingly similar to that of EF-3 in Ca^{2+}-bound recoverin and calmodulin. The RMS deviations of the 116 main chain atoms of EF-3 are 0.66 Å in comparing

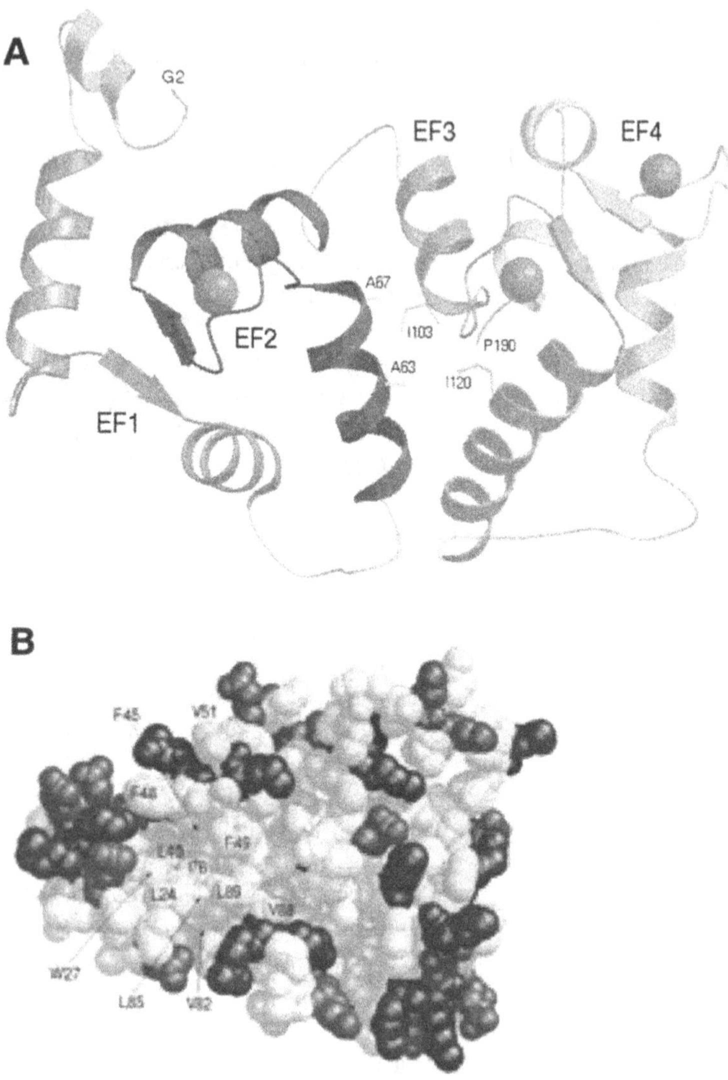

Figure 6. Schematic ribbon diagram (A) and space-filling representation (B) of the structure of unmyristoylated GCAP-2 with three bound Ca^{2+}. The side-chain atoms of residues at the domain interface (Ala 63, Ala 67, Ile 103, and Ile 120) are indicated in "A". Hydrophobic, negatively charged, and positively charged residues of the exposed hydrophobic patch are highlighted in "B". Solvent-exposed hydrophobic residues from EF-1 and EF-2 are indicated.

GCAP-2 with recoverin, and 0.80 Å in comparing GCAP-2 with calmodulin. The structure of EF-4 is quite different from that of recoverin. In recoverin, the second residue in the loop (Lys 161) forms a salt bridge with residue 12 (Glu 171) that disables Ca^{2+} binding. In GCAP-2, the second residue of the EF-4 loop (E159) is negatively charged and cannot form a salt bridge that would impede Ca^{2+} binding.

Furthermore, residues 1 and 3 of the EF-4 loop (Asp 158 and Asn 160) contain oxygen atoms in their side chains that can ligate Ca^{2+}, in contrast with the corresponding residues of recoverin (Gly 160 and Lys 162). Thus, Ca^{2+} binds to EF-4 much as it does to EF-2 and EF-3.

GCAP-2 has a solvent-exposed, hydrophobic surface formed by residues from EF-1 and EF-2 (Fig. 6B). The exposed patch of hydrophobic residues is formed by the clustering of several aromatic side chains (Trp 27, Phe 31, Phe 45, Phe 48, Phe 49, and Tyr 81), and several aliphatic residues (Leu 24, Leu 40, Ile 76, Val 82, Leu 85, and Leu 89). These exposed hydrophobic residues are highly conserved in members of the family[8] and form a similar nonpolar patch in Ca^{2+}-bound recoverin.[18,23] The exposed hydrophobic patch of GCAP-2 may serve a role in regulating guanylyl cyclase. Recent site-directed mutagenesis studies reveal that many of these exposed residues are important in the cyclase interaction.[37,38] In particular, replacement of residues 78-110 (the exiting helix of EF-2) with corresponding residues of neurocalcin results in a chimeric protein that fails to inhibit guanylyl cyclase, but activates it at high Ca^{2+} levels. Also, the replacement of residues in EF-1 (residues 24-49) with the corresponding residues of neurocalcin renders the chimera completely inactive.

The structure of GCAP-2 near the amino-terminus (residues 2-18) appears different from that of recoverin. There is virtually no sequence similarity between recoverin and the GCAPs in this region. The orientation of the amino-terminal helix is different in recoverin and GCAP-2. This helix in recoverin extends close to the inter-domain linker, whereas it interacts primarily with the entering helix of EF-1 in GCAP-2. We note, however, that these apparent structural differences in the amino-terminal region between recoverin and GCAP-2 may result from the very low precision of our structure in this region because of dynamical disordering. Deletion of the amino-terminal region by mutagenesis has no effect on the function of GCAP-2,[38] consistent with our finding that this region is structurally disordered.

The GCAP-2 structure is likely to be similar to that of GCAP-1 (40% sequence identity), GCAP-3 (35% identity) and GCIP (37% identity), since the overall main chain structure appears so similar to recoverin (2.2 Å RMS deviation; 30% identity). Most of the hydrophobic residues in the hydrophobic core and in the exposed patch are highly conserved. Also conserved are the residues that ligate Ca^{2+} in the EF-hand loops. Interestingly, important residues in the entering helix of EF-2 at the domain interface (Ala 57, Ala 63, and Ala 67) are not conserved. Other structurally important and non-conserved residues include Asn 74, Leu 79, Thr 93, His 95 and Thr 100. Considerable differences are also found in the amino-terminal (residues 2-18) and carboxyl-terminal (residues 191-204) regions. These differences suggest that the interaction and/or orientation between the N-terminal and C-terminal domains might be different in GCAP-1, GCAP-3 and GCIP. Indeed, a point mutation at the domain interface causes very different phenotypes in GCAP-1 and GCAP-2. The mutation (Y99C) causes GCAP-1 to be constitutively active,[39,40] resulting in autosomal dominant cone dystrophy.[41] In contrast, the corresponding mutation in GCAP-2 (Y104C) does not alter its Ca^{2+}-sensitivity and is not linked with cone degeneration.[39]

In summary, the overall main chain structure of GCAP-2 is similar to that of Ca^{2+}-bound recoverin except for structural differences near the amino-terminus (residues 2-18) and the binding of Ca^{2+} to EF-4. We see an exposed hydrophobic patch of residues belonging to EF-1 and EF-2 that may play an important role in regulating guanylyl cyclase. Our next goal is to solve the structure of Ca^{2+}-free GCAP-2, a formidable challenge because of its lower stability and solubility, to fully elucidate the Ca^{2+}-induced structural changes that enable GCAP-2 to activate guanylyl cyclases in the absence of Ca^{2+}

ACKNOWLEDGMENTS

This work was supported by grants to J.B.A from the NIH, and to M.I. from the Medical Research Council of Canada. J.B.A. is a recipient of a Beckman Foundation Young Investigator Award. MI. is a Howard Hughes Medical Institute international research scholar.

REFERENCES

1. Dizhoor AM, Ray S, Kumar S et al. Recoverin: a calcium sensitive activator of retinal rod guanylate cyclase. Science 1991; 251:915-918.
2. Palczewski K et al. Molecular cloning and characterization of retinal photoreceptor guanylyl cyclase-activating protein. Neuron 1994; 13:395-404.
3. Dizhoor AM, Lowe DG, Olshevskaya EV et al. The human photoreceptor membrane guanylyl cyclase is present in outer segments and is regulated by calcium and a soluble activator. Neuron 1994; 12:1345-1352.
4. Gorczyca WA et al. Guanylyl cyclase activating protein. A calcium-sensitive regulator of phototransduction. J Biol Chem 1995; 270:22029-22036.
5. Haeseleer F et al. Molecular characterization of a third member of the guanylyl cyclase-activating protein subfamily. J Biol Chem 1999; 274:6526-6535.
6. Li N et al. Guanylate-cyclase-inhibitory protein is a frog retinal Ca2+-binding protein related to mammalian guanylate-cyclase-activating proteins. European J Biochem 1998; 252:591-9.
7. Moncrief ND, Kretsinger RH, Goodman M. Evolution of EF-hand calcium-modulated proteins. J. Mol. Evol. 1990; 30:522-562.
8. Ames JB, Tanaka T, Stryer L et al. Portrait of a myristoyl switch protein. Curr Opin Struct Biol 1996; 6:432-8.
9. Hendricks KB, Wang BQ, Schnieders EA et al. Yeast homologue of neuronal frequenin is a regulator of phosphatidylinositol-4-OH kinase. Nature Cell Biol 1999; 1:234-241.
10. Dizhoor AM et al. The NH2 terminus of retinal recoverin is acylated by a small family of fatty acids. J Biol Chem 1992; 267: 16033-6.
11. Olshevskaya EV, Hughes R, Hurley JB et al. Calcium binding, but not a calcium-myristoyl switch, controls the ability of guanylyl cyclase-activating protein GCAP-2 to regulate photoreceptor guanylyl cyclase. J Biol Chem 1997; 272: 14327-14333.
12. Zozulya S, Stryer L. Calcium-myristoyl protein switch. Proc Natl Acad Sci USA 1992; 89:11569-73.
13. Dizhoor AM et al. Role of the acylated amino terminus of recoverin in Ca(2+)-dependent membrane interaction. Science 1993; 259: 829-32.
14. Ladant D. Calcium and membrane binding properties of bovine neurocalcin expressed in Escherichia coli. J Biol Chem 1995; 270: 3179-3185.

15. Kobayashi M, Takamatsu K., Saitoh S et al. Myristoylation of hippocalcin is linked to its calcium-dependent membrane association properties. J Biol Chem 1993; 268: 18898-904
16. Tanaka T, Ames JB, Harvey TS et al. M. Sequestration of the membrane-targeting myristoyl group of recoverin in the calcium-free state. Nature 1995; 376:444-7.
17. Ames JB, Tanaka T, Ikura M et al. Nuclear magnetic resonance evidence for Ca(2+)-induced extrusion of the myristoyl group of recoverin. J. Biol. Chem. 1995; 270:30909-13.
18. Ames JB et al. Molecular mechanics of calcium-myristoyl switches. Nature 1997; 389: 198-202.
19. Chen CK, Inglese J, Lefkowitz, RJ et al. Ca(2+)-dependent interaction of recoverin with rhodopsin kinase. J Biol Chem 1995; 270: 18060-18066.
20. Klenchin VA, Calvert PD, Bownds MD. Inhibition of rhodopsin kinase by recoverin. Further evidence for a negative feedback system in phototransduction. J Biol Chem 1995; 270:16147-16152.
21. Erickson MA et al. The effect of recombinant recoverin on the photoresponse of truncated rod photoreceptors. 1998; Proc Natl Acad Sci USA 95:6474-9.
22. Gray-Keller MP, Polans AS, Palczewski K et al. The effect of recoverin-like calcium-binding proteins on the photoresponse of retinal rods. 1993; Neuron 10:523-531.
23. Flaherty KM, Zozulya S, Stryer L et al. Three-dimensional Structure of Recoverin, a Calcium Sensor in Vision. Cell 1993; 75:709-716.
24. Babu YS, Bugg CE, Cook WJ. Structure of calmodulin refined at 2.2 A resolution. J Mol Biol 1988; 204:191-204.
25. Herzberg O, James MN. Refined crystal structure of troponin C from turkey skeletal muscle at 2.0 Å resolution. J Mol Biol 1988; 203:761-779.
26. Ames JB, Hamasaki N, Molchanova T. Structure and calcium binding studies of a recoverin mutant (E85Q) in an allosteric intermediate state. Biochemistry 2002; 41:5776-5787.
27. Ames JB, Porumb T, Tanaka T et al. Amino-terminal myristoylation induces cooperative calcium binding to recoverin. J Biol Chem 1995; 270:4526-4533.
28. Matsuda S, Hisatomi O, Ishino T et al. The role of calcium binding sites in S-modulin function. J Biol Chem 1998; 273:20223-20227.
29. Permyakov SE et al. Effects of mutations in the calcium binding sites of recoverin on its calcium affinity. Protein Engineering 2000; 13:783-790.
30. Alekseev AM et al. Obtaining and characterization of EF-hand mutants of recoverin. FEBS Letts. 1998; 440:116-118.
31. Zhang M, Tanaka T et al. Calcium-induced conformational transition revealed by the solution structures of apo calmodulin. Nat Struct Biol 1995; 2:758-767.
32. Kuboniwa H et al. Structure of calcium-free calmodulin. Nat Struct Biol 1995; 2:768-776.
33. Yap KL, Ames JB, Swindells MB et al. Diversity of conformational states and changes within the EF-hand protein superfamily. Proteins 1999; 37:499-507.
34. Fernandez EJ, Abad-Zapatero C, Olsen KW. Crystal structure of Lysb182-Lysb182 crosslinked hemoglobin: A possible allosteric intermediate. J Mol Biol 2000; 296:1245-1256
35. Schumacher MA. et al. Allosteric intermediates indicate R2 is the liganded hemoglobin end state. Proc. Natl Acad Sci USA 1997; 94:7841-7844.
36. Dizhoor AM, Hurley JB. Inactivation of EF-hands makes GCAP-2 a constitutive activator of photoreceptor guanylyl cyclase by preventing a calcium-induced transition. J Biol Chem 1996; 271:19346-19350.
37. Ermilov AN, Olshevskaya EV, Dizhoor AM. Instead of binding calcium, one of the ef-hand structures in guanylyl cyclase activating protein-2 is required for targeting photoreceptor guanylyl cyclase. J Biol Chem 2001; 276:48143-48148.
38. Olshevskaya EV, Ermilov AN, Dizhoor AM. Dimerization of guanylyl cyclase-activating protein. J Biol Chem 1999; 274:25583-25587.
39. Dizhoor AM, Boikov SG Olshevskaya EV. Constitutive activation of photoreceptor guanylate cyclase by Y99C mutant of GCAP-1. Possible role in causing human autosomal dominant cone degeneration. J Biol Chem 1998; 273:17311-17314.

40. Sokal I et al. GCAP1 (Y99C) mutant is constitutively active in autosomal dominant cone dystrophy. Molecular Cell 1998; 2:129-133.
41. Payne AM et al. A mutation in GCAP1 in an autosomal dominant cone dystrophy pedigree mapping to a new locus on chromosome 6p21.1. Human Mol Gen 1998; 7:273-277.
42. Ferrin T, Huang C, Jarvis L et al. The MIDAS display system. J Mol Graph 1998; 6:13-27.
43. Kraulis PJ. Molscript a program to produce both detailed and schematic plots of protein structures. J Appl Crystallogr 1991; 24:946-950.
44. Bacon DJ, Anderson WF. A fast algorithm for rendering space-filling molecule pictures. J Mol Graph 1988; 6:219-220.

TARGET RECOGNITION OF GUANYLATE CYCLASE BY GUANYLATE CYCLASE-ACTIVATING PROTEINS

Karl-Wilhelm Koch

ABSTRACT

Guanylate cyclase-activating proteins (GCAPs) control the activity of membrane bound guanylate cyclases in vertebrate photoreceptor cells. They form a permanent complex with guanylate cyclase 1 (ROS-GC1) at low and high Ca^{2+}-concentrations. Five different target regions of GCAP-1 have been identified in ROS-GC1 at rather distant sites. These findings could indicate a multipoint attachment site for GCAP-1 or, alternatively, the presence of transient binding sites with short contact to GCAP-1. In addition some data are consistent with the operation of one or more transducer units , that represent regulatory regions without being direct binding sites. A permanent ROS-GC1/GCAP-1 complex is physiologically significant, since it allows a very short response time of cyclase activity when the intracellular Ca^{2+}-concentration changes. Thereby, activation of cyclase participates in speeding up the recovery of the photoresponse after illumination and restores the circulating dark current.

INTRODUCTION

Light sensitivity of vertebrate photoreceptor cells is controlled by multiple feedback loops that include Ca^{2+}-dependent and Ca^{2+}-independent mechanisms [1-3] (see also other chapters in this book). A significant contribution to Ca^{2+}-dependent feedback mechanisms is conferred by the Ca^{2+}-sensitive regulation of two membrane bound guanylate cyclases, which are synonymously termed ROS-GC1, retGC1, GC-E

Institut für Biologische Informationsverarbeitung 1, Forschungszentrum Jülich, D-52425 Jülich, Germany.

or ROS-GC2, retGC2, GC-F.[4] Regulation of cyclases by Ca^{2+} is indirect and mediated by small EF-hand Ca^{2+}-binding proteins which are named guanylate cyclase-activating proteins or GCAPs of which three isoforms have been identified, GCAP-1, GCAP-2 and GCAP-3.[5,6] So far most work on the expression profile of GCAPs and on their regulatory properties has been done with GCAP-1 and GCAP-2. For example rods of transgenic mice that do not express GCAP-1 and GCAP-2 (GCAPs-/-) show a higher light sensitivity than rods from wildtype mice.[7] Rod cells from mutant mice also show a slowed time course of their flash response and slower kinetics of the recovery phase of a flash response. Guanylate cyclase activity of transgenic GCAPs-/- mice is normal, but any Ca^{2+}-dependent regulation is absent. These results are consistent with previous biochemical observations, that GCAPs have a strong effect on regulation of cyclase activity over the physiologically relevant range of free [Ca^{2+}] from 600 nM in the dark adapted state (low guanylate cylase activity) and 50 nM after illumination (high guanylate cyclase activity).[4-6]

GCAPs belong to a family of neuronal calcium sensor proteins (NCS) that show a wide distribution among species and within the nervous system. NCS proteins can be grouped into five subfamilies named recoverins, VILIPs, frequenins, Kchips and GCAPs.[8,9] In contrast to calmodulin that is ubiquitously expressed and can regulate a large variety of proteins NCS proteins seem to have a specialized function and target to a small set of proteins. This article discusses aspects concerning target recognition and protein-protein interactions of GCAPs when they regulate cyclases.

COMPLEX OF GUANYLATE CYCLASE AND GCAP

When guanylate cyclase activities are assayed in vitro, GCAP-1 and GCAP-2 activate ROS-GC1 severalfold at low free [Ca^{2+}] and inhibit the enzyme at higher [Ca^{2+}] $\geq 1 \mu M$. ROS-GC2 is activated by GCAP-2 and not by GCAP-1,[10] which indicates that GCAPs and cyclases differ in their regulatory features. The physiological meaning of this different targeting however is unclear. GCAPs form a direct complex with cyclases and results from different experimental approaches are consistent with this model (see below). One method to demonstrate a direct interaction is chemical crosslinking. Proteins of interest in an existing complex react with bifunctional small organic molecules resulting in one or more crosslinked products of higher molecular mass. Most crosslinkers react with free amino or sulfhydryl groups in proteins. The presence of two (or more) proteins in a crosslinked product can then be detected by immunoblotting. Crosslinking of GCAP-1 and ROS-GC1 had been demonstrated by this approach.[11,12]

CROSSLINKING OF GCAP-1

A typical result of a crosslinking experiment is shown in Fig. 1. The homobifunctional water insoluble crosslinker Disuccinimidyl suberate (DSS) was incubated with a suspension of rod outer segments containing ROS-GC1 and endogeneous GCAP-1. Proteins were separated by SDS-PAGE, electroblotted and

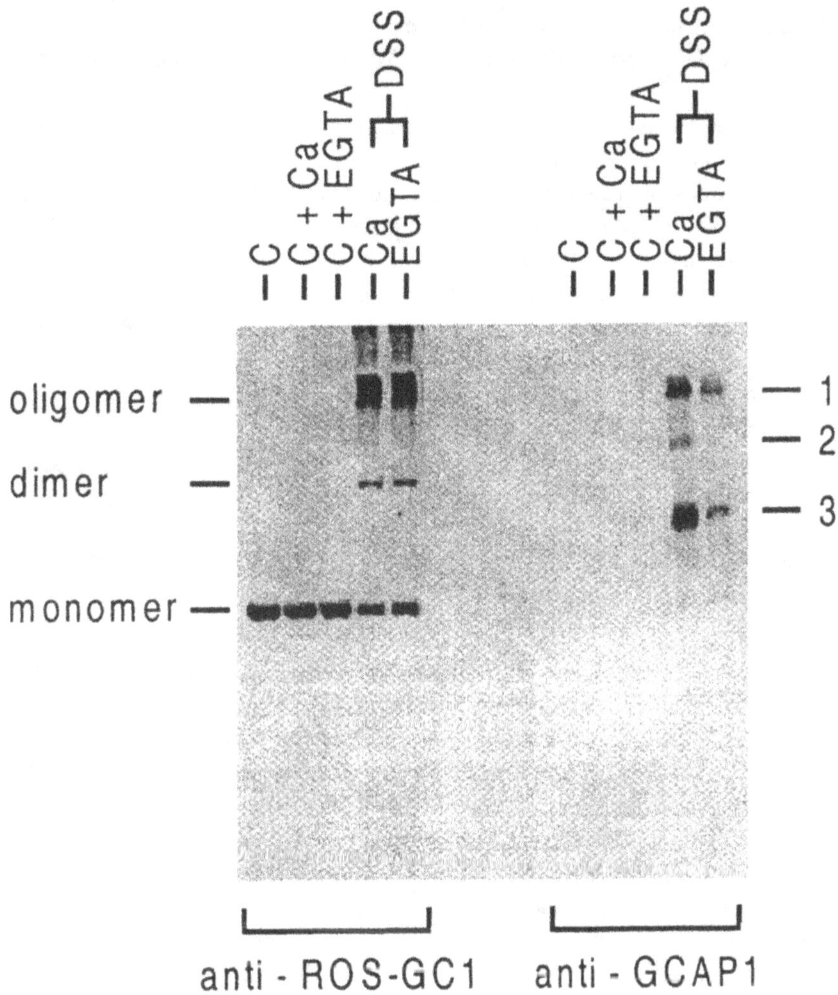

Figure 1. Crosslinking of ROS-GC1 and GCAP-1. A suspension of purified bovine rod outer segments was incubated with 1 mM DSS, a homobifunctional crosslinker (Pierce), in the presence of 1 mM Ca^{2+} or 1 mM EGTA. Control incubations were done without addition of crosslinker (lanes C) and without crosslinker but with either Ca^{2+} or EGTA. Incubations were stopped by the addition of 50 mM Tris-HCl pH 8.0 and samples were analyzed by electrophoresis (7.5 % acrylamide gel). The gel was blotted and the blot was probed by anti-ROS-GC1 (left side) and anti-GCAP-1 (right side) specific antibodies. The anti-ROS-GC1 antibody recognized a monomer, a dimer and an oligomer of ROS-GC1.[25] The anti-GCAP-1 antibody recognized GCAP-1,[31] which can not be seen on a 7.5 % gel, and three bands above the ROS-GC1 monomer.

detected by specific antibodies against GCAP-1 and ROS-GC1. The results show several features of a cyclase/GCAP complex:

1. ROS-GC1 forms a dimer at 230 kDa and an oligomer that barely enters the gel.
2. Three bands are visible in the molecular range above 100 kDa, that were immunoreactive for a GCAP-1 antibody in the presence of crosslinker, but were absent in control incubations without crosslinker. One of these bands (see arrows indicating band 1) co-migrate with the oligomer ROS-GC1 band which indicates a crosslinked product of ROS-GC1 and GCAP-1.
3. Crosslinking was observed in the presence and absence of Ca^{2+}, which indicates that the complex of cyclase and GCAP-1 exists under high and low $[Ca^{2+}]$. In fact, crosslinking was more efficient at high $[Ca^{2+}]$ as can be seen in the more intense labelling of band 1 in the presence of Ca^{2+} by the anti-GCAP-1 antibody.
4. Crosslinking was performed by a water insoluble substance, which points to a ROS-GC1/GCAP-1 contact site in the vicinity of the membrane.

INTERACTION OF GUANYLATE CYCLASES AND GCAPs AT LOW AND HIGH $[Ca^{2+}]$

The existence of a guanylate cyclase/GCAP complex at low and high $[Ca^{2+}]$ as revealed by chemical crosslinking was also inferred from other experiments. Ca^{2+}-loaded GCAP-2 and to a lesser degree also Ca^{2+}-loaded GCAP-1 inhibit ROS-GC activity leading to a lower basal activty at micromolar free $[Ca^{2+}]$.[13,14] Under these conditions ROS-GC1 activity is lower than in washed membrane preparations where no GCAP is present. Furthermore, Ca^{2+}-loaded GCAP-1 and GCAP-2 can compete with Ca^{2+}-insensitive mutants of GCAPs, that activate ROS-GC constitutively in a Ca^{2+}-independent way.[13,15,16] Co-immunoprecipitation studies had shown that GCAP-1, ROS-GC1 and tubulin form a complex at high and low $[Ca^{2+}]$.[17] Thus it seems, that dissociation of GCAP from the guanylate cyclase/GCAP complex is not triggered by a change in $[Ca^{2+}]$. However, the complex is not very tight, since GCAPs can easily be washed off from rod outer segment membranes. This is different from the regulatory features of some calmodulin/target complexes, that cannot be disrupted by complexing Ca^{2+}. Examples are phosphorylase b kinase and the inducible form of nitric oxide synthase (NOS-II), which show strong binding to Ca^{2+}-calmodulin and Ca^{2+}-free calmodulin (apocalmodulin).[18]

In summary the current working hypothesis is that GCAPs exist in a permanent complex with ROS-GCs. A change in the intracelluar $[Ca^{2+}]$ triggers a conformational change in GCAP leading to a rearrangement within the complex. In consequence ROS-GC is transformed into a state of higher catalytic activity. A first step to decipher this switch mechanism has been to identify regions in ROS-GC1 that directly interact with GCAP-1 or represent critical sites for regulation. Since all published work on this specific matter so far was done on ROS-GC1 and GCAP-1, I restrict the following discussion on this protein pair.

TARGET REGIONS IN ROS-GC1

Three different groups have identified a total of 5 different sites in ROS-GC1 that are thought to represent target regions of GCAP-1. These sites are located in rather distant regions in the cytoplasmic part of the enzyme, two are in the juxtamembrane domain (JMD), two are in the kinase homolgy domain (KHD) and one is in the cyclase catalytic domain (CCD) (Fig. 2). Lange et al[19] identified two regions in the juxtamembrane domain (445M-456L; 503L-522I) by screening a synthetic peptide library and by site directed mutagenesis of hypothetic regulatory regions. The screening of the peptide library was restricted to the ROS-GC1 domain between amino acids 437R-552G. The library was screened with peptides overlapping by 10 out of 12 amino acids. This region was chosen because in a prior study, deletion- and hybrid-analysis of ROS-GC1 had revealed that this domain was pivotal in the GCAP modulation of the cyclase.[12] GCAP1 regulatory regions identified in the study by Lange et al gave a hint to explain the molecular causes of the rod dystrophy Leber's congenital amaurosis (LCA1) where the patients are born blind or soon become blind after birth.[20] In these patients there is a point mutation in the ROS-GC1 gene between the region 503L-522I. Biochemical studies on heterologously expressed ROS-GC1 mutants show that this LCA mutant has completely lost its sensitivity to modulation by GCAP-1.[21]

Krylov and Hurley[11] found two regions in KHD, 680S-699E and 724Q-738A by a different experimental strategy. They first stabilized ROS-GC1/GCAP-1 complexes by homobifunctional crosslinking of sulfhydryl groups in spatially proximate cysteines. These complexes were digested with trypsin in a second step and then analyzed by mass spectrometry to find crosslinked peptides. Candidate regions in ROS-GC1 and GCAP-1 were identified by assigning peaks in the mass spectrum. Synthetic peptides of candidate regions were then used in peptide competition studies.

Sokal et al[22] identified a region in the catalytic domain, 966G-983G, which corresponds to 910G-927G in the mature protein (Fig. 2). In this study the authors screened the whole cytoplasmic domain with peptides of 15 amino acids in length with 8 amino acids overlapping by a similar competition assay as Lange et al.[19] Interestingly, the region in the CCD shows a 2-3 fold higher affinity for GCAP-2 than for GCAP-1 as revealed by peptide titration experiments. Thus, the identified region in this study may represent the GCAP-2 site in ROS-GC1. This interpretation is consistent with the results of earlier studies, which demonstrated by the use of deletion and chimeric mutants that the GCAP-2 modulated domain in ROS-GC1 resides at its C-terminus.[23,24]

APPARENT AFFINITIES OF TARGET REGIONS IN ROS-GC1

Peptide competition experiments were used in all three studies to estimate apparent IC_{50} values (concentration of peptide at which inhibition is halfmaximal). All

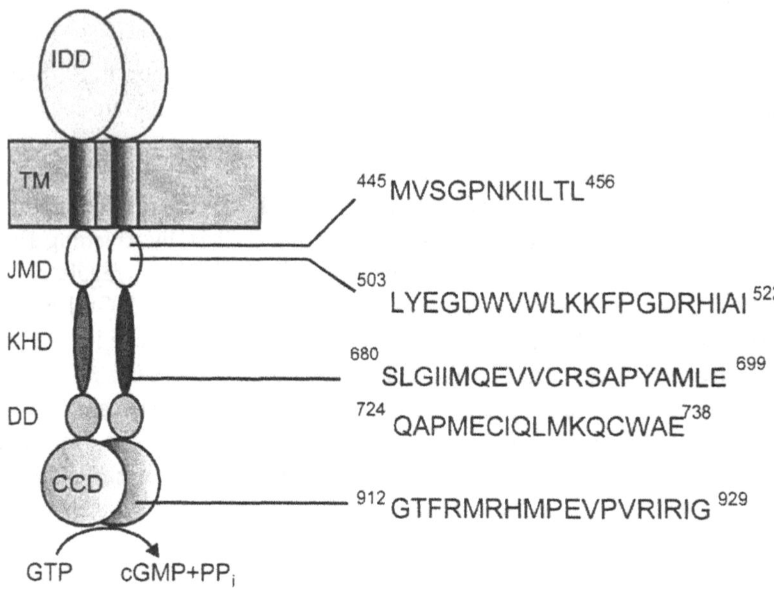

Figure 2. Target regions of GCAP-1 in a topographical model of ROS-GC1. Denoted regions are the intradiscal domain (IDD), the transmembrane domain (TM) within the disk membrane, the juxtamembrane domain (JMD), the kinase homology domain (KHD), the dimerization domain (DD) and the cyclase catalytic domain (CCD). Peptide sequences of identified target regions are indicated at the corresponding domains (see text for details).

reported values are within the same order of magnitude, the two peptides in the JMD showed an IC_{50} between 100 and 260 μM, the two peptides in the KHD had an IC_{50} between 30 and 80 μM and the peptide of the CCD had an IC_{50} of 290 μM. If the identified regions represent true binding sites, then each site alone would bind GCAP-1 only with rather low affinity. Activation of ROS-GC1 by GCAP-1 is halfmaximal at 0.8 μM in enzymatic assays.[19,23] The apparent dissociation constant of GCAP-1 and ROS-GC1 then is expected to be < 0.8 μM. All reported IC_{50} values are two to three orders of magnitude higher. Although the three studies do not rely solely on peptide competition studies and include site–directed mutagenesis, peptide-affinity column chromatography and crosslinking, they do not give a conclusive answer about right or wrong target sites so far. However, a large cytoplasmic fragment (aa733-1054) of ROS-GC1 that includes three of the identified target regions (two in the KHD and one in the CCD) exhibits a small, but measurable cyclase activity. Interestingly this fragment is not stimulated by GCAP-1, but can still be activated by GCAP-2 at low $[Ca^{2+}]$.[25] In addition this fragment can be activated by S100b, a small Ca^{2+}-binding protein that stimulates ROS-GC1 activity severalfold at micromolar $[Ca^{2+}]$. These experiments show that many properties regarding structural integrity are retained in the ROS-GC1 fragment aa733-1054, but that necessary regions for GCAP-1 recognition are not present.

MODEL OF GCAP-1 MEDIATED ACTIVATION OF ROS-GC1

The conflicting results on target regions in ROS-GC1 could indicate a multipoint attachment site for GCAP-1.[19] Taking into account that all peptides are rather weak inhibitors in competition assays it is reasonable that several regions in ROS-GC1 are needed to form together a site of higher affinity for GCAP-1. These sites could well be located in distant regions of the molecule. The different results could also reflect the existence of several transient binding sites. These sites might have a short contact with GCAP-1, when GCAP-1 changes its conformational state due to binding or release of Ca^{2+}. It is also possible that some regions are important for the switch from a low activity state to a high activity state in the cyclase as this has been proposed for the region 445M-456L in the JMD, that was assigned as transducer unit.[19] Such a site must not necessarily be a direct binding site, but it could represent an important regulatory unit within the cytoplasmic part of the cyclase that is necessary to transmit the Ca^{2+}-signal from GCAP-1 to the catalytic center of the cyclase.

PHYSIOLOGICAL SIGNIFICANCE OF A PERMANENT ROS-GC1/GCAP-1 COMPLEX

It is somehow surprising that GCAP-1 and ROS-GC1 form a complex that is not disrupted by a change in free [Ca^{2+}]. Why does GCAP-1 bind to ROS-GC1 in both its Ca^{2+}-loaded and its Ca^{2+}-free form? When a photoreceptor cell is excited by light, the cytoplasmic [Ca^{2+}] decreases and the guanylyl cyclase activity increases with a time constant of approx. 200 msec.[26] Therefore, the activation of ROS-GC1 participates significantly in speeding up the recovery of the photoresponse after saturating light flashes and restores the circulating dark current.[1] A permanent complex of GCAP-1 and ROS-GC1 would immediately respond to changes in cytoplasmic [Ca^{2+}] after illumination. No additional (time consuming) diffusion step is necessary. Dissociation of Ca^{2+} from the EF-hand Ca^{2+}-binding sites in GCAP-1 would then induce a conformational change in GCAP-1 that is directly transmitted to the cyclase catalytic domain. If ROS-GC1 and the built-in Ca^{2+}-sensor GCAP-1 operate in the time frame of a photoresponse, the Ca^{2+}-sensing of GCAP-1 must also fulfill at least two requirements: 1. the dissociation of Ca^{2+} from the EF-hand Ca^{2+}-binding sites must be fast enough (less than 200 msec) and 2. the conformational change in GCAP-1 must be sufficient to trigger a severalfold increase in the catalytic turnover rate of the cyclase.

AFFINITIES OF GCAP-1 FOR Ca^{2+}

According to a systematic study by Renner et al [27] relative affinities of Ca^{2+} among different groups of Ca^{2+}-binding proteins can be predicted. If EF-hands contain a Glu at the ninth position, they release Ca^{2+} with slower dissociation rates than

EF-hands that contain either an amino acid with a shorter side chain (Asp) or an uncharged side chain (Gln). Ca^{2+}-buffer proteins like parvalbumin have slower dissociation rates of Ca^{2+} than Ca^{2+}-trigger like calmodulin or troponin C (for review ref. 28). All three functional EF-hands in GCAP-1 (and also in GCAP-2 and GCAP-3) contain either an Asp or a Thr/Ser at the ninth position. GCAPs fit therefore perfectly into the group of Ca^{2+}-trigger proteins with rapid off kinetics (mean residence time of Ca^{2+} would be a few milliceconds). A recent study reported an intermediate affinity of EF-hand 3 for Ca^{2+} in GCAP-1.[29] With an apparent K_D of 2.9 μM and the assumption that the on-rate of Ca^{2+} to all three EF-hands is under control of diffusion [30] (k_1=2 x 10^8 $M^{-1}s^{-1}$; K_D=k_{-1}/k_1) the authors obtained an off-rate of k_{-1}=5.8 x 10^{-3} s^{-1}, which would result in a time constant of ~2 ms. Thus, dissociation of Ca^{2+} from EF-hand 3 is fast enough to trigger a conformational change in GCAP-1 and subsequent activation of ROS-GC1.

GCAP-1 probably undergoes a large conformational change upon binding Ca^{2+}-ions. This change has been previously monitored by different methods, e.g., tryptophan fluorescence spectroscopy, Ca^{2+}-induced shifts in electrophoretic mobility and limited proteolysis.[30-32] This conformational change could be sensed by ROS-GC1 at one or more of the defined sequence motifs (s. above) and then transmitted to the catalytic center to increase the turnover rate of cGMP synthesis.

DIMERIZATION OF ROS-GC1

Efficient catalysis of GTP to cGMP in ROS-GC1 requires the formation of a correct dimer interface. Thus, the catalytically active form of ROS-GC1 is a dimer. The dimerization domain DD that is located between the KHD and CCD forms a coiled-coil domain and is critical for a correct dimerization of the whole protein [25,33-35] (Fig.2). Mutations in the DD in the ROS-GC1 gene are linked to a cone-rod dystrophy (CORD6) and cause a severe distortion of dimer formation and GC activity.[25] Concomitant with an impaired dimer formation is a decrease in basal, i.e., unstimulated cyclase activity. However some CORD6 related mutations cause a hypersensitivity to activation by GCAP-1.[25,35] It seems that binding of GCAP-1 can compensate for some loss of catalytic efficiency when the dimer formation is distorted. Binding of GCAP-1 seems to control formation of a correct dimer interface in the CCD without interaction with the DD. It is however unclear how GCAP-1 can control this step, whether it binds to the CCD directly or whether it triggers the correct conformational change by binding to a distant site. To resolve this point conflicting results on target and binding sites as outlined above need to be clarified by further experiments.

CONCLUSION

In summary the following sequence of events is plausible when ROS-GC1 is activated by GCAP-1 (Fig. 3). In the unstimulated state Ca^{2+}-loaded GCAP-1 binds to a site in ROS-GC1 that is close to the membrane (see interpretation of crosslinking

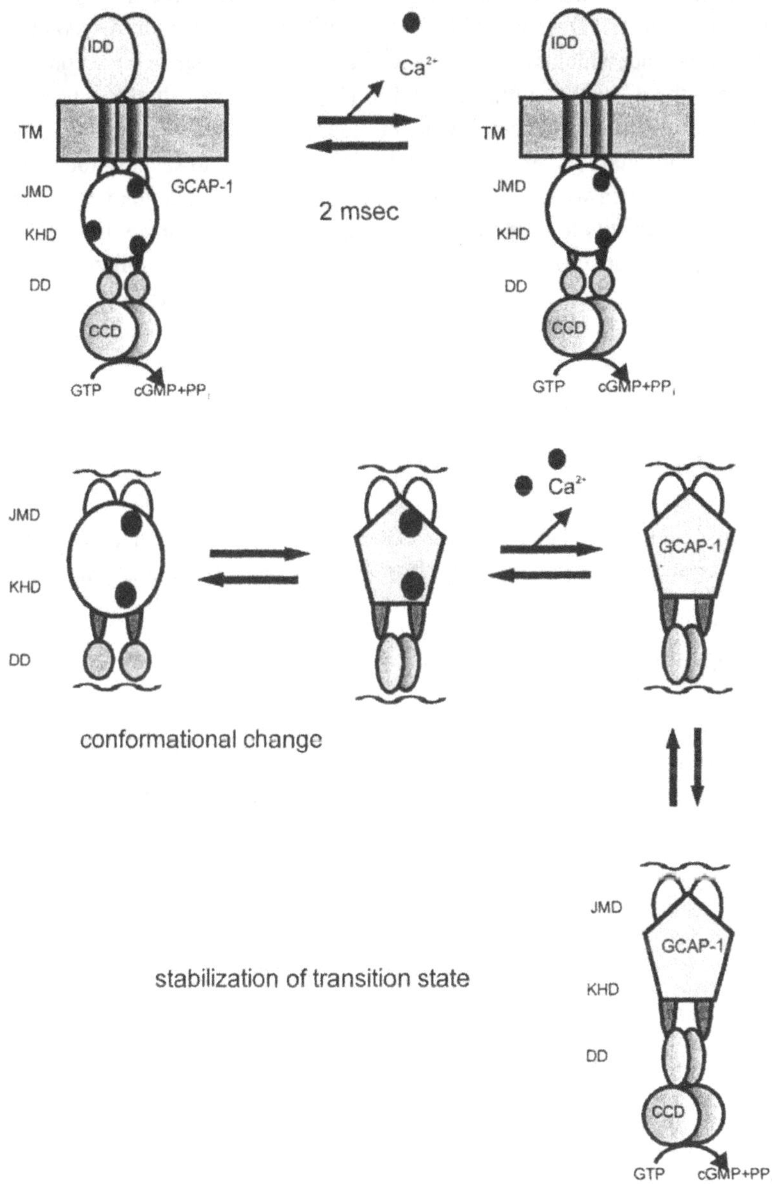

Figure 3. Model of activation of ROS-GC1 by GCAP-1. Decrease of cytoplasmic [Ca^{2+}] leads first to dissociation of Ca^{2+} from EF-hand 3 in GCAP-1. This induces a conformational change in GCAP-1 and subsequently in ROS-GC1, which facilitates a reorientation in the DD. Formation of a correct dimer interface then increases the catalytic efficiency. Further release of Ca^{2+} from EF-hand 2 and 4 (EF-hand 1 is nonfunctional) might then stabilize this transition state of cGMP synthesis.

experiment in Fig. 1), which could be ^{503}L-^{522}I in the JMD and/or ^{680}S-E^{699} and ^{724}Q-A^{738} in the KHD. Decrease of cytoplasmic [Ca^{2+}] is sensed by GCAP-1 first on its third EF-hand, which rapidly loses its Ca^{2+}-ion and induces a large conformational change in GCAP-1 around EF-hand 3. This change leads first to a reorientation of the DD coiled-coil helices,[35] which influences the conformation of the CCD monomers in the catalytic center. GCAP-1 might be bound to several noncontiguous binding sites even when it undergoes a conformational change after dissociation of Ca^{2+}. The affinity to every single binding site probably changes and Ca^{2+}-free GCAP-1 might be transferred to a different location where it can stabilize the active state of ROS-GC1 probably by stabilization of the transition state during catalysis, but this has to be tested experimentally. GCAP-2 probably binds to different sites in ROS-GC1. It is therefore reasonable to speculate that a different sequence of events takes place when GCAP-2 activates ROS-GC1.

ACKNOWLEDGMENT

I thank D. Höppner-Heitmann for excellent technical assistance. Work from this laboratory was supported by grants from the Deutsche Forschungsgemeinschaft (DFG).

REFERENCES

1. Pugh EN Jr, Nikonov S, Lamb TD. Molecular mechanisms of vertebrate photoreceptor light adaptation. Curr Opinion Neurobiol 1999; 9:410-418.
2. Fain GL, Matthews HR, Cornwall MC et al. Adaptation in vertebrate photoreceptors. Physiol Rev 2001; 81:117-151.
3. Burns ME, Baylor DA. Activation, deactivation, and adaptation in vertebrate photoreceptor cells. Annu Rev Neurosci 2001; 24:779-805.
4. Koch KW, Duda T, Sharma RK. Photoreceptor specific guanylate cyclases in vertebrate phototransduction. Molec Cell Biochem 2002; 230:97-106.
5. Palczewski K, Polans AS, Baehr W et al. Ca^{2+}-binding proteins in the retina: structure, function, and the etiology of human visual diseases. BioEssays 2000; 22:337-350.
6. Dizhoor AM, Hurley JB. Regulation of photoreceptor membrane guanylyl cyclases by guanylyl cyclase activator proteins. Methods 1999; 19:521-531.
7. Mendez A, Burns ME, Sokal I et al. Role of guanylate cyclase-activating proteins (GCAPs) in setting the flash sensitivity of rod photoreceptors. Proc Natl Acad Sci USA 2001; 98:9948-9953.
8. Braunewell KH, Gundelfinger ED. Intracellular neuronal calcium sensor proteins: a family of EF-hand calcium-binding proteins in search of a function. Cell Tissue Res 1999; 295:1-12.
9. Burgoyne RD, Weiss JL. The neuronal calcium sensor family of Ca^{2+}-binding proteins. Biochem J 2001; 353:1-12.
10. Haeseleer F, Sokal I, Li N et al. Molecular characterization of a third member of the guanylyl cyclase-activating protein subfamily. J Biol Chem 1999; 274:6526-6535.
11. Krylov DM, Hurley JB. Identification of proximate regions in a complex of retinal guanylyl cyclase 1 and guanylyl cyclase-activating protein-1 by a novel mass spectrometry-based method. J Biol Chem 2001; 276:30648-30654.
12. Duda T, Goraczniak R, Surgucheva I et al. Calcium modulation of bovine photoreceptor guanylate cyclase. Biochemistry 1996; 35:8478-8482.

13. Dizhoor AM, Hurley JB. Inactivation of EF-hands makes GCAP-2 (p24) a constitutive activator of photoreceptor guanylyl cyclase by preventing a Ca^{2+}-induced „activator-to-inhibitor" transition. J Biol Chem 1996; 271:19346-19350.
14. Otto-Bruc A, Buczylko J, Surgucheva I et al. Functional reconstitution of photoreceptor guanylate cyclase with native and mutant forms of guanylate cyclase-activating protein 1. Biochemistry 1997; 36:4295-4302.
15. Dizhoor AM, Boikov SG, Olshevskaya EV. Constitutive activation of photoreceptor guanylate cyclase by Y99C mutant of GCAP-1. J Biol Chem 1998; 273:17311-17314.
16. Sokal I, Li N, Surgucheva I et al. GCAP-1 (Y99C) mutant is constitutively active in autosomal dominant cone dystrophy. Molec Cell 1998; 2:129-133.
17. Schrem A, Lange C, Beyermann M et al. Identification of a domain in guanylyl cyclase-activating protein 1 that interacts with a complex of guanylyl cyclase and tubulin in photoreceptors. J Biol Chem 1999; 274:6244-6249.
18. Jurado LA, Chockalingam PS, Jarret HW. Apocalmodulin. Physiol Rev 1999; 79:661-682.
19. Lange C, Duda T, Beyermann M et al. Regions in vertebrate photoreceptor guanylyl cyclase ROS-GC1 involved in Ca^{2+}-dependent regulation by guanylyl cyclase-activating protein GCAP-1. FEBS Lett 1999; 460:27-31.
20. Perrault I, Rozet JM, Calvas P et al. Retinal-specific guanylate cyclase gene mutations in Leber's congenital amaurosis. Nature Genet 1996 14:461-464.
21. Duda T, Venkataraman V, Goraczniak R et al. Functional consequences of a rod outer segment membrane guanylate cyclase (ROS-GC1) gene mutation linked with Leber's congenital amaurosis. Biochemistry 1999; 38:509-515.
22. Sokal I, Haeseleer F, Arendt A et al. Identification of a guanylyl cyclase-activating protein-binding site within the catalytic domain of retinal guanylyl cyclase 1. Biochemistry 1999; 38:1387-1393.
23. Krishnan A, Goraczniak RM, Duda T et al. Third calcium-modulated rod outer segment membrane guanylate cyclase transduction mechanism. Molec Cell Biochem 1998; 178:251-259.
24. Goraczniak RM, Duda T, Sharma RK. Calcium modulated signaling site in type 2 rod outer segment membrane guanylate cyclase (ROS-GC2). Biochem Biophys Res Commun 1998; 245:447-453.
25. Duda T, Venkataraman V, Jankowska A et al. Impairment of the rod outer segment membrane guanylate cyclase dimerization in a cone-rod dystrophy results in defective calcium signaling. Biochemistry 2000; 39:12522-12533.
26. Calvert PD, Ho TW, LeFebvre YM et al. Onset of Feedback reactions underlying vertebrate rod photoreceptor light adaptation. J Gen Physiol 1998; 111:39-51.
27. Renner M, Danielson MA, Falke JJ. Kinetic control of Ca(II) signaling: Tuning the ion dissociation rates of EF-hand Ca(II) binding sites. Proc Natl Acad Sci USA 1993; 90:6493-6497.
28. Falke JJ, Drake SK, Hazard AL et al. Molecular tuning of ion binding to calcium signaling proteins. Quart Rev Biophys 1994; 27:219-290.
29. Hwang JY, Schlesinger R, Koch KW. Calcium-dependent cysteine reactivities in the neuronal calcium sensor guanylate cyclase-activating protein 1. FEBS Lett 2001; 508:355-359.
30. Sokal I, Otto-Bruc AE, Surgucheva I et al. Conformational changes in guanylyl cyclase-activating protein 1 (GCAP1) and its tryptophan mutants as a function of calcium concentration. J Biol Chem 1999; 274:19829-19837.
31. Frins S, Bönigk W, Müller F et al. Functional characterization of a guanylyl cyclase-activating protein from vertebrate rods. J Biol Chem 1996; 271:8022-8027.
32. Rudnicka-Nawrot M, Surgucheva I, Hulmes JD et al. Changes in biological activity and folding of guanylate cyclase-activating protein 1 as a function of calcium. Biochemistry 1998; 37:248-257.
33. Yang RB, Garbers DL. Two eye guanylyl cyclases are expressed in the same photoreceptor cells and form homomers in preference to heteromers. J Biol Chem 1997; 272:13738-13742.

34. Liu Y, Ruoho AE, Rao VD et al. Catalytic mechanism of the adenylyl and guanylyl cyclases: Modeling and mutational analysis. Proc Natl Acad Sci USA 1997; 94:13414-13419.
35. Ramamurthy V, Tucker C, Wilkie S et al. Interactions within the coiled-coil domain of RetGC-1 guanlyly cyclase are optimized for regulation rather than for high affinity. J Biol Chem 2001; 276:26218-26229.

MOUSE MODELS TO STUDY GCAP FUNCTIONS IN INTACT PHOTORECEPTORS

Ana Mendez and Jeannie Chen

ABSTRACT

In photoreceptor cells cGMP is the second messenger that transduces light into an electrical response. Regulation of cGMP synthesis by Ca^{2+} is one of the key mechanisms by which Ca^{2+} exerts negative feedback to the phototransduction cascade in the process of light adaptation. This Ca^{2+} feedback to retinal guanylyl cyclases (Ret-GCs) is conferred by the guanylate cyclase-activating proteins (GCAPs). Mutations in GCAP1 that disrupt the Ca^{2+} regulation of Ret-GCs in vitro have been associated with severe human vision disorders. This chapter focuses on recent data obtained from biochemical and electrophysiological studies of GCAP1/GCAP2 knockout mice and other GCAP transgenic mice, addressing:

1. the quantitative aspects of the Ca^{2+}-feedback to Ret-GCs in regulating the light sensitivity and adaptation in intact rods;
2. functional differences between GCAP1 and GCAP2 in intact rod photoreceptors; and
3. whether GCAP mutants with impaired Ca^{2+} binding lead to retinal disease in vivo by constitutive activation of Ret-GCs and elevation of intracellular cGMP, as predicted from in vitro studies.

INTRODUCTION

Visual excitation in retinal photoreceptor cells is mediated by a cascade that leads to the enzymatic hydrolysis of cGMP and the subsequent closure of cGMP-gated channels in the plasma membrane. Recovery of the dark state requires the resynthesis

---

The Mary D. Allen Laboratory for Vision Research, Doheny Eye Institute, and [2]Department of Ophthalmology and [3]Cell and Neurobiology, Keck School of Medicine, University of Southern California, Los Angeles, CA 90089-9112, U.S.A.

of cGMP, which is catalyzed by particulate (membrane-associated) guanylate cyclases (Ret-GCs). Ret-GC activity has long been known to be stimulated with high cooperativity by the decrease in cytosolic [Ca^{2+}] that follows light exposure.[1] This cyclase stimulation has been proposed to be the major mechanism by which photoreceptors adjust their sensitivity according to background illumination.[2,3] The soluble regulators that confer Ca^{2+}-sensitivity to Ret-GCs have been cloned and identified as 23-kDa Ca^{2+}-binding proteins from the calmodulin superfamily,[4-6] the guanylate cyclase-activating proteins (GCAPs).

Two isoforms of Ret-GCs are expressed in mammalian retinas, Ret-GC1 and Ret-GC2.[7,8] Ret-GC1 localizes to the outer segments and synaptic terminals of rods and cones,[9-12] whereas the distribution of Ret-GC2 within photoreceptors is less well known. Although Ret-GC1 appears to be indispensable for phototransduction in human rods and cones,[13] both Ret-GC1 and Ret-GC2 are expressed,[14] and probably functional[15] in murine rod outer segments. In photoreceptor cells, Ret-GCs are transmembrane proteins oriented in the disc membranes so that the "extracellular domain" is intradiscal, and are regulated intracellularly by the GCAPs.

Three isoforms of mammalian GCAPs, namely GCAP1, GCAP2 and GCAP3 have been isolated.[5,6,16] GCAP1 and GCAP2 have both been localized to photoreceptors in numerous studies,[4-6,17,18] although reports of the distribution of GCAPs between rods and cones in different species have been more ambiguous.[17-19] In the human, monkey and bovine species, GCAP1 immunoreactivity appears stronger in the cone outer segments, whereas GCAP2 immunostaining is seen in the entire region of rods and cones.[20] GCAP3 expression seems to be limited to humans[16] and zebrafish,[21] localizing to all cone types in human retinas.[21] Whether individual GCAPs might play distinct roles in phototransduction is not known.

Extensive biochemical analyses have been done to characterize the molecular properties and biological activity of each GCAP. The extent and the mechanism by which GCAPs modulate Ret-GC activity in a bidirectional way in response to [Ca^{2+}] is now better understood, thanks to a thorough mutational approach defining functional domains in both GCAPs and Ret-GC molecules.[22-29] Both GCAP1 and GCAP2 possess three high affinity EF hand motifs that bind Ca^{2+}. They appear to have similar Ca^{2+}-dependence and cooperativity for GC activation by Ca^{2+}, with reported EC50s for the inhibitory effect of Ca^{2+} of ~580 nM for GCAP1 and ~250 nM for GCAP2.[4,5,30,31] Consistent with the ability of GCAPs to activate and inhibit Ret-GCs, Ret-GCs and GCAPs appear to bind to each other independently of [Ca^{2+}].[25,32,33] Therefore, it has been proposed that GCAPs are constitutively bound to Ret-GCs, and that Ca^{2+} binding to the GCAP.Ret-GC complexes induces structural changes that stabilize the inactive state relative to the active state of the Ret-GC catalytic dimer[31,32] (see Fig. 1). Mutations in the Ret-GC1 or GCAP1 genes that have been shown to alter this fine-tuning of Ret-GC activity by Ca^{2+} in vitro have been linked to autosomal dominant cone or cone-rod dystrophies in human patients.[30,31,34-38]

The cloning of the GCAP proteins and the characterization of their gene structure[17,39] have opened up the possibility of developing mouse models to study the normal functions of this Ca^{2+}-regulation of Ret-GCs by GCAPs in the context of

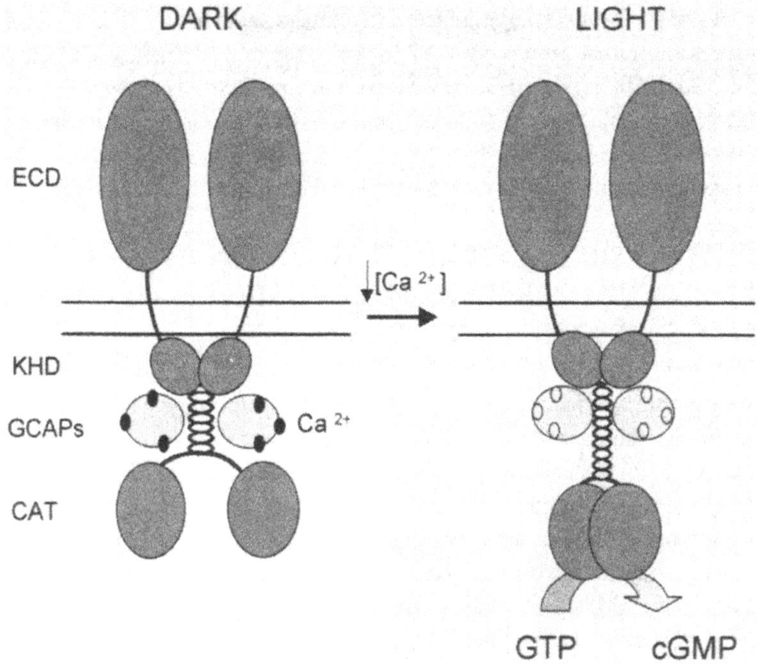

Figure 1. Model showing the regulation of Ret-GCs by GCAPs. Particulate Ret-GCs are functional as dimers. GCAPs are thought to constitutively bind to Ret-GCs, and regulate cyclase activity by structural changes caused by the binding or dissociation of Ca^{2+}. The Ca^{2+}-loaded form of GCAPs (predominant in resting photoreceptors in the dark) would stabilize the inactive conformation. Dissociation of Ca^{2+} from the GCAPs in response to light would promote dimerization of the cyclase catalytic domains, and cGMP synthesis. ECD, extracellular domain; KHD, kinase homology domain; CAT, catalytic domain.

intact, living rods. It has also made it possible to study the mechanism by which malfunctioning GCAPs lead to disease in vivo. This chapter describes new insights gained from mouse models developed to study the functions of the GCAP proteins in vivo. We have divided this chapter into three sections. In the first section we review the characterization of the GCAP1/GCAP2 knockout mice, in which the Ca^{2+}-feedback to Ret-GCs is specifically abolished. We discuss how disrupting this feedback loop affects the kinetics and the sensitivity of the light response in dark-adapted rods and how it affects the ability of rod photoreceptors to adapt to light. In the second section we discuss the effect of restoring GCAP2 in the GCAP1/GCAP2 double knockout mice as a first step toward determining whether GCAP1 and GCAP2 have distinct functions in regulating the light response in rods. Individual restoration of GCAP1 and GCAP2 expression in GCAP1/GCAP2 knockout mice might show functional differences that would only be revealed when Ret-GC regulation is analyzed in intact cells under real time kinetics. The third section of this chapter summarizes the characterization of transgenic mice developed to study the mechanism by which mutant forms of GCAPs with impaired Ca^{2+}-binding (and in vitro altered Ca^{2+}-sensitivity) might lead to retinal disease in vivo. These transgenic mice express a mutant form of GCAP2 that is incapable of binding Ca^{2+}. In vitro characterization of this mutant GCAP2 shows it to constitutively activate Ret-GCs.[23] Because several mutations in the GCAP1 gene disrupt Ca^{2+} binding, these transgenic mice may reveal the molecular basis of the pathology caused by mutations in the GCAP1 gene that have been linked to autosomal dominant cone dystrophy.[34,35,37]

FUNCTIONS OF THE GCAP PROTEINS IN INTACT PHOTORECEPTORS: EFFECTS OF THE Ca^{2+}-FEEDBACK TO Ret-GCs ON THE LIGHT RESPONSE

The vertebrate visual system can operate in a wide range of light intensities due, in part, to the capacity of rod and cone photoreceptors to light adapt. Rod photoreceptors, for instance, are specialized for the detection of dim light. In rods, activation of one rhodopsin molecule by a single photon gives rise to a measurable electrical response.[40] This is due to a huge amplification of the signal in the transduction process, achieved in two successive steps: the activation of hundreds of transducin molecules (and cGMP-PDE catalytic subunits) by a single photoactivated rhodopsin during its lifetime (120 s^{-1} per R*), and the hydrolysis of thousands of cGMP molecules by each activated PDE holomer, whose catalytic efficiency is reflected in a K_{cat}/K_m value that exceeds 10^8 M^{-1} s^{-1}.[41] This extraordinary signal amplification would lead to rod saturation at very dim illumination if the sensitivity of rods to light were fixed. But rods have the capacity to adjust their sensitivity in the presence of a background light, making it smaller as the background light is brighter. By shifting the response-intensity function toward brighter light stimuli, photoreceptors extend the range of light intensities in which they can operate.

It has been known since the late 1980s that light adaptation in vertebrate rods and cones can be abolished by slowing or preventing movements of Ca^{2+} across the plasma membrane.[42,43] These studies established that Ca^{2+} is the major second messenger modulating transduction during light adaptation. Several steps in the transduction cascade are now known to be sensitive to Ca^{2+}, responding to the light-driven decrease in $[Ca^{2+}]$ by either reducing the gain or speeding the termination of the light response. The light-activated PDE activity is reduced in response to a decrease in $[Ca^{2+}]$[44-47] through an effect that may be mediated by recoverin.[48-50] In addition, a decrease in $[Ca^{2+}]$ is thought to cause calmodulin to dissociate from the β subunit of the cGMP-gated channel, increasing the channel's sensitivity for cGMP.[51-53] As mentioned, a decrease in $[Ca^{2+}]$ stimulates cGMP synthesis by Ret-GCs[1,2,54,55] through the action of the GCAPs, speeding restoration of cGMP and recovery of the dark current. Light-driven changes in $[Ca^{2+}]$ could also affect the response kinetics by affecting the phosphorylation state of RGS9-1;[56] or by acting on as yet unidentified targets.

The dependence of several of the transduction steps on Ca^{2+} has made it difficult in the past to assess the individual contributions of these mechanisms to the overall process of adaptation in vivo, where conditions designed to minimize Ca^{2+} dynamics would affect them all. More recently, a quantitative characterization of each of these pathways was undertaken in truncated salamander rods, by manipulation of the internal ionic conditions and introduction of inhibitors by dialysis to dissect the pathway under study.[2,3,46,53] These studies concluded that over most of the range of background light intensities, the most important mechanism regulating the sensitivity adjustment to light was the Ca^{2+}-dependent modulation of the guanylyl cyclase.[3]

In this section we discuss the effect of disrupting the GCAPs mediated Ca^{2+}-feedback loop to Ret-GCs in intact vertebrate photoreceptors. Based on a mouse genetics approach, this analysis presents the advantage that ionic conditions and concentration of soluble components are under true physiological conditions. The analysis of the effect of disrupting the Ca^{2+}-feedback loop to Ret-GCs on the light response of intact, living rods, represents one of the most direct assessments of the functions of the GCAP proteins in vivo.

Development of a Mouse Model to Assess the Contribution of Ca^{2+}-Regulation of Ret-GCs in the Light Response

Given that both GCAP1 and GCAP2 appear to be present in murine rod outer segments[17] and that they behave similarly in vitro in their ability to stimulate Ret-GCs, a genetic approach intended to abolish Ca^{2+}- regulation of Ret-GC activity in rods required targeting the expression of both genes simultaneously. The fact that GCAP1 and GCAP2 genes are arranged in a tail-to-tail array made it possible to disrupt both genes in embryonic stem cells using a single homologous recombination step.[57] The recombination strategy outlined in Figure 2A led to the removal of exons 2 to 4 of GCAP1 and exons 3 and 4 of GCAP2 in the disrupted allele. Heterozygous mice for the disrupted allele showed the corresponding reduction of expression of GCAP1 and GCAP2 in the retina, whereas homozygous knockout mice showed no expression at all (Fig. 2C). Importantly, the expression level of proteins directly involved in cGMP turnover, such as Ret-GC1, Ret-GC2, and PDE catalytic and inhibitory subunits, were unchanged (Fig. 2C). Immunoblot analysis of several other phototransduction proteins and expression profiling of ~ 11000 murine genes and expressed sequence tags similarly showed little, if any changes resulting from the absence of GCAPs. Consistent with the lack of expression changes, the morphology of GCAPs-/- retinas was largely normal (Fig. 2D). Removing GCAP1 and GCAP2 effectively abolished Ca^{2+} regulation of rod Ret-GC activity, as shown by the absence of Ca^{2+}-dependent GC activity throughout the entire physiological range of $[Ca^{2+}]$ in rod outer segments from GCAPs-/- mice (Fig. 2E). These results show that disruption of GCAPs expression by gene targeting was highly specific and did not lead to compensatory changes in the expression of other photoreceptor genes or

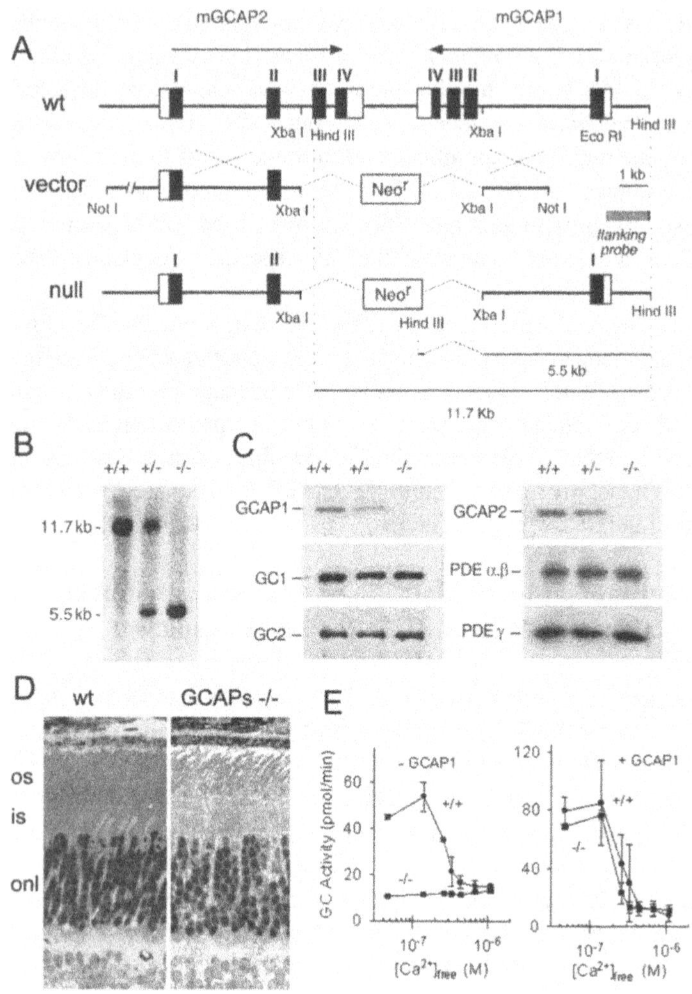

Figure 2. Disruption of GCAP1 and GCAP2 expression in mice. A, Genomic organization of GCAP1 and GCAP2 and targeting strategy. By homologous recombination, a 6.3-kb genomic DNA fragment comprising GCAP1 exons 2 to 4 and GCAP2 exons 3 and 4 is replaced by a neomycin resistance cassette. The resulting recombinant loci can be distinguished from the wildtype loci by Southern hybridization. B, Southern blot analysis of GCAPs+/+, +/-, and -/- mouse genomic DNA. C, Immunoblot analysis of GCAPs+/+, +/-, and -/- retinal homogenates, with polyclonal Abs recognizing the indicated proteins. D, Light micrographs of retinal sections from 3-month-old GCAPs+/+ and -/- mice: os, outer segment; is, inner segment; onl, outer nuclear layer. E, Ca^{2+}-titration of GC activity in rod outer segment (ROS) extracts from wildtype and GCAPs-/- mice in the absence (left) or presence (right) of 3 μM recombinant GCAP1. The left panel shows that the GC activity in GCAPs-/- ROS does not respond to changes in Ca^{2+}. The right panel shows that Ret-GCs were present in similar amounts in wildtype and GCAPs-/- ROS preparations, as assessed by addition of excess recombinant GCAP1. Error bars represent standard errors, from three independent assays. Reprinted from: Mendez A, Burns ME, Sokal I et al. Role of guanylate cyclase-activating proteins (GCAPs) in setting the flash sensitivity of rod photoreceptors. Proc Natl Acad Sci USA 2001; 98: 9948-9953. "Copyright (2001) National Academy of Sciences, USA."

disruption of retinal morphology. These mice therefore constitute an ideal model to study the functions of Ca^{2+}-feedback to Ret-GCs in intact photoreceptors.

Basal Ret-GC Activity in Rods Lacking GCAP1 and GCAP2

One of the first questions addressed in the GCAP1/GCAP2 knockout mice was whether GCAP1 and GCAP2 regulate the basal activity of Ret-GCs in dark-adapted rods. This can be done by comparing the free [cGMP] in wildtype and GCAPs-/- rods, assuming that the dark PDE activity is not affected by the absence of GCAPs. The free [cGMP] in the outer segments is proportional to the outer segment light-sensitive membrane current. This is because the light-sensitive current is co-operatively regulated by free [cGMP], obeying a Hill relation in which the equilibrium between the free cGMP and the channels is established very rapidly, within less than 3 milliseconds.[58,59] Therefore, the instantaneous magnitude of the light-sensitive membrane current reflects the free [cGMP] in the outer segment.

There were no significant differences in the dark current values of GCAPs-/- and wildtype rods (ref. 57, Table 1). Assuming that the PDE basal activity hasn't changed in GCAPs-/- mice, this result suggests that the dark [cGMP] is similar in GCAPs-/- and wildtype rods, implying that both wildtype and GCAPs-/- rods have similar dark (basal) levels of GC activity. Consistent with this result, no significant differences were observed in the total levels of cGMP in wildtype and GCAPs-/- retinas, when total (bound and free) cGMP levels were determined in whole retinas by radioimmunoassay (data not shown).

In conclusion, removal of GCAPs had little effect on the size of the inward dark current (and therefore, basal GC activity) or the physiological levels of cGMP in resting photoreceptors. Because GCAPs have dual actions to activate and inhibit Ret-GCs, there would be a certain value of $[Ca^{2+}]$ for which the GC activity is the same in the presence or absence of GCAPs. This would be ~400 nM for GCAP2[25,60] and ~1 μM for GCAP1.[29] The fact that the basal GC activity is similar in wildtype and GCAPs-/- mice might be explained if at the free $[Ca^{2+}]$ of resting rods some of the GCAP molecules are locally activating Ret-GC activity while others are inhibiting it, so that the steady state activity is unchanged in the presence or the absence of GCAPs.

Effect of GCAPs Deletion on Dark-Adapted Flash Responses

It has been known since the 1980s that slowing the rate of change of $[Ca^{2+}]$ by incorporation of Ca^{2+} buffers such as BAPTA has profound effects on the dark-adapted response of photoreceptors.[61-63] Ca^{2+}-feedback to the transduction cascade severely limits the response to light. But what is the effect of abolishing Ca^{2+} regulation of Ret-GCs in the dark-adapted light response?

Figure 3 shows representative flash families from a GCAPs-/- and a wildtype rod. For dim flashes of comparative strength (see figure legend) GCAPs-/- responses rose for a longer time, to a larger peak amplitude than wildtype responses, and took

Table 1. Characteristics of dark-adapted GCAPs-deficient and wildtype mouse rods

	Dark current (pA)	Flash sensitivity (pA ϕ^{-1} μm^2)	I_o‡ (ϕ μm^{-2})	Single photon response amplitude (pA)	Time to peak (ms)	Integration time (ms)
GCAPs +/+	12.3 ± 0.5 (28)	0.13 ± 0.02 (15)	88.0 ± 7.0 (26)	0.50 ± 0.05 (25)	114 ± 5 (27)	240 ± 14 (27)
GCAPs +/-	12.2 ± 1.0 (20)	0.31 ± 0.06 (18)	46.9 ± 7.4 (20)	0.89 ± 0.16 (18)	254 ± 23 (18)	500 ± 44 (18)
GCAPs -/-	13.3 ± 0.8 (42)	0.78 ± 0.10 (14)	11.3 ± 0.8 (33)	2.31 ± 0.20 (41)	315 ± 13 (42)	589 ± 19 (42)
GCAPs -/- GCAP2+	10.2 ± 0.9 (20)	0.45 ± 0.13 (16)	33.1 ± 6.9 (20)	1.17 ± 0.30 (16)	228 ± 19 (18)	374 ± 36 (18)

Error bars represent SEM. Number of cells is indicated in parentheses.
‡ - Flash strength required to elicit a half-maximal response.
Reprinted from: Mendez A., Burns ME, Sokal I et al. Role of guanylate cyclase-activating proteins (GCAPs) in setting the flash sensitivity of rod photoreceptors. Proc Natl Acad Sci USA 2001; 98: 9948-9953.

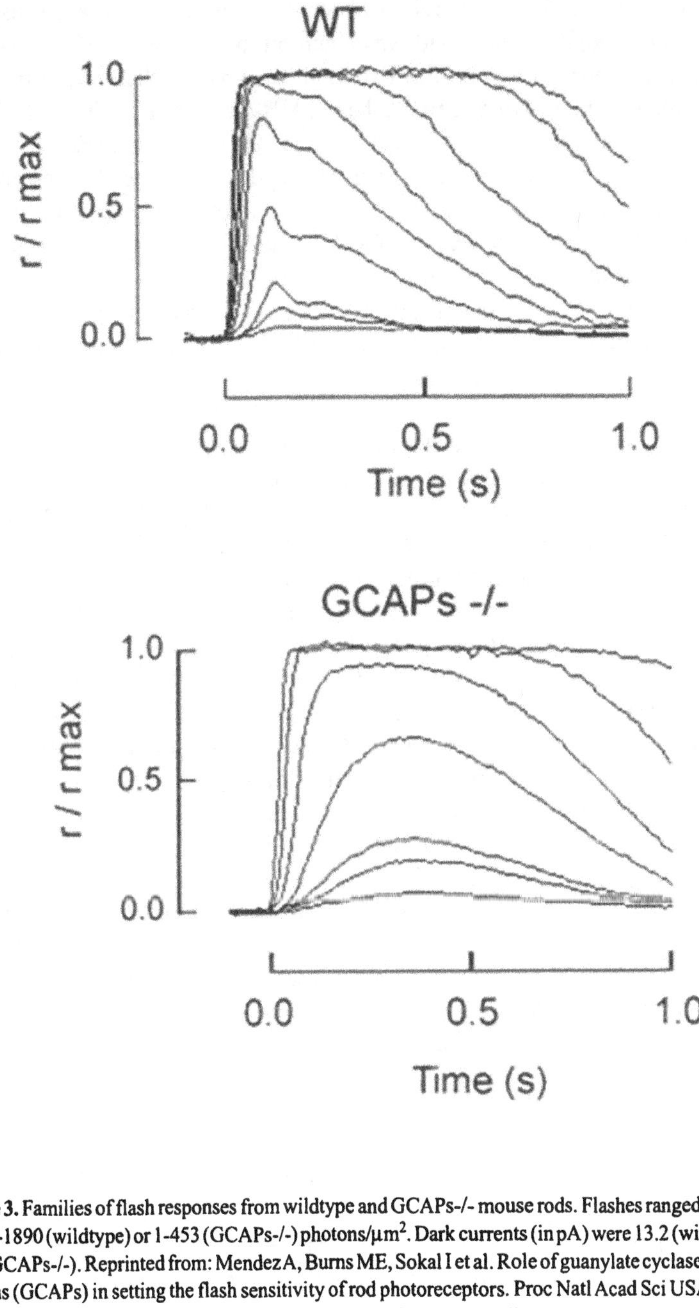

Figure 3. Families of flash responses from wildtype and GCAPs-/- mouse rods. Flashes ranged in strength from 4-1890 (wildtype) or 1-453 (GCAPs-/-) photons/μm^2. Dark currents (in pA) were 13.2 (wildtype) and 14.4 (GCAPs-/-). Reprinted from: Mendez A, Burns ME, Sokal I et al. Role of guanylate cyclase-activating proteins (GCAPs) in setting the flash sensitivity of rod photoreceptors. Proc Natl Acad Sci USA 2001; 98: 9948-9953. "Copyright (2001) National Academy of Sciences, USA."

longer to recover. The flash sensitivity (the amplitude of the linear range response divided by the strength of the flash) was about 6 times higher in GCAPs-/- rods. Another index of sensitivity, the flash strength required to elicit a half-maximal response (Io), was about 8-fold lower in the GCAPs-/- rods (ref. 57, Table 1). These results indicate that Ca^{2+} feedback to Ret-GCs via GCAP1 and/or GCAP2 greatly limits the flash sensitivity of wildtype dark-adapted rods. These changes are also reflected in the single photon response (Fig. 4). Heterozygous knockout rods, expressing about half the normal levels of GCAP1 and GCAP2 (GCAPs+/-) had flash sensitivities and response kinetics between those of knockout and wildtype rods (Fig. 4), indicating that GCAPs are not present in excess in wildtype rod outer segments.

Figure 4 shows that Ca^{2+}-feedback to Ret-GCs is sensed at the single photon level. This fact implies that the local closure of cGMP-channels triggered by photoactivation of a single rhodopsin molecule is enough to cause a local drop in [Ca^{2+}] that is sensed by Ret-GCs during the time-course of the single photon response (on the order of a few hundred milliseconds). The speed of this Ca^{2+}-feedback loop might derive from the close proximity of the cGMP-gated channels, the Na^+/Ca^{2+},K+ exchanger and the Ret-GCs at the rim of the discs. These proteins may be assembled in a macromolecular complex that may also include cGMP-PDE and the ATP-binding cassette transporter (ABCR) by interactions with the multivalent glutamic-acid-rich proteins (GARPs).[64] Fluctuations of [Ca^{2+}] would be greatest near the plasma membrane at the site of Ca^{2+} entry and extrusion. By rapidly responding to local changes in [Ca^{2+}] (through a high Hill coefficient[1] and rapid conformational change[65]) GCAPs transmit a local drop in [Ca^{2+}] during the photoresponse to a quick and robust stimulation of Ret-GC activity that limits the amplitude and shortens the duration of the light response.

Interestingly, the Ca^{2+} feedback to Ret-GCs that boosts cGMP synthesis imposes a constraint on the amplitude of the light response even in mouse models with defects in other biochemical processes governing inactivation. The loss of rhodopsin phosphorylation by RK in several mouse models,[66-68] for instance, results in constant rhodopsin's catalytic activity, but causes only a 2-3 fold increase in single photon response amplitude, in contrast to the 5-fold increase seen in GCAPs-/- mice. These results indicate that the amplitude of the responses in all these mouse models is limited by replenishment of cGMP by stimulated Ret-GCs, and stress the strength of the feedback loop mediated by the GCAP proteins.

Effect of GCAPs Deletion on Light Adaptation of Rod Photoreceptors

As mentioned above, [Ca^{2+}] fluctuations are sensed at several enzymatic steps in the phototransduction cascade, and each of these steps could potentially contribute to the photoreceptor's ability to light adapt. Therefore, a complete understanding of the molecular mechanisms behind this important photoreceptor cell attribute requires a quantitative assessment of each of these steps. This can only be accomplished in a system where each of the pathways can be specifically disrupted without affecting Ca^{2+}-feedback to the other enzymatic reactions. As we have demonstrated,

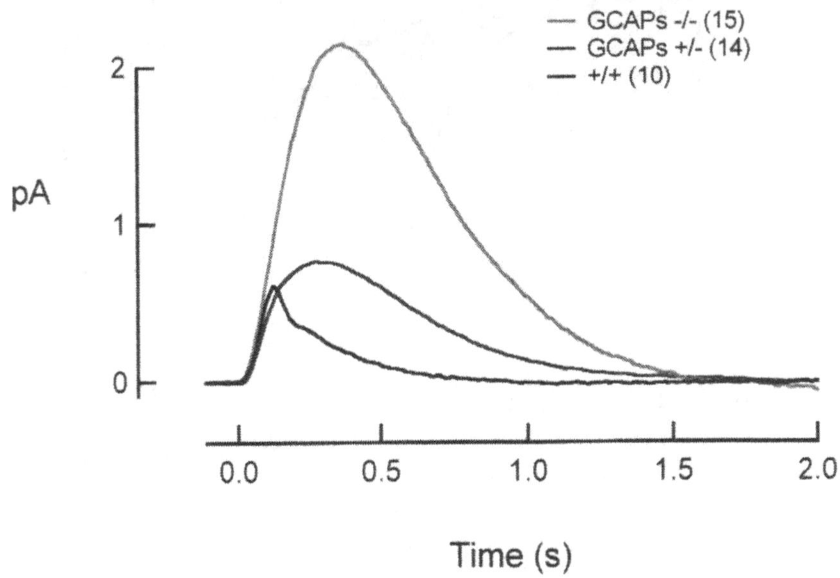

Figure 4. Mean single photon responses from GCAPs+/+, +/- and -/- rods. Flashes were delivered at t=0. Reprinted from: Mendez A, Burns ME, Sokal I et al. Role of guanylate cyclase-activating proteins (GCAPs) in setting the flash sensitivity of rod photoreceptors. Proc Natl Acad Sci USA 2001; 98: 9948-9953. "Copyright (2001) National Academy of Sciences, USA."

the GCAPs-/- rod fulfills this requirement. Measurements of flash sensitivities of wildtype and GCAPs-/- rods in the presence of background lights of increasing intensities showed the gradual desensitization obeying the Weber-Fechner law for wildtype rods (filled symbols, Fig. 5), but a much steeper decline of sensitivity in GCAPs-/- rods (open symbols, Fig. 5). However, adaptation was not completely absent in GCAPs-/- rods. The dependence of incremental sensitivity on background intensity for a GCAPs-/- rod lacking any adaptation to background light was theoretically predicted by assuming an exponential saturation relation and making use of the GCAPs-/- dark-adapted response parameters (thin trace, Fig. 5). A comparison between the experimental results and this theoretical prediction indicated that GCAPs stimulation of GC activity accounts for only one half of the factor by which the rods' operating range of light intensities is extended. The remainder is mediated by other mechanisms (discussed in ref. 57).

Thus, Ret-GCs activation by GCAPs in wildtype rods served to increase the incremental flash sensitivity in the presence of bright steady light by restoring current to rods that would otherwise have saturated. It is interesting to note that rod photoreceptors, specialized for the detection of dim illumination and extremely sensitive to light, could be considerably more sensitive to light in the absence of the GCAP proteins. However, in the absence of these proteins the range of light intensities in which the rods could operate would be severely limited.

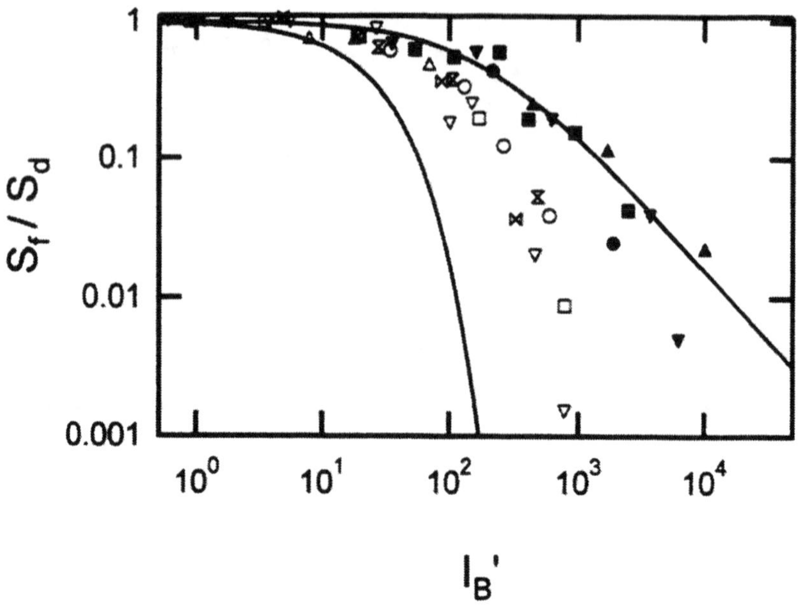

Figure 5. Relative incremental flash sensitivity as a function of normalized background light intensity in wildtype (solid symbols) and GCAPs-/- rods (open symbols). Bold curve fitted to the wildtype data is the Weber-Fechner relation; the thin curve is the relation expected if the superposition of the elementary responses of GCAPs-/- rods exponentially saturated the cell (see text). Reprinted from: Mendez A, Burns ME, Sokal I et al. Role of guanylate cyclase-activating proteins (GCAPs) in setting the flash sensitivity of rod photoreceptors. Proc Natl Acad Sci USA 2001; 98: 9948-9953. "Copyright (2001) National Academy of Sciences, USA."

EFFECT OF RESTORING GCAP2 IN GCAPs-/- PHOTORECEPTORS

Introduction

While the mechanism of modulation of Ret-GCs by GCAPs is now better understood, less is known about the significance of multiple Ret-GCs and GCAPs. In vitro, GCAPs display some preferences toward Ret-GCs. For example, GCAP1 stimulates Ret-GC1 more efficiently than Ret-GC2, while GCAP2 and GCAP3 effectively stimulate both Ret-GC1 and Ret-GC2. Ret-GC1 and GCAP1 appear to be important for cone and rod function in humans. Mutations in the Ret-GC1 gene have been linked to Leber's congenital amaurosis type I[13] and to autosomal dominant cone-rod dystrophy (adCORD),[69] and defects in the GCAP1 gene have been linked to adCORD.[34,35,37] A naturally occurring null mutation in Ret-GC1 in chickens, which have cone-dominated retinas, leads to the loss of cones and rods.[70] However, disruption of Ret-GC1 expression in mice leads to the degeneration of cones primarily,[15] indicating that Ret-GC2 may substitute for the loss of Ret-GC1 in murine rods.

It is not clear why photoreceptors express two GCAP proteins with very similar biochemical behavior. The ratio of GCAP1 to GCAP2 in bovine retinas has been estimated at between 3:1 and 4:1 (reference 18, and references herein), and the GC activity stimulation attributable to GCAP2 has been estimated at about 30%.[18] Based on this data, the question was raised of whether GCAP2 would have a physiological role in phototransduction.[18] The compartmentalization of GCAP2 differs from that of GCAP1 in rods and cones of retinas from higher species.[18-20] GCAP1 appears to be more abundantly expressed in the cone outer segments of human, monkey and bovine retinas, whereas GCAP2 immunoreactivity, even if weaker in the outer segments, is seen in the whole region of rods and cones.[20]

As a starting point to address functional differences and redundancy of GCAPs in the photoreceptors, we generated a transgenic mouse line that expressed GCAP2 in the GCAPs-/- background.

GCAP2 is Functional in Rods and Partially Rescues the GCAP1/ GCAP2 Knockout Phenotype

GCAP2 is first expressed in murine retinas at postnatal day 9, paralleling rod opsin expression (Fig. 6). Transgenic mice expressing GCAP2 in rod photoreceptors were generated by expressing the bovine GCAP2 cDNA under the control of the rod opsin promoter (Fig. 7A). Several lines were obtained, and a line in which heterologous GCAP2 expression was about 2-fold higher than endogenous GCAP2 expression was selected for the study (Fig. 7B, see figure legend). This line was bred to GCAPs-/- mice to establish a line in which GCAP2 was expressed in rod photoreceptors in the absence of GCAP1. Heterologous GCAP2 in the GCAPs-/-GCAP2+ line showed the same pattern of localization as endogenous GCAP2 in wildtype retinas (Fig. 7C); and in vitro GC assays using retinal homogenates showed that heterologous GCAP2 fully restored stimulation of Ret-GC activity at low [Ca^{2+}] (Fig. 7D).

GCAPs-/-GCAP2+ transgenic mice were analyzed by single cell recordings. Although GCAP2 expression seemed to vary from cell to cell (see reference 57, summarized here in Fig. 8 legend), the mean single photon response from GCAPs-/-GCAP2+ rods was between those of wildtype and GCAPs-/- responses (Fig. 8A). This result indicates that GCAP2 expression alone partially restores the dark flash sensitivity in GCAPs-/- rods. Furthermore, a correlation between the dark flash sensitivity and the time course of the decline of the Na^+/Ca^{2+}, K^+ exchange current (thought to reflect the level of expression of GCAP2 in individual cells, see ref. 57), could be established for individual rods. In general, cells with time constants of the Na^+/Ca^{2+},K+ exchanger close to that of wildtype rods (and presumably high GCAP2 expression) had nearly normal dark flash sensitivities. Cells with briefer exchanger time constants (and presumably lower GCAP2 expression) had higher flash

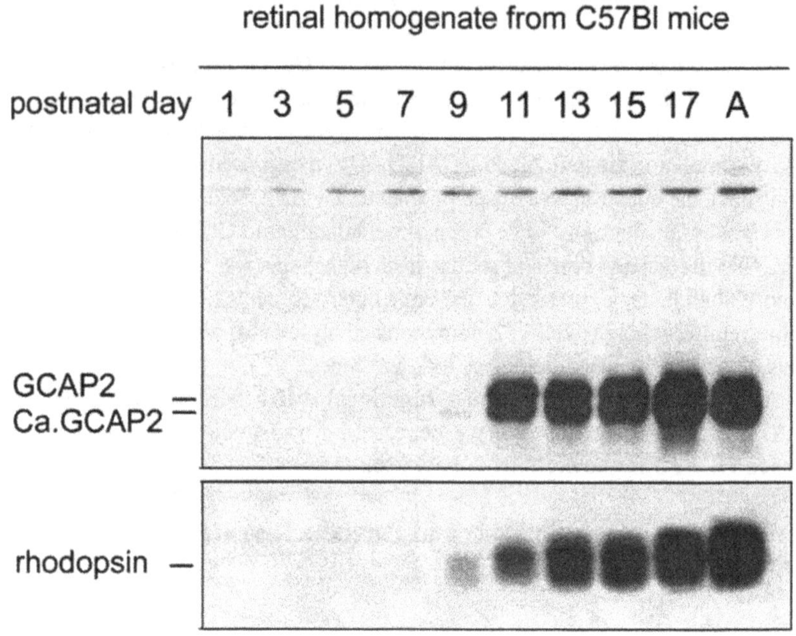

Figure 6. Analysis of GCAP2 expression in the retina during postnatal development. Immunoblot analysis of retinal homogenates from C57Bl mice at 1, 3, 5, 7, 9, 11, 13, 15, 17 and 30 (A, adult) postnatal days, with a polyclonal Ab against GCAP2 (p24ΔN, ref. 5) (top panel); or anti-rhodopsin mAb 1D4 (bottom panel). Retinal homogenates were standardized per total protein, so that equal amounts of retinal protein were loaded per lane. GCAP2 expression is detected at postnatal day 9, paralleling rhodopsin expression.

sensitivities, approaching those of GCAPs-/- (Fig. 8B). These results indicate that GCAP2 has the capacity to stimulate GC activity in vivo.

In addition, responses from GCAPs-/-GCAP2+ rods to saturating flashes showed that GCAP2 alone can maximally stimulate GC activity. After a bright flash, in which the [Ca^{2+}] drops to the minimum level, the time at which a rod comes out of saturation is determined by the light-triggered PDE activity (assumed to be the same in wildtype, GCAPs-/- and GCAPs-/-GCAP2+ rods for a flash of the same strength), and the maximal cyclase activity. The similarity of wildtype and GCAPs-/-GCAP2+ responses to bright flashes indicated that GCAP2 alone was capable of producing maximal GC activation (Fig. 8C).

Taken together, these results show that GCAP2 is active in rod outer segments, partially rescuing the GCAPs-/- phenotype, and that GCAP2 alone is capable of maximally activating Ret-GC activity under low intracellular [Ca^{2+}].

Although GCAP2 was capable of maximally activating Ret-GC activity and restoring sensitivity, it was not sufficient to restore the normal kinetics of the photoresponse. Responses to subsaturating flashes lacked the rapid initial recovery that follows the peak of the response in wildtype rods.[57] This result suggests that GCAP1 is responsible for the characteristic rapid recovery, indicating that GCAP1

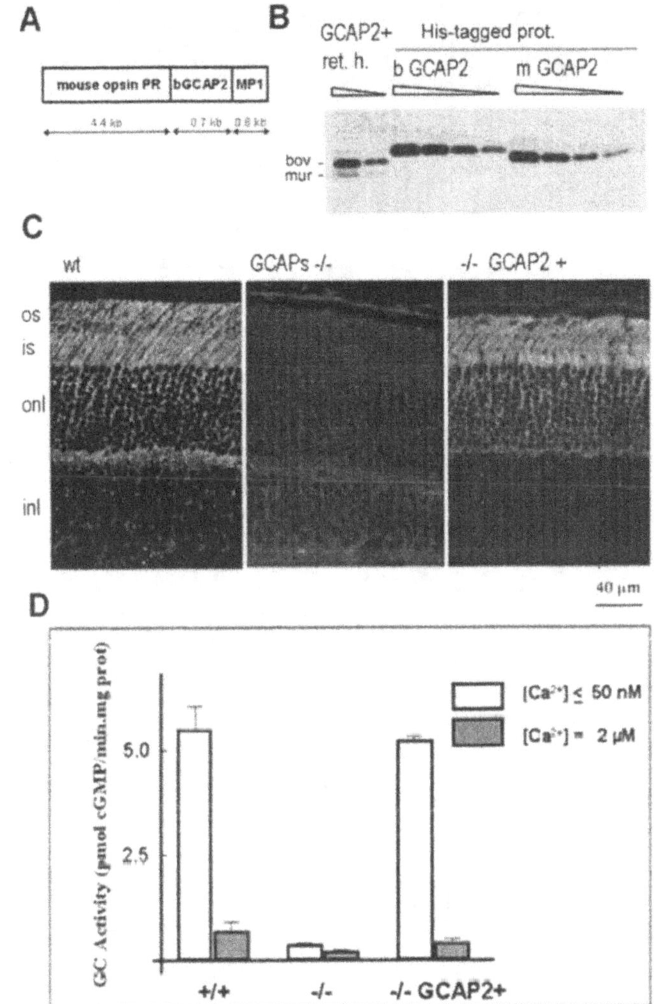

Figure 7. Restoration of GCAP2 expression in GCAPs-/- retinas. A, Construct used to generate transgenic mice expressing bovine GCAP2 in the retina. A 5.7-kb fusion gene was obtained by assembling the 4.4-kb mouse opsin promoter, the 0.7-kb bovine GCAP2 cDNA, and the mouse protamine 1 polyadenylation signal. B, Estimation of the transgene expression level by immunoblot. Fractions corresponding to 1/40 and 1/80 of a retina from mice expressing both the endogenous mGCAP2 and heterologous bGCAP2 (3 aa different in size) were separated by SDS/PAGE (lanes 1 and 2) and compared to fixed amounts of bovine and murine recombinant His-tagged GCAP2 (120 to 15 ng, two-fold dilutions). Proteins were detected with polyclonal Ab p24ΔN, raised against the bovine form of GCAP2.[5] Heterologous GCAP2 expression was estimated to be 2-fold higher than the endogenous GCAP2. C, Immunolocalization of bGCAP2 in the GCAPs-/-GCAP2+ retina. D, GC activity in retinal homogenates from wildtype, GCAPs-/- and GCAPs-/-GCAP2+ mice at low (< 50 nM) or high (2 μM) free [Ca^{2+}]. Values are the average of three independent experiments. Error bars represent the standard deviation. Reprinted from: Mendez A, Burns ME, Sokal I et al. Role of guanylate cyclase-activating proteins (GCAPs) in setting the flash sensitivity of rod photoreceptors. Proc Natl Acad Sci USA 2001; 98: 9948-9953. "Copyright (2001) National Academy of Sciences, USA."

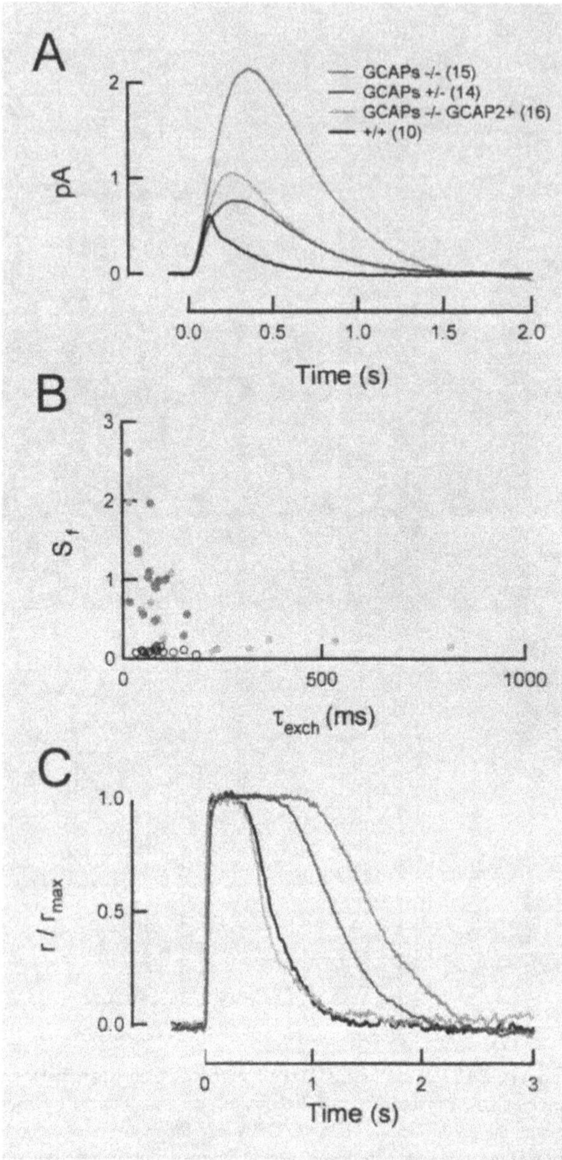

Figure 8. GCAP2 activates GC activity and partially rescues the GCAPs-/- phenotype. A. Mean single photon response of GCAPs-/-GCAP2+ rods compared to GCAPs+/+, +/- and -/- responses. B, Dark-adapted flash sensitivity ($S_F{}^D$) and the time constant of the Na+/Ca2+, K+ exchange current in wildtype, GCAPs-/ - and GCAPs-/-GCAP2+ rods. The rate of Ca^{2+} extrusion in GCAPs-/-GCAP2+ rods was highly variable, suggesting that the expression of GCAP2 varied widely in different cells. C, Representative responses of wildtype and GCAPs-deficient rods to a bright flash (560 photons/μm²). Traces are averages of two responses. Dark currents (in pA) were 13.1 (wildtype), 12.4 (GCAPs+/-), 13.3 (GCAPs-/-), and 10.8 (GCAPs-/-GCAP2+). The time constant of Na+/Ca2+, K+ exchange in this GCAPs-/-GCAP2+ rod was slightly longer than that of wildtype rods. Reprinted from: Mendez A, Burns ME, Sokal I et al. Role of guanylate cyclase-activating proteins (GCAPs) in setting the flash sensitivity of rod photoreceptors. Proc Natl Acad Sci USA 2001; 98: 9948-9953. "Copyright (2001) National Academy of Sciences, USA."

might act faster in stimulating Ret-GCs. This functional difference might result from differences in the rates of Ca^{2+} binding and dissociation to each GCAP. The rate of Ca^{2+} release by GCAP1 is very rapid (k_{-1} = 72 s^{-1}; ref. 65) and would be consistent with an early role of GCAP1 in the recovery phase of the photoresponse, while the Ca^{2+} dissociation rate for GCAP2 has not yet been determined. In addition, as the EC50s for the inhibitory effect of Ca^{2+} were estimated to be ~580 nM for GCAP1 and ~250 nM for GCAP2, GCAP1 would be expected to sense a decrease in [Ca^{2+}] earlier than GCAP2. Alternatively, this difference in kinetics may be imparted by the topology of GCAPs and Ret-GCs near the plasma membrane where [Ca^{2+}] changes first occur. As mentioned above, the proximity of cGMP-gated channels, the Na^+/Ca^{2+},K+ exchanger and Ret-GCs at the rim of the disc near the plasma membrane may play a role in the rapid feedback to resynthesize cGMP. In this regard, the concentration of each GCAP near this complex would affect the speed of the feedback. Future experiments will be needed to determine the mechanisms that underlie the differential effects of GCAP1 and GCAP2 on the time course of the flash response.

GCAPs AND DISEASE: MOUSE MODELS EXPRESSING FORMS OF GCAPS WITH IMPAIRED Ca^{2+}-BINDING PROPERTIES

Three mutations in the gene encoding GCAP1 have been linked to autosomal dominant cone dystrophy, a disease characterized by the loss of color vision and central visual acuity with retention of peripheral sight.[34,35,37] The first mutation characterized, Y99C,[34] impairs Ca^{2+} binding at EF3 and has been shown to alter Ca^{2+} sensitivity of GCAP1 in in vitro assays. The mutant protein causes normal stimulation of Ret-GC1 at low [Ca^{2+}], but this stimulation persists as the [Ca^{2+}] increases even above 1 μM,[30,38] leading to constitutive activation of the cyclase over the physiological range of [Ca^{2+}]. Another mutation, causing an E155G substitution within the EF4 domain, was shown to alter Ca^{2+} sensitivity of GCAP1 in the same way.[37] Both Y99C-GCAP1 and E155G-GCAP1 stimulate Ret-GCs constitutively in the presence of wildtype GCAP1, consistent with the dominant phenotype of the disease. A model has been proposed in which constitutive synthesis of cGMP in vivo would increase the steady-state level of cGMP and lead to cell death. Increased cGMP levels have been associated with cell death in a number of animal models with mutations in the genes encoding cGMP-PDE subunits.[71-73]

Consistent with this interpretation, substitutions at Arg838 in Ret-GC1 that have been associated with autosomal dominant cone rod dystrophy (adCORD) result in constitutive synthesis of cGMP in vitro, even at the highest physiological [Ca^{2+}] concentrations.[31,36] As in the case of GCAP1 mutations, the Arg838-substituted RetGC-1 mutants exhibit a dominant biochemical phenotype that would explain the characteristics of the disease. Arg838 was proposed to be critical for maintaining the normal structure of the α-helical coiled-coil domain, which could regulate the strength of the dimerization of the catalytic domains and consequently the Ca^{2+}-sensitivity.[31]

Despite the logic of the predicted association between elevated levels of cGMP and photoreceptor cell death in these cone or cone-rod dystrophies, some questions remain. A third mutation in GCAP1, leading to a P50L substitution, exhibits a marked variability in phenotype, ranging from mild abnormalities of macular function to severe cone-rod dystrophy.[35] Contrary to the other mutants, however, P50L-GCAP1 activates Ret-GC with the same Ca^{2+}-sensitivity as the wildtype protein in vitro.[74,75] Interestingly, P50L-GCAP1 showed a reduced thermal stability when compared to wildtype GCAP1, observed both by circular dicroism and by in vitro GC assays.[75] Expression of a thermally unstable GCAP1 in human photoreceptors, therefore, may be the underlying cause of this dominantly inherited disease.

To investigate how mutations affecting Ca^{2+} binding in GCAPs may lead to retinal degeneration in vivo, a mutant form of GCAP2 with inactivated EF hands (E80Q/E116Q/D158N; or $EF(2;3;4)^-$ GCAP2, ref. 23), which leads to constitutive activation of Ret-GCs in vitro over the whole physiological range of $[Ca^{2+}]$,[23] was expressed in the photoreceptor cells of transgenic mice. Because this GCAP2 mutant shows a biochemical behavior similar to that of Y99C-GCAP1,[30,38] this transgenic animal model is expected to show a similar phenotype to the human disease.

GCAP2(EF2;3;4)⁻ Expression in Transgenic Mice Causes Retinal Dysplasia and Retinal Degeneration

Four independent transgenic lines were established that express the bovine GCAP2(E80Q/E116Q/D158N) cDNA under the control of the mouse rod opsin promoter. Two of the lines expressed the mutant protein to higher than the endogenous GCAP2 levels (lines A and B), whereas the other two expressed mutant GCAP2 to lower than the endogenous levels (lines C and D) (A. Mendez, A. Dizhoor and J. Chen, unpublished results). Lines could be ordered by the level of expression of the transgene as follows: line B > line A > lines C and D.

Transgenic mice from lines A and B showed a progressive retinal degeneration, the time course and severity of which correlated with the level of expression of the transgene (Fig. 9). Mice from line B, the line with the highest level of expression, manifested a retinal dysplasia as early as 20d of age. Initially, a loss of the integrity of the outer limiting membrane was seen, with occasional invaginations of the outer nuclear layer into the inner retina (Fig. 10B). At this early age, however, retinas maintained the 10-12 normal rows of nuclei at the outer nuclear layer, and about the normal rod outer segment and inner segment length. By 30d of age, disorganization and shortening of the outer segments was apparent, and the outer nuclear layer showed whirls and rosettes over the whole retinal section in some cases. By 2.5 months of age, the outer nuclear later was reduced to 4-5 rows of nuclei (Fig. 9, line B). Mice from line A, which expressed the mutant form of GCAP2 to approximately four times the endogenous GCAP2 levels, showed a slower progression of the disease. Disorganization of the outer limiting membrane and occasional rosette formation

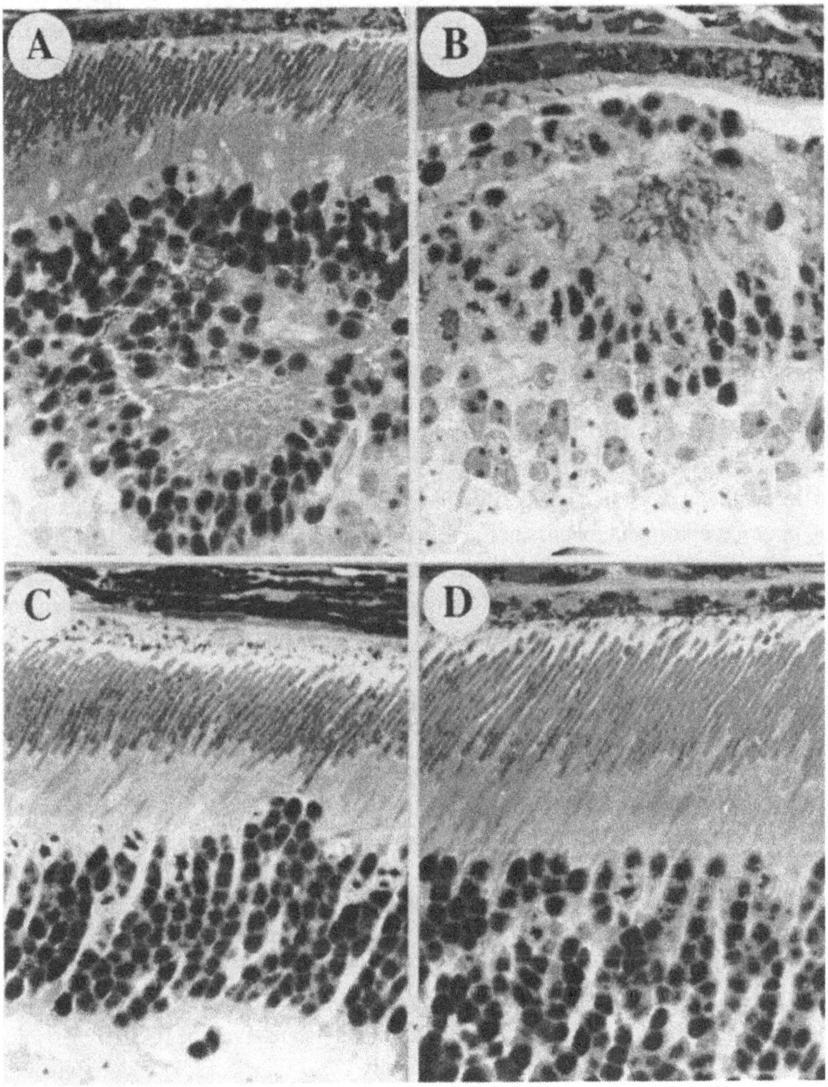

Figure 9. Retinal degeneration in EF(2;3;4)⁻GCAP2 transgenic mice. The bovine E80Q/E116Q/D158N GCAP2 cDNA was expressed under the control of the mouse opsin promoter, and four transgenic lines were obtained that differed in the levels of transgene expression as follows: Line B > Line A > Lines C & D. The retinal degeneration in these lines correlated with the level of expression of the transgene, as shown in the light micrographs of 1-μm retinal sections from 2.5-month-old mice reared in cyclic light. A, B, C & D are transgene-positive mouse retinas from lines A, B, C & D, respectively.

were seen early in time in some retinal sections, but the outer nuclear layer thickness was still 8-9 rows by 2.5 months of age (Fig 9). Mice from lines C and D showed normal retinal morphology up to one year of age.

In contrast to bGCAP2(E80Q/E116Q/D158N), expression of wildtype bGCAP2 to two-fold the endogenous levels did not cause retinal degeneration or signs of rosette formation in a control line of transgenic mice (data not shown), suggesting that the degeneration observed in mice expressing bGCAP2(E80Q/E116Q/D158N) most probably results from the physical properties of this particular mutant form of GCAP2 that does not bind Ca^{2+}.

The Severity of Retinal Degeneration Correlates with Transgene Expression, but not with the Levels of Total cGMP

Historically, the hypothesis that elevated levels of cGMP are toxic for the photoreceptor cell originated from the rd/rd mouse model, in which autosomal recessive mutations in the gene encoding the beta subunit of cGMP-PDE result in massive photoreceptor cell death.[72,76,77] In rd/rd mouse retinas, the deficiency of cGMP-PDE activity results in 4-5 -fold higher total cGMP levels than normal at 14-15 postnatal days. This peak of [cGMP] preceeds the fast degeneration of all the rod photoreceptors that occurs in the third postnatal week.[77,78] Other animal models of retinal degeneration with imbalances in cGMP metabolism, due to a deficiency in cGMP-PDE (e.g., the Irish Setter dog, ref. 71) or in Ret-GC1 (the rd/rd chicken, ref. 70) show either abnormally high or abnormally low total [cGMP] in the retina. In these animal models deregulation of cGMP metabolism is reflected in changes in the total [cGMP] in the photoreceptors that preceeds photoreceptor cell death.

If the loss of photoreceptor cells in transgenic mouse retinas expressing bGCAP2(E80Q/E116Q/D158N) was due to constitutive activation of the cyclase in rods, one reasonable expectation would be that an increase in the levels of total cGMP might precede the onset of the degeneration. Another predictable outcome would be that the time course and severity of the degeneration should be affected by the lighting conditions in which the mice are reared, accelerating or worsening when mice are raised in constant darkness, and slowing down when mice are exposed to constant light. This latter prediction is based on the fact that PDE activity increases with background light intensity. Therefore, constitutive activation of the cyclase should translate to the higher increase in the steady-state levels of cGMP when mice are raised under constant darkness, when only basal PDE activity is counteracting cGMP synthesis. However, in the presence of background light, PDE activity can be expected to overcome the cyclase activity.

The first prediction was tested by measuring total cGMP in freshly dissected retinas by radioimmunoassay (RIA), at time points that precede the retinal degeneration in each transgenic line. For mice from line B (showing the earliest and more severe phenotype), the selected time points were 20 and 30 postnatal days, as in this time window the disorganization of the outer nuclear layer becomes apparent (Fig. 10B), and loss of outer nuclei rows begins to be measurable (Fig. 10A). Mice from line B were raised under constant darkness and sacrificed at 20 or 30 postnatal days. Whole litters were processed at each timepoint, so that transgene-positive mice could be directly compared to transgene-negative littermate controls. One eye of each

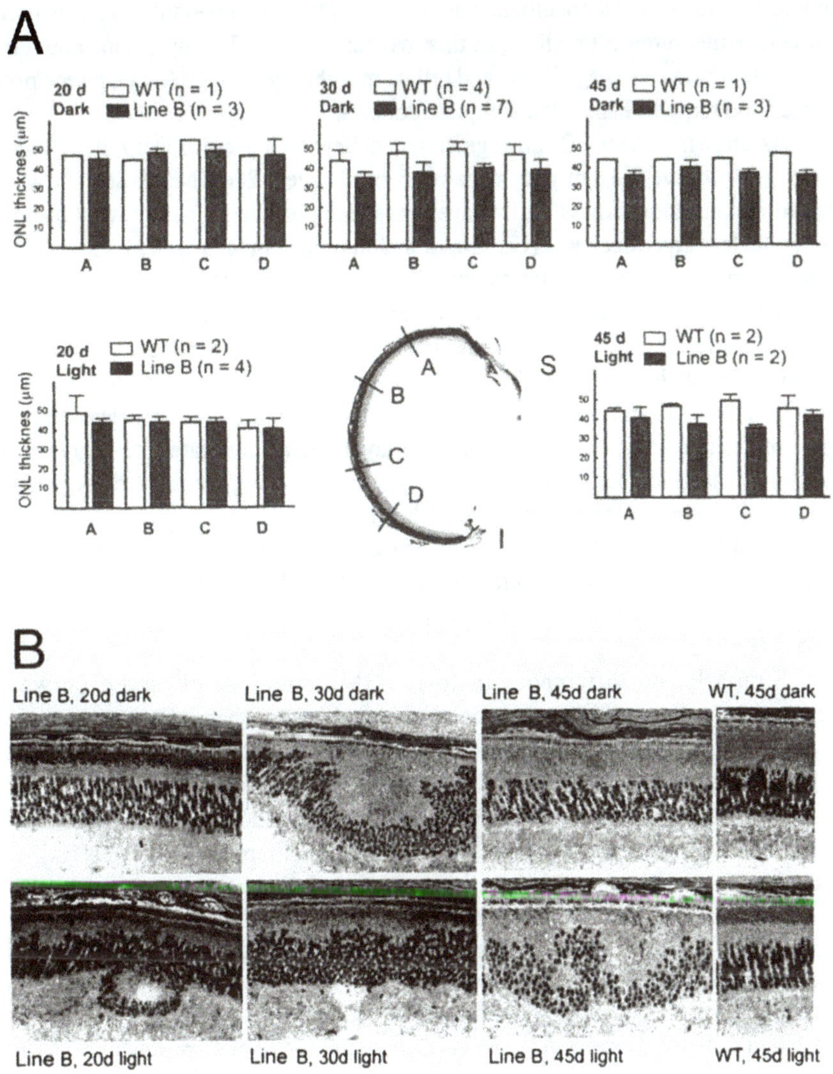

Figure 10. Changes in retinal morphology in transgenic mice from line B at 20, 30 and 45 postnatal days, when raised in constant darkness or constant light (1500 lux). A. The outer nuclear layer thickness was measured in transgene-positive-mouse retinas and littermate control retinas at four distinct regions: A, B, C and D, as shown schematically. For morphometry, the Slide Book 3.0 software (Intelligent Imaging Innovations, Inc., Denver, CO) was used. S, superior; I, inferior. The number of measurements is indicated in parentheses for each histogram. Error bars represent standard deviation, except for determinations with n=2, in which they show the range of the measurements. B, Representative retinal sections of transgene-positive mice from line B, raised in constant darkness or under constant light exposure, at either 20, 30 or 45 days of age. Retinal dysplasia was a common feature whether the mice were raised in darkness or constant light. Retinal sections from wildtype mice at 45d are shown as a control.

mouse was fixed for morphological analysis, and the corresponding eye was used for cGMP measurement by RIA. In this manner, individual comparisons could be established between cGMP levels and retinal morphology. For RIA, mice were processed and samples handled in complete darkness.

As shown in Table 2, no significant differences were observed between transgene-positive and transgene-negative control mice at 20d when these mice were raised in constant darkness. At this age, transgene positive retinas showed normal outer nuclear layer thickness (Fig. 10A), and inner and outer segment length (data not shown), indicating that retinas were not degenerated. However, it should be pointed out that disorganization of the outer limiting membrane and rosette formation were already incipient in some retinas. These signs of the pathology, shown in Figure 10B in the 20d light-reared-mouse retina, were present in some of the transgenic retinas and absent in others within the same litter. Nevertheless, the degree of disorganization of the outer nuclear layer did not correlate with higher total cGMP levels in the corresponding eye (e.g., mice showing the more disorganized retinas did not show the higher cGMP values).

At 30 days, the levels of total cGMP were actually lower in the transgene-positive mice than in their littermate controls, most probably reflecting the loss of a percentage of rod photoreceptors (see measurement of rod outer nuclear layer thickness at 30d, Fig.10A).

Similarly, no differences in total cGMP could be observed between transgene-positive and littermate control mice of line A, when mice were raised in cyclic light for 2 months, and dark-adapted for 20d prior to the experiment (Table 2); nor in mice from lines C or D at later ages (data not shown).

These data show that even when the experimental conditions were optimized to identify changes in the levels of total cGMP, no major differences in the total [cGMP] could be detected in transgene-positive versus control mouse retinas.

The second prediction was tested by raising mice from line B under constant light exposure (1,500 lux), and analyzing their retinal morphology at 20, 30, and 45 postnatal days. The outer nuclear layer thickness was measured at four points along the retinal section. No difference was observed between transgene positive mice raised in constant darkness or raised in constant light at 20 or 45 postnatal days (Fig. 10A). In addition, no differences in retinal morphology were observed when 2-month cyclic-light-reared mice from line A were placed for 20d in constant light, or under constant light exposure (data not shown).

Taken together, these results show that overexpression of bGCAP2(E80Q/E116Q/D158N) in transgenic mice did not lead to major changes in the retinal content of cGMP, or differences in retinal morphology depending on the light-rearing conditions.

Several scenarios would explain these results. First, that bGCAP2(E80Q/E116Q/D158N) causes constitutive activation of Ret-GC activity resulting in the 5-12 fold increase in cGMP synthesis in the dark state as predicted from in vitro studies; but that the cGMP hydrolysis rate by basal PDE is accelerated to adjust to the new darkness equilibrium. In this case, a change in the cGMP turnover rate wouldn't

Table 2. Total cGMP in whole retinas from mice expressing EF (2;3;4)$^-$ GCAP2

Sample	Total cGMP* (fmol/μg prot)
EF(2;3;4)$^-$GCAP2 Line B 20d D	308.94 ± 47.29 (n=3)
wildtype littermate controls	290.41 ± 55.08 (n=3)
EF(2;3;4)$^-$GCAP2 Line B 30d D	285.5 ± 54.18 (n=3)
wildtype littermate controls	390.64 ± 34.56 (n=2)
EF(2;3;4)$^-$GCAP2 Line B 20d L	39.08 ± 17.88 (n=3)
wildtype littermate controls	27.30 ± 8.9 (n=3)
EF(2;3;4)$^-$GCAP2 Line B 30d L	44.18 ± 16.47 (n=3)
wildtype littermate controls	41.79 ± 25.12 (n=3)
EF(2;3;4)$^-$GCAP2 Line A, 2m CL & 20d D	282.6 ± 34.54 (n=4)
wildtype littermate controls	299.7 ± 22.09 (n=4)

* Transgene positive mice and littermate controls were processed in complete darkness, under an infrared setting. For RIA, the TRK500 cyclic GMP [^{3}H] assay system from Amersham (UK) was used. The number of determinations per litter is indicated in parenthesis. Standard deviation is indicated for n > 3, and the range of measurements for n = 2. d = day; D = dark; L = light; CL = cyclic light.

necessarily be reflected in bulk [cGMP], but could result in small changes in free [cGMP]. Since the light-sensitive current is cooperatively regulated by free [cGMP] with a Hill coefficient of ~3, a small increase in free [cGMP] would result in a much bigger difference in the light-sensitive conductance and therefore in Ca^{2+} influx, which could be toxic for the cell (see refs. 79 and 80 , and references herein).

A second possibility is that bGCAP2(E80Q/E116Q/D158N) is affecting some other cellular processes unrelated to cGMP synthesis, e.g., cell adhesion at the outer limiting membrane, by interacting with an as yet unidentified target. Finally, the possibility cannot be excluded that bGCAP2(E80Q/E116Q/D158N) in transgenic lines kills the cells nonspecifically due to an overexpression artifact.

Future experiments will be oriented toward the expression of this mutant form of GCAP2 in the GCAPs-/- background to address whether these mutant proteins localize to rod outer segments, and whether they are active in vivo. The question remains open, therefore, as to whether similar mutations affecting Ca^{2+} binding in the GCAP1 protein cause the loss of cones through constitutive activation of Ret-GC, or whether other mechanisms are involved.

CONCLUSION

In vivo, the GCAP proteins mediate a Ca^{2+}-feedback loop to Ret-GCs that greatly limits the amplitude and duration of the photoresponse of dark-adapted rods, but also serves to extend the rod's operational range of light intensities during background light adaptation. The biophysical properties of GCAPs (their k_{on} and k_{off} for

Ca^{2+}, their Ca^{2+}-sensitivity and the cooperativity of GC activation by Ca^{2+}) confer speed and robustness to this feedback loop. During the photoresponse, when there is a decrease in the free [cGMP] and some cGMP-channels close, the GCAPs sense the resulting decrease in [Ca^{2+}] very quickly, transducing it to a robust stimulation of Ret-GC activity that restores free cGMP and accelerates recovery. During background adaptation, this feedback loop prevents rod's saturation by restoring current to rods that are exposed to ambient illumination. In vivo, GCAP2 is functional in rod outer segments, and has the capacity to maximally stimulate Ret-GC activity when [Ca^{2+}] drops to a minimum. However, responses to dim flashes of rods expressing GCAP2 alone lack the rapid initial decline that follows the peak in wildtype responses, suggesting that GCAP1 may function earlier in the photoresponse—and be responsible for this early phase of the recovery—, and GCAP2 regulation of GCs may proceed more slowly and/or require a higher intensity of light. Future experiments are needed to confirm this subtle difference in GCAP1 and GCAP2 kinetic behavior. Finally, the in vivo mechanism by which mutant forms of GCAP1 lead to cone photoreceptor death remains to be determined.

ACKNOWLEDGMENTS

The electrophysiological analysis of GCAPs-/- and GCAPs-/-GCAP2+ mice was done by Dr. M. E. Burns in Dr. D. A. Baylor's laboratory at Stanford University. Financial support to J. C. was provided by the National Eye Institute (NEI) (EY 12703).

REFERENCES

1. Koch KW, Stryer L. Highly cooperative feedback control of retinal rod guanylate cyclase by calcium ions. Nature 1988; 334:64-66.
2. Koutalos Y, Nakatani K, Tamura T et al. Characterization of guanylate cyclase activity in single retinal rod outer segments. J Gen Physiol 1995; 106:863-890.
3. Koutalos Y, Yau KW. Regulation of sensitivity in vertebrate rod photoreceptors by calcium. Trends Neurosci 1996; 19:73-81.
4. Gorczyca WA, Polans AS, Surgucheva IG et al. Guanylyl cyclase activating protein. A calcium-sensitive regulator of phototransduction. J Biol Chem 1995; 270:22029-22036.
5. Dizhoor AM, Olshevskaya EV, Henzel WJ et al. Cloning, sequencing, and expression of a 24-kDa Ca(2+)-binding protein activating photoreceptor guanylyl cyclase. J Biol Chem 1995; 270:25200-25206.
6. Palczewski K, Subbaraya I, Gorczyca WA et al. Molecular cloning and characterization of retinal photoreceptor guanylyl cyclase-activating protein. Neuron 1994; 13:395-404.
7. Shyjan AW, de Sauvage FJ, Gillett NA et al. Molecular cloning of a retina-specific membrane guanylyl cyclase. Neuron 1992; 9:727-737.
8. Lowe DG, Dizhoor AM, Liu K et al. Cloning and expression of a second photoreceptor-specific membrane retinal guanylyl cyclase (RetGC), RetGC-2. Proc Natl Acad Sci USA 1995; 92:5535-5539.
9. Liu X, Seno K, Nishizawa Y et al. Ultrastructural localization of retinal guanylate cyclase in human and monkey retinas. Exp Eye Res 1994; 59:761-768.

10. Dizhoor AM, Lowe DG, Olshevskaya EV et al. The human photoreceptor membrane guanylyl cyclase, RetGC, is present in outer segments and is regulated by calcium and a soluble activator. Neuron 1994; 12:1345-1352.
11. Koch KW. Purification and identification of photoreceptor guanylate cyclase. J Biol Chem 1991; 266:8634-8637.
12. Hallett MA, Delaat JL, Arikawa K et al. Distribution of guanylate cyclase within photoreceptor outer segments. J Cell Sci 1996; 109:1803-1812.
13. Perrault I, Rozet JM, Calvas P et al. Retinal-specific guanylate cyclase gene mutations in Leber's congenital amaurosis. Nat Genet 1996; 14:461-464.
14. Yang RB, Garbers DL. Two eye guanylyl cyclases are expressed in the same photoreceptor cells and form homomers in preference to heteromers. J Biol Chem 1997; 272:13738-13742.
15. Yang RB, Robinson SW, Xiong WH et al. Disruption of a retinal guanylyl cyclase gene leads to cone-specific dystrophy and paradoxical rod behavior. J Neurosci 1999; 19:5889-5897.
16. Haeseleer F, Sokal I, Li N et al. Molecular characterization of a third member of the guanylyl cyclase-activating protein subfamily. J Biol Chem 1999; 274:6526-6535.
17. Howes K, Bronson JD, Dang YL et al. Gene array and expression of mouse retina guanylate cyclase activating proteins 1 and 2. Invest Ophthalmol Vis Sci 1998; 39:867-875.
18. Otto-Bruc A, Fariss RN, Haeseleer F et al. Localization of guanylate cyclase-activating protein 2 in mammalian retinas. Proc Natl Acad Sci USA 1997; 94:4727-4732.
19. Cuenca N, Lopez S, Howes K et al. The localization of guanylyl cyclase-activating proteins in the mammalian retina. Invest Ophthalmol Vis Sci 1998; 39:1243-1250.
20. Kachi S, Nishizawa Y, Olshevskaya E et al. Detailed localization of photoreceptor guanylate cyclase activating protein-1 and -2 in mammalian retinas using light and electron microscopy. Exp Eye Res 1999; 68:465-473.
21. Imanishi Y, Li N, Sowa ME et al. Characterization of retinal guanylate cyclase-activating protein 3 (GCAP3) from zebrafish to man. Eur J Neurosci 2002; 15:63-78.
22. Duda T, Goraczniak R, Surgucheva I et al. Calcium Modulation of Bovine Photoreceptor Guanylate Cyclase. Biochemistry 1996; 35:8478-8482.
23. Dizhoor AM, Hurley JB. Inactivation of EF-hands makes GCAP-2 (p24) a constitutive activator of photoreceptor guanylyl cyclase by preventing a Ca2+-induced "activator-to-inhibitor" transition. J Biol Chem 1996; 271:19346-19350.
24. Laura RP, Dizhoor AM, Hurley JB. The membrane guanylyl cyclase, retinal guanylyl cyclase-1, is activated through its intracellular domain. J Biol Chem 1996; 271:11646-11651.
25. Laura RP, Hurley JB. The kinase homology domain of retinal guanylyl cyclases 1 and 2 specifies the affinity and cooperativity of interaction with guanylyl cyclase activating protein-2. Biochemistry 1998; 37:11264-11271.
26. Krylov DM, Niemi GA, Dizhoor AM et al. Mapping sites in Guanylyl Cyclase Activating Protein-1 required for regulation of photoreceptor membrane Guanylyl Cyclases. J Biol Chem 1999; 274:10833-10839.
27. Krylov DM, Hurley JB. Identification of proximate regions in a complex of retinal guanylyl cyclase 1 and guanylyl cyclase-activating protein-1 by a novel mass spectrometry-based method. J Biol Chem 2001; 276:30648-30654.
28. Olshevskaya EV, Boikov S, Ermilov A et al. Mapping functional domains of the guanylate cyclase regulator protein, GCAP2. J Biol Chem 1999; 274:10823-10832.
29. Rudnicka-Nawrot M, Surgucheva I, Hulmes JD et al. Changes in biological activity and folding of guanylate cyclase-activating protein 1 as a function of calcium. Biochemistry 1998; 37:248-257.
30. Dizhoor AM, Boikov SG, Olshevskaya EV. Constitutive activation of photoreceptor guanylate cyclase by Y99C mutant of GCAP-1. Possible role in causing human autosomal dominant cone degeneration. J Biol Chem 1998; 273:17311-17314.
31. Ramamurthy V, Tucker C, Wilkie SE et al. Interactions within the coiled-coil domain of RetGC-1 guanylyl cyclase are optimized for regulation rather than for high affinity. J Biol Chem 2001; 276:26218-26229.

32. Olshevskaya EV, Ermilov AN, Dizhoor AM. Dimerization of guanylyl cyclase-activating protein and a mechanism of photoreceptor guanylyl cyclase activation. J Biol Chem 1999; 274:25583-25587.
33. Yu H, Olshevskaya E, Duda T et al. Activation of retinal guanylyl cyclase-1 by Ca2+-binding proteins involves its dimerization. J Biol Chem 1999; 274:15547-15555.
34. Payne AM, Downes SM, Bessant DAR et al. A mutation in guanylate cyclase activator 1A (GUCA1A) in an autosomal dominant cone dystrophy pedigree mapping to a new locus on chromosome 6p21.1. Hum Mol Genet 1998; 7:273-277.
35. Downes SM, Holder GE, Fitzke FW et al. Autosomal dominant cone and cone-rod dystrophy with mutations in the guanylate cyclase activator 1A gene-encoding guanylate cyclase activating protein-1. Arch Ophthalmol 2001; 119:96-105.
36. Wilkie SE, Newbold RJ, Deery E et al. Functional characterization of missense mutations at codon 838 in retinal guanylate cyclase correlates with disease severity in patients with autosomal dominant cone-rod dystrophy. Hum Mol Genet 2000; 9:3065-3073.
37. Wilkie SE, Li Y, Deery EC et al. Identification and functional consequences of a new mutation (E155G) in the gene for GCAP1 that causes autosomal dominant cone dystrophy. Am J Hum Genet 2001; 69:471-480.
38. Sokal I, Li N, Surgucheva I et al. GCAP1(Y99C) mutant is constitutively active in autosomal dominant cone dystrophy. Mol Cell 1998; 2:129-133.
39. Surguchov A, Bronson JD, Banerjee P et al. The human GCAP1 and GCAP2 genes are arranged in a tail-to-tail array on the short arm of chromosome 6 (p21.1). Genomics 1997; 39:312-322.
40. Baylor DA, Lamb TD, Yau KW. Responses of retinal rods to single photons. J Physiol (Lond) 1979; 288:613-634.
41. Leskov IB, Klenchin VA, Handy JW et al. The gain of rod phototransduction: reconciliation of biochemical and electrophysiological measurements. Neuron 2000; 27:525-537.
42. Matthews HR, Murphy RL, Fain GL et al. Photoreceptor light adaptation is mediated by cytoplasmic calcium concentration. Nature 1988; 334:67-69.
43. Nakatani K, Yau KW. Calcium and light adaptation in retinal rods and cones. Nature 1988; 334:69-71.
44. Pepperberg DR, Cornwall MC, Kahlert M et al. Light-dependent delay in the falling phase of the retinal rod photoresponse. Vis Neurosci 1992; 8:9-18.
45. Rispoli G, Sather WA, Detwiler PB. Visual transduction in dialysed detached rod outer segments from lizard retina. J Physiol (Lond) 1993; 465:513-537.
46. Koutalos Y, Nakatani K, Yau KW. The cGMP-phosphodiesterase and its contribution to sensitivity regulation in retinal rods. J Gen Physiol 1995; 106:891-921.
47. Matthews HR. Actions of Ca2+ on an early stage in phototransduction revealed by the dynamic fall in Ca2+ concentration during the bright flash response. J Gen Physiol 1997; 109:141-146.
48. Kawamura S, and Murakami M. Calcium-dependent regulation of cyclic GMP phosphodiesterase by a protein from frog retinal rods. Nature 1991; 349:420-423.
49. Kawamura S. Rhodopsin phosphorylation as a mechanism of cyclic GMP phosphodiesterase regulation by S-modulin. Nature 1993; 362:855-857.
50. Klenchin VA, Calvert PD, Bownds MD. Inhibition of rhodopsin kinase by recoverin. Further evidence for a negative feedback system in phototransduction. J Biol Chem 1995; 270:16147-16152.
51. Hsu YT, Molday RS. Modulation of the cGMP-gated channel of rod photoreceptor cells by calmodulin. Nature 1993; 361:76-79.
52. Hsu YT, Molday RS. Interaction of calmodulin with the cyclic GMP-gated channel of rod photoreceptor cells. Modulation of activity, affinity purification, and localization. J Biol Chem 1994; 269:29765-29770.
53. Nakatani K, Koutalos Y, and Yau KW. Ca2+ modulation of the cGMP-gated channel of bullfrog retinal rod photoreceptors. J Physiol (Lond) 1995; 484:69 76.

54. Lolley RN, Racz E. Calcium modulation of cGMP synthesis in rat visual cells. Vision Res 1982; 22:1481-1486.
55. Kawamura S, Murakami M. Regulation of cGMP levels by guanylate cyclase in truncated frog rod outer segments. J Gen Physiol 1989; 94:649-668.
56. Hu G, Jang GF, Cowan CW et al. Phosphorylation of RGS9-1 by an endogenous protein kinase in rod outer segments. J Biol Chem 2001; 276:22287-22295.
57. Mendez A, Burns ME, Sokal I et al. Role of guanylate cyclase-activating proteins (GCAPs) in setting the flash sensitivity of rod photoreceptors. Proc Natl Acad Sci USA 2001; 98:9948-9953.
58. Cobbs WH, Pugh EN Jr. Kinetics and components of the flash photocurrent of isolated retinal rods of the larval salamander, Ambystoma tigrinum. J Physiol (Lond) 1987; 394:529-572.
59. Karpen JW, Zimmerman AL, Stryer L et al. Gating kinetics of the cyclic-GMP-activated channel of retinal rods: flash photolysis and voltage-jump studies. Proc Natl Acad Sci USA 1988; 85:1287-1291.
60. Dizhoor AM, Hurley JB. Regulation of photoreceptor membrane guanylyl cyclases by guanylyl cyclase activator proteins. Methods 1999; 19:521-531.
61. Lamb TD, Matthews HR, Torre V. Incorporation of calcium buffers into salamander retinal rods: a rejection of the calcium hypothesis of phototransduction. J Physiol (Lond) 1986; 372:315-349.
62. Matthews HR, Torre V, Lamb TD. Effects on the photoresponse of calcium buffers and cyclic GMP incorporated into the cytoplasm of retinal rods. Nature 1985; 313:582-585.
63. Nikonov S, Engheta N, Pugh EN Jr. Kinetics of recovery of the dark-adapted salamander rod photoresponse. J Gen Physiol 1998; 111:7-37.
64. Korschen HG, Beyermann M, Muller F et al. Interaction of glutamic-acid-rich proteins with the cGMP signalling pathway in rod photoreceptors. Nature 1999; 400:761-766.
65. Sokal I, Otto-Bruc AE, Surguchcva I et al. Conformational changes in guanylyl cyclase-activating protein 1 (GCAP1) and its tryptophan mutants as a function of calcium concentration. J Biol Chem 1999; 274:19829-19837.
66. Chen J, Makino CL, Peachey NS et al. Mechanisms of rhodopsin inactivation in vivo as revealed by a COOH-terminal truncation mutant. Science 1995; 267:374-377.
67. Chen CK, Burns ME, Spencer M et al. Abnormal photoresponses and light-induced apoptosis in rods lacking rhodopsin kinase. Proc Natl Acad Sci USA 1999; 96:3718-3722.
68. Mendez A, Burns ME, Roca A et al. Rapid and reproducible deactivation of rhodopsin requires multiple phosphorylation sites. Neuron 2000; 28:153-164.
69. Kelsell RE, Gregory-Evans, K, Payne AM et al. Mutations in the retinal guanylate cyclase (RetGC-1) gene in dominant cone-rod dystrophy. Hum Mol Genet 1998; 7:1179-1184.
70. Semple-Rowland SL, Lee NR, Van Hooser JP et al. A null mutation in the photoreceptor guanylate cyclase gene causes the retinal degeneration chicken phenotype. Proc Natl Acad Sci USA 1998; 95:1271-1276.
71. Suber ML, Pittler SJ, Qin N et al. Irish setter dogs affected with rod/cone dysplasia contain a nonsense mutation in the rod cGMP phosphodiesterase beta-subunit gene. Proc Natl Acad Sci USA 1993; 90:3968-3972.
72. Bowes C, Li T, Frankel WN et al. Localization of a retroviral element within the rd gene coding for the beta subunit of cGMP phosphodiesterase. Proc Natl Acad Sci USA 1993; 90:2955-2959.
73. Tsang SH, Burns ME, Calvert PD et al. Role for the target enzyme in deactivation of photoreceptor G protein in vivo. Science 1998; 282:117-121.
74. Sokal I, Li N, Verlinde C et al. Ca^{2+}-binding proteins in the retina: from discovery to etiology of human disease. Biochim Biophys Acta 2000; 1498:233-251.
75. Newbold RJ, Deery EC, Walker CE et al. The destabilization of human GCAP1 by a proline to leucine mutation might cause cone-rod dystrophy. Hum Mol Genet 2001; 10:47-54.

76. Pittler SB, Baehr W. Identification of a nonsense mutation in the rod photoreceptor cGMP phosphodiesterase beta-subunit gene of the rd mouse. Proc Natl Acad Sci USA 1991; 88:8322-8326.
77. Farber DB, Lolley RN. Cyclic guanosine monophosphate: elevation in degenerating photoreceptor cells of the C3H mouse retina. Science 1974; 186:449-451.
78. Farber DB. From mice to men: the cyclic GMP phosphodiesterase gene in vision and disease. The Proctor Lecture. Invest Ophthalmol Vis Sci 1995; 36:263-275.
79. Fain GL, Lisman JE. Light, Ca2+, and Photoreceptor Death: New Evidence for the Equivalent-Light Hypothesis from Arrestin Knockout Mice. Invest Ophthalmol Vis Sci 1999; 40:2770-2772.
80. Palczewski K, Polans AS, Baehr W et al. Ca(2+)-binding proteins in the retina: structure, function, and the etiology of human visual diseases. Bioessays 2000; 22:337-350.

CALCIUM-DEPENDENT ACTIVATION OF GUANYLATE CYCLASE BY S100b

Ari Sitaramayya

ABSTRACT

Calcium concentration in the dark-adapted retinal rod outer segment is in the 200 to 600 nM range, and the guanylate cyclase of rod outer segments is thought to be activated in response to a fall in calcium concentration triggered by light. Calcium-binding proteins that mediate such activation, i.e., activation in the absence of or presence of low nanomolar calcium concentrations, have been identified and termed GCAPs (Guanaylate Cyclase Activating Proteins). In the course of our search for GCAP-like proteins in bovine retina, we isolated a protein fraction that stimulated rod outer segment cyclase activity at calcium concentrations higher than those in dark-adapted rod outer segments. We purified the protein responsible for this calcium-dependent stimulation of cyclase activity and found it to be of 6-7 kDa molecular weight as judged by electrophoresis under denaturing conditions and about 40 kDa by gel filtration analysis. Maximum stimulation of cyclase activity was observed at 3-4 micromolar concentration of the protein. It required about 1.5 micromolar free calcium concentration for half-maximal activation of the enzyme. Partial amino acid sequencing of peptide fragments of the activator suggested that the protein was identical with S100b, a previously described calcium-binding protein. Further characterization with antibody specific for S100b supported this possibility. However, the protein isolated in our laboratory and termed CD-GCAP (Calcium-Dependent Guanylate Cyclase Activator Protein) was found to differ significantly from commercially available S100b in the magnitude and calcium dependence of cyclase activation. It was also found to be inactivated by hydroxylamine while S100b was resistant. Investigation into these differences showed that

Eye Research Institute, Oakland University, Rochester, MI 48309, U.S.A.

purification methods had a significant influence on the properties of the activator, producing a less active (S100b) or more active (CD-GCAP) protein, but that it was, otherwise, one and the same protein. We conclude from this study that rod outer segment guanylate cyclase, unlike any cyclase known so far, is capable of activation by two different types of calcium-binding proteins, one that activates in response to a decrease in calcium concentration, and the other, described here, which activates in response to an increase in calcium-concentration. We hypothesize that this cyclase and others like it will be colocalized with one or the other type of activator depending upon the physiological requirement, i.e., activation in response to decreasing or increasing calcium concentration.

INTRODUCTION

In 1978 Cohen and colleagues reported that the cyclic GMP content of mouse retina increased dramatically when dark-adapted tissue was transferred from a normal to a calcium-free medium.[1] Since both media contained an inhibitor of phosphodiesterase, it was hypothesized that the observed increase in cyclic GMP was mainly due to increased synthesis of the cyclic nucleotide by guanylate cyclase. Four years later, Lolley and Racz made a similar observation in studies on rat retina: whole retinas, retinal homogenates or lysed rod outer segments (ROS) were found to produce more cyclic GMP in low-calcium than in normal medium.[2] They concluded that lowering intracellular calcium below micromolar level would stimulate synthesis of cyclic GMP in retinal rod outer segments.

Guanylate cyclase activity in rod outer segments (ROS-GC) was mostly membrane-associated. Its activity in isolated membrane preparations was, however, not dramatically influenced by changes in calcium concentration suggesting the involvement of a soluble protein. The evidence for the existence of such a soluble protein was provided by Koch and Stryer in 1988 in studies on bovine rod outer segment membranes: cyclase activity in the membranes was unaffected by changing the calcium concentration, but addition of a soluble extract of rod outer segments increased the activity at 20 nanomolar, but not at 144 nanomolar calcium.[3] The report by Koch and Stryer stimulated efforts to identify the soluble protein which increased the activity of the membrane cyclase at low calcium concentrations.

Membrane guanylate cyclases known at the time were regulated either by peptide hormones in the extracellular space, or by the intracellular calcium-binding protein calmodulin in response to an increase in calcium concentration.[4-6] ROS-GC was unaffected by hormones or calmodulin, and was, thus, unique. However, attempts at purification, characterization, and cloning of the rod outer segment guanylate cyclase have revealed that the enzyme was structurally similar to other known membrane guanylate cyclases and that it had two isoforms, referred to as ROS-GC 1 and 2 (or Ret GC1 and 2).[7-9,10]

The search for proteins that would stimulate rod outer segment guanylate cyclase at nanomolar calcium concentrations resulted in the identification of three proteins called Guanylate cyclase activating proteins (GCAPs).[11-13] Efforts in my laboratory to identify a GCAP-like protein from bovine retina produced an unexpected result: we found a soluble protein fraction of retina that stimulated ROS-GC when calcium concentration was raised to micromolar level. We have referred to this activity as "high calcium activator." This paper describes the isolation and characterization of the protein responsible for this activity, and its eventual identification as S100b.

DETECTION OF "HIGH CALCIUM ACTIVATOR" IN CRUDE RETINAL EXTRACT

In studies aimed at obtaining a crude soluble fraction which could stimulate ROS-GC at nanomolar calcium concentrations, washed bovine ROS membranes were used as the source of ROS-GC. For routine measurements, low (nominal nanomolar) calcium concentration was obtained by the addition of 2 mM EGTA to the cyclase assay mixture. To serve as a high calcium control in which ROS-GC was not expected to be activated, 1 mM $CaCl_2$ was added to the assay mixture in the absence of EGTA. As shown in Figure 1, ROS-GC activity in the membranes was similar at nanomolar and high calcium concentrations indicating that the membranes were washed free of soluble activator(s). The presence of activators was investigated in a retinal soluble fraction, which was obtained by heating the supernatant of a retinal homogenate at 75°C (3 min) in the presence of 5 mM $CaCl_2$, clarification by centrifugation, and dialysis against 10 mM Tris, pH 8.0. This soluble fraction had no cyclase activity by itself either at low or high calcium concentration. However, in its presence, ROS-GC activity of the washed membranes was higher both at low and high calcium concentrations (Fig. 1). This observation indicated that the soluble fraction contained the activator expected to stimulate cyclase at nanomolar calcium concentrations (GCAP), and, in addition, an unknown factor that stimulated cyclase at high calcium concentrations. With the discovery of GCAPs that stimulate cyclase at nanomolar calcium concentrations already announced,[11,12] we attempted to characterize the factor responsible for activation of cyclase at high calcium concentration.

PURIFICATION OF THE FACTOR RESPONSIBLE FOR ACTIVATION AT HIGH CALCIUM CONCENTRATION

The crude extract prepared as described above from bovine retinas was processed successively through ion-exchange chromatography on DEAE-Sepharose CL-6B, hydrophobic chromatography on Phenylsepharose CL-4B, and gel filtration by HPLC on a Biosep-sec 2000 column.[14] The factor responsible for activation at high calcium concentration eluted as a single peak after each chromatography. The apparent molecular weight of the active protein eluting from the gel filtration column was 40 kDa. Electrophoresis of the material on a 10-20% gradient SDS-polyacrylamide gel showed a single protein band at 6-7 kDa (Fig. 2), suggesting

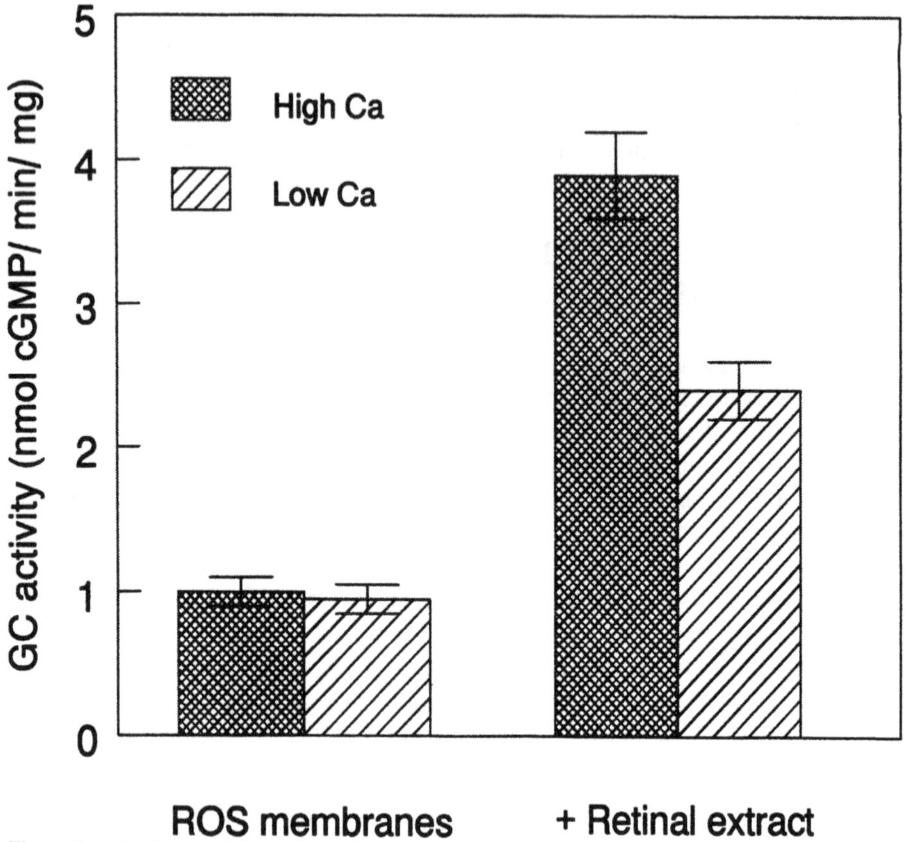

Figure 1. Activation of ROS membrane GC by crude retinal extract. Reprinted with permission from Meth. Enzymol., 315:730-742, 2000. Copyright, Academic Press.

that the native protein (40 kDa) existed as a multimer. To insure that this small molecular weight protein was indeed the factor responsible for activation of the cyclase at high calcium concentrations, the 6-7 kDa protein band was extracted from the acrylamide gel, regenerated and subjected to gel filtration by HPLC.[15] Analysis of the fractions showed that the cyclase-stimulating activity eluted again with a peak corresponding to 40 kDa, suggesting that the protein appearing at 6-7 kDa in SDS-PAGE was indeed the activator.

DOSE AND CALCIUM DEPENDENCE OF THE ACTIVATION

The 40 kDa protein activated ROS-GC activity in a dose-dependent fashion (Fig. 3). With 0.1 or 1.0 mM calcium in the assays, cyclase activity was stimulated to about 2400 % of the control in the presence of 4 μM activator. Half-maximal stimulation was reached at about 1 μM concentration. Calcium dependence of the

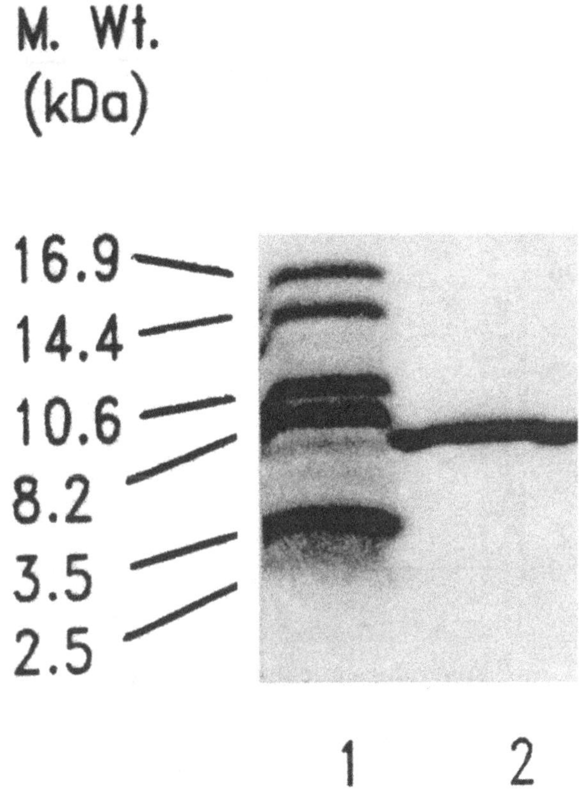

Figure 2. Electrophoresis of the purified activator. Lane 1, molecular weight standards; Lane 2, four µg of purified activator. Reprinted with permission from Biochemistry 34:14279-13283, 1995. Copyright American Chemical Society.

activation was measured at saturating concentration (12 µM) of the activator and free calcium concentrations between 0.5 µM and 1000 µM, obtained by the use of calcium-dibromo BAPTA buffers (Fig. 4). The results showed that half-maximal activation was achieved at about 1.5 µM free calcium concentration. However, upon further analysis of the data, it was observed that calcium dependence was biphasic (Fig. 4); half maximal calcium concentrations for the two phases were at 0.3-0.4 µM and 3.5 µM. The corresponding maximal activation obtained was about 1200% and 1800% of control, respectively. [Calcium dependence studies were carried out at 25°C whereas the dose-dependence measurements, which showed a maximal, 2400% of the control activity, were done at 30°C.]

Since this activator required calcium for the stimulation of cyclase, unlike GCAPs which activated the enzyme when freed of calcium, it was referred to as CD-GCAP or calcium-dependent GCAP.

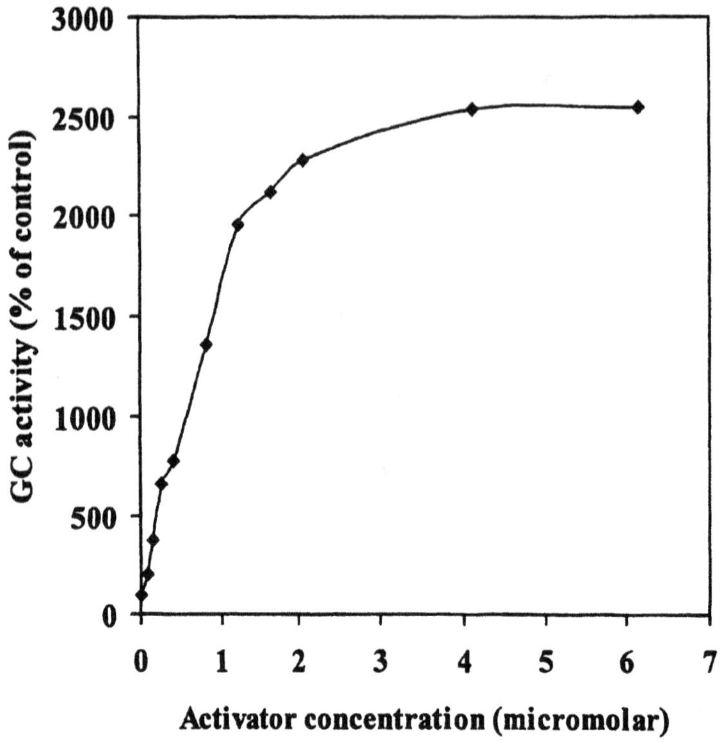

Figure 3. Dependence of GC activation on activator concentration.

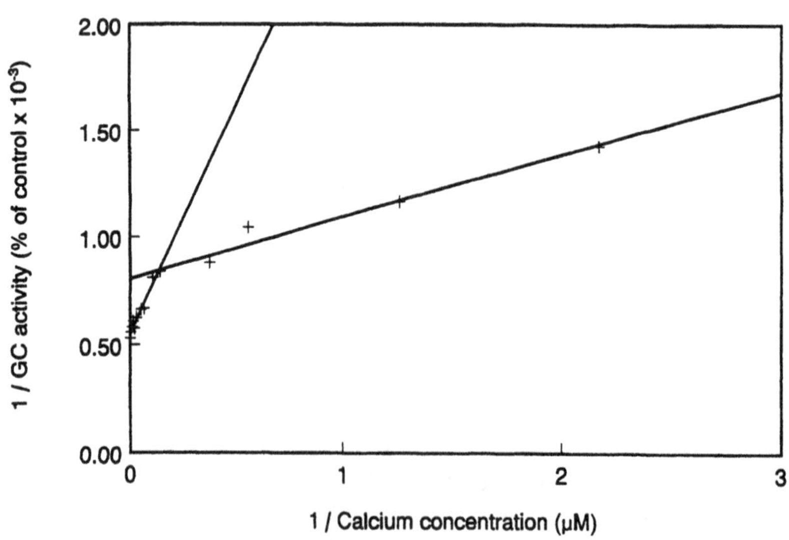

Figure 4. Calcium dependence of ROS-GC activation by CD-GCAP.

IDENTIFICATION OF CD-GCAP AS S100b

Partial amino acid sequences of tryptic fragments of CD-GCAP had identical matches in the sequence of bovine brain S100b, a calcium binding protein. CD-GCAP reacted with an antibody against S100b, but not with one against the closely related S100A1 protein (Figure 5). A previous report on S100b indicated that the protein migrated as a 7 kDa protein on SDS-PAGE, just like CD-GCAP.[17] Mass spectrometric analysis of CD-GCAP and commercially available S100b showed similar mass, about 10,570 daltons. Though S100b was generally considered to exist as a dimer with an apparent molecular weight of about 21 kDa, HPLC gel filtration revealed that both S100b and CD-GCAP eluted as 40 kDa proteins under our experimental conditions. Together, these observations indicated that CD-GCAP was identical to S100b.

CD-GCAP and commercially obtained S100b, however, exhibited differences in the activation of ROS-GC. At saturating calcium concentrations CD-GCAP activated cyclase by about 24 fold while the activity with S100b never reached above 8 fold. The calcium concentration required for half-maximal activation by S100b was about 30 μM, more than an order of magnitude higher than that required for CD-GCAP. Also, incubation with 2 M hydroxylamine degraded CD-GCAP and nearly completely eliminated its cyclase-stimulating ability, while it had little or no effect on S100b. These differences appeared to be due to differences in the isolation of the proteins: the commercially obtained S100b was apparently isolated through a procedure involving incubation with zinc followed by zinc affinity chromatography on phenylsepharose,[18] whereas CD-GCAP was isolated following treatment with Ca^{2+} at 75°C. Tightly bound zinc in the S100b preparation could have changed the conformation of the protein in a way that resulted in resistance to hydroxylamine and reduced its ability to activate cyclase. To test this possibility, purified S100b was subjected to the purification protocol of CD-GCAP to isolate S100b→CD-GCAP, and purified CD-GCAP was subjected to the protocol of S100b to isolate CD-GCAP→S100b.[19] The proteins thus isolated, S100b→CD-GCAP and CD-GCAP→S100b, behaved like CD-GCAP and S100b, respectively, in the magnitude and calcium sensitivity of the activation of cyclase, and in their ability to withstand treatment with hydroxylamine, indicating that the purification procedures were indeed responsible for the differences between them. CD-GCAP and S100b were, otherwise, one and the same protein. S100A1, another S100 protein, also activated ROS-GC, though less vigorously than S100b.[20]

THE SIGNIFICANCE OF CYCLASE ACTIVATION BY S100b

Since the discovery of a "high calcium activator" of ROS-GC was unexpected, it remained to be investigated whether the activator (S100b) and cyclase were colocalized in ROS or elsewhere in retina and whether calcium concentrations in vivo ever rose to the level favorable for the activation of cyclase.

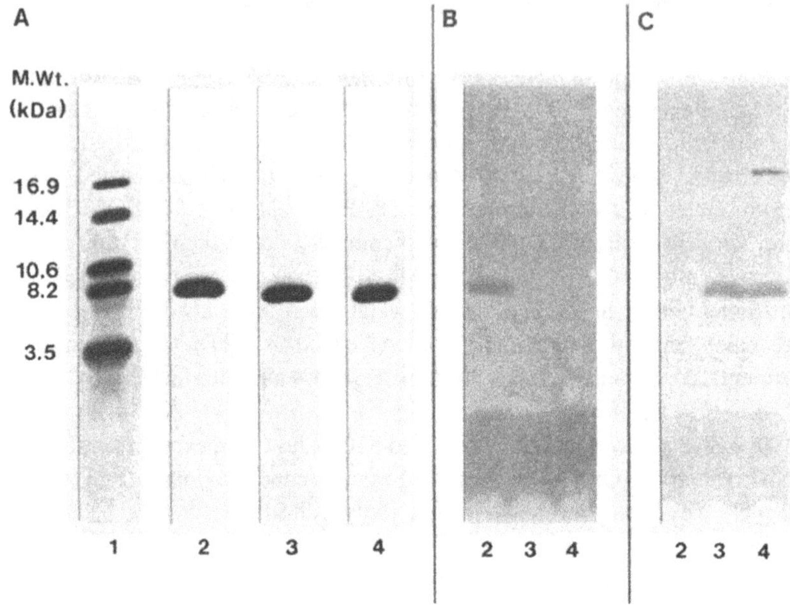

Figure 5. Lane 1, molecular weight standards; Lane 2, S100 A1; Lane 3, S100b; Lane 4, CD-GCAP. The set of lanes in A is stained for protein, in B for reaction with antibody against S100A, and in C for reaction with antibody against S100b. Reprinted with permission from Biochemistry 36:14159-14166, 1997. Copyright American Chemical Society.

CD-GCAP was initially isolated from the soluble fraction of whole retina. Since it activated the cyclase of the ROS, we attempted to determine if CD-GCAP was present in ROS. At the time, our method of preparation of crude CD-GCAP had involved extraction of ROS soluble fraction, heating it for 3 min at 75°C in 5 mM Ca^{2+}, clarification by centrifugation, and dialysis. Such a preparation had contained little or no ability to activate ROS-GC and hence, we had concluded that ROS did not contain the high calcium activator. However, it has since been found that heat treatment of the soluble fraction denatured and removed about 90% of the S100b. Therefore, the presence of S100b was tested again, this time directly in the crude ROS soluble fraction by Western blotting, and it was found that ROS did contain the protein, though at a low level. Our estimation was that a bovine retina contained about 250 μg of S100b of which 2.7 μg was in ROS (A. Margulis and A. Sitaramayya, unpublished observations).

Rambotti et al investigated the presence of S100b in bovine retina by immunofluorescence analysis.[21] They too found the protein in the ROS, and not surprisingly, at a much lower concentration than in inner nuclear and ganglion cell layers. In addition, Rambotti et al observed stimulation of cyclase activity in the disc membranes of photoreceptor cells when sections of dark-adapted retina were incubated with S100b or S100A1.

The physiological significance of the activation of ROS-GC by S100b remains uncertain since calcium concentration in the dark-adapted ROS is 200 to 600 nM and known to decrease in response to light,[10] whereas activation of cyclase by S100b is favored at higher calcium concentrations. Even so, S100b does stimulate the cyclase significantly at the calcium concentrations in dark-adapted ROS.[16] The physiological role of this activation remains unknown.

ROS-GC-like cyclases and S100 proteins are present elsewhere in the retina and in other tissues,[21-24] and it is likely that they contribute to the increase in cyclic GMP synthesis in response to an increase in calcium concentration. It was recently reported that neurocalcin, another calcium-binding protein, can also activate ROS-GC in a calcium-dependent fashion.[25]

From a structural point of view it is fascinating that ROS-GC has the ability to respond to one type of calcium-binding protein at low calcium concentrations and another at higher concentrations. It is reasonable to assume that in most cases this cyclase and similar cyclases would be colocalized with one or the other type of calcium-binding protein depending upon whether the stimulation of cyclase activity is required in response to a decrease or an increase in calcium concentration.

ACKNOWLEDGMENTS

The work described here was performed by Nikolay Pozdnyakov, Akiko Yoshida and Alexander Margulis, and supported by grants from the National Institutes of Health (EY 07158 and EY 05230).

REFERENCES

1. Cohen AI, Hall IA, Ferrendelli JA. Calcium and cyclic nucleotide regulation in incubated mouse retinas. J Gen Physiol 1978; 71:595-612.
2. Lolley RN, Racz E. Calcium modulation of cyclic GMP synthesis in rat visual cells. Vision Res 1982; 22:1481-1486.
3. Koch KW, Stryer L. Highly cooperative feedback control of retinal rod guanylate cyclase by calcium ions. Nature 1988; 334:64-66.
4. Garbers DL, Koesling D, Schultz G. Guanylyl cyclase receptors. Mol Biol Cell 1994; 5:1-5.
5. Kakiuchi S, Sobue K, Yamazaki R et al. Ca2+-dependent modulator proteins from Tetrahymena pyriformis, sea anemone, and scallop and guanylate cyclase activation. J Biol Chem 1981; 256:19-22.
6. Klumpp S, Schultz JE. Characterization of a Ca2+-dependent guanylate cyclase in the excitable ciliary membrane from Paramecium. Eur J Biochem 1982; 124:317-324.
7. Shyjan AW, de Sauvage FJ, Gillett NA et al. Molecular cloning of a retina-specific membrane guanylyl cyclase. Neuron 1992; 9:727-737.
8. Lowe DG, Dizhoor AM, Liu K et al. Cloning and expression of a second photoreceptor-specific membrane retina guanylyl cyclase (RetGC), RetGC-2. Proc Natl Acad Sci USA 1995; 92:5535-5539.
9. Goraczniak RM, Duda T, Sitaramayya A et al. Structural and functional characterization of the rod outer segment membrane guanylate cyclase. Biochem J 1994; 302 (Pt 2):455-461.
10. Pugh EN, Jr., Duda T, Sitaramayya A et al. Photoreceptor guanylate cyclases: A review. Biosci Rep 1997; 17:429-473.

11. Dizhoor AM, Lowe DG, Olshevskaya EV et al. The human photoreceptor membrane guanylyl cyclase, RetGC, is present in outer segments and is regulated by calcium and a soluble activator. Neuron 1994; 12:1345-1352.
12. Gorczyca WA, Gray-Keller MP, Detwiler PB et al. Purification and physiological evaluation of a guanylate cyclase activating protein from retinal rods. Proc Natl Acad Sci USA 1994; 91:4014-4018.
13. Haeseleer F, Sokal I, Li N et al. Molecular characterization of a third member of the guanylyl cyclase- activating protein subfamily. J Biol Chem 1999; 274:6526-6535.
14. Pozdnyakov N, Yoshida A, Cooper NG et al. A novel calcium-dependent activator of retinal rod outer segment membrane guanylate cyclase. Biochemistry 1995; 34:14279-14283.
15. Pozdnyakov N, Yoshida A, Cooper NG et al. A novel calcium-dependent activator of retinal rod outer segment membrane guanylate cyclase. Biochemistry 1995; 34:14279-14283.
16. Pozdnyakov N, Goraczniak R, Margulis A et al. Structural and functional characterization of retinal calcium-dependent guanylate cyclase activator protein (CD-GCAP): Identity with S100beta protein. Biochemistry 1997; 36:14159-14166.
17. Kligman D, Marshak D. Purification and characterization of a neurite extension factor from bovine retina. Proc Natl Acad Sci USA 1985; 82:7136-7139.
18. Baudier J, Holtzscherer C, Gerard D. Zinc-dependent affinity chromatography of the S100b protein on phenyl- Sepharose. A rapid purification method. FEBS Lett 1982; 148:231-234.
19. Pozdnyakov N, Goraczniak R, Margulis A et al. Structural and functional characterization of retinal calcium-dependent guanylate cyclase activator protein (CD-GCAP): Identity with S100beta protein. Biochemistry 1997; 36:14159-14166.
20. Margulis A, Pozdnyakov N, Sitaramayya A. Activation of bovine photoreceptor guanylate cyclase by S100 proteins. Biochem Biophys Res Commun 1996; 218:243-247.
21. Rambotti MG, Giambanco I, Spreca A et al. S100B and S100A1 proteins in bovine retina: Their calcium-dependent stimulation of a membrane-bound guanylate cyclase activity as investigated by ultracytochemistry. Neuroscience 1999; 92: 1089-1101.
22. Liu X, Seno K, Nishizawa Y et al. Ultrastructural localization of retinal guanylate cyclase in human and monkey retinas. Exp Eye Res 1994; 59:761-768.
23. Cooper N, Liu L, Yoshida A et al. The bovine rod outer segment guanylate cyclase, ROS-GC, is present in both outer segment and synaptic layers of the retina. J Mol Neurosci 1995; 6:211-222.
24. Venkataraman V, Nagele R, Duda T et al. Rod outer segment membrane guanylate cyclase type 1-linked stimulatory and inhibitory calcium signaling systems in the pineal gland: Biochemical, molecular, and immunohistochemical evidence. Biochemistry 2000; 39:6042-6052.
25. Kumar VD, Vijay-Kumar S, Krishnan A et al. A second calcium regulator of rod outer segment membrane guanylate cyclase, ROS-GC1: Neurocalcin. Biochemistry 1999; 38:12614-12620.

THE ROLE OF CADHERINS IN Ca^{2+}-MEDIATED CELL ADHESION AND INHERITED PHOTORECEPTOR DEGENERATION

Hanno Bolz[1], Jan Reiners[2], Uwe Wolfrum[2] and Andreas Gal[1]

Cadherins are Ca^{2+}-binding, transmembrane proteins involved in cell adhesion. Recently, three cadherin molecules, cadherin-23, protocadherin-15, and cadherin-3, were found to be defective in various human diseases, many of them with photoreceptor degeneration and/or sensorineural hearing loss as major features such Usher syndrome type 1D (USH1D), USH1F, and hypotrichosis with juvenile macular dystrophy. The process, by which mutations lead to photoreceptor degeneration is still not fully understood. Data from the inner ear phenotype of USH1 mouse models suggest that loss of cell adhesion is a crucial event.

CADHERINS: FEATURES AND FUNCTIONS

Cadherins are Ca^{2+}-binding, transmembrane proteins involved in cell adhesion. The cadherin superfamily consists presently of more than 300 proteins (in vertebrates alone) and represents, with the immunoglobulin-type molecules, a major group of cell adhesion molecules both in vertebrates and invertebrates. In human, more than 80 different cadherins have been identified to date. Some members of the cadherin protein family gained special attention due to their involvement in different forms of cancer.[1] The identification of mutations in three cadherin genes in patients with various sensory disorders, most of them with photoreceptor degeneration as a major feature, revealed the importance of cadherins for retinal integrity. This Chapter will focus on the cadherin molecules implicated in human retinal disease.

Common to all cadherins are the multiple (5 to 34) cadherin domains (extracellular domains, EC), tandemly repeated, ~100-amino-acid stretches connected by

---

[1]Institut für Humangenetik, Universitätsklinikum Hamburg-Eppendorf, Hamburg, Germany.
[2]Institut für Zoologie, Johannes Gutenberg-Universität Mainz, Germany.

~10-amino-acid linker regions. ECs contain the evolutionarily highly conserved, negatively charged DXD, DRE, and DXNDN motifs that mediate Ca^{2+}-dependent homophilic binding between cadherin molecules (Fig. 1). Cadherins can be grouped based on the number and sequence of their ECs (as to the latter by comparing the first N-terminal EC, called EC1) and other domains, e.g., the cytoplasmic domain which may provide information on interacting partners of members of a given subfamily (for overview of phylogenetic classification, see ref. 2). Presently, members of the cadherin superfamily fall into six groups according to size, number and feature of domains, function, and binding partners. We distinguish classical cadherins, desmosomal cadherins, protocadherins, cadherins with tyrosine-kinase domains (pointing towards a role of cadherins in signal transduction pathways), Fat-like cadherins (large cadherins with similarity to Fat, a cadherin with 34 ECs), and seven-pass transmembrane cadherins (Flamingo)[3,4] (Fig. 2).

The so-called classical cadherins possess five ECs (Fig. 1). In addition, classical cadherins share a similar intracellular peptide sequence that gives clue to interacting partners. Indeed, all classical cadherins interact with the actin cytoskeleton via catenins. The intracellular domain is thought to play a regulatory role for the adhesive state of the extracellular domains.[5] The Ca^{2+}-binding pocket is formed by residues of EC1 and EC2 and those of the linker region (Fig. 3a). Each cadherin dimer associates with six Ca^{2+} ions primarily via the residues of the linker region between EC1 and EC2, whereas the interaction with amino acids from the ECs is also required (Fig. 3b). Ca^{2+} -binding seems to provide the molecule with a rigid and proteolysis-resistant arrangement of the ECs.[6]

The structural basis of cell adhesion and molecular interaction has been studied so far mainly for classical cadherins. The cadherin fold consists of a seven-strand β-sheet. It seems that cadherins of the same cell, by lateral association, form parallel *cis*-dimers which are today considered "building blocks" for lateral clustering and thereby form the basis for stable cell adhesion.[7] *Trans*-dimerization results from contacts between the N-terminal ECs of cadherin molecules from the two opposite cells involved.[8] More recent studies suggested the possibility of a variable degree of anti-parallel overlap, to an extent where all five ECs may overlap.[9] Remarkably, the predicted distance (20-25 nm) at maximal overlap between two classical cadherin molecules of adjacent membranes corresponds well to the cell-cell distance at adherens junctions, at which those cadherins play an important role, supporting the assumption of a possible overlap to a greater extent than just the N-terminal ECs.

Cadherins mediate two types of adhesive contacts: Firstly, oligomers of *trans*-dimers form diffuse adhesive contacts between neighboring cells. Secondly, a greater density of adhesive contacts is reached in clusters of *trans*-dimers that are found in specialized adherens junctions such as the zonula adherens. Although homophilic *cis*-binding seems typical, heteromeric binding has also been observed.[10]

Cadherins are also implicated in determining cell polarity, that is particularly important in highly organized tissue structures such as epithelial layers.[11] Initial cell-cell contacts are stabilized by classical cadherins (and strengthened by their link to cytoskeleton), that appear in high density in adherens junctions as this

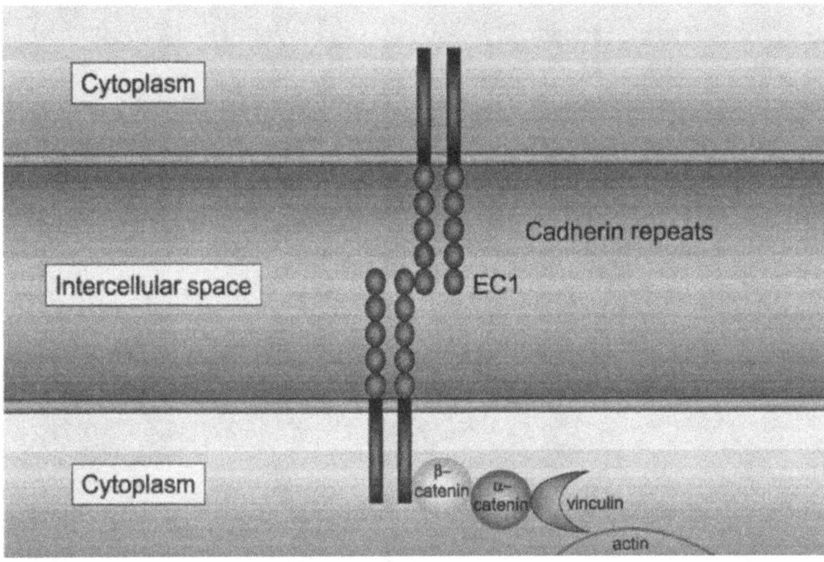

Figure 1. Cartoon of classical cadherins. Two opposite cells are shown. Five extracellular cadherin domains (interspersed linker regions are not shown) of each cadherin molecule protrude from the cells, form dimers with the horizontally neighbouring molecule and bridge the intercellular space by interacting laterally with dimers of the other cell (see text for details). By the cytoplasmic tail, classical cadherins are linked to the actin cytoskeleton via α- and β-catenins.

contact broadens across neighbouring membranes. Moreover, cadherins seem to play a role in recruiting sec6/8, a multiprotein complex that targets exocytic vesicles with specific molecular components to selected docking sites on the (basal lateral) plasma membrane, thereby establishing epithelial apical-basal polarity.[12] The importance of cadherin-mediated cell adhesion for polarity and movement during embryonic morphogenesis is highlighted by the observation of left-right asymmetry after disruption of N-cadherin function during chick gastrulation.[13]

Many cadherins show strong expression in neural tissues. Protocadherins of the diverse Pcdhα/CNR group show specific expression in selected brain areas/tissues and function as receptors for the extracellular matrix molecule reelin, an interaction that is assumed to be crucial for positioning of neuronal sub-populations in the cortex.[14] Particularly noteworthy in view of the focus of this book is the physiological role of N-cadherin that was shown to be required for axonal outgrowth and guidance in the retina.[15] The finding that numerous cadherins localize to synapses in structures resembling epithelial adherens junctions suggests that cadherins may be important in building complex neural networks. Cadherin-mediated adhesion seems, in turn, to be influenced by synaptic activity, suggesting a role of cadherin both in synapse plasticity and activity modulation,[16] hence making cadherins excellent candidates for being involved in long term potentiation (LTP) at synapses.[6]

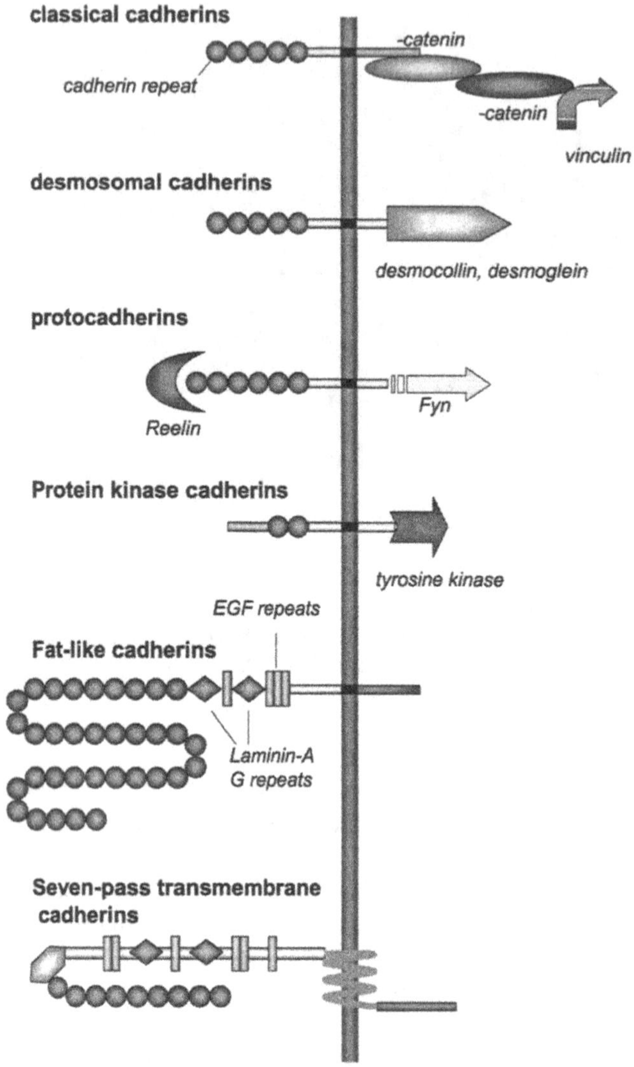

Fig. 2

Figure 2. The cadherin superfamily. Schematic overview of representative members of the cadherin superfamily. See text for details.

A remarkable feature of protocadherin genes is their genomic organization in clusters, strikingly resembling clustering of the immunoglobulin genes in the mammalian genome, and giving rise to a large variety of similar but distinct transcripts from each cluster that could be involved in formation and reorganization of synaptic connections in nervous tissues (overview in ref. 17). A recent review about the role of cadherins in both embryonic and neural morphogenesis was published by Tepass et al.[4]

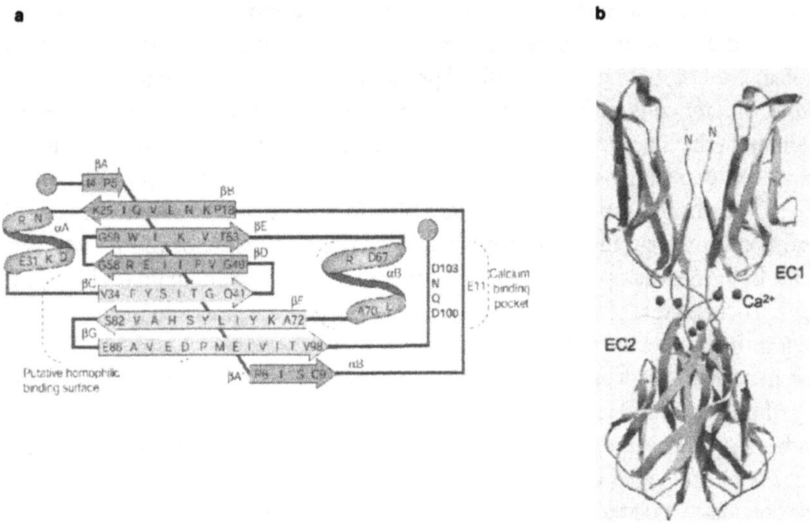

Figure 3. Structure of the cadherin domain and *cis*-dimer formation (based on analysis of EC1 of mouse E-cadherin). (a) Schematic topology of the amino-terminal EC1. βA, βA', βB, βE and βD (green/dark grey), and βC, βF and βG (yellow/light grey) form β-sheets. α-helices are on left and right. Dotted lines indicate the putative homophilic binding surface and the Ca^{2+}-binding pocket. (b) Each cadherin *cis*-dimer associates with six Ca^{2+} ions primarily via the residues of the linker region between EC1 and EC2 (Reprinted by permission from Tepass et al.: Cadherins in embryonic and neural morphogenesis. Fig. 1a, Nat Rev Mol Cell Biol 2000; 1(2):91-100; copyright 2001 Macmillan Magazines Ltd.).

ROLE OF CADHERINS IN HUMAN DISEASE

Prior to the elucidation of the role of cadherins in sensory disorders, members of this protein family were mainly known to be implicated in malignancies. E-cadherin mutations were found in invasive gastric cancer and various other neoplasms, mainly of the digestive system. The role of cadherins in cancerogenesis is thought to be related to impaired cell adhesion, which, in turn, leads to a higher degree of invasiveness.

ROLE OF CADHERINS IN HUMAN RETINA

The expression (and its variability during development) as well as the localization of various cadherins was documented in several publications (see ref. 3 and references therein). The particular importance of three cadherins, cadherin-23, protocadherin-15, and cadherin-3, in retinal function has been revealed only recently by the identification of cadherin gene mutations in autosomal recessive disorders, that all show retinal pathology as a common feature.

Recent studies in mouse and human showed that cadherin-23 is expressed in a variety of tissues, including the neurosensory epithelia of the inner ear and the retina.[18-20] Indirect immunofluorescence on the murine retina with an antiserum generated against the cytoplasmic domain at the C-terminus of the human protein indicates that cadherin-23 is localized primarily in two distinct compartments of the photoreceptor cells, at the synapse and the inner segment (Fig. 4). Anti-cadherin-23 staining in the inner segment is rather diffuse and not restricted to membranes. This may be due to staining of de novo synthesized protein in diverse cellular compartments. The bright staining of the outer plexiform layer of the retina suggests that cadherin-23 is a prominent component of the ribbon synapse of rod photoreceptor cells. In the nervous system, both sides of synaptic junctions contain highly specialized structures that promote rapid and efficient signal transmission from pre-synaptic terminal to post-synaptic membrane (for review see refs. 21 and 22). While the complex cytomatrix of post-synaptic density is thought to be important for clustering of post-synaptic receptors, the numerous structural elements at the pre-synaptic button may be necessary for exocytosis of synaptic vesicles at the pre-synaptic active zone. Molecular analysis of ribbon synapses demonstrated that these specialized synapses, that transmit signals both from auditory hair cells and photoreceptor cells, exhibited an even higher complexity in their composition.[23] The detection of cadherin-23 at the ribbon synapse of the photoreceptor cells by immunofluorescence suggests that cadherin-23 is required for the proper function of ribbon synapses, perhaps by forming adhesive contacts. It is commonly accepted that cell-cell adhesion molecules of the two synaptic sides may interact with each other via their extracellular domains protruding into the extracellular space of the synaptic cleft keeping components of the cytomatrix of both synaptic membranes well-organized.[21]

Cadherin-23 Mutations in Usher Syndrome Type 1D

Usher syndrome (USH) is an autosomal recessive disorder characterized by sensorineural hearing loss and early onset visual impairment due to retinitis pigmentosa (RP), a degenerative disease of photoreceptors (overview in ref. 24). Three USH subtypes are distinguished according to the degree of clinical symptoms. Usher syndrome type 1 (USH1) is the most severe form with profound congenital deafness, vestibular dysfunction, and early onset RP, and is a common cause of deaf-blindness in developed countries.[25] In addition to the clinical differences between the different subtypes, USH is also heterogeneous genetically. To date, seven loci have been mapped for USH1 (*USH1A-USH1G*), whereas four of the underlying gene defects have been identified (overview in refs. 26, 36).

Using a positional candidate approach to identify the USH1D gene mapped previously to the long arm of chromosome 10, a novel member of the cadherin gene superfamily, *CDH23*, was identified. *CDH23* encodes a protein of 3,354 amino acids (referred to analysis of human retinal mRNA) with a single transmembrane domain and 27 cadherin repeats, and is expressed in a wide range of tissues, including the cochlea and retina. Mutations in *CDH23* were shown to underlie both USH1D[18] and an autosomal recessive non-syndromic form of deafness (*DFNB12*, ref. 19).

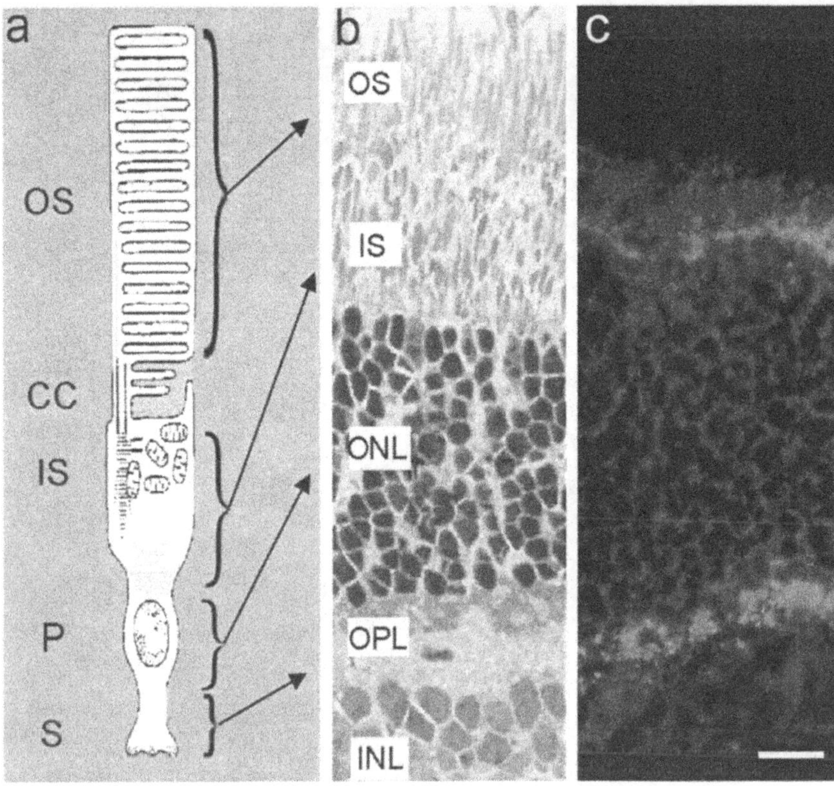

Figure 4. Localization of cadherin-23 in the retina by immunofluorescence. (a) Diagram of a vertebrate rod photoreceptor cell. Outer segment (OS), connecting cilium (CC), inner segment (IS), perikaryon with nucleus (P), synaptic terminal (S). (b) Light microscopic image of a semi-thin section through the mouse retina. In the outer plexiform layer (OPL), rod photoreceptor cells are connecting to the 2nd order retinal neurons (bipolar cells, horizontal cells) via ribbon synapses. Outer nuclear layer (ONL), inner plexiform layer (IPL). (c) Indirect immunofluorescence in a longitudinal cryosection through the mouse retina. Note prominent anti-cadherin-23 immunofluorescence (Alexa488, Molecular Probes) present in the ONL at the synapses of photoreceptor cells. Additional staining is found in the IS of photoreceptor cells. Bar: 8 μm

All but one of the *CDH23* mutations identified in USH1D patients occurred in portions of the gene that encode the extracellular part of the protein. The only mutation affecting the cytoplasmic domain reported so far was found in an atypical case of USH1 with mild retinal phenotype. As shown in Figure 5, for all but one of the *USH1D* mutations, a truncated gene product is predicted. Missense mutations have been found only in USH1 patients with more 'severe' mutations on the other allele (compound heterozygosity) or, if homozygous, in patients with atypically mild retinal phenotype.[18] In contrast, all disease relevant changes identified in patients with non-syndromic deafness (*DFNB12*) were missense mutations.[19] These observations suggest that the inner ear function is already sensitive to 'minor' changes caused by

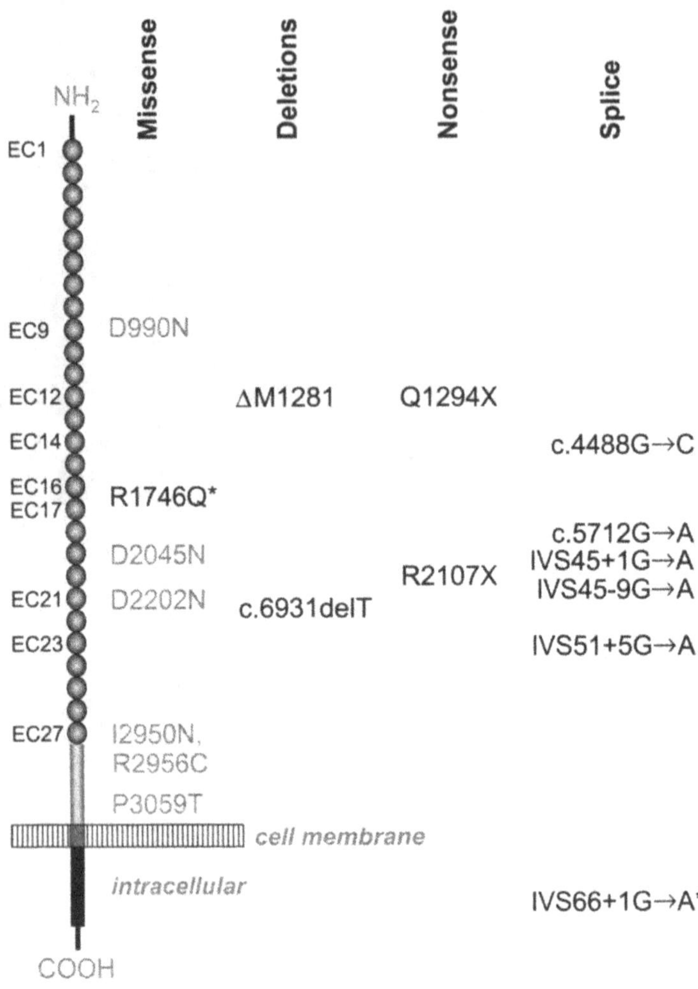

Figure 5. Cartoon of the molecular structure of human *CDH23*, with the position of disease-relevant mutations (on the right) and ECs (on the left). *CDH23* mutations causing non-syndromic deafness (DFNB12) are in grey letters whereas mutations that lead to additional retinal affection (USH1D) are given in black letters. A star indicates mutations associated with mild retinal phenotype. Note that autosomal-recessive DFNB12 is caused by missense mutations only, whereas in USH1D, all mutations but R1746Q (that was found in homozygous state only in cases of atypical USH1), predict a truncated gene product.

missense mutations, whereas heavily reduced or absent protein function result, in addition, in retinal impairment. A summary of *CDH23* mutations in patients with DFNB12 and USH1D is shown in Figure 5. Of the 24 disease alleles identified to date, a high proportion (58.5%) predict aberrant splicing. In a panel of 52 USH1 patients, *CDH23* mutations accounted for about 10% of cases.[27]

The orthologous murine gene, *cdh23,* is mutated in *waltzer* (*v*), a mouse model for USH1D.[18] All *waltzer* mutants analysed to date have *cdh23* mutations, for which a loss of function of the gene product is predicted.[28-30] As the *v* mouse presents no obvious retinal pathology, only the inner ear morphology was investigated in greater detail. Stereocilia of hair cells in *v* mutants are heavily disorganized suggesting that cdh23 may have a "cross-linking" function for stereocilia in hair bundle formation.[20] Clearly, it is unknown at present whether the reported defects in development and morphogenesis result from impaired Ca^{2+}-dependent adhesion or from loss of other yet unknown functions of cdh23, or both.

The retina is a complex system of neuronal cells connected with each other and arranged in a well-defined order. Defects in cell adhesion may impair both development and maintenance of this architecture. In contrast to the inner ear symptoms, that seem to be due to developmental defects impairing the structure of stereociliae, retinitis pigmentosa in USH1D patients (the morphological basis of which has not been yet investigated) is unlikely to result from a comparable mechanism. It is possible that *CDH23* mutations affect rather the maintenance of the retinal structure than its development. Yet, the precise roles of CDH23 in the human retina still await to be determined. Based on our preliminary results on histological localization of cadherin-23 in the mature mammalian retina, it is tempting to speculate that retinitis pigmentosa in patients with Usher syndrome type 1D may in part result from a functional impairment of the ribbon synapse of photoreceptor cells, and hence a defect in signal transmission. It will be interesting to see whether the cadherin-23 deficient *waltzer* mice exhibit defects in synaptic function and whether cadherin-23 and the products of the other Usher 1 genes are assembled at the photoreceptor synapse to a protein complex as recently suggested by Petit (2001) for the stereovilli of the mechanosensitive hair cells.

Protocadherin-15 Mutations in Usher Syndrome Type 1F

USH1F, with the disease locus near the *USH1D* locus on chromosome 10, represents a condition clinically indistinguishable from the other USH1 syndromes. In a positional cloning approach, mutations in a novel cadherin gene, *PCDH15*, were identified both in USH1F families[31,32] and the corresponding mouse model *ames waltzer (av).*[33] For all mutations identified in human studies, a truncated PCDH15 protein, and therefore a loss of function is predicted. *PCDH15* is expressed in retina, brain, cochlea, lung, and kidney. As in the mouse models for USH1B and USH1D (*shaker1* and *waltzer*, respectively), *av* mice have inner ear defects but no retinal degeneration. Nonetheless, electroretinography in different *shaker1* mice showed, compared to unaffected mice, a weaker response to light, with reduced a- and b-waves documenting an abnormal physiological situation.[34] Clearly, the same could also be true for the cadherin mutants.

Cadherin-3 Mutations in Hypotrichosis with Juvenile Macular Dystrophy

Sprecher et al[35] reported the identification of a protein-truncating 1 bp deletion in the *CDH3* gene encoding P-cadherin in patients with hair loss and progressive macular degeneration leading to early onset visual handicap. CDH3 belongs to classical cadherins and the gene has previously been shown to be expressed in the retinal pigment epithelium. The 981delG mutation predicts a protein lacking three of the five extracellular domains, the transmembrane domain, and the cytoplasmic tail. The mutation may therefore define a functional null allele. Of note, loss of CDH3 in mice does not cause hair or retinal abnormalities, which might be due to expression of other cadherins in the affected tissues and/or functional redundancy.

CONCLUSION

To date, several of the mutant proteins of the cadherin superfamily has been discovered to be causal for various retinal diseases. Three cadherin molecules, though expressed in several tissues throughout the body, are implicated in distinct retinal (and, as in USH, cochlear) phenotypes: CDH23, PCDH15, and CDH3. Most likely, the list is far from being complete. The process, by which mutations lead to photoreceptor degeneration is still not fully understood. Data from the inner ear phenotype of USH1 mouse models suggest that loss of cell adhesion is a crucial event. More experimental work is needed to investigate the functions of cadherins involved in retinopathies, also with respect to other putative functions than adhesion, such as signal transduction.

ACKNOWLEDGMENT

The experimental work carried out in the authors' laboratory and described in this study was financially supported by The Foundation Fighting Blindness, FAUN-Stiftung, Deutsche Forschungsgemeinschaft (Wo548/4), and the Forschung Contra Blindheit-Initiative Usher Syndrome.V. The authors thank Dr. Irene Huber and Boris Reidel for helpful experimental support.

REFERENCES

1. Christofori G, Semb H. The role of the cell-adhesion molecule E-cadherin as a tumour-suppressor gene. Trends Biochem Sci 1999; 24(2):73-76.
2. Nollet F, Kools P, van Roy F. Phylogenetic analysis of the cadherin superfamily allows identification of six major subfamilies besides several solitary members. J Mol Biol 2000; 299(3):551-572.
3. Angst BD, Marcozzi C, Magee AI. The cadherin superfamily: diversity in form and function. J Cell Sci 2001; 114(Pt 4):629-641.
4. Tepass U, Truong K, Godt D et al. Cadherins in embryonic and neural morphogenesis. Nat Rev Mol Cell Biol 2000; 1(2):91-100.

5. Gumbiner BM. Cell adhesion: The molecular basis of tissue architecture and morphogenesis. Cell 1996; 84(3):345-357.
6. Stevens CF. A million dollar question: does LTP = memory? Neuron 1998; 20(1):1-2.
7. Brieher WM, Yap AS, Gumbiner BM. Lateral dimerization is required for the homophilic binding activity of C-cadherin. J Cell Biol 1996; 135(2):487-496.
8. Shapiro L, Fannon AM, Kwong PD et al. Structural basis of cell-cell adhesion by cadherins. Nature 1995; 374(6520):327-337.
9. Sivasankar S, Brieher W, Lavrik N et al. Direct molecular force measurements of multiple adhesive interactions between cadherin ectodomains. Proc Natl Acad Sci USA 1999; 96(21):11820-11824.
10. Shan WS, Tanaka H, Phillips GR et al. Functional cis-heterodimers of N- and R-cadherins. J Cell Biol 2000; 148(3):579-590.
11. Yeaman C, Grindstaff KK, Nelson WJ. New perspectives on mechanisms involved in generating epithelial cell polarity. Physiol Rev 1999; 79(1):73-98.
12. Grindstaff KK, Yeaman C, Anandasabapathy N et al. Sec6/8 complex is recruited to cell-cell contacts and specifies transport vesicle delivery to the basal-lateral membrane in epithelial cells. Cell 1998; 93(5):731-740.
13. Garcia-Castro MI, Vielmetter E, Bronner-Fraser M. N-Cadherin, a cell adhesion molecule involved in establishment of embryonic left-right asymmetry. Science 2000; 288(5468):1047-1051.
14. Senzaki K, Ogawa M, Yagi T. Proteins of the CNR family are multiple receptors for Reelin. Cell 1999; 99(6):635-647.
15. Riehl R, Johnson K, Bradley R et al. Cadherin function is required for axon outgrowth in retinal ganglion cells in vivo. Neuron 1996; 17(5):837-848.
16. Yagi T, Takeichi M. Cadherin superfamily genes: functions, genomic organization, and neurologic diversity. Genes Dev 2000; 14(10):1169-1180.
17. Hamada S, Yagi T. The cadherin-related neuronal receptor family: A novel diversified cadherin family at the synapse. Neurosci Res 2001; 41(3):207-215.
18. Bolz H, von Brederlow B, Ramirez A et al. Mutation of CDH23, encoding a new member of the cadherin gene family, causes Usher syndrome type 1D. Nat Genet 2001; 27(1):108-112.
19. Bork JM, Peters LM, Riazuddin S et al. Usher syndrome 1D and nonsyndromic autosomal recessive deafness DFNB12 are caused by allelic mutations of the novel cadherin-like gene CDH23. Am J Hum Genet 2001; 68(1):26-37.
20. Di Palma F, Holme RH, Bryda EC et al. Mutations in Cdh23, encoding a new type of cadherin, cause stereocilia disorganization in waltzer, the mouse model for Usher syndrome type 1D. Nat Genet 2001; 27(1):103-107.
21. Garner CC, Nash J, Huganir RL. PDZ domains in synapse assembly and signalling. Trends Cell Biol 2000; 10(7):274-280.
22. Garner CC, Kindler S, Gundelfinger ED. Molecular determinants of presynaptic active zones. Curr Opin Neurobiol 2000; 10(3):321-327.
23. Dick O, Hack I, Altrock WD et al. Localization of the presynaptic cytomatrix protein Piccolo at ribbon and conventional synapses in the rat retina: comparison with Bassoon. J Comp Neurol 2001; 439(2):224-234.
24. Keats BJ, Corey DP. The usher syndromes. Am J Med Genet 1999; 89(3):158-166.
25. Boughman JA, Vernon M, Shaver KA. Usher syndrome: definition and estimate of prevalence from two high-risk populations. J Chronic Dis 1983; 36(8):595-603.
26. Petit C. Usher syndrome: From genetics to pathogenesis. Annu Rev Genomics Hum Genet 2001; 2:271-297.
27. von Brederlow B, Bolz H, Janecke A et al. Identification and in vitro expression of novel CDH23 mutations of patients with Usher syndrome type 1D. Hum Mutat 2002;in press.
28. Wada T, Wakabayashi Y, Takahashi S et al. A point mutation in a cadherin gene, Cdh23, causes deafness in a novel mutant, Waltzer mouse niigata. Biochem Biophys Res Commun 2001; 283(1):113-117.

29. Wilson SM, Householder DB, Coppola V et al. Mutations in Cdh23 cause nonsyndromic hearing loss in waltzer mice. Genomics 2001; 74(2):228-233.
30. Di Palma F, Pellegrino R, Noben-Trauth K. Genomic structure, alternative splice forms and normal and mutant alleles of cadherin 23 (Cdh23). Gene 2001; 281(1-2):31-41.
31. Ahmed ZM, Riazuddin S, Bernstein SL et al. Mutations of the protocadherin gene PCDH15 cause Usher syndrome type 1F. Am J Hum Genet 2001; 69(1):25-34.
32. Alagramam KN, Yuan H, Kuehn MH et al. Mutations in the novel protocadherin PCDH15 cause Usher syndrome type 1F. Hum Mol Genet 2001; 10(16):1709-1718.
33. Alagramam KN, Murcia CL, Kwon HY et al. The mouse Ames waltzer hearing-loss mutant is caused by mutation of Pcdh15, a novel protocadherin gene. Nat Genet 2001; 27(1):99-102.
34. Libby RT, Steel KP. Electroretinographic anomalies in mice with mutations in Myo7a, the gene involved in human Usher syndrome type 1B. Invest Ophthalmol Vis Sci 2001; 42(3):770-778.
35. Sprecher E, Bergman R, Richard G et al. Hypotrichosis with juvenile macular dystrophy is caused by a mutation in CDH3, encoding P-cadherin. Nat Genet 2001; 29(2):134-36.
36. Mustapha M, Chouery E, Torchard-Pagnez D et al. A novel locus for Usher syndrome type I, USH1G, maps to chromosome 17q24-25. Hum Genet 2002; 110(4):348-350.

GUANYLATE CYCLASE ACTIVATING PROTEINS, GUANYLATE CYCLASE AND DISEASE

Richard J. Newbold,[1] Evelyne C. Deery,[1] Annette M. Payne,[2] Susan E. Wilkie,[3] David M. Hunt[3] and Martin J. Warren[1]

ABSTRACT

A range of cone and cone-rod dystrophies (CORD) have been observed in man, caused by mutations in retinal guanylate cyclase 1 (RetGC1) and guanylate cyclase activating protein 1 (GCAP1). The CORD causing mutations in RetGC1 are located at a mutation "hot spot" within the dimerisation domain, where R838 is the key residue. Three disease causing mutations have been found in human GCAP1, resulting in cone or cone-rod degeneration. All three mutations are dominant in their effect although the mechanism by which the P50L mutation exerts its influence remains unclear although it might act due to a haplo-insufficiency, arising from increased susceptibility to protease activity and increased thermal instability. In contrast, loss of Ca^{2+} sensitivity appears to be the main cause of the diseased state for the Y99C and E155G mutations. The cone and cone-rod dystrophies that are caused by mutations in RetGC1 or GCAP1 arise from a perturbation of the delicate balance of Ca^{2+} and cGMP within the photoreceptor cells and it is this disruption that is believed to cause cell death. The diseases caused by mutations in RetGC1 and GCAP1 prominently affect cones, consistent with the higher concentrations of these proteins in cone cells.

[1]School of Biological Sciences, Queen Mary, University of London, Mile End Road, London, E1 4NS, UK. [2]Institute of Cancer Genetics and Pharmacogenomics, Department of Biological Sciences, Brunel University, Uxbridge, Middlesex UB8 3PH, UK. [3]Institute of Ophthalmology, University College London, 11-43 Bath Street, London, EC1V 9EL, UK.

INTRODUCTION

To date three disease causing mutations have been found in human guanylate cyclase activating protein 1, resulting in cone degeneration. For the Y99C and E155G mutations the loss of Ca^{2+} sensitivity appears to be the main cause of the diseased state. All three mutations are dominant in their effect although the mechanism by which the P50L mutation exerts its influence remains unclear although it is possible that it acts due to a haplo-insufficiency, which results in increased susceptibility to protease activity and increased instability.

The mutations in human retinal guanylate cyclase 1 that cause dominant cone-rod dystrophy (CORD) are located in a different region of the protein (dimerisation domain) than other mutations which produce the more severe recessive phenotype described as Leber's Congenital Amaurosis. The active form of RetGC1 is believed to be a dimer with the two RetGC1 molecules interacting via a coiled coil motif that enables dimerisation, see Figure 3. The CORD mutations are found at a mutation "hot spot" within the dimerisation domain, where R838 is the key residue. The leading theory is that altered Ca^{2+} sensitivity, caused by perturbation of the normal dimerisation process, yields an imbalance of cellular Ca^{2+} and cGMP concentrations which results in photoreceptor cell degeneration.

RetGC1, GCAP1 and Phototransduction

The phototransduction cascade and the dark state recovery can be thought of as an extremely complicated but highly responsive method of modulating [cGMP]. Photoactivation of rhodopsin leads to the replacement of GDP by GTP on transducin which enables transducin to activate phosphodiesterase (PDE) when it displaces two inhibitory subunits. The activated PDE then hydrolyses cGMP to 5'-GMP. The reduction in [cGMP] closes the gated cation channels in the plasma membrane that are normally held in the open position by bound cGMP. The cell then hyperpolarizes and recovers. Photo activated rhodopsin is subsequently phosphorylated and binds arrestin, with retinal re-associating with the opsin in the 11-*cis* configuration. In addition, PDE re-associates with its inhibitory subunits preventing further hydrolysis of cGMP. With closed cation channels, the [$Ca^{2+}{}_{cellular}$] falls and retinal guanylate cyclase is activated and produces cGMP. As the concentration of cGMP rises the channels are reopened and the normal dark potential of the cell returns. It is therefore the cellular concentration of cGMP that is crucial for the maintenance of Ca^{2+} homeostasis.[1-3]

The key enzyme involved in cGMP production in cone photoreceptor cells is RetGC1. Synthesis of cGMP, in the cell, by RetGC is stimulated by a guanylate cyclase activating protein (GCAP), with GCAP1 implicated as the main activating species in cone cells. Thus, GCAP1 and RetGC1 are crucial components of the cone cells and their role is highlighted in Figure 1. As well as RetGC1, there is also a second closely related guanylate cyclase, RetGC2, which has not yet been associated with any eye diseases. GCAP1 is a member of the EF-hand super-family of

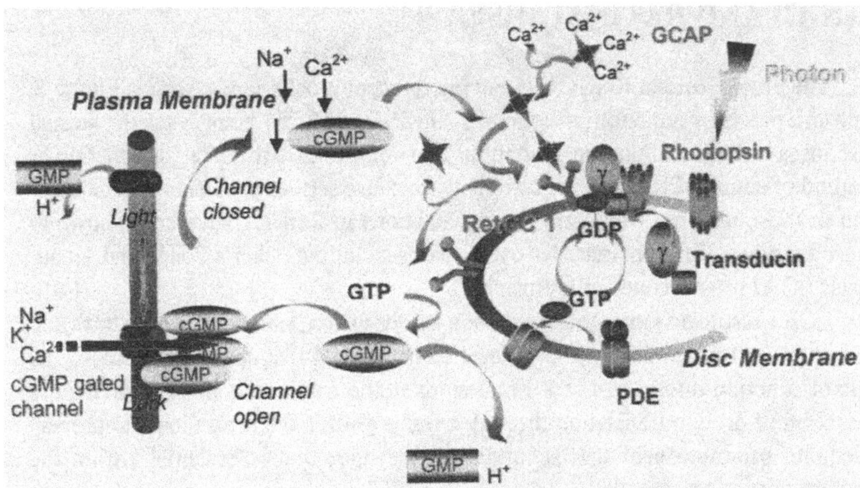

Figure 1. A cartoon representation of the phototransduction cascade.

calcium binding proteins, and three guanylate cyclase activator proteins (GCAP1, GCAP2 and GCAP3) are present in man. These proteins stimulate RetGCs, increasing [cGMP]. No mutations of GCAP2,[4] despite an in-depth search[5] nor of GCAP3[6] have yet been demonstrated to cause any retinal abnormalities. The roles of RetGC2, GCAP2 and GCAP3[7] have not been as clearly defined as those of RetGC1 and GCAP1 although it appears that RetGC2 is not involved in the phototransduction cascade in cone cells and is limited to activity within rods.[8]

Cone and Cone-Rod Dystrophies

Cone dystrophies are a subgroup of the inherited chorioretinal dystrophies that are characterized by progressive loss of colour vision, photophobia, visual acuity and central visual field, coupled with preservation of rod function. By contrast in cone-rod dystrophy abnormal rod function may be part of the initial symptoms but rod involvement may be less severe, or occur later than the cone dysfunction. As the disease progresses the retinal pigment epithelium (RPE) may take on a granular appearance followed by central atrophy.

Autosomal dominant cone–rod dystrophy (adCORD) is a distinct type of chorioretinal disease causing initial degeneration of cone photoreceptors followed by loss of rods.[9,10] The disease is generally characterized by an early loss of visual acuity and colour discrimination, associated with loss of cones, followed by nyctalopia and progressive peripheral field loss as the rods subsequently degenerate.[11]

Numerous mutations in a range of proteins have been identified as causing autosomal dominant cone and cone-rod dystrophies, including mutations in peripherin/human retinal degeneration slow (*RDS*)[12], retinal guanylate cyclase;[9,10] the cone-rod homeobox (*Crx*)[13,14]; and three disease causing mutations in GCAP.[15-17]

RetGC1 AND RETINAL DISEASE

The pivotal role of RetGC1 within photoreceptor cells can be seen in Figure 1, with the precise regulation of cellular [cGMP] and [Ca^{2+}] being vital for normal responses to light stimulation. Mutations that compromise the catalytic function or method of action of RetGC1 are also likely to cause serious problems to cone function in those patients who carry them. Mutations in RetGC1 have been shown to cause a range of different inherited eye diseases including Leber's Congenital Amaurosis (LCA1)[18] and cone-rod dystrophy.[9,10]

LCA1 results in either total or drastic loss of vision, at or shortly after birth and to date seven mutations have been detected in RetGC1 that cause this disease. The loss of function due to mis-sense mutations in the catalytic domain or premature cessation of protein translation through a frame shift, have been shown to prevent adequate production of this secondary messenger that is essential within the phototransduction cascade.[19]

The missense mutations within RetGC1 that cause cone and cone-rod dystrophies are found in the dimerisation domain. With RetGC1 shown to be active as a dimer,[20,21] disruption of the dimerisation processes would be expected to affect the rate of cGMP synthesis. The first reported cone-dystrophy-causing mutations in RetGC1 were discovered within the dimerisation domain at positions 837 and 838.[9] A double mis-sense mutation E837D/R838S[22] was found in the original CORD6 family, with a second missense mutation, R838C, in three other families with dominant cone-rod dystrophy. The cone-rod dystrophy in the CORD6 family displays an early onset, with loss of central vision reported before seven years of age, early loss of visual acuity and colour discrimination, with peripheral field loss by the fourth decade. In addition, a marked photophobia is present, particularly when dark adapted. Electroretinography showed no detectable cone responses early in disease, with progressive abnormality of rod responses appearing later.[22]

A separate affected family[10] was found to have three consecutive mis-sense mutations: E837D, R838C and T839M, located within the molecule's highly conserved dimerisation domain. All affected individuals displayed an early cone dysfunction characterised by decreased vision acuity, with severe colour dyschromatopsia and photophobia, during the first decade of life. Electrophysiological testing revealed marked loss of cone function, with rod function relatively well preserved. During the second and third decades, visual acuity decreased dramatically and by the fourth decade, peripheral visual field loss and progressive night blindness were observed.

An additional mutation in the dimerisation domain has also been reported: R838H.[23] Later comparisons of families with a variety of mutations in the dimerisation domain of RetGC1 showed that the R838C and R838H mutations showed milder phenotypes than did those observed for carriers of E837D and R838S. Lifelong poor vision in bright light was observed, with major reduction in visual acuity not occurring until after their late teens for most subjects. Fundus abnormalities were confined to the central macula, and increasing central atrophy was noted with age. Electrophysiological testing revealed a marked loss of cone function with only minimal rod involvement, even in older subjects. There was also some variation in disease severity within the three families with the R838C mutation.[24,21]

Biochemical Role of RetGC1

A schematic diagram highlighting the various segments of the protein is shown in Figure 2. The key areas of the protein are the catalytic domain (where mutations have been shown to be responsible for LCA[18]) and the dimerisation domain (where cone-rod dystrophy-causing mutations are located[9,10]). In addition, there is a membrane binding segment, a kinase-like domain and an extracellular domain (located in the lumen of the photoreceptor disks). A schematic diagram showing the basal and activated, dimerized forms of RetGC1 is shown in Figure 3.

The site of the interaction between GCAP1 and RetGC1 has been probed using replacement of the predicted interaction domain with a corresponding sequence from a GCAP-insensitive GC.[26] This showed that the region between β-strand 3 and α-helix 4 is required for stimulation of RetGC1 activity by GCAP.

The dimerisation domain, is predicted to be an amphipathic α-helical coil and this region has been described as a mutation 'hot spot' region[23] with R838C, R838H, E837D/R838S and E837D/R838C/T839M mutations all being found in this area.

GCAPs and the guanylate cyclases have been shown to form stable complexes at high and low $[Ca^{2+}]$[27,28] and this would appear to be consistent with GCAP's dual role of activating GC activity at low $[Ca^{2+}]$ whilst inhibiting it at high $[Ca^{2+}]$.[29] However, the binding sites on GCAP1 remain to be identified and will probably remain elusive until a high resolution structure of GCAP1 is obtained. In addition any conformational changes that occur when RetGC1 dimerizes will be impossible to characterize without the acquisition of technically challenging, high resolution structural data.

RetGC1's ROLE IN CONE AND CONE-ROD DYSTROPHIES

Mutations in RetGC1 produce disparate disease phenotypes depending on their nature and position within the protein's primary sequence: mis-sense mutations in the catalytic domain and out of frame-induced truncations result in Leber's Congenital Amaurosis,[14] whereas missense mutations in the dimerisation domain produce the less severe cone-rod dystrophies.[9,10]

An important point to note is the dominant nature of the RetGC1-related cone-rod degenerations. By comparison, in several reported cases of LCA only null alleles in homozygotes are deleterious[18] indicating that a reduction of functional RetGC1 to half the level normally found does not cause a significant visual impairment.[30] Hence, for LCA, it is the dramatic loss of functional RetGC1 that causes such a severe phenotype and this cannot be compensated by RetGC2. With the dimerisation domain mutations RetGC1 remains functional but it is the altered Ca^{2+} sensitivity that appears to be the disease-determining factor.

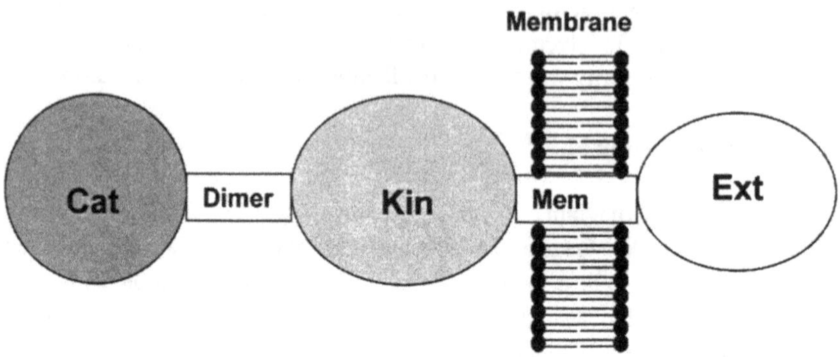

Figure 2. A model of RetGC1 showing the catalytic (Cat), dimerisation (Dimer), kinase (Kin), membrane binding (Mem) and extracellular (Ext) domains of the protein.

RetGC1 activity can be stimulated by either GCAPs or Mn^{2+}/Triton X-100.[31] Using recombinant RetGC1 produced by transient transfection of mammalian cells, wild type RetGC1 was shown to be 2-3 times more active than mutant R838C.[30] Stimulation of the enzyme, as assessed by cGMP production, revealed that mutant RetGC1 was stimulated by GCAP1. Higher affinity for GCAP1 was observed for R838C than for WT RetGC1 and the mutant also required higher concentrations of Ca^{2+} for inactivation than WT enzyme.

The importance of the identity of the amino acid residue at position 838 has also been assessed by production of the naturally occurring disease causing mutants and some artificial mutant forms.[32] The overall catalytic activities of the RetGC1 species are shown in Figure 4. Basal activity and Mn^{2+}-stimulated activity of all RetGC1 mutants was reduced compared with the WT enzyme. Analysis of initial rates of cGMP formation for the mutants indicated similar or higher V_{max} values than for WT enzyme but lower K_m values, with R838H showing an apparent 5-fold greater affinity for GCAP1 than wild type RetGC1. Ca^{2+} titrations of wild type RetGC1 and the R838C and R838H mutants are shown in Figure 5. For wild type RetGC1, the free Ca^{2+} concentration for half-maximal activity was estimated to be 420nM, whereas R838C mutant was shifted to 530nM and R838H shifted to 800nM. The mutants displayed a significant 'tail' of elevated activity at high $[Ca^{2+}]$, with R838C, which shows the smallest shift, retaining >16% of maximal activity. Computer prediction of the ability of the dimerisation domain to form a coiled-coil structure (using Coils[33]) was used to assess the possible effects of the mutations upon the extent of this domain. Using the Coils algorithm, the coiled-coil structure for wild type RetGC1, is predicted to extend for four turns, ending at R838. The three disease causing mutations are predicted to have extended coiled coil motifs.[32]

The striking similarities between the effect of the R838C RetGC1 mutation and that of the Y99C mutation of GCAP1, in causing shifts in GCAP1 Ca^{2+} sensitivity, imply that the two diseases may have a similar molecular etiology. Ramamurthy et al also suggested that the R838C mutant protein displays increased sensitivity to

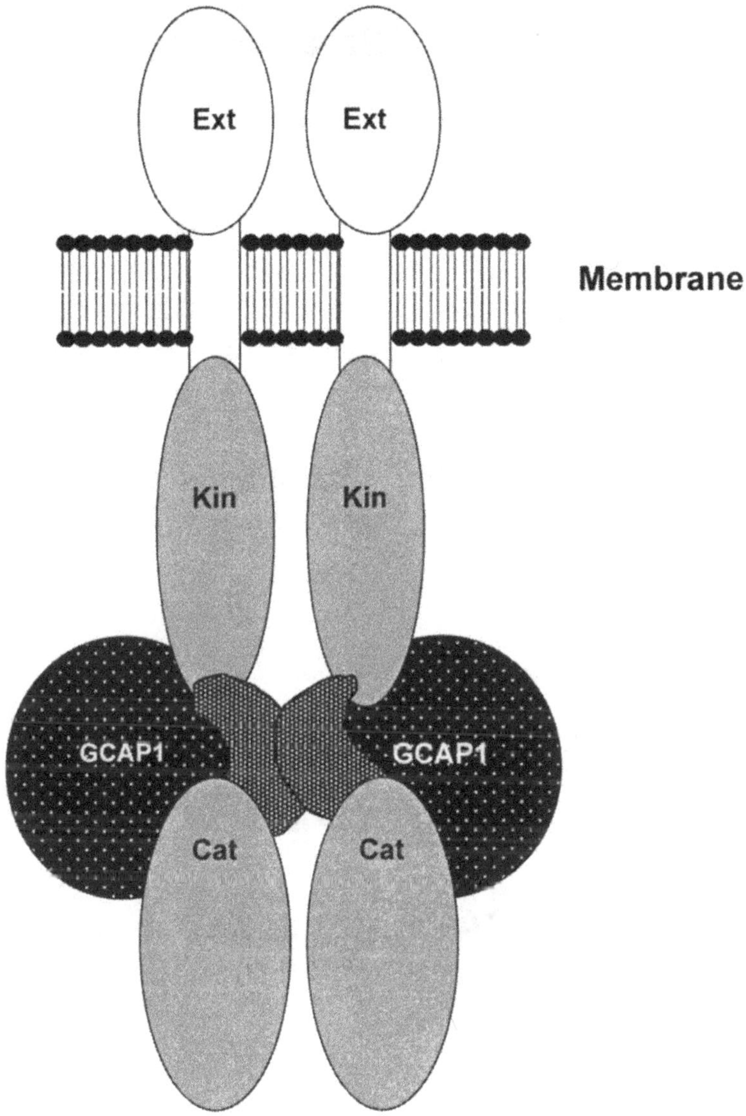

Figure 3. A model of the dimerisation of RetGC1 to the active form in the presence of Ca^{2+}-free GCAP1. The catalytic, kinase, and extracellular domains of RetGC1 are labelled Cat, Kin and Ext respectively.

activation by GCAP1.[21] Thus, the increased sensitivity of R838C was suggested to accelerate synthesis of cGMP, resulting in higher [cGMP] in dark adapted photoreceptors, with concomitant alteration in the balance of [Ca^{2+}], enzyme activity, and cGMP modulated channel activity leading to cone and rod degeneration.

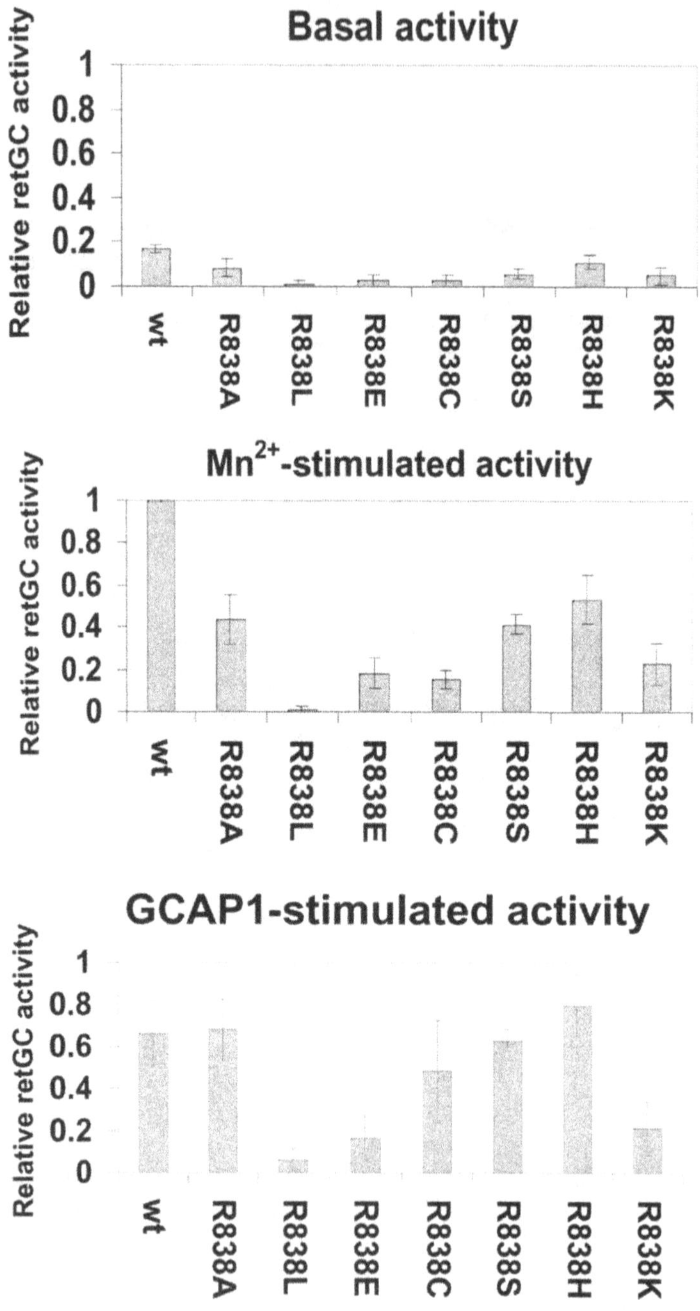

Figure 4. Relative guanylate cyclase activities of WT RetGC1 and RetGC1 mutant forms (Diagram adapted from Wilkie et al 2000[28]).

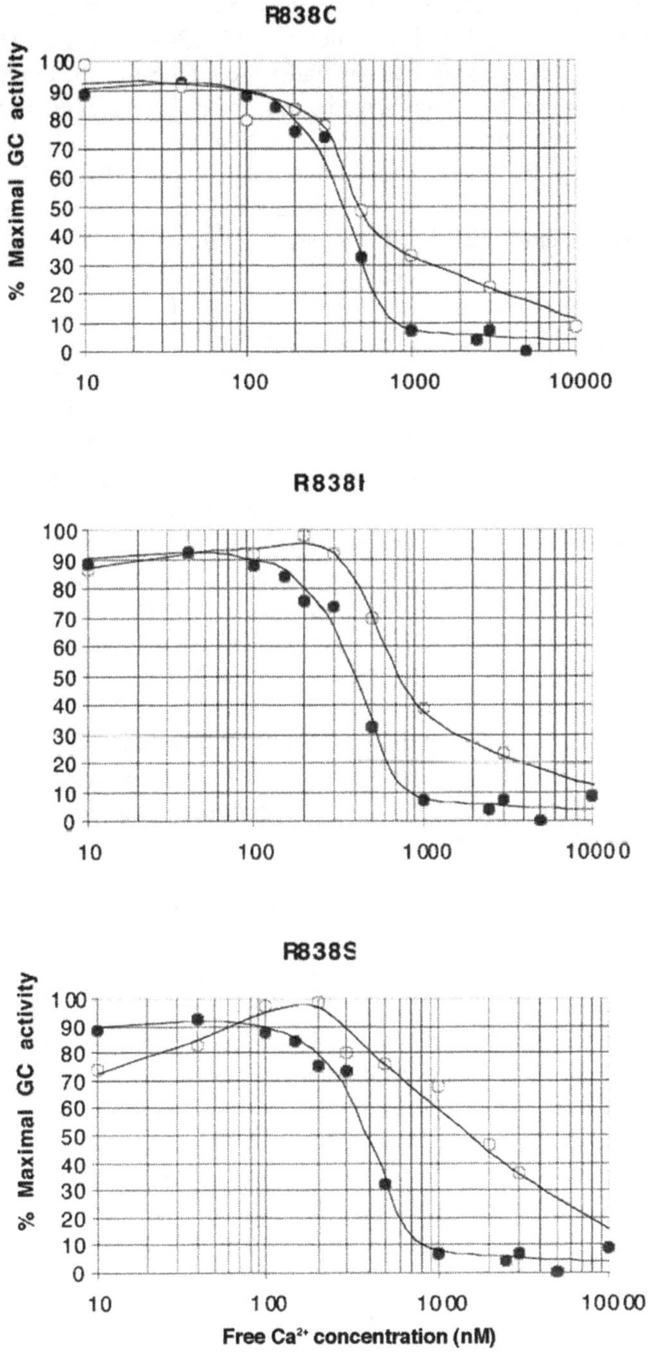

Figure 5. Ca^{2+} sensitivity curves for the guanylate cyclase activities of WT RetGC1 RetGC1 (closed circles) and RetGC1 mutants (open circles). Figure adapted from Wilkie et al 2000.[28]

The RetGC1 mutations associated with dominant cone-rod dystrophy are in conserved residues in the dimerization domain of RetGC-1. The role of this domain in membrane GC regulation is not fully understood, reflecting the limited information available for this large, membrane bound protein. The disease-causing R838C mutation produces a catalytically active homo-dimer, as is seen for the wild type enzyme. However, the mutant is easier to activate, via GCAP1, and harder to turn off. The stabilization of the active conformation of RetGC1 due to stronger interaction between the individual RetGC1 molecules, promoted by the mutation in the dimerisation domain, results in altered cellular levels of cGMP and Ca^{2+} and degeneration of the photoreceptor cells.

By contrast the work by Duda et al on the RetGC1 mutants (E837D, R838C, and E837D/R838C/T839M) did not reveal altered Ca^{2+} sensitivity.[34,35] The study of the single mutation RetGC1s showed that E837D results in no change in basal activity compared with WT RetGC1 and identical, 5-fold activation with GCAP1, and no change in Ca^{2+} sensitivity. By contrast, R838C showed reduced basal activity (1/4 of WT) and 25-fold increased activity in the presence of GCAP1.[34] The triple mutation also showed reduced basal activity (1/4 of WT), and the increased activity in the presence of GCAP1 (18-fold) was insufficient to compensate for the lowered basal level. The reduced activity was estimated to result in only 35% of the cGMP production observed for WT RetGC1. Again, no change in sensitivity to Ca^{2+} was observed. However, cross-linking studies indicated that the mutant's dimerisation ability had increased by 60-70% over that seen for wild type RetGC1.[35] The reduction in intrinsic cyclase activity in those RetGC1s containing the R838C substitution was proposed to result in insufficient cGMP production for proper function of the phototransduction machinery.[34,35] The authors suggested that in the triple mutant this disruption occurred because the mutant cyclase is unable to change from its monomeric to the dimeric form, which was supported by cross-linking experiments that showed a 60-70% reduction in dimerisation. Reduction of cyclase activity in the triple mutant was predicted to result in closure of almost all the cGMP-gated channels, thereby lowering [Ca^{2+}free] and resulting in a false light adapted state for the photoreceptors. The shift of the response light curve would result in photoreceptors perceiving illumination without being in a state of illumination. Therefore under bright illumination, low levels of cGMP synthesis would result in permanent photo-excitation and hyperpolarization.[35]

Ramamurthy et al investigated the role of amino acid residue R838 within the dimerisation domain, the only amino acid residue that is consistently substituted in all cases of RetGC1-based autosomal dominant cone-rod dystrophy (adCORD) mutations.[21,24] The roles of the individual amino acid substitutions in the triple (E837D/R838C/T839M) mutant were assessed by production of single mutations for each amino acid residue. Although the T839M mutation was shown to result in a 2-fold increase in basal activity and sensitivity to GCAP1, it was shown that only substitutions at position 838 resulted in altered Ca^{2+} sensitivity, which was proposed as the sole cause of RetGC1-produced adCORD. The model produced for wild type RetGC1 predicted that the functional protein is a dimer with the two sub-units forming a coiled coil in the dimerisation domain. In the wild type protein, the amino acid residues after R838 splay apart, whereas the R838C and R838S mutants are predicted to show continuations of the coiled coil region beyond amino acid residue 838. Further artificial mutations (R838E and R838D were shown to result in a RetGC1 that displayed 50% activity even at [Ca^{2+}] of 27μM, whereas the conservative mutation, R838K, which modelling predicted might maintain the wild-type hydrogen bonding network, produced an enzyme with unaltered Ca^{2+} sensitivity.[17]

The results and conclusions reached in Duda et al[35] conflict with those subsequently presented by Tucker et al.[36] In the later paper, the authors found that the triple mutant reduced the cyclase activity of RetGC1, in agreement with Duda et al, but they suggested that the effect on intrinsic activity is insufficient to explain the disease phenotype. Ramamurthy et al proposed that it is the shifted Ca^{2+} sensitivity not the altered intrinsic activity that is linked to the adCORD mutations.[21] This argument was further developed to produce a hypothesis where the dimerisation strength of the coiled-coil domain must be fine-tuned for proper regulation by Ca^{2+}. In contrast to the findings of Duda et al,[34,35] it was suggested that the mutations enhance dimerisation, i.e., they stabilize the active state and hence higher Ca^{2+} concentrations are required to overcome the increased stability of the active state. The shifted Ca^{2+} sensitivity in the mutants is accompanied by an apparent increase in sensitivity to GCAP-1 in the absence of Ca^{2+}. The authors state that it is unlikely that this represents increased affinity of RetGC1 for GCAP1 but instead suggest that when GCAP1 binds to RetGC1 in the absence of Ca^{2+} the active state is stabilized by the mutations so that abnormally high activity is detected.

Ramamurthy et al identify R838 as the key residue in the disease state given that all of the reported adCORD mutations in RetGC-1 include a substitution at R838.[21] Hydrogen bonding and charge interactions involving Arg838 were implicated as the main determining factor in the strength of the coiled-coil interaction, which establishes the Ca^{2+} sensitivity of RetGC-1. The mutations at position 838 were suggested to dominantly enhance cGMP synthesis even at the highest Ca^{2+} concentrations. Molecular dynamics simulations predicted that mutations at position 838 would disrupt a network of salt bridges, allowing an extension of coiled-coil structure. The position of R838 at the edge of the coiled coil motif is similar to the positions of other deleterious mutations observed at the ends of other coiled coils for proteins such as keratin[37] and spectrin.[38] Hence, it was suggested that inhibition of the mutant cyclase activity in dark adapted photoreceptor cells requires a higher Ca^{2+} concentration, because dissociation of the dimer is a less favorable process. Some mutants retain 15-30% of maximal activity even at 30μM Ca^{2+} and hence this was predicted to result in constitutively accelerated cGMP synthesis in darkness in human photoreceptor cells, leading to dominant cone-rod degeneration via methods already discussed in this review.

The differences between the results obtained by Duda et al[34,35] on the one hand and Ramamurthy et al[21] and Wilkie et al[32] on the other are puzzling and may reflect the different constructs used by the different groups. The work by Ramamurthy et al and Wilkie et al provide a simple model for the regulation of RetGCs by GCAPs and Ca^{2+}, where Ca^{2+} binds to GCAP1 in the dark state, producing a conformational change which pulls apart the polypeptide chains of the catalytic dimer, thereby reducing cGMP synthesis. When $[Ca^{2+}]$ falls after light stimulation, Ca^{2+}-free GCAP1 activates RetGC1 forcing the catalytic domains closer together.[21]

GCAP1 AND RETINAL DISEASE

Three mutations in human guanylate cyclase activating protein 1 (GCAP1) have been shown to be responsible for cone and cone-rod dystrophies.[15-18] These degenerative retinal diseases show a variety of symptoms, of varying severity, that vary from patient to patient within the same family.[16] Given GCAP1's key role in modulation of the phototransduction cascade and the high concentration of GCAP1 in cone cells[39-41] it is not surprising that mutations within this key phototransduction cascade protein result in serious problems for those who carry them.

Disease Presentation

The first detected mutation in GCAP1, which was also the first report of an identified gene defect causing cone dystrophy, was Y99C.[15] The sufferers displayed a variability of expression of the disease, with the first reported symptom of reduced visual acuity with associated loss of colour vision between the ages of 20 and 40 years. Prior to visual loss, changes in the retinal pigment epithelium (RPE) and the macula were observed with development of central atrophy. Visual field testing showed central loss but preservation of peripheral visual fields, even in late disease with significant generalized loss of cone function shown by reduction in photopic electroretinograms (ERG).[15] However, even at later stages of the disease, the peripheral visual field is preserved, indicating that the rod cells are largely unaffected.[16]

Patients suffering from retinal degeneration caused by P50L, the second mutation in GCAP1 to be discovered, showed a wider range of symptoms than seen for Y99C, even within the same family.[16] Three sufferers, all from the same family, each exhibited a different phenotype: minimal macular involvement with mild symptoms; a cone dystrophy, with the same phenotype as seen in the Y99C mutation; and a moderately severe cone-rod dystrophy. Age of onset of disease ranged from the late-20s to the mid-50s, and symptoms include granular changes in the RPE, photophobia and reduced central vision. However, in the limited available pedigree there was also an example of accompanying loss of rod function and nyctalopia.

Investigation of the genetic inheritance of the E155G mutation also showed an autosomal dominant pattern of inheritance. E155G is similar to Y99C, in that it causes a severe phenotype. However, the mean age at onset of disease (16 years), as detected by reduced visual acuity, colour vision defects and photophobia is noticeably earlier in patients with E155G than in patients with Y99C.[17]

Biochemical Basis of the Role of GCAP1

As a vital regulatory component of the phototransduction cascade, it is not surprising that mutations in GCAP1 affect photoreceptor function. GCAPs are Ca^{2+}-binding proteins that activate the membrane bound photoreceptor guanylate cyclases RetGC1 and RetGC2 with GCAP1 directly activating RetGC1 and GCAP2 shown to activate RetGC1 and RetGC2.[29,42-44]

GCAP1 is a member of the Ca^{2+}-binding EF-hand protein family that includes calmodulin[45] and recoverin.[46] In contrast to the behavior of other EF–hand proteins, which are stimulated by high Ca^{2+} levels, GCAP1 only acts as an activator at low levels of Ca^{2+}, typically when the $[Ca^{2+}]$ drops below ~300nM.[44] GCAP1 has four EF-hand motifs but only three of these are functional, and able to bind Ca^{2+}, the fourth having changed from a Ca^{2+}-binding motif into a site involved in the direct interaction between GCAP1 and RetGC1,[47] see Figure 6. Similarly the ancestral EF1-hand in GCAP2 has also lost its Ca^{2+} binding capability but is involved in the interaction of GCAP2 with the cyclase.[48]

A crystal structure of recoverin in the Ca^{2+}-free form has been elucidated,[49] as have NMR structures of recoverin in the Ca^{2+}-free[46] and Ca^{2+}-bound states,[50] and a NMR structure of GCAP2.[51] Currently, recoverin provides the only model for the conformational changes that EF-hand proteins may undergo on Ca^{2+} binding. The NMR structures of myristoylated recoverin in the Ca-free and Ca-bound states reveal considerable changes in the tertiary structure on Ca^{2+} binding, including a dramatic change in the position of the myristoyl group. The myristoyl group in GCAP1 is believed to help target the protein to the membrane and its absence has been observed to prevent GCAP1 activation of RetGC1.[52] A pictorial representation of a model of GCAP1 (unmyristoylated for clarity) is shown in Figure 7, and highlights the positions of the EF-hands and the three mutations in GCAP1 discussed in this review.

ANALYSIS AND DISCUSSION OF THE EFFECTS OF GCAP1 MUTATIONS

When the Y99C mutation was discovered in a cone dystrophy family it was predicted to have a significant effect on the structure of GCAP1.[15] As a member of the EF-hand containing super-family of Ca^{2+} binding proteins, GCAP1 was observed to display a structural similarity to calmodulin, recoverin and calcineurin B at the amino acid level. The degree of similarity between these proteins was declared to be particularly striking in the EF3-hand, which is adjacent to tyrosine 99, and it was suggested that this region would display a highly conserved topography, see Figure 8.[17] The mutation in GCAP1 changes the coding of tyrosine 99 to cysteine, a position which in the other members of the protein super-family is also occupied by an aromatic amino acid, either tyrosine or phenylalanine. Study of the structures of calmodulin,[53,54] recoverin[49] and calcineurin B[55] showed that the large

```
                                    Ancestral EF1
         10         20         30         40         50         60
          |          |          |          |          |          |
GNVMEGKSV EELSSTECHQ WYKKFMTECP SGQLTLYEFR QFFGLKNLSP SASQYVEQMF

     EF2                                         EF3
         70         80         90        100        110        120
          |          |          |          |          |          |
ETFDFNKDGY IDFMEYVAGL SLVLKGKVEQ KLRWYFKLYD VDGNGCIDRD ELLTIIQAIR

                                    EF4
        130        140        150        160        170        180
          |          |          |          |          |          |
AINPCSDTTM TAEEFTDTVF SKIDVNGDGE LSLEEFIEGV QKDQMLLDTL TRSLDLTRIV

        190        200
          |          |
RRLQNGEQDE EGADEAAEAA G
```

Figure 6. Amino acid sequence of human guanylate cyclase activating protein 1 (GCAP1), Accession Number P43080. The positions of the three disease causing mutations (P50L, Y99C and E155G) are in bold and surrounded by a box. The four EF-hands (three functional Ca^{2+}-binding motifs and one nonfunctional ancestral motif are indicated by underlining and bold text.

aromatic side chain fits into a hydrophobic pocket which forms part of the hydrophobic core of the C-terminal domain of the proteins. Hence, changing the tyrosine side chain to a considerably smaller cysteine (van der Waals volume of Cys = 86Å^3 as compared to Tyr = 141Å^3)[56] would be expected to result in a disruption of the hydrophobic packing below the EF3-hand. It was suggested that the mutation could affect the conformation of the C-terminal domain and consequently prevent GCAP1 from binding Ca^{2+} at least in the EF3-hand. Consequently, this deleterious mutation in GCAP1 was predicted to lead to an aberrant change in the concentration of cGMP.[15]

The primary method by which the effects of the mutations in GCAP1 have been assessed is measurement of guanylate cyclase activation.[60] Assay analysis of WT and Y99C GCAP1 suggested activation of RetGC1 by Y99C GCAP1 that was not properly inactivated at high levels of Ca^{2+}.[61] Cyclase activity in the presence of Y99C remained at 60% of the maximal level even at Ca^{2+} concentrations over ten times those required to inactivate the wild type protein, see Figure 9A. In addition, mixing Y99C GCAP1 and wild type GCAP1 in stoichiometric quantities showed that the presence of half the quantity of mutant GCAP1 would still produce considerable (~30%) activation of RetGC1, preventing full recovery.

The Y99C mutant lacks the Ca^{2+} sensitivity observed for the WT protein, and is unable to act as a Ca^{2+} switch. In addition, it was suggested that the mutant protein interferes with the remaining wild-type GCAP, preventing GCAP1 from being switched off. This would leave GCAP1 permanently active and RetGC1 would become constitutively active within the physiological range of free Ca^{2+} concentrations.[62] These experimental observations were suggested to support the inference

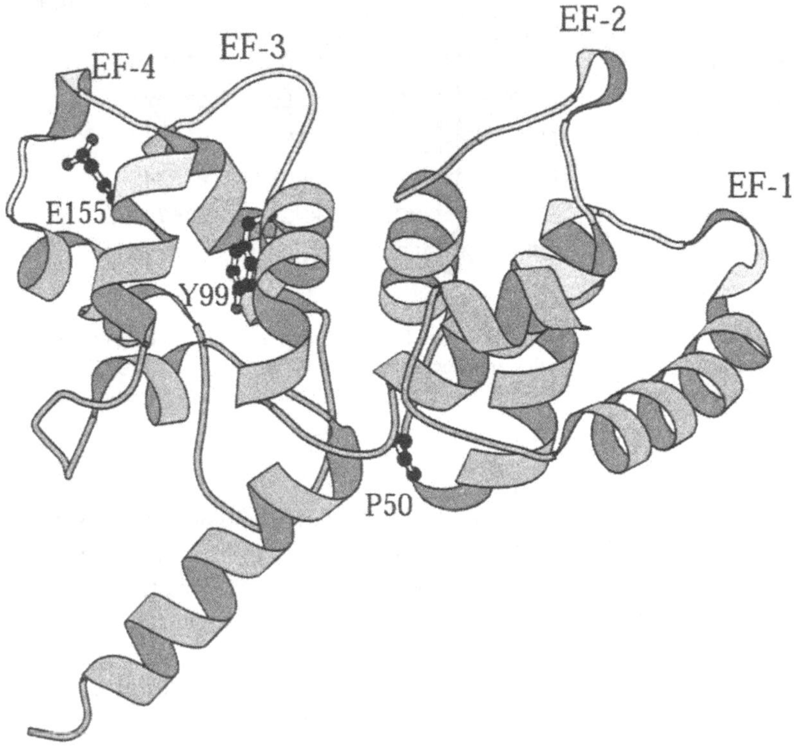

Figure 7. Model of GCAP1 indicating the relative positions of the EF-hands (shown in yellow) and the three disease causing mutations in GCAP1: P50L, Y99C and E155 (shown in ball and stick form).

that the cone degeneration associated with the Y99C mutation in GCAP1 is a result of constitutive activation of cGMP synthesis and the subsequently elevated levels of cytoplasmic cGMP. The concept of interference by Y99C of the normal function of wild type GCAP1 was undoubtedly derived from the hypothesis that GCAP1 activates RetGC1 as a dimer. However, no completely compelling evidence for the existence of an activating dimeric species has been shown and a recent paper by the same group indicated that the activating species is monomeric and that the dimerized form is inactive.[47] Hence the activation of Ret GC1 at high [Ca^{2+}] observed in mixed samples of Y99C and wild type proteins is probably solely due to the constitutive activation provided by Y99C.

The discovery of the disease-causing E155G mutation in GCAP1 was particularly interesting as this amino acid residue is located within the EF4-hand, with the glutamate side chain predicted to act as a bidentate ligand in Ca^{2+} binding. This represented the first disease-associated mutation in GCAP1, or any neuron-specific Ca^{2+}-binding protein within an EF-hand domain that directly coordinates Ca^{2+}. Within functional EF-hands, an invariant glutamate residue is found at position 12, i.e., that

	EF3	Accession Number
human GCAP1	LRWYFKL**Y**DVDGNGCIDRDE	**P43080**
human GCAP2	LKWTFKI**Y**DKDGNGCIDRLE	**Q9UMX6**
human GCAP3	LKWYFKL**Y**DADGNGSIDKNE	**O95843**
human recoverin	LEWAFSL**Y**DVDGNGTISKNE	**P35243**
human calcineurin	LRFAFRI**Y**DMDKDGYISNGE	**P06705**
human vislip-1	LNWAFNM**Y**DLDGDGKITRVE	**P28677**

Figure 8. Pile up of amino acid sequences around EF3 for GCAP1 and closely related EF-superfamily proteins showing the conserved bulky aromatic amino acid residues adjacent to EF3. The conserved tyrosine residue adjacent to EF3 is highlighted in bold and within a box, with the EF-hand shown by underling.

normally occupied by E155 in GCAP1, and this has been shown to be essential for Ca^{2+} coordination.[62] The substitution of Glu155 in EF4 by a glycine effectively reduces the number of coordinating oxygen ligands from seven to five and has a drastic effect on Ca^{2+} binding. A model of the EF4 motif based on the NMR structure of GCAP2[51] shows the role of E155, see Figure 10.

Analysis using the RetGC1 activation assay showed that the mutant protein activates the cyclase at low $[Ca^{2+}]$ but fails to inactivate at high $[Ca^{2+}]$, resulting in constitutive activation of guanylate cyclase similar to that observed for Y99C, see Figure 9C.[17] Mixing E155G GCAP1 with wild-type (WT) protein in stoichiometric amounts also showed activation of RetGC1 even at high $[Ca^{2+}]$.

The E155G mutation, like the Y99C mutation, results in constitutive activation of RetGC1 at physiologically high $[Ca^{2+}{}_{free}]$, although the severity of the disease is described as being greater than that produced by Y99C.[17] There is also an earlier age of onset of disease for patients with the E155G mutation. The E155G mutation also results in a reduced apparent affinity of the protein for RetGC1, an effect not observed with the Y99C mutant. Both these effects have been shown to persist in the presence of wild-type GCAP1 and thus act in a dominant manner. Constitutive activation of RetGC1 would be expected to result in a persistently elevated level of cGMP in the cell, above that required to keep the cGMP-gated cation channels open, resulting in high $[Ca^{2+}{}_{free}]$. The authors suggest that the mechanism of the ensuing photoreceptor degeneration is unclear, but it is known that persistently elevated Ca^{2+} levels tend to disrupt the membrane potential of the mitochondrial outer membrane, leading to release of cytochrome *c*, with subsequent caspase activation and apoptosis.[63] As already discussed for Y99C, elevated cGMP levels, arising from mutations in the phosphodiesterase gene, result in cell death[58,59] and a similar mechanism was proposed to be the cause of cone cell death for carriers of the E155G mutation.

Concentrations of both RetGC1 and GCAP1 are higher in cone photoreceptors than in rod photoreceptors.[41,63] Hence, a mutation in GCAP1 that affects the cyclase

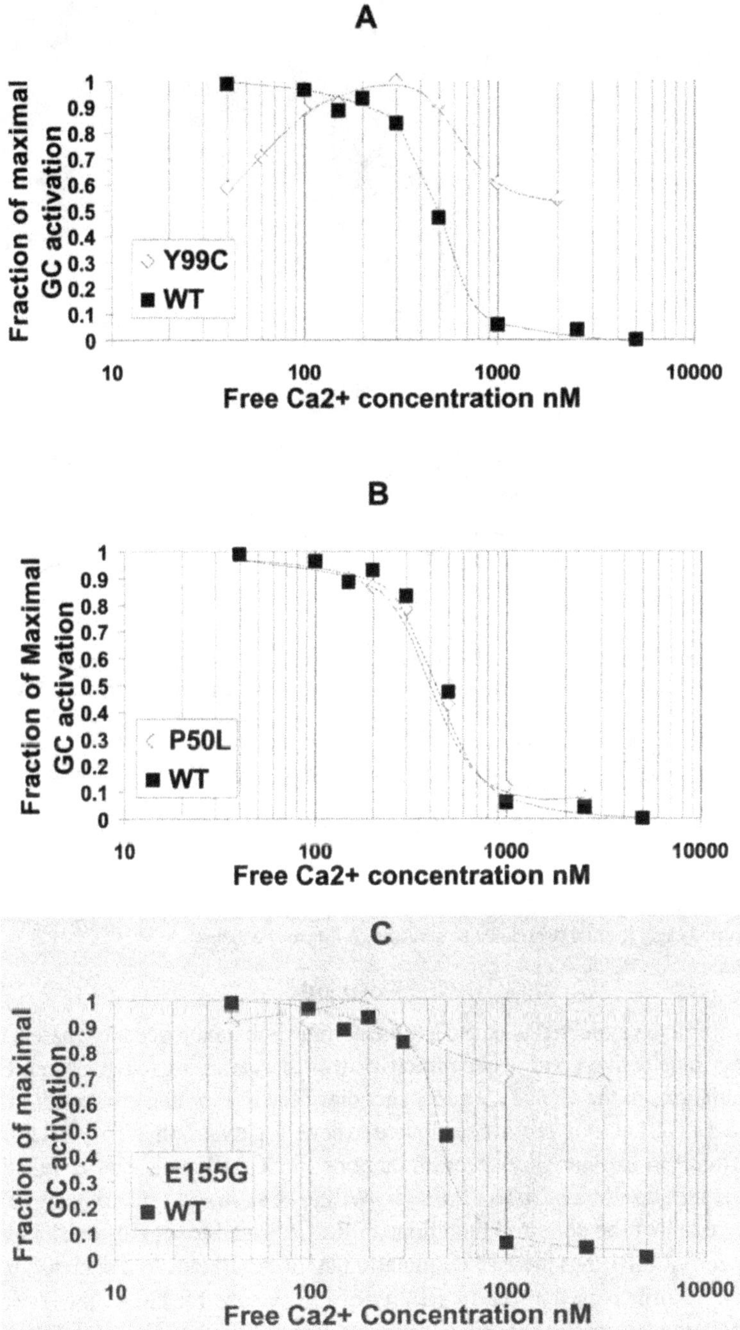

Figure 9. Fraction of maximal guanylate cyclase activation (GTP → cGMP) assessed by the guanylate cyclase activation assay at varying concentrations of Ca^{2+}_{free} for the Y99C, P50L and E155G mutants compared with WT GCAP1.

Figure 10. The role of E155 in EF4. The structure around EF4 from the NMR-derived coordinates of bovine GCAP2[46] where the bidentate ligand, E155 is rendered in dark grey in 'stick' format, the Ca^{2+} ion in a lighter grey in 'spacefill' representation, the remainder of the structure is shown in 'ribbon' format with the E and F-helices shown in light grey.
EF4 is highly conserved and it is likely that the conformation of this motif will be similar for human GCAP1.

		EF4 155
Bovine	GCAP2	DENGDGQLSLNE
Human	GCAP1	DVNGDGELSLEE

Figure created using Rasmol (written by Roger A.Sayle, Glaxo-Wellcome)

activity might be expected to cause a more serious imbalance in cones than in rods, resulting in selective cone degeneration with preservation of rod function. However, mutations in the *GUCY2D* gene, encoding RetGC1, which also result in constitutive RetGC1 activity, have been linked to cone-rod dystrophy.[9,10,23,32] It is therefore difficult to explain why severe mutations in GCAP1 with similar functional consequences do not also affect the rods. Wilkie et al suggest that one significant difference is that the cone-rod mutations in RetGC1 lead to an increased apparent affinity for GCAP1, but the E155G mutation in GCAP1 leads to a decrease in this property.[17] An increased apparent affinity of the cyclase for GCAP1 will tend to exacerbate the failure to abolish cyclase activation at high levels of $[Ca^{2+}{}_{free}]$, but a reduced apparent affinity should compensate for it. Hence it was suggested that the overall severity of the effect of the E155G mutation might be lessened in comparison with the RetGC1 mutations, with the sparing of the rods.

The roles of individual amino acid residues within EF-hands and the function of bovine GCAP1, have been probed by site directed mutagenesis of key residues within EF-hands.[64] Substitution of glutamate with aspartate at position 12 of the three functional EF motifs in GCAP1 altered the affinity of the domains for Ca^{2+} with minimal disruption to their structure. This study demonstrated that only EF3 and EF4 contribute to the Ca^{2+}-dependent inhibition of RetGC1 by GCAP1. The mutants with a single substitution, E111D or E155D, resulted in ~50% inhibition of cyclase activity at a $[Ca^{2+}_{free}]$ of 500 μM, but double mutants (E111D and E155D) and triple mutants (E75D, E111D, and E155D) activated cGMP synthesis in a manner completely independent of $[Ca^{2+}_{free}]$. In contrast, the results presented by Wilkie et al[17] reveal that a considerably more significant substitution by glycine has a drastic effect on Ca^{2+} binding, with <25% inhibition of the cyclase activity at a Ca^{2+} concentration of 500 μM, undoubtedly reflecting greater disruption of the topology within the affected EF-hand.

The identity of P50L as a disease-causing mutation found in GCAP1 was unexpected as proline 50 is not located in a region of the protein that is known to be functionally important. The proline residue at position 50 in GCAP1 should be present on an external loop, in the Ca^{2+}-free form, which is not spatially close to any of the Ca^{2+} binding motifs. However, the Ca^{2+} bound form of recoverin undergoes a huge conformational change, as observed by comparison with the Ca^{2+}-free.[50,51] The proline in recoverin, that is equivalent to proline 50 of GCAP1, moves from a peripheral loop position to a key central role within the molecule, forming potential Van der Waals contacts, between two helices, which form the boundary between the two halves of the recoverin molecule.[50] Therefore, the identity of residue 50 in GCAP1 could be important if GCAP1 undergoes a similar change in its tertiary confirmation upon Ca^{2+} binding. Alignment of published GCAP1 sequences also shows that the proline residue is conserved in species as diverse as man,[65] zebrafish[66] and leopard frog,[67] although intriguingly the equivalent residue in the pufferfish (*Fugu*) is a leucine.[68] The general conservation of this residue is in marked contrast to other residues in close proximity to the proline, see Figure 11. Both the homology modelling of recoverin and the general conservation of the proline at position 50 in GCAP1 suggest that this residue may play a structural role within the protein, which may not affect the activity of the molecule but may affect its long term stability within the cell. Proline is well known as a helix breaker and is commonly found in turns (as it is here in wild type GCAP1) and hence its replacement by leucine, which is the third most stabilizing residue within an α-helix[56] would be expected to yield considerable destabilizing conformational changes to the native GCAP1 fold. The presence of a leucine at position 50 in *Fugu* GCAP1 may not result in vision problems for a fish with a body temperature considerably lower than that of man.

An assessment of the ability of P50L to stimulate RetGC1 was carried out using native bovine rod outer segments (ROS) and recombinant bovine and human RetGC1 as sources of guanylate cyclase (GC) activity.[69] It was found that P50L GCAP1 activates photoreceptor guanylate cyclase at $[Ca^{2+}]$<100 nM and inhibits it at micromolar concentrations. P50L GCAP1 activated photoreceptor guanylate cyclase

	50	Accession Number
Human	FFGLKNLSPSASQYVEQMFE	P43080
Cow	FFGLKNLSPWASQYVEQMFE	P46065
Mouse	FFGLKNLSPSASQYVEQMFE	P43081
Chicken	FFGLKNLSPSANKYVEQMFE	P79880
Frog	FFGLKNLSPASNQYIEQMFD	O73761
Fugu	FFGLRGLDPEANAYIEQMFR	
Zebrafish	FFGLRGLDPKANAYIEQMFR	Q90WX4

Figure 11. Pile up of GCAP1 sequences around Proline 50. Proline (and leucine) 50 are highlighted in bold type. Conserved residues are shaded in dark grey and conservative substitutions in light grey.

to the same relative extent as wild type GCAP1, indicating that Ca^{2+} dependent stimulation of guanylate cyclase by P50L GCAP1and WT GCAP1 is indistinguishable. In addition to stimulation of guanylate cyclase by Ca^{2+}-free GCAP, guanylate cyclase activity was also inhibited by the Ca^{2+}-loaded form of GCAP1. In the same study, the Ca^{2+} binding parameters of P50L GCAP1 were obtained by equilibrium dialysis, measuring the uptake of $^{45}Ca^{2+}$. The results showed that the apparent dissociation constants for Ca^{2+} binding in WT and P50L GCAPs are almost identical (5.2 and 5.6 μM for GCAP1 and P50L GCAP1, respectively). However, the number of occupied Ca^{2+} binding sites in the mutant GCAP1 is significantly reduced (2.9 sites for wild type vs. 1.4 sites for P50L). This result is consistent with the presence of three functional EF-hands in the native polypeptide, with the reduction to less than two functional binding sites in P50L indicating impaired Ca^{2+} binding for the P50L form.[69]

It was suggested that the EF2-hand was likely to be affected by this mutation as this site is nearest to proline 50.[69] In mutants, of other EF-hand proteins, disabling Ca^{2+}-binding to the EF2 loop has been shown to be unimportant for Ca^{2+} dependent stimulation of guanylate cyclase.[64] However, reduced Ca^{2+}-binding ability could lead to changes in cone photoreceptor Ca^{2+} homeostasis and thus trigger cone degeneration. However, the cause of the dominant character of the P50L mutation in vivo is unclear because the activity profiles obtained for enzymatic assays of P50L GCAP1 appear to be similar to WT GCAP1. Two possible hypotheses were put forward: a proposal that the mutant protein interferes with uncharacterized

Ca^{2+}-dependent processes of phototransduction leading to the photoreceptor dystrophy, and alternatively that the P50L mutation may interfere with dimerisation of the native GCAP1, depleting the amount of GCAP in the outer segments. However, the latter suggestion is at odds with the argument progressed by the same group in a subsequent paper.[47]

The analysis of P50L by Newbold et al also revealed no alteration in Ca^{2+} sensitivity compared with wild type.[70] In addition, the level of activity was similar to that observed for the wild type and mixtures of wild type and P50L GCAP1 also behaved in a similar manner to wild type alone, see Figure 9B. In order to explore this alternative disease causing mechanism, Newbold et al[70] examined the sensitivity of the mutant protein to trypsin proteolysis and thermal degradation.[70] Treatment of purified P50L, Y99C and WT protein showed that both the Ca^{2+}-free and Ca^{2+}-rich forms of P50L GCAP1 were considerably more susceptible to trypsin action than wild type or the Y99C mutant form, with addition of Ca^{2+} providing some protection against digestion for all forms, see Figure 12. Assessment of the thermal stability of the P50L form by CD spectroscopy showed that it lost structure at a significantly lower temperature than either WT GCAP1 or the Y99C-form (transition midpoint temperature (T_m) ~25°C lower), as shown in Figures 13 and 14. Hence, it was suggested that the dependence of the disease on the stability of the protein might help explain the phenotypic variability associated with the P50L form of cone dystrophy, since this is likely to be influenced by other factors.[70]

It was proposed that the lower stability of P50L could result in reduced cellular concentration, which would in turn alter the $[Ca^{2+}]$ within the photoreceptor cell. Such disruption of the fine balance of cellular ion levels could set in motion a series of events resulting in the observed cone/rod dystrophy phenotype.[70] The P50L mutation produces a similar, although milder, retinal degenerative effect compared with the other two mutations.[16] This is consistent with a reduced level of P50L GCAP1 within the cell resulting in a slower rate of cell loss.

CONCLUSION

To date three disease causing mutations have been found in human guanylate cyclase activating protein 1, resulting in cone or cone-rod degeneration. For the Y99C and E155G mutations the loss of Ca^{2+} sensitivity appears to be the main cause of the diseased state. All three mutations are dominant in their effect although the mechanism by which the P50L mutation exerts its influence remains unclear although it is possible that it acts due to a haplo-insufficiency, arising from increased susceptibility to protease activity and increased thermal instability.

The mutations in human retinal guanylate cyclase 1 that cause cone-rod dystrophy (CORD) are located in a different region of the protein (dimerisation domain) to other mutations that cause the more severe phenotype described as Leber's Congenital Amaurosis. The CORD mutations are found at what appears to be a mutation "hot spot" where R838 is the key residue. The leading theory is that altered Ca^{2+}

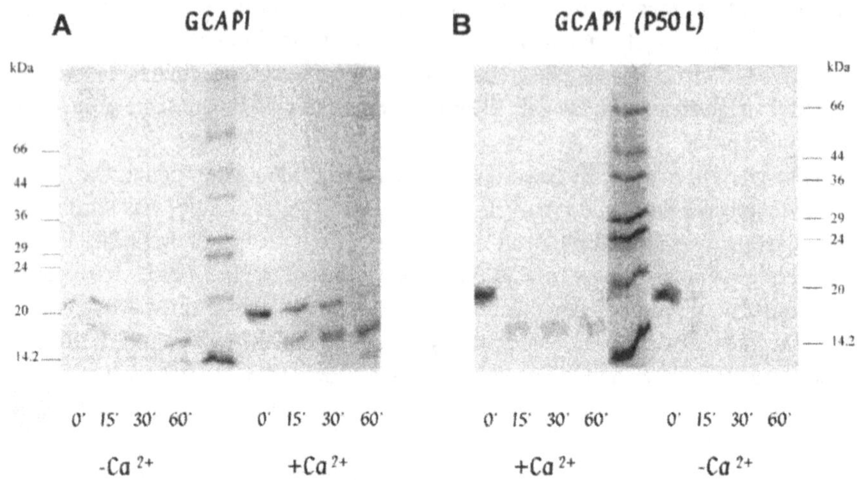

A. Wild Type

Left to right

1 - No calcium T = 1 min
2 - No calcium T= 15 min
3 - No calcium T= 30 min
4 - No calcium T= 60 min
5 - Dalton VII markers (66, 45, 36, 29, 24, 20.1, 14.2kDa)
6 - 1mM Ca^{2+} T=1 min
7 - 1mM Ca^{2+} T=15 min
8 - 1mM Ca^{2+} T=30 min
9 - 1mM Ca^{2+} T=60 min

B. P50L

Left to right

1 - Dalton VII markers (66, 45, 36, 29, 24, 20.1, 14.2kDa)
2 - No calcium T = 1 min
3 - No calcium T= 15 min
4 - No calcium T= 30 min
5 - No calcium T= 60 min
6 - 1mM Ca^{2+} T=1 min
7 - 1mM Ca^{2+} T=15 min
8 - 1mM Ca^{2+} T=30 min
9 - 1mM Ca^{2+} T=60 min

Figure 12. Digestion of WT GCAP1 and P50L by trypsin in the presence and absence of calcium analysed by SDS-PAGE. The P50L shows greater susceptibility to proteolysis in the presence of trypsin. Both proteins show greater resistance to proteolysis in the presence of Ca^{2+}.

sensitivity, caused by perturbation of the normal dimerisation process, yields an imbalance of cellular Ca^{2+} and cGMP concentrations which results in photoreceptor cell degeneration.

The cone and cone-rod dystrophies that are caused by mutations in RetGC1 or GCAP1, arise from a perturbation of the delicate balance of Ca^{2+} and cGMP within

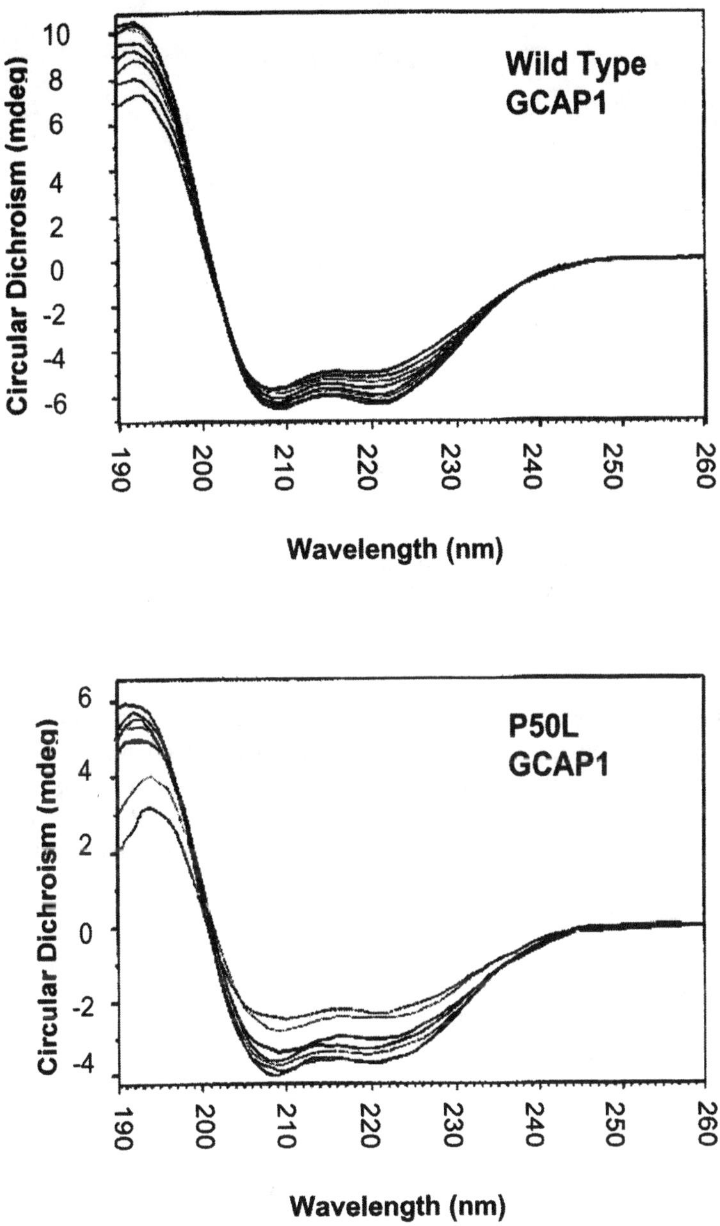

Figure 13. CD Spectra of A - wild type GCAP1, measured at 20-90°C
B - P50L GCAP1, measured at 20-90°C
The P50L mutant shows more rapid breakdown of secondary structure at elevated temperatures compared with wild type GCAP1. Diagram adapted from spectra shown in Newbold et al 2001.[62]

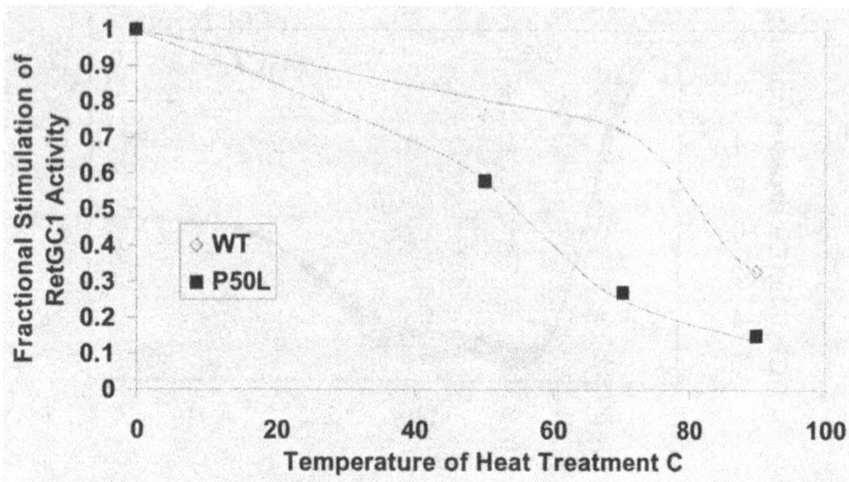

Figure 14. Fractional stimulation of RetGC1 by GCAP1 after heat treatment. The ability of P50L to stimulate RetGC1 is greatly reduced after 5 minutes at elevated temperatures when compared to WT GCAP1.

the photoreceptor cells and it is this disruption that is believed to cause cell death. The diseases caused by mutations in RetGC1 and GCAP1 prominently affect cones, consistent with the higher concentrations of these proteins in cone cells. However, the precise interpretation of the mechanism involved in GCAP1-RetGC1 interaction and the possible perturbations caused by the mutations remain to be elucidated.

ACKNOWLEDGMENTS

The Wellcome Trust
Figure 1 was created by Caroline Graham, QMUL.
Figure 7 was created by Prof. R. W. Pickersgill, QMUL.

REFERENCES

1. Kaupp UB, Koch K-W. Role of cGMP and Ca^{2+} in vertebrate photoreceptor excitation and adaptation. Annu Rev Physiol 1992; 54:153-175.
2. Polans A, Baehr W, Palczewski K. Turned on by Ca^{2+}! The physiology and pathology of Ca(2+)-binding proteins in the retina. Trends Neurosci 1996; 19:547-554.
3. Dizhoor AM, Hurley JB. Regulation of photoreceptor membrane guanylyl cyclases by guanylyl cyclase activator proteins. Methods 1999; 19:521-531.
4. Surguchov A, Bronson JD, Banerjee P et al. The human GCAP1 and GCAP2 genes are arranged in a tail-to-tail array on the short arm of chromosome 6 (p21.1). Genomics 1997; 39:312-322.

5. Payne AM, Downes SM, Bessant DA et al. Genetic analysis of the guanylate cyclase activator 1B (GUCA1B) gene in patients with autosomal dominant retinal dystrophies. J Med Genet 1999; 36:691-693.
6. Haeseleer F, Sokal I, Li N et al. Molecular characterization of a third member of the guanylyl cyclase-activating protein subfamily. J Biol Chem 1999; 274:6526-6535.
7. Imanishi Y, Li N, Sokal I et al. Characterization of retinal guanylate cyclase-activating protein 3 (GCAP3) from zebrafish to man. Eur J Neurosci 2002; 15:63-78.
8. Yang, R-B, Robinson SW, Xiong W-H et al. Disruption of a Retinal Guanylyl Cyclase Gene Leads to Cone-Specific Dystrophy and Paradoxical Rod Behavior. J Neurosci 1999; 19:5889-5897.
9. Kelsell RE, Gregory-Evans K, Payne AM et al. Mutations in the retinal guanylate cyclase (RETGC-1) gene in dominant cone-rod dystrophy. Human Molecular Genetics 1998; 7:1179-1184.
10. Perrault I, Rozet JM, Gerber S et al. A retGC-1 mutation in autosomal dominant cone-rod dystrophy. Am J Hum Genet 1998; 63:651-654.
11. Moore AT, Evans K Molecular genetics of central retinal dystrophies Aust N.Z. J Ophthalmology 1996; 24:189-98.
12. Nakazawa M, Kikawa E, Chida Y et al. Asn244His mutation of the peripherin/RDS gene causing autosomal dominant cone rod degeneration. Hum Mol Genet 1996; 3:1195–1196.
13. Freund CL, Gregory-Evans CY, Furukawa T et al. Cone-rod dystrophy due to mutations in a novel photoreceptor-specific homeobox gene (CRX) essential for maintenance of the photoreceptor. Cell 1997; 91:543-553.
14. Swain PK, Chen S, Wang QL. Mutations in the cone-rod homeobox gene are associated with the cone-rod dystrophy photoreceptor degeneration. Neuron 1997; 19:1329-1336.
15. Payne AM, Downes SM, Bessant DA et al. A mutation in guanylate cyclase activator 1A (GUCA1A) in an autosomal dominant cone dystrophy pedigree mapping to a new locus on chromosome 6p21.1. Hum Mol Genet 1998; 7:273-277.
16. Downes SM, Holder GE, Fitzke FW et al. Autosomal dominant cone and cone-rod dystrophy with mutations in the guanylate cyclase activator 1A gene-encoding guanylate cyclase activating protein-1. Arch Ophthalmol 2001; 119:96-105.
17. Wilkie SE, Li Y, Deery EC et al. Identification and functional consequences of a new mutation (E155G) in the gene for GCAP1 that causes autosomal dominant cone dystrophy. Am J Hum Genet 2001; 69:471-480.
18. Perrault I, Rozet JM, Calvas P et al. Retinal specific guanylate cyclase gene mutations in Leber's congenital amaurosis. Nature Genet 1996; 14:461–464.
19. Duda T, Venkataraman V, Goraczniak R et al. Functional consequences of a rod outer segment membrane guanylate cyclase (ROS-GC1) gene mutation linked with Leber's congenital amaurosis. Biochemistry 1999; 38:509-515.
20. Yu H, Olshevskaya E, Duda T et al. Activation of retinal guanylyl cyclase-1 by Ca2+-binding proteins involves its dimerization. J Biol Chem 1999; 274:15547-55.
21. Ramamurthy V, Tucker C, Wilkie SE et al. Interactions within the coiled-coil domain of RetGC-1 guanylyl cyclase are optimized for regulation rather than for high affinity. J Biol Chem 2001; 276:26218-26229.
22. Gregory-Evans K, Kelsell RE, Gregory-Evans CY et al. Autosomal dominant cone–rod retinal dystrophy (CORD6) from heterozygous mutation of *GUCY2D*, which encodes retinal guanylate cyclase. Ophthalmology 2000; 107:55-61.
23. Weigell-Weber M, Fokstuen S, Torok B et al. Codons 837 and 838 in the retinal guanylate cyclase gene on chromosome 17p: hot spots for mutations in autosomal dominant cone-rod dystrophy? Arch Ophthalmol 2000; 118:300.
24. Payne AM, Morris AG, Downes SM et al. Clustering and frequency of mutations in the retinal guanylate cyclase (GUCY2D) gene in patients with dominant cone-rod dystrophies. J Med Genet 2001; 38:611-614.

25. Downes SM, Payne AM, Kelsell RE et al. Autosomal dominant cone-rod dystrophy with mutations in the guanylate cyclase 2D gene encoding retinal guanylate cyclase-1 Arch. Ophthalmology 2001; 119:1667-1673.
26. Sokal I, Haeseleer F, Arendt A. Identification of a guanylyl cyclase-activating protein-binding site within the catalytic domain of retinal guanylyl cyclase. Biochemistry 1999; 38:1387-1393.
27. Gorczyca WA, Polans AS, Surgucheva IG et al. Guanylate cyclase activating protein: a calcium-sensitive regulator of phototransduction. J Biol Chem 1995; 270:22029-22036.
28. Yu H, Olshevskaya E, Duda T et al. Activation of retinal guanylyl cyclase-1 by Ca2+-binding proteins involves its dimerization. J Biol Chem 1999; 274:15547-15555.
29. Palczewski K, Subbaraya I, Gorczyca WA et al. Molecular cloning and characterization of retinal photoreceptor guanylyl cyclase-activating protein. Neuron 1994; 13:395-404.
30. Payne AM, Downes SM, Bessant DA et al. A mutation in guanylate cyclase activator 1A (GUCA1A) in an autosomal dominant cone dystrophy pedigree mapping to a new locus on chromosome 6p21.1. Hum Mol Genet 1998; 7:273-277.
31. Garbers DL, Dyer E, Hardman JG. Effects of cations on guanylate cyclase of sea urchin sperm. J Biol Chem 1975; 250:382-387.
32. Wilkie SE, Newbold RJ, Deery EC et al. Functional characterization of missense mutations at codon 838 in retinal guanylate cyclase correlates with disease severity in patients with autosomal dominant cone-rod dystrophy. Hum Mol Genet 2000; 9:3065-3073.
33. Lupas A, Van Dyke M, Stock J. Predicting Coled Coils from Protein Sequences. Science 1991; 252:1162-1164.
34. Duda T , Krishnan A, Venkataraman V et al. Mutations in the rod outer segment membrane guanylate cyclase in a cone-rod dystrophy cause defects in calcium signaling. Biochemistry 1999; 38:13912-13919.
35. Duda T, Venkataraman V, Jankowska A et al. Impairment of the rod outer segment membrane guanylate cyclase dimerization in a cone-rod dystrophy results in defective calcium signaling. Biochemistry 2000; 39:12522-12533.
36. Tucker CL, Woodcock SC, Kelsell RE et al. Biochemical analysis of a dimerization domain mutation in RetGC-1 associated with dominant cone-rod dystrophy. Proc Natl Acad Sci USA 1999; 96:9039-9044.
37. Coetzer TL, Sahr K, Prchal J et al. Four different mutations in codon 28 of alpha spectrin are associated with structurally and functionally abnormal spectrin alpha I/74 in hereditary elliptocytosis. J Clin Invest 1991; 88:743-749.
38. Hovnanian A, Pollack E, Hilal L et al. A missense mutation in the rod domain of keratin 14 associated with recessive epidermolysis bullosa simplex. Nature Genetics 1993; 3:327-332.
39. Otto-Bruc A, Fariss RN, Haeseleer F et al. Localization of guanylate cyclase-activating protein 2 in mammalian retinas. Proc Natl Acad Sci USA 1997; 94:4727-32.
40. Cuenca N, Lopez S, Howes K et al. The localization of guanylyl cyclase-activating proteins in the mammalian retina. Invest Ophthalmol Vis Sci 1998; 39:1243-1250.
41. Kachi S, Nishizawa Y, Olshevskaya E et al. Detailed localization of photoreceptor guanylate cyclase activating protein-1 and -2 in mammalian retinas using light and electron microscopy. Exp Eye Res 1999; 68:465-473.
42. Dizhoor AM, Lowe DG, Olshevskaya EV et al. The human photoreceptor membrane guanylyl cyclase, RetGC, is present in outer segments and is regulated by calcium and a soluble activator. Neuron 1994; 12:1345-1352.
43. Dizhoor AM, Olshevskaya EV, Henzel WJ et al. Cloning, sequencing, and expression of a 24-kDa Ca(2+)-binding protein activating photoreceptor guanylyl cyclase. J Biol Chem 1995; 270:25200-25206.
44. Lowe DG, Dizhoor AM, Liu K et al. Cloning and expression of a second photoreceptor specific membrane guanylyl cyclase (RetGC), RetGC-2. Proc Nat Acad Sci. USA 1995; 92:5535-5539.
45. Ikura M, Clore GM, Gronenborn AM et al. Solution structure of a calmodulin-target peptide complex by multidimensional NMR. Science 1992; 256:632-638.

46. Tanaka T, Ames JB, Harvey TS et al. Sequestration of the membrane-targeting myristoyl group of recoverin in the calcium-free state. Nature 1995; 376:444-447.
47. Sokal I, Li N, Klug CS et al. Calcium-sensitive regions of GCAP1 as observed by chemical modifications, fluorescence, and EPR spectroscopies. J Biol Chem 2001; 276:43361-43373.
48. Ermilov AN, Olshevskaya EV, Dizhoor AM. Instead of Binding Calcium, One of the EF-hand Structures in Guanylyl Cyclase Activating Protein-2 is Required for Targeting Photoreceptor Guanylyl Cyclase. J Biol Chem 2001; 276:481-438.
49. Flaherty KM, Zozulya S, Stryler L et al. Three-dimensional structure of recoverin, a calcium sensor in vision. Cell 1993; 75:709–716.
50. Ames JB, Ishima R, Tanaka T et al. Molecular mechanics of calcium-myristoyl switches. Nature 1997; 389:198-202.
51. Ames JB, Dizhoor AM, Ikura M et al. Three-dimensional structure of guanylyl cyclase activating protein-2, a calcium-sensitive modulator of photoreceptor guanylyl cyclases. J Biol Chem 1999; 274:19329-19337.
52. Frins S, Bonigk W, Muller F et al. Functional characterization of a guanylyl cyclase-activating protein from vertebrate rods. Cloning, heterologous expression, and localization. J Biol Chem 1996; 271:8022-8027.
53. Kretsinger RH, Rudnick SE, Weissman LJ. Crystal structure of calmodulin. J Inorg Biochem 1986; 28:289-302.
54. Babu YS, Bugg C, Cook WJ. Structure of calmodulin refined at 2.2A resolution. J Mol Biol 1988; 204:191–204.
55. Griffiths JP, Kim JL, Kim EE et al. X-Ray structure of calcineurin inhibited by the immunophilin-immuno-suppressant FKBP12-FK506 complex. Cell 1995; 82:507–522.
56. Miller S, Janin J, Lesk AM et al. Interior and surface of monomeric proteins. J Mol Biol 1987; 196:641-56.
57. Semple-Rowland SL, Gorczyca WA et al. Expression of GCAP1 and GCAP2 in the retinal degeneration (rd) mutant chicken retina. FEBS Lett 1996; 385:47–52.
58. Pittler SJ, Baehr W. Identification of a nonsense mutation in the rod photoreceptor cGMP phosphodiesterase beta-subunit gene of the rd mouse. Proc Natl Acad Sci USA 1991; 88:322-326.
59. Farber DB, Danciger JS, Aguirre G. The beta subunit of cyclic GMP phosphodiesterase mRNA is deficient in canine rod-cone dysplasia 1. Neuron 1992; 9:349–356.
60. Laura RP, Dizhoor AM, Hurley JB. The membrane guanylyl cyclase retinal guanylyl cyclase-1 is activated through its intracellular domain. J Biol Chem 1996; 271:11646-11651.
61. Sokal I, Li N, Surgucheva I et al. GCAP1 (Y99C) mutant is constitutively active in autosomal dominant cone dystrophy. Mol Cell 1998; 2:129-133.
62. Kretsinger RH. Structure and evolution of calcium-modulated proteins. Crit Rev Biochem. 1976; 8:119-174.
63. Liu X, Seno K, Nishizawa Y et al. Ultrastructural localization of retinal guanylate cyclase in human and monkey retinas. Exp Eye Res 1994; 59:761-768.
64. Rudnicka-Nawrot M, Surgucheva I, Hulmes JD et al. Changes in biological activity and folding of guanylate cyclase-activating protein 1 as a function of calcium. Biochemistry 1998; 37:248-257.
65. Subbaraya I, Ruiz CC, Helekar BS et al. Molecular characterization of human and mouse photoreceptor guanylate cyclase-activating protein (GCAP) and chromosomal localization of the human gene. J Biol Chem 1994; 269:31080-31089.
66. Taylor MR, Niemi GA, Tucker et al. Zebrafish phototransduction genes. Unpublished results, 2001; Accession AY050501.
67. Ning L, Fariss RN, Zhang, K et al. Guanylate-cyclase-inhibitory protein is a frog retinal Ca^{2+}-binding protein related to mammalian guanylate-cyclase-activating proteins. Eur J Biochem 1998; 252:591-599.

68. Wilkie SE, Stinton I, Cottrill P et al. Characterisation of two genes for guanylate cyclase activator protein (GCAP1 and GCAP2) in the Japanese Pufferfish, Fugu rubripes. (submitted) 2002.
69. Sokal I, Li N, Verlinde CL et al. Ca^{2+}-binding proteins n the retina: from discovery to etiology of human disease. Biochim Biophys Acta 2000; 1498:233-251.
70. Newbold RJ, Deery EC, Walker CE et al. The destabilization of human GCAP1 by a proline to leucine mutation might cause cone-rod dystrophy. Hum Mol Genet 2001; 10:47-54.

USING MUTANT MICE TO STUDY THE ROLE OF VOLTAGE-GATED CALCIUM CHANNELS IN THE RETINA

Sherry L. Ball[1,2,3] and Ronald G. Gregg[4]

ABSTRACT

Neuronal voltage-gated calcium channels (VGCCs) are critical to numerous cellular functions including synaptogenesis and neurotransmitter release. Mutations in individual subunits of VGCCs are known to result in a wide array of neurological disorders including episodic ataxia, epilepsy, and migraines. The characterization of these disorders has focused on channel function within the brain. However, a defect in the retina-specific α_{1F} subunit of an L-type VGCC results is a loss of visual sensitivity or the incomplete form of X-linked congenital stationary night blindness (CSNB2). Based on the electroretinographic phenotype of these patients this channel type is localized to the axon terminal of photoreceptor cells and results in a loss of signal transmission from photoreceptors to bipolar cells. A mouse with a deletion of the β_2 subunit of VGCCs in the central nervous system was recently shown to have a similar phenotype as CSNB2 patients. The identification of the role of VGCCs in this disorder highlights the potential association of other VGCC mutations with retinal disorders. The study of the role of these channels in normal retinal function may also be elucidated by the characterization of retinal structure and visual function in the numerous knockout, transgenic, and naturally occurring mouse mutants currently available.

[1]Research Service, Cleveland VA Medical Center, 10701 East Boulevard, Cleveland, OH 44106. [2]Cole Eye Institute, Cleveland Clinic Foundation, 9500 Euclid Avenue, Cleveland, OH 44195. [3]Department of Psychology, Case Western Reserve University. [4]Department of Biochemistry & Molecular Biology and Ophthalmology & Visual Sciences, University of Louisville, Louisville, KY 40202

INTRODUCTION

The characterization of animals with ion channel mutations provides useful information regarding their role in vivo and provides systems in which to study the pathogenesis of channelopathies.[1] In fact, an extensive literature has developed where animal models have provided information regarding how abnormalities in a specific channel type including water channels, receptors, and ion channels such as K^+, Na^+, Cl^- and Ca^{2+} alter the function of systems.

A wide variety of neurological disorders have now been linked to voltage-gated calcium channel (VGCC) subunit defects.[2,3,4,5,6,7,8,9] All VGCCs are composed of an α_1 pore-forming subunit, a β and α_2/δ, and possibly γ-subunits. Figure 1 illustrates the topological organization of these subunits. Each VGCC type is defined according to its pharmacological sensitivities, kinetics and specific α_1 subunit. To date, ten α_1 (A-S), four β (1-4), three α_2/δ and eight γ subunits have been identified.[10,1,12] Additionally, for many subunits there are multiple splice variants, which create the potential for considerable functional diversity.

NEURONAL Ca^{2+} CHANNELS

In the central nervous system, VGCCs are required for neurotransmitter release and they are present at two general synapse types, those exhibiting fast synaptic transmission and those that release neurotransmitter in a tonic manner. Fast synapses contain primarily the P/Q and N- type channels in various ratios, which are composed of the α_{1A} and the α_{1B} subunits, respectively. Ribbon synapses present in the cochlea and retina release neurotransmitter tonically. They contain L-type VGCCs, which are composed of α_{1D}, α_{1F}, and possibly α_{1C} subunits. In addition to synaptic transmission, as regulators of both membrane potential and intracellular Ca^{2+} levels, VGCCs serve diverse cellular functions such as neurite outgrowth and synaptogenesis.[13,14] Thus, they are located on the cell soma as well as neuronal processes.[15]

VGCCs in the Retina

The retina contains two synaptic or plexiform layers. The outer plexiform layer (OPL) contains two types of synapses that connect photoreceptors to bipolar and horizontal cells. Rod photoreceptors have a ribbon synapse that is characterized by a large electron dense structure extending from the synaptic contact into the rod terminal, and four invaginating post-synaptic processes, two each from bipolar and horizontal cells.[16] This "ribbon" is thought to be involved in the tonic release of neurotransmitter that occurs in the dark at this synapse.[17,18] The cones also have ribbon synapses, however, each cone terminal has multiple sets of invaginating processes each from bipolar and horizontal cells. In addition, cone terminals also have conventional synapses with horizontal cells. The inner plexiform layer (IPL)

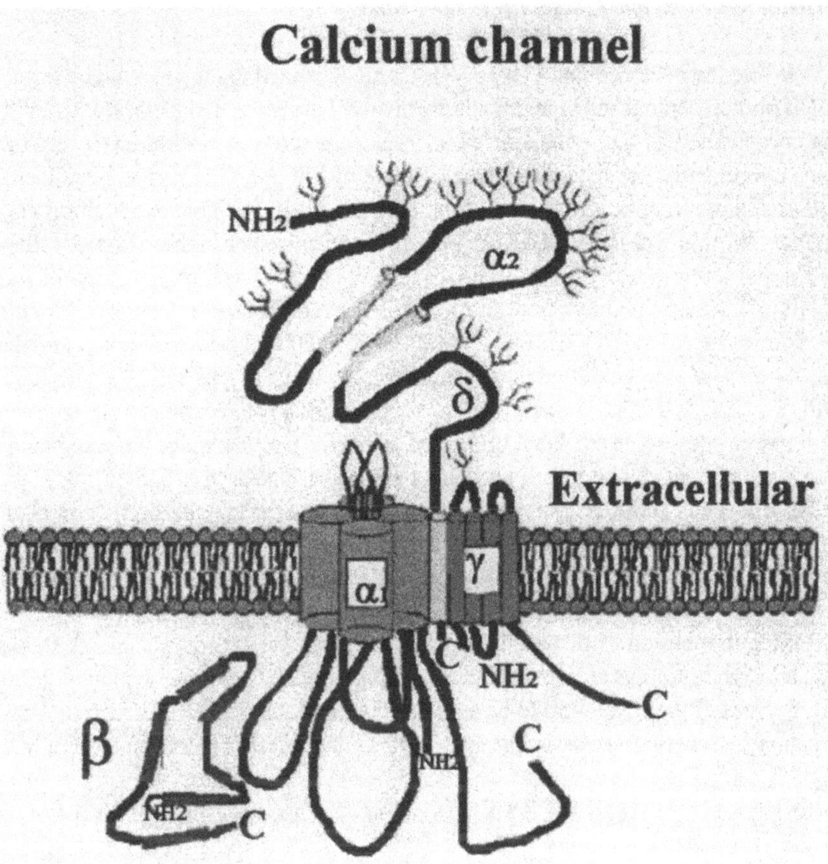

Figure 1. Topological organization of the three core subunits comprising a voltage-gated calcium channel. The pore forming transmembrane α_1 subunit is composed of 4 homologous domains each containing 6 transmembrane segments (not shown). The α_1 subunit interacts with the auxillary β subunit at the intracellular connecting loop and $\alpha_2\delta$ auxillary subunits. Diagram courtesy of Dr. Kevin Campbell.

contains synapses involving bipolar, amacrine and ganglion cells. Bipolar cells have ribbon synapses, although structurally they are much smaller than the ribbons found in photoreceptor terminals.

The types of VGCCs have not been studied extensively in all retinal cell types. The best studied are the slowly-inactivating L-type channels that control the tonic release of neurotransmitter creating a graded potential at the axon terminal in both photoreceptors and bipolar cells.[13,19,20] Although the role of P/Q and N-type channels in the retina is unclear, it is possible that these channel types are utilized at conventional synapses present in both the OPL and IPL. In addition to synaptic localization, VGCCs such as T-type are located on the cell bodies of neurons.[21]

Distribution of α_1 and β Subunits in the Retina.

While the evidence that L-type VGCCs are localized to the presynaptic terminal of photoreceptors and bipolar cells are involved in synaptic transmission is solid, the composition of these channels is under intense scrutiny. Both the α_{1C} and the α_{1D} subunit and more recently, the α_{1F} subunit of L-type VGCCs have been localized to the presynaptic terminal of photoreceptors.[13,22,23,24,25] One explanation may be that multiple types of VGCCs are present at synaptic sites in the retina. A differential distribution of the L-type channel subunit α_{1D}, between photoreceptors was noted in the cone-dominant tree shrew retina. The rod photoreceptors as well as the blue cones lacked the α_{1D} subunit.[22] In rat, differential label between the α_{1F} and the pan α_1 antibody suggests that there are different types of L-type VGCCs within the OPL and IPL, which may correspond to rod and cone photoreceptors.[25] In addition to synaptic sites, some α_{1F} label was noted within the photoreceptor inner segments and faint label in cell bodies of both inner and outer nuclear layers. Staining in the OPL, IPL, and Müller cells has also been noted using the α_{1C} antibody.[13] In addition, there exists a great deal of evidence suggesting the presence of other channel types throughout the retina although their pattern of expression remains undetermined.

The distribution of β subunits is even less clear. Western blot analysis shows that all four β subunits are expressed in the retina (Fig. 2).[26] Indirect evidence showing the loss of the β_2 subunit to be associated with changes in OPL structure[27] and function[26] strongly suggests the presence of this subunit at the photoreceptor terminal.

ASSESSMENT OF RETINAL FUNCTION IN MUTANT MICE

When light is presented to an otherwise steady state retina the resulting changes in ionic current or evoked potential can be measured from the surface of the cornea. This response, the electroretinogram (ERG), represents the temporally summed activity of retinal cells in response to a light stimulus and is composed of two main waveforms. The most prominent waveform, the b-wave, is generated by the activation of bipolar cells.[28] Under some conditions, the b-wave is preceded by a negative wave representing the hyperpolarization of photoreceptors referred to as the a-wave.[29] The ERG is useful as a means to study synaptic transmission between photoreceptors and bipolar cells and thus Ca^{2+} channel function at the photoreceptor axon terminal. This method has been well characterized in rodents and is often used in animal studies as it is noninvasive and can be readily compared to data from human studies.

IMPACT OF VGCC SUBUNIT MUTATIONS ON RETINAL FUNCTION

For many of the VGCC α_1 and auxiliary subunits either natural mutations have been identified, or gene targeting has been used to produce null alleles (Table 1).

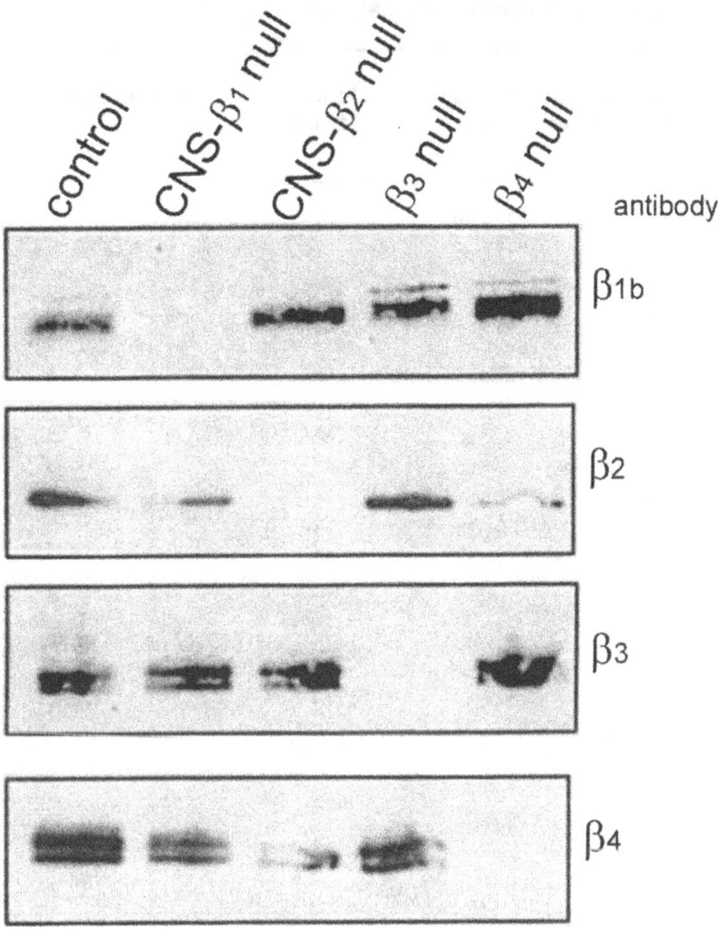

Figure 2. Western blots showing β subunit expression in the retina of control and the four β subunit knockout mouse lines. All four β subunits are expressed in the retina and the expected subunit is absent from each of the mutant mouse lines. Modified from Ball et al, 2000.

Mutations in the α_{1A} subunit genes of P/Q type VGCCs have been identified in lines of mutant mice, such as *rocker*,[30] *tottering*,[31] and *leaner*,[32] which display neurological symptoms similar to those of human patients with transient motor abnormalities.[33,34] Based on labeling with specific radioligands and an analysis of mRNA, N-and P/Q type channels utilizing the α_{1B} and α_{1A} subunit were reported to be present in the retina.[35,36] However, little is known regarding their molecular structure and function and, until now, no information was available regarding visual function in these mice. Additionally, within the past year neurological deficits have been described in 3 knockout mice, in which the pore forming α_{1G}[37], α_{1E} [38] and α_{1D}[39]

Table 1. Mice with Calcium Channel Mutations

Subunit	Human Chromosome Location*	Current Type	Mouse Mutant	Identification of Mutant Reference	Retinal Phenotype
α_{1A}	19p13.2	P- and/or Q-type	*rocker*	30	?
			tottering	31	
			leaner	32	?
α_{1B}	9q34	N-type	knockout	48	?
α_{1C}	12p13.3	L-type	inactivation	49	?
α_{1D}	3p14.3	L-type	knockout	39	?
α_{1E}	1q25-31	R-type	knockout	50	?
α_{1F}	Xp11.23	L-type	(CNS-β_2-/-)	26	CSNB2
α_{1G}	17q22	T-type	knockout	37	?
α_{1H}	16p13.3	T-type			
α_{1I}	22q13.1	T-type			
α_{1S}	1q32	L-type	*mdg*	51	None
β_1	17q21-22		knockout	43	None
β_2	10p12		knockout	26	CSNB2
β_3	12q13		knockout	52	Alteration in visual axonal pathway
β_4	2q22-23		*lethargic*	53	None
α_2/δ-2	7q21-22				?
α_2/δ-2	3p21.3		*ducky*	54	?
γ_1	17q23				
γ_2	22q13		*stargazer*	55	?

* All chromosome location information was obtained at the following web site: http://www.ncbi.nlm.nih.gov/LocusLink/.

subunit corresponding to T-type, R-type and L-type channels respectively, were targeted. Again, although similar channel types have been shown to be present in the retina making it likely that the mutant mice will also posses retinal dysfunction, visual function has not been assessed.

Of the various animals with mutations in α_1 subunits, only mutations in the human α_{1F} have been associated with a retinal disease phenotype. A recent study examining retinal function in mice with mutations in the β subunit of VGCCs provides the first mouse model for a retinal disorder linked to a Ca^{2+} channel mutation. In this study, mice lacking all four β subunits were examined using electroretinography, only the CNS-β_2 null mouse showed a detectable phenotype.[26] There are no reports describing visual system structure or function in mice with mutations in the other auxiliary α_2/δ- and γ- subunits. Mutations in the α_{1F} subunit in humans and the β_2 subunit in mice produce the same ERG phenotype as found in the incomplete form of X- linked congenital stationary night blindness (CSNB2), indicating that these subunits are likely to make up the major L-type channel at the photoreceptor terminals. Mutations in the β_3 subunit have been shown to have an impact on organization of the lateral geniculate nucleus.[40]

α_{1F} Subunit Mutations Cause CSNB2

A mouse with a mutation in the α_{1F} subunit has not yet been reported, however, mutations in this subunit in humans cause CSNB2.[5,6,41] CSNB2 is a retinal disorder resulting in decreased visual acuity and a profound loss of visual sensitivity under both scotopic and photopic conditions.[42] ERG testing of these patients shows that CSNB2 is associated with a 'negative' ERG, where the amplitude of the b-wave is selectively reduced, indicating a loss of synaptic transmission between photoreceptors and bipolar cells.[42] The loss of post-receptoral electrical activity in CSNB2 patients suggests that VGCCs composed of the α_{1F} subunit are required for normal neurotransmitter release at the photoreceptor terminal. In support of this, the α_{1F} subunit has been localized to the OPL and IPL of the rat retina.[25] The α_{1F} subunit has not yet been expressed in vitro and shown to be sensitive to dihyropyridines (DHP) and therefore has not been clearly classified as L-type. However, sequence analyses indicates it is likely to have the DHP binding site and the phenotype also is consistent with the absence of the photoreceptor VGCC, which is known to be an L-type channel. A second isoform of the α_{1F} subunit also is expressed on the cell bodies of the photoreceptors. Currently the function of this isoform is unknown, although it could be involved in regulation of gene expression.

α_{1A} Mutant Mice

In addition to L-type VGCCs the P/Q type are dispersed throughout the brain and are especially prominent in cerebellar Purkinje and granule cells. Because this subunit is expressed in the retina we have begun to assess retinal function in the α_{1A} mutant mice by recording light-adapted ERGs. Figure 3 shows that according to this recording method, retinal function appears to be intact. However, this preliminary analysis needs to expanded to more mice, and extended to dark-adapted conditions before final conclusions can be drawn.

β_2 Subunit Mutations Result in a CSNB2-Like Phenotype

Although the α_1 subunit primarily determines channel characteristics, β subunits are known to play an important role in channel assembly and can modulate channel activity through interaction with the α_1 subunit. The discovery that mutations in the α_{1F} gene were responsible for CSNB2 indicates that a single α_1 subunit is present as the primary L-type channels at photoreceptor terminals. This indicated that there also may be a single β subunit expressed in photoreceptors. Mutations in the β_3 and β_4 (*lethargic*) subunit genes are viable. However, mutations in the β_1[43]

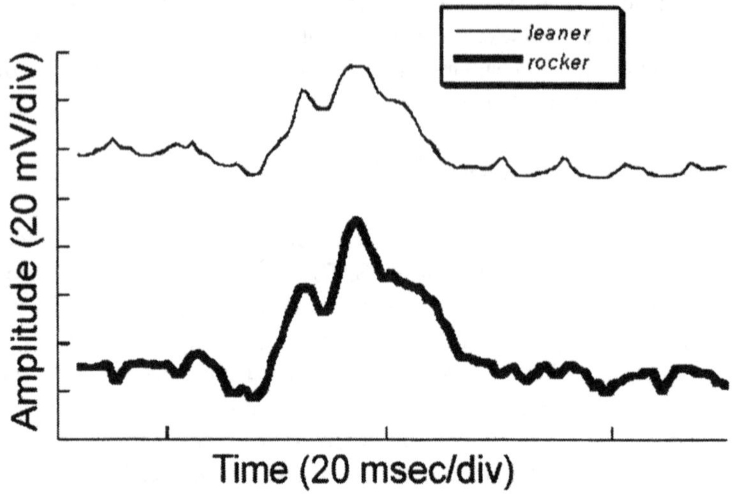

Figure 3. Light adapted electroretinograms from *leaner* (thin trace) and *rocker* (thick trace) mice.

and β_2[26] subunits produced a lethal phenotype. Mice carrying the β_1 mutation were rescued by expressing the subunit under the control of a skeletal muscle specific promoter. Mice carrying the β_2 mutation were rescued by expressing the subunit under the control of a cardiac muscle specific promoter. Both the β_1 and β_2 rescued mice are viable and lack the expected subunit in the brain and retina. Figure 2 confirms the loss of each subunit in the retina of each respective mutant mouse.

Retinal structure and visual function were examined in each of the β subunit mutants by ERG, histology, and a behavioral task. The ERGs and the histology were normal in the β_1, β_3, and β_4 null mice. However, CNS-specific deletion of the β_2 gene had profound effects on visual function. Figure 4 shows ERGs using stimulus conditions that represent primarily rod-mediated activity.[28,44] Two major ERG components found in normal mice were noted in the CNS-β_1, β_3 and β_4 null mice (Fig. 4). At all intensities, the ERG included a positive polarity b-wave and higher frequency oscillatory potentials representing the summed activity of bipolar cells and other inner retinal neurons.[28,45] At higher intensities, the ERG also included a negative polarity a-wave, which is generated by the light-induced closure of cation channels along the photoreceptor outer segments[46]. In contrast to the results in the CNS-β_1, β_3 and β_4 null mice, the b-wave responses of CNS-β_2 null mice were reduced in amplitude at all flash intensities, although the a-wave component appeared unaffected.

When stimulus conditions designed to examine the cone system were used, a similar pattern was observed. Namely, the cone ERGs of CNS-β_1, β_3 and β_4 null mice were comparable to those of normal animals (Fig. 5), while the cone ERGs of CNS-β_2 null mice were markedly reduced in amplitude (Fig. 5). To determine if the two mouse cone types[47] were differently affected by β subunit deletion, the function

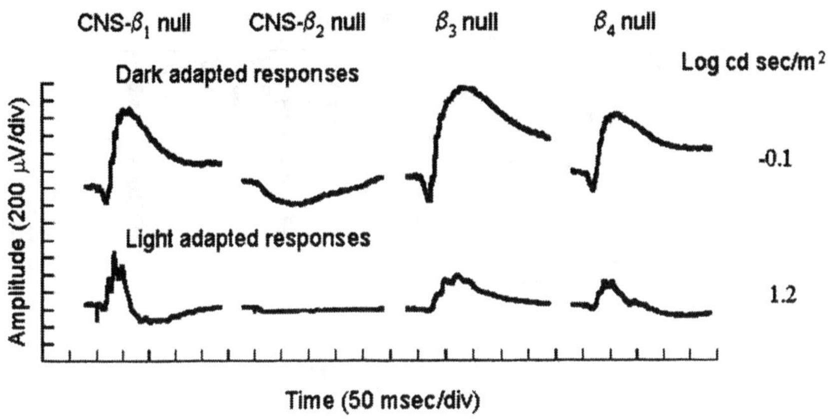

Figure 4. Dark (top trace) and light (bottom trace) adapted ERGs from each β-null mouse in response to a stimulus flash of −0.1 and 1.2 and log cd sec/m^2 respectively.

of each was examined using appropriate spectral stimuli.[44] Figure 5A shows representative ERGs recorded in response to either a 400 nm or a 500 nm stimulus flash for each of the four mutant mouse lines. Distinct responses of positive polarity were obtained from CNS-β_1-, β_3- and β_4-null mice, which were similar to responses obtained from wild-type animals. Under both stimulus conditions, the CNS-β_2 null responses were negative in polarity and markedly reduced in amplitude (Fig. 5B).

These ERG results indicate that CNS-β_2 null mice possess normal photoreceptor function, however, there is an abnormality in transmission between photoreceptors and bipolar cells in these animals. While the amplitude of the b-wave of the ERG in the CNS-β_2 null mice is markedly decreased there is a small-amplitude, but reproducible, b-wave-like late positive component.[26] The corneal positive component was always smaller in amplitude and displayed much slower kinetics than a normal b-wave but suggests the presence of some post receptoral activity.[26] A thorough ERG analysis of the CNS-β_2 null mice indicates they represent a phenocopy of CSNB2.[26] CSNB2 in humans was initially called incomplete CSNB because of a small amplitude b-wave with slow kinetics could be detected[42] while the complete form (CSNB1) resulted in a total loss of post receptoral components. The same comparison is true of the ERG in the CNS-β_2 null mice versus the *nob* mouse, the animal model for CSNB1.[48] There are at least two explanations for this remnant activity in the CNS-β_2 null mouse. First, there may be a low level of expression of the α_{1F} subunit either alone or it may partner with one of the other β subunits. Second, there may be a low level of expression of another Ca^{2+} channel in the photoreceptors, which is normally present at this synapse.

Immunohistochemical analysis of the retinas of the CNS-β_2 null mice show they lack expression of the α_{1F} subunit in the OPL,[26] which further supports the idea that they are a model for CSNB2. In addition to an abnormal ERG, the CNS-β_2 null mice show a thinning of the OPL that is associated with a loss of the ribbon

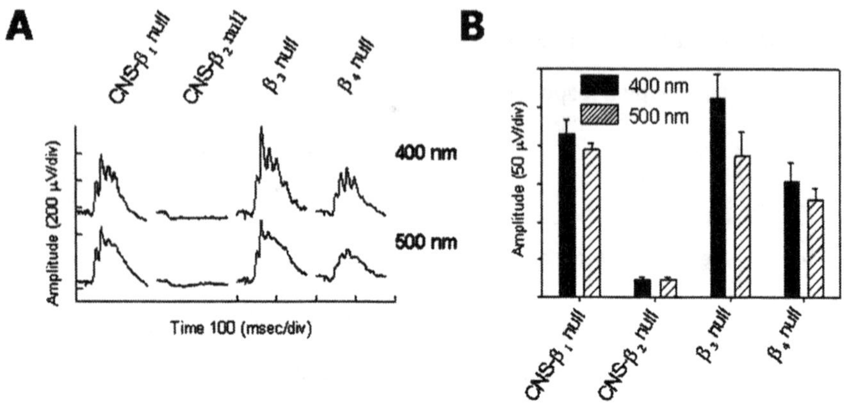

Figure 5. A) Cone ERG waveforms recorded from each β-null mouse in response to a flash of 400 nm or 500 nm light preceded by a rod-desensitizing conditioning flash. B) Amplitude of the cone ERGs obtained to a 400 nm or 500 nm stimulus flash. Both responses were significantly reduced in CNSβ2-null mice ($P<0.05$). Data points indicate the average (±s.d.) for 3-4 mice.

synapses.[27] Behavioral studies confirmed that the CNS-β_2 null mice have a loss of visual sensitivity, but that under normal lighting conditions can perform a visually cued task at a normal level[26] similar to CSNB2 patients.[42] These data indicate that the predominant L-type channel in the photoreceptor terminals is most likely a complex between the α_{1F} and β_2 subunit and presumably an α_2/δ subunit.

CONCLUSIONS AND FUTURE DIRECTIONS

Current evidence suggests the α_{1F} and the β_2 subunit are paired to form the L-type VGCC at the photoreceptor axon terminal. Ca^{2+} channels exist in virtually all retinal cell types and presumably play diverse roles in information processing throughout the retina. Thus, defects in Ca^{2+} channel function can potentially lead to a variety of retinal disorders. It is likely that many human retinal eye diseases may involve Ca^{2+} channel mutations. By studying visual function in mice with Ca^{2+} channel mutations it is likely that additional animal models for retinal diseases will be identified.

REFERENCES

1. Ashcroft FM. Ion channels and disease. San Diego: Academic Press, 2000.
2. Ptacek LJ, George AL, Griggs RC et al. Identification of a mutation in the gene causing hyperkalemic periodic paralysis. Cell 1991; 67:1021-1027.
3. Jurkat-Rott K, Lehmann-Horn F, Elbaz A et al. A calcium channel mutation causing hypokalemic periodic paralysis. Hum Mol Genet 1994; 3:1415-9.
4. Ophoff RA, Terwindt GM, Vergouwe MN et al. Familial hemiplegic migraine and episodic ataxia type-2 are caused by mutations in the Ca^{2+} channel gene CACNL1A4. Cell 1996; 87:543-552.

5. Bech-Hansen NT, Naylor MJ, Maybaum TA. Loss-of-function mutations in a calcium-channel α_1-subunit gene in Xp11.23 cause incomplete X-linked congenital stationary night blindness. Nat Genet 1998; 19:264-267.
6. Strom TM, Nyakatura G, Apfelstedt-Sylla E et al. An L-type calcium-channel gene mutated in incomplete X-linked congenital stationary night blindness. Nat Genet 1998; 19:260-263.
7. Friend KL, Crimmins D, Phan TG et al. Detection of a novel missense mutation and second recurrent mutation in the CACNA1A gene in individuals with EA-2 and FHM. Hum Genet 1999; 105:261-265.
8. Escayg A, De Waard M, Lee DD et al. Coding and noncoding variation of the human calcium-channel β_4-subunit gene CACNB4 in patients with idiopathic generalized epilepsy and episodic ataxia. Am J Hum Genet 2000; 66:1531-1539.
9. Ducros A, Denier C, Jourtel A. The clinical spectrum of familial hemiplegic migraine associated with mutations in a neuronal calcium channel. N Engl J Med 2001; 345:17-24.
10. Catterall WA. Structure and regulation of voltage-gated Ca^{2+} channels. Annu Rev Cell Dev Biol 2000; 16:521-555.
11. Ertel EA, Campbell KP, Harpold MMA. The nomenclature of voltage-gated calcium channels. Neuron 2000; 25:533-535.
12. Burgess DL, Gefrides LA, Poreman PJ et al. A cluster of three novel Ca^{2+} channel γ subunit genes on chromosome 19q13.4: evolution and expression profile of the γ subunit gene family. Genomics 2001; 71:339-350.
13. Nachman-Clewner M, St. Jules R, Townes-Anderson E. L-type calcium channels in the photoreceptor ribbon synapse: localization and role in plasticity. J Comp Neurol 1999; 415:1-16.
14. Singer JH, Mirotznik RR, Mirotznik RR et al. Potentiation of L-type calcium channels reveals nonsynaptic mechanisms that correlate spontaneous activity in the developing mammalian retina. J Neurosci 2001; 21:8514-22.
15. DeWaard M, Gurnett CA, Campbell KKP. Stuctural and functional diversity of voltage-activated calcium channels. Ion channels 1996; 4:41-87.
16. Dowling JE, Boycott BB. Organization of the primate retina: electron microscopy. Proc R Soc London 1966; 166:80-111.
17. von Gersdorff H. The ribbon synapse: Versatile signal transducers. Neuron 2001; 29:7.
18. Morgans CW. Neurotransmitter release at ribbon synapses in the retina. Immunol Cell Biol 2000; 78:442-6.
19. DeVries S, Baylor D. Synaptic circuitry of the retina and olfactory bulb. Cell 1993; 72:139-149.
20. Tachibana M, Okada T, Arimura T et al. Dihydropyrinine-sensitive calcium current mediates neurotransmitter release from bipolar cells of the goldfish. J Neurosci 1993; 13:2898-2909.
21. Pan Z-H. Differential expression of high- and tow types of low-voltage-activated calcium currents in rod and cone bipolar cells of the rat retina. J Neurophysiol 2000; 83:513-527.
22. Morgans CW. Calcium channel heterogeneity among cone photoreceptors in the tree shrew retina. Eur J Neurosci 1999; 11:2989-2993.
23. Yagi T, Macleish PR. Ionic conductances of monkey solitary cone inner segments. J Neurophysiol 1994; 71:656-665.
24. de la villa P, Vaquero CF, Kaneko A. Two types of calcium currents of the mouse bipolar cells recorded in the retinal slice preparation. Eur J Neurosci 1998; 10:317-323.
25. Morgans CW. Localization of the α_{1F} calcium channel subunit in the rat retina. Invest Ophthalmol & Vis Sci 2001; 42:2414-2418.
26. Ball SL, Powers PA, Shin H-S et al. The β_2 subunit of voltage-gated Ca^{2+} channels is essential for normal synaptic transmission between photoreceptors and the inner retina. Invest Ophthal & Vis Sci 2002; in press.
27. Read DS, Ball SL, Pardue MT et al. Photoreceptor L-type voltage-dependant Ca^{2+} channels are required for formation and/or maintenance of ribbon synapses in the OPL. Invest Ophthal Vis Sci 2001; 42:S365.

28. Robson JG, Frishman LJ. Response linearity and kinetics of the cat retina: The bipolar cell component of the dark-adapted electroretinogram. Vis Neurosci 1995; 12:837-850.
29. Hood D, Birch DG. Assessing abnormal rod photoreceptor activity with the a-wave of the electroretinogram: applications and methods. Doc Ophthalmol 1997; 92:253-267.
30. Zwingman T, Neumann PE, Noebels JL et al. Rocker is a new variant of the voltage-dependent calcium channel gene Cacna1a. J Neurosci 2001; 21:1169-1178.
31. Doyle J, Ren X, Lennon G et al. Mutations in the Cacnl1a4 calcium channel gene are associated with seizures, cerebellar degeneration, and ataxia in *tottering* and *leaner* mutant mice. Mamm Genome 1997; 8:113-120.
32. Fletcher CF, Lutz CM, O'Sullivan TN et al. Absence epilepsy in *tottering* mutant mice is associated with calcium channel defects. Cell 1996; 87:607-617.
33. Sidman RL, Green MC, Appel SH. Catalog of the neurological mutants of the mouse. Harvard UP: Cambridge, 1965.
34. Noebels JL, Sidman RL. Inherited epilepsy: spike-wave and focal motor seizures in the mutant mouse *tottering*. Science 1979; 204:1334-1336.
35. Melena J, Osborne NN. Voltage-dependent calcium channels in the rat retina: involvement in NMDA-stimulated influx of calcium. Exp Eye Res 2001; 72:393-401.
36. Kamphius W, Hendriksen H. Expression patterns of voltage-dependent calcium channel alpha 1 subunits (alpha 1A-alpha 1E) mRNA in rat retina. Brain Res Mol Brain Res. 1998; 55:209-20.
37. Kim D, Song I, Keum S et al. Lack of the burst firing of thalmocortical relay neurons and resistance to absence seizures in mice lacking alpha 1G T-type Ca^{2+} channels. Neuron 2001; 31:35-45.
38. Wilson SM, Toth PT, Oh SB et al. The status of voltage-dependent calcium channesl in α_{1E} knockout mice. J Neurosci 2000; 20:8566-8571.
39. Platzer J, Engel J, Schrott-Fischer A et al. Congenital deafness and sinoatrial node dysfunction in mice lacking class D L-type Ca^{2+} channels. Cell 2000; 102:89-97.
40. Cork RJ, Namkung Y, Shin H-S et al. Development of the visual pathway is disrupted in mice with a targeted disruption of the calcium chanvel β3-subunit gene. J Comp Neurol 2001; 440:177-191.
41. Nakamura M, Ito S, Terasaki H et al. Novel CACNA1F mutations in Japanese patients with incomplete congenital stationary night blindness. Invest Ophthalmol Vis Sci 2001; 42:1610-1616.
42. Miyake Y, Yagasaki K, Horiguchi M et al. Congenital stationary night blindness with negative electroretinogram. A new classification. Arch Ophthalmol. 1986; 104:1013-1020.
43. Gregg RG, Messing A, Strube C et al. Absence of the β subunit (*cchb1*) of the skeletal muscle dihydropyridine receptor alters expression of the α_1 subunit and eliminates excitation-contraction coupling. Proc Natl Acad Sci USA 1996; 93:3961-13966.
44. Xu X, Quiambao AB, Roveri L et al. Degeneration of cone photoreceptors induced by expression of the *Mas1* oncogene. Exp Neurol 2000; 163:207-219.
45. Wachmeister L. Some aspects of the oscillatory response of the retina. Prog Brain Res 2001; 131:465-74.
46. Lamb TD. Gain and kinetics of activation in the G-protein cascade of phototransduction. Proc Natl Acad Sci USA 1996; 93:566-570.
47. Jacobs GH, Neitz J, Deegan 2d JF. Retinal receptors in rodents maximally sensitive to ultraviolet light. Nature 1991; 353:655-656.
48. Pardue MT, McCall MA, LaVail MM et al. A naturally occurring mouse model of X-linked congenital stationary night blindness. Invest Ophthalmol Vis Sci 1998; 39:2443-2449.

CALDENDRINS IN THE INNER RETINA

Constanze I. Seidenbecher, C. Reissner and Michael R. Kreutz

ABSTRACT

Caldendrin is the first member of a novel family of Ca^{2+}-binding proteins (CaBPs). Its unique two-domain structure is composed of a calmodulin-homologous C-terminus and an unrelated N-terminal part. The latter is thought to mediate the tight association of caldendrin with the subsynaptic cytoskeleton. Caldendrin is expressed in forebrain regions with a laminar cytoarchitecture as well as in the inner retina where it is localized to OFF cone bipolar and a subset of amacrine and ganglion cells. In addition, caldendrin is prominently present in processes and synapses of the inner plexiform layer. Thus, caldendrin-immunoreactivity is displayed by subpopulations of most retinal cell classes, with the exception of glial cells. Caldendrin is most likely involved in dendritic Ca^{2+}-signaling, one of the functions of its close relative, calmodulin. However, several lines of evidence suggest that due to its unique properties caldendrin might not merely substitute for calmodulin. It is speculated that either the specific enrichment in cellular micro-compartments like the postsynaptic cytomatrix, the unique two-domain structure or the altered distribution of surface charges renders caldendrin specific for distinct binding partners or certain Ca^{2+}-triggered signaling events.

INTRODUCTION

The retina is a highly specialized part of the central nervous system that has evolved to receive, transduce and integrate visual signals. This is accomplished by a variety of sophisticated cell types that are equipped with unique sets of proteins to fulfill these cellular functions (for review see Masland[1]). Because of the central role

AG Molecular of Plasicity, Department of Neurochemistry/Molecular Biology, Leibnitz Institute for Neurobiology, 39118 Magdeburg, Germany.

that Ca^{2+} ions play as second messengers in the entire phototransduction process the retina is a particularly rich source of intracellular Ca^{2+}-binding proteins (CaBPs). Accordingly, numerous studies have been published concerning the expression, localization and possible function of this multitude of different CaBPs in the retina. The great variety of retinal CaBPs is believed to reflect the cells' needs to accurately sense actual Ca^{2+} concentrations, to buffer local Ca^{2+} overload, to couple plasma membrane voltage-gated Ca^{2+} channels to intracellular Ca^{2+} stores and to transduce the Ca^{2+} signals to other compartments of the cell, e.g., the nucleus. Finally, most CaBPs exhibit a rather restricted expression pattern and therefore CaBPs are frequently used as markers for special retinal cell populations in functional studies (Andressen et al[2]). However, the reasons for the existence of a multitude of CaBPs and their functional implications for intracellular Ca^{2+} signaling still remain elusive.

Among the retinal CaBPs there is a subpopulation predominantly or exclusively expressed in the outer retina and a second group that occurs in the inner retina as well as in other parts of the central nervous system (CNS). Whereas members of the first group mainly function in the primary phototransduction process within photoreceptor cells, the latter ones apparently fulfill functions in Ca^{2+}-mediated signaling which seem to occur basically in the same way both in retinal neurons and nerve cells of the forebrain, e.g., hippocampal pyramidal cells. Accordingly, the expression of members of two sub-families of the neuronal Ca^{2+}-sensors (NCS; for review see Burgoyne & Weiss[3]), namely recoverins and guanylate cyclase-activating proteins (GCAPs; for review see Palczewski et al[4]), is largely restricted to retinal photoreceptor cells. In contrast, proteins of the frequenin and visinin-like protein (VILIP) subfamilies are found in distinct sets of neurons in brain and retina (for review see Polans et al,[5] Braunewell & Gundelfinger[6]), where their expression is mainly restricted to the inner retinal layers (Lenz et al;[7] Nakano et al;[8] Bastianelli et al;[9] De Raad et al;[10] Reynolds et al[11]).

Caldendrin, a neuronal CaBP cloned from rat brain by our laboratory, clearly belongs to the second group because it is expressed both in the inner part of the retina and in the gray matter areas of the brain. The protein's name derives from its ability to bind Ca^{2+} and its exclusive localization in the somato-dendritic compartment of mature neurons (Seidenbecher et al[12]).

Caldendrin is a member of the EF hand family of neuronal CaBPs that displays an interesting two-domain primary structure with four EF hand motifs in its C-terminal half forming a CaM-like three-dimensional structure and an unrelated N-terminal half with a rather flexible shape. Interestingly, the caldendrin gene is alternatively spliced. Shorter variants lacking the N-terminal part of caldendrin are only detectable in the retina and the cerebellum (Haeseleer et al;[13] Laube et al[14]).

Within neurons, caldendrin localizes both to the cytosolic and to the cytoskeletal compartments with a clear enrichment in the subsynaptic cytoskeleton of the post-synaptic density (PSD). Both its two-domain structure and the differential association with subcellular compartments suggest that caldendrin serves functions different from those of known neuronal Ca^{2+} sensors and Ca^{2+} buffers.

THE MOLECULAR STRUCTURE OF CALDENDRIN

Caldendrin displays a unique primary structure with a bipartite organization of the molecule into a very basic N-terminus (P_I 11.5) and a rather acidic C-terminus (P_I 4.6) resulting in an overall P_I of 7.1. Interestingly, in the splice variants caldendrin-S1 (S-CaBP1) and -S2 (L-CaBP1) the much shorter N-terminal part can not neutralise the acidic C-terminus resulting in a clearly lower overall P_I of 4.7 and 4.8, respectively.

The N-terminal half of caldendrin has a unique primary structure with no homology to other polypeptide sequences in public databases. The N-terminus is predicted to start with a short hydrophobic helix comprising residues with rather small side chains (PHD prediction according to Rost and Sander[15]). Although the isoform caldendrin-S1 harbors an N-terminal helical region, this helix has an amphipathic character and is primarily composed of amino acid residues with large side chains. Furthermore, both shorter isoforms are myristoylated at their N-terminus. Therefore, the amino-terminal regions of the three caldendrins are unrelated (Fig. 1B). The flexible Arg- and Pro-rich regions of caldendrin match several PROSITE entries (http://www.expasy.org/prosite/), which were used to build the alignment in Fig. 1A.

As shown there, the entire C-terminal half is highly homologous to calmodulin (CaM), the ancestral member of the family of EF-hand Ca^{2+}-binding proteins. In fact, apart from the CaM-like protein caldendrins are the closest relatives of CaM identified so far.

Crystal structures with and without bound peptide ligands or Ca^{2+} ions have been solved for CaM and troponin C. According to these structures we have modeled caldendrin and the CaBP structures. Their topology is described as a dumb-bell shape formed by two "halfes" or "lobes" connected by a flexible helical linker. That dumb-bell shape is stabilized by the 4 EF hand helix-loop-helix motifs. Troponin C has quite low Ca^{2+}-affinities in EF-hands 1 and 2.

Calmodulin chelates helical peptides of various length enclosing the helix symmetrically with both "opened" halfes which results in an overall compact globular shape. Troponin C does not show a comparable compact form and remains in the elongated dumb-bell shape with each half binding to different peptides. The structures show that most of the conserved residues are part of the large hydrophobic folding unit, which is built by each of the two CaM-like halves. From the crystal structures with bound peptides it is known, that at least one half is involved in peptide binding.

Although the C-terminal caldendrin structure determined by modeling perfectly fits the CaM structure a different surface of the molecules has to be considered because most structure-determining residues are part of the interior hydrophobic cavity within both molecules but the outer surface is covered by residues that clearly differ between CaM and caldendrin. Furthermore, caldendrin as well as most of the CaBPs harbor a longer hinge region between EF-hands 2 and 3 than CaM possibly enhancing the flexibility of the two halfs (Haeseleer et al [13]).

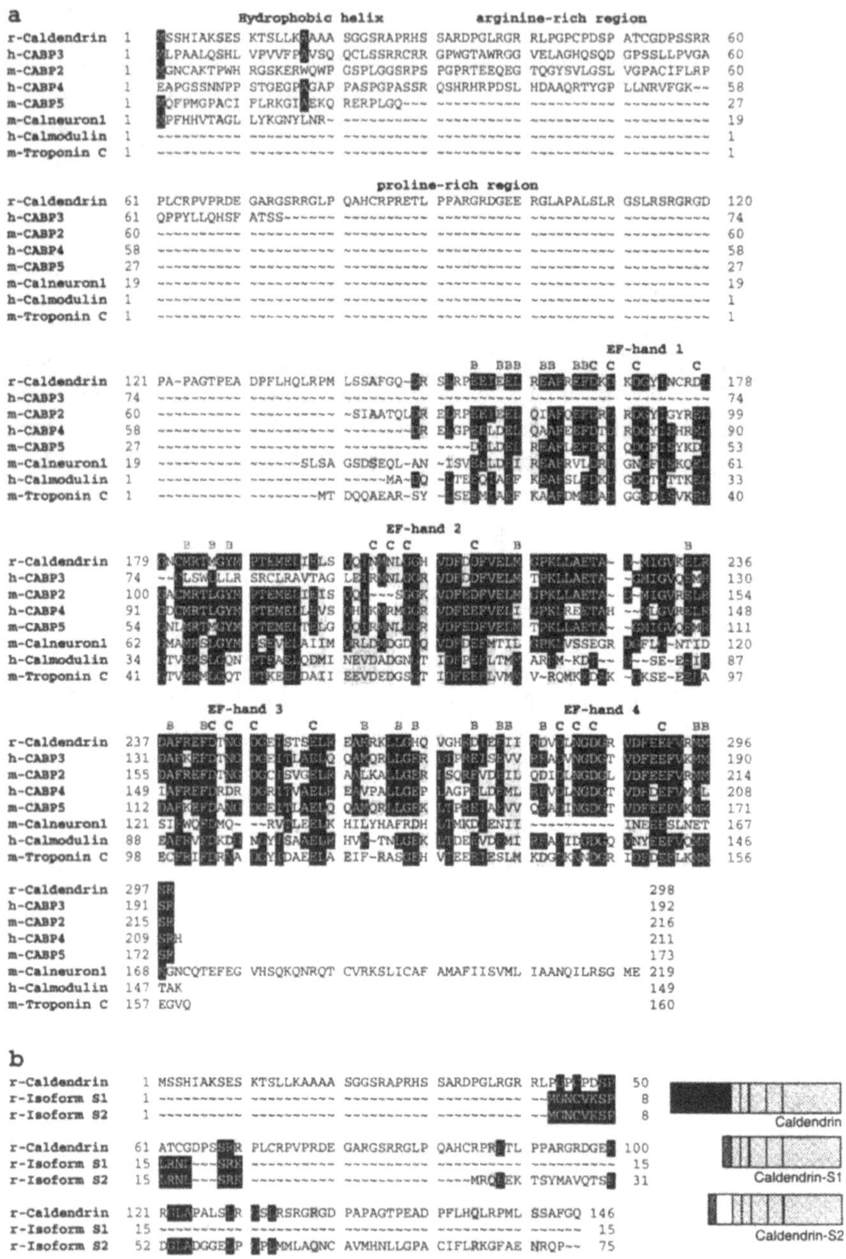

Figure 1. See figure legend next page.

Figure 1A. Multiple alignment of rat caldendrin with structural relatives containing a comparable calmodulin-like fold. r- rat, h- human, m- mouse sequences. Residues identical in at least 50% of the sequences are shaded in black, those with PAM30 similarity in gray. Residues which do bind Ca^{2+} in the reference calmodulin structure are marked with C, while residues in the vicinity from the peptide binding site (4Å) as shown from several crystal structures (PDB entries: 1CDM,1WDC,) are marked with B. Deviations at the C or B positions may be interpreted as adaptations to specific functional requirements for each protein. The alignment has been optimized by superposition of crystal structures (1CDM,1NCX) and model structures of caldendrin and the Ca^{2+}-binding proteins. Structural data are available at synProt (http://www.synprot.de/). The phylogenetic tree has been calculated on the basis of the alignment shown using program TreeTop from GeneBee (http://www.genebee.msu.su). Figure 1B. Alignment of the N-terminal parts of the three rat caldendrin splice variants. In the schematic drawing the black bar represents the exon encoding the caldendrin-specific N-terminus. Gray bars depict the alternative start in the splice isoforms S1 and S2 and the white bar represents the alternatively spliced exon in S2. The 5 exons forming the common C-terminal part are drawn as hatched bars.

The comparison between CaM and caldendrins/CaBPs shows that evolutionary modifications of one or more of the EF hands presumably disturb EF-hand Ca^{2+}-binding, but keep the overall structural topology of the molecules conserved. Troponin C and CaBP3, for instance, have no functional EF-hands 1 and 2, whereas caldendrin and all CaBPs lack a Ca^{2+}-binding EF-hand 2 and calneuron1 has no functional EF-hands 3 and 4.

Ca^{2+} is chelated by acidic amino acids (D, E) or Asn contained in the EF hand loop. In troponin C and caldendrin single point mutations of the type [Asp|Asn]->[Gly|Ser] lead to inactivity of EF-hands 1 and 2, respectively. In the CaBPs 3, 4 and 5, positive charges have been introduced into EF-hand 2 and the consequence is inactivation of Ca^{2+}-binding. Finally, calneuron1 represents an example for inactivation of EF-hands 3 and 4 by shortening the loop of the helix-loop-helix motif by 10 residues which leads to the deletion of three Ca^{2+}-coordinating residues. The peptide binding site shows rather minor deviations with respect to the described EF hand modifications.

In a next step, we generated phylogenetic trees for the full-length alignment, for all of the CaM-like halves, for small fragments, and for those residues which represent the conserved structural core. Because of the statistically equal distribution of homologous vs. non-homologous residues, nearly identical trees were generated by all methods mentioned above. The tree in Figure 1 was calculated for the alignment shown. Based on the CaM structure a structural core was derived and all residues not fitting the template structure were deleted. It is noteworthy, that this tree suggests caldendrin to be most distant to all other CaBPs including CaM, when a gap-less alignment of the structural core is performed. However, phylogenetic considerations rule out an ancestral position for caldendrin, because the gene must have evolved relatively late during vertebrate evolution (it is not present in the *Drosophila* and *C. elegans genomes*), whereas the CaM gene is present in all known eukaryotic organisms (Rogers and Strehler[16]).

The results show that the singularity of the caldendrin surface is intrinsic and independent from insertions or deletions and we therefore speculate that this is probably due to adaptations of its surface to a specific localization and function in neurons of higher vertebrates.

BIOCHEMICAL FEATURES OF CALDENDRIN

Many CaBPs are either soluble cytoplasmatic proteins or membrane-associated via a myristoylation. Caldendrin, however, occurs as both, a soluble pool, easily extractable with physiological buffer and a poorly soluble cytoskeleton-associated pool. On SDS gels loaded with soluble proteins caldendrin migrates as a single band at 33 kDa. Insoluble cytoskeletal protein preparations contain a second caldendrin band at 36 kDa that corresponds to the full-length in vitro translation product. Although it is not yet completely understood how this dichotomy is mediated, we could demonstrate that phosphorylation may play an important role. Caldendrin harbors 7 putative Ser phosphorylation sites for PKC and in situ phosphorylation experiments reveal to caldendrin as a PKC substrate. On two-dimensional protein gels it became obvious that the 33 and 36 kDa isoforms differ in their P_I: whereas the 36 kDa isoform displays a P_I of approx. 6.8-6.9 the 33 kDa variant separates into a chain of spots within the range of 5.9-6.8 possibly indicating phosphorylation by PKC or other kinases as a means of release from the cytoskeleton. A specialization of the neuronal cytoskeleton is the subsynaptic cytomatrix termed postsynaptic density (PSD). We have demonstrated earlier that caldendrin is a component enriched in PSD preparations extracted from rat brain and this seems also to be the case for retinal caldendrin. As depicted in Figure 2, crude synaptic protein preparations from rat retina contain higher amounts of caldendrin immunoreactivity as compared to retinal homogenates. Especially the 36 kDa isoform seems to be particularly abundant in the synaptic fraction. Interestingly, the smaller caldendrin splice variant S2 (L-CaBP1) migrating as a faint band at 28 kDa appears not to be enriched, suggesting that the unique N-terminus of the long isoform is responsible for its restrictive localization and its tight association with the synaptic cytomatrix.

A similar fractionation was performed for the neuronal Ca^{2+}-sensor-1 protein (NCS-1; Reynolds et al[11]). This study showed that NCS-1 equally partitions between the cytosolic and detergent-soluble membrane fractions but is not detectable in the detergent-insoluble membrane skeleton fraction where caldendrin is enriched.

DISTRIBUTION OF CALDENDRIN IN THE FOREBRAIN

Caldendrin expression is clearly restricted to neurons. In situ hybridization experiments and immunostaining have shown that caldendrin transcripts and protein are preferentially found in CNS regions with a laminar cytoarchitecture, i.e., cerebral and cerebellar cortices, hippocampus, olfactory bulb and the retina. Within the cerebral cortex the expression is highly heterogenous, with highest levels of caldendrin in layer III of the piriform and layers II/III of the somatosensory cortex. The segregated cortical distribution to areas which represent the most important primary sensory systems of the rodent brain may reflect different requirements for

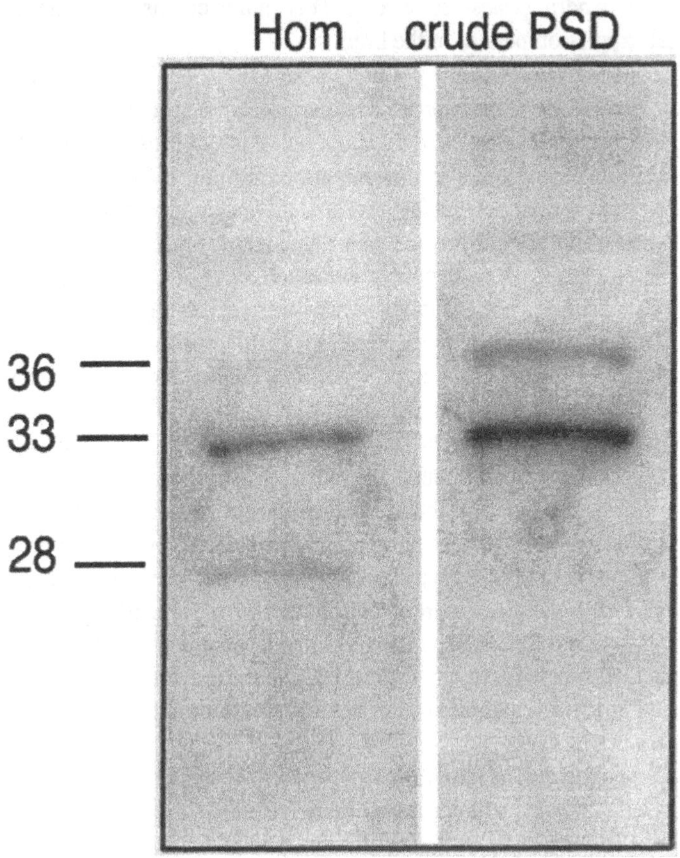

Figure 2. Western blot of 25 μg total homogenate (Hom) or a PSD-enriched fraction (crude PSD, for preparation details see Smalla et al[29]) extracted from rat retina and separated on a 5-20% Laemmli gel. The blot was probed with a 1:500 diluted polyclonal rabbit anti-caldendrin antiserum (Seidenbecher et al[12]).

dendritic Ca^{2+}-signaling in these neurons (Laube et al[14]). Thus, caldendrin immunoreactivity is present in the somato-dendritic compartment of a subpopulation of mainly principal neurons in the brain and it appears to be a component of a discrete subset of mature excitatory synapses.

DISTRIBUTION OF CALDENDRIN IN THE RETINA

Several studies describe in detail the distribution of caldendrin immunoreactivity in the rodent retina (Wässle et al,[17] Menger et al,[18] Haverkamp and Wässle,[19] Haeseleer et al[13]). All these publications concurrently revealed a predominant localization of the protein to the inner retina. Müller glial cells, the main retinal glia cells, seem to be devoid of caldendrin (Haeseleer et al[13]). This data is consistent with the exclusively neuronal expression pattern in the brain.

Hence the expression of caldendrin is not unique for a neuronal cell type with identified function. In accordance to the expression pattern described for other CaBPs, caldendrin-immunoreactivity was displayed by subpopulations of most retinal cell classes, with the exception of horizontal cells where no immunoreactivity was detected. Immunostainings of retinal flat mounts revealed that caldendrin is restricted to a small subpopulation of fluorogold-prelabeled retinal ganglion cells (RGC) with no staining of axonal compartments in the optic nerve layer. Surprisingly caldendrin immunoreactivity was most frequently found in amacrines, a cell type with neuritic processes of variable length and no clear differentiation into axonal and dendritic compartments. Amacrines, however, exhibit arborizations of their neurites, which clearly resemble the arborization pattern of dendritic trees and also express the protein components of postsynaptic structures. Thus, although in amacrines the differentiation of dendrites and axons is ambiguous, the retinal distribution of caldendrin immunoreactivity is consistent with the somato-dendritic localization found in rat brain (Seidenbecher et al[12]). Double labeling experiments suggest that caldendrin expression is not restricted to a single amacrine cell subtype but instead might be the CaBP with the probably most widespread distribution in rat amacrines of the inner nuclear layer (INL) identified to date. Interestingly, in contrast to glycinergic amacrines of the INL the displaced glycinergic amacrines were frequently caldendrin immunonegative. The functional significance of this differentiation, however, is at present unknown.

In line with this heterogeneous expression pattern in rat amacrines an exact co-localization with a distinct retinal neurotransmitter system or excitatory vs. inhibitory neurons could also not be found. A detailed analysis of the mouse retina by Haverkamp & Wässle[19] pointed out that exclusively bipolar cells with a rather large cell body in the center of the INL and a morphology resembling that of OFF-cone bipolar cells are caldendrin-immunopositive. Rod and ON cone bipolar cells, on the other hand, are apparently not labeled. Moreover, according to double-labeling experiments the immunopositive neurons are not glycinergic. Whether this unique staining pattern reflects distinct requirements for the transduction of Ca^{2+}-signals is still unknown and it has to be mentioned in this regard that considerable differences in the expression pattern of CaBPs have been found between different species.

COMPARISON WITH OTHER CaBPs

The localization of a variety of CaBPs belonging to the CaM-troponin-C family has been investigated in the rat retina by several laboratories. Apart from recoverin, visinin and the GCAPs (Braunewell and Gundelfinger;[6] Polans et al[5]) most proteins supposed to have a triggering function upon Ca^{2+}-binding as well as the Ca^{2+}-buffer proteins have been mainly localized to the inner retina. Accordingly, caldendrin is also present in these inner retinal layers. Although the caldendrin distribution overlaps in total or partially with that of hitherto characterized CaBPs there is no complete congruence.

In the rod-dominated retina of the rat parvalbumin prominently labels AII-amacrine cells (Wässle et al[20]), which are the most important amacrines in the rod signal pathway. Significantly fewer widefield amacrines and RGCs are stained. Calretinin labels many amacrines and RGCs (Pasteels et al [21]). Calbindin immunocytochemistry labels the axon-bearing B-type horizontal cells (Peichl and Gonzalez-Soriano[22]), small amacrines and a subpopulation of RGCs (Bastianelli et al[9], Hamano et al,[23] Pasteels et al[21]). Co-distribution of caldendrin and calbindin immunoreactivity was found only in a minority of bipolar and amacrine cells of the INL. A substantial number of calbindin positive cells exhibited no caldendrin-staining and vice versa. Furthermore, horizontal cells, which abundantly contain calbindin, seem not to express caldendrin. Calbindin immunoreactive cells of the ganglion cell layer (GCL) were frequently immunopositive for caldendrin, but the staining intensity for caldendrin was considerably lower. In retinae from calbindin-null mutant (-/-) mice the caldendrin expression pattern is not changed as holds true for all retinal CaBPs examined in this study (Wässle et al[17]).

CaBP5, a cognate member of the caldendrin family, is detectable in rod and cone bipolar cell perikarya and fine processes including dendrites in the outer plexiform layer (OPL) and axons ramifying in the inner plexiform layer (IPL ; Haeseleer et al[13]). Despite the close relation to caldendrin the distribution of both molecules differs significantly. Furthermore, caldendrin immunoreactivity does not seem to be congruent with that of CaM (Pochet et al[24]). Although CaM is generally discussed as a ubiquitously expressed molecule it shows a quite distinct expression pattern in the retina. CaM was found to be below the detection limit in photoreceptors but was detected in horizontal and prominently in amacrine, bipolar and retinal ganglion cells. In addition, in the mouse retina the optic nerve layer is strongly CaM immunoreactive. Thus, although the C-terminus of caldendrin is highly similar to CaM the different distribution in the retina suggests that both proteins have cell-specific functions.

CALDENDRIN LOCALIZATION AT SYNAPSES

Caldendrin immunoreactivity is tightly associated with the cortical cytoskeleton, enriched in the PSD fraction and associates late during development with the synaptic cytomatrix in the brain (Laube et al[14]).

The retina comprises two plexiform layers where synaptic contacts are organized, the OPL and the IPL. Caldendrin prominently stains the IPL (see Fig. 3) but is absent from the OPL where synaptic contacts between photoreceptor cells and bipolar and horizontal cells reside. In contrast to calbindin and calretinin, which are also present in the IPL caldendrin labeling appears not as a clear stratification but more diffuse. Codistribution with the presynaptic marker synaptophysin suggests a localization at chemical synapses.

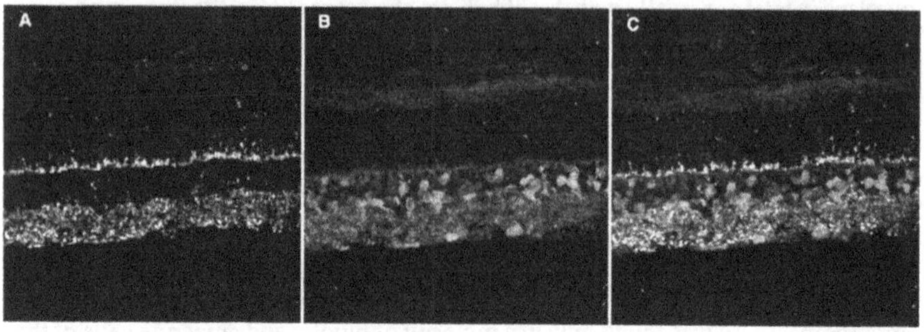

Figure 3. Double immunofluorescence labeling in the IPL and OPL of the rat retina with polyclonal rabbit anti-caldendrin antiserum (A) and mouse anti-synaptophysin (B). Double immunofluorescence confocal laserscans show a large overlap of synaptophysin and caldendrin fluorescence signals in the IPL (C).

CALDENDRIN EXPRESSION DURING RETINAL DEVELOPMENT

Caldendrin transcript levels were differentially regulated in different cell types during the postnatal development of the retina. In the inner neuroblast layer (INB) where amacrines reside, caldendrin transcripts were present at postnatal day P1. In contrast, ganglion cells and displaced amacrines were devoid of hybridization signals at time points P1 and P5, indicating that displaced amacrines in the GCL do not express caldendrin mRNA at this early developmental stage. This is consistent with the generally lower expression level of caldendrin RNA in displaced amacrines compared to amacrines in the INL in the adult retina. However, the functional significance of the lack of expression in displaced amacrines during early development remains unclear.

It is noteworthy that the early detection of caldendrin mRNA in amacrines of the INL is in contrast to the developmental expression pattern in neuronal cells of the forebrain. Preliminary studies in brain suggest that only minor amounts of caldendrin mRNA and protein are present in the early postnatal brain (Seidenbecher et al[12]; Kreutz and Seidenbecher, unpublished observations) as compared to adult levels. The amount of caldendrin transcripts increases during the second postnatal week coinciding with dendrite maturation and synaptogenesis. This pattern fits with the developmental localization of hybridization signals in bipolar and ganglion cells but not with that detected in amacrines. Thus, it seems that in amacrines caldendrin might play a role in the development of neurite processes before functional synapses have formed.

IS THERE A SPECIFIC FUNCTION FOR CALDENDRIN IN DENDRITIC Ca^{2+}-SIGNALING?

Previous work has demonstrated an exclusive association of caldendrin with the somato-dendritic compartment of brain neurons (Seidenbecher et al[12]). The distribution pattern in retinal sections is consistent with this localization. Also in accordance with the brain data a substantial amount of caldendrin immunoreactivity was found in the particulate and crude synaptic protein fractions of the retina. In addition, no staining of the optic nerve layer was detected in retinal flat mounts. We therefore assume that a significant fraction of caldendrin immunoreactivity is also associated with the somato-dendritic cytoskeleton in the retina. Its precise role in dendritic Ca^{2+}-signaling, however, remains still enigmatic.

Several attempts have been made to unravel the cellular functions of caldendrin/CaBP1 by analyzing the interactions with known CaM targets. Yamaguchi et al[25] reported a truncated C-terminal caldendrin fragment termed calbrain. It is unclear, however, whether calbrain occurs in nature. In their studies this fragment comprising EF-hands 3 and 4 can compete with CaM for Ca^{2+}/calmodulin-dependent kinase II (CaMKII) binding and was shown to reduce CaMKII activation and autophosphorylation. For the N-terminal caldendrin-S1 (S-CaBP1) splice variant the interaction with CaMKII was tested by Haeseleer et al[13] demonstrating a weak stimulation of CaMKII but no binding to the CaM-binding peptide of CaMKII. The different results may be explained by the use of a small artificial fragment in the Yamaguchi study. The stimulating effect of caldendrin-S1 on CaMKII, however, is rather weak as compared to CaBP2 or CaM. Furthermore, in contrast to CaM or CaBPs 2 and 5 no significant influence of caldendrin-S1 on the activity of the Ca^{2+}-dependent phosphatase calcineurin could be measured (Sokal et al[26]).

Interestingly, an inhibiting effect of caldendrin-S1 on the G-protein coupled receptor kinase GRK5 was reported in vitro (Haeseleer et al[13]). This finding could be of particular importance because GRK5 was recently demonstrated to play a role in the SAP90/PSD95-mediated anchoring of receptors at synapses (Hu et al[27]).

A further functional implication for caldendrin was hypothesized in the Ca^{2+}-independent inhibition of $Ca_v2.1$ channels (Lee et al[28]). However, these channels are discussed as presynaptic and caldendrin expression is more or less confined to postsynaptic sites. Therefore, the physiological significance of this finding is questionable.

Thus, although caldendrin like its relative CaM is most probably involved in local Ca^{2+}signaling rather than the regulation of intracellular Ca^{2+} concentrations, several lines of evidence suggest that due to its unique properties caldendrin may not merely serve as a CaM substitute under certain conditions. In most cells CaM has a several fold higher intracellular concentration as compared to caldendrin. However, we speculate that either the restrictive localization in cellular micro-compartments like the postsynaptic cytomatrix, its unique two-domain structure or its distinct molecular surface renders caldendrin specific for particular binding partners or certain synaptic Ca^{2+}-triggered events. In any case cellular functions mediated by caldendrin are obviously required only in a subset of retinal neurons and future studies with knock-out mice may help to identify the impact of caldendrin on dendritic Ca2+ signaling.

ACKNOWLEDGMENT

The authors are grateful to Prof. E.D. Gundelfinger for helpful criticism of the manuscript, Dr. W. Zuschratter for advice in confocal microscopy and to C. Borutzki and M. Marunde for expert technical assistance. This work was supported by the DFG (Kr1879/2-1, Se952/2-1, 2-2), the Fonds der Chemischen Industrie and the Land Sachsen-Anhalt (FKZ: 2508A/0086 and 3004A/0088H).

REFERENCES

1. Masland RH. The fundamental plan of the retina. Nat Neurosci 2001; 4(9):877-886.
2. Andressen C, Blümcke I, Celio MR. Calcium-binding proteins: selective markers of nerve cells. Cell Tissue Res 1993; 271(2):181-208.
3. Burgoyne RD, Weiss JL. The neuronal calcium sensor family of Ca^{2+}-binding proteins. Biochem J 2001; 353(Pt 1):1-12.
4. Palczewski K, Polans AS, Baehr W et al. $Ca^{(2+)}$-binding proteins in the retina: structure, function, and the etiology of human visual diseases. Bioessays 2000; 22(4):337-350.
5. Polans A, Baehr W, Palczewski K. Turned on by Ca^{2+}! The physiology and pathology of Ca(2+)-binding proteins in the retina. Trends Neurosci 1996; 19(12):547-554.
6. Braunewell KH, Gundelfinger ED. Intracellular neuronal calcium sensor proteins: a family of EF-hand calcium-binding proteins in search of a function. Cell Tissue Res 1999; 295(1):1-12.
7. Lenz SE, Henschel Y, Zopf D et al. VILIP, a cognate protein of the retinal calcium binding proteins visinin and recoverin, is expressed in the developing chicken brain. Brain Res Mol Brain Res 1992; 15(1-2):133-140.
8. Nakano A, Terasawa M, Watanabe M et al. Neurocalcin, a novel calcium binding protein with three EF-hand domains, expressed in retinal amacrine cells and ganglion cells. Biochem Biophys Res Commun 1992; 186(3):1207-1211.
9. Bastianelli E, Takamatsu K, Okazaki K et al. Hippocalcin in rat retina. Comparison with calbindin-D28k, calretinin and neurocalcin. Exp Eye Res 1995; 60(3):257-266.
10. De Raad S, Comte M, Nef P et al. Distribution pattern of three neural calcium-binding proteins (NCS-1, VILIP and recoverin) in chicken, bovine and rat retina. Histochem J 1995; 27(7):524-535.
11. Reynolds AJ, Bartlett SE, Morgans C. The distribution of neuronal calcium sensor-1 protein in the developing and adult rat retina. Neuroreport 2001; 12(4):725-728.
12. Seidenbecher CI, Langnaese K, Sanmarti-Vila L et al. Caldendrin, a novel neuronal calcium-binding protein confined to the somato-dendritic compartment. J Biol Chem 1998; 273(33):21324-21331.
13 Haeseleer F, Sokal I, Verlinde CL et al. Five members of a novel $Ca^{(2+)}$-binding protein (CABP) subfamily with similarity to calmodulin. J Biol Chem 2000; 275(2):1247-1260.
14. Laube G, Seidenbecher CI, Richter K et al. The neuron-specific Ca^{2+}-binding protein caldendrin: gene structure, splice isoforms and expression in rat brain. Mol Cell Neurosci (in press).
15. Rost B, Sander C. Prediction of protein secondary structure at better than 70% accuracy. J Mol Biol 1993; 232(2):584-599.

16. Rogers MS, Strehler EE. Calmodulin. In: Celio MR, Pauls TL, Schwaller B, eds. Guidebook to the calcium-binding proteins. Oxford Univ Press, 1996.
17. Wässle H, Peichl L, Airaksinen MS et al. Calcium-binding proteins in the retina of a calbindin-null mutant mouse. Cell Tissue Res 1998; 292(2):211-218.
18. Menger N, Seidenbecher CI, Gundelfinger ED et al. The cytoskeleton-associated neuronal calcium-binding protein caldendrin is expressed in a subset of amacrine, bipolar and ganglion cells of the rat retina. Cell Tissue Res 1999; 298(1):21-32.
19. Haverkamp S, Wässle H. Immunocytochemical analysis of the mouse retina. J Comp Neurol 2000; 424(1):1-23.
20. Wässle H, Grunert U, Rohrenbeck J. Immunocytochemical staining of AII-amacrine cells in the rat retina with antibodies against parvalbumin. J Comp Neurol 1993; 332(4):407-420.
21. Pasteels B, Rogers J, Blachier F et al. Calbindin and calretinin localization in retina from different species. Vis Neurosci 1990; 5(1):1-16.
22. Peichl L, Gonzalez-Soriano J. Morphological types of horizontal cell in rodent retinae: a comparison of rat, mouse, gerbil, and guinea pig. Vis Neurosci 1994; 11(3):501-517.
23. Hamano K, Kiyama H, Emson PC et al. Localization of two calcium binding proteins, calbindin (28 kD) and parvalbumin (12 kD), in the vertebrate retina. J Comp Neurol 1990; 302(2):417-424.
24. Pochet R, Pasteels B, Seto-Ohshima A et al. Calmodulin and calbindin localization in retina from six vertebrate species. J Comp Neurol 1991; 314(4):750-762.
25. Yamaguchi K, Yamaguchi F, Miyamoto O et al. Calbrain, a novel two EF-hand calcium-binding protein that suppresses Ca^{2+}/calmodulin-dependent protein kinase II activity in the brain. J Biol Chem 1999; 274(6):3610-3616.
26. Sokal I, Li N, Verlinde CL et al. $Ca^{(2+)}$-binding proteins in the retina: from discovery to etiology of human disease(1). Biochim Biophys Acta 2000; 1498(2-3):233-251.
27. Hu LA, Chen W, Premont RT et al. G protein-coupled receptor kinase 5 regulates beta1-adrenergic receptor association with PSD-95. J Biol Chem 2001; 77.
28. Lee A, Westenbroek RE, Haeseleer F et al. Inhibition of CAV2.1 Ca^{2+} channels by a neuronal Ca^{2+} binding protein. Soc Neurosci Abstr 2001; 27(1):14.2.
29. Smalla KH, Matthies H, Langnäse K et al. The synaptic glycoprotein neuroplastin is involved in long-term potentiation at hippocampal CA1 synapses. Proc Natl Acad Sci USA 2000; 97(8):4327-4332.

CALCIUM CHANNELS AT THE PHOTORECEPTOR SYNAPSE

Steven Barnes[1] and Melanie E.M. Kelly[2]

ABSTRACT

Presynaptic Ca^{2+} channels mediate early stages of visual information processing in photoreceptors by facilitating the release of neurotransmitter and by receiving modulatory input that alters transmission. Two types of L-type Ca^{2+} channels, composed of $\alpha 1F$ and $\alpha 1D$ subunits and having similar biophysical and pharmacological properties, appear to form the principle voltage-dependent Ca^{2+} influx pathways in rods and cones, respectively. The role played by these channels in neurotransmitter release at these graded potential, non-spiking synapses, has been well described. The channels mediate sustained glutamate release in darkness where the cells rest at potentials near –40 mV, and signal increases in light intensity as the cells hyperpolarize negative to this value. Synaptic modulation and integration mediated by these channels has not yet been as fully described but appears to involve GABA, nitric oxide (NO), glutamate, and dopamine. Ca^{2+} permeable cyclic nucleotide gated (CNG) channels appear to have supporting roles at the photoreceptor output synapse and may transduce NO signals from other cells by either directly permitting Ca^{2+} influx or by providing depolarizing influences that gate voltage dependent Ca^{2+} channels.

INTRODUCTION

Animals see when features of the neural image first formed in the photoreceptors are extracted and encoded in the successive and parallel stages of visual information processing. The output synapse of rod and cone photoreceptors is an obvious crucial link between the transduction of light and the neural processing of that

Departments of [1]Physiology & Biophysics, [2]Pharmacology and [1,2]Ophthalmology, Dalhousie University, Halifax, Nova Scotia, B3H 4H7 signal. At this first synapse there is communication not only from the photoreceptors to second order bipolar and horizontal cells, but retrograde, feedback communication from second order cells to photoreceptors in return. Photoreceptor presynaptic Ca2+ c^{han}nels are key elements in this bi-directional transfer of visual information.

This chapter reviews three fundamental aspects of the photoreceptor presynaptic terminal: 1) the function of voltage-gated Ca^{2+} channels in transmission of the light response from the rod spherules and cone pedicles, the presynaptic terminals of the photoreceptors, 2) the presynaptic function of non-voltage gated Ca^{2+} permeable channels in the photoreceptor terminals, and 3) the role of Ca^{2+} and Ca^{2+} permeable channels in mediating the synaptic signals received from second order cells.

Physiology of Transmission at the Photoreceptor Synapse

The neurons of the outer retina, e.g., rod and cone photoreceptors, horizontal cells, and bipolar cells, do not generate action potentials (but see Kawai et al, 2001).[1] Instead, they signal via graded and relatively small changes in their membrane potentials.[1a] For photoreceptors, the range of membrane potential excursions is limited to values from around –40 mV in darkness, to –60 or –70 mV in bright light. For rod photoreceptors, clipping of hyperpolarizing signals greater than about 5 - 10 mV is known to occur at the synapse,[2] while for cones, evidence suggests a greater dynamic range for the synapse (up to 30 mV).[3] This is surprising, considering that for both types of photoreceptors, biophysically similar but molecularly distinct voltage-gated Ca^{2+} channels mediate the Ca^{2+} dependent synaptic release of glutamate.[4] The apparent Ca^{2+} channel activation threshold at about –45 mV suggests a poor alignment between the range of potentials over which the photoreceptors operate and the voltage range over which presynaptic Ca^{2+} channels activate. However, it must be recalled that there is no true threshold to ion channel activation (such as there is for an action potential), and that at potentials negative to the apparent threshold, the open probability for the channels falls off exponentially to nearly imperceptible levels.

As at most other synapses in the nervous system, glutamate is released when the membrane is depolarized and Ca^{2+} channels are activated.[5,6] Postsynaptic responses in horizontal cells and hyperpolarizing ("off") bipolar cells are mediated by a diversity of ionotropic glutamate-gated channels. Light-evoked photoreceptor hyperpolarizations become depolarizations in "on" bipolar cells, the inversion being mediated by a metabotropic glutamate receptor (the APB-receptor), which affects membrane potential via CNG channels.[7]

Synaptic transmission is strikingly sensitive to Ca^{2+} signal modulation at the rod and cone terminals. Since a narrow range of Ca^{2+} channel activation is utilized for synaptic release, high gain by the processes translating membrane potential into glutamate release is required. The first consequence of this is that a steep exponential relationship between the presynaptic and postsynaptic potentials arises.[8] The second consequence is that small changes in Ca^{2+} channel activation translate into large changes of synaptic efficacy.[2,9] The third consequence, especially important considering the likely role of CNG channels at the synapse, is that all processes resulting in Ca^{2+} influx and/or membrane potential changes alter synaptic function dramatically.[10]

Receptive Field Formation and Adaptation at the Photoreceptor Synapse

Ganglion cells, which carry the output of the retina, characteristically have center-surround antagonistic receptive fields. In some ganglion cells, light falling in the center of the receptive field evokes excitation, while light falling on concentric surrounding regions evokes a response of opposite polarity due to lateral inhibition. Other ganglion cell receptive fields have inhibitory centers and excitatory surrounds, while for some others, the color of the light determines the sign of the response. These receptive field organizations support some of the basic principles of retinal processing: contrast enhancement and color opponency. It is known that center-surround antagonistic receptive fields are first formed through presynaptic integration at the cone photoreceptor synaptic terminal.

Another major emphasis of research at the photoreceptor synapse derives from the fact that for the visual processing cascade to function optimally, neural signals must be maximized for all levels of light adaptation.[1a] The visual system operates over an extremely broad range (10^{10}) of stimulus intensities. The system achieves this remarkable operating range by exploiting the dynamic range at numerous, cascaded stages where adaptation occurs. Several adaptation stages have been described in rods and cones, beginning of course in the light transducing outer segments. Resetting of the operating range of the photoreceptor output synapse is another such stage, but the process occurs via unknown mechanisms.

The relative strength of cone versus rod synaptic input to second order cells can be modulated.[11] By altering the strength of the rod and cone output synapses inversely, an effective means for broadening the operating range of the retina arises. Most vertebrate retinas are duplex: rod photoreceptors and associated circuits mediate photon-sparse night vision, whereas cone photoreceptors and associated circuitry mediate information-rich day vision. Duplex retinas share circuitry for the rod and cone systems wherein some retinal neurons in the pathway, downstream from photoreceptors, may take part in both night and day vision. Thus, during the light-dark switch, retinal signaling pathways undergo reconfiguration to facilitate the rod or cone domination of the duplex system. A major stage of adaptation in this scheme is at the photoreceptor synapse, where bias is applied to cone synaptic function during the day and to rod synaptic function during the night.

VOLTAGE-GATED Ca^{2+} CHANNELS AT THE PRESYNAPTIC TERMINALS OF THE PHOTORECEPTORS

Photoreceptors and other non-spiking neurons communicate with graded potentials and their synapses mediate the continuous release of neurotransmitter. A prominent and specialized morphological feature at the presynaptic membrane of such cells, the synaptic ribbon, is responsible for guiding the transport of vesicles to the release site. Voltage-gated Ca^{2+} channels, as well as other Ca^{2+} permeable channels such as CNG channels, mediate the influx of Ca^{2+} required for vesicle fusion. A great deal is known about the pharmacology and biophysics of the rod and cone L-type Ca^{2+} channels. First recorded in rods, the properties of these channels have been remarkably consistent in all preparations examined and share features with high-voltage-activated, dihydropyridine sensitive, non-inactivating Ca^{2+} channels described throughout the nervous system.[12-19]

However, the Ca^{2+} channels of rod and cone photoreceptors can be distinguished from mainstream cardiovascular L-type Ca^{2+} channels by their relatively weak interactions with dihydropyridines and the partial and reversible block by the N-type blocker ω-conotoxin.[18,20] The pharmacological characterization most closely resembling this profile is that of α1D subunit-containing Ca^{2+} channels.[21]

In humans expressing a mutated L-type Ca^{2+} channel α1F subtype, there is no visual function of the rod system. Genetic analysis of small populations affected with Congenital Stationary (e.g., non-progressive) Night Blindness, type 2 (CSNB2), revealed mutations in the gene coding for the α1F L-type Ca^{2+} channel forming subunit.[22,23] Electroretinography revealed that these individuals had full function of rod photoreceptors in transduction but a complete lack of transmission of this signal to second order cells. Since CSNB2 is a rod disease, the identification of the gene product responsible suggests that the rod Ca^{2+} channel is composed of α1F subunits. In support of this, additional evidence suggests that rod Ca^{2+} channels, which share most biophysical and pharmacological properties with cone Ca^{2+} channels,[18,20] are composed of α1F Ca^{2+} channel subunits.[22,23] In situ hybridization localized α1F subunit mRNA in rod photoreceptors[23] and antibodies directed against α1F Ca^{2+} channel subunits labeled rod synaptic terminals.[24]

Immunohistochemistry has not yet allowed firm conclusions to be drawn about the molecular identity of the Ca^{2+} channels in cones. Some cones in mammalian retina label with antibodies directed against α1D subunits,[19,25] while cones (and rods) in salamander retina label with antibodies directed against rat brain α1C subunits, although the specificity of these antibodies in salamander was not established.[26]

Modifiers of Ca^{2+} Channel Activity

Several models of Ca^{2+} channel modulation at this synapse have been described. Protons, divalent cations, and anions have each been investigated for their Ca^{2+} channel modulating activity and all have been shown to have profound effects upon the gating and permeation of photoreceptor channels. Protons are known to alter ion channel gating via several mechanisms, including direct block of the conducting pore, surface charge effects, and binding to channel proteins. In rods and cones, protons produce changes in Ca^{2+} channel gating and conductance by all of these mechanisms. Acidic conditions reduce Ca^{2+} channel activity and basic conditions increase it.[27] The consequences to synaptic transmission are that synaptic efficacy is increased as pH is increased.[9,28] It has been suggested that protons liberated during vesicular release from photoreceptors produce a rapid auto-inhibition of further release via Ca^{2+} channel inhibition.[29]

Divalent cations including Ca^{2+}, Mg^{2+}, Cd^{2+}, Co^{2+}, and Ni^{2+} have all been shown to exert important modulatory effects on photoreceptor presynaptic Ca^{2+} channels, and to effect synaptic transmission in both expected and unexpected manners. Ca^{2+} channel blocking divalent cations (Cd^{2+}, Co^{2+}, and Ni^{2+}), as expected, block synaptic transmission. Surprisingly, the block by low concentrations of these divalents can be reversed by lowering Ca^{2+} levels, since this action reduces the screening of membrane surface charge and increases Ca^{2+} channel activation.[30,31] Inversely, as Ca^{2+} concentrations increase, surface charge effects dominate over a narrow range of the foot of activation curve for Ca^{2+} channels and reduce Ca^{2+} channel activity.[32]

Anion modulation of Ca^{2+} current voltage dependence and amplitude in salamander photoreceptors has also been reported. Reductions of extracellular Cl^- suppress the activity and conductance of Ca^{2+} channels and this leads to reduction of synaptic efficacy.[33,34]

At the cone synapse in particular, the invaginating horizontal cell processes require that extracellular current flow to the dendrite tips must follow a significant and necessarily resistive path to the glutamate gated channels. Byzov put forward a model of electrical feedback, in which the large extracellular flow of current to post-synaptic glutamate-gated channels in horizontal cell dendrites, which serve as current sinks, produced voltage drops in the extracellular space.[35] Extracellular voltage drops affect the voltage-dependent gating of Ca^{2+} channels equally as much as intracellular voltage changes. Byzov postulated that the post-synaptic horizontal cell response itself modulates presynaptic release via this auto-inhibitory feedback loop. His ingenious model of ephaptic interaction has since been modified and stands as the cornerstone of further attempts, described later in this chapter, to describe synaptic feedback to cones.

OTHER Ca^{2+} PERMEABLE CHANNELS IN THE PHOTORECEPTOR TERMINALS

Another type of ion channel that is well known for having significant Ca^{2+} permeablility is the CNG channel of the outer segments of rods and cones. While mostly permeable to monovalent cations, the Ca^{2+} permeability of these channels underlies their role in providing a negative feedback on the production of cGMP. In addition to their high-density expression in the outer segments of photoreceptors, the inner segment membranes of rods also contain CNG channels, albeit at a much lower density.[36] CNG channels are capable of supporting exocytosis and glutamate release at cone synaptic terminals.[4,10] CNG channel activity was also shown to be stimulated by NO, which acts via guanylate cyclase to produce cGMP.[10] The preponderance of sites of NO synthesis in the retina, the demonstrated production of NO in the outer retinal layers during darkness, and the strong actions of NO synthase inhibitors on synaptic release from photoreceptors, suggested a role for CNG channels in synaptic signaling at these synapses.

THE ROLE OF Ca^{2+} PERMEABLE CHANNELS IN FEEDBACK FROM SECOND ORDER CELLS

Retinal structure is dominated by lateral networks of neurons that mediate processing of the neural image. At the outer plexiform layer, photoreceptor to bipolar cell transmission is modified by laterally interacting horizontal cells. It is known that basic features of the ganglion cell receptive field structure, e.g., simple center-surround antagonism, are present in bipolar cells.[37,38] Morphological studies of the outer plexiform layer support electrophysiological data: bipolar cells sample many photoreceptors directly (accounting for the center region) and, via horizontal cells, sample photoreceptors over broad lateral regions (accounting for the surround component of the receptive field). Integration of center and surround signals occurs presynaptic to bipolar cells at the cone presynaptic terminal.[39] The first study of cones showed that surround illumination decreases the hyperpolarizing response of a cone to direct illumination.[3] The mechanism of inhibitory feedback to cones has remained elusive.[40]

GABA as a Feedback Transmitter

An early hypothesis offered for the synaptic mechanisms of feedback was based on histochemical and biochemical experiments showing that some classes of horizontal cells synthesize and release GABA.[41] Electrophysiological evidence suggested that GABA is a feedback transmitter and acts at cone $GABA_A$ receptors[42] which form Cl channels.[43,44] When cones are hyperpolarized strongly by bright light, the effect of surround inhibition, as recorded in bipolar cells, diminishes.[45] Others demonstrated a conductance increase to Cl^- in cones during surround illumination[46,47] later suggested to be secondary to a conductance increase involving Ca^{2+} ions.[48] However, failure to block surround inhibition in cones with $GABA_A$ antagonists has been demonstrated.[49] These diverse observations complicated the model in which GABA is the sole feedback neurotransmitter acting via $GABA_A$ receptors. The results can also be interpreted in terms of a Ca^{2+} channel modulation that would vanish at hyperpolarized levels simply because the Ca^{2+} channels are not activated at this negative potential.

Nitric Oxide as a Feedback Transmitter and Adaptation Signal

Evidence from a variety of cells and tissues has demonstrated the important roles of nitric oxide (NO) as both a second messenger and a neurotransmitter, which mediates diverse physiological functions.[50] Recent discoveries suggest that NO may not only be integral to visual processing and retinal function, but may also be a contributor to the underlying pathology associated with diseases such as retinal ischemia and glaucoma.[51]

Three common isoforms of the enzyme nitric oxide synthase (NOS), which converts L-arginine to NO and L-citrulline, are found in the retina and its vasculature:

Constitutive isoforms, denoted as cNOS, are Ca^{2+}/calmodulin dependent and include eNOS (endothelial) and nNOS (neural). A third, Ca^{2+}-independent, inducible isoform, iNOS, is up regulated by immunological signals. NO can activate guanylate cyclase, which produces cGMP as a second messenger, but additional actions include activation of ADP-ribosyltransferase and direct nitrosylation of proteins.

NOS-immunoreactivity or diaphorase staining (histochemical indicator of NOS activity) has been found in every retina examined.[52,53] NOS is found commonly in the specialized glia of the retina, the Müller cells, in subpopulations of horizontal and amacrine cells, and with less consistency in a variety of neurons.[54,55] Detailed studies of several species find evidence for subcellular distributions of cNOS in rod and cone photoreceptors, bipolar cells, Müller cells, amacrine cells, and ganglion cell axon-ensheathing astrocytes, as well as in both the inner and outer plexiform layers. Prominent sites of NOS localization in the outer plexiform layer are within horizontal cell synaptic processes. Such localization supports a role for NO as a feedback neurotransmitter carrying surround signals, which are integrated in horizontal cells, back to the cone photoreceptors. In addition, NOS is present in the sclerad region of Müller cells that envelop the photoreceptor inner segments, potentially supporting a role for NO as an adaptation molecule in the outer retina.

Demonstrated actions of NO on retinal cells are consistent with a role in light-dark adaptation.[51] For example, whereas voltage-gated Ca^{2+} channels responsible for rod synaptic release are potentiated by NO,[56] evidence suggested that those in cones are suppressed.[57] If NO were a dark-adaptation messenger in the outer retina, an increase in NO would strengthen rod signaling while decreasing that of cones.

Another broadly recognized action of NO in photoreceptors[4] is the activation of cyclic nucleotide-gated (CNG) channels; the voltage-independent, Ca^{2+}-permeable non-specific cation channels best known for their role in phototransduction. While CNG channel activation by NO may have ramifications ranging from phototransduction to synaptic modulation at numerous pre- and post-synaptic sites, this mechanism of signaling raises the issue of whether NO signals darkness or light, whether it signals broadly or locally, and whether its actions as a neuromodulator occur on a fast-or slow-timescale. It has been suggested that this type of mechanism operates to facilitate neurotransmitter release at the photoreceptor terminal.[10] These studies have shown that activation of CNG channels in synaptic terminals elicits transmitter release from cones. This model of neuromodulation at the photoreceptor terminals suggests that nitric oxide is the transmitter responsible for increasing or decreasing neurotransmitter release from photoreceptors in response to other retinal stimuli, either by way of modulating voltage dependent Ca^{2+} channels[56] or activation of Ca^{2+}-permeable CNG channels.[10]

Synaptic Modulation by Dopamine

Dopamine is a classic neuromodulator with a well-established role in the retina in light-dark adaptation.[58,59] Dopamine has specific actions on virtually all types of neurons of the retina including photoreceptors, horizontal cells, and bipolar cells.

Release of dopamine from subclasses of amacrine and interplexiform cells of the inner retina results in increased cone input to horizontal cells and decreased rod input.[11,60] In salamander, both D1 and D2 dopamine receptor subtypes have been localized to the photoreceptors.[11] D2 receptors appear to have a role in decreasing rod synaptic output by suppressing the activity of the primary signal shaping ionic conductance in the rod inner segment, the hyperpolarization-activated, non-selective cation conductance I_h.[61] I_h is responsible for shaping the characteristic peak-plateau sag in the hyperpolarizing rod response to light.[62] By suppressing I_h, dopamine would have the effect of diminishing rod output since rods would be more strongly hyperpolarized. In modulatory roles of this nature, dopamine seems to mediate retinal adaptation to increasing levels of illumination. However, dopamine acting via D2 receptors can decrease the activity of Ca^{2+} channels of cones while increasing that in rods.[63] Due to the plethora of actions of dopamine on the postsynaptic cells (targets: glutamate receptors, Ca^{2+} channels, potassium channels, gap junction channels), as well as other effects on the presynaptic photoreceptors (targets: h-current, Na-K ATPase, gap junctions), the overall actions of dopamine on synaptic transmission may be different than the effect of dopamine on the presynaptic Ca^{2+} channels alone.

Synaptic Modulation by Glutamate

Both rod and cone photoreceptors release glutamate in a manner graded with their degree of depolarization. Evidence that possible glutamate autoreception inhibits glutamate release, mediated by a type of metabotropic glutamate receptor (mGluR8) has been provided.[64] Metabotropic glutamate receptors signal via G-proteins, and in the example described in rod photoreceptors, this signaling couples glutamate reception to reduction of intracellular Ca^{2+} levels, an apparent inhibitory feedback mechanism that could serve to protect the cells receptive to glutamate at this synapse to damaging levels of the excitotoxic neurotransmitter.

Synaptic Modulation through an Ephaptic Interaction

The electric feedback model of Byzov (section "Modifiers of Ca^{2+} Channel Activity") was modified after the identification of putative hemi-gap junction channels in the dendrites of horizontal cells.[65] In this model (shown in Fig. 1), hemi-gap junction channels serve as the current sinks instead of the post-synaptic glutamate-gated channels in Byzov's model. This simplifies understanding of the model since little or no conductance change need occur in these elements to gate the horizontal cell response. As in the original postulation of Byzov, extracellular current flow to these hemi-gap junction channel current sinks produces an extracellular voltage drop, and this causes an apparent shift of the presynaptic Ca^{2+} channel activation curve.

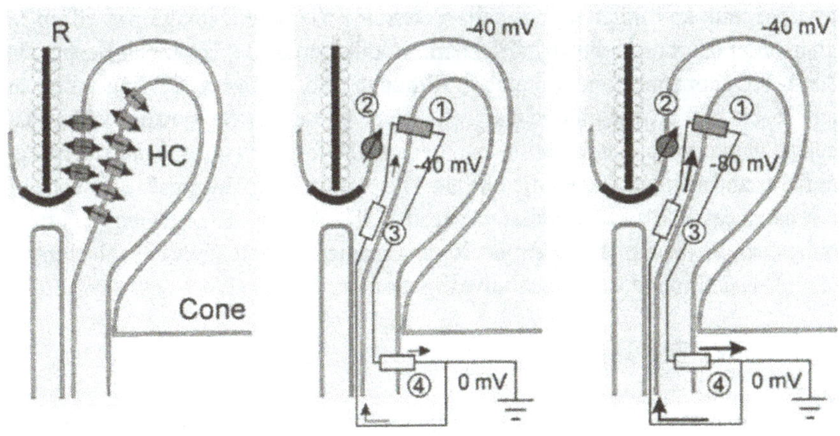

Figure 1. The modified Byzov electric feedback model describing an ephaptic interaction between horizontal cells and photoreceptors. In a), the hemi gap junction channels in the postsynaptic horizontal cell membrane (HC) are juxtaposed with voltage-gated Ca^{2+} channels adjacent to the synaptic ribbon (R) in the presynaptic cone membrane (Cone). In b), the electrical equivalent circuit under conditions of darkness is abbreviated: 1) shows the hemi gap junction channels, 2) the presynaptic Ca^{2+} channels, 3) the interstitial resistance that is responsible for the extracellular voltage drop that leads to the ephaptic interaction, and 4), the input resistance of the horizontal cell. Cones and horizontal cells are depolarized to –40 mV. In c) the horizontal cell has been hyperpolarized to –80 mV as if it were responding to bright surround illumination. This increased hyperpolarization drives a larger current into the hemi gap junction channels, which means that a larger current flows across the resistance of the extracellular space (larger arrow). In this example shown, a doubling of the horizontal cell membrane potential would double the extracellular current flow, and this would double the voltage drop in the extracellular space outside the presynaptic Ca^{2+} channels. The voltage drop is negative with respect to ground, and has been estimated at a few millivolts.[65] This leads to greater Ca^{2+} channel activity and enhanced transmitter release since an extracellular negative potential is equivalent to the cone membrane being depolarized by a few millivolts. This figure is reproduced from an original by Kamermans et al. which is posted on the world wide web at http://www.sciencemag.org/cgi/content/full/292/5519/1178/DC1.

CONCLUSION

This chapter reviewed the biophysical and physiological mechanisms of photoreceptor synaptic function and modulation as they pertain to the formation of receptive fields and adaptation. Voltage-gated L-type Ca^{2+} channels appear to be the principle ion channel type responsible for Ca^{2+} entry into photoreceptors. Important features of these channels, as well as how signal processing occurs at the photoreceptor synapse, remain intact among all vertebrates studied such that descriptions of retinal function in lower vertebrates and nocturnal mammals can be extrapolated to other species, including humans.

The presynaptic Ca^{2+} channels are L-type and appear to be composed of 1F subunits in rods and 1D subunits in cones. Modulatory synaptic interactions between photoreceptors and other retinal cells determine the neural image passed on to higher-order neurons in the visual system. Modulation of Ca^{2+} channels occurs via GABA, NO, dopamine and glutamate. Although information processing occurs at many levels in the vertebrate visual system, the photoreceptor synapse is the first locus at which synaptic integration can mediate the formation of specialized receptive fields and adaptation-specific output. The photoreceptor synapse is also one of the most accessible and well characterized CNS synapse and will permit further biophysical, molecular and pharmacological manipulation that will enable continued high-resolution physiological investigation.

ACKNOWLEDGMENTS

The authors' work was supported with operating funds (SB) and salary awards (SB, MEMK) from the Canadian Institutes of Health Research and a NSERC research grant (MEMK).

REFERENCES

1. Kawai F, Horiguchi M, Suzuki H et al. Na^+ Action Potentials in Human Photoreceptors. Neuron 2001; 30:451-458.
1a. Dowling JE. The Retina: An approachable part of the brain. Cambridge: Belknapp Press, 1987.
2. Attwell D. The photoreceptor output synapse. Progress in Retinal Research 1990; 9:337-362.
3. Baylor DA, Fuortes MGF, O'Bryan PM. Receptive fields of cones in the retina of the turtle. J Physiol 1971; 214:265-294.
4. Rieke F, Schwartz EA. A cGMP-gated current can control exocytosis at cone synapses. Neuron 1994; 13:863-873.
5. Copenhagen DR, Jahr CE. Release of endogenous excitatory amino acids from turtle photoreceptors. Nature 1989; 341:536-539.
6. Schmitz Y, Witkovsky P. Dependence of photoreceptor glutamate release on a dihydropyridine-sensitive Ca^{2+} channel. Neurosci 1997; 78:1209-16.
7. Nawy S. The metabotropic receptor mGluR6 may signal through G(o), but not phosphodiesterase, in retinal bipolar cells. J Neurosci 1999; 19:2938-2944.
8. Attwell D, Borges S, Wu S et al. Signal clipping by the rod output synapse. Nature 1987; 328:522-524.
9. Barnes S, Merchant V, Mahmud F. Modulation of transmission gain by protons at the photoreceptor output synapse. Proceedings of the National Academy of Science (U.S.A.) 1993; 90:10081-10085.
10. Savchenko A, Barnes S, Kramer RH. Cyclic nucleotide-gated channels in synaptic terminals mediate feedback modulation by nitric oxide. Nature 1997; 390:694-698.
11. Hare WA, Owen WG. Similar effects of carbachol and dopamine on neurons in the distal retina of the tiger salamander. Vis Neurosci 1995; 12: 443-455.
12. Bader CR, Bertrand D, Schwartz EA. Voltage-activated and calcium-activated currents studied in solitary rod inner segments from the salamander retina. J Physiol 1982; 331:253-284.
13. Corey DP, Dubinsky JM, Schwartz EA The calcium current in inner segments of rods from the salamander (Ambystoma tigrinum) retina. J Physiol 1984; 354:557-575.
14. Maricq, AV, Korenbrot JI. Calcium and calcium-dependant chloride currents generate action potentials in solitary cone photoreceptors. Neuron 1988; 1:503-515.
15. Barnes S, Hille B. Ionic channels of the inner segment of tiger salamander cone photoreceptors. J Gen Physiol 1989; 94:719-743

16. Lasater EM, Witkovsky P. The calcium current of turtle cone photoreceptor axon terminals. Neuroscience Research Supplement 1991; 15:S165-S173.
17. Yagi T, Macleish PR. Ionic conductances of monkey solitary cone inner segments. J Neurophysiol 1994; 71:656-65.
18. Wilkinson MF, Barnes S. The dihydropyridine-sensitive Ca^{2+} channel subtype in cone photoreceptors. J Gen Physiol 1996; 107:621-630.
19. Taylor WR, Morgans C. Localization and properties of voltage-gated calcium channels in cone photoreceptors of Tupaia belangeri. Vis Neurosci 1998; 15:541-52.
20. Kourennyi DK, Barnes S. Depolarization-induced calcium channel facilitation in rod photoreceptors is independent of G-proteins and phosphorylation. J Neurophysiol 2000; 84:133-138.
21. Williams ME, Feldman, DH, McCue AF et al. Structure and functional expression of alpha1, alpha2, and beta subunits of a novel human neuronal calcium channel. Neuron 1992; 8:71-84.
22. Bech-Hansen NT, Naylor MJ, Maybaum TA et al. Loss-of-function mutations in a calcium-channel alpha1-subunit gene in Xp11.23 cause incomplete X-linked congenital stationary night blindness. Nat Genet 1998; 19:264-7
23. Strom TM, Nyakatura G, Apfelstedt-Sylla E et al. An L-type calcium-channel gene mutated in incomplete X-linked congenital stationary night blindness. Nat Genet 1998; 19:260-263.
24. Morgans CW. Localization of the alpha(1F) Calcium channel subunit in the rat retina. Invest Ophth Vis Sci 2001; 42:2414-2418.
25. Morgans CW. Calcium channel heterogeneity among cone photoreceptors in the tree shrew retina. Eur J Neurosci 2000; 11:2989-2993.
26. Nachman-Clewner M, St. Jules R, Townes-Anderson E. L-type calcium channels in the photoreceptor ribbon synapse: Localization and role in plasticity. J Comp Neurol 1999; 415:1-16.
27. Barnes S. Modulation of vertebrate retinal function by pH. In: K Kaila, BR Ransom, eds. pH and Brain Function. New York: John Wiley & Sons Inc., 1998.
28. Harsanyi K, Mangel SC. Modulation of cone to horizontal cell transmission by calcium and pH in the fish retina. Vis Neurosci 1993; 10:81-91.
29. DeVries SH. Exocytosed protons feedback to suppress the Ca^{2+} current in mammalian cone photoreceptors. Neuron 2001; 32:1107-1117.
30. Piccolino M, Byzov AL, Kurenny DE et al. Low-calcium-induced enhancement of chemical synaptic transmission from photoreceptors to horizontal cells in the vertebrate retina. Proceedings of the National Academy of Science (USA) 1996; 93:2302-2306.
31. Piccolino M, Vellani V, Rakotobe LA et al. Manipulation of synaptic sign and strength with divalent cations in the vertebrate retina: Pushing the limits of tonic, chemical neurotransmission. Eur J Neurosci 1999; 11:1-5.
32. Baldridge WK, Kurennyi DK, Barnes S. Calcium-sensitive calcium influx in photoreceptor inner segments. J Neurophysiol 1998; 79:3012-3018.
33. Thoreson WB, Nitzan R, Miller RF. Reducing extracellular Cl- suppresses dihydropyridine-sensitive Ca2+ currents and synaptic transmission in amphibian photoreceptors. J Neurophysiol 1997; 77:2175-90.
34. Thoreson WB, Stella SL. Anion modulation of calcium current voltage dependence and amplitude in salamander rods. Biochem Biophys Acta 2000; 1464:142-50.
35. Byzov AL, Shura-Bura TM. Electric feedback mechanism in the processing of signals in the outer plexiform layer of the retina. Brain Res 1986; 26:33-44.
36. Watanabe S, Matthews G. Regional distribution of cGMP-activation ion channels in plasma membrane of rod photoreceptors. J Neurosci 1988; 8:2334-2337.
37. Werblin FS, Dowling JE. Organization of the retina of the mudpuppy, Necturus maculosus. II. Intracellular recording. J Neurophysiol. 1969; 32:339-355.
38. Cook PB, McReynolds JS. Lateral inhibition in the inner retina is important for spatial tuning of ganglion cells. Nature Neurosci 1998; 1:711-719.
39. Kaneko A, Shimazaki H. Synaptic transmission from photoreceptors to bipolar and horizontal cells in the carp retina. Cold Spring Harbor Symp Quant Biol 1976; 40:537-546.

40. Burkhardt DA. Synaptic feedback, depolarization, and color opponency in cone photoreceptors. Vis Neurosci 1993; 10:981-9.
41. Lam DMK. Biosynthesis of gamma-aminobutyric acid by isolated axons of cone horizontal cells in the goldfish retina. Nature 1975; 254:345-347.
42. Murakami M, Shimoda Y, Nakatani K et al. GABA-mediated negative feedback from horizontal cells to cones in carp retina. Jap J Physiol 1982; 32:911-926.
43. Kaneko A, Tachibana M. Effects of gamma-butyric acid on isolated cone photoreceptors of the turtle retina. J Physiol 1986; 373:443-461.
44. Wu SM. Effects of gamma-butyric acid on cones and bipolar cells of the tiger salamander retina. Brain Res 1986; 365:70-77.
45. Skrzypek J, Werblin F. Lateral interactions in the absence of feedback to cones. J Neurophysiol 1983; 49:1007-1017.
46. O'Bryan PM. Properties of the depolarizing synaptic potential evoked by peripheral illumination in cones of the turtle retina. J Physiol 1973; 235:207-223.
47. Lasansky A. Synaptic action mediating cone responses to annular illumination in the retina of the larval tiger salamander. J Physiol 1981; 310:205-14.
48. Piccolino M, Gerschenfeld HM. Activation of a regenerative calcium conductance in turtle cones by peripheral stimulation. Proc R Soc London B 1978; 201:309-315.
49. Thoreson WB, Burkhardt DA. Effects of synaptic blocking agents on the depolarizing responses of turtle cones evoked by surround illumination. Vis Neurosci 1990 5:571-83.
50. Snyder SH. Nitric oxide and neurons. Curr Opin Neurobiol 1992; 2:323-327.
51. Kelly MEM, Barnes S. Physiology and pathophysiology of nitric oxide in the retina. The Neuroscientist 1997; 3:357-360.
52. Peretz MTR, Larsson B, Alm P et al. Localisation of neuronal nitric oxide synthase-immunoreactivity in rat and rabbit reitnas. Exp Brain Res 1995; 104:207-217.
53. Yamamoto R, Bredt DS, Snyder SH et al. The localization of nitric oxide synthase in the rat eye and related cranial ganglia. Neuroscience 1993; 54:189-200.
54. Liepe BA, Stone C, Koistanaho J et al. Nitric oxide synthase in Muller cells and neurons of salamander and fish retina. J Neurosci 1994; 14:7641-7654
55. Blute TA, Mayer B, Eldred WD. Immunocytochemical and histochemical localization of nitric oxide synthase in the turtle retina. Vis Neurosci 1997; 14:717-729.
56. Kurenny DE, Moroz LL, Turner RW et al. Modulation of ion channels in rod photoreceptors by nitric oxide. Neuron 1994; 13:315-324.
57. Kurennyi DE, Baldridge W, Hart J et al. Differential action of the nitric oxide donor S-nitrosocysteine on calcium channels in cone versus rod photoreceptors. Soc Neurosci Abstr 1997; 23:449.
58. Morgan IG, Boelen MK. A retinal light-dark switch: a review of the evidence. Vis Neurosci 1996; 13:399-409.
59. Djamgoz MB, Hankins MW, Hirano J et al. Neurobiology of retinal dopamine in relation to degenerative states of the tissue. Vis Res 1997; 37:3509-3529.
60. Krizaj D, Witkovsky P. Effects of submicromolar concentrations of dopamine on photoreceptor to horizontal cell communication. Brain Res 1993; 627:122-128.
61. Akopian A, Witkovsky P. D2-dopamine receptor mediated inhibition of a hyperpolarization-activated current in rod photoreceptors. J Neurophysiol 1996; 76:1828-1835.
62. Barnes S. After transduction: Response shaping and control of transmission by ion channels of the photoreceptor inner segment. Neuroscience 1994; 58:447-459.
63. Stella SL, Thoreson WB. Differential modulation of rod and cone calcium currents in tiger salamander retina by D2 dopamine receptors and cAMP. Eur J Neurosci 2000; 12:3537-48.
64. Koulen P, Kuhn R, Wassle H et al. Modulation of the intracellular calcium concentration in photoreceptor terminals by a presynaptic metabotropic glutamate receptor. PNAS 1996; 96:9909-9914
65. Kamermans M, Fahrenfort I, Schultz K et al. Hemichannel-mediated inhibition in the outer retina. Science 2001; 292:1178-1180.

ON BIPOLAR CELLS: FOLLOWING IN THE FOOTSTEPS OF PHOTOTRANSDUCTION

Malcolm M. Slaughter and Gautam B. Awatramani

ABSTRACT

The electrical signals resulting from phototransduction are decomposed by bipolar cells and then encoded into spike trains by ganglion cells. The signal decomposition by bipolar cells includes formation of ON and OFF pathways and separation of tonic and phasic signals. The decomposition is accomplished by post-synaptic receptors in the ON and OFF bipolar cells. This chapter focuses on these phenomena in ON bipolar cells and the role of metabotropic glutamate receptors in these processes.

INTRODUCTION

Photoreceptor neurotransmitter release drives three types of neurons: the ON and OFF bipolar cells and the horizontal cells. The bipolar cells are interneurons linking photoreceptors with ganglion cells. Photoreceptors and ganglion cells are clearly necessary: the former for phototransduction and the latter for spike encoding. The need for bipolar cells is less apparent. There is an order of magnitude more photoreceptors than bipolar cells, and another order of magnitude less ganglion cells. Perhaps this amount of convergence is developmentally unwieldy in a single step, necessitating an interneuron. However, bipolar cells are far more than simple conduits, they decompose the photoreceptor's signal, parsing it among specialized classes of ganglion cell.

A prominent example of this decomposition is the formation of ON and OFF responses in two types of bipolar cell. This distinction is created by the sign-inverting

Department of Physiology and Biophysics, University at Buffalo, Buffalo, New York, 14214, U.S.A.

metabotropic glutamate receptors in ON bipolar cells and the sign-conserving ionotropic glutamate receptors in OFF bipolar cells. Within these two sets of glutamate receptors there are further distinctions that yield transient and sustained responses. In the OFF bipolar cells this seems to be established by employing AMPA and kainate receptors with different kinetics. In the ON bipolar cells, a calcium positive feedback to the signal transduction pathway may produce phasic or tonic responses.

SIGNAL TRANSDUCTION AT THE ON BIPOLAR CELL SYNAPSE

From the perspective of signal transduction, the ON bipolar cell is a simile for the photoreceptor. The ON bipolar neuron receives glutamatergic signals at its synapse with photoreceptors, but this 'excitatory' neurotransmitter hyperpolarizes the ON bipolar cell.[1] Like the effect of light in the outer segments of photoreceptors, glutamate acts to close a cationic channel that is constitutively open in the absence of a stimulus (Fig. 1). In one case the stimulus is light, in the other it is glutamate, but in both cases the stimulus evokes a hyperpolarization produced by channel closing. Furthermore, the light-regulated and synaptic channels have similar permeation properties, including significant calcium permeability. And, as in photoreceptors, elimination of calcium alters the apparent 'adaptive' mechanisms in the ON bipolar cells.[2,3] These similarities have encouraged efforts to extend the analogy between photoreceptor and ON bipolar cell signal transduction. Following the clues provided by phototransduction, the early postsynaptic second messenger candidate in ON bipolar cells was cGMP, and several studies seemed to support this suspicion.[4,5]

Closer inspection has revealed some differences in the transduction pathway of photoreceptors and ON bipolar cells and raised doubts about the parallelism. It was reasonable to hypothesize that glutamate activated a G-protein linked receptor in ON bipolar cells that activated a transducin-like G-protein leading to enhanced activation of phosphodiesterase. However, the G-protein found in the ON bipolar cell is different, a Go.[6,7] Furthermore, inhibitors of phosphodiesterase do not block the light response of ON bipolar cells[8] and inhibitors of cGMP do not block the light response. Thus, while the cGMP photoreceptor transduction pathway may be important at the ON bipolar cell synapse, the function may be to modulate the main, but still undiscovered, cascade.

A NOTE ABOUT METHODS

Most of the experiments described in this chapter were performed on the retina of the tiger salamander, *Ambystoma tigrinum*, an aquatic amphibian. The preparation was the retinal slice, which was examined under infrared illumination and stimulated by full field light generated by an LED. Electrical signals were recorded using the whole-cell patch technique, in either current clamp or voltage clamp mode. Drug application was by superfusion, thus affecting the entire retinal slice uniformly.

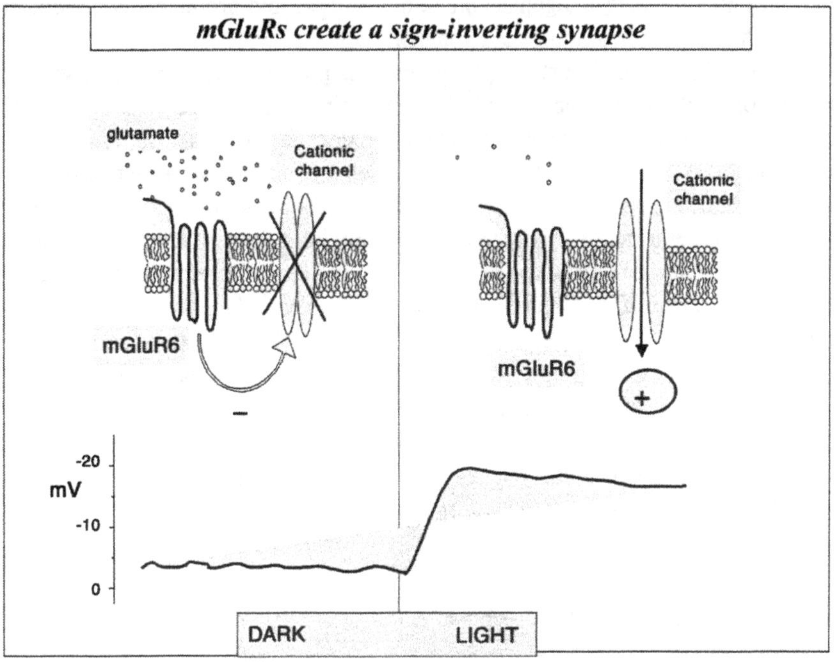

Figure 1. The photoreceptor and the ON bipolar cell have opposite responses to light. In the dark, glutamate released by depolarized photoreceptors activates a metabotropic glutamate receptor (mGluR6) in the ON bipolar cell, resulting in the closing of a cationic channel and the hyperpolarization of the cell. In the light, the photoreceptor hyperpolarizes, less glutamate is released, causing the ON bipolar cell channels to open and the cell to depolarize. The top part of the figure shows a cartoon of the relationship between glutamate release and channel conductance regulated by the ON bipolar cell metabotropic glutamate receptor (mGluR6). The bottom shows the change in ON bipolar cell membrane voltage in dark and light.

THE ON BIPOLAR CELL'S SYNAPTIC RECEPTORS

The synapse between photoreceptors and ON bipolar cells is a very unusual one, utilizing mechanisms that might be unique in the central nervous system. The ON bipolar cell possesses one kind of metabotropic glutamate receptor, mGluR6, and this receptor is exquisitely localized to the postsynaptic region at the tip of ON bipolar cell dendrites.[9] Full activation of this receptor closes channels on the bipolar cell. Several standard agonists of metabotropic glutamate receptors (mGluRs) produce the same effect, including the one first identified two decades ago: 2-amino-4-phosphonobutyrate, although its acronym has evolved from APB back then to AP4 now.[10] Application of mGluR agonists eliminates the light response and appears to close the same channels that are shut by the photoreceptor transmitter. This conclusion is based on comparing the current-voltage relationship of agonists with that of receptor activation by the endogenous neurotransmitter.[11] Furthermore, metabotropic glutamate antagonists block the synapse and produce a conductance increase that is the mirror image of agonists.[31]

The ON bipolar cell postsynaptic membrane possesses ionotropic as well as metabotropic glutamate receptors.[12] In fish retina, ON bipolar cells use both ionotropic and metabotropic receptors at the synapse. When AP4 is applied to saturate the metabotropic receptor, fish ON bipolar cells still exhibit synapse responses.[13] This is not true in terrestrial vertebrates, where AP4 eliminates the synaptic response of ON bipolar cells and ON light responses throughout the retina.[14,15] While most transmitters open a conductance that shunts cell responses, glutamate agonists increase the input resistance of the ON bipolar cell. This would make any other synaptic inputs even easier to detect. Thus, the effect of AP4 indicates that no other glutamate receptor mediates signal transduction at this synapse in amphibians through mammals. The significance of ionotropic glutamate receptors in ON bipolar cells in terrestrial vertebrate retina remains an enigma. It is tempting to speculate that they serve to modulate the metabotropic receptor, possibly by regulating calcium influx (see below).

TWO TYPES OF BIPOLAR CELLS: FIDELITY AT THE SYNAPSE

The OFF bipolar neuron, which is complementary to the ON bipolar cell, is driven by a conventional, synaptic ionotropic glutamate receptor. The OFF bipolar cell, having a sign-conserving synapse, follows the potential changes of photoreceptors. The ON bipolar cell reverses the polarity of the photoreceptor signal. The advantages of having two bipolar cells of opposite polarity reading the output of photoreceptors are probably varied, but one important function is mechanistic.[16] In the case of the OFF bipolar cell, as photoreceptors release more and more glutamate the incremental response of the OFF bipolar cell declines, following a hyperbolic curve (Fig. 2). This is not unique to the OFF bipolar cell, it is the inescapable result of progressively driving the potential of the postsynaptic cell closer and closer to the synaptic reversal potential. Thus, as more channels open both the cell's input resistance and the ionic driving force continuously decline. In the dark, the OFF bipolar neuron is relatively close to its synaptic reversal potential and is relatively insensitive to small changes in photoreceptor transmitter release. Under the same dark conditions, the ON bipolar cell is maximally sensitive (Fig. 2). It has most of its synaptic channels closed and is relatively far from its synaptic reversal potential. It is estimated that the opening of the first synaptic channel produces a 1 mV depolarization of the ON bipolar cell in amphibian retina.[17] Under dim light conditions, the synaptic gain at the ON bipolar synapse in dogfish retina is 135 fold compared to 20 fold at the OFF bipolar synapse.[18] Thus, the sensitivity of the ON bipolar cell is very high in the dark, which might explain why the mammalian rod system relies so heavily on the ON bipolar cells.

The responsiveness at the bipolar cell synapse is more complex than can be explained by channel driving force and cell resistance. The transmitter event at the

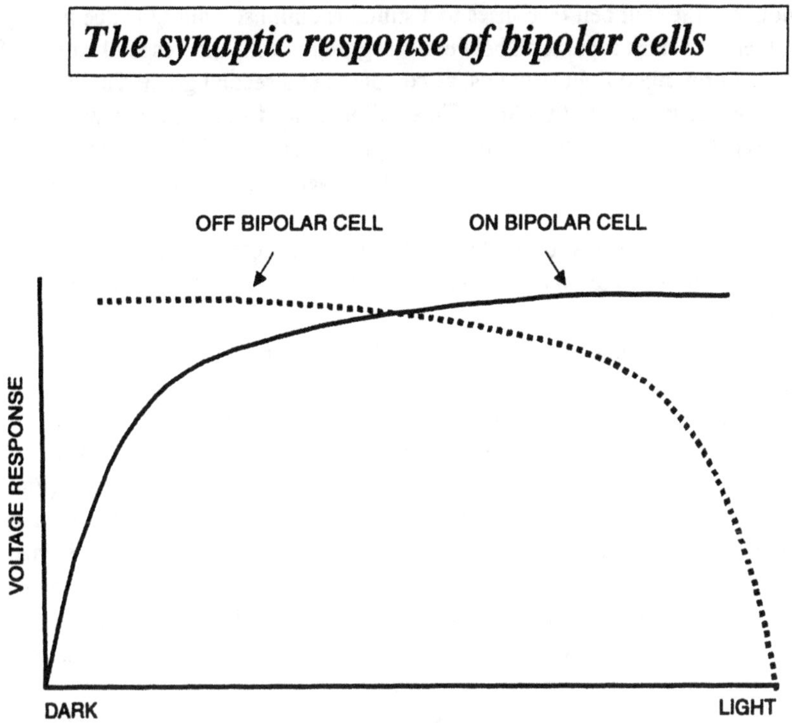

Figure 2. The synaptic responses of ON bipolar cells are most responsive in the dark and OFF bipolar cells in the light. Based on theoretical properties of a synapse, the sign-inverting ON bipolar cell responds with high gain (rate of change of voltage with light) to small increments of light from a dark baseline, while the OFF bipolar cell responds with high gain to small decrements of light from a baseline of bright light.

ON bipolar cell synapse has a duration that is about ten-fold longer than at the OFF bipolar cell synapse. This is expected at a metabotropic synapse and might complement the temporal integration of phototransduction that is characteristic of rods. The ON bipolar cell synapse also transmits a narrower range of frequencies, which is also characteristic of rod signals and mesopic responses.[19] Nevertheless, Yang and Wu (1993) found that OFF bipolar cells in amphibian retina have a lower threshold than ON bipolar cells.[20] The explanation is unclear.

Overall, the ON and OFF bipolar cells ensure the fidelity of the photoreceptor signal throughout its dynamic range. The importance of this dual system is exemplified by the maintenance of ON and OFF responses throughout the visual system.

TRANSIENT AND SUSTAINED RESPONSES IN GANGLION CELLS

Another prominent physiological property that is evident throughout the visual system is the dichotomy in temporal responses to a prolonged light stimulus.

Although all retinal cell responses to a stimulus diminish with time due to adaptation, there is a stark separation between one group of cells that respond only briefly, even to a prolonged light stimulus, as compared to a second group that have a persistent response to the stimulus.[21] This delineation of transient and sustained responses rides on top of the ON and OFF signals, producing ON sustained, ON transient, OFF sustained, and OFF transient cells, as well as the signal doubling, transient ON-OFF cells.

Unlike sustained responses that originate in photoreceptors, transient responses were first found and are most evident in ganglion cells. These are the neurons that get input from bipolar cells, so it has been hypothesized that some mechanism at this bipolar-ganglion cell synapse creates the transient response. And a number of potential mechanisms have been found (Fig. 3). For example, bipolar cells receive negative feedback from amacrine cells. This feedback is delayed and somewhat slow, generated largely by several classes of GABAergic amacrine cell, which activate numerous subtypes of GABA receptor ON bipolar cell axon terminals (Fig. 4). One is the metabotropic $GABA_BR$, which suppresses the output of bipolar cells through a delayed inhibitory feedback. The time-delayed suppression of synaptic calcium channels mediated by this G-protein linked receptor can lead to a transient output signal, even if the bipolar cell itself has a sustained light response. Naturally, this would have to be coupled with a sustained system that would maintain the inhibition, but the outlines of such a network have been developed.[22,23] One possible function of this system is the encoding of movement across the retina. $GABA_CRs$ are also present at the presynaptic axon terminal of bipolar cells, and again delayed activation could yield a transient output signal.[24,25] Indeed, when the $GABA_CR$ is blocked, the responses of ganglion cells do become less transient. Transient signals associated with this feedback pathway may be involved in edge detection.[26]

In addition to these feedback circuits, mechanisms that control the kinetics of bipolar cell transmitter release may also contribute to the formation of transient and sustained signals (Fig. 3). The output of bipolar cells is mediated by two types of calcium channel. One, the L-type, is well designed to produce a sustained signal, but the accompanying T-type is rapidly inactivating and would only be able to contribute to a transient signal.[27,28] Thus, intrinsic properties of bipolar cells and a selective circuitry lead to generation of transient responses.

There are also postsynaptic mechanisms that contribute to formation of transient signals in ganglion cells. Bipolar cells release glutamate, which activates AMPA receptors on ganglion cells. In transiently responding ganglion cells these AMPA receptors rapidly desensitize, so that even a sustained output from a bipolar cell would yield a transient response in these ganglion cells.[29] Furthermore, the intrinsic membrane properties of transient and sustained ganglion cells appear to differ, and this could also contribute to the difference in temporal response properties of these ganglion cells.[30]

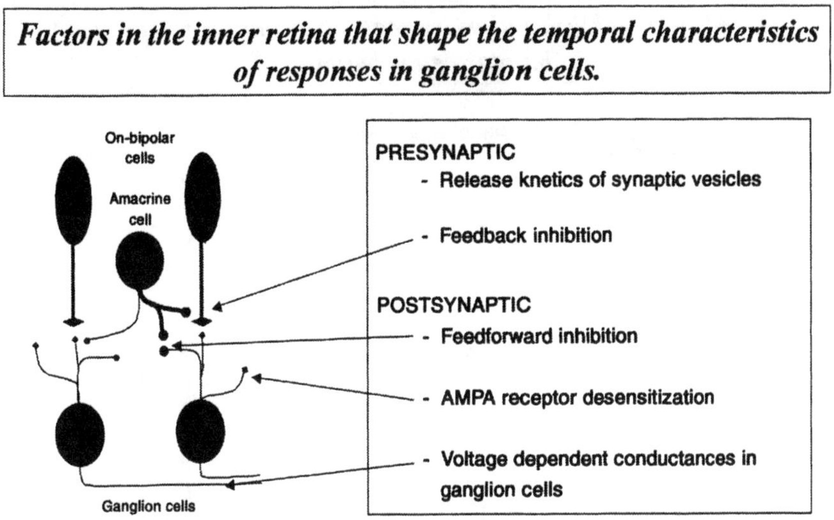

Figure 3. Properties of the bipolar-to-ganglion cells synapse can convert a sustained signal in bipolar cells to a transient signal in ganglion cells. Both presynaptic and postsynaptic mechanisms have been shown to influence this transition, as described in text.

TRANSIENT AND SUSTAINED BIPOLAR CELLS

While all these factors contribute to properties of transient and sustained signals, the origin of transient signals occurs at the input, not at the output, synapse of bipolar cells.[31] Thus, the synapse between photoreceptors and bipolar cells determines which cells are transient and which are sustained (Fig. 5). For technical reasons we examined this phenomenon in the ON bipolar cells. These neurons, which have some striking similarities to photoreceptors, have a synaptic mechanism that creates transient responses in some and sustained responses in other ON bipolar cells. Calcium is implicated in formation of the transient signal, analogous to adaptation in photoreceptors. However, unlike photoreceptors, calcium appears to produce a positive, rather than a negative, feedback in the ON bipolar cells.

In reaching these conclusions, we were aided by the discovery of a selective antagonist of the photoreceptor-ON bipolar cell synapse. This permitted the parsing of ON bipolar cell responses into two groups and the examination of their distinctive mechanisms. Although the metabotropic glutamate receptor at ON bipolar cell dendrites is a member of a small family of proteins, referred to as Group III metabotropic glutamate receptors, it has proven difficult to find an antagonist. A careful analysis has shown that many of the anticipated antagonists were ineffective or nonspecific.[32] However, we stumbled upon one, RS-α-cyclopropyl-4-phosphonophenylglycine (CPPG), that was suitable for analysis of the ON bipolar synapse. In a series of experiments to analyze this antagonist, we found that it

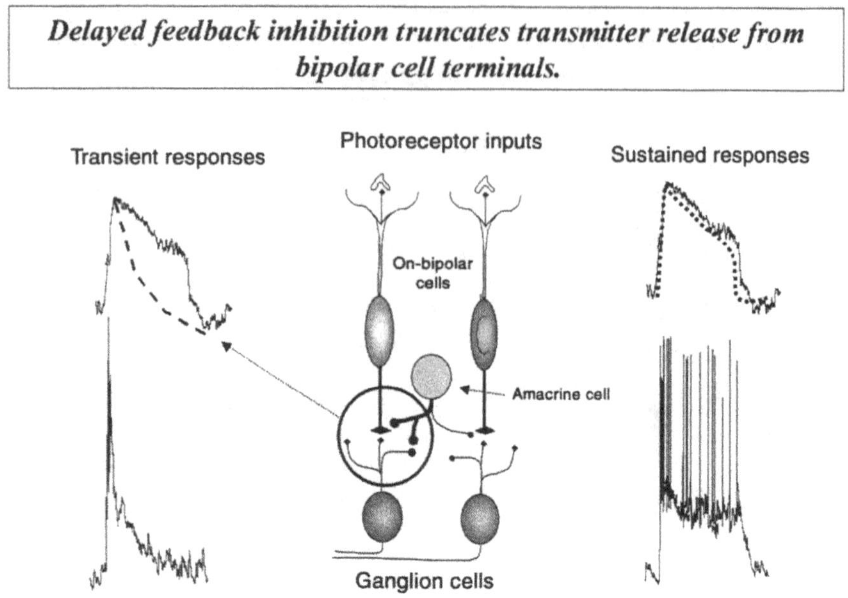

Figure 4. Delayed amacrine cell inhibition of bipolar cells contributes to transient responses in ganglion cells. Several pathways have been demonstrated, principally GABAergic feedback that activates $GABA_B$ or $GABA_C$ receptors on a select group of bipolar cells. In this model, bipolar cells would have sustained light responses (upper left and right traces) but amacrine cell feedback would truncate the synaptic output of a subset of bipolar cells (dotted line in upper right bipolar cell). This would result in transient and sustained ganglion cell responses (lower left and right traces).

blocked the light response of ON bipolar cells, that it blocked the b-wave of the ERG (which is produced by ON bipolar cells) and that it blocked AP4, the selective agonist of this receptor. Thus, it was an effective and selective blocker of photoreceptor input to ON bipolar cell synaptic receptors.[31,33] A disadvantage was that it was not very potent.

Bipolar cell light responses are considered to follow those of photoreceptors, which are sustained responding cells. But it is well known that the photoreceptor response to light is not a tonic, square wave response. Similarly, bipolar cell light responses exhibit an early peak followed by a sag. Over the years there have been numerous reports of electrophysiological recordings from bipolar cells showing that some bipolar cells sagged faster than others. But it was concluded that the sag was variable and covered a spectrum of kinetics. However, in whole-cell recordings in the amphibian retinal slice, we found that the kinetics of the sag fell into two distinct and nonoverlapping groups. In one set of ON bipolar cells, the light response sagged with a time constant of about 4 seconds, while the other set sagged with a time constant of approximately 400 milliseconds. In voltage-clamp experiments, the current sag was faster for both groups of cells, but the basic distinction between the two

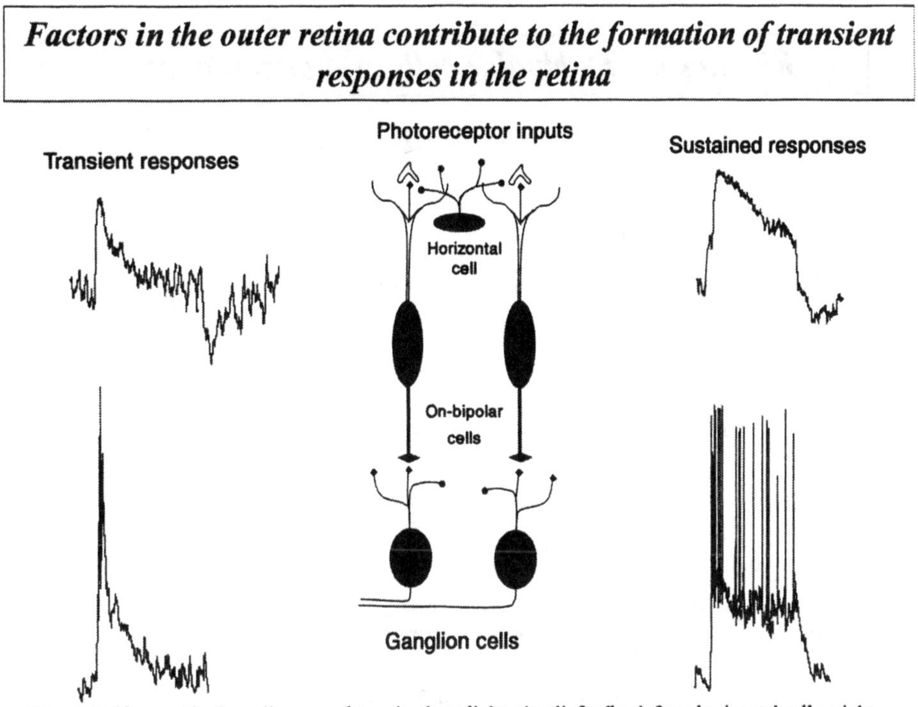

Figure 5. If some bipolar cells respond transiently to light stimuli, feedback from horizontal cells might account for this response. Horizontal cell feedback could make the output of photoreceptors, or the response of bipolar cells, transient. This feedback would have to be selective to affect one group of ON bipolar cells (upper left trace, but not upper right trace), leading to transient ganglion cell responses (lower left trace).

groups remained. In the group of ON bipolar cells with a rapidly decaying response, the light response decayed almost completely back to the baseline voltage (the potential in the dark). In the ON bipolar cells with a slow decay, the steady state response (after several seconds) was approximately midway between the peak response and the dark potential (Fig. 6, top). This provided the first clue that there might be two sets of bipolar cells that create the sustained and transient responses in ganglion cells.[31]

Bipolar cells and horizontal cells read the output of photoreceptors and share information at the ribbon synapse.[34] Yet the horizontal cell response is very sustained (Fig. 6, bottom). This contrasts with the two types of responses in ON bipolar cells, suggesting that the intrinsic properties of ON bipolar cells create the transient/sustained distinction.

We then examined the effect of CPPG, the metabotropic glutamate receptor antagonist, on the light response and membrane voltage of ON bipolar cells. Since bipolar cells are hyperpolarized by glutamate, we anticipated that CPPG would depolarize the ON bipolar cell. More specifically, we expected CPPG to depolarize these neurons to the peak of their light response, since this point presumably represented the response to no glutamate at the synapse (Fig. 7). Surprisingly, CPPG never produced this effect. Instead, CPPG depolarized ON bipolar cells close to the

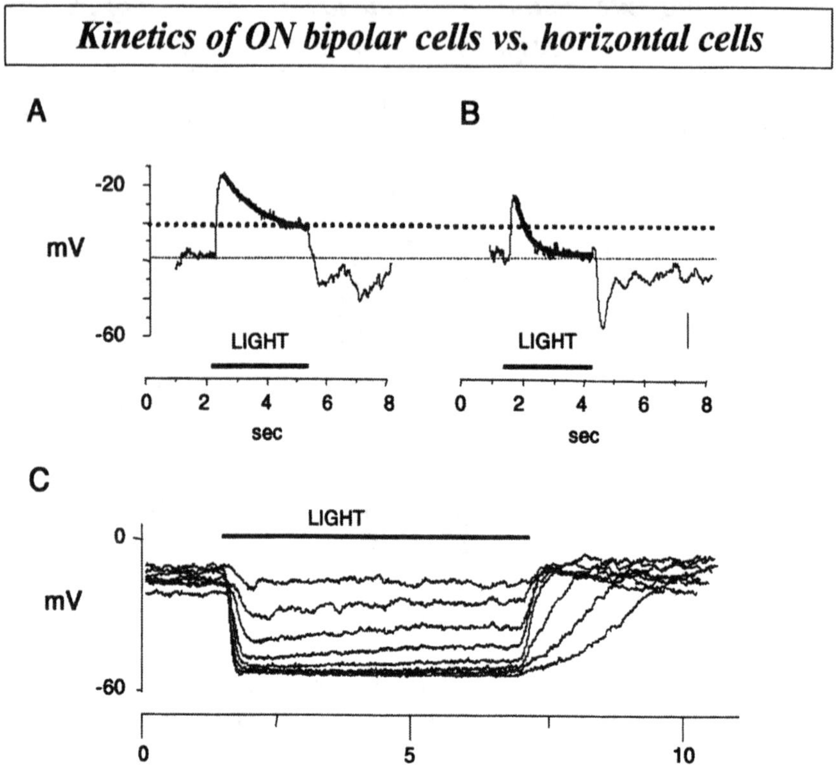

Figure 6. Bipolar cells and horizontal cells have different kinetics. In the retinal slice preparation, ON bipolar cells responded either tonically (A) or phasically (B) to a prolonged light stimulus. The dark line superimposed on the two traces is a single exponential fit to the sag of the light response. The sag was about an order of magnitude different in the two groups of ON bipolar cells. But horizontal cells only responded tonically (C) to a prolonged light stimulus.

same steady state voltage observed during a prolonged light stimulus. Thus, CPPG produced a greater depolarization in sustained ON bipolar cells, while depolarizing transient ON bipolar cells very little. Thus, when synaptic input was blocked, the two types of ON bipolar cell responded very differently (Fig. 8). In both cases, the light response was blocked. But transient ON bipolar cells remained near the dark membrane potential (Fig. 8, left) while sustained ON bipolar cells were depolarized by CPPG.This suggested that intrinsic properties of the two sets of neuron were dissimilar.

But possibly it was not the ON bipolar cells that differed, perhaps the circuitry differed. It has been demonstrated that bipolar cells receive inhibitory feedback and this feedback could make some bipolar cells more transient (Fig. 4). Thus, while CPPG freed the ON bipolar cell from the influence of photoreceptors, it did not eliminate the feedback from amacrine cells or horizontal cells. To distinguish between the intrinsic model and the feedback model, we tested the action of cadmium.

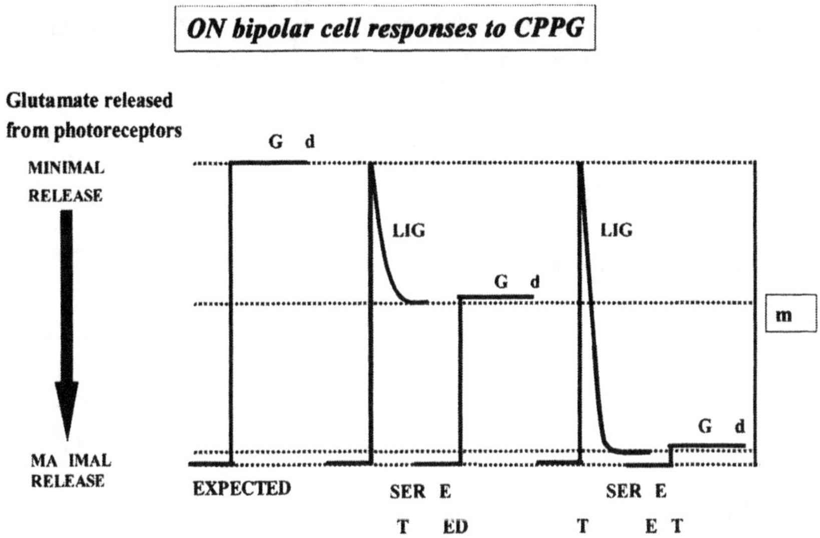

Figure 7. CPPG, or cadmium, produced different effects in transient and sustained ON bipolar cells. Both of these agents block the photoreceptor signal to bipolar cells and it was anticipated that they would fully depolarize all ON bipolar cells (Expected response at left). Instead, sustained bipolar cells were partially depolarized by either CPPG or cadmium ; transient bipolar cells were very slightly depolarized by CPPG or cadmium. In both cases, these synaptic blockers depolarized ON bipolar cells close to the steady state voltage evoked by a prolonged light stimulus.

This divalent cation will block calcium-dependent synaptic transmission, thus suppressing not only photoreceptor input to bipolar cells, but also amacrine cell and horizontal cell input. If the transient and sustained ON bipolar cell responses were due to intrinsic cell properties, then cadmium should produce the same effect as CPPG. If the feedback model was correct, then cadmium should depolarize both types of ON bipolar cell to the same voltage: the peak of the light response. Cadmium had the same effects as CPPG, supporting the conclusion that there were intrinsic differences in transient and sustained ON bipolar cells.[31] Thus, when photoreceptors were exposed to light and stopped releasing neurotransmitter, one set of ON bipolar cells had a steady state response that was significantly more positive than the dark potential, while the synaptic channels closed in the other set of ON bipolar cells and it settled back near the dark membrane potential.

An obvious question is why does the light stimuli produce a peak response in ON bipolar cells that sags back to a steady-state voltage while both CPPG and cadmium only produce a depolarization to the level of the steady-state response. The answer is probably that the peak light response involves a set of fast events, that rapidly fades. Our drug application is comparatively slow and does not evoke this fast response.

Two separate channels, originating in bipolar cells, generate transient and sustained responses.

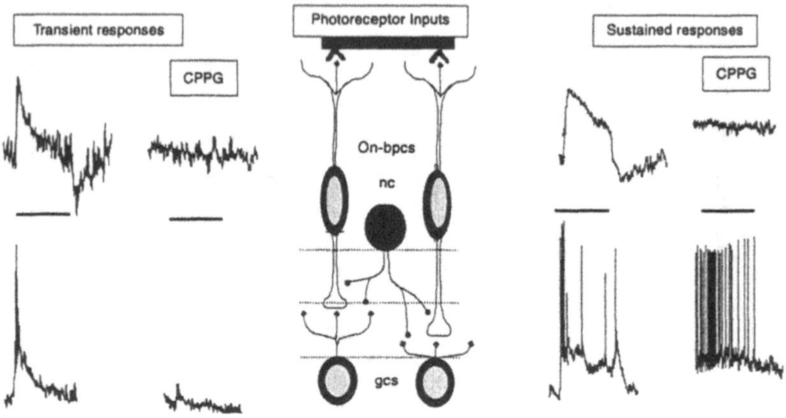

Figure 8. CPPG's effects on transient and sustained ON bipolar cells correlated with actions on transient and sustained ganglion cells. On the left side of the figure, CPPG suppressed the light response of a transient ON bipolar cell while producing only a small steady state depolarization (top two traces on left side of figure). The effect of CPPG on a dark membrane potential and light response of a transient ganglion cell was very similar to the action on the ON bipolar cell (bottom row, left). On the right side of the figure, CPPG suppressed the response of a sustained ON bipolar cell and depolarized the steady state potential (top two traces on right). Correspondingly, CPPG suppressed the light response of a sustained ON ganglion cell, and depolarized the cell producing a train of spikes (bottom left pair of traces). The dark bars below the bipolar cell traces indicate the timing of light stimuli. Abbreviations in central diagram are ON bipolar cell (On-bpcs), amacrine cell (ac), and ganglion cell (gcs).

DOES CALCIUM INFLUENCE TRANSIENT SIGNALS?

To look more closely at this question, we turned again to the behavior of the photoreceptor. The light response of the photoreceptor sags and this is due to a calcium feedback system that is important in adaptation. Calcium influx has also been associated with a similar process in bipolar cells[2,3]. The analogy seemed suggestive, so we performed experiments that were similar to the adaptation studies in photoreceptors. The hypothesis was that the depolarization of the ON bipolar cell led to an influx of calcium and this produced a negative modulation of the synaptic conductance and thus a sag in the light response. In the experiments up to this point, EGTA was used as the calcium buffer in the internal solution. If this were changed to BAPTA, the buffering would be faster. The anticipation was that, as in photoreceptors, BAPTA would block calcium feedback and eliminate the sag in the light response. Also, CPPG or cadmium should then depolarize the ON bipolar to the peak voltage of the light response. Again we were surprised because BAPTA eliminated the peak, instead of removing the sag, in the light response (Fig. 9). When 0.2

Calcium influx contributes to the transient response

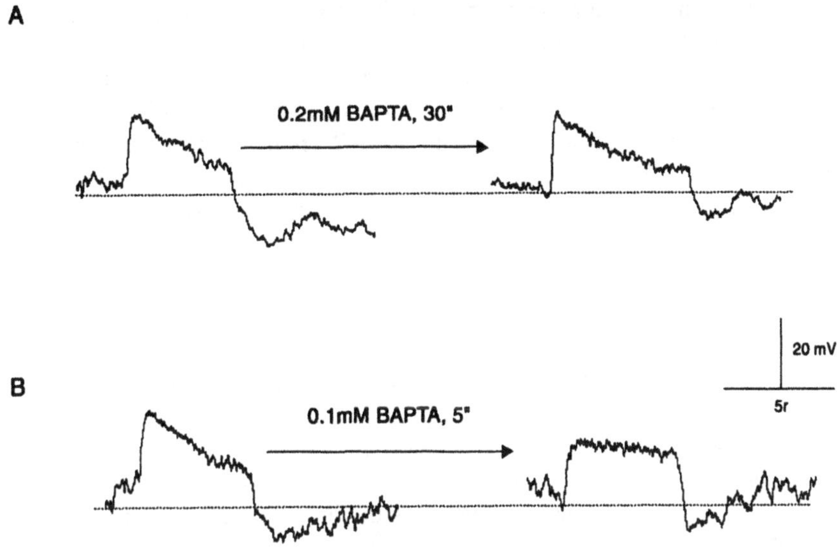

Figure 9. A calcium positive-feedback mechanism may contribute to transient responses of ON bipolar cells. When the ON bipolar cell was dialyzed with 0.2 mM BAPTA (in pipette, whole-cell recording mode), then the bipolar cell light response showed a sag even after 30 minutes of recording (top set of traces). However, in another ON bipolar cell that was dialyzed with 1 mM BAPTA, there was a sag in the light response in the first minute of recording, but five minutes later the light response of the bipolar cell was sustained and the transient portion of the ON bipolar cell light response was suppressed.

mM BAPTA was included in the whole-cell recording pipette, then the ON bipolar cell light response continued to show a sag. But when 1 mM BAPTA was used to increase calcium buffering, this sag was blocked. This was exactly the opposite of our expectation. It implicates calcium in the formation of the transient response but suggests that calcium is providing positive feedback, not negative feedback.

This possibility is intriguing because transient ganglion cell responses have a small dynamic range as compared to sustained ganglion cells. A positive feedback system would have this effect because the response would quickly saturate. While we have not done enough experiments to substantiate this model, the working hypothesis is that one type of ON bipolar cell initially responds to photoreceptor input with a boosted response, but then that positive feedback fades and the response of the cell becomes small. In contrast, sustained ON bipolar cells may not possess that positive feedback system (or at least to a far smaller degree), so they do not display a large sag and have a larger dynamic range.

DIFFERENT BIPOLAR CELLS DRIVE TRANSIENT AND SUSTAINED GANGLION CELLS

With this information, we returned to the original question: Do two types of ON bipolar cell produce the transient and sustained responses in ganglion cells? If this model is correct, it predicts that CPPG will have opposing effects on the two kinds of ganglion cell. If transient bipolar cells drive transient ganglion cells exclusively, or primarily, then CPPG should block the ON response but not depolarize the cell. In contrast, CPPG should not only eliminate the ON response of sustained ON ganglion cells, but it should also depolarize them. This is what was observed[31] (Fig. 8, bottom left and right, respectively).

CONCLUSION

In summary, the metabotropic glutamate receptors at the dendritic tips of ON bipolar cells make possible an array of properties that are at the foundation of vision. They permit the sign opposing ON and OFF responses that facilitate high gain responses in darkness versus light. In addition, these receptors trigger the formation of transient and sustained signals, which provide high gain and broad dynamic range. The ON and OFF bipolar cells utilize metabotropic and ionotropic receptors, respectively, to extract unique information from the photoreceptors. The temporal properties of ON bipolar cells seem to be conveyed by one receptor, but perhaps different transduction pathways. The OFF bipolar cells use different types of ionotropic receptors to accomplish a similar end.[35] The transient and sustained dichotomy that is formed in the outer plexiform layer is then modulated by a variety of mechanisms in the inner plexiform layer. Thus, the complexities of phototransduction are accompanied by equally complex decompositions at the first synapse in retina.

ACKNOWLEDGMENTS

This work was supported by the National Eye Institute, grant # EY05725.

REFERENCES

1. Murakami M, Otsuka T, Shimazaki H. Effects of aspartate and glutamate on the bipolar cells in the carp retina. Vision Res 1975; 15:456-458.
2. Shiells RA, Falk G. A rise in intracellular Ca2+ underlies light adaptation in dogfish retinal 'on' bipolar cells. J Physiol 1999; 514:343-350.
3. Nawy S. Regulation of the ON bipolar cell mGluR6 pathway by Ca2+. J Neurosci 2000; 20:4471-4479.
4. Nawy S, Jahr CE. cGMP-gated conductance in retinal bipolar cells is suppressed by the photoreceptor transmitter. Neuron 1991; 7:677-683.
5. Shiells RA, Falk G. Responses of rod bipolar cells isolated from dogfish retinal slices to concentration-jumps of glutamate. Vis Neurosci 1994; 11:1175-1183.

6. Vardi N. Alpha subunit of Go localizes in the dendritic tips of ON bipolar cells. J Comp Neurol 1998; 395:43-52.
7. Dhingra A, Lyubarsky A, Jiang M et al. The light response of ON bipolar neurons requires G[alpha]o. J Neurosci 2000; 20:9053-9058.
8. Nawy S. The metabotropic receptor mGluR6 may signal through G(o), but not phosphodiesterase, in retinal bipolar cells. J Neurosci 1999; 19:2938-2944.
9. Nomura A, Shigemoto R, NakamuraY et al. Developmentally regulated postsynaptic localization of a metabotropic glutamate receptor in rat rod bipolar cells. Cell 1994; 77:361-369.
10. Slaughter MM, Miller RF. 2-amino-4-phosphonobutyric acid: a new pharmacological tool for retina research. Science 1981; 211:182-185.
11. Tian N, Slaughter MM. Pharmacological similarity between the retinal APB receptor and the family of metabotropic glutamate receptors. J Neurophysiol 1994; 71:2258-2268.
12. Morigiwa K, Vardi N. Differential expression of ionotropic glutamate receptor subunits in the outer retina. J Comp Neurol 1999; 405:173-184.
13. Nawy S, Copenhagen DR. Multiple classes of glutamate receptor on depolarizing bipolar cells in retina. Nature 1987; 325:56-58.
14. Slaughter MM, Miller RF. 2-amino-4-phosphonobutyric acid: a new pharmacological tool for retina research. Science 1981; 211:182-185.
15. Bloomfield SA, Dowling JE. Roles of aspartate and glutamate in synaptic transmission in rabbit retina. I. Outer plexiform layer. J Neurophysiol. 1985; 53:699-713.
16. Tian N, Slaughter MM. Functional properties of a metabotropic glutamate receptor at dendritic synapses of ON bipolar cells in the amphibian retina. Vis Neurosci 1995; 12:755-765.
17. Tian N, Slaughter MM. Functional properties of a metabotropic glutamate receptor at dendritic synapses of ON bipolar cells in the amphibian retina. Vis Neurosci 1995; 12:755-765.
18. Ashmore JF, FalkG. The single-photon signal in rod bipolar cells of the dogfish retina. J Physiol. 1980; 300:151-166.
19. Ashmore JF, Copenhagen DR. Different postsynaptic events in two types of retinal bipolar cell. Nature 1980; 288:84-86.
20. Yang XL, Wu SM. Synaptic transmission from rods to rod-dominated bipolar cells in the tiger salamander retina. Brain Res 1993; 613:275-280.
21. Werblin FS, Dowling JE. Organization of the retina of the mudpuppy, Necturus maculosus. II. Intracellular recording. J Neurophysiol 1969; 32:339-355.
22. Maguire G, Maple B, Lukasiewicz P et al. Gamma-aminobutyrate type B receptor modulation of L-type calcium channel current at bipolar cell terminals in the retina of the tiger salamander. Proc Natl Acad Sci USA 1989; 86:10144-10147.
23. Werblin F, Maguire G, Lukasiewicz P et al. Neural interactions mediating the detection of motion in the retina of the tiger salamander. Vis Neurosci 1988; 1:317-329.
24. Lukasiewicz PD, Werblin FS. A novel GABA receptor modulates synaptic transmission from bipolar to ganglion and amacrine cells in the tiger salamander retina. J Neurosci 1994; 14:1213-1223.
25. Lukasiewicz PD, Maple BR, Werblin FS. A novel GABA receptor ON bipolar cell terminals in the tiger salamander retina. J Neurosci 1994; 14:1202-1212.
26. Jacobs AL, Werblin FS. Spatiotemporal patterns at the retinal output. J Neurophysiol 1998; 80:447-451.
27. Pan ZH. Differential expression of high- and two types of low-voltage-activated calcium currents in rod and cone bipolar cells of the rat retina. J Neurophysiol 2000; 83:513-527.
28. Pan ZH. Voltage-activated Ca2+ channels and ionotropic GABA receptors localized at axon terminals of mammalian retinal bipolar cells. Vis Neurosci 2001; 18:279-288.
29. Lukasiewicz PD, Lawrence JE, Valentino TL. Desensitizing glutamate receptors shape excitatory synaptic inputs to tiger salamander retinal ganglion cells. J Neurosci 1995; 15:6189-6199.
30. Mobbs P, Everett K, Cook A. Signal shaping by voltage-gated currents in retinal ganglion cells. Brain Res 1992; 574:217-223.

31. Awatramani GB, Slaughter MM. Origin of transient and sustained responses in ganglion cells of the retina. J Neurosci 2000; 20:7087-7095.
32. Thoreson WB, Gottesman J, Jane DE et al. Two phenylglycine derivatives antagonize responses to L-AP4 in ON bipolar cells of the amphibian retina. Neuropharmacol 1997; 36:13-20.
33. Awatramani GB, Slaughter MM. Origin of transient and sustained responses in ganglion cells of the retina. J Neurosci 2000; 20:7087-7095.
34. Dowling JE, Werblin FS. Organization of retina of the mudpuppy, Necturus maculosus. I. Synaptic structure. J Neurophysiol 1969; 32:315-338.
35. DeVries SH. Bipolar cells use kainate and AMPA receptors to filter visual information into separate channels. Neuron 2000; 28:847-856.

Ca^{2+} REGULATION OF *DROSOPHILA* PHOTOTRANSDUCTION

Joseph E. O'Tousa

ABSTRACT

Drosophila vision research has benefited from simultaneous application of genetic, molecular, electrophysiological analyses. The work establishes an essential role of Ca^{2+} in regulation of phototransduction. Many different proteins are the targets of Ca^{2+} regulation and these proteins act at multiple steps of the process. These targets include proteins involved in the rhodopsin cycle, proteins responsible for intermediate steps of phototransduction, and the TRP and TRPL light-gated channels. The regulation of these phototransduction components by Ca^{2+} occurs in three different ways. First, the presence of Ca^{2+}/calmodulin-binding sites in phototransduction-mediating proteins places these proteins under Ca^{2+} control. Second, a protein kinase C regulated by Ca^{2+} phosphorylates proteins to modulate their activity. Finally, some proteins contain Ca^{2+}-binding sites and apparently are directly regulated by Ca^{2+}. Here I review the photoreceptor proteins regulated by Ca^{2+}, and summarize current views on the roles of these proteins in the *Drosophila* photoresponse.

INTRODUCTION

The phototransduction cascade of invertebrate photoreceptor culminates in cell depolarization in large part due to the influx of Ca^{2+}. The increased levels of intracellular Ca^{2+} participate in two relevant activities of the photoreceptor during the light response. First, Ca^{2+} regulates intermediate processes controlling kinetics and magnitude of the cellular response to light. Second, Ca^{2+}-controlled activation of

Department of Biological Sciences, University of Notre Dame, Notre Dame, IN 46556, U.S.A.

recovery processes restores the photoreceptor to the dark condition enabling additional light responses. Additional information on *Drosophila* and invertebrate phototransduction can be found in recent reviews.[1-4]

Ca^{2+} CONTROL OF THE RHODOPSIN PHOTOCYCLE

The photoreceptors of both vertebrates and invertebrates contain similar visual pigments, or rhodopsins, consisting of a seven-transmembrane protein bound to a retinal chromophore. Some insects, including *Drosophila*, utilize 9-hydroxy-11-cis-retinal in place of 11-cis-retinal found in vertebrate pigments. The most profound difference recognized between the behavior of invertebrate and vertebrate rhodopsins is the rhodopsin cycle. In invertebrates, light activation of rhodopsin converts the photopigment into a stable photoproduct, metarhodopsin. Light carries out the reverse process, converting all-trans-retinal of metarhodopsin back to the 11-cis-retinal of rhodopsin. This process differs in a fundamental way from the vertebrate rhodopsin cycle, in which newly formed metarhodopsin releases the all-trans-retinal chromophore, and then the chromophore is transported from the photoreceptors to retinal pigment epithelium (RPE) cells. Following a series of conversion steps within the RPE cells, the chromophore is returned again to the photoreceptors and rhodopsin is reconstituted.

Figure 1 shows the steps involved in the invertebrate rhodopsin cycle. Despite the stability of the invertebrate chromophore-protein linkage, the initial steps are similar to the vertebrate rhodopsin cycle. The metarhodopsin form is rapidly phosphorylated,[5] and complexes with arrestin. The protein possessing rhodopsin kinase activity has not been identified in Drosophila. Two candidate G protein coupled receptor kinases have been described,[6] and surprisingly the Drosophila genome project failed to identify additional candidates.[7] As in vertebrates, arrestin binding terminates the ability of metarhodopsin to activate G protein resulting in a rapid termination of the light response.[8, 9] At this point, in invertebrates only, light converts the metarhodopsin back to rhodopsin. The action of the rdgC phosphatase then removes the phosphates on rhodopsin to complete the cycle and bring rhodopsin back to its initial state.

Ca^{2+} Control of Arrestin Phosphorylation

Within the rhodopsin photocycle, there are two recognized mechanisms of Ca^{2+} regulation. The first of these is control of arrestin action. Two distinct gene products, Arr1 and Arr2, are found in *Drosophila* photoreceptors; Arr2 plays the major role.[9] Both arrestins are rapidly phosphorylated during the light response by a Type II Ca^{2+}/calmodulin-dependent kinase.[10,11] Recent studies of arrestin mutants suggest that Ca^{2+}-controlled arrestin phosphorylation is not required for arrestin binding to metarhodopsin, but rather to render bound arrestin competent for release from metarhodopsin.[12] Further, this model suggests that the rapid photoreceptor degeneration seen in situations in which intracellular Ca^{2+} levels do not rise and hence

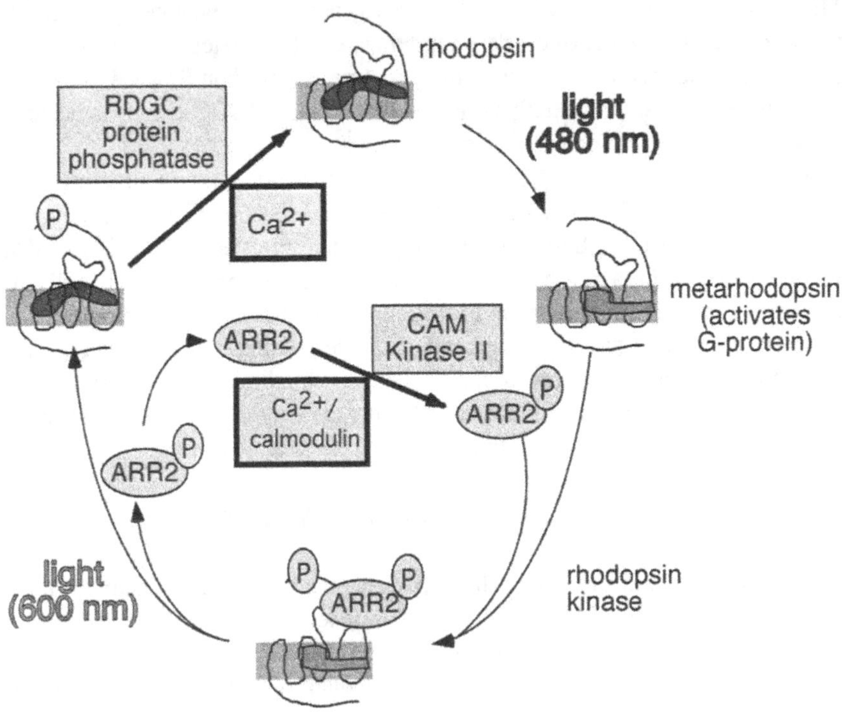

Figure 1. Ca^{2+} regulation of the *Drosophila* rhodopsin cycle. From top, moving clockwise, rhodopsin is activated by light to the active metarhodopsin state capable of activating phospholipase C. The major form of rhodopsin in the *Drosophila* retina is activated by blue light, with other rhodopsin showing distinct spectral properties.[69] The chromophore is shown in different conformations within the diagram to signify that rhodopsin contains 11-cis-retinal whereas metarhodopsin contains all-trans-retinal. Metarhodopsin activity is rapidly quenched by the combined action of rhodopsin phosphorylation and arrestin binding. Although the gene product responsible for rhodopsin phosphorylation is not known, there are two candidate G protein-coupled receptor kinases expressed in the retina.[6] There are two arrestins expressed in the *Drosophila* retina;[9] the predominant ARR2 arrestin is shown in the diagram. During the photoresponse, both arrestins are rapidly phosphorylated via a Ca^{2+}/calmodulin dependent protein kinase. For the major rhodopsin, the photoconversion of the phosphorylated, arrestin-bound, metarhodopsin back to the rhodopsin form occurs via longer wavelength light. ARR2 dissociates from rhodopsin and is dephosphorylated by an unknown protein phosphatase to return to its dark state. Phosphorylated rhodopsin is restored to its original state by action of the Ca^{2+}-dependent RDGC protein phosphatase.

arrestin is not phosphorylated is due to excessive phosphorhodopsin-arrestin complexes triggering photoreceptor cell death.[13,14]

The *Drosophila rdgC* Gene is a Novel Protein Phosphatase

Many *Drosophila* genes playing roles in phototransduction exhibit light-dependent retinal degeneration.[2] Among these is the *rdgC* mutant. The striking aspect of the *rdgC* mutant phenotype is that prior to degeneration the photoreceptors generate a normal light response. Under dark conditions, or when rhodopsin activity is compromised, the photoreceptors retain normal structure and function. This led to the proposal that, despite the substantially normal photoresponse, RDGC activity is required for the cell to recover from rhodopsin stimulation.[15]

Cloning of the *rdgC* gene showed that the encoded protein is a serine/threonine protein phosphatase. Three classic Ser/Thr protein phosphatase families were previously recognized. Description of additional phosphates related to RDGC has led to general agreement that a fourth subfamily, rdgC/PP5, be recognized.[16, 17] The RDGC subgroup is distinguished from the PP5 group on the basis of a carboxy-terminal regulatory domain (Figure 1). The prominent feature of this domain is the repeated EF hand motifs found in calmodulin and other Ca^{2+}-binding proteins. The recognition of these motifs led to the hypothesis that the carboxy-domain acts as a regulatory domain allowing Ca^{2+} levels to control the phosphatase activity.

Ca^{2+} Regulation of *rdgC* Family of Protein Phosphatases

Experimental data supporting the hypothesis that Ca^{2+} controls activity of the RDGC phosphatase family has been obtained in several studies. The first study [8] showed that poor rhodopsin dephosphorylation was observed in cell extracts from the *Drosophila* retinal degeneration C (*rdgC*) mutant relative to the wild-type extracts. In the wild-type extracts, the dephosphorylation was enhanced by Ca^{2+}, but not dependent on calmodulin. A second important result is that phosphorylated rhodopsin is a major substrate of RDGC in this in vitro assay. A subsequent study confirmed that rhodopsin is a major substrate for rdgC under in vivo conditions.[18]

Regarding other RDGC family members, only the phosphatase activity of the human RDGC family member PPEF-1 has been tested and shown to be stimulated by Ca^{2+}.[19] There are more biochemical analysis waiting to be done on these enzymes, as other family members have not been tested, and furthermore in no case are there data showing that Ca^{2+} exerts its effect on RDGC activity by binding to the carboxy EF-hand containing domain.

Non-*Drosophila* RDGC Orthologs Are Involved in Other Cellular Processes

The clear role of *Drosophila* RDGC in rhodopsin phosphorylation and retinal degeneration led to speculation that vertebrate homologs are also involved in regulation of G protein-coupled receptors. To test the role of PPEF-1 and PPEF-2, which are expressed in the vertebrate retina, knockout mice were constructed and analyzed.[20] These mutants neither show defects in rhodopsin phosphorylation nor trigger retinal degeneration. Thus, the gene products responsible for Ca^{2+}-activated rhodopsin phosphatase activity[21] remain unidentified. These results are remarkable, because despite the high sequence identity shared with RDGC, the PPEF-1 or the

PPEF-2 phosphatases do not act on rhodopsin. The in vivo substrates of these proteins, as well as the C. elegans PPEF,[22] remain unidentified; it is possible that they act on non-rhodopsin members of the G protein-coupled receptor family.

Ca^{2+} REGULATION OF INTERMEDIATE STEPS OF THE PHOTOTRANSDUCTION CASCADE

There are fundamental differences in the photransduction cascade of vertebrates and invertebrates. In vertebrates, rhodopsin triggers a G protein, transducin, that activates a cGMP phosphodiesterase. Increased cGMP hydrolysis leads to closure of the cGMP-gated channels and hyperpolarization of the photoreceptor cell. In *Drosophila* and other invertebrates, rhodopsin activates a distinct G-protein class member, Gq, able to stimulate a different effector, phospholipase C. The essential role of each these proteins is well documented, however the subsequent steps remain under investigation.[4, 23] One major impetus in the work is the widespread use of phospholipase C signaling in vertebrates and invertebrates, with *Drosophila* phototransduction representing an excellent experimental system to elucidate molecular processes underlying these events.

Analyses of individual phototransduction components implicate Ca^{2+} control of many different intermediate steps (Fig. 2). The large Ca^{2+} influx during the light response [24-26] creates a situation in which Ca^{2+} feedback can modulate components of the photoresponse.[3] During these intermediate steps, Ca^{2+} may also activate enzymes that restore substrates depleted during the photoresponse.

The INAD Scaffolding Protein

An unexpected recent finding is that key components of the invertebrate phototransduction cascade are located within a larger complex. The central component of this complex is the INAD protein. INAD is considered a "scaffolding protein" because it contains no intrinsic activity but rather consists of binding sites, referred to as PDZ domains, for other phototransduction proteins. The first *Drosophila inaD* mutant characterized specifically affected binding to the TRP channel.[27] Subsequent analyses recognize five PDZ domains within INAD. There is good agreement that TRP, PLC, and PKC proteins are bound within the INAD complex;[28, 29] additional reports implicate other proteins.[1]

The importance of INAD complex to the photoresponse is indicated by the phototransduction defects seen in *inaD* mutants.[30] The possibility that Ca^{2+} regulates organization of the complex comes from two reports. One report shows that INAD retains consensus protein kinase C phosphorylation sites and is phosphorylated in a Ca^{2+} dependent fashion.[31] It has also been reported that INAD will directly bind calmodulin,[32] providing a second mechanism for Ca^{2+} regulation of the complex.

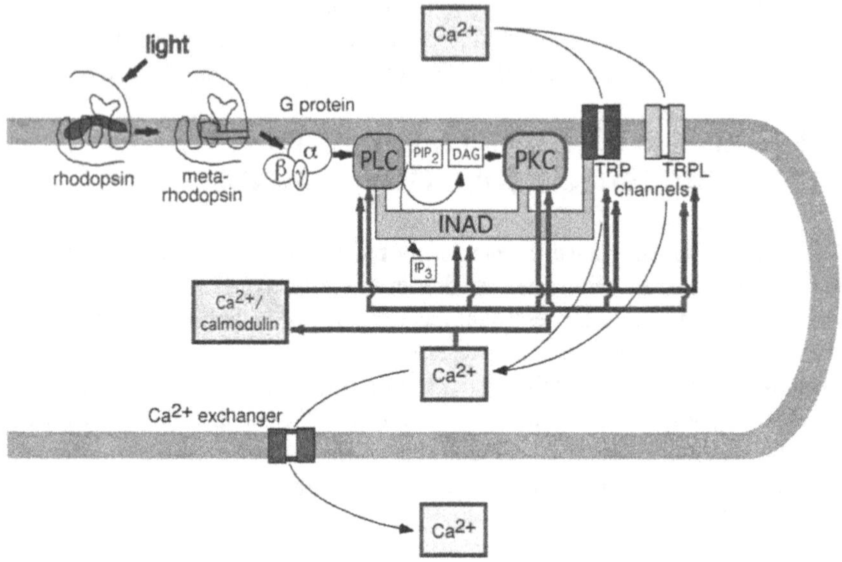

Figure 2. Potential Ca^{2+}-regulated components of *Drosophila* phototransduction. Transduction occurs by step-wise activation of rhodopsin, G protein, and phospholipase C (PLC). PLC creates inositol trisphosphate (IP_3) and diacylglycerol (DAG) from PIP_2 (phosphatidylinositol 4,5-bisphosphate). Subsequent steps leading to opening of the TRP and TRPL channels are not understood. The Ca^{2+} entry from the TRP and TRPL channels has the potential to regulate the photransduction cascade at multiple levels. DAG and Ca^{2+} cooperate in activating protein kinase C (PKC). Identified PKC substrates are PLC, TRP and INAD. INAD is a scaffolding protein that restrains PKC, PLC and TRP in a multiprotein complex. Ca^{2+} also complexes to calmodulin. Identified Ca^{2+}/calmodulin-binding proteins are PLC, INAD and TRP/TRPL. Not shown in the figure is the regulation of NINAC by Ca^{2+}/calmodulin and PKC, and possible regulation of the RDGB by Ca^{2+}. Finally, removal to the Ca^{2+} from the cell likely occurs via activity of a Ca^{2+} exchanger.

Ca^{2+} Control of Phospholipase C

Phospholipase C, PLC, is the target of the rhodopsin/G protein cascade in invertebrate photoreceptors. An in vitro study shows Ca^{2+} stimulates *Drosophila* PLC at low concentrations (up to 1 μM), but has an inhibitory effect at higher (5 μM) concentrations.[33] During the light response, intracellular Ca^{2+} levels rise above 200 μM.[34] Thus, the increased intracellular Ca^{2+} levels during the photoresponse may rapidly attenuate PLC activity. This mechanism may contribute to the brief refractory period observed in invertebrate photoreceptors.[3]

Two mechanisms have been proposed for Ca^{2+} control of PLC activity. The first is Ca^{2+}/calmodulin-binding to PLC. In *Limulus*, the light response and PLC activity is effectively inhibited by action of calmodulin antagonists.[35] These results led to the proposal that there is a calmodulin-like domain within PLC that mediates Ca^{2+} sensitivity of the enzyme. The second proposed mechanism is that phosphorylation of PLC via a Ca^{2+} dependent process controls PLC activity. Data in support of this proposal come from experiments demonstrating that PKC inhibitors block phototransduction at the level of G protein or PLC activity.[36]

Ca^{2+} Regulation of Protein Kinase C

The control of protein kinase C activities by Ca^{2+} and DAG, both of which rise during the light response, is well-documented.[37] *Drosophila* photoreceptors express a unique PKC,[38] and photoreceptors lacking this PKC activity show a defect in deactivation and light adaptation of the photoresponse.[39,40] There are three known targets of this PKC, and there is evidence that each of these contribute to the mutant phenotype. These are the TRP[41] channel, the INAD scaffolding protein[31] and the NINAC p174 protein.[42]

A Ca^{2+}-Binding Domain in the RDGB Phosphatidylinositol Transfer Protein

rdgB mutants are characterized by a rapid light-dependent degeneration of photoreceptors. The *rdgB* gene encodes a large membrane-associated protein containing phosphatidylinositol transfer protein structure within its amino terminal domain. This domain retains phosphatidylinositol transfer activity essential for the rescue of the mutant phenotype.[43] The RDGB domain immediately adjacent to the phosphatidylinositol transfer domain is labeled the Ca^{2+}-binding domain due to its ability to sequester Ca^{2+} in vitro.[44] Photoreceptors lacking rdgB activity do not replenish PIP_2 during the light response, with evidence that both decreased synthesis and increased hydrolysis of PIP_2 contribute to this effect.[45] These data implicate RDGB activity in regeneration of PIP_2, and suggest the possibility that Ca^{2+} stimulates this activity.

NINAC is Major Calmodulin-Binding Protein of the Photoreceptor

ninaC mutants exhibit slow termination of the photoresponse. Cloning of the *ninaC* gene showed that the gene encodes two different proteins, NINAC p174 and NINAC p134, and these localize to distinct regions of the photoreceptor. Both of these are hybrid proteins containing serine/threonine kinase domain coupled to a myosin heavy chain head.[46] The NINAC proteins have two calmodulin-binding sequences; these act to both localize and stabilize calmodulin within the photoreceptor cell.[47] Further, mutation of one or both of these calmodulin-binding sites slows the termination of the photoresponse.[48] Thus, the calmodulin-binding sites in NINAC are critical for proper levels of calmodulin and proper photoresponse. These results have led to the proposal that the major deficits in the photoresponse are attributable to low steady state levels of calmodulin in these cells.[49]

There are likely additional roles for NINAC in mediating the photoresponse. Recent work has implicated PKC phosphorylation in the function of NINAC.[42] Also, the NINAC proteins retain a protein kinase domain of unknown function. The analysis of *ninaC* transgenes lacking the protein kinase domain suggest that the kinase domain plays only a minor role in the altered photoresponse of *ninaC* mutants, and is not required to prevent *ninaC*-induced retinal degeneration.[47]

Calphotin is a Ca^{2+}-Binding Protein

Calphotin, first identified as the protein target of a monoclonal antibody, is expressed exclusively within *Drosophila* photoreceptors. Within photoreceptors, the protein is spatially restricted to the cytoplasmic domain directly adjacent to the photosensitive rhabdomeres. The Calphotin protein contains a 600 amino acids stretch containing 9% negatively charged residues. This region is thought to mediate Ca^{2+}-binding observed in gel overlay experiments. The spatial distribution and Ca^{2+}-binding properties of the protein suggests that it may play a role in sequestering Ca^{2+} and limiting its diffusion into different subcellular regions.[50,51] The description of Calphotin mutants further suggests that the protein is required for proper rhabdomere development and in photoreceptor cell survival.[52]

Ca^{2+} Exchangers in the Photoreceptor

As in other cell types, Ca^{2+} exchangers likely remove the large amount of intracellular Ca^{2+} entering the photoreceptor during the light response. However, there is limited evidence on molecular components and mechanisms responsible for these processes. Release of caged Ca^{2+} within the photoreceptor triggers a rapid efflux of Ca^{2+} that is dependent on external Na^{+}.[53] These results suggest a role for an electrogenic Na^{+}/Ca^{2+} exchanger in Ca^{2+} extrusion from photoreceptors. Two *Drosophila* genes encoding Ca^{2+}/Na^{+} exchangers, *Calx*[54] and *Nckx30C*[55] have been characterized. Although both are expressed in photoreceptors, the Nckx30C mode of action is considered more appropriate for the major role in Ca^{2+} efflux during the photoresponse recovery.[55]

Other Ca^{2+}-Binding Proteins in the *Drosophila* Retina

Additional retinal proteins potentially regulated by Ca^{2+} have been identified in biochemical screens using calmodulin as bait. These include two Ca^{2+}/calmodulin dependent protein kinases and one Ca^{2+}/calmodulin dependent protein phosphatase.[56] The roles and modes of regulation of these in the photoreceptor function have not been determined. Another retinal protein with calmodulin-binding sites is the *Drosophila* ryanodine receptor[57] active in Ca^{2+} release from endoplasmic reticulum membranes. Functional tests suggest calmodulin regulates this activity, and, further, that the ryanodine receptor plays a role in the photoresponse.[49]

Ca^{2+}/CALMODULIN CONTROL OF THE LIGHT-GATED TRP AND TRPL CHANNELS

The TRP and TRPL channels are the channels responsible for the light-activated conductance of *Drosophila* photoreceptors. A third channel component, TRP-γ, may form heteromultimers with TRPL,[58] but TRP-γ must act in an accessory role as TRPL forms functional channels on its own,[59] and all light-induced currents are abolished in the *trpl;trp* double mutant.[60]

Ca^{2+} regulation of the TRP and TRPL channels has both facilitation and inactivation components.[61-63] Ca^{2+}/calmodulin likely mediates some aspects of this Ca^{2+} regulation. TRPL was first identified as a calmodulin-binding protein, and contains two recognized calmodulin-binding sites.[64] TRPL proteins engineered to lack each of the calmodulin-binding sites show that these sites are critical for normal inactivation under in vivo conditions.[65] Calmodulin-deficient mutants also show poor inactivation of the TRPL channels. In the case of TRP, there is one putative calmodulin-binding site;[66] further work is required to document its role in TRP regulation.

A second possible Ca^{2+} dependent mechanism in TRP and TRPL control is PKC phosphorylation. In the larger fly *Calliphora*, TRP channel is phosphorylated by a Ca^{2+}- and phorbol ester-dependent process, implicating PKC as the enzyme responsible for TRP phosphorylation.[41] The delayed repolarization of the TRP channels seen in the PKC mutant is consistent with PKC phosphorylation playing a role in channel inactivation.[67] It is also known that TRPL channels contain a domain that, at least in vitro, serves as a substrate for PKC and PKA phosphorylation.[68] Thus, TRPL may also be controlled by PKC-mediated phosphorylation.

SUMMARY

Invertebrate phototransduction has proven to be a challenging experimental problem. Our current understanding of the process remains incomplete, notably the mechanism responsible for triggering opening of the TRP and TRPL channels, and the influx of Ca^{2+} into the cell, is not understood. It is clear, however, that the combined use of sophisticated genetic, molecular, electrophysiological analyses in *Drosophila* have successfully addressed a large number of these questions. This body of work recognizes the key role of Ca^{2+} regulation. Ca^{2+} regulates many aspects of phototransduction, including proteins involved in the rhodopsin cycle, proteins responsible for intermediate steps of phototransduction, and the TRP and TRPL channels themselves.

ACKNOWLEDGMENTS

The National Eye Institute of the National Institutes of Health supports vision research in the author's laboratory.

REFERENCES

1. Montell C. Visual transduction in *Drosophila*. Annu Rev Cell Dev Biol 1999; 15:231-68.
2. Minke B, Hardie RC. Genetic dissection of *Drosophila* phototransduction. In: Stavenga DG, DeGrip WJ, Pugh ENJ, eds. Handbook of Biological Physics. Amsterdam: Elsevier, 2000:449-525.
3. Hardie RC. Phototransduction in *Drosophila melanogaster*. J Exp Biol 2001; 204:3403-9.
4. Hardie RC, Raghu P. Visual transduction in *Drosophila*. Nature 2001; 413:186-93.
5. Bentrop J, Paulsen R. Light-modulated ADP-ribosylation, protein phosphorylation and protein binding in isolated fly photoreceptor membranes. Eur J Biochem 1986; 161:61-7.
6. Cassill JA, Whitney M, Joazeiro CA et al. Isolation of *Drosophila* genes encoding G protein-coupled receptor kinases. Proc Natl Acad Sci USA 1991; 88:11067-70.
7. Fortini ME, Skupski MP, Boguski MS, Hariharan IK. A survey of human disease gene counterparts in the *Drosophila* genome. J Cell Biol 2000; 150:F23-30.
8. Byk T, Bar-Yaacov M, Doza YN et al. Regulatory arrestin cycle secures the fidelity and maintenance of the fly photoreceptor cell. Proc Natl Acad Sci USA 1993; 90:1907-11.
9. Dolph PJ, Ranganathan R, Colley NJ et al. Arrestin function in inactivation of G protein-coupled receptor rhodopsin in vivo. Science 1993; 260:1910-6.
10. Kahn ES, Matsumoto H. Calcium/calmodulin-dependent kinase II phosphorylates *Drosophila* visual arrestin. J Neurochem 1997; 68:169-75.
11. Matsumoto H, Kurien BT, Takagi Y et al. Phosrestin I undergoes the earliest light-induced phosphorylation by a calcium/calmodulin-dependent protein kinase in *Drosophila* photoreceptors. Neuron 1994; 12:997-1010.
12. Alloway PG, Dolph PJ. A role for the light-dependent phosphorylation of visual arrestin. Proc Natl Acad Sci USA 1999; 96:6072-7.
13. Alloway PG, Howard L, Dolph PJ. The formation of stable rhodopsin-arrestin complexes induces apoptosis and photoreceptor cell degeneration. Neuron 2000; 28:129-38.
14. Kiselev A, Socolich M, Vinos J et al. A molecular pathway for light-dependent photorecptor apoptosis in *Drosophila*. Neuron 2000; 28:139-52.
15. Steele F, O'Tousa JE. Rhodopsin activation causes retinal degeneration in *Drosophila rdgC* mutant. Neuron 1990; 4:883-90.
16. Andreeva AV, Evans DE, Hawes CR et al. PP7, a plant phosphatase representing a novel evolutionary branch of eukaryotic protein Ser/Thr phosphatases. Biochem Mol Biol Int 1998; 44:703-15.
17. Cohen PT. Novel protein serine/threonine phosphatases: variety is the spice of life. Trends Biochem Sci 1997; 22:245-51.
18. Vinos J, Jalink K, Hardy RW et al. A G protein-coupled receptor phosphatase required for rhodopsin function. Science 1997; 277:687-90.
19. Huang X, Honkanen RE. Molecular cloning, expression, and characterization of a novel human serine/threonine protein phosphatase, PP7, that is homologous to *Drosophila* retinal degeneration C gene product (*rdgC*). J Biol Chem 1998; 273:1462-8.
20. Ramulu P, Kennedy M, Xiong WH et al et al. Normal light response, photoreceptor integrity, and rhodopsin dephosphorylation in mice lacking both protein phosphatases with EF-hands (PPEF-1 and PPEF-2). Mol Cell Biol 2001; 21:8605-14.
21. Kutuzov MA, Bennett N. Calcium-activated opsin phosphatase activity in retinal rod outer segments. Eur J Biochem 1996; 238:613-22.
22. Ramulu P, Nathans J. Cellular and subcellular localization, N-terminal acylation, and calcium binding of *Caenorhabditis elegans* protein phosphatase with EF- hands. J Biol Chem 2001; 276:25127-35.
23. Scott K, Zuker C. TRP, TRPL and trouble in photoreceptor cells. Curr Opin Neurobiol 1998; 8:383-8.

24. Ranganathan R, Bacskai BJ, Tsien RY, Zuker CS. Cytosolic calcium transients: spatial localization and role in *Drosophila* photoreceptor cell function. Neuron 1994; 13:837-48.
25. Hardie RC. INDO-1 measurements of absolute resting and light-induced Ca^{2+} concentration in *Drosophila* photoreceptors. J Neurosci 1996; 16:2924-33.
26. Peretz A, Suss-Toby E, Rom-Glas A et al. The light response of *Drosophila* photoreceptors is accompanied by an increase in cellular calcium: effects of specific mutations. Neuron 1994; 12:1257-67.
27. Shieh BH, Zhu MY. Regulation of the TRP Ca^{2+} channel by INAD in *Drosophila* photoreceptors. Neuron 1996; 16:991-8.
28. Huber A, Sander P, Gobert A et al. The transient receptor potential protein (Trp), a putative store- operated Ca^{2+} channel essential for phosphoinositide-mediated photoreception, forms a signaling complex with NorpA, InaC and InaD. Embo J 1996; 15:7036-45.
29. Tsunoda S, Sierralta J, Sun Y et al. A multivalent PDZ-domain protein assembles signalling complexes in a G- protein-coupled cascade. Nature 1997; 388:243-9.
30. Tsunoda S, Zuker CS. The organization of INAD-signaling complexes by a multivalent PDZ domain protein in *Drosophila* photoreceptor cells ensures sensitivity and speed of signaling. Cell Calcium 1999; 26:165-71.
31. Huber A, Sander P, Paulsen R. Phosphorylation of the InaD gene product, a photoreceptor membrane protein required for recovery of visual excitation. J Biol Chem 1996; 271:11710-7.
32. Xu XZ, Choudhury A, Li X, Montell C. Coordination of an array of signaling proteins through homo- and heteromeric interactions between PDZ domains and target proteins. J Cell Biol 1998; 142:545-55.
33. Running Deer JL, Hurley JB, Yarfitz SL. G protein control of *Drosophila* photoreceptor phospholipase C. J Biol Chem 1995; 270:12623-8.
34. Oberwinkler J, Stavenga DG. Calcium transients in the rhabdomeres of dark- and light-adapted fly photoreceptor cells. J Neurosci 2000; 20:1701-9.
35. Richard EA, Ghosh S, Lowenstein JM, Lisman JE. Ca^{2+}/calmodulin-binding peptides block phototransduction in *Limulus* ventral photoreceptors: evidence for direct inhibition of phospholipase C. Proc Natl Acad Sci USA 1997; 94:14095-9.
36. Dabdoub A, Payne R. Protein kinase C activators inhibit the visual cascade in *Limulus* ventral photoreceptors at an early stage. J Neurosci 1999; 19:10262-9.
37. Tanaka C, Nishizuka Y. The protein kinase C family for neuronal signaling. Annu Rev Neurosci 1994; 17:551-67.
38. Schaeffer E, Smith D, Mardon G, Quinn W, Zuker C. Isolation and characterization of two new *Drosophila* protein kinase C genes, including one specifically expressed in photoreceptor cells. Cell 1989; 57:403-12.
39. Hardie RC, Peretz A, Pollock JA, Minke B. Ca^{2+} limits the development of the light response in *Drosophila* photoreceptors. Proc R Soc Lond B Biol Sci 1993; 252:223-9.
40. Smith DP, Ranganathan R, Hardy RW, Marx J, Tsuchida T, Zuker CS. Photoreceptor deactivation and retinal degeneration mediated by a photoreceptor-specific protein kinase C. Science 1991; 254:1478-84.
41. Huber A, Sander P, Bahner M, Paulsen R. The TRP Ca^{2+} channel assembled in a signaling complex by the PDZ domain protein INAD is phosphorylated through the interaction with protein kinase C (ePKC). FEBS Lett 1998; 425:317-22.
42. Li HS, Porter JA, Montell C. Requirement for the NINAC kinase/myosin for stable termination of the visual cascade. J Neurosci 1998; 18:9601-6.
43. Milligan SC, Alb JG, Jr., Elagina RB, Bankaitis VA, Hyde DR. The phosphatidylinositol transfer protein domain of *Drosophila* retinal degeneration B protein is essential for photoreceptor cell survival and recovery from light stimulation. J Cell Biol 1997; 139:351-63.
44. Vihtelic TS, Goebl M, Milligan S, O'Tousa JE, Hyde DR. Localization of *Drosophila* retinal degeneration B, a membrane- associated phosphatidylinositol transfer protein. J Cell Biol 1993; 122:1013-22.

45. Hardie RC, Raghu P, Moore S et al. Calcium influx via TRP channels is required to maintain PIP2 levels in *Drosophila* photoreceptors. Neuron 2001; 30:149-59.
46. Montell C, Rubin GM. The *Drosophila ninaC* locus encodes two photoreceptor cell specific proteins with domains homologous to protein kinases and the myosin heavy domain head. Cell 1988; 52:757-772.
47. Porter JA, Montell C. Distinct roles of the *Drosophila NinaC* kinase and myosin domains revealed by systematic mutagenesis. J. Cell Biol. 1993; 122:601-612.
48. Porter JA, Minke B, Montell C. Calmodulin binding to *Drosophila NinaC* required for termination of phototransduction. Embo J 1995; 14:4450-9.
49. Arnon A, Cook B, Gillo B et al. Calmodulin regulation of light adaptation and store-operated dark current in *Drosophila* photoreceptors. Proc Natl Acad Sci USA 1997; 94:5894-9.
50. Martin JH, Benzer S, Rudnicka M, Miller CA. Calphotin: A *Drosophila* photoreceptor cell calcium-binding protein. Proc Natl Acad Sci USA 1993; 90:1531-5.
51. Ballinger DG, Xue N, Harshman KD. A *Drosophila* photoreceptor cell-specific protein, calphotin, binds calcium and contains a leucine zipper. Proc Natl Acad Sci USA 1993; 90:1536-40.
52. Yang Y, Ballinger D. Mutations in calphotin, the gene encoding a *Drosophila* photoreceptor cell-specific calcium-binding protein, reveal roles in cellular morphogenesis and survival. Genetics 1994; 138:413-21.
53. Hardie RC. Photolysis of caged Ca^{2+} facilitates and inactivates but does not directly excite light-sensitive channels in *Drosophila* photoreceptors. J Neurosci 1995; 15:889-902.
54. Schwarz EM, Benzer S. Calx, a Na-Ca exchanger gene of *Drosophila melanogaster*. Proc Natl Acad Sci USA 1997; 94:10249-54.
55. Haug-Collet K, Pearson B, Webel R et al. Cloning and characterization of a potassium-dependent sodium/calcium exchanger in *Drosophila*. J Cell Biol 1999; 147:659-70.
56. Xu XZ, Wes PD, Chen H et al. Retinal targets for calmodulin include proteins implicated in synaptic transmission. J Biol Chem 1998; 273:31297-307.
57. Hasan G, Rosbash M. *Drosophila* homologs of two mammalian intracellular Ca(2+)-release channels: identification and expression patterns of the inositol 1,4,5- triphosphate and the ryanodine receptor genes. Development 1992; 116:967-75.
58. Xu XZ, Chien F, Butler A et al. TRPgamma, a *Drosophila* TRP-related subunit, forms a regulated cation channel with TRPL. Neuron 2000; 26:647-57.
59. Hardie RC, Reuss H, Lansdell SJ, Millar NS. Functional equivalence of native light-sensitive channels in the *Drosophila trp301* mutant and TRPL cation channels expressed in a stably transfected Drosophila cell line. Cell Calcium 1997; 21:431-40.
60. Niemeyer BA, Suzuki E, Scott K et al. The *Drosophila* light-activated conductance is composed of the two channels TRP and TRPL. Cell 1996; 85:651-9.
61. Hardie RC, Raghu P. Activation of heterologously expressed *Drosophila* TRPL channels: Ca^{2+} is not required and InsP3 is not sufficient. Cell Calcium 1998; 24:153-63.
62. Zimmer S, Trost C, Wissenbach U et al. Modulation of recombinant transient-receptor-potential-like (TRPL) channels by cytosolic Ca^{2+}. Pflugers Arch 2000; 440:409-17.
63. Henderson SR, Reuss H, Hardie RC. Single photon responses in *Drosophila* photoreceptors and their regulation by Ca^{2+}. J Physiol 2000; 524 Pt 1:179-94.
64. Phillips AM, Bull A, Kelly LE. Identification of a *Drosophila* gene encoding a calmodulin-binding protein with homology to the trp phototransduction gene. Neuron 1992; 8:631-42.
65. Scott K, Sun Y, Beckingham K, Zuker CS. Calmodulin regulation of *Drosophila* light-activated channels and receptor function mediates termination of the light response in vivo. Cell 1997; 91:375-83.
66. Chevesich J, Kreuz AJ, Montell C. Requirement for the PDZ domain protein, INAD, for localization of the TRP store-operated channel to a signaling complex. Neuron 1997; 18:95-105.
67. Hardie RC, Peretz A, Suss-Toby E et al. Protein kinase C is required for light adaptation in *Drosophila* photoreceptors. Nature 1993; 363:634-7.

68. Warr CG, Kelly LE. Identification and characterization of two distinct calmodulin-binding sites in the Trpl ion-channel protein of *Drosophila melanogaster*. Biochem J 1996; 314:497-503.
69. Salcedo E, Huber A, Henrich S et al. Blue- and green-absorbing visual pigments of *Drosophila*: Ectopic expression and physiological characterization of the R8 photoreceptor cell-specific Rh5 and Rh6 rhodopsins. J Neurosci 1999; 19:10716-26.

SIMULTANEOUS ROLES FOR Ca^{2+} IN EXCITATION AND ADAPTATION OF *LIMULUS* VENTRAL PHOTORECEPTORS

John E. Lisman,[1] Edwin A. Richard,[1] Sridhar Raghavachari[1] and Richard Payne[2]

ABSTRACT

The ventral photoreceptors of *Limulus* have been one of the main preparations for the study of invertebrate phototransduction. The study of ventral photoreceptors has revealed that they have remarkable performance characteristics, most notably the very large amplification of the transduction process. This amplification is critically dependent upon the coupling of photoactivated rhodopsin to the phosphoinositide cascade, resulting in the release of Ca^{2+} from intracellular stores. The consequent elevation of Ca^{2+} within the photoreceptor's cytosol is amongst the most rapid and dramatic known to be activated by the phosphoinositide cascade. This review summarizes the evidence that intracellular Ca^{2+} is a key regulator of transduction in *Limulus* photoreceptors. The mechanisms that regulate Ca^{2+} as well as the possible targets of the action of Ca^{2+} are reviewed. Ca^{2+} elevation is critical for triggering both excitation and adaptation processes in the photoreceptor. The question of how a single second messenger can produce these two opposing effects is of obvious interest and is a topic dealt with throughout this review.

[1]Department of Biology and Volen Center for Complex Systems, Brandeis University, Waltham, MA 02454, U.S.A. [2]Department of Biology, University of Maryland, College Park, MD 20742, U.S.A.

ABBREVIATIONS

A lobe	a rhabdomeral lobe
BAPTA	N,N'-(1,2-ethanediylbis(oxy-2,1-phenylene))bis(N-carboxymethyl))-glycine
cGMP	guanosine 3',5'-cyclic monophosphate
CPA	cyclopiazonic acid
DAG	diacylglycerol
EGTA	ethylenediaminetetraacetic acid
ER	endoplasmic reticulum
HEDTA	N-(2-hydroxyethyl)ethylenediaminetriacetic acid
$InsP_3$	inositol-1,4,5-trisphosphate
$InsP_3R$	$InsP_3$-receptor protein
PI	phosphoinositide
PIP_2	phosphatidylinositol (4,5) bisphosphate
PKC	protein kinase C
PLC	phospholipase C
R lobe	rhabdomeral lobe
SER	smooth endoplasmic reticulum

INTRODUCTION

The ventral photoreceptors of *Limulus* and the compound eyes of *Drosophila* have been the two main preparations for the study of invertebrate phototransduction. The *Limulus* ventral photoreceptors were originally chosen as a model because of their large size (>200 μm). This facilitates insertion of multiple electrodes and makes it possible to voltage-clamp the cells and inject compounds of interest in the cytoplasm. The study of ventral photoreceptors has revealed that they have remarkable performance characteristics, most notably the very large amplification of the transduction process. Amplification refers to the amount of charge that is carried across the plasma membrane as a result of excitation by a single photon. In *Limulus* photoreceptors, this gain can be directly measured because the single photon response, termed a quantum bump, is easily recorded. Quantum bumps can be over 10 mV in amplitude and involve the passage of millions of ions.[1] Under voltage-clamp, the peak current can be over 1 nA and appears to be generated over several square microns of membrane surface, containing hundreds of microvilli. By contrast, in the smaller *Drosophila* photoreceptors or vertebrate rods, a single photon event involves a maximum current of only about 10 pA[2] and excitation may be confined to a single microvillus or disc respectively.[3] *Limulus* photoreceptors achieve this large amplification in only 100-200 ms, much faster than in amphibian rods. A further remarkable feature of these cells is their broad dynamic range. Whereas vertebrate rods work over about a 4 log unit range of light intensity before saturating, *Limulus* photoreceptors work over 8.[4,5] They achieve this range through a strong adaptation process that reduces amplification at high light intensity. The large dynamic range of these cells obviates the need for the dual system of photoreceptors (rods and cones) used to achieve a large dynamic range in the vertebrate eye. Understanding the adaptation processes that make possible the wide dynamic range of *Limulus* photoreceptors is a second interesting reason to study these cells.

It is now clear that intracellular Ca^{2+} is a key intracellular regulator of transduction in *Limulus* photoreceptors. Initially it appeared that Ca^{2+} was involved primarily in adaptation processes. It was shown that Ca^{2+} is elevated during light,[6] that artificially elevating Ca^{2+} desensitizes the cells,[7] and that light adaptation can be blocked by introducing high intracellular concentrations of Ca^{2+} buffers.[8] Subsequent work, however, revealed that Ca^{2+} was also involved in the excitation process.[9,10] This conclusion followed from the discovery that certain protocols for elevating intracellular Ca^{2+} produce a large inward current that closely mimics the currents elicited by light. Furthermore, by carefully monitoring the effect of increasing the intracellular Ca^{2+} buffering capacity of the cells, it was concluded the Ca^{2+} elevation is likely to be one of the steps in the excitation cascade.[11,12] Thus, according to our current understanding, Ca^{2+} elevation is critical for triggering both excitation and adaptation. The interplay between these two processes forms the central theme of this review.

An important advance in the study of *Limulus* photoreceptors was the discovery that the light-induced elevation in intracellular Ca^{2+} is produced by the activation of phospholipase C (PLC)[13,14]It was further shown that the inositol-1,4,5-trisphosphate ($InsP_3$) resulting from PLC activity releases Ca^{2+} from intracellular stores.[15,16] Indeed this system was one of the earliest systems for demonstrating the Ca^{2+} releasing role for this pathway. *Limulus* remains an important system for the study of this signaling pathway because of the precision with which the input (light) can be controlled both spatially and temporally, and because it is has proven possible to directly measure the resulting Ca^{2+} elevation with high temporal and spatial resolution.

The study of *Limulus* photoreceptors continues to be an exciting field because some of the key processes of phototransduction in these cells remain to be identified. Some elements of the excitation cascade are clear, in particular the role of G-protein, PLC, $InsP_3$ and Ca^{2+}. However, it is less certain which second messenger leads to the opening of the light-dependent channels. cGMP is the best current candidate, but more work needs to be done to be certain about this. Similarly, the molecular pathways by which Ca^{2+} elevation leads to adaptation remain uncertain. This review will summarize the information that has been learned about the role of Ca^{2+} in excitation and adaptation in *Limulus* photoreceptors, but also will attempt to identify the important questions that still need to be addressed. The interested reader may consult several previous reviews on this topic.[17-19]

MECHANISMS FOR CHANGES IN INTRACELLULAR Ca^{2+}

The Phosphoinositide Cascade

A large body of evidence now demonstrates that the phosphoinositide (PI) pathway links the absorption of light to the activation of ion channels in the plasma membrane of invertebrate photoreceptors (review ref. 19). The PI pathway is a mechanism for releasing intracellular messengers upon the activation of a receptor protein, using inositol phospholipids as a substrate. Because activation of the PI pathway elevates cytosolic Ca^{2+} in most cells, it is an attractive candidate for the visual cascade in microvillar photoreceptors. In the case of *Limulus* photoreceptors, the receptor protein is rhodopsin and the cascade is localized to the membrane of microvilli that cover the plasma membrane of the light-sensitive region of the photoreceptor. Activated rhodopsin catalyzes the exchange of GTP for GDP bound to the alpha subunit of a heterotrimeric GTP-binding protein of the G_q sub-family, which in turn activates phospholipase C. PLC cleaves phosphatidylinositol (4,5) bisphosphate (PIP_2), a minor membrane phospholipid, into a lipid messenger, diacylglycerol (DAG), and the water-soluble messenger, $InsP_3$. The major known target of DAG is protein kinase C (PKC). DAG also can be metabolized into fatty acids, which might serve as downstream messengers. The best-characterized target of $InsP_3$ is a Ca^{2+} channel, the $InsP_3R$ protein ($InsP_3R$), located in the membrane of endoplasmic reticulum (ER). $InsP_3$ therefore releases Ca^{2+} from intracellular stores within the ER. Like many signal transduction cascades, the PI pathway is capable of great signal amplification. One photoisomerized rhodopsin molecule may activate several GTP-binding proteins, many PIP_2 molecules may be hydrolyzed during the active lifetime of one PLC molecule and thousands of Ca^{2+} ions may enter the cytosol through a single opening of the $InsP_3R$.

Rapid Light-Induced Ca^{2+} Release from Stores

The *Limulus* ventral photoreceptor (Fig. 1) is a highly polarized cell divided into two lobes, analogous to the inner and outer segments of vertebrate retinal photoreceptors.[20] The rhabdomeral (R) lobe bears microvilli on its plasma membrane and is therefore light-sensitive. The light-insensitive arhabdomeral (A) lobe contains the cell's nucleus. An axon projects from the A lobe towards the animal's central nervous system. Physiologically relevant measurements of cytosolic Ca^{2+} must be made as close as possible to the site at which the elevation of Ca^{2+} is initiated. Recently, confocal fluorescent light microscopy has enabled the measurement of light-induced elevation of Ca^{2+} in ventral photoreceptors at spots within 4 μm of the microvillar membrane in the R lobe (Fig. 2).[21] In darkness, ventral photoreceptors maintain free Ca^{2+} at 200-600 nM.[22] Following a very bright flash, Ca^{2+} begins to rise after a latent period of approximately 20 ms. Thereafter, Ca^{2+} rises at an initial rate of 1-2 mM/s to reach a peak of 100-200 μM within 200 ms. For less intense

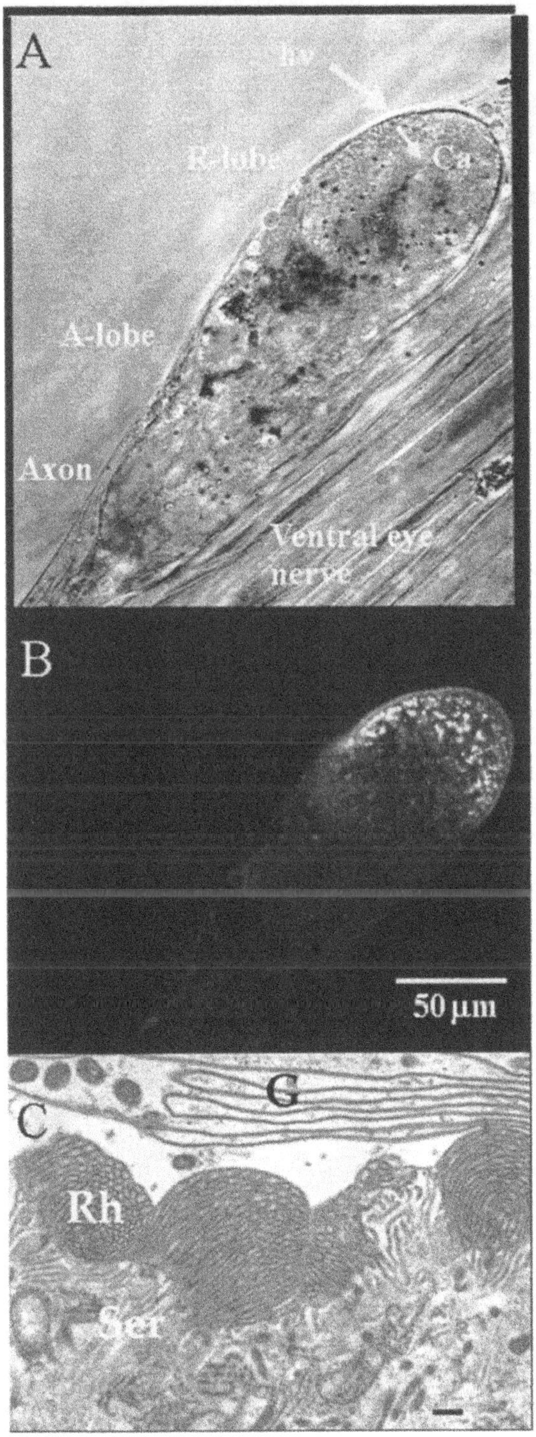

Figure 1. A)Light micrograph of a *Limulus* ventral photoreceptor attached to the ventral eye nerve. Light (hv) is absorbed by the microvillar membrane of the rhabdomeral (R) lobe, resulting in the release of Ca^{2+} ions (Ca) into the cytosol. The light-insensitive arhabdomeral (A) lobe contains the cell nucleus. The axon extends along the ventral eye nerve and terminates in the optic lobe. Photomicrograph courtesy of Dr. A. Dabdoub. B) Immunofluorescent labeling of same photoreceptor by an anti-rhodopsin antibody reveals that rhodopsin is localized to the R-lobe. (For details see ref. 42) Photomicrograph courtesy of Dr. A. Dabdoub. C) Electron micrograph of a section through the plasma membrane of the R-lobe of another ventral photoreceptor showing the whorls of microvilli that form the rhabdom (Rh), the underlying cisternae of smooth endoplasmic reticulum (SER) within the cytosol and glia (G) external to the photoreceptor. Scale bar = 0.5 μm. From reference 108.

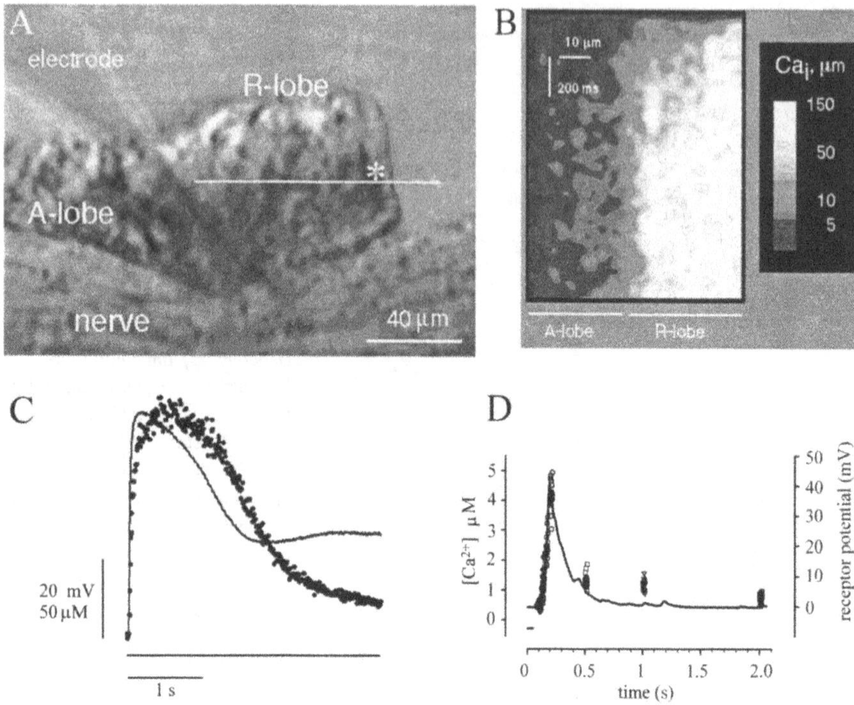

Figure 2. Measurement of light-induced elevation of Ca^{2+} in ventral photoreceptors using fluorescent Ca^{2+}-indicator dyes and both spot and line scan modes of a confocal microscope. A) Ventral photoreceptor showing positions of the confocal spot (asterisk) and the line scan where the fluorescence of an injected Ca^{2+}-indicator dye was measured (The dye used was a mixture of Calcium Green-5N and ANTS). B) Line scan of dye fluorescence, calibrated as $Ca^{2+}{}_i$, showing temporal progression of light-induced elevation of Ca^{2+} from the edge of the R-lobe (right-hand edge of scan) towards the A-lobe. The laser beam scanned the line shown in (A) every 4 ms and the resulting lines of fluorescence data were stacked to create the image. The laser beam started scanning across the cell at a time indicated by the top-most line of the image. C) Measurement of Ca^{2+} (circles) during intense illumination, delivering approximately 10^8 effective photons per second to a spot at the edge of the R-lobe (asterisk in A) The electrical response to light (a depolarization of the membrane potential) is also shown (line). Measurement of Ca^{2+} (circles) at a spot close to the edge of the R-lobe and representative electrical response (line) recorded from another cell following a dim flash, delivering approximately 50 effective photons per second. From references 21 and 23.

flashes, the peak elevation of Ca^{2+} is graded with flash intensity, increasing from 2 µM to more than 140 µM as light intensity increases from 10 effective photons to 10,000 effective photons. Over the same intensity range, the latency of the Ca^{2+} signal drops from 140 to 20 ms.[23] For the dimmest flashes so far investigated, these concentrations represent ~600 free Ca^{2+} ions per effective photon generated within the confocal measurement volume of ~5 μm^3.

Very high levels of cytosolic Ca^{2+} are reached only transiently upon illumination. Following a dim or moderate intensity short light flash, Ca^{2+} falls close to its resting value within 1 s, although a small lingering elevation may persist for up to a minute afterwards (Fig. 2D). Even during sustained intense illumination, Ca^{2+} falls within 5 s to a sustained "plateau" elevation of less than 20 μM.[21] As with the initial transient, the level of Ca^{2+} during sustained illumination is graded with intensity, but with a much-reduced sensitivity, so that a 1000-fold increase in intensity results in only a 3- to 4-fold increase in Ca^{2+} [22] This decline in sensitivity is probably a result of adaptation, and not depletion of Ca^{2+} stores, since release of Ca^{2+} from stores is subject to negative feedback control by the level of cytosolic Ca^{2+} (see below).

In contrast to the detailed knowledge about the amplitude and latency of the Ca^{2+} signal, the spatial spread has only been investigated in detail following intense illumination. Not surprisingly, the light-induced Ca^{2+} signal is initiated at the outer edge of R lobe, beneath the microvillar membrane, and rapidly spreads into the interior of the R lobe.[21] Within 500 ms of the beginning of a bright flash, Ca^{2+} concentrations above 50 μM can be measured 20-30 μm deep into the R lobe, consistent with a diffusion coefficient of 10^{-6}-10^{-5} cm^2/s, a value[24] consistent with $InsP_3$ being the messenger for Ca^{2+} release. In contrast to the rapid spread of very high Ca^{2+} elevations throughout the R lobe, only small and slow light-induced elevations are detected in the A lobe. Confocal images confirm earlier work using aequorin and Ca^{2+}-sensitive electrodes and indicate that the light-induced elevation of Ca^{2+} barely penetrates the A lobe, resulting in increases following intense illumination of less than 1 μM.

Confocal fluorescence microscopy has not so far been applied to the lateral spread of the Ca^{2+} signal across the microvillar membrane. Preliminary measurements, made using the photoprotein aequorin, indicate that elevation of Ca^{2+} is confined to within 20 μm of a bright spot of illumination.[25] If lateral diffusion of the Ca^{2+} signal is similar to the spread into the interior of the R lobe, elevation of Ca^{2+} could spread several microns across the interior surface of the microvillar membrane during the first 10 ms after release. Thus a single photon might generate hundreds of free Ca^{2+} spread over several square microns of membrane surface.

Transient $InsP_3$-Induced Release of Ca^{2+} from the SER

The large transient elevation of cytosolic Ca^{2+} during the first few hundred milliseconds of the light response is unaltered by removal and chelation of extracellular Ca^{2+}, but is severely reduced by agents expected to deplete Ca^{2+} stores. The light-induced elevation of Ca^{2+} therefore arises from the release of Ca^{2+} ions from intracellular stores. The R lobe of the ventral photoreceptor contains a dense meshwork of smooth endoplasmic reticulum (SER), which can accumulate Ca^{2+} ions in an ATP-dependent manner.[26,27] Consistent with a role in light-induced Ca^{2+} release, extensions of SER are tightly juxtaposed to the base of microvilli, within 100 nm.[27] The SER meshwork in the R lobe is continuous with a less dense meshwork in the A lobe. The cytoplasm of both the R lobe and, to a lesser extent, the A lobe are immunoreactive when probed with an anti- $InsP_3R$ antibody (Fig. 3D).[28]

The SER therefore is a store from which $InsP_3$ can release Ca^{2+}. Micro-injections of $InsP_3$, or photolysis of caged $InsP_3$, rapidly release Ca^{2+} from the R lobe in

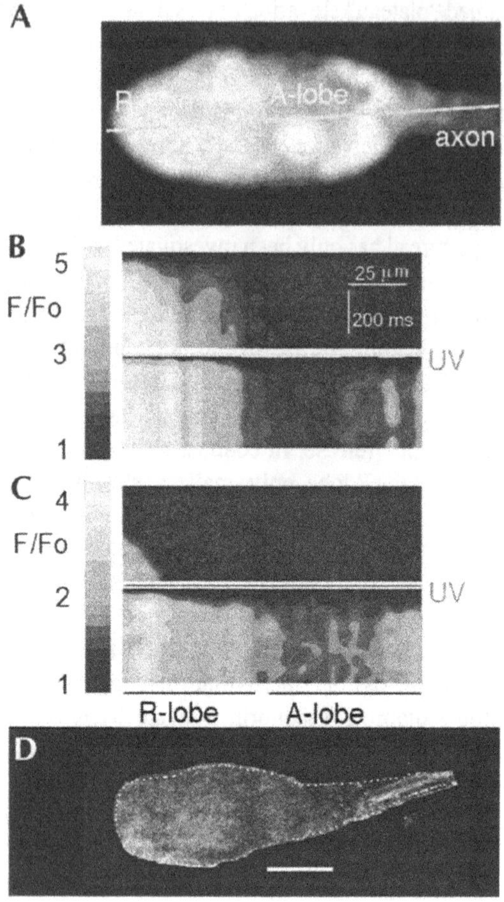

Figure 3. Photolysis of caged $InsP_3$ initiates Ca^{2+} release throughout *Limulus* ventral photoreceptors. (A) Fluorescence micrograph of a ventral photoreceptor loaded with 10 mM GDPβS, 1 mM fluo-3 (fluorescent Ca^{2+}-indicator dye) and 10 mM caged $InsP_3$. The laser beam scanned along the green line every 4 ms and the resulting lines of fluorescence data were stacked to create the images below.
(B) (upper frame) Fluorescence recorded during a line scan using a 488nm laser beam, which excited the photoreceptor through the physiological mechanism. The scan shows the temporal progression of the physiological elevation of Ca^{2+} from the edge of the R-lobe membrane (left hand edge of scans) into the A-lobe and axon (right hand side of scans). The laser beam started scanning across the cell at a time indicated by the first line of the image. (upper edge of frame). (lower frame) A 10 ms flash from a 364 nm UV laser was superimposed at the beginning of the scan. The resulting photolysis of caged $InsP_3$ induced fast Ca^{2+} release throughout the cell.
(C) To exclude any contribution of Ca^{2+} influx, a different cell was kept in darkness for 30 min in artificial sea water containing 1 mM EGTA instead of 10 mM Ca^{2+}. This increased the latency of response to the 488 nm laser via the physiological mechanism (upper frame) but did not alter the latency or pattern of Ca^{2+} elevation following photolysis of $InsP_3$ (lower frame).
(D) Immunofluorescent labeling of another photoreceptor by an anti-$InsP_3$ receptor antibody suggests that the $InsP_3$ receptor is more concentrated in the R-lobe than in the A-lobe, a distribution similar to that of $InsP_3$-induced Ca^{2+} release, as shown in B and C. From references 28 and 29.

darkness (Fig. 3A-C).[15,16] Photolysis of caged $InsP_3$ by UV light delivered to a spot beneath the microvillar membrane results in local elevations of Ca^{2+} that are comparable in magnitude and rate of rise to those elicited by bright visible light (Fig. 4). However, the latency of the Ca^{2+} signal that follows illumination by visible light is ~30 ms longer than that of the response to the release of caged $InsP_3$, the difference being presumably the time required for light to activate the PI pathway.[29] Light-induced Ca^{2+} release is inhibited by heparin, an $InsP_3R$ antagonist[11] and by cyclopiazonic acid (CPA), which depletes ER Ca^{2+} stores by inhibiting the Ca^{2+}-ATP-ase.[21,23,30] There is therefore convincing evidence that the light-induced release of Ca^{2+} from internal stores is mediated by the PI pathway acting on $InsP_3Rs$ in SER that are closely juxtaposed to the microvillar membrane. A number of approaches, including the application of ryanodine and caffeine, have failed to indicate any role for Ca^{2+} release by ryanodine receptors (R. Payne, unpublished observations).

Injection of $InsP_3$ or photolysis of caged $InsP_3$ can also release Ca^{2+} from the A lobe (Fig. 3B and C), but with a much slower time course and smaller peak amplitude, consistent with the reduced $InsP_3R$ immunoreactivity and ER density found there.[28] Thus the inability of light to greatly elevate Ca^{2+} in the A lobe may result from a reduced density of $InsP_3R$ and SER in the A lobe as well as the dilution of $InsP_3$ as it diffuses away from the microvillar membrane.

Feedback Inhibition by Ca^{2+} of $InsP_3$-Sensitive Ca^{2+} Release

A striking feature of both light and $InsP_3$-induced Ca^{2+} release is its transience. Cytosolic Ca^{2+} levels above 20 μM are only sustained for a few seconds. This transience is part of an adaptation mechanism that desensitizes the PI pathway and protects the Ca^{2+} stores from depletion. Prior exposure to bright light desensitizes subsequent light-induced Ca^{2+} release.[31] Adaptation of the PI cascade is not unique to ventral photoreceptors. $InsP_3$-induced Ca^{2+} release is, in many cells, a transient phenomenon. In *Limulus* ventral photoreceptors, as in some other cells, part of the explanation for this transience lies in the inhibition of $InsP_3R$ by micromolar Ca^{2+}, providing negative feedback control.[32,33]

Following an injection of 100 mM $InsP_3$ into a ventral photoreceptor, the large initial transient elevation of Ca^{2+}, lasting less than 1 s, is followed by a small lingering elevation of less than 5 μM which declines over 10-60 s.[34] Immediately following the first injection of $InsP_3$, the cell is completely desensitized to a second injection (Fig. 5). However, as the lingering elevation of Ca^{2+} declines, sensitivity recovers. A pulse injection of a low (1 mM) concentration of the "slow" Ca^{2+} chelator EGTA during this period knocks down the lingering elevation of Ca^{2+} and restores sensitivity to $InsP_3$. In contrast, rapid injection of a pressure-driven pulse of 1 mM Ca^{2+} solution prior to an injection of $InsP_3$ results in desensitization. It therefore appears that the lingering elevation of Ca^{2+} causes desensitization of the $InsP_3$-induced Ca^{2+}-release mechanism. $InsP_3$-induced Ca^{2+} release appeared to be extremely sensitive to Ca^{2+}-2 μM elevations of Ca^{2+} were associated with almost complete inhibition.

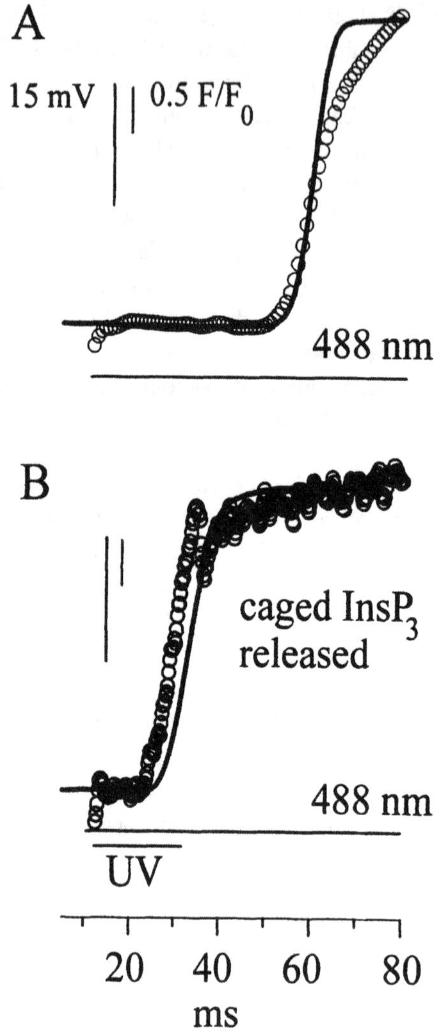

Figure 4. Timing of Ca^{2+} release by caged $InsP_3$. Photoreceptors were loaded with fluo-3 or Ca^{2+} Green-5N, caged $InsP_3$ and GDP-βS. 488nm and UV laser beams were focused onto the edge of the R-lobe. A) Membrane potential (solid line) and fluo-3 fluorescence (dots) recorded during illumination by the 488nm laser. Laser stimulation began at the beginning of the fluorescence trace. B) Effect of superimposing a 20 ms duration, UV flash and so releasing caged $InsP_3$. From reference 29.

The result is surprising in view of the ability of $InsP_3$ and light to elevate Ca^{2+} much beyond the micromolar range. One obvious resolution is a delay in the onset of feedback desensitization. At room temperature, the onset of desensitization is delayed by 30-40 ms after the first injection of $InsP_3$, with desensitization approaching 100% only after 200 ms.[35]

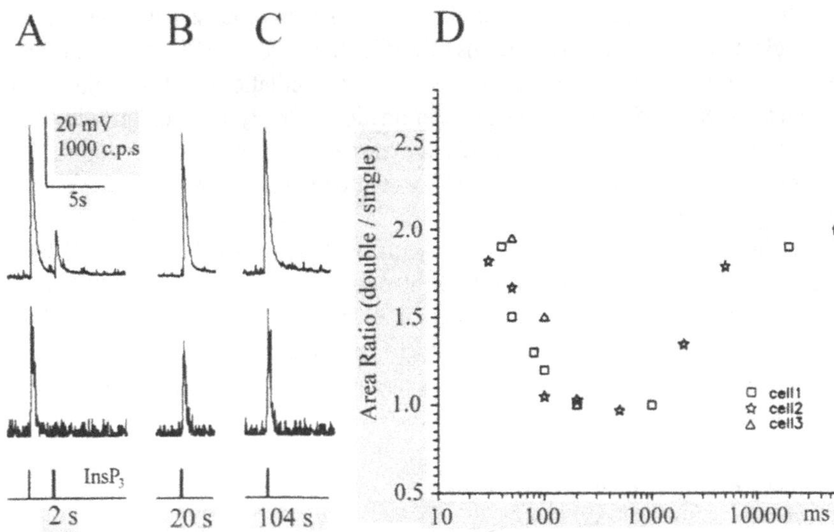

Figure 5. Feedback inhibition of $InsP_3$-induced Ca^{2+} release. A-C) Membrane potential (upper traces) and luminescence of the Ca^{2+}-indicator aequorin (middle traces) indicating Ca^{2+} release and depolarization following pulsed pressure injection of 100 μM $InsP_3$ into the R-lobe of a ventral photoreceptor (lower traces). A) Responses to a pair of $InsP_3$ injections delivered 2 s apart. The electrical response to the second injection is much diminished, while the aequorin signal is undetectable. B-C) The response to a second injection recovers as the interval between injections is increased to 20 and 104 s. D) Timing of the onset and recovery of feedback inhibition. The area under the electrical response to two successive injections of $InsP_3$, relative to the area under the response to a single injection, is plotted as a function of the interval between the injections. After a delay of approximately 30 ms, the ratio falls to 1 indicating a delayed total desensitization of the response to the second injection. Recovery takes approximately 1 minute. From reference 35.

Facilitation of $InsP_3$-Sensitive Ca^{2+} Release by Positive Feedback

In other cells it has been directly shown that feedback inhibition of $InsP_3R$ by Ca^{2+} is just one aspect of regulation. There is also a facilitative effect at low concentrations of Ca^{2+}, creating a "bell-shaped" dependency of $InsP_3$-induced Ca^{2+} release on Ca^{2+} concentration.[36] An analysis of the dynamics of responses to $InsP_3$ analogs has provided evidence for a positive feedback mechanism that governs $InsP_3$-induced Ca^{2+} release in *Limulus* photoreceptors.[37] Slow, sustained injections of poorly hydrolysable $InsP_3$ analogs result in dramatic bursts of Ca^{2+} release after a long latency that can only be modeled by positive as well as negative feedback control of the $InsP_3R$ by released Ca^{2+}. In addition, the model $InsP_3R$ requires co-operative binding of $InsP_3$ and Ca^{2+}. The same mechanism may underlie facilitation of responses to dim light and injections of $InsP_3$ by sub-threshold backgrounds levels of poorly hydrolysable $InsP_3$ analogs.

The bell-shaped dependency on Ca^{2+} has also been proposed as a mechanism that might create sustained oscillations of Ca^{2+} activated by the PI pathway in some cells (review ref. 38). In *Limulus* photoreceptors, oscillations of Ca^{2+} release are only rarely seen following very bright illumination or $InsP_3$ injection (provided the $InsP_3$ is purely the 1,4,5 isomer). $InsP_3$ is expected to be hydrolysed within a few seconds of its production,[39] so that $InsP_3$ would be gone before sensitivity returns following the initial transient release of Ca^{2+}. However, injection of poorly hydrolysable $InsP_3$ analogs does result in oscillatory bursts of Ca^{2+} release.[33,40] Heavy metal ions, which inhibit $InsP_3$ hydrolysis, also induce oscillatory responses to InsP3 injection and light flashes.[40] Regulation of the $InsP_3R$ by Ca^{2+} is therefore likely to involve cooperatively, positive feedback and negative feedback. This complexity may account for the latency, the explosive rise and occasional oscillations of light-induced Ca^{2+} release.

Ca^{2+} Buffering and Sequestration

The Ca^{2+} buffering capacity of the cytosol has been estimated by the method of stimulating reverse sodium Ca^{2+} exchange and then comparing influx of Ca^{2+} determined from Na^+-Ca^{2+} exchange currents with the elevation of cytosolic Ca^{2+} under the same conditions. 99.6 % of Ca^{2+} ions are buffered or sequestered.[41] The relevance of this number for the light response is unclear since it represents a spatially averaged, steady-state analysis of events throughout the cell. The times taken to sequester or buffer transient elevations of Ca^{2+} in the spaces beneath the microvillar membrane are unknown. The molecules responsible for Ca^{2+} buffering are also unclear. Calmodulin is present at high levels at the microvillar membrane[42] and may be expected to locally buffer intracellular Ca^{2+}.

Sequestration of Ca^{2+} by the SER allows it to act as a sink as well as a source of cytosolic free Ca^{2+}. The Ca^{2+} ATP-ase in the SER membrane recovers released Ca^{2+} from the cytosol. It takes hundreds of bright flashes to completely deplete SER Ca^{2+} stores in the virtual absence of extracellular Ca^{2+}, a testimonial to the efficiency of this re-uptake, the large amount of stored Ca^{2+} and the relative impermeability of the plasma membrane to Ca^{2+}.[10] An inhibitor of Ca^{2+}-ATP-ases in the ER, CPA, greatly accelerates store depletion, resulting in loss of detectable light-induced Ca^{2+} release within 30-40 minutes, even in darkness.[21]

Mitochondria are also prominent in R lobe. While there is no direct evidence for mitochondrial Ca^{2+} uptake, Ca^{2+} released by light is able to trigger a burst of mitochondrial oxidation,[43] so that some Ca^{2+} must presumably enter the mitochondria. However, the significance of mitochondrial uptake for regulating cytosolic Ca^{2+} is unknown.

The degree to which these buffering and re-sequestration mechanisms, might be responsible for restoring cytosolic Ca^{2+} to normal after a single flash of light has been investigated. Following a flash, Ca^{2+}, monitored from the entire R lobe using the indicator dye arsenazo III, showed fast and slow declining phases of approximately equal amplitude.[44] The fast phase was complete within 2 s and was not dependent on cell membrane potential, consistent with the buffering of released

Ca^{2+}, re-sequestration into ER or uptake by mitochondria. The slow phase, which lasted for tens of seconds, could be reversed at positive membrane potentials, consistent with extrusion of Ca^{2+} from the cell by electrogenic forward Na^+-Ca^{2+} exchange (see below, ref. 44).

Sustained Elevation of Ca^{2+} is a Balance Between Light-Activated Influx of Extracellular Ca^{2+} Ions and Efflux via the Na^+/Ca^{2+} Exchanger

During the first few seconds of the light response, removal of extracellular Ca^{2+} has very little effect on the light response.[45] Negative feedback control of Ca^{2+} release will, in the short term, tend to stabilize Ca^{2+} levels at a value determined by the concentration of $InsP_3$ (and hence the photon flux), compensating for reduced Ca^{2+} influx by increasing Ca^{2+} release. However, release of Ca^{2+} from the SER cannot determine cytosolic Ca^{2+} in the long term because the plasma membrane contains pumps and exchangers that remove Ca^{2+}. Steady state levels of Ca^{2+} can only be the result of a balance between Ca^{2+} influx and efflux across the plasma membrane. A steady elevation of Ca^{2+} that can be maintained for hours in the light-adapted state requires influx of Ca^{2+} from the extracellular space. Ca^{2+} may enter the cell either by Ca^{2+} permeation through light-sensitive channels or through voltage-activated Ca^{2+} channels.[46] In an effort to determine Ca^{2+} influx through light-sensitive channels, Hsiao and Payne,[47] used Mn^{2+} as a surrogate for extracellular Ca^{2+}, on the assumption that it would permeate non-selective cation channels as well as Ca^{2+}. Influx of Mn^{2+} was measured from its ability to quench the dye fura-2. A Mn^{2+} influx rate of 0.09-0.9 μM/s was observed during a 10 s flash of light, through a light-activated, but not a voltage-activated, pathway. A comparable influx of Ca^{2+} would be, as expected, insignificant compared to the initial rate of light-induced Ca^{2+} release from stores. However, this influx is sufficient to compensate for the loss of Ca^{2+} due to Na^+/Ca^{2+} exchange activity during prolonged illumination, estimated to reach a maximum rate of approx. 1 μM/s.[41] The permeability of Ca^{2+} through the light-dependent channels is relatively small; the bulk of the light-dependent current is carried by Na^+.[48] Low permeability may limit the role in light adaptation of entering Ca^{2+}. This contrasts sharply with the high selectivity for divalent over monovalent cations of the light-dependent current in *Drosophila*.[49] In these cells external $[Ca^{2+}]$ has the major role in both light-induced elevation of cytosolic Ca^{2+} and in light adaptation (review ref. 3).

The major mechanism that has been identified for Ca^{2+} efflux is Na^+/Ca^{2+} exchange across the plasma membrane. Forward, electrogenic, Na^+/Ca^{2+} exchange appears to be critical in maintaining cytosolic Ca^{2+}. Reducing or reversing Na^+/Ca^{2+} exchange, by reducing extracellular sodium or by depolarization results in an elevation of Ca^{2+} in both light- and dark-adapted cells.[41,44] The resting level of Ca^{2+} measured using fluorescent indicators, 200-600 nM, is in the range calculated to be maintained by the exchanger at equilibrium.[41] The localization of Na^+/Ca^{2+} exchanger within the cell and whether it is restricted to the R lobe, is not known.

ROLE OF Ca^{2+} IN EXCITATION

In this section we summarize the growing evidence that the light-induced elevation of Ca^{2+} described in the previous section is necessary and sufficient for excitation. The general criteria for making conclusions of this kind are clear: the Ca^{2+} elevation must precede the activation of the light-induced conductance, the conductance-increase must be mimicked by artificially elevating Ca^{2+} and the light-induced conductance should be blocked when the light-induced elevation of Ca^{2+} is blocked. Generally, the evidence indicates that these requirements are met, but there remains some ambiguity with regard to one issue: it is possible that in addition to a light-dependent conductance that requires Ca^{2+} elevation there may be a second that does not.

Ca^{2+} Elevation Activates the Light-Dependent Conductance

Early experiments involving the slow iontophoretic injection of Ca^{2+} indicated that Ca^{2+} could dramatically adapt the cell,[7] but did not reveal a strong component of excitation. Evidence for this excitatory component came when brief pulses of solutions were injected. Pulsed pressure injections of solutions containing 1-2 mM Ca^{2+} into the R lobe of ventral photoreceptors activated up to 20 nA of a cation conductance in the plasma membrane with a similar reversal potential, sodium dependence and outward rectification to that activated by light (Fig. 6A).[9] Release of Ca^{2+} ions via $InsP_3$ activates the same conductance, indicating that concentrations of Ca^{2+} generated via the endogenous Ca^{2+} release pathway are sufficient.[13,14] The current, typically 5-20 nA in amplitude following a pulse of 100 μM $InsP_3$, generates a transient depolarization of the photoreceptor lasting for less than 1 s. The coupling between the elevation of Ca^{2+} and the depolarization of the photoreceptor is rapid. Depolarization follows $InsP_3$-induced Ca^{2+} release after 2.5 ± 3.3 ms, while photolysis of caged Ca^{2+} (o-nitrophenyl EGTA) at the edge of the R lobe activates current within 1.8 ± 0.7 ms (Fig. 6B).[29]

A quite different method for elevating Ca^{2+} was by injection of buffered Ca^{2+} solutions. These have the advantage of giving an indication of the quantitative levels of Ca^{2+} required. One approach was to load cells with mM levels of Ca^{2+} buffer to "set" varying levels of free Ca^{2+}.[12] When a 100 or 200 mM 5,5'-dibromo-BAPTA solution with free Ca^{2+} set to resting levels (0.5 μM), was injected into *Limulus* ventral photoreceptors there was no effect on resting membrane properties. That intracellular Ca^{2+} levels were being buffered could be seen in a progressive decrease of light sensitivity and slowing of the residual light response as dibromo-BAPTA levels were increased in the cell. Free Ca^{2+} levels during sustained illumination have been measured using Ca^{2+}-selective electrodes and indicator dyes as being between 3 and 30 μM.[50] When cell Ca^{2+} was buffered to 5 μM, a sustained inward current of approximately 1 nA was observed.[12] The amplitude of this current was comparable to that produced by a low-to-moderate intensity adapting background light. That free Ca^{2+} had been set at 5 μM could be inferred from several criteria. The first injections each rapidly induced inward current increases of a

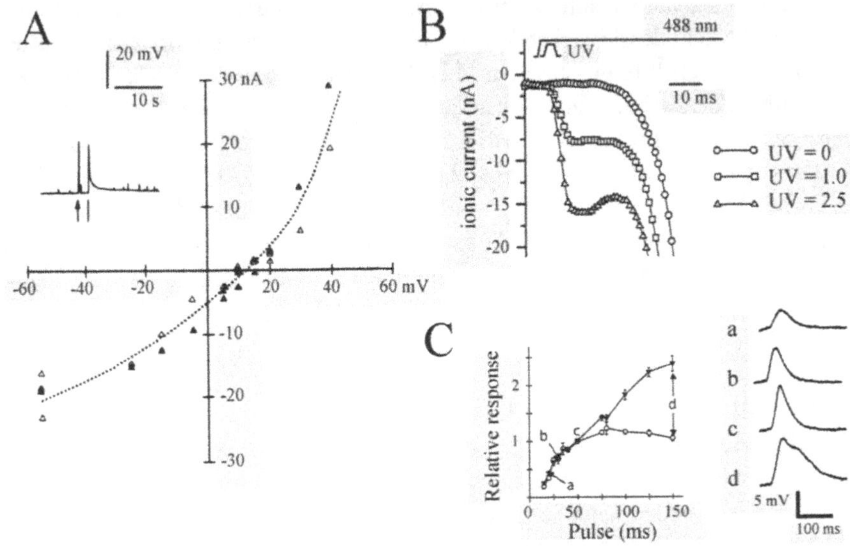

Figure 6. Characteristics of current induced by elevation of Ca^{2+} in ventral photoreceptors by various means. (A) Inset. Membrane depolarization elicited by light flashes (arrow) and a pulsed- pressure injection of 2 mM calcium aspartate into a ventral photoreceptor. Graph. Voltage-clamp currents underlying the electrical responses to light (solid symbols) and Ca^{2+}-injections (open symbols) are plotted as a function of membrane potential. Both currents show a similar dependence on membrane potential and reverse between +10 and +20 mV. From reference 9. B) Voltage-clamp currents elicited by photolysis of caged Ca^{2+} (o-nitrophenyl EGTA). Ca^{2+} was released by flashes from a UV laser superimposed upon a step of light from a 488 nm laser, with UV laser intensities as indicated. Both lasers were focused on a spot at the edge of the R-lobe of a ventral photoreceptor. The UV laser flashes elicited an early inward current (triangles and squares), that is graded with UV intensity and begins within 3 ms of the onset of the UV flash, clearly distinguishable from the later physiological response to the 488 nm light alone (circles). For details see reference 29. C) Membrane depolarizations induced by intracellular injection of Ca^{2+}/HEDTA buffer solution. Response peak amplitude (open circles) and areas (solid triangles) are plotted against pressure pulse length. **a-d** match amplitude and pulse duration (left) to voltage trace (right). Bars represent the standard error of the measurements. Responses were normalized by dividing response amplitude and area by those of concurrent 50 ms pressure pulse responses. From reference 51.

fraction of a nA. Eventually additional injections had little additional effect consistent with reaching the free Ca^{2+} concentration set by the buffer. The effects of elevated Ca^{2+} levels on dark membrane properties and the kinetics of a flash response were stable for at least 30' after injection. If Ca^{2+} is an intracellular second messenger for excitation, the response should increase in parallel with Ca^{2+} concentration. This was tested using dibromo-BAPTA buffered to 45 μM free Ca, a level representative of peak Ca^{2+} increases measured in response to bright light. With this high Ca^{2+} solution only a few injections were need to produce a sustained inward current of 1 nA. Further injections increased the inward current proportionally upwards of 5 nA. At about this level, a rapid and irreversible increase of current to as much as 100 nA ended the experiments.

A complementary result can be obtained by causing transient, focal elevations of Ca^{2+} in the light-sensitive rhabdomeric lobe using low concentrations of the buffer HEDTA (Fig. 6C).[51] In this case 2 mM HEDTA was used with free Ca^{2+} estimated at 8.5 μM. Focal stimulation was not only a result of diffusion from the site of injection, but also due to loss of buffering by HEDTA. HEDTA is poorly selective for Ca^{2+} over Mg^{2+} and inside the cell bound Ca^{2+} should be replaced with endogenous Mg^{2+}. With the Ca^{2+}/HEDTA buffered solution the peak depolarization increased to a limit near 10 mV over approximately a three-fold range in pressure pulse duration (typically 25 to 75 ms with the electrodes used). This limit likely reflected the local establishment of free Ca^{2+} near the 8 μM level set by the electrode solution.

Light-Induced Ca^{2+} Release Can Be Detected Before the Electrical Response

The above experiments indicate that micromolar concentrations of Ca^{2+}, released from internal stores by $InsP_3$, can activate a large inward current through the plasma membrane within a few milliseconds. It follows that if light-induced Ca^{2+} release is to act similarly, then a component of the photocurrent must be initiated a few milliseconds after Ca^{2+} is released. Early measurements of Ca^{2+} changes did not support this relationship—no Ca^{2+} release could be observed prior to the electrical response—indeed delays of 10-25 ms being reported.[6,52,53] There are several possible explanations for this failure. The slowness of some Ca^{2+} indicators and the use of optical measurements that average the Ca^{2+} concentration in the entire photoreceptor may have diminished the detection of Ca^{2+} release initiated directly beneath the plasma membrane. The technical problems have recently been reduced by the use of confocal microscopy to both excite ventral photoreceptors at a spot as close as possible to their microvillar membrane and to measure Ca^{2+} release using fluorescent Ca^{2+}-indicator dyes at that spot.[22,23,29]

At all light intensities, the Ca^{2+} signal detected by confocal fluorescence microscopy is highly correlated with that of the electrical response (Fig. 7).[23] The time for Ca^{2+} to exceed 2 μM is approximately equal to that for the receptor potential to exceed 8 mV (mean difference; 2.2 ± 6.4 ms). This correlation is unaltered when successive responses to the same intensity exhibit stochastic variation in their latency of up to 20 ms, indicating that a shared stochastic process generates both signals. Removal of extracellular Ca^{2+} has no effect on this correlation, eliminating the possibility that it arises from influx of Ca^{2+} ions through the light-sensitive conductance. Given the evidence above, it is more plausible that the correlation arises from the initiation of inward current by Ca^{2+} released from intracellular stores. In this case, one should expect to be able to detect the Ca^{2+} signal first. The question of "which event occurs first" is difficult to address, since two signals with different noise levels are compared, but two approaches have been adopted. The first is to define a detection criterion of 2 standard deviations above the initial mean value of each signal. By this criterion, the Ca^{2+} signal led the electrical response by up to 5

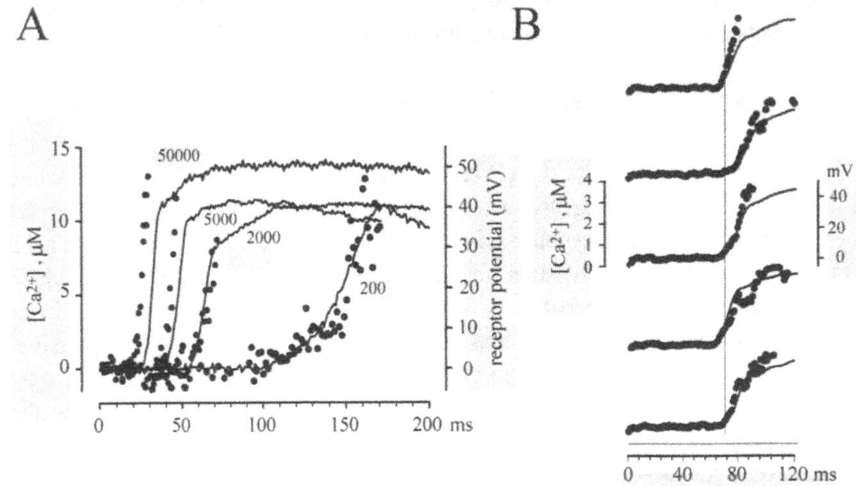

Figure 7. Tight correlation between the time of onset of confocally-measured Ca^{2+} signals and the electrical response of ventral photoreceptors. A) Rising phases of receptor potentials (solid lines) and reconstructed elevations of Ca^{2+} (symbols) recorded from cells following flashes that delivered 50000, 5000, 2000 and 200 effective photons. The flashes were delivered at the origin of the time axis. B) Receptor potentials (lines) and estimated elevations of Ca^{2+} (symbols) recorded from a photoreceptor filled with fluo-4 during 5 successive continuous steps of illumination (bar below traces). The latencies of both the Ca^{2+} signals and the electrical responses varied to the same extent from flash-to-flash, either side of the vertical time marker. From reference 23.

ms in recordings from spots within about 1/3 of cells examined. The lag in other cells could indicate the presence of an early Ca^{2+}-independent component of the response present in some cells, but it is difficult to be sure of this. The possibility exists that these lags result from the time for released Ca^{2+} to diffuse to measuring spots. In any case, the Ca^{2+} signal only ever lagged the electrical response by a few milliseconds. Given the rapidity with which Ca^{2+} can elicit an inward currrent (1-3 ms; see above) and the 50 ms rise-time of the photocurrent, this timing appears to be sufficient for released Ca^{2+} ions to contribute to the activation of the photocurrent during the rising edge of the response to light.

The second approach is to develop a kinetic model that predicts the photocurrent from the Ca^{2+} signal.[23] Variation in the latency of light responses makes it difficult to define an accurate relationship, but normalizing the time-base for each response to the time at which photocurrent is first detected allows the approximate fitting of a model to mean data from several cells. A first-order, linear process with a sensitivity of inward current to cytosolic Ca^{2+} of 1.75 $nA.\mu M^{-1}$ and a time constant of 5 ms could transform the resulting Ca^{2+} signals into photocurrent. This model provides a first glimpse of the mechanism that might link Ca^{2+} to the activation of channels in the plasma membrane. The process is linear, relatively insensitive to Ca^{2+}-mediated adaptation, and delays the electrical response relative to cytosolic

Ca^{2+} by a few milliseconds. The linearity and delay are consistent with the results of experiments on Ca^{2+}-activated inward currents, summarized above.

Ca^{2+} Elevation is Required to Produce the Normal Light Response

The first indication that Ca^{2+} elevation might be necessary for excitation came from experiments involving prolonged removal of extracellular Ca^{2+}.[10,50] Under these conditions, the light response initially gets larger. Moreover, if cells are placed in zero Ca^{2+} and not illuminated for long periods, they retain their ability to produce large responses when illumination is finally given. However, if illumination is repeatedly given under these conditions, the cells gradually loose their responsiveness altogether. Responsiveness can be restored by reintroduction of Ca^{2+}. Direct measurements of the light induced Ca^{2+} elevation show that the transition to the unresponsive state is accompanied by a loss of the light-induced Ca^{2+} elevation.[10] These results suggest that if Ca^{2+} is repeatedly released by light from intracellular stores, some of this Ca^{2+} is extruded out of the cell and therefore does not refill intracellular stores. Normally, this extrusion would be compensated by entry of Ca^{2+} from the extracellular space through voltage-dependent Ca^{2+} channels [54] and the light-dependent conductance.[45,47,55] Because this does not occur if there is no external Ca^{2+}, the internal stores eventually become depleted. When this depletion is sufficient to prevent light-induced release, excitation of the cells is greatly reduced (Fig. 8).[21,23] As mentioned before, blocking light-induced release by other means (heparin)[11,56] also greatly reduces excitation.

More direct evidence for a role of Ca^{2+} in excitation came from experiments using Ca^{2+} buffers and it was these experiments that led to the proposal that the light-induced Ca^{2+} elevation is a required component of the excitation cascade.[8,11,12] Upon cumulative injection of Ca^{2+} buffers, the responses to dim or moderate intensity lights are progressively reduced. This reduction can be large, reaching over 100-fold. The main uncertainty regarding these experiments comes from the interpretation the residual responses when moderate or bright intensity stimuli are used. It is important here to clarify what is meant by the terms "moderate" or "bright," because the quantitative considerations are quite unlike other types of experiments (e.g., pharmacological) where dose is varied. In pharmacological experiments a high dose might mean 10 times the IC_{50}. In *Limulus* photoreceptors, the cells are responsive over such a wide range that a moderate light might be 10,000 times brighter than a dim one and a bright light could be 100 million times brighter than a dim one. Because the non-saturated response to brief stimuli varies nearly linearly with intensity, this extremely broad range of input introduces interpretational problems. Suppose that a drug is introduced that blocks a transduction step by 99%, an effect that would be considered highly robust in most fields. However in experiments on *Limulus* it is not unusual to simply raise the intensity 100 fold, a change that can largely restore the normal response. Indeed a compound that greatly reduces the response to dim lights might produce no reduction at all to a saturating intensity light sufficient to completely open the light-dependent conductance. If the light

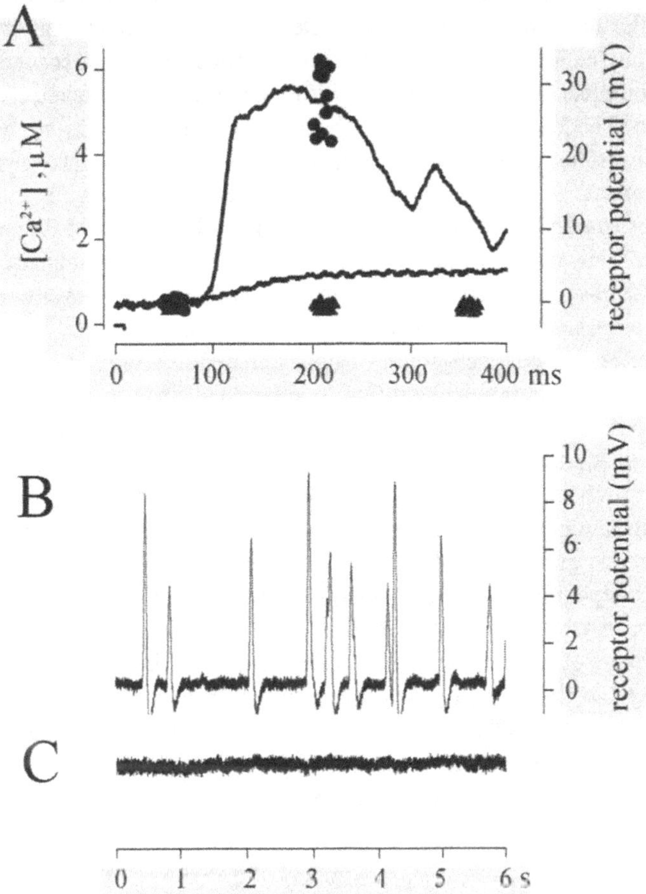

Figure 8. Depletion of Ca^{2+} stores by treatment with cyclopiazonic acid greatly desensitizes ventral photoreceptors and eliminates detection of single quantal events. A) Representative receptor potentials (solid lines) and elevations of Ca^{2+} (symbols) recorded from a photoreceptor following a flash that delivered ~100 effective photons, when the cell was bathed for 10 min in 0 Ca^{2+}-ASW (circles) and 30 min after switching the bathing solution to 0 Ca^{2+}-ASW containing 100 µM cyclopiazonic acid (CPA; triangles). B) Single photon signals (bumps), recorded during dim continuous illumination, when the cell was bathed for 10 min in 0 Ca^{2+}-ASW and (C) 30 min after switching the bathing solution to 0 Ca^{2+}-ASW containing 100 µM CPA. From reference 23.

intensity is more than 100-fold over the intensity sufficient for saturation, this 99% effective inhibitor might produce no observable effects on membrane current. For this reason, conclusions should be based on how a substance affects the response to dim stimuli and the lack of effect on the response to bright light must be interpreted with caution.

Shin et al[12] specifically studied the large residual response that could be observed after injecting Ca^{2+} buffer and after raising the light intensity to moderate levels. They concluded that this "residual" response could be accounted for by a lack of complete buffering rather than by a second Ca^{2+}-independent component of excitation. The key experiment was to monitor the effects of further buffer injection. If the residual response was Ca^{2+}-independent, these additional injections should have no effect. On the other hand, if the residual response was due to incomplete Ca^{2+} stabilization by the first injection, additional buffer should further slow and reduce Ca^{2+} elevation and thereby decrease excitation. Such a cumulative reduction was in fact observed; there was not sign of a component of the response that could not be progressively reduced by increasing buffer concentration.

Other pharmacological studies have attempted to address the problem of the "residual" response by blocking Ca^{2+} release rather than by buffering it. The PI pathway has been interrupted by inhibitors of PLC,[5, 56-58] by inhibitors of the $InsP_3R$[11] and by depletion of Ca^{2+} stores.[21,23] All of these interventions desensitized the response to flashes of light 40- to 1000-fold, eliminating the detection of single photon signals, but none succeeded in eliminating the light response. In one study, Ca^{2+} elevation was monitored within a few micrometers of the microvillar membrane following depletion of Ca^{2+} stores by the Ca^{2+}-ATP-ase inhibitor CPA.[21] Following depletion, very bright, prolonged flashes were still capable of generating saturating photocurrents of several hundred nA, albeit with slow rise times, in the absence of any detectable elevation of $[Ca^{2+}]$. These studies prevent the universal acceptance of the view that Ca^{2+} release exclusively couples the PI pathway to the activation of channels in the plasma membrane. It still seems possible that a slow pathway of excitation remains in the absence of detectable light-induced elevation of Ca^{2+}. A definitive proof of this pathway and its interaction with released Ca^{2+} awaits a clearer understanding of the ion channels mediating each component of the response.

Molecular Basis for the Late Stages in Transduction by Which Ca^{2+} Activates the Light-Dependent Conductance

The light-dependent channels responsible for generation of the light response have been characterized in cell-attached patches.[59,60] They have a reversal potential near zero, a maximal open state conductance of 40 pS and various subconductance states. Their mean open time at resting potential is in the millisecond range, but can be considerably larger at positive voltages. The finding that similar channels can be activated by the application of cGMP to the cytoplasmic side of excised patches is one of the strongest lines of evidence in favor of the hypothesis that a rise in cGMP activates the light-dependent conductance.[61]

Additional lines of evidence support a cyclic nucleotide-gated photocurrent. The cyclic-nucleotide gated channel blocker l-*cis*-diltiazem was indeed found to decrease the light-dependent conductance.[62] Further consistent with this hypothesis is the finding that the conductance can be activated by cGMP injection into living cells[63,64] or by inhibiting phosphodiesterase.[65,66] Recent evidence strengthens this hypothesis by showing that an antibody to a putative cGMP channel clone from the *Limulus* genome specifically labels the rhabdom.[67] Furthermore, inhibiting guanylate cyclase with a variety of agents greatly reduces the light response.[68] The cyclase involved appears to be NO-insensitive, suggesting a particulate rather than soluble form.

The major difficulty with the cGMP hypothesis has been the inability to measure a reproducible elevation in cGMP during light.[65,69-71] However, such changes were difficult to measure even in vertebrate photoreceptors due to the large background of cGMP bound to proteins and contained in non-transducing regions.[72] Thus, the failure to detect light-induced cGMP elevation in invertebrate photoreceptors cannot be considered disproof of the cGMP hypothesis.

The available data places strong constraints on the possible mechanisms involved in the late stages of transduction. First, in the experiments on excised patches, solutions containing high Ca^{2+} were unable to activate channels whereas application of cGMP to the same patches did produce channel activation.[61] This makes it unlikely that the light dependent channels are Ca^{2+} channels directly activated by the light-induced rise in Ca^{2+}. Given the recent result that the light response was inhibited by guanylate cyclase inhibitors,[68] a strong working hypothesis is that Limulus photoreceptors contain a Ca^{2+}-activated guanylate cyclase. Evidence for regulation of cyclases by Ca^{2+} have been obtained in vertebrate rods[73-76] and ciliates,[77,78] in neither of these cases is the Ca^{2+} detector incorporated directly into the cyclase, but is rather a member of the EF-hand super-family of Ca^{2+} binding proteins. The study of guanylate cyclase and its activation is a promising area for further research.

According to current models, the light-dependent current in *Drosophila* flows through some combination of three related ion channel proteins, TRP, TRPL, and TRPγ (for a recent review see ref. 3). Evolutionary parsimony based on morphology and kinship lead to the inference that these channel proteins might underlie the light-dependent current in *Limulus*. A definitive refutation of this possibility is not possible sans reconstitution studies of the light-dependent conductance in *Limulus*. However the body of positive evidence for a cGMP-gated conductance is unchallenged by evidence for a member of the TRP family of ion channels. As an example, diacylglycerol[79] and polyunsaturated fatty acids[80] activate the *Drosophila* TRP and TRPL conductances. None of these compounds induced depolarization of *Limulus* photoreceptors.[81]

INTRACELLULAR Ca^{2+} ELEVATION PRODUCES LIGHT ADAPTATION

In this section we review the evidence that the light-induced elevation of Ca^{2+} is necessary and sufficient to produce light adaptation of *Limulus* photoreceptors. The first evidence that intracellular Ca^{2+} might powerfully down regulate the photoresponse came from experiments in which the Na^+ pump was inhibited.[82] Inhibition led to a powerful reduction in the light response that was far greater in magnitude than could be expected to result from the moderate elevation in intracellular Na^+. Importantly, the reduction in the response did not occur when if extracellular Ca^{2+} was removed. Our current understanding of this is that Ca^{2+} removal across the plasma membrane requires a large Na^+ gradient to drive Na^+/Ca^{2+} exchange; if this gradient is even partially reduced, the level of intracellular Ca^{2+} rises.[41] It is this elevation of Ca^{2+} that dramatically reduces the light response.

The clear evidence that elevation of intracellular Ca^{2+} can powerful adapt the cell has come from several different methods used to raise intracellular Ca^{2+}. Interestingly, these methods raised Ca^{2+} slowly. Perhaps for this reason that the most dramatic effect of Ca^{2+} elevation is adaptation, rather than the excitation that dominates when Ca^{2+} is rapidly injected (see above). The first method used to elevate Ca^{2+} was slow iontophoretic injection from an intracellular electrode.[7] This produced a progressive desensitization of the light response that recovered over a time course of minutes after the end of the injection. Further evidence for the role of Ca^{2+} came from the pressure injection of solutions containing 1-2 mM Ca^{2+} into ventral photoreceptors.[9] Besides inducing membrane depolarization, these injections caused an immediate desensitization to light. As in the case of ionotophoretic injection of Ca^{2+}, the light response recovered over time.

The quantitative relationship between Ca^{2+} and light-adaptation was studied by Levy and Fein[50] using Ca^{2+}-sensitive microelectrodes. They found that steady-state Ca^{2+} levels increased monotonically with the log intensity of a sustained adapting light. While the absolute levels of resting Ca^{2+} and the light-dependent increase varied somewhat between cells, a 2-fold decrease in light sensitivity was roughly correlated with a 200-400 nM increase in Ca^{2+} levels.

Ca^{2+} Elevation is Necessary for Light Adaptation

To test the hypothesis that elevation of Ca^{2+} produces light adaptation, Lisman and Brown[8] injected the Ca^{2+} buffer EGTA into the cytoplasm of photoreceptors. Under control conditions, the adapting light produced a large initial transient response followed by a smaller maintained "plateau". After the adapting light was turned off, the response to the subsequent test flash was greatly reduced as the intensity of the adapting light was increased, indicating that light adaptation had occurred. When this experiment was repeated after buffer injection, the response to the adapting light no longer made the transition from transient to plateau and the response to the subsequent test flash was relatively unaffected by the adapting light intensity. The interpretation given these findings was that both the transition from the transient to plateau amplitudes and the reduction of the subsequent test flash were indicative of adaptation. By both measures, adaptation was blocked by Ca^{2+} buffer.

How Does Ca^{2+} Both Excite and Adapt the Cell?—Delayed Negative Feedback

The ability of Ca^{2+} to both excite and adapt the cell is not unexpected. One of the earliest quantitative model of invertebrate phototransduction[83] proposed that an element that mediates excitation also feeds back to reduce the gain of the visual cascade. The surprising concept that emerges from the above studies is that it seems to require a higher concentration of Ca^{2+} to excite the cell than to adapt it. The curious situation existed in which a slow increase in Ca^{2+} by ionotophoretic injection was thought to lead to adaptation, but not excitation; while a light-induced rapid increase in Ca^{2+}, albeit in the presence of EGTA, led to excitation, but not adaptation. The explanation for such results follows from consideration of the spatial spread of excitation and adaptation processes. *Limulus* photoreceptors begin to adapt when less than 100 photons/s are effective over the approximately 10^6 microvilli in the R lobe.[84,85] Assuming tight packing of the 0.07 um diameter microvilli,[20] Ca^{2+} concentrations that are effective in adapting the cell must therefore spread over the bases of 10^4 microvilli or about 4 μm from the site of photoisomerization and will therefore be subject to buffering by the cell's cytoplasm. The adaptation mechanism must therefore be extremely sensitive to Ca^{2+}. However, excitation can be confined to the site of Ca^{2+} release and requires only that Ca^{2+} released from the SER cross the less than 100 nm gap between the SER and the microvilli.[20,86] High local Ca^{2+} concentrations can be generated in this space while $InsP_3R$ channels are open, so that excitation need not be as sensitive to Ca^{2+}. Ionophoretic injection of Ca^{2+} succeeded in inducing light adaptation by producing a low, global increase in Ca^{2+}. EGTA failed to block excitation in *Limulus* photoreceptors due to the slow binding of Ca^{2+} by EGTA. Adler et al [87] demonstrated that EGTA, but not the faster buffer 5,5'-dibromo-BAPTA, was inadequate to prevent Ca^{2+}-dependent transmitter release at the squid giant synapse. More recently Naraghi and Neher[88] have modeled the effects of EGTA on Ca^{2+} gradients in the vicinity of a Ca^{2+} channel and propose that EGTA has little effect on free Ca^{2+} concentration within 100 nm of the channel mouth. The Ca^{2+} elevations necessary for light adaptation in *Limulus* photoreceptors spread 40 times farther from the channel pore and are effectively blocked by EGTA.

Possible Sites for Feedback Inhibition by Ca^{2+} of the Enzymes of the PI Cascade

Adaptation can be modeled as a negative feedback process[89] delayed by approximately 200 ms from the onset of excitation. This process can be seen in the time course of the transient to sustained levels of response to a step increase in light intensity. A major site of adaptation is probably desensitization caused by feedback inhibition of $InsP_3$-induced Ca^{2+} release, described above. As regards other possible sites of adaptation, responses to successive injections of Ca^{2+} do not show the dramatic desensitization of the resulting inward current or depolarization that accompanies successive injections of $InsP_3$.[15,51] Therefore there does not seem to be a major site of adaptation downstream of Ca^{2+} release. Upstream of Ca^{2+} release, there are several possible sites for feedback inhibition and it seems logical that cells would also reduce $InsP_3$ production as they desensitize the Ca^{2+} release mechanism.

The reactions that terminate the PI cascade have been extensively investigated in *Drosophila* photoreceptors. Mutants lacking the Ca^{2+}-binding protein, calmodulin,[90,91] as well as those lacking an eye specific protein kinase C,[92,93] are defective in inactivation of the cascade. Since both of these proteins are activated by Ca^{2+}, they represent possible sites for feedback adaptation during the light response.

In *Limulus* photoreceptors, calmodulin is clearly present and localized at the microvillar membrane,[42] but little progress has been made with understanding its function. Calmodulin-binding peptides desensitize *Limulus* photoreceptors,[51] but may act directly by binding to a calmodulin-like domain of PLC, rather than by interfering with the interaction of endogenous calmodulin with a protein in the PI pathway. Nevertheless, the presence of this domain hints at a possible involvement of calmodulin in PLC regulation. Light adaptation also leads to a decrease in the lifetime of rhodopsin activity.[94] An attractive possibility is the calmodulin-reguated phosphorylation of arrestin, the protein whose binding is the decisive event in arresting rhodopsin activity. Arrestin phosphorylation in *Limulus* photoreceptors is mediated by CaM-kinase II[95] and the calmodulin-dependent phosphatase, calcineurin.[96] Calcineurin inhibitors desensitize[97] and CaM-kinase II inhibitors increase the sensitivity of *Limulus* ventral photoreceptors,[98] but whether these effects are due to changes in arrestin phosphorylation has not been determined.

Protein kinase C activators also profoundly desensitize light-induced Ca^{2+} release and the photocurrent, but not the response to injected $InsP_3$.[99] This suggests that a PKC modulates a step in the PI pathway that is upstream of the $InsP_3R$. Since the natural activators of PKC, DAG and Ca^{2+}, are both generated by the PI pathway, PKC-induced desensitization of the PI pathway is an attractive feedback mechanism. There is, however, no evidence yet of the target protein nor is their evidence for a role for PKC in light adaptation, since protein kinase C inhibitors have not so far been shown to significantly relieve light adaptation. Instead, the desensitization observed following the activation of PKC may be a prelude to long-term structural changes in the microvillar membrane that are observed in sections fixed for electron microscopy following intense illumination[100] or PKC activation.[101]

SUMMARY

The *Limulus* ventral photoreceptor is an excellent model system for examining the G protein-coupled receptor release of Ca^{2+} from the sub-rhabdomeric endoplasmic reticulum by means of G_q, phospholipase C, and the $InsP_3$ receptor. This Ca^{2+} signal acts as the central control for this signal transduction pathway. One consequence of Ca^{2+} elevation is the rapid opening of the light-dependent conductance in the immediate vicinity of photon absorption. Evidence was presented supporting the conclusion that the final steps of the transduction pathway take place near to the $InsP_3$ receptor and are activated by the high Ca^{2+} levels reached near the pore mouth, a volume tightly bounded by the base of the microvilli and the sub-rhabdomeric endoplasmic reticulum membrane. These experiments were done with the confocal scan perpendicular to the plasma membrane. This allowed the spread of the Ca^{2+}

signal to be followed into the interior of the R lobe and onwards into the A lobe. A complementary experiment would be to look at the spread of the rise in Ca^{2+} along the base of the microvilli. Confinement of the Ca^{2+} signal around the base of the microvillus could underlie the high gain observed in this transduction pathway and the linear summation of single photon responses at low light intensities. The simplest hypothesis consistent with current evidence is that Ca^{2+} release leads to an increase of cGMP concentration and opening of cyclic nucleotide-gated channels. However a determination of how Ca^{2+} raises cGMP levels and reconstitution of the light-dependent conductance will be needed before this model can be fully accepted. A second consequence of the initial phase of Ca^{2+} elevation is a transient facilitation of release, most likely by positive feedback onto the $InsP_3$ receptor. The third consequence is a slower, long-range decrease of the response of the photoreceptor to subsequent excitation by light. Negative feedback by Ca^{2+} may occur at each of the early steps in excitation: rhodopsin activation of G_q, G_q activation of phospholipase C, and release of Ca^{2+} by the $InsP_3$ receptor.

The idea that Ca^{2+} both excites and adapts *Limulus* photoreceptors may at first seem odd, but the following simple calculations show that a transduction cascade organized in this way is in fact quite plausible. The release of Ca^{2+} (L) is set equal to three stages of gain times the light intensity (I) expressed in units of Ca^{2+}/s/photon (Fig. 9A). These stages can be thought of as G_q-PLC/rhodopsin, $InsP_3/G_q$-PLC, and $[Ca^{2+}]_i/InsP_3$. Positive feedback onto the $InsP_3R$ is only significant at low light intensities and will be ignored for now. For simplicity we assume that the maximal gains at each stage are the same (g) and that the gain of each stage is inversely proportional to $[Ca^{2+}]^2$ thereby providing negative feedback. A Hill coefficient of near 2 is typical for Ca^{2+} binding to the highly cooperative low- and high-affinity sites of calmodulin.

$$L = I\{gK_m^{\,2} / (K_m^{\,2} + [Ca^{2+}]^2)\}^3 \tag{1}$$

Let Ca^{2+} at steady state be set by the rate of Ca^{2+} release (L) and sequestration ($k_{-1}Ca^{2+}$).

$$I\{gK_m / (K_m + [Ca^{2+}]^2/K_m)\}^3 = k_{-1}[Ca^{2+}] \tag{2}$$

Solving for I gives

$$I = k_{-1}\, g^{-3}\, \{[Ca^{2+}] + 3K_m^{-2}\, [Ca^{2+}]^3 + 3K_m^{-4}\, [Ca^{2+}]^5 + K_m^{-6}[Ca^{2+}]^7\} \tag{3}$$

Since the $[Ca^{2+}]^7$ term dominates, this equation can be reduced to

$$[Ca^{2+}]\ a\ I^{1/7} \tag{4}$$

That at moderate to high light intensities steady-state $[Ca^{2+}]$ is proportional to $I^{1/7}$ is in reasonable agreement with $[Ca^{2+}]$ during constant illumination.[50]

This simple model has been expanded to include positive feedback by Ca^{2+} onto the $InsP_3R$ at low $[Ca^{2+}]$[36,37] and run as a dynamic simulation (Fig. 9B). The

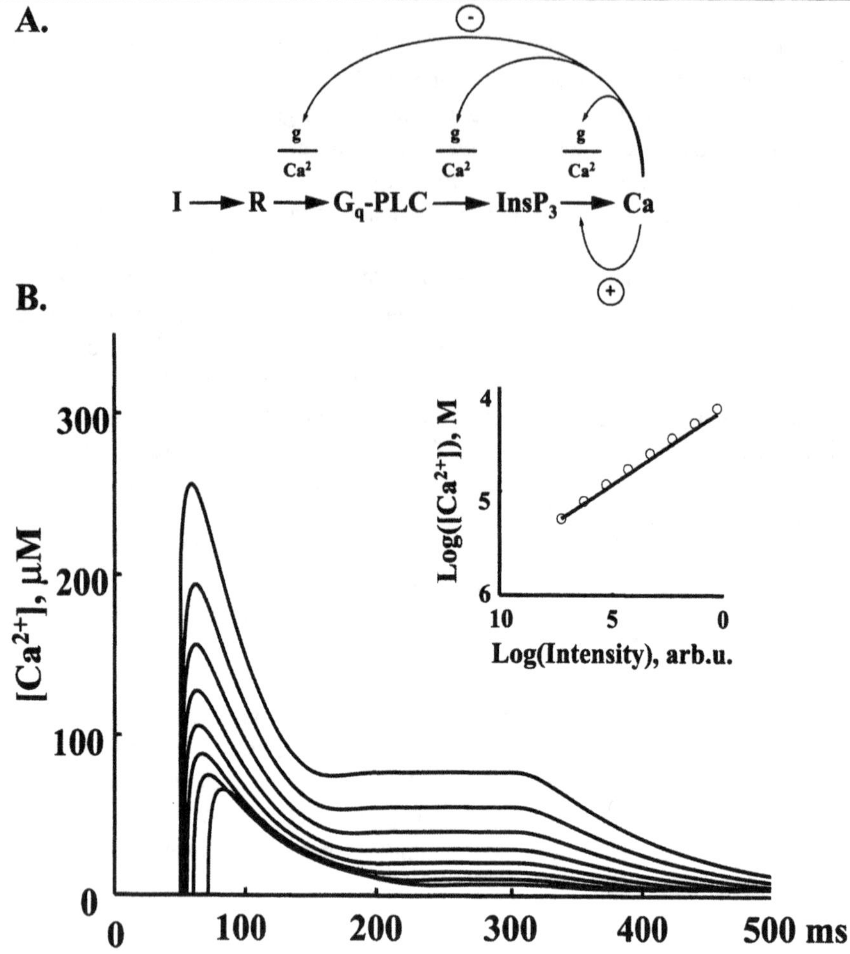

Figure 9. Model for control of Ca^{2+} release during *Limulus* phototransduction. A) Light (I) photoisomerizes rhodopsin (R). Active rhodopsin catalyzes the formation of G protein (G_q). The G_q/PLC complex catalyzes the release of $InsP_3$ from membrane lipid. $InsP_3$ binds to the $InsP_3$ receptor thereby releasing Ca^{2+} (Ca) from intracellular stores. The gain between stages (g) is inversely proportional to the square of cytoplasmic Ca^{2+} (Ca^2). Low levels of cytoplasmic Ca^{2+} enhance release by the $InsP_3$ receptor. Not shown are the first-order loss of R, G_q/PLC, $InsP_3$, and Ca. B) Simulation of the Ca^{2+} feedback model qualitatively replicates the time course and intensity dependence of [Ca^{2+}] in the R lobe. A constant intensity stimulus (arbitrary units) starts at 50 ms and ends at 300 ms. Intensity was increased 10-fold over the curve below. In the inset Ca^{2+} levels at 300 ms (open circles) are plotted against stimulus intensity. The line has a slope of 1/7.

elements of the initial Ca^{2+} signal seen in the photoreceptors, namely a decrease in response latency as light intensity increases and a rapid rise in Ca^{2+} to a peak value of hundreds of micromolar,[23] can be seen in the traces. As calculated above, the plateau steady-state Ca^{2+} increases approximately 10 fold over seven orders of magnitude increase in light intensity. Comparison with Fig. 2C indicates that for the highest light intensity, the model somewhat underestimates the peak-to-plateau decline in Ca^{2+}, but it does provide a qualitative explanation for many of the essential features of the light-induced elevation of Ca^{2+}

During a rapid change in light intensity, the negative feedback of adaptation must inherently lag the direct effect of Ca^{2+} on excitation. While the feedback regulation of each gain step is instantaneous in this model, the fact that Ca^{2+} elevation at the late stage of the cascade responds with a delay with respect to earlier changes is sufficient to generate the dynamics of negative feedback.[89] The steep dependence of adaptation on Ca^{2+} will facilitate spread of the negative feedback signal away from the points of photon absorption even in the absence of Ca^{2+} release or excitation in the surrounding area. Not included in this model are the relationships between Ca^{2+}, cGMP, and membrane conductance. Experimental evidence indicates that the waveforms of changes of membrane potential closely follow those of bursts of $InsP_3$-induced Ca^{2+} release.[33,102]

The role and mechanisms of increases of Ca^{2+} can be compared between the two arthropods, *Limulus* and *Drosophila*. In *Drosophila*, photoisomerization of caged Ca^{2+} was unable to excite cells[103] and genetic removal of the $InsP_3$ receptor had no effect on phototransduction.[104] These results argue against a necessary role for elevation of Ca^{2+} during excitation. Ca^{2+} does enter through the light-dependent conductance and has roles in facilitation, termination, and adaptation of the light response.[105,106] Some of these roles for Ca^{2+} such as CaMKII-dependent phosphorylation of arrestin appear likely to be similar between *Limulus*[95] and *Drosophila*,[107] but others are likely to be quite different. Recent models for *Drosophila* transduction[3] feature the complete excitation pathway taking place not only within a single microvillus, but possibly within individual macromolecular assemblies, the so-called transducisome or signaplex. The cost may be a greatly reduced amplification of the light signal. The average response to a single photon in *Drosophila* is only 10 pA, one-hundredth the size of the *Limulus* response. While transduction may be faster, reduced gain dictates a smaller photoreceptor to ensure rapid depolarization. Indeed, in addition to greater gain, the horseshoe crab has a larger photoreceptor containing 100 times more rhodopsin. Shrinking the phototransduction process and photoreceptor size may underlie the fly's ability to transduce a light signal in 10's of ms. Such speed may be necessary for making the right-or-flight decision involved in finding a mate or fleeing a predator. The fly photoreceptor may have lost the ability to utilize the $InsP_3$ receptor for added gain and control in favor of faster activation of TRP family ion channels. *Limulus*, a slow moving animal often active under conditions of low light intensity, can use the multiple advantages of size, additional gain, and feedback to maximize photon capture.

REFERENCES

1. Millecchia R, Mauro A. The ventral photoreceptor cells of *Limulus*. II. The basic photoresponse. J Gen Physiol 1969; 54:310-330.
2. Scott K et al. Gq alpha protein function in vivo: genetic dissection of its role in photoreceptor cell physiology. Neuron 1995; 15:919-927.
3. Hardie RC, Raghu P. Visual transduction in *Drosophila*. Nature 2001; 413(6852):186-193.
4. Lisman JE, Brown JE. Light induced changes of sensitivity in *Limulus* ventral photoreceptors. J Gen Physiol 1975; 66:473-488.
5. Brown JE, Coles JA. Saturation of the response to light in *Limulus* ventral photoreceptors. J Physiol 1979; 296:373-392.
6. Brown JE, Blinks JR. Changes in intracellular free calcium concentration during illumination of invertebrate photoreceptors. Detection with aequorin. J Gen Physiol 1974; 64:643-665.
7. Lisman JE, Brown JE. The effects of intracellular iontophoretic injection of calcium and sodium ions on the light response of *Limulus* ventral photoreceptors. J Gen Physiol 1972; 59:701-719.
8. Lisman JE, Brown JE. Effects of intracellular injection of calcium buffers on light adaptation in Limulus ventral photoreceptors. J Gen Physiol 1975; 66:489-506.
9. Payne R, Corson DW, Fein A. Pressure injection of calcium both excites and adapts *Limulus* ventral photoreceptors. J Gen Physiol 1986; 88:107-126.
10. Bolsover SR, Brown JE. Calcium ion, an intracellular messenger of light adaptation, also participates in excitation of *Limulus* photoreceptors. J Physiol Lond 1985; 364:381-393.
11. Frank TM, Fein A. The role of the inositol phosphate cascade in visual excitation of invertebrate microvillar photoreceptors. J Gen Physiol 1991; 97:697-723.
12. Shin J, Richard EA, Lisman JE. Ca^{2+} is an obligatory intermediate in the excitation cascade of *Limulus* photoreceptors. Neuron 1993; 11:845-855.
13. Brown JE et al. myo-inositol polyphosphate may be a messenger for visual excitation in *Limulus* photoreceptors. Nature 1984; 311:160-163.
14. Fein A et al. Photoreceptor excitation and adaptation by inositol 1,4,5-trisphosphate. Nature 1984; 311:157-160.
15. Payne R et al. Excitation and adaptation of *Limulus* ventral photoreceptors by inositol 1,4,5 triphosphate result from a rise in intracellular calcium. J Gen Physiol 1986; 88:127-142.
16. Brown JE, Rubin LJ. A direct demonstration that inositol-trisphosphate induces an increase in intracellular calcium in *Limulus* photoreceptors. Biochem Biophys Res Commun 1984; 125:1137-1142.
17. Richard EA, Sampat P, Lisman JE. Distinguishing between roles for calcium in *Limulus* photoreceptor excitation. Cell Calcium 1995; 18:331-341.
18. Payne R. Dynamics of the release of calcium by light and inositol 1,4,5-trisphosphate in *Limulus* ventral photoreceptors. In: Hidalgo C et al, eds. Transduction in Biological Systems. New York: Plenum, 1990:9-25.
19. Nasi E, del Pilar Gomez M, Payne R. Phototransduction mechanisms in microvillar and ciliary photoreceptors of invertebrates. In: Stavenga DG, de Grip WJ, Pugh EN Jr, eds. Molecular Mechanisms in Visual Transduction. New York: Elsevier Science, 2000:389-448.
20. Calman BG, Chamberlain SC. Distinct lobes of *Limulus* ventral photoreceptors. II. Structure and ultrastructure. J Gen Physiol 1982; 80:839-862.
21. Ukhanov K, Payne R. Light activated calcium release in *Limulus* ventral photoreceptors as revealed by laser confocal microscopy. Cell Calcium 1995; 18:301-313.
22. Ukhanov KY et al. Measurement of cytosolic Ca^{2+} concentration in *Limulus* ventral photoreceptors using fluorescent dyes. J Gen Physiol 1995; 105:95-116.
23. Payne R, Demas J. Timing of Ca^{2+} release from intracellular stores and the electrical response of *Limulus* ventral photoreceptors to dim flashes. J Gen Physiol 2000; 115(6):735-748.
24. Allbritton NL, Meyer T, Stryer L. Range of messenger action of calcium ion and inositol 1,4,5- trisphosphate. Science 1992; 258(5089):1812-1815.

25. Payne R, Fein A. Localized adaptation within the rhabdomeral lobe of *Limulus* ventral photoreceptors. J Gen Physiol 1983; 81(5):767-769.
26. Feng JJ et al. Three-dimensional organization of endoplasmic reticulum in the ventral photoreceptors of *Limulus*. J Comp Neurol 1994; 341(2):172-183.
27. Payne R et al. The localization of calcium release by inositol trisphosphate in *Limulus* photoreceptors and its control by negative feedback. Philos Trans R Soc Lond B Biol Sci 1988; 320:359-379.
28. Ukhanov K et al. Putative inositol 1,4,5-trisphosphate receptor localized to endoplasmic reticulum in *Limulus* photoreceptors. Neuroscience 1998; 86(1):23-28.
29. Ukhanov K, Payne R. Rapid coupling of calcium release to depolarization in *Limulus* polyphemus ventral photoreceptors as revealed by microphotolysis and confocal microscopy. J Neurosci 1997; 17:1701-1709.
30. Dorlochter M, Yuan W, Stieve H. Effects of calcium and cyclopiazonic acid on the photoresponse in the *Limulus* ventral photoreceptor. Zeitschrift fur Naturforschung 1999; 54c:446-455.
31. Maaz G, Stieve H. The correlation of the receptor potential with the light induced transient increase in intracellular calcium-concentration measured by absorption change of arsenazo III injected into *Limulus* ventral nerve photoreceptor cell. Biophys Struct Mech 1980; 6:191-208.
32. Payne R, Flores TM, Fein A. Feedback inhibition by calcium limits the release of calcium by inositol trisphosphate in *Limulus* ventral photoreceptors. Neuron 1990; 4(4):547-55.
33. Payne R, Potter BV. Injection of inositol trisphosphorothioate into *Limulus* ventral photoreceptors causes oscillations of free cytosolic calcium. J Gen Physiol 1991; 97:1165-1186.
34. Levy S, Payne R. A lingering elevation of Ca_i accompanies inhibition of inositol 1,4,5 trisphosphate-induced Ca release in *Limulus* ventral photoreceptors. J Gen Physiol 1993; 101:67-84.
35. Levitan I, Hillman P, Payne R. Fast desensitization of the response to $InsP_3$ in *Limulus* ventral photoreceptors. Biophys J 1993; 64:1354-1360.
36. Bezprozvanny I, Watras J, Ehrlich BE. Bell-shaped calcium-response curves of Ins(1,4,5)P_3- and calcium-gated channels from endoplasmic reticulum of cerebellum. Nature 1991; 351:751-754.
37. Levitan I et al. Facilitation of the responses to injections of inositol 1,4,5-trisphosphate analogs in *Limulus* ventral photoreceptors. Biophysical Journal 1994; 67:1161-1172.
38. Keizer J et al. InsP3-induced Ca^{2+} excitability of the endoplasmic reticulum. Mol Biol Cell 1995; 6(8):945-951.
39. Wood SF, Szuts EZ, Fein A. Metabolism of inositol 1,4,5-trisphosphate in squid photoreceptors. J Comp Physiol [B] 1990; 160(3):293-298.
40. Vallet AM, Fein A. A role for hydrolysis of inositol 1,4,5-trisphosphate in terminating the response to inositol 1,4,5-trisphosphate and to a flash of light in *Limulus* ventral photoreceptors. Brain Res 1997; 768:91-101.
41. O'Day PM, Gray-Keller MP. Evidence for electrogenic Na^+/Ca^{2+} exchange in *Limulus* ventral photoreceptors. J Gen Physiol 1989; 93:473-494.
42. Battelle BA et al. Immunocytochemical localization of opsin, visual arrestin, myosin III, and calmodulin in *Limulus* lateral eye retinular cells and ventral photoreceptors. J Comp Neurol 2001; 435(2):211-225.
43. Fein A, Tsacopoulos M. Activation of mitochondrial oxidative metabolism by calcium ions in *Limulus* ventral photoreceptor. Nature 1988; 331:437-440.
44. Deckert A, Stieve H. Electrogenic Na(+)-Ca2+ exchanger, the link between intra- and extracellular calcium in the *Limulus* ventral photoreceptor. J Physiol Lond 1991; 433:467-482.
45. Lisman JE. Effects of removing extracellular Ca^{2+} on excitation and adaptation in *Limulus* ventral photoreceptors. Biophys J 1976; 16:1331-1335.
46. Lisman JE, Fain GL, O'Day PM. Voltage-dependent conductances in *Limulus* ventral photoreceptors. J Gen Physiol 1982; 79:187-209.

47. Hsiao HS, Payne R. Light-induced Mn^{2+} influx in *Limulus* ventral photoreceptors. J Comp Physiol [A] 1998; 183(2):193-202.
48. Brown JE, Mote MI. Ionic dependence of reversal voltage of the light response in *Limulus* ventral photoreceptors. J Gen Physiol 1974; 63(3):337-350.
49. Hardie RC, Minke B. The *trp* gene is essential for a light-activated Ca^{2+} channel in *Drosophila* photoreceptors. Neuron 1992; 8:643-651.
50. Levy S, Fein A. Relationship between light sensitivity and intracellular free Ca concentration in *Limulus* ventral photoreceptors. A quantitative study using Ca-selective microelectrodes. J Gen Physiol 1985; 85:805-841.
51. Richard EA et al. Ca^{2+}/calmodulin-binding peptides block phototransduction in *Limulus* ventral photoreceptors: Evidence for direct inhibition of phospholipase C. Proc Natl Acad Sci USA 1997; 94:14095-14099.
52. Payne R, Flores TM. The latency of the response of *Limulus* photoreceptors to inositol trisphosphate lacks the calcium-sensitivity of that to light. J Comp Physiol A 1992; 170:311-316.
53. Stieve H, Benner S. The light-induced rise in cytosolic calcium starts later than the receptor current of the *Limulus* ventral photoreceptor. Vision Research 1992; 32:403-416.
54. O'Day PM, Lisman JE, Goldring M. Functional significance of voltage-dependent conductances in *Limulus* ventral photoreceptors. J Gen Physiol 1982; 79:211-232.
55. Stommel G et al. The light-stimulated cytosolic calcium transient in *Limulus* ventral nerve photoreceptors: two components in the rising phase. Z Naturforsch [C] 1996; 51(1-2):101-112.
56. Faddis MN, Brown JE. Intracellular injection of heparin and polyamines. Effects on phototransduction in *Limulus* ventral photoreceptors. J Gen Physiol 1993; 101:909-931.
57. Nagy K, Contzen K. Inhibition of phospholipase C by U-73122 blocks one component of the receptor current in *Limulus* photoreceptor. Vis Neurosci 1997; 14:995-998.
58. Johnson EC, Gray-Keller MP, O'Day PM. Rescue of excitation by inositol following Li^{+}-induced block in *Limulus* ventral photoreceptors. Vis Neurosci 1998; 15(1):105-112.
59. Bacigalupo J, Lisman JE. Single-channel currents activated by light in *Limulus* ventral photoreceptors. Nature 1983; 304:268-270.
60. Bacigalupo J, Chinn K, Lisman JE. Ion channels activated by light in *Limulus* ventral photoreceptors. J Gen Physiol 1986; 87:73-89.
61. Bacigalupo J et al. Light-dependent channels from excised patches of *Limulus* ventral photoreceptors are opened by cGMP. Proc Natl Acad Sci USA 1991; 88:7938-7942.
62. Nagy K. Inhibition of the first component of the receptor current in *Limulus* photoreceptor. Neuroreport 1994; 5:847-849.
63. Johnson EC, Robinson PR, Lisman JE. Cyclic GMP is involved in the excitation of invertebrate photoreceptors. Nature 1986; 324:468-470.
64. Feng JJ, Frank TM, Fein A. Excitation of *Limulus* photoreceptors by hydrolysis-resistant analogs of cGMP and cAMP. Brain Res 1991; 552:291-294.
65. Brown JE, Kaupp UB, Malbon CC. 3',5'-cyclic adenosine monophosphate and adenylate cyclase in phototransduction by *Limulus* ventral photoreceptors. J Physiol1984; 353:523-539.
66. Johnson EC, O'Day PM. Inhibitors of cyclic-GMP phosphodiesterase alter excitation of *Limulus* ventral photoreceptors in Ca^{2+}-dependent fashion. J Neurosci 1995; 15:6586-6591.
67. Chen FH et al. A cGMP-gated channel subunit in *Limulus* photoreceptors. Visual Neuroscience 2001; 18(4):517-526.
68. Garger A, Richard EA, Lisman JE. Inhibitors of guanylate cyclase inhibit phototransduction in *Limulus* photoreceptors. Visual Neuroscience 2001; 18(4):625-632.
69. Schmidt JA, Farber DB. Light-induced changes in cAMP levels in *Limulus* photoreceptors. Biochem Biophys Res Commun 1980; 94(2):438-42.
70. Brown JE, Faddis M, Combs A. Light does not induce an increase in cyclic-GMP content of squid or *Limulus* photoreceptors. Exp Eye Res 1992; 54:403-410.
71. Dorlochter M, de Vente J. Cyclic GMP in lateral eyes of the horseshoe crab *Limulus*. Vision Res 2000; 40(27):3677-3684.

72. Gillespie PG, Beavo JA. cGMP is tightly bound to bovine retinal rod phosphodiesterase. Proc Natl Acad Sci USA 1989; 86:4311-4315.
73. Gorczyca WA et al. Purification and physiological evaluation of a guanylate cyclase activating protein from retinal rods. Proc Natl Acad Sci USA 1994; 91(9):4014-4018.
74. Palczewski K et al. Molecular cloning and characterization of retinal photoreceptor guanylyl cyclase-activating protein. Neuron 1994; 13(2):395-404.
75. Dizhoor AM et al. Cloning, sequencing, and expression of a 24-kDa Ca^{2+}-binding protein activating photoreceptor guanylyl cyclase. J Biol Chem 1995; 270(42):25200-25206.
76. Pozdnyakov N et al. A novel calcium-dependent activator of retinal rod outer segment membrane guanylate cyclase. Biochemistry 1995; 34(44):14279-14283.
77. Schultz JE, Klumpp S. Lanthanum dissociates calmodulin from the guanylate cyclase of the excitable ciliary membrane from *Paramecium*. FEMS Microbiol Letts 1982; 13:303-306.
78. Klumpp S, Schultz JE. Characterization of a Ca^{2+}-dependent guanylate cyclase in the excitable ciliary membrane from *Paramecium*. Eur J Biochem 1982; 124:317-324.
79. Estacion M, Sinkins WG, Schilling WP. Regulation of *Drosophila* transient receptor potential-like (TrpL) channels by phospholipase C-dependent mechanisms. J Physiol 2001; 530(Pt 1):1-19.
80. Chyb S, Raghu P, Hardie RC. Polyunsaturated fatty acids activate the *Drosophila* light-sensitive channels TRP and TRPL. Nature 1999; 397:255-259.
81. Fein A, Cavar S. Divergent mechanisms for phototransduction of invertebrate microvillar photoreceptors. Vis Neurosci 2000; 17(6):911-917.
82. Brown JE, Lisman JE. An electrogenic sodium pump in *Limulus* ventral photoreceptor cells. J Gen Physiol 1972; 59:720-733.
83. Fuortes MGF, Hodgkin AL. Changes in time scale and sensitivity in the ommatidia of *Limulus*. J Physiol 1964; 172:239-263.
84. Lisman JE, Bering H. Electrophysiological measurement of the number of rhodopsin molecules in single *Limulus* photoreceptors. J Gen Physiol 1977; 70:621-633.
85. Lisman JE. An electrophysiological investigation of the ventral eye of the horseshoe crab *Limulus polyphemus*. [Ph.D.]. Massachusetts Institute of Technology, 1971.
86. Payne R, Fein A. Inositol 1,4,5 trisphosphate releases calcium from specialized sites within *Limuluss* photoreceptors. J Cell Biol 1987; 104:933-937.
87. Adler EM et al. Alien intracellular calcium chelators attenuate neurotransmitter release at the squid giant synapse. J Neurosci 1991; 11:1496-1507.
88. Naraghi M, Neher E. Linearized bu$^{\text{ffered Ca2+}}$ diffusion in microdomains and its implications for calculation of $[Ca^{2+}]$ at the mouth of a calcium channel. J Neurosci 1997; 17(18):6961-6973.
89. Grzywacz NM, Hillman P, Knight BW. Response transfer functions of *Limulus* ventral photoreceptors: Interpretation in terms of transduction mechanisms. Biol.Cybern. 1992; 66:429-435.
90. Arnon A et al. Calmodulin regulation of light adaptation and store-operated dark current in *Drosophila* photoreceptors. Proc Natl Acad Sci USA 1997; 94:5894-5899.
91. Scott K et al. Calmodulin regulation of *Drosophila* light-activated channels and receptor function mediates termination of the light response in vivo. Cell 1997; 91:375-383.
92. Hardie RC et al. Ca^{2+} limits the development of the light response in *Drosophila* photoreceptors. Proc R Soc Lond B Biol Sci 1993; 252:223-229.
93. Smith DP et al. Photoreceptor deactivation and retinal degeneration mediated by a photoreceptor-specific protein kinase C. Science 1991; 254:1478-1484.
94. Richard EA, Lisman JE. Rhodopsin inactivation is a modulated process in *Limulus* photoreceptors. Nature 1992; 356:336-338.
95. Calman BG et al. Calcium/calmodulin-dependent protein kinase II and arrestin phosphorylation in *Limulus* eyes. J Photochem Photobiol B Biol 1996; 35:33-44.

96. Ellis DZ, Edwards SC. Characterization of a calcium/calmodulin-dependent protein phosphatase in the *Limulus* nervous tissue and its light regulation in the lateral eye. Vis Neurosci 1994; 11:851-860.
97. Kass L et al. Inhibition of the calcineurin-like protein phosphatase activity in *Limulus* ventral eye photoreceptor cells alters the characteristics of the spontaneous quantal bumps and the light-mediated inward currents, and enhances arrestin phosphorylation. Vis Neurosci 1998; 15(6):1039-1049.
98. Kass L, Bray WO. Kinetic model for phototransduction and G-protein enzyme cascade: Understanding quantal bumps during inhibition of CaM-KII or PP2B. J Photochem Photobiol B 1996; 35(1-2):105-113.
99. Dabdoub A, Payne R. Protein kinase C activators inhibit the visual cascade in *Limulus* ventral photoreceptors at an early stage. J Neurosci 1999; 19:10262-10269.
100. Herman KG. Light-stimulated rhabdom turnover in *Limulus* ventral photoreceptors maintained in vitro. J Comp Neurol 1991; 303(1):11-21.
101. Jinks RN, White RH, Chamberlain SC. Dawn, diacylglycerol, calcium, and protein kinase C—The retinal wrecking crew. A signal transduction cascade for rhabdom shedding in the *Limulus* eye. J Photochem Photobiol B 1996; 35:45-52.
102. Corson DW, Fein A. Inositol 1,4,5-trisphosphate induces bursts of calcium release inside *Limulus* ventral photoreceptors. Brain Res 1987; 423:343-346.
103. Hardie RC. Photolysis of caged Ca^{2+} facilitates and inactivates but does not directly excite light-sensitive channels in *Drosophila* photoreceptors. J Neurosci 1995; 15:889-902.
104. Acharya JK et al. $InsP_3$ receptor is essential for growth and differentiation but not for vision in *Drosophila*. Neuron 1997; 18:881-887.
105. Ranganathan R et al. A *Drosophila* mutant defective in extracellular calcium-dependent photoreceptor deactivation and rapid desensitization. Nature 1991; 354:230-232.
106. Hardie RC. Whole-cell recordings of the light induced current in dissociated *Drosophila* photoreceptors: evidence for feedback by calcium permeating the light-sensitive channels. Proc R Soc Lond B 1991; 245:203-210.
107. Matsumoto H et al. Phosrestin I undergoes the earliest light-induced phosphorylation by a calcium/calmodulin-dependent protein kinase in *Drosophila* photoreceptors. Neuron 1994; 12:997-1010.
108. Dabdoub A, Payne R, Jinks RN. Protein kinase C-induced disorganization and endocytosis of photosensitive membrane in *Limulus* ventral photoreceptors. J Comp Neurol 2002; 442:217-225.

CALCIUM HOMEOSTASIS IN FLY PHOTORECEPTOR CELLS

Johannes Oberwinkler

ABSTRACT

In fly photoreceptor cells, two processes dominate the Ca^{2+} homeostasis: light-induced Ca^{2+} influx through members of the TRP family of ion channels, and Ca^{2+} extrusion by Na^{+}/Ca^{2+} exchange. Ca^{2+} release from intracellular stores is quantitatively insignificant. Both, the light-activated channels and the Ca^{2+}-extruding exchangers are located in or close to the rhabdomeric microvilli, small protrusions of the plasma membrane. The microvilli also contain the molecular machinery necessary for generating quantum bumps, short electrical responses caused by the absorption of a single photon. Due to this anatomical arrangement, the light-induced Ca^{2+} influx results in two separate Ca^{2+} signals that have different functions: a global, homogeneous increase of the Ca^{2+} concentration in the cell body, and rapid but large amplitude Ca^{2+} transients in the microvilli. The global rise of the Ca^{2+} concentration mediates light adaptation, via regulatory actions on the phototransduction cascade, the voltage-gated K^{+} channels and small pigment granules controlling the light intensity. The local Ca^{2+} transients in the microvilli are responsible for shaping the quantum bumps into fast, all-or-nothing events. They achieve this by facilitating strongly the phototransduction cascade at early stages of the light response and subsequently inhibiting it. Many molecular targets of these feedback mechanisms have been identified and characterized due to the availability of numerous *Drosophila* mutant showing defects in the phototransduction.

Department of Neurobiophysics, University of Groningen, The Netherlands.

INTRODUCTION

Fly photoreceptor cells are one of the best-studied model systems for receptor-mediated Ca^{2+} entry, a ubiquitous signal transduction pathway found in many cell types (for reviews, see refs. 1-3). Generally, after the stimulation of a receptor, which in fly photoreceptor cells is rhodopsin, this pathway leads the opening of channels permeable for Ca^{2+}, through the sequential activation of G_q and phospholipase C (PLC). It is still unclear, how precisely PLC activity causes the channels to open. Also, the molecular nature of the Ca^{2+}-permeable channels has not been identified in most systems. In photoreceptor cells of the fruitfly *Drosophila melanogaster*, however, the Ca^{2+}-influx channels are known to be at least partly encoded by the *trp* and *trpl* gene.[4,5] Highlighting the role of *Drosophila* phototransduction as a model system, the mammalian homologues of these genes are amongst the best candidates for the elusive channels opened during receptor-activated Ca^{2+}-entry (for reviews, see refs. 6-9). In heterologous expression systems, the seven mammalian genes most closely related to the *Drosophila trp* and *trpl* (the so-called "classical" TRP-channels) are all activated by increased PLC-activity. The PLC cleaves the minor membrane phospholipid phospho-inositol-diphosphate (PIP_2) into inositol-1,4,5-triphosphate ($InsP_3$) and diacylglycerol (DAG). Both of these reaction-products have been implicated in the molecular mechanism leading to the activation of the members of the TRP-channel family in separate hypotheses (Fig. 1). The first hypothesis assumes that DAG or its metabolites, the polyunsaturated fatty acids, directly activate the channels.[10,11] In the alternative scenario, the so-called "store-operated" or "capacitative" Ca^{2+} entry, $InsP_3$ causes Ca^{2+} release from the endoplasmic reticulum (ER) and the resulting depletion of Ca^{2+} from the stores causes the channels to open, either through direct coupling of $InsP_3$-receptors and the channels located in the plasma-membrane or by an—yet to be identified—diffusible messenger (for review see refs. 2,6).

Whatever the exact mechanism of channel activation, the activation of receptors leads to an influx of Ca^{2+} through the activated channels, and consequently to an increase of the cytosolic free Ca^{2+} concentration (Ca_i). The rise of Ca_i can be quite inhomogeneous, especially in the vicinity of open Ca^{2+} channels where the local Ca^{2+} concentration can be very high, up into the millimolar range.[12] The spatial extent of these local Ca^{2+} signals, often referred to as Ca^{2+} microdomains, is very small (in the range of tens of nanometers) and depends on the concentration, kinetics and diffusion coefficient of the Ca^{2+} buffers present in the cytosol.[13,14] Ca^{2+} microdomains can form without any compartmentalization or other morphological specialization of the cell. Nevertheless, morphological compartments of the cytosol, connected to the cell body only via narrow passages, can help to isolate local Ca^{2+} signals from each other, as shown e.g., in dendritic spines.[15,16] The local Ca^{2+} signal in these microdomains, often is of functional importance as it may regulate the Ca^{2+} channels, or other proteins in its vicinity, in ways that can not be understood from the global Ca^{2+} concentration.[17,18]

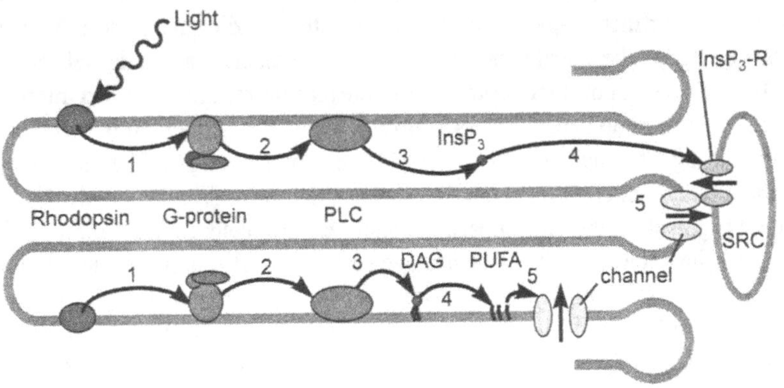

Figure 1. The two main hypotheses about the signal transduction pathway leading to the opening of members of the TRP channel family. Activation of the receptor (in the case of phototransduction rhodopsin) leads to the activation of a PLC via a G_q-protein (steps 1 and 2). The PLC splits PIP_2 into $InsP_3$ (step 3A) and DAG (step 3B). a) In the first hypothesis (capacitative Ca^{2+} entry), $InsP_3$ binds to $InsP_3$-receptors located on the ER (step 4A). In fly photoreceptor cells, the subrhabdomeric cisternae (SRC) are part of the ER and located very close to the rhabdomere. The subsequent Ca^{2+} release from the ER leads to a reduction of the Ca^{2+} concentration and that causes the plasma membrane Ca^{2+} channels to open (step 5A). In the alternative hypothesis, either DAG directly,[11] or its metabolites (step 4B), the polyunsaturated fatty acids (PUFAs)[10] open the plasma membrane Ca^{2+} channels (step 5B).

The evidence reviewed here suggests that in fly photoreceptors, both global and local Ca^{2+} signals with strongly differing properties exist. The unique morphology of these cells is very important for shaping the different Ca^{2+} signals spatially and temporally. Global and local Ca^{2+} signals have entirely different roles in fly photoreceptors, and are both indispensable for proper functioning of these cells. Fly phototransduction therefore is a highly informative system to study how different Ca^{2+} signals bring about their action on the physiology. Additionally, the availability of a large amount of mutants in *Drosophila* aids in identifying and characterizing the proteins responsible for the processes underlying the regulatory actions of the different Ca^{2+} signals.

THE SPECIAL ANATOMY OF THE FLY PHOTORECEPTOR CELLS

General Organization of Fly Photoreceptor Cells

Fly photoreceptor cells have large, elongated cell bodies. In *Drosophila*, they are about 80 μm long and 4-5 μm wide, while in the blowfly *Calliphora vicina*, they can be up to 250 μm long and 10 μm wide (for review, see ref. 19). Insect photoreceptors are polarized cells, meaning that the cell membrane has two distinct compartments, apical and basolateral, both extending along the entire length of the cell body. The apical membrane contains the rhabdomere, a densely packed, highly ordered stack of small protrusions of the plasma-membrane called microvilli. Optically, the rhabdomere works as a wave-guide, trapping the light that enters into it and guiding it through its entire length (for review, see ref. 20). The membrane of the rhabdomeric microvilli contains high concentrations of rhodopsin, the light absorbing pigment. In their distal part, the photoreceptor cells contain small pigment granules (diameter: ca 200 nm) that move close to the rhabdomere when the cells are stimulated with bright light.[21] Located close to the rhabdomere, the pigment granules absorb a fraction of the light travelling just outside the wave-guide (the so-called boundary wave). The pigment granules therefore act as an "intracellular pupil" diminishing the effective light intensity (for reviews, see refs. 22,23). This mechanism only comes into play at fairly high light intensities[24,25] and avoids saturation of the light response, thereby guaranteeing a high signal-to-noise ratio, even when the other light adaptation mechanisms do not suffice.[24]

At the base of the rhabdomere the so-called subrhabdomeric cisternae (SRCs, often also called SMCs, sub-microvillar cisternae) are located that are part of the ER.[26-28] Fly photoreceptor cells also contain many mitochondria, to meet the high metabolic demand of this cell type.[29,30] The mitochondria are located mainly in the distal part and close to the basolateral membrane of the photoreceptors.[31]

The Microvilli are Extremely Tiny

The microvilli are round, elongated tubes, between 0.8 and 1.6 μm long and only 60 nm in diameter, including the plasma-membrane.[26,32] The volume of an individual microvillus therefore is tiny, enabling small ion fluxes across its membrane to cause major changes of the local ion concentrations.[33] This is especially important in the case of calcium, as the (resting) Ca^{2+} concentration typically is much lower than the concentration of other cations. For the dimensions given above (and subtracting two times 5 nm for the thickness of the membrane), adding one single calcium ion works out to increase the total Ca^{2+} concentration inside a microvillus by 0.5 - 1 μM. This figure might even be an underestimate as it does not take into account actin filaments running along the entire length of the microvilli[34,35] and other proteins (see below), that further reduce the volume of the microvilli. The actin filaments extend through the narrow neck of the microvilli (diameter ca 35 nm) that connects them with the membrane of the cell body.[26]

The Microvilli Contain the Molecular Machinery to Produce Bumps

Despite their small size, the microvilli contain almost all molecules that are thought to be involved in the phototransduction pathway (Fig. 1). Often these molecules are present in high concentrations; for example, a microvillus contains approximately 1000 rhodopsins,[36] 1250 calmodulins (in *Drosopila*),[37] 100 copies of the proteins TRP, NORPA, INAC and INAD each (in the blowfly *Calliphora*),[38] and up to 20 G-proteins (in the housefly *Musca domestica*).[39]

In response to light stimulation, fly photoreceptor cells generate discrete events of electrical activity, the so-called quantum bumps, that are evoked by the absorption of a single photon.[40-43] When measuring the membrane potential, the bumps can be seen under dim illumination as small depolarizations of at most a few millivolt.[40,41] Under voltage-clamp conditions the peak amplitude of bumps is on average 9 pA in dark-adapted *Drosophila* photoreceptors, which corresponds to about 15 channels being open simultaneously.[43]

At higher light intensities, the bumps overlap and produce a macroscopic light response with characteristic noise properties. This electrical noise still reflects the properties of the underlying unitary events, the bumps, that can be recovered employing the so-called shot-noise analysis.[24,44-48] Illuminating fly photoreceptors with relatively bright, long lasting (>100 ms) light stimuli, the membrane potential typically depolarizes within a few milliseconds and, after having reached a peak, repolarizes to a relatively stable plateau that is maintained as long as the light intensity does not change (Fig. 2A). The quantitative values for both peak and plateau thereby depend on the light intensity (Fig. 2). This phenomenon, which is a characteristic manifestation of light adaptation, is called peak-to-plateau transition and reflects the decrease of the size of the bumps induced by the prolonged light-stimulus.[47,49] The peak-to-plateau transition can also be observed under voltage-clamp conditions.[42,50]

It is generally assumed, but has not yet been rigorously tested, that a bump is generated by a single microvillus, i.e., that a microvillus corresponds to a "transduction unit".[24,51] Under bright illumination, a photoreceptor as a whole can generate as much as 10^6 bumps per second.[24,46,48] Since there are ca $3*10^4$ (*Drosophila*) - 10^5 (*Calliphora*) microvilli per photoreceptor cell, this suggests that each microvillus can sustain tens of bumps per second, a figure which seems plausible.[51-54]

Extracellular Space in the Retina

The photoreceptors in the fly retina are arranged in groups of eight in a ring-like fashion forming an ommatidium. Each ommatidium is surrounded by glia cells, leaving only a very narrow extracellular cleft between photoreceptor and glia cells. In flies (but not in all other insects, see section "Calcium Signals in the Photoreceptor Cells of Another Insect, the Honeybee drone"), there is a fairly large, extracellular space in the middle of the ommatidium, surrounded by the photoreceptors. This inner space, called central matrix or intraommatidial cavity, is separated from the rest of the extracellular space by tight-junctions that connect the ring of photoreceptors. These tight-junctions act as diffusion barriers for large molecules, but not, or only incompletely, for La^{3+},[55] and therefore presumably also not for other ions such as Ca^{2+}. For the purpose of Ca^{2+} homeostasis, the extracellular space in the fly retina might therefore be viewed as a single, interconnected interstitium, allowing diffusion between its different compartments. Ca^{2+} entry from the extracellular space

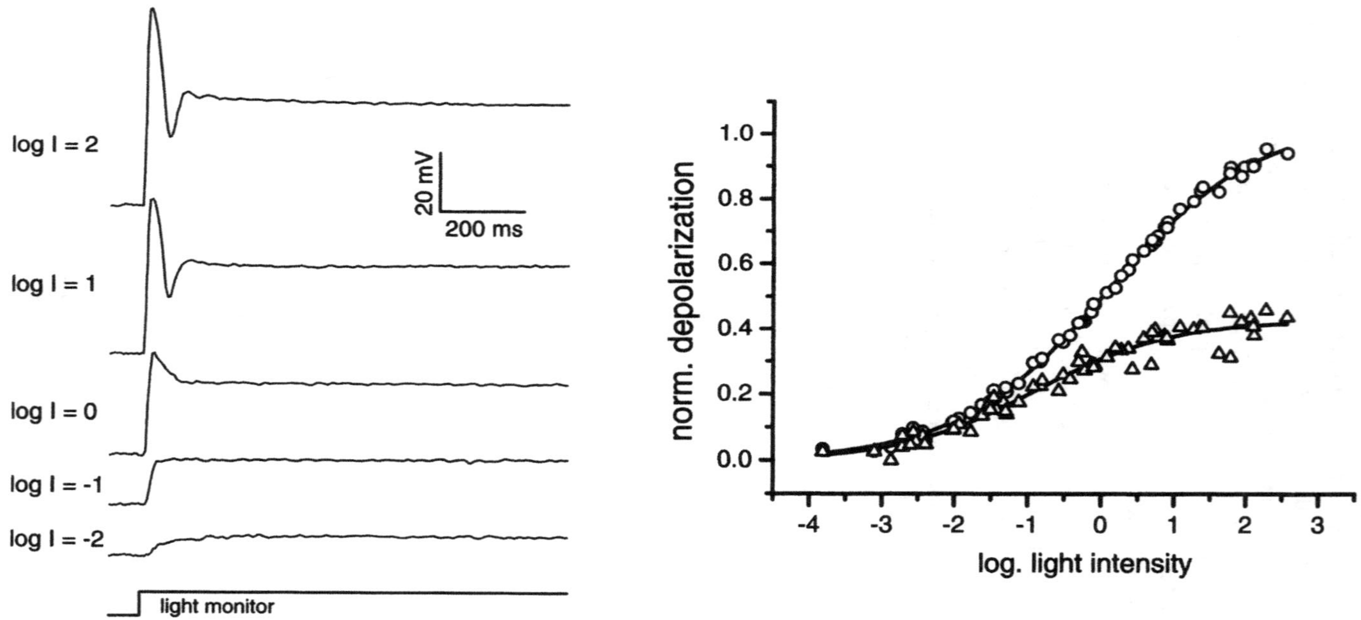

Figure 2. The light response of fly photoreceptor cells. A) The changes of the membrane potential due to illumination of various intensities (as indicated). The light response is graded with the light intensity. At the onset of illumination with intermediate or strong light intensities, the light response shows a peak and then stabilizes on a plateau value after sometimes having shown an undershoot or oscillations. B) The amplitude of the peak (circles) and the plateau (triangles) plotted versus the light intensity (in logarithmic units). Data redrawn from ref. 108.

into the photoreceptor cells therefore results in a measurable decrease in the extracellular Ca^{2+} concentration (Ca_o).[56,57]

PHYSIOLOGICAL AND MOLECULAR COMPONENTS INVOLVED IN CALCIUM HOMEOSTASIS

Calcium Influx is The Major Source of Calcium During Light-Induced Increase of the Cytosolic Calcium Concentration

The first indications that Ca^{2+} influx provides an important contribution to Ca^{2+} homeostasis in fly photoreceptors came from studies in which Ca_o was manipulated. Removing extracellular Ca^{2+}, as well as chelating the intracellular Ca^{2+} with EGTA, reduces the peak-to-plateau transition of the light response,[58] and abolishes the light-induced migration of the intracellular pupillary pigment granules in photoreceptor cells.[59] The observed effects imply intracellular changes to the physiology caused by the changes in Ca_o, thus suggesting that Ca^{2+} ions permeate the light-activated conductance. Early on, it was also noted that the light-induced Ca^{2+} influx was diminished in flies carrying mutations in the *trp* gene. In those flies, the movement of the pupillary pigment granules is strongly reduced or even absent.[60-62] Increasing Ca_o restores the movement of the pigment granules, at least partially.[61] Furthermore, at the onset of light stimulation Ca_o decreases,[56,57,63] indicating that the photoreceptor cells take up Ca^{2+}. Again, this decrease in Ca_o is strongly diminished in flies carrying mutations in the *trp* gene compared to wild-type flies.[64,65]

From the early 90's onwards, patch-clamping *Drosophila* photoreceptor cells after isolating the ommatidia has been the technique of choice to investigate Ca^{2+} influx.[42,50] The advantage of this technique is that it allows controlling the membrane voltage and the composition of the extra- and intracellular fluids with great precision. Varying the driving force for Ca^{2+} entry by changing the voltage across the plasma-membrane demonstrated that the kinetics of the light response slows down strongly when the cell is depolarized and the Ca^{2+} influx is reduced accordingly.[42,50] In the absence of extracellular Ca^{2+}, the kinetics of the light response is slow, regardless of the holding potential.[42,50] Besides confirming the important role of Ca^{2+} influx, this finding demonstrates that Ca^{2+} influx results in positive and negative feedback to the light-response (see section "Local Calcium Transients in the Rhabdomere Shape the Form of the Quantum Bumps").

Using flies with mutations of the *trp* gene and measuring the reversal potential of the light-induced current at different extracellular Ca^{2+} concentrations, Hardie and coworkers showed that the light-activated channels in *trp* mutants are ca 10 times less permeable for Ca^{2+} as compared to wild-type flies.[66,67] This explains why in these mutants the light-induced Ca^{2+} influx into photoreceptor cells is reduced. It also shows that the *trp* gene codes for a membrane channel, or a subunit thereof.

The patch-clamp technique also allows to dialyze the interior of the photoreceptors with fluorescent Ca^{2+}-indicator dyes, making it possible to directly visualize the light-induced increase in Ca_i. Light stimulation was thus shown to produce very rapid and large increases in Ca_i. Removing extracellular Ca^{2+} greatly reduces or even abolishes this light-induced increase of Ca_i.[65,68-71] Again, the light-induced rise of Ca_i was greatly reduced, but not abolished, in *Drosophila trp* mutants.[65,68] The combined evidence from these studies has thoroughly established that Ca^{2+} influx through the light-activated channels is the predominant way to increase Ca_i in fly photoreceptor cells, and that TRP is a part of the Ca^{2+} influx channels.

Calcium Influx Occurs via Two Different Types of Channels Located in the Microvilli

The sequence of the *trp* gene[4] shows some weak homology with the sequence of voltage-gated Ca^{2+} channels, suggesting that the *trp* gene itself codes for a Ca^{2+} channel, or a subunit thereof.[5] Flies carrying the *trp* mutation, however, are not blind, indicating that other proteins contribute to the light-activated conductance. This role is filled in by TRPL, a protein with ca 40% homology to TRP,[5] which is required for the remaining light-induced activity in *trp* mutant flies, as *trp/trpl* double mutants are blind.[72]

The noise properties of the light-activated currents of wild-type flies indicate that, physiologically, only two types of channels open in response to light.[67] Each of these channel types is eliminated separately by a null mutation in either the *trp* or the *trpl* gene, respectively, thus allowing their isolation and characterization in vivo.[67] The channels depending on the *trp* gene product (from now on called TRP channels) are highly selective to Ca^{2+}, while the channels dependent on the *trpl* gene product (the TRPL channels) are rather unselective cation channels, but nevertheless show considerable permeability for Ca^{2+}.[66,67] TRP channels have a much smaller single channel conductance (ca 4 pS) compared to TRPL channels (35 pS).[67,73] TRP proteins, on the other side, are about 10 times more abundant compared to TRPL proteins.[74,75] TRP channels are subject to a pronounced, voltage-dependent block by Mg^{2+}, while TRPL channels are not.[76] Importantly, TRP and TRPL channels are regulated in different ways by Ca^{2+} (see section "Calcium Signals in the Photoreceptor Cells of Another Insect, the Honeybee drone").

The sequence homology with voltage-gated Ca^{2+} channels suggests that 4 individual TRP proteins assemble to form a functional channel.[5] Proteins from the TRP-family (including the vertebrate homologues), however, are notorious for being capable of forming heteromultimers (for reviews, see refs. 9,77). The subunit composition of the two physiologically identified channel types in *Drosophila* photoreceptors has not yet been unequivocally determined. Besides the TRPL protein, TRPL channels have been proposed to contain TRPγ, a third TRP homologue.[78] TRPL channels may also contain TRP proteins, although their channel properties clearly do not depend on the presence of TRP. The evidence that TRP may be part of

TRPL channels comes from heterologous expression system, where formation of TRP/TRPL heteromultimers have been reported.[74,79] Also TRPL has been shown to co-immunoprecipitate with TRP from retinal tissue of *Drosophila*, indicating that TRP/TRPL multimerization takes place also in vivo.[74] TRP channels, on the other hand might contain almost only TRP protein as TRP is available in much higher concentrations than TRPL.[74,75] However, INAF, a protein of unknown function, might be a regulatory subunit of TRP channels, because TRP channels are almost inactive in flies lacking INAF.[80]

Fast spatial imaging of *Drosophila* photoreceptor cells filled with Ca^{2+} indicator dye showed that, a few milliseconds after the onset of light stimulation, the Ca^{2+} concentration increases in the rhabdomere, and only later in the cell body,[69] a finding that later has been repeated for the photoreceptor cells of the larger blowflies (*Calliphora vicina*, Fig. 3).[81] This implies that the Ca^{2+}-influx channels are located in or very close to the rhabdomere. That this is indeed the case has been shown immunohistochemically for the channel-proteins TRP and TRPL.[4,38,72,82,83] Consequently, the rise of Ca_i in the cell body is delayed and slower with respect to the increase of Ca_i in the rhabdomere (Fig. 4), as Ca^{2+} reaches the cell body only by diffusion through the narrow neck of the microvilli.

Calcium Release from Intracellular Stores is Quantitatively Insignificant

In most cell types, the production of $InsP_3$ leads to Ca^{2+} release through the $InsP_3$-receptors ($InsP_3$-R) located on the ER (for review, see e.g., ref. 84). $InsP_3$ is produced in fly photoreceptor cells due to the action of the PLCβ NORPA,[39,85-88] suggesting that light-induced Ca^{2+} release from intracellular stores could occur in fly photoreceptor cells. The subrhabdomeric cisternae (SRCs), belonging to the ER, are a good candidate for the Ca^{2+}-releasing store, as they are located very close to the microvilli,[26-28] where NORPA is located.[89] $InsP_3$-Rs[90-93] and ryanodine receptors (RyR)[90,94] are expressed in fly photoreceptor cells, albeit only weakly. In flies, RyR proteins are dispersed throughout the photoreceptor cell body, and only sparsely expressed on the SRCs.[94]

Although all the molecular and structural requirements for light-induced Ca^{2+}-release are therefore met, initial reports indicated that there is no such Ca^{2+} release in *Drosophila* photoreceptor cells.[68,69,71] Hardie estimated that Ca^{2+} release—if occuring at all—increases Ca_i by less than 20 nM.[71] Additionally, increasing the intracellular concentration of $InsP_3$ experimentally (by releasing caged $InsP_3$ in blind flies) did not lead to a significant increase of Ca_i.[95]

The $InsP_3$-Rs and RyRs are each encoded by only one gene in the Drosophila genome and mutant photoreceptor cells lacking one of these genes have been generated.[92,93,96] Photoreceptor cells without $InsP_3$-Rs[92,93] display normal photoresponses, as do photoreceptor cells without RyRs.[96] Importantly, they also show normal light-adaptation behavior (peak-to-plateau transition), indicating that the Ca^{2+} homeostasis is not disturbed.

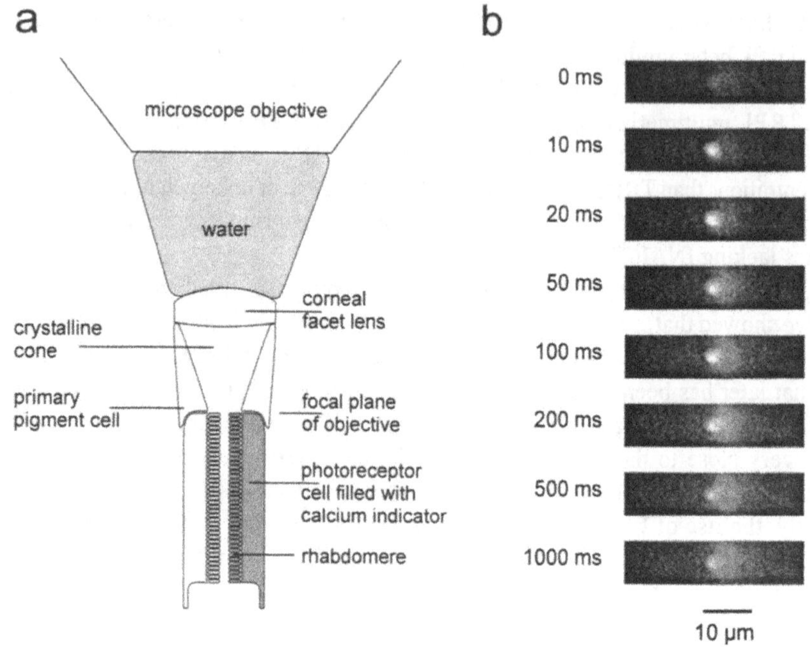

Figure 3. The light induced Ca^{2+} increase occurs first in the rhabdomere, and then in the rest of the cell body. A) Schematic drawing to illustrate the method used to image a cross-section of a photoreceptor in the intact eye of the white-eyed blowfly mutant *chalky*. In these mutants (but not in the wild type, see Fig. 7A), the primary pigment cells are basically transparent. A drop of water between the microscope objective and the cornea largely eliminates the optical properties of the cornea,[21] allowing to directly image the distal tips of the photoreceptor cells, one of which was previously filled with a Ca^{2+} indicator dye. B) Images of a photoreceptor cell filled with the Ca^{2+} indicator dye Oregon Green 5N at various times after the onset of light stimulation (as indicated). The rhabdomere shows a strong increase of the fluorescence intensity already after 10 ms, while it takes longer before the Ca^{2+} indicator reports an increase of the Ca^{2+} concentration in the cell body. Adapted from ref. 81.

More recently, however, Minke and coworkers reported that light-induced Ca^{2+} release in *Drosophila* can be elicited under carefully chosen conditions.[97-99] Applying M5, a Ca^{2+}/calmodulin blocker, increased this Ca^{2+} release.[97,98] It is still controversially discussed whether light-induced Ca^{2+} release actually occurs, and whether the reduction of Ca^{2+} content in the SRCs, which may accompany the Ca^{2+} release, is relevant to phototransduction.[99,100] For the Ca^{2+} homeostasis under physiological conditions it is quite evident, however, that the contribution of Ca^{2+} release is quantitatively negligible compared to the massive influx of Ca^{2+} occurring in the presence of extracellular Ca^{2+}.

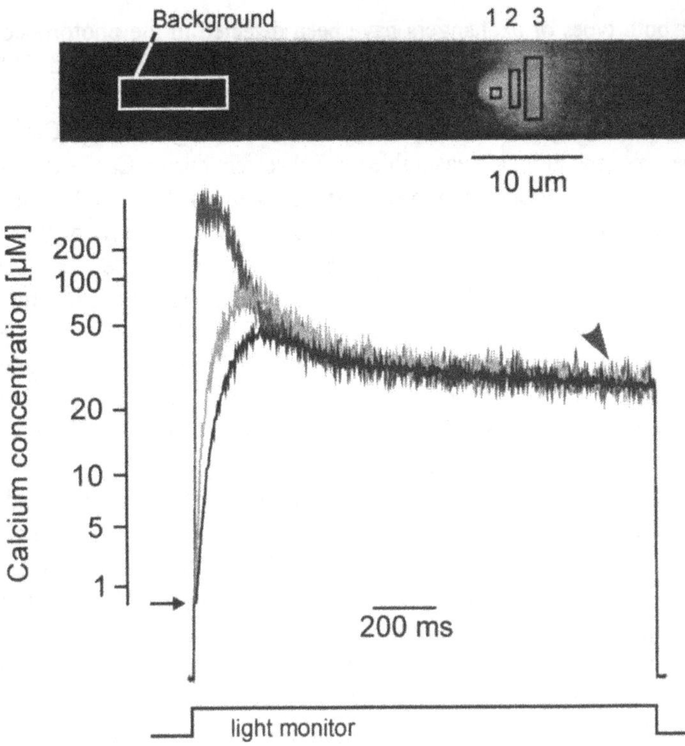

Figure 4. Quantitative analysis of the data shown in Fig. 3. A) Three region of interest were defined in the cross-section of the photoreceptor cell. Region Nr. 1 corresponds to the rhabdomere, region Nr. 2 to a region of the cell body close to the rhabdomere, while region Nr. 3 is on the opposite part of the cell body. Background values were obtained from the indicated region. B) The time course of the fluorescence intensity normalized to the first reading after the light was turned on (black, horizontal arrow). From the normalized fluorescence intensity the Ca^{2+} concentration was calculated.[135] Shortly after turning on the light, the Ca^{2+} concentration in the rhabdomere (traces Nr. 1) increases strongly to values exceeding 200 μM, while the Ca^{2+} concentration in the cell body (traces Nr. 2 and 3) increases much slower and only reaches lower values. After ca 100 ms, the Ca^{2+} concentration in the rhabdomere declines, and reaches after ca 500 ms the same value as found in the cell body (arrowhead). The Ca^{2+} concentration in the photoreceptor cell is thus homogeneous in the steady state. Modified from ref. 81.

Calcium Extrusion

The strong Ca^{2+} influx into fly photoreceptors is balanced by the activity of Na^+/Ca^{2+}-exchanger proteins. Other types of Ca^{2+} extrusion mechanisms have not yet been positively identified. Plasma-membrane Na^+/Ca^{2+}-exchangers exist as two different types. The first type, found e.g., in the mammalian heart, has a stoichiometry of 3 Na^+ vs 1 Ca^{2+}. The second type, found e.g., in rod photoreceptor cells, exchanges 4 Na^+ for 1 Ca^{2+} and 1 K^+ (for a review, see ref. 101). Interestingly, transcripts for both types of exchangers have been detected in the photoreceptor cells of *Drosophila*.[102,103] The cardiac type exchanger protein has been termed CALX in *Drosophila*, while the rod-type exchanger has been designated NCKX30.

Both types of exchangers move one net electrical charge against each Ca^{2+} ion transported thereby generating a measurable inward current during Ca^{2+} extrusion. Under voltage-clamp conditions, the inward current due to Na^+/Ca^{2+} exchange can be directly observed in *Drosophila* photoreceptor cells.[104,105] When recording the membrane potential of a photoreceptor in the intact eye, the Na^+/Ca^{2+} exchange activity can be observed as a small, Na^+-dependent afterdepolarization, lasting for up to 10 seconds after the end of a light stimulus.[106-108] As the Na^+/Ca^{2+} exchangers do not use ATP as an energy source, the electrochemical gradients of Na^+ and Ca^{2+} determine the efficiency of the Ca^{2+} removal from the cytosol. Accordingly, changing these gradients by removing Na^+ from the extracellular solution or by increasing Ca_o, has been shown to increase Ca_i reversibly.[109]

The subcellular location of CALX and NCKX30 proteins has not yet been studied with immunohistochemical methods, but physiological evidence indicates that strong Na^+/Ca^{2+} exchange activity is localized in the microvilli. In blowflies, it was found that Ca_i is elevated homogeneously in all parts of the cell during strong illumination (Fig. 4).[81] Modeling the Ca^{2+} diffusion in the photoreceptor cell bodies indicates that this is only possible when Ca^{2+} influx and Ca^{2+} extrusion are closely co-localized, as otherwise rather strong gradients of Ca_i would be expected (Fig. 5).[81] Interestingly, in another invertebrate, the squid *Loligo pealei*, Na^+/Ca^{2+} exchange activity was demonstrated in vesicles obtained from the rhabdomeric membranes.[110]

Calcium Uptake into Intracellular Organelles

Although Ca^{2+}-release from intracellular stores is at best of minor quantitative importance (see section "Calcium Release from Intracellular Stores in Quantitatively Insignificant"), Ca^{2+}-uptake into such stores could still play an important role in helping to reduce the cytosolic Ca^{2+}. The ER[111,112] and the mitochondria[113,114] are generally believed to play such a role.

Ca^{2+} uptake into ER requires the action of a Ca^{2+} pump (SERCA) that consumes ATP while transporting Ca^{2+} into the ER. The only gene encoding for a SERCA (Ca-P60A) in *Drosophila* has been shown to be transcribed in photoreceptor cells.[115] Hypomorphic mutants of this gene show severe defects during development of the eyes.[116] In the blowfly *Calliphora*, ATP-dependent Ca^{2+} uptake into the SRCs has been experimentally demonstrated.[27] Blocking the SERCA (with drugs such as thapsigargin or CPA) induces a measurable increase in Ca_i in *Drosophila* photoreceptor cells within seconds, indicating that the Ca^{2+} pumps are continuously active.[117] The quantitative contribution of ATP-dependent Ca^{2+}-uptake to the Ca^{2+} homeostasis, however, has not yet been determined.

Up to now, there is no evidence for a mitochondrial role in Ca^{2+} homeostasis in fly photoreceptor cells. Mitochondria in insect photoreceptors, however, react very quickly upon light stimulation with an increase in activity, faster than the metabolic

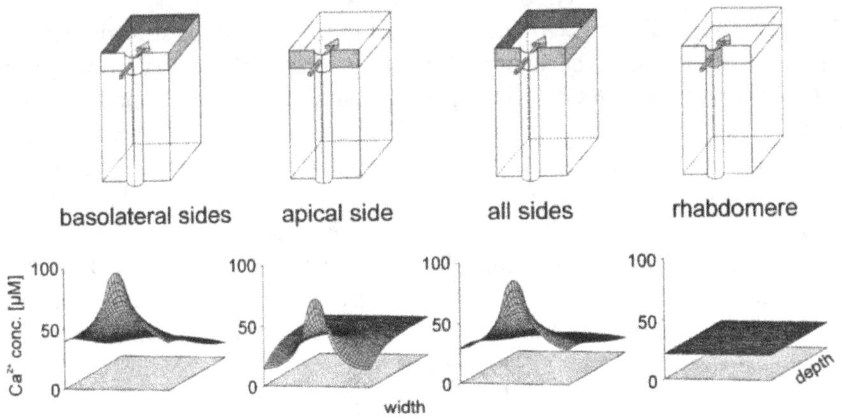

Figure 5. Close co-localization of Ca^{2+}-influx and Ca^{2+} extrusion leads to a homogeneous distribution of the Ca^{2+} concentration. Modeling Ca^{2+} diffusion in a photoreceptor cell shows that the steady state distribution of the Ca^{2+} concentration depends strongly on the distance between the location of Ca^{2+} influx and Ca^{2+} extrusion. Only Ca^{2+} diffusion in the cross-section of the photoreceptor cell needs to be considered, as the Ca^{2+} influx is likely to be similar along the length of the photoreceptor cell. The geometry was further simplified by assuming a square cross-section. Placing the Ca^{2+} extrusion (as indicated by the shaded areas in the schematic drawings) on the basolateral sides (A), the apical side excluding the rhabdomere (B), all of the membrane except the rhabdomere (C) and only to the rhabdomere (D) demonstrates that large gradients exist in the distribution of Ca_i under conditions A) - C). Only when assuming co-localization of Ca^{2+} influx and extrusion (D), the homogeneous distribution of Ca_i is obtained, that is consistent with the results shown in Figure 4B.[81]

demand necessitates.[118,119] The light-induced rise of Ca_i and subsequent Ca^{2+} entry into the mitochondria have been hypothesized to play a role in this rapid mitochondrial stimulation.[23] In the ventral photoreceptors of *Limulus*, evidence for Ca^{2+}-dependent activation of mitochondria exists.[120]

Calcium Buffering

Two different approaches have been taken to estimate the Ca^{2+} buffering capacity of fly photoreceptors. In *Drosophila*, Hardie compared the light-induced current with the increase of Ca_i caused by this current.[70] After calculating the fractional Ca^{2+} current from the light-induced current, he found that less than 0.2% of the Ca^{2+} entering the cell appeared as free Ca^{2+}. This estimate however does not differentiate between Ca^{2+} extrusion (by the Na^{+}/Ca^{2+} exchangers) and Ca^{2+} binding to buffer molecules or uptake into intracellular stores. Since Ca^{2+} extrusion is a fairly rapid and powerful process (see section "Calcium Extrusion"), almost certainly less than 99.8% of Ca^{2+} is buffered or taken up into intracellular stores in *Drosophila* photoreceptor cells.

In *Calliphora*, estimates exist for the extracellular decrease of extracellular Ca^{2+} in the retina[56] and of the corresponding increase in Ca_i,[108] both as a function of stimulating light intensity. By matching these two functions, the effective Ca^{2+}-buffering could be determined, and was found to decrease strongly with increasing light-intensity. This indicates that the predominant Ca^{2+}-buffering system in *Calliphora* has a high Ca^{2+}-binding affinity and that it can be saturated. It was concluded that the Ca^{2+} buffering capacity of *Calliphora* photoreceptor cells is equivalent to a Ca^{2+} buffer with a K_d of 770 nM, available at a concentration of 18 μM.[108]

Most likely, the Ca^{2+} buffering capacity is not homogeneously distributed in the cell body. For example, the Ca^{2+}-binding protein calmodulin is found preferentially in the rhabdomeric microvilli, where it reaches a concentration of 0.5 mM.[37] Calmodulin on its own therefore yields a buffering capacity of 2 mM in the microvilli, as each calmodulin molecule can bind up to 4 Ca^{2+} ions. Calmodulin is also found in the cell body, but at lower concentrations.[37] The microvilli have a very high surface to volume ratio, leading to very high effective concentrations of phospholipids. Especially the negatively charged phospholipids (like phospatidylserine) bind Ca^{2+} and might therefore act as low affinity, immobile Ca^{2+} buffers in the microvilli.[33]

Another Ca^{2+}-binding protein, Calphotin, is localized to a cytosolic region just below the rhabdomeres of *Drosophila* photoreceptors, potentially creating a special compartment with high Ca^{2+} buffering capacity in the cell body.[121,122] While mutants of the calphotin gene are available and show defects in the morphology of rhabdomeres and the cornea,[123] they have not been analyzed with respect to the Ca^{2+} buffering capacity of the photoreceptor cells.

The pigment granules abundantly present in wild-type (but not in white-eyed mutants) show a high content of Ca^{2+} in electron-probe X-ray analysis in flies as well as in other invertebrates.[124-126] They might therefore play a role in Ca^{2+} homeostasis. Comparing changes of the extracellular Ca^{2+} concentration in *Calliphora* wild-type and *chalky* mutants, in which the pigment granules are absent, Sandler and Kirschfeld however concluded that the pigment granules are not of major importance to the Ca^{2+} homeostasis.[63]

Calcium Diffusion

Changes of Ca_i in the cell body are brought about by diffusion of Ca^{2+} ions between the microvilli, where influx and extrusion take place, and the cell body. The microvillar neck is likely to constitute a considerable diffusion barrier, due to its narrow geometry[26] and the fact that actin filaments running through it further diminishing the space available for free diffusion.[35] A quantitative understanding of this process is not yet possible, since neither the diffusion coefficient of Ca^{2+}, nor the quantitative characteristics of Ca^{2+} buffering have been determined in fly photoreceptor cells in sufficient detail.

Inside the microvilli, the diffusion of Ca^{2+} is essentially one-dimensional. Due to the elongated shape of the microvilli, the average time needed for a Ca^{2+} ion to move the distance across the microvilli is more than 500 times less than the time needed to diffuse along the length of the microvillus.[33] This indicates that Ca^{2+}

gradients can exist along the length of the microvillus, but not in its cross-section. During a bump, high-affinity Ca^{2+} buffers in the microvilli, such as calmodulin, become saturated very quickly, and Ca^{2+} is then available in large excess over the buffers (see section "Calcium signals During Bumps can be Estimated by Biophysical Modeling"). Ca^{2+} buffers therefore have only minor effects on the Ca^{2+} homeostasis and on the effective diffusion of Ca^{2+} in the microvilli during a bump.[33]

FORM AND FUNCTION OF THE CALCIUM SIGNALS

Global Calcium Signals are Homogeneous Elevations of the Cytosolic Calcium Concentration

The resting Ca^{2+} concentration in patch-clamped, unstimulated photoreceptor cells of *Drosophila* is 0.16 µM,[70] not much different from other cell types. Strong light stimulation, however, quickly leads to a remarkable increase of the global Ca^{2+} concentration, up to tens of micromoles per liter.[69,70] These high concentrations of Ca^{2+} are also obtained in the intact eye of blowflies (*Calliphora vicina* mutant *chalky*).[108] In this preparation it was possible to determine how Ca_i depends on the stimulating light intensity. The steady-state Ca^{2+} concentration was found to be proportional to the square root of the stimulating light intensity.[108] Consistent with the findings in *Drosophila*, the maximal Ca^{2+} concentration induced by the brightest intensities was ca 20 µM (see also Fig. 4).

Directly imaging blowfly photoreceptor cells filled with a Ca^{2+} indicator dye showed that the Ca^{2+} concentration in a cross-section of the cell body is homogeneous during the steady-state (Fig. 4).[81] This is a direct consequence of co-localizing Ca^{2+} influx and Ca^{2+} extrusion (see section "Calcium Extrusion"). Another consequence of the co-localization of Ca^{2+} influx and Ca^{2+} extrusion is that Ca^{2+} removal from the rhabdomere is faster than in the rest of the cell (Fig. 6).[81] It is however difficult to estimate the time course of Ca^{2+} removal under natural conditions when using cells filled with Ca^{2+} indicators. The main obstacle is that the Ca^{2+} indicator dye adds additional Ca^{2+} buffering capacity, which, at least in blowflies, quickly leads to changes of the Na^+/Ca^{2+} currents during Ca^{2+} removal as witnessed by the afterdepolarization after a light-stimulus (see section "Calcium Extrusion").[108] Another approach to estimate the time necessary for complete Ca^{2+} removal is to monitor the restoration of sensitivity after an adapting light stimulus. Full restoration of sensitivity of a strongly light adapted photoreceptor cell takes 1-2 min.[127,128] Although a very indirect measure, it still yields an upper estimate of the time necessary to fully remove Ca^{2+}, as an increased Ca_i causes the cells to be less sensitive.

The Global Calcium Signal Mediates Light Adaptation

At least three different mechanisms work together to change the cell's physiology when exposed to an increased background light intensity: (1) Reduction of the

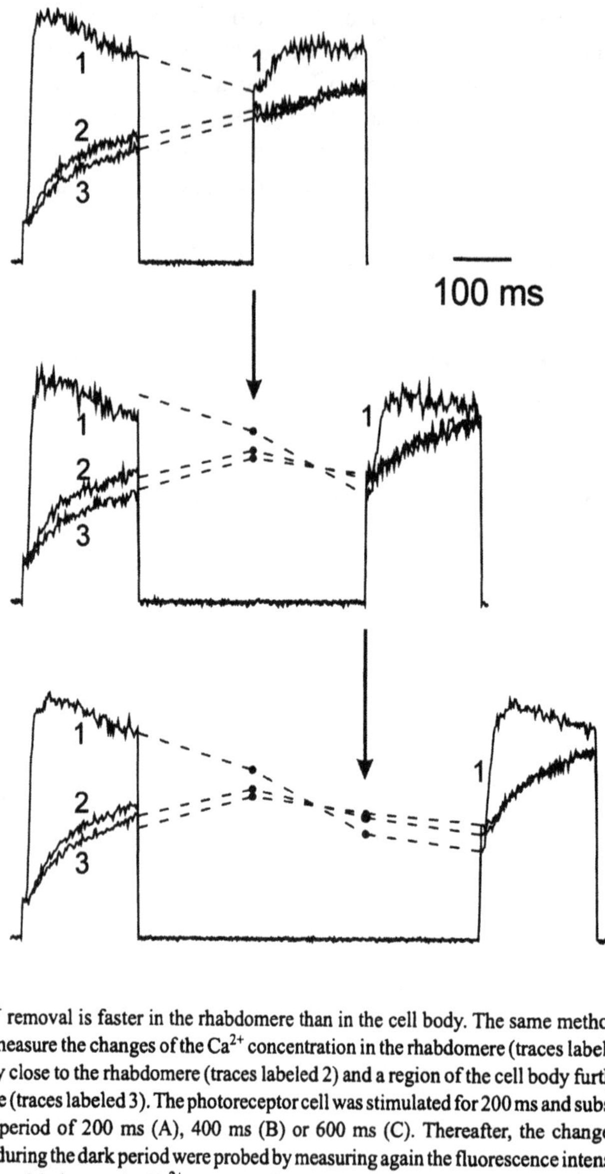

Figure 6. Ca^{2+} removal is faster in the rhabdomere than in the cell body. The same method as in Figure 4 was used to measure the changes of the Ca^{2+} concentration in the rhabdomere (traces labeled 1), a region of the cell body close to the rhabdomere (traces labeled 2) and a region of the cell body further away from the rhabdomere (traces labeled 3). The photoreceptor cell was stimulated for 200 ms and subsequently dark adapted for a period of 200 ms (A), 400 ms (B) or 600 ms (C). Thereafter, the changes of the Ca^{2+} concentration during the dark period were probed by measuring again the fluorescence intensity. A) During the first 200 ms of darkness, the Ca^{2+} concentration in the rhabdomere declines, while the Ca^{2+} concentration in the cell body still increases. B) and C) During longer periods of darkness, the Ca^{2+} concentration in the cell body also declines. In the rhabdomere, however, the decrease of the Ca^{2+} concentration is faster, and the Ca^{2+} concentration in the rhabdomere even becomes lower than in the cell body. From ref. 81.

light-induced change in conductance per absorbed photon. When voltage-clamping the photoreceptor cells, this translates directly to a reduction in bump amplitude. (2) Activation of voltage-gated K^+ channels. These channels reduce the resistance of the membrane, and therefore reduce the depolarization caused by a given light-induced current.[129] In *Drosophila* photoreceptor cells, at least two different voltage-gated K^+ currents have been described: a quickly inactivating A-current, dependent on expression of the *shaker* gene, and a slowly inactivating delayed rectifier current.[130,131] In other fly species, the composition of voltage-gated K^+ channels can be different.[132] For example, blowfly (*Calliphora vicina*) photoreceptors only express delayed rectifier K^+ currents.[129] (3) Reduction of the light intensity in the rhabdomeres due to the movement of small pigment granules towards the rhabdomere. The net effect of these three mechanisms is that the photoreceptor reacts less sensitively to light, and thus retains its ability to faithfully report relative changes in the light intensity. The crucial role of Ca^{2+} in light adaptation is highlighted by the fact that all three of the aforementioned mechanisms have been reported to depend on Ca^{2+}, or to be regulated by Ca^{2+}.

At the onset of illumination, the peak-to-plateau transition reflects the very rapid re-adjustment of the sensitivity due to the first two processes (reduced light-activated conductance and voltage-gated K^+-currents). The peak-to-plateau transition is greatly reduced or eliminated when no Ca^{2+} influx takes place.[42,50,52,58] After entering the photoreceptor cell through the light-activated channels, Ca^{2+} diffuses into the cell body, and importantly, also into neighboring microvilli that might not have been stimulated. The neighboring microvilli desensitize due to this diffusional Ca^{2+} entry. The causal role of Ca^{2+} was demonstrated by injecting the Ca^{2+} buffer EGTA into photoreceptor cells; this manipulation greatly reduces the peak-to-plateau transition.[58,105] The intracellular diffusion of the adaptation signal in the form of Ca^{2+} allows the cell to adapt already to background light levels at which only a fraction of the microvilli are stimulated per second. The spatial spread of the adaptation signal was directly demonstrated in isolated housefly ommatidia (*Musca domestica*).[133] Quantitatively, the increase of the Ca^{2+} concentration in the nonstimulated microvilli will be at best as large as the global increase in Ca_i, possibly even lower due to the Ca^{2+} extruding action of the Na^+/Ca^{2+} exchangers in the microvilli. These values are much smaller than the local Ca^{2+} transient in a stimulated microvillus (see below). Relatively modest increases in Ca_i in the rhabdomeric region therefore bring about gain reduction and thus result in light adaptation.

Ca^{2+}/calmodulin regulates voltage-gated K^+-channels in *Drosophila* via a calcium/calmodulin-dependent kinase (CaMK).[134] Applying pharmacological agents, which are supposed to block calmodulin or the CaMK, severely reduced the amplitude of both, the delayed rectifier current, and to a lesser extent, the *shaker*-dependent A-current.[134] This resulted in larger light responses and slowed deactivation phases.[134] The exact subcellular localization of the voltage-gated K^+-channels is unknown, but at least a portion of it is found in the basolateral membranes.[129] Probably, therefore, these channels, and the CaMK regulating them, sense and respond to the global Ca^{2+} signals.

The movement of the pigment granules is also dependent on Ca^{2+} influx, since it is inhibited by removing extracellular Ca^{2+}.[59] In flies carrying a mutation in the *trp* gene (and therefore having a reduced Ca^{2+} entry) the movement of the pigment granules is largely reduced or even absent, but can to some extent be restored by increasing the extracellular Ca^{2+} concentration.[61]

Local Calcium Signals in The Microvilli are Fast Transients of Large Amplitude

The Ca^{2+} concentration in the rhabdomere can be measured in the intact eye of wild-type blowflies by exploiting the natural optics of these eyes (Fig. 7A).[135] At the onset of light stimulation, the Ca^{2+} concentration in the rhabdomere displays a very fast Ca^{2+} transient, lasting about 100 ms when the cells were previously dark-adapted (Figs. 4B and 7B).[135] Comparing signals from fluorescent Ca^{2+} indicators with different affinities for Ca^{2+}, it was estimated that the Ca^{2+} concentration rises above 200 μM during the peak of the Ca^{2+} transient (Fig. 5B). Alternatively, the cross-section of a Ca^{2+} indicator-filled cell can be imaged using high frame rates.[81] This technique directly demonstrates how the kinetics of the Ca^{2+} concentration depends on the compartment of the cell (Fig. 4). In light-adapted photoreceptor cells, the rhabdomeric Ca^{2+} transient is smaller and faster than that in the dark-adapted situation.[135] In strongly adapted cells, the Ca^{2+} transient can be as short as 30 ms (Fig. 8). The rhabdomeric Ca^{2+} signal thus adapts to the background light-level, as does the electrical response. This is to be expected, because the Ca^{2+} transient and the electrical response share a common underlying cause, the opening of the light-activated channels.

However, the rhabdomeric Ca^{2+} signal, as measured with fluorescent Ca^{2+} indicators, does not always behave as expected from the electrical response of the cells. In insect photoreceptor cells, the first detectable increase of Ca_i has consistently been reported to lag the electrical signal measured in the same cell by a few milliseconds (*Apis*,[136] *Drosophila*,[70] *Calliphora*[135]). The cause for this delay has not yet satisfactorily been clarified, but it is possible that the Ca^{2+}-indicator dyes do not have access to the vicinity of the channels. Additionally, strong local buffering close to the channels could limit the detectability of the first Ca^{2+} influx. Contrary to the situation in insects, an increase of Ca_i prior to the electrical response can be measured in *Limulus polyphemus*.[137,138] In this species, Ca^{2+} release from stores triggered by $InsP_3$ is a well-established[139,140] and important step in the generation of the light response (for review, see ref. 141 and Chapter 31).

Calcium Signals During Bumps can be Estimated by Biophysical Modeling

Measuring the Ca^{2+} concentration with fluorescent Ca^{2+} indicators necessitates high light intensities and thus strong stimulation of the cells. To obtain estimates of the changes of the Ca^{2+} concentration induced by stimuli as weak as single photons, modeling based on the measurable characteristics of bumps has proven to be a useful approach.[33]

Assuming the validity of the Goldman-Hodgkin-Katz equation[142] for the TRP and TRPL channels, and using the measured permeability ratios,[67] the fractional Ca^{2+} current was calculated to be roughly 50% of the total current during a bump

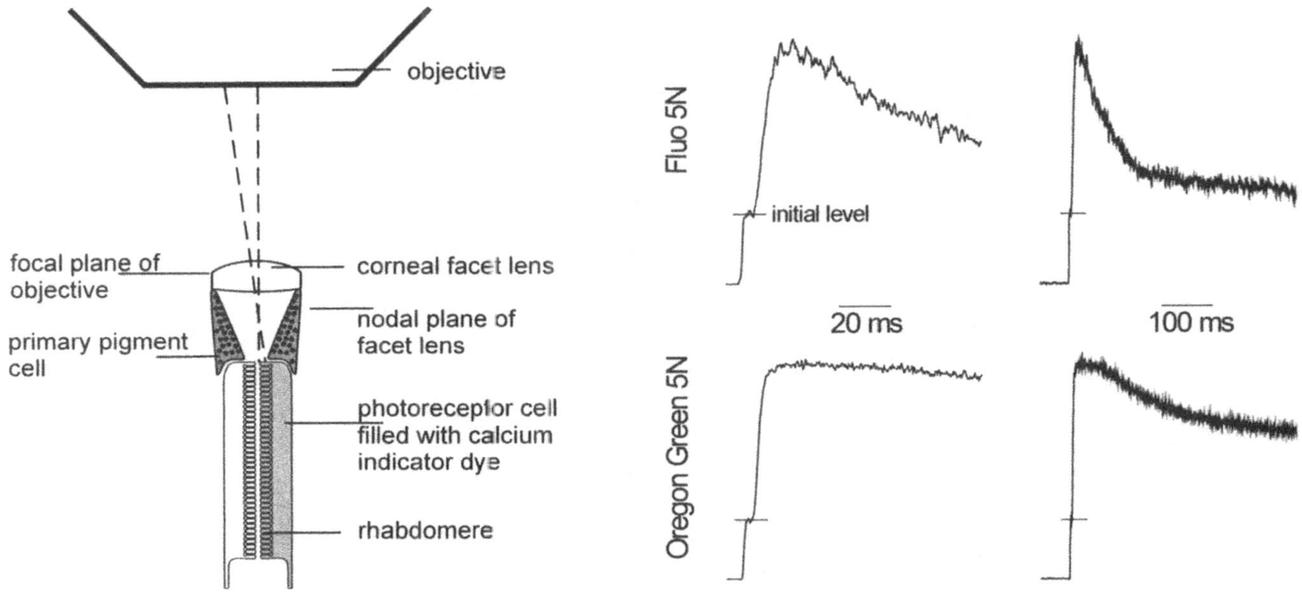

Figure 7. The Ca^{2+} transient in the microvilli exceeds 200 μM. A) In wild type flies, the primary pigment cells are densely filled with light-absorbing pigment granules, shielding the photoreceptor cells from stray light. They ensure that only light that enters the corneal facet lens in an angle of ca 2 degrees reaches the rhabdomere, thus endowing the photoreceptor cells with high spatial acuity. Light emanating from a fluorophore inside the photoreceptor cells therefore only can leave the eye on the same path in the opposite direction. Hence, a very large fraction of the fluorescence leaving the photoreceptor cells stems from the rhabdomere. B) Fluorescence intensity measured from wild type photoreceptor cells filled with Oregon Green 5N (lower panels) or Fluo 5N (upper panels; the same data are plotted at two different time scales. Fluo 5N ($K_d = 90$ μM) has a lower affinity for Ca^{2+} than Oregon Green 5N ($K_d = 20$ μM); at Ca^{2+} concentrations of ca 200 μM Oregon Green 5N therefore saturates, while Fluo 5N does not. The traces obtained with Oregon Green 5N show that the fluorescence rises after the onset of the light stimulation to a plateau, at which it stays for ca 50 ms. The traces obtained with Fluo 5N do not show this plateau. This demonstrates that the indicator Oregon Green 5N becomes saturated during the Ca^{2+} transient, indicating that the Ca^{2+} concentration in the rhabdomere exceeds 200 μM.[135]

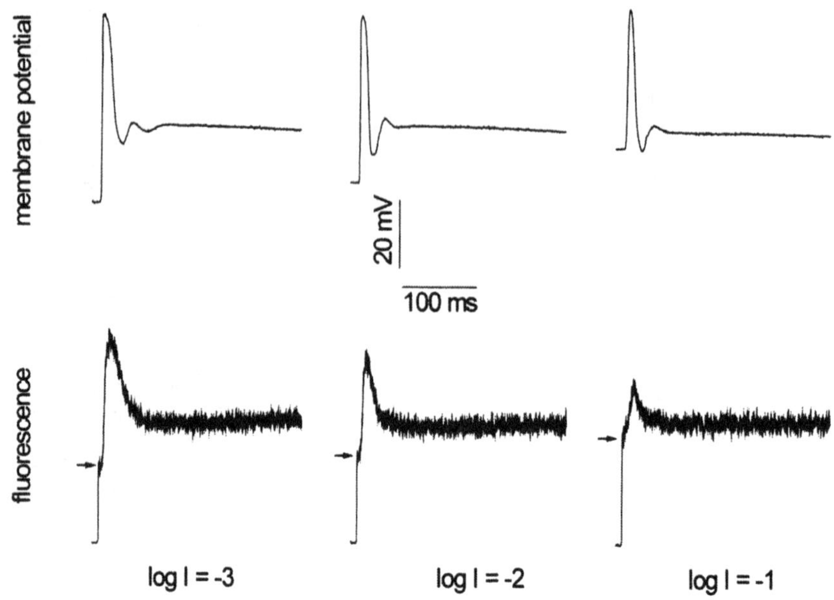

Figure 8. Light adaptation shortens the duration and diminishes the size of the Ca^{2+} transient in the rhabdomere. Membrane potential (upper panels) and fluorescence from the rhabdomere (lower panels) were recorded simultaneously from the same cell (filled with the Ca^{2+} indicator Oregon Green 5N) under three different adaptation conditions. Increasing the background light intensity, reduces the peak amplitude of the Ca^{2+} transient, and shortens the duration of the Ca^{2+} transient (as well as the initial peak of the membrane potential). This demonstrates that the Ca^{2+} transients in the rhabdomere adapt to the background light.[135]

(Fig. 9).[33] Combining this with a diffusion model for Ca^{2+} ions, and assuming that the channels that open during a bump are homogeneously distributed in the membrane of a single microvillus, the spatio-temporal profile of the Ca^{2+} concentration inside the microvilli could be calculated (Fig. 10).[33] The estimated peak of the Ca^{2+} concentration is ca 20 mM, e.g., more than 100,000 times the resting Ca^{2+} concentration. The kinetics of the Ca^{2+} concentration is very rapid, and the Ca^{2+} concentration returns to baseline levels in ca 100 ms. The conclusion that the Ca^{2+} concentration goes into the millimolar range is not significantly influenced by taking Ca^{2+} buffers into account, such as calmodulin (Fig. 11) or the negatively charged phospholipids of the membrane. This rather surprising finding is due to the tiny volume of a single microvillus compared to the sizable Ca^{2+} influx. Under these conditions, even relative high concentrations of Ca^{2+} buffers quickly become fully saturated. The Ca^{2+} concentration thus increases until influx and diffusional loss into the cell body are equal. Diffusion in the microvillus, however, occurs only in one dimension (see section "Calcium Diffusion") and is therefore much slower than unrestricted diffusion in three dimensions. This is an important reason why the Ca^{2+} influx during a single bump is capable of increasing the Ca^{2+} concentration in the microvillus

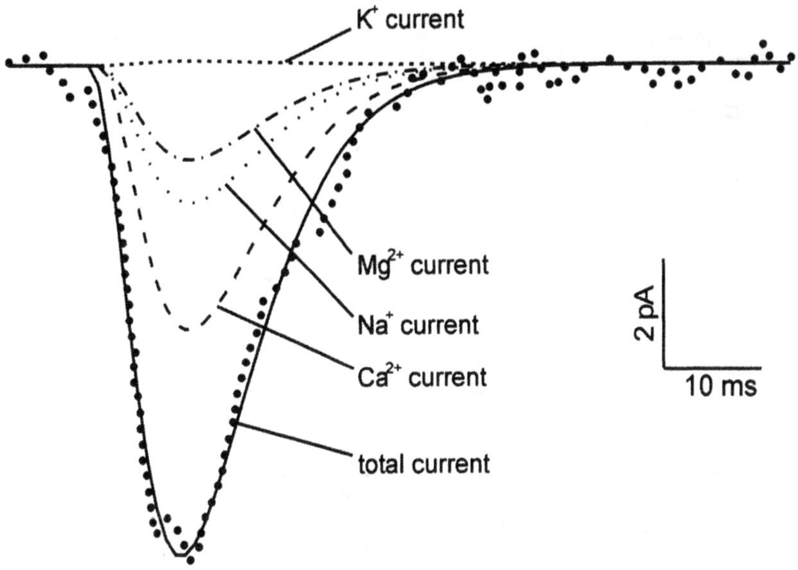

Figure 9. Ca^{2+} carries approximately 50% of the current during a bump. Under voltage clamp, a bump is a short, transient inward current having on average an amplitude of ca 9 pA. The curve labeled "total current" is a gamma function fitted to the average waveform of measured bumps (dots).[186] Using the Goldman-Hodgkin-Katz equation and the known permeability ratios for the light-activated conductance,[67] the total current can be divided into the fractional currents carried by Na^+, K^+, Ca^{2+} and Mg^{2+}. The fractional currents for these cations are also plotted. The fractional Ca^{2+} current is the largest current and is responsible for about 50% of the total current. Adapted from ref. 33.

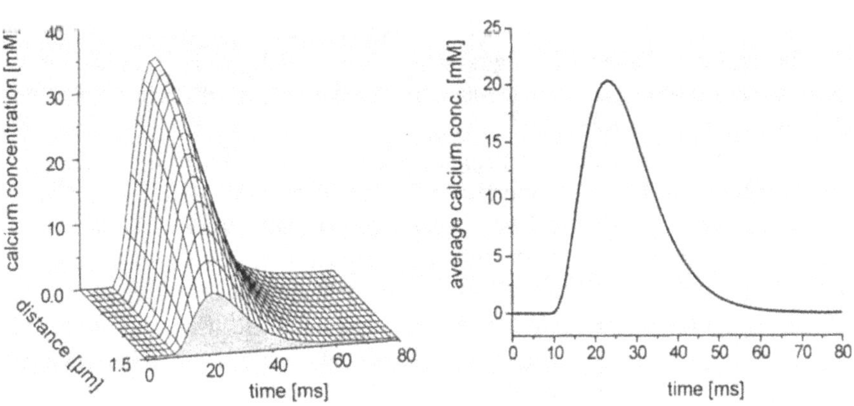

Figure 10. A) The spatio-temporal Ca^{2+} concentration profile during a bump. Combining the fractional Ca^{2+} current (Fig. 9) with a diffusion model for Ca^{2+} ions and assuming that the Ca^{2+} influx channels are homogeneously distributed along the microvilli, the Ca^{2+} concentration profile was calculated. Anywhere in the microvillus, the Ca^{2+} concentration reaches more than 10 mM. B) The spatially averaged Ca^{2+} concentration in the microvillus during a bump. Adapted from ref. 33.

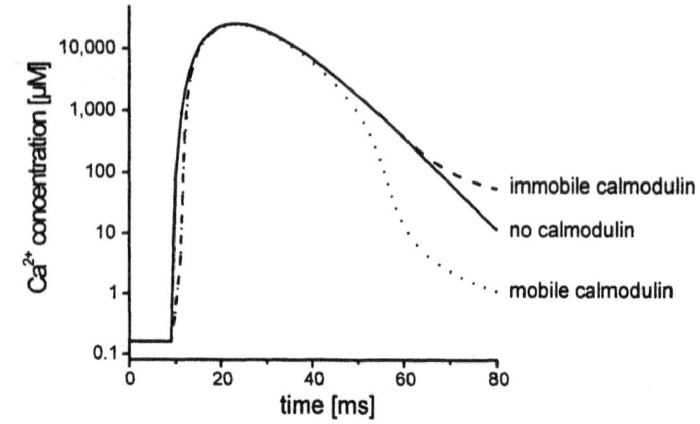

Figure 11. Adding high concentrations of a high-affinity calcium buffer does not change the peak values of the Ca^{2+} concentration during bump. Calmodulin is present at very high concentrations (0.5 mM) in the microvilli,[37] each of which can bind four Ca^{2+} ions. Calmodulin is thus likely to constitute the major Ca^{2+} binding substance in these structures. However, the calculated peak Ca^{2+} concentration does not change by taking 2 mM Ca^{2+} binding sites into accound during the calculations. The enormous excess of the free Ca^{2+} simply saturates the Ca^{2+} binding sites of calmodulin. It is thereby also not important, whether calmodulin is assumed to be mobile or not.[33]

to extreme levels. In effect, the geometry of the microvilli causes the Ca^{2+} microdomains, which typically build at the mouth of open Ca^{2+} channels, to enlarge and to spread across the entire microvillus.

Notably, during a single bump, the estimated peak Ca^{2+} concentration in the stimulated microvillus is at least 10 times higher than the measured peak concentration during bright-light stimulation. This somewhat paradoxical quantitative difference is readily explained. High light-intensities rapidly depolarize the photoreceptor cells (in the absence of voltage-clamp), thereby reducing the driving force for Ca^{2+} and causing a reduced Ca^{2+} entry.

In conclusion, both fluorometric measurements and modeling indicate that photon absorption leads to a Ca^{2+} transient in the stimulated microvillus. The Ca^{2+} transient lasts for about 100 ms in dark-adapted photoreceptors, and is much faster (<40 ms) when the photoreceptor is strongly light adapted. The peak of the Ca^{2+} concentration reaches at least 200 µM in dark-adapted cells, but can be as high as 20 mM, depending on the membrane potential. Hence, the Ca^{2+} concentration in the microvilli reaches values that in other systems are only found in Ca^{2+} microdomains, very close to the Ca^{2+} channels.

Local Calcium Transients in the Rhabdomere Shape the Form of Quantum Bumps

The Ca^{2+} transients in the microvilli are crucial for shaping the quantum bumps into rapid, distinct events and to increase their amplitude.[43] The exceptionally fast kinetics of fly quantum bumps are achieved by Ca^{2+} exerting both positive and negative feedback on the phototransduction cascade.[42,50,104] These feedback mechanisms have predominantly been studied on responses to short light flashes, rather than directly on bumps. However, as flash responses are a superposition of many bumps,[43] they accurately reflect the properties of the single bumps. When Ca^{2+} influx is reduced, which can be achieved by voltage-clamping to depolarized values or by reducing Ca_o, the light response has much slower kinetics and a reduced amplitude.[42,50,52,67] This indicates that both, positive feedback, increasing the amplitude, and negative feedback, speeding up the termination phase, are Ca^{2+}-dependent processes. Using caged Ca^{2+} to rapidly manipulate the Ca^{2+} concentration, Hardie showed that the type of feedback occurring critically depends on the precise timing with respect to the light response, and on the type of the channel.[104] Ca^{2+} exerts positive feedback only during the rising phase of the light response, later on it inhibits the light response. The situation is the opposite in mutants missing the TRP channels. Early increases in Ca^{2+} downregulate the light response, but later on an increase in Ca_i enhances the light response; this enhancement, however, is only minor.[104] In photoreceptor cells from *trp* mutant flies, Ca^{2+} influx downscales the light response, while in wild-type cells the facilitation by positive feedback is very prominent.[67] Under physiological conditions therefore, an early positive feedback dominates, strongly boosting the activity of the TRP channels.

Both types of feedback react very rapidly to changes in the Ca^{2+} concentration, i.e., within a few milliseconds. This implies that they play a major role during individual bumps. Also quantitatively, both feedbacks occur at Ca^{2+} concentrations that are well within the range of Ca_i-values occurring during a bump. The positive feedback on TRP channels occurs at Ca^{2+} concentration of 20 μM but not at 1 μM, while the negative feedback is already very strong for Ca^{2+} concentrations of 1 μM,[104] values that are much lower than the Ca^{2+} concentration found in the microvilli during bumps. This might explain why Ca^{2+}-dependent feedback on the phototransduction cascade is very robust under conditions where Ca_o is reduced. Henderson et al showed that the extracellular Ca^{2+} concentration needs to be reduced to below 200 μM (from ca 1.5 mM under natural conditions) before the positive feedback breaks down.[43] At extracellular Ca^{2+} levels between 50 and 200 μM, the positive feedback just begins to work. A bump then shows a relatively slow rising phase (termed "creep"), but it still develops to almost full size and terminates normally.[43] When extracellular Ca^{2+} is entirely absent, the bumps are very slow, both in their activation and termination phase, and have a much reduced amplitude.[43] The overall picture is therefore that during a quantum bump fast positive feedback requires a Ca^{2+} concentration as high as 20 μM and only acts on TRP channels, while the slower negative feedback works already at elevations of Ca^{2+}

around 1 μM and acts on both, TRP and TRPL channels. Together, these feedback mechanisms ensure that a quantum bump is a relatively large and rapid all-or-nothing event similar to an action potential.[54]

The Phototransduction Machinery becomes Refractory after a Quantum Bump

The phototransduction cascade becomes refractory for a short period after having generated a bump, in line with the analogy between action potentials and quantum bumps.[51-53] Initially this was shown with protocols employing three consecutive intense flashes in *Calliphora* photoreceptor cells at various intervals.[51] It was found that the second flash did not have any effects when it was delivered within 50-100 ms of the first flash. Importantly, this effect was only observed when the first flash is so intense that all microvilli absorb at least one photon, indicating that all microvilli need to be inactivated independently. This result was interpreted in favor of the hypothesis stating that a microvillus corresponds to a transduction unit.[51]

Experiments on fly mutants with a reduced calmodulin concentration in the photoreceptor cells support the notion of a refractory period after the generation of a bump.[52,53] In these mutants stimulated rhodopsins do not get inactivated and can stimulate the phototransduction cascade much longer than usual. As a consequence, after the absorption of a single photon, the photoreceptor cells in these mutants produce an electrical activity that resembles a train of quantum bumps.[52,53] It takes 100-200 ms before a new quantum bump is generated. The molecular mechanisms of the refractory period are unknown, but probably the increase of the Ca^{2+} concentration in the microvillus during the bump is an important factor, given the powerful negative feedback that Ca^{2+} exerts on the phototransduction cascade.

Local Calcium Transients Prevent Depletion of PIP_2

Ca^{2+} influx via the TRP channels is instrumental for maintaining high PIP_2 levels in the rhabdomeres during illumination.[143] In flies missing the TRP channels, PIP_2 levels decrease strongly under continuous bright light stimulation, showing that Ca^{2+} influx through TRPL channels alone is not sufficient to maintain PIP_2 levels. The decline in PIP_2 concentration can be rescued by inhibiting the Na^+/Ca^{2+} exchangers (by removing extracellular Na^+), thus demonstrating the causal role of a strongly increased Ca^{2+} concentration during light stimulation in preventing PIP_2 depletion.[143] Two ideas have been advance to explain PIP_2 depletion in the absence of Ca^{2+} influx through TRP channels. The first assumes that the large Ca^{2+} transients inhibits PLC activity (see section "Phospholipae Cβ (NORPA)") and that the close vicinity of TRP and NORPA proteins in necessary for this. The alternative hypothesis is that the increased Ca^{2+} influx through the TRP channels stimulates PIP_2 resynthesis.[143]

THE MOLECULAR TARGETS OF CALCIUM FEEDBACK IN THE MICROVILLI

Ca^{2+} regulates many proteins involved in the phototransduction cascade, either directly or indirectly through binding to calmodulin. Changes of the Ca^{2+} concentration therefore influence the phototransduction cascade at many different points. The following section is intended to give a brief summary of the known regulatory actions of Ca^{2+} on proteins of the phototransduction cascade. A recent review on the more general functions of these (and other proteins) in phototransduction and retinal degeneration can be found in ref. 144. Fig. 12 schematically summarizes the established Ca^{2+}-dependent regulatory processes in the microvilli.

Calmodulin

Calmodulin is highly enriched in the rhabdomeric microvilli, up to a concentration of 0.5 mM.[37] This enrichment depends on the presence of a particular isoform (called p174) of the NINAC protein. This protein contains two calmodulin binding domains. The rhabdomeric concentration of calmodulin is severely reduced in flies without $NINAC^{p174}$ protein, or in flies expressing a mutated form of $NINAC^{p174}$ lacking one, or both, calmodulin binding sites.[37,145]

There is only one gene in *Drosophila* encoding for calmodulin,[146,147] and the quantitative Ca^{2+} binding characteristics of the corresponding calmodulin protein have been determined in detail.[148,149] As typically found for this class of proteins, the *Drosophila* calmodulin has four Ca^{2+} binding sites, two with high affinity for Ca^{2+} (K_d = 1.25 μM and 5 μM) and two with low affinity (K_d = 14 μM and 25 μM).[148] The binding of Ca^{2+} to these binding sites occurs within milliseconds.[148,149] All of the Ca^{2+} binding sites on the calmodulins in the rhabdomere are therefore rapidly saturated during a bump.[33]

Calmodulin in turn has been shown to bind to six different proteins involved in the phototransduction cascade (Fig. 12B): TRP, TRPL, INAD, NINAC, an unidentified Cam-kinase II, and the rhodopsin phosphatase RDGC. Calmodulin is therefore an important intermediate that relays the information of changes in Ca_i to a multitude of other proteins. Eight additional putative calmodulin-binding proteins are enriched in the retina.[150] Most of them have a role in synapse functioning, but none is known to make a contribution to phototransduction.[150]

Flies in which the calmodulin content of the rhabdomere is reduced or in which a calmodulin binding protein is affected, generally show an abnormally prolonged deactivation phase after the cessation of a light stimulus.[37,52,53,145] This indicates that calmodulin plays a crucial part in the negative feedback mechanisms that lead to rapid termination of the light response, but calmodulin has also been implicated in regulating light adaptation. Reducing the rhabdomeric calmodulin concentration by mutating NINAC, or applying the calmodulin inhibitor M5, decreases the peak-to-plateau transition.[98,145] Exposing photoreceptors to M5 for a prolonged duration has the additional effect of abolishing the light response.[98]

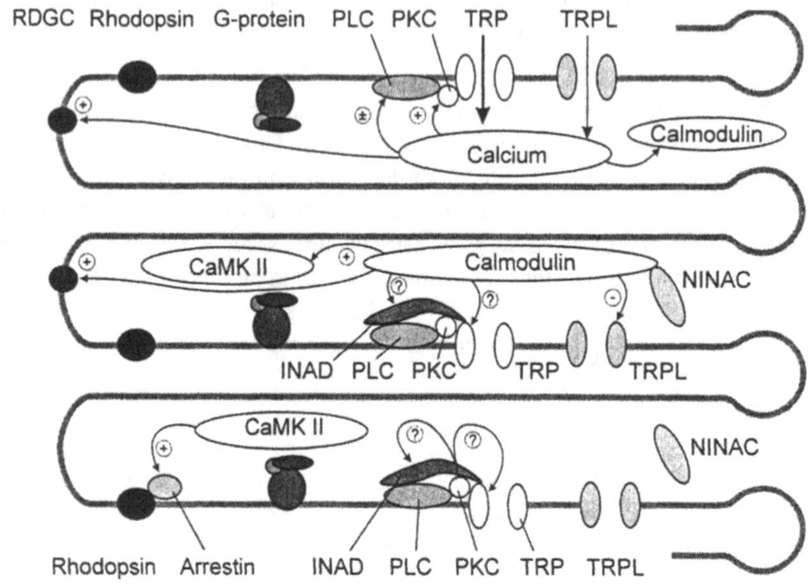

Figure 12. Schematic drawing of the known regulatory actions of Ca^{2+}. A) Ca^{2+} flowing into the microvilli through the TRP and (to a lesser extend) the TRPL channels is assumed to directly bind to and regulate the PLC encoded by NORPA, the PKC INAC, the rhodopsin kinase RDGC and calmodulin. B) Calmodulin, in turn has been shown to interact with a CaMK II, RDGC, INAD and the channel proteins TRP and TPL. It also binds to NINAC, an unconventional myosin. C) The Ca^{2+} regulated phosphorylations in the microvillus. The CaMK II, activated by Ca^{2+}/calmodulin phosphorylates arrestin, while the PKC INAC, directly activated by Ca^{2+} and DAG phosphorylates INAD and TRP. NINAC also has a kinase domain, but its substrate has not yet been determined. A + sign indicates activation, a - sign inhibition, a +/- sign both facilitation and inhibition, and a question mark uncharacterized interactions. See section "The Molecular Targets of Calcium Feedback in the Microvilli" for details and references.

TRP and TRPL Channels

TRP and TRPL channels are strongly modulated by a change of the Ca^{2+} concentration. As pointed out above (section "Local Calcium Transients in the Rhabdomere Shape the Form of Quantum Bumps), light-induced TRP channel activity first is enhanced by an increase of Ca_i at early stages of the light response and later it is inhibited.[104] In contrast, TRPL channel activity is mainly downregulated by Ca^{2+}.[67,104] Two lines of evidence indicate that Ca^{2+} acts directly on the channels or on a subunit closely associated with the channels. The action of Ca^{2+} is very rapid, taking at most 2-3 ms, for both positive and negative feedback.[104] Responses on this time scale are only possible when the components involved do not have to diffuse over appreciable distances. Secondly, the Ca^{2+}-dependent downregulation of TRP- and TRPL-channel activity also occurs when the channels are activated by metabolic stress, i.e., the lack of sufficient ATP.[104,151-153] The PLCβ (NORPA) is not required for the activation of TRP- and TRPL-channels by metabolic stress,[153] indicating that under this condition the TRP and TRPL channels become uncoupled from the upstream phototransduction cascade. However, both Ca^{2+}-dependent facilitation and inactivation of TRP channels still occur, indicating that these regulations take at least partially place downstream of PLC activation.[104,152]

The sequence of TRP shows one calmodulin binding site, while TRPL has two.[5,154-156] The interaction between TRPL and calmodulin has been studied by removing one or both of the TRPL calmodulin binding sites.[52] Such TRPL channels deactivate much slower than wild-type channels.[52] The action of calmodulin depends on Ca^{2+} entry through the TRPL channels, because both wild-type TRPL channels and mutated TRPL display the same slow inactivation without Ca^{2+} entry, demonstrating that Ca^{2+} acts through calmodulin to inhibit TRPL channels in vivo.[52]

INAD

The INAD protein consists almost in its entirety of 5 PDZ protein-protein interaction domains. In fly photoreceptors at least three other proteins, TRP, the PLCβ NORPA and the protein kinase (PKC) INAC bind strongly to INAD.[38,155,157,158] When INAD is absent, or when the binding of any of those proteins to INAD is impaired, these proteins are not localized to the microvilli anymore.[158-161] Other proteins, like TRPL, rhodopsin, and NINAC may bind less strongly, or only under certain circumstances to INAD.[155,162,163] The proper localization of these proteins to the rhabdomere takes place even in the absence of functional INAD.[160]

INAD function was studied by disrupting INAD interaction with its binding partners. The *inaD*^{P215} allele has a point mutation in the third PDZ domain abolishing TRP binding to INAD.[157] Flies with this mutation show abnormally prolonged termination phases of the light response.[157,164] The prolonged termination phase is due to the failure of individual bumps to terminate normally.[43] In another approach, the *norpA* encoded PLCβ,[159,165] the protein kinase C INAC[166] and the unconventional myosin NINAC[163] were modified so that they do not bind to INAD any longer. Disruption of INAD binding to NORPA and INAC drastically reduces their concentration in the rhabdomere, resulting in a phenotype very similar to severe *norpA* or *inaC* alleles.[159,166] For example, flies with interrupted NORPA-INAD binding are essentially blind.[159] Finally, INAD null mutant flies have an extremely reduced light response, consistent with the finding that TRP, NORPA and PKC are essentially absent from the rhabdomeres in these mutants.[158] The observed phenotypes of the various mutants with defective INAD protein interactions thus can—at least qualitatively—be explained by the mislocalizations of INAD binding partners. This argues that INAD functions mainly as a structural protein that acts as a scaffold to tie together various other proteins in a supramolecular complex.

Two lines of evidence suggest however that INAD also plays a more dynamic, Ca^{2+}-regulated role. Calmodulin binds to INAD in vitro, in a Ca^{2+}-dependent way.[155] The calmodulin binding region of INAD has been determined to lie between PDZ1 and PDZ2.[162] Secondly, INAD is phosphorylated by INAC in a Ca^{2+}-dependent way.[167,168] Whether these two potential regulatory mechanisms interfere with and modulate the phototransduction cascade is not yet known.

Phospholipase Cβ (NORPA)

The PLCβ encoded by *norpA* is the crucial intermediate between the G-protein coupled rhodopsins and the light-activated channels (see Introduction). NORPA cleaves the minor membrane lipid PIP_2 into $InsP_3$ and diacylglycerol (DAG).[39,85-88] Flies carrying a mutation in the *norpA* gene that reduces its enzyme activity have a much smaller light response.[87,88] In severe cases, the flies are blind, which is reflected in the name given to the mutation: no receptor potential A (norpA).[169,170] Partially purified NORPA[171] or NORPA in membrane preparations[172] was used to determine that the enzyme activity depends on the Ca^{2+} concentration in a bell-shaped fashion. At Ca^{2+} concentration around 1-10 μM the phospholipase activity was maximal, while further increasing the Ca^{2+} concentrations lead to declining enzyme activity. Interestingly, the same bell-shaped Ca^{2+}-dependence was found for the PLC activity from squid photoreceptor membranes.[173] If Ca^{2+} regulates NORPA similarly in vivo during a bump, the bell-shaped Ca^{2+} regulation will initially cause a positive and then a negative feedback on the phospholipase C activity and thereby on the phototransduction cascade. Ca^{2+} influx through the TRP channels is especially well suited to provide the Ca^{2+} for such a feedback, because TRP channels are tightly coupled to the NORPA proteins via INAD.[38,158] Accordingly, it was recently proposed that TRP-dependent Ca^{2+} influx is necessary to downregulate PLC activity in order to avoid PIP_2 depletion during prolonged illumination (see section "Local Calcium Transients Prevent Depletion of PIP_2").[143]

It has also been reported that calmodulin binding proteins can inhibit PLC activity in vitro.[174] They also diminish the light response in vivo in *Limulus* ventral photoreceptor cells. Calmodulin itself can inhibit PLC activity in vitro by 30%.[174] Whether calmodulin regulation of PLC activity also occur in vivo in fly photoreceptors is not known.

Protein Kinase C (INAC)

The *inaC* gene codes for an eye-specific protein kinase C (PKC).[175,176] PKCs generally are activated by both DAG and Ca^{2+}. INAC is therefore a prime candidate for mediating Ca^{2+}-dependent feedback on the light-response. Accordingly, in *inaC* mutant flies, the termination phase of the light response is characteristically prolonged,[50,166,176,177] resembling those obtained from wild-type cells in the absence of extracellular Ca^{2+}. Without extracellular Ca^{2+} there is no difference between wild-type and *inaC* null mutants, firmly establishing that INAC causes Ca^{2+}-dependent negative feedback on the light response.[50,177] The abnormal termination of the light response is caused by defective inactivation of individual quantum bumps.[177] The quantum bumps of *inaC* mutant photoreceptor thus last longer and, at low to intermediate light intensities, overlap more than wild-type bumps. This in turn causes the light response to dim illumination to be somewhat larger in flies lacking INAC compared to wild-type flies.[68] In addition, the Ca^{2+} influx through the prolonged quantum bumps is larger, which is reflected in an augmented increase in Ca_i during intense light stimulation and an augmented decrease of Ca_o.[68] This additional Ca^{2+} influx is carried mainly by TRP channels, as the Ca^{2+} homeostasis in *trp;inaC* double mutants is not much different from their *trp* controls.[68] Additionally, flies without INAC fail to show some characteristic effects of light-adaptation. The peak-to-plateau transition is absent and the photoreceptors do not change their sensitivity due to background illumination (no shift in the response/intensity curve).[177] Instead, the light response of *inaC* photoreceptor cells declines slowly to (almost) baseline under intense illumination.[177]

Ca^{2+} homeostasis in *inaC* mutant flies was also studied by investigating the pupil response. It was found that the pupil response is activated at slightly lower light intensities compared to wild-type flies and that a light flash induced a longer lasting pupil response.[62] Considering that the pupil response is a Ca^{2+}-dependent process, these results are fairly straightforwardly explained by considering the electrophysiological phenotype, i.e., enhanced response to weak light stimulation and delayed response termination. Interestingly however, the pupil response in flies lacking INAC did not decay to baseline during prolonged illumination—as the light response does. Also, the pupil in INAC deficient flies does not open entirely during prolonged dark adaptation.[62] These findings are not readily explained from the electrophysiologically determined phenotype and it is therefore unclear how the lack of INAC causes these effects.

INAC is, together with TRP and NORPA, part of the INAD-dependent signalling complex.[38,158] The close spatial proximity is thought to promote rapid and efficient phosphorylation of TRP and INAD by INAC. INAC phosphorylates INAD and TRP in a Ca^{2+}-dependent manner in vitro.[167,168,178] Phosphorylation of INAD and TRP is very fast, as more than 50% of the maximal amount of phosphorylation is obtained already after 1 s.[178] Ca^{2+}-dependent phosphorylation of TRP occurs also in intact retinas.[178] Contrary to TRP and INAD, the PLCβ NORPA is not—or only slightly—phosphorylated by INAC.[168,178] Although the consequence of the phosphorylations of TRP and INAD have not been elucidated directly, for example by genetically removing the relevant phosphorylation sites, the available data strongly suggest that they are key elements of Ca^{2+}-dependent negative feedback on the phototransduction cascade.

NINAC

Through alternative splicing, the *ninaC* locus codes for two different proteins that are expressed in *Drosophila* photoreceptor cells.[179,180] The protein designated $NINAC^{p132}$ is localized to the cell body while $NINAC^{p174}$ is exclusively found in the rhabdomeres.[181,182] In agreement with this subcellular localization, only *ninaC* mutants without or with an altered $NINAC^{p174}$ manifest an electrophysiological phenotype.[182]

NINAC codes for an unconventional myosin with a kinase domain.[180] Additionally, NINAC^{p174} contains two calmodulin binding sites, two sites for phosphorylation by a serine/threonine kinase and a domain that allows binding to INAD.[145,163,183] NINAC^{p174} has been exceptionally well studied by deleting each of these domains separately, thereby showing that they are all of importance to the function of the protein. For the purpose of Ca^{2+} signaling however, the only well-established function of NINAC is to ensure a very high (0.5 mM) concentration of calmodulin in the rhabdomeres.[37,145]

The myosin domain is important for the correct subcellular localization of NINAC^{p174} to the rhabdomere.[184] NINAC binds to actin in vitro[185] and it is therefore believed that the myosin domain serves to anchor NINAC to the actin filaments that run through the microvilli. Accordingly, NINAC^{p174} is not concentrated to the rhabdomeres any longer when it lacks the myosin domain, but also found in the cell bodies.[184] Although not yet demonstrated directly, the absence of the myosin domain might therefore have an impact on the localization of calmodulin. The calmodulin concentration in the rhabdomere also decreases when only one, or both, of the calmodulin binding sites of NINAC^{p174} are mutated.[145] The NINAC mutants that result in reduced calmodulin content in the rhabdomere generally have an abnormally prolonged termination phase of the photoresponse.[37,145,182,186] Individual bumps in these mutants also fail to terminate normally,[186] but NINAC^{p174} has additionally been implicated in modulating the latency period between light absorption and the generation of a bump.[187] Both, the form of the bumps and the latency distribution therefore might contribute to the prolonged termination after strong illumination. It seems likely that at least part of the prolonged termination phenotype of *ninaC* mutants can be attributed to the reduced calmodulin content of the rhabdomere and its effect on arrestin phosphorylation (see below).

Besides the defects during the termination phase, flies missing NINAC^{p174} have a reduced peak-to-plateau transition when stimulated with strong light; this might indicate that NINAC plays a role in light-adaptation.[145,184] Deletion of the NINAC kinase domain,[182] the phosphorylation sites[183] or the site that mediates interaction of NINAC and INAD[163] also causes the mutant flies to have abnormal light responses, but neither Ca^{2+} nor calmodulin has yet been implicated in the regulation of these processes.

Arrestin

After a rhodopsin has absorbed a photon and activated a G-protein it is quickly turned off. Failure to do so would allow the rhodopsin to continuously activate the phototransduction cascade, severely compromising the temporal acuity of the photoreceptor. In fly photoreceptors, the first mechanism to inactivate rhodopsin is binding of arrestin to rhodopsin.[188-190] Two different arrestins are expressed in the photoreceptor cells,[191-194] both of which can bind to and thereby inactivate rhodopsin.[188] Arrestin2 (Arr2; also called phosrestin1)[195] is far more abundant than arrestin1, but even both proteins together are significantly less abundant than rhodopsin.[188] When metarhodopsin (the activated form of rhodopsin) is generated in excess of the available arrestin, the light response does not terminate (this phenomenon has been termed PDA, prolonged depolarizing afterpotential, for review see refs. 36,144). In flies, which have a reduced arrestin concentration (*arr2* mutant flies), this phenomenon occurs at lower metarhodopsin concentrations compared to wild types.[188,190] Also, when *arr2* mutant photoreceptor cells absorb only a single photon, they may produce more than one bump, sometimes even a train of bumps, indicating that rhodopsin inactivation is impaired.[52]

After arrestin binds to activated rhodopsin, it promotes the phosphorylation of rhodopsin.[196] For the subsequent liberation of arrestin from rhodopsin, arrestin itself needs to be phosphorylated.[197] Both arrestins are phosphorylated,[198,199] but Arr2 phosphorylation occurs much faster than the phosphorylation of arrestin1.[195,200] The phosphorylation of Arr2 depends on a calmodulin-dependent kinase type II (CaMKII)[192,200-202] and therefore on the presence of calmodulin. Consequently, mutants with a reduced calmodulin content in the microvilli (hypomorphic *cam* alleles and NINAC^{p174} deficient flies) will have a reduced concentration of free Arr2, as Arr2 stays bound to the membrane.[197] These mutants display a similar phenotype as *arr2* mutants (e.g., delayed response termination and multiple bumps after the absorption of a single photon).[52] The prolonged deactivation phase of *ninaC* mutants, which have a reduced calmodulin content in the rhabdomere,[37,145] can—at least qualitatively—be understood from the resulting shortage of free Arr2. The CaMKII that phosphorylates Arr2 has not yet been genetically identified, although a CaMKII that binds to calmodulin has been reported to be expressed in the brain and the retina of Drosophila.[150,203]

Phosphorylation of arrestins is light-dependent[192,195] and depends in vitro on the presence of Ca^{2+}.[191,204] It seems therefore that light-induced Ca^{2+} influx leads to an increased binding of Ca^{2+} to calmodulin which in turn stimulates a CaMKII to phosphorylate Arr2.

RDGC and Other Phosphatases

The various phosphorylation reactions that are catalyzed by the various kinases, the PKC INAC (Section 5.5), an unidentified PKA,[187] an unidentified CaMKII,[201] NINAC (Section 5.4) and the unidentified rhodopsin kinase, require the presence of phosphatases for reversing the phosphorylations. Up to now, only the rhodopsin phosphatase RDGC has been identified molecularly.[204-206] Dephosphorylation of rhodopsin depends on the presence of RDGC.[204,206] The dephosphorylation of rhodopsin is a Ca^{2+}-dependent process,[189,204] and from the primary structure of the RDGC gene it has been inferred that it contains Ca^{2+} binding domains (so-called EF-hand domains).[205]

In addition to the proposed direct binding of Ca^{2+} to RDGC, RDGC binds to calmodulin, which stimulates the phosphatase activity.[207] Calmodulin binding to

RDGC depends at least partially on Ca^{2+}, and flies expressing a genetically modified RDGC that does not bind calmodulin show an increased level of rhodopsin phosphorylation.[207]

Electrophysiologically, flies lacking RDGC have a severely prolonged termination phase of the light response.[206,207] This phenotype results from overphosphorylated rhodopsin, as it does not occur in flies that express a truncated rhodopsin without the phosphorylation sites.[206] How the phosphorylation of rhodopsin causes the prolonged inactivation phase is not known. Flies expressing RDGC without the calmodulin binding site have a similar, albeit less severe, electrophysiological phenotype as flies without any RDGC.[207]

Contrary to the rhodopsin phosphatase RDGC, the phosphatases that dephosphorylate INAD and TRP have not yet been identified. They are, however, enriched in the cytosol of retinal cells and their enzymatic activity is not dependent on divalent cations.[168] Likewise, the dephosphorylation of arrestin has been reported to be unaffected by Ca^{2+} levels.[204]

CALCIUM SIGNALS IN THE PHOTORECEPTOR CELLS OF ANOTHER INSECT, THE HONEYBEE DRONE

Besides the flies, the only other insect species in which Ca^{2+} homeostasis of photoreceptor cells has been investigated is the honeybee drone (*Apis mellifera* male). Honeybee drones are—like flies—highly visual animals with photoreceptors that perform excellently under natural conditions.[208-210] The Ca^{2+} homeostasis in this preparation is sufficiently different compared to the situation in flies to warrant a brief comparison.

Only global Ca^{2+} signals have been measured with fluorescent Ca^{2+} indicator dyes from drone photoreceptor cells.[136] They are qualitatively similar to those measured in flies, showing a large increase in Ca_i developing rapidly after the onset of illumination. Quantitatively, the resting Ca^{2+} concentration in drones is ca 92 nM (160 nM in *Drosophila*), while the light-induced increase in Ca_i does not exceed 1 μM, which is considerably smaller than the increase seen in flies (see section "Global Calcium Signals are Homogeneous Elevations of the Cytosolic Calcium Concentration").[70,108] The latency between the first detectable electrical response and the first increase in Ca_i is comparable to those measured in flies, e.g., a few milliseconds.[136]

Compared to the tiny SRCs of *Drosophila* photoreceptor cells, the SRCs in drone photoreceptor cells are large and well developed.[211,212] They take up Ca^{2+} in an ATP dependent way.[213] The Ca^{2+} sequestering process works at half-maximal speed at cytosolic Ca^{2+} concentration of ca 600 nM, allowing this process to dynamically respond to the physiologically occurring range of Ca^{2+} concentrations.[136,213] Contrary to the situation in *Drosophila*, strong evidence exists that the ER in drone photoreceptor cells releases Ca^{2+} in response to $InsP_3$.[213] The Ca^{2+} dependence of the $InsP_3$-induced Ca^{2+} release displays a bell-shaped curve,[213] as is typically found for $InsP_3$-Rs.[214] Additionally, Ca^{2+} can be released from drone photoreceptor ER by the so-called Ca^{2+}-induced Ca^{2+} release.[215] The physiological trigger for this Ca^{2+} release is an increase in Ca_i. As mentioned above, this type of Ca^{2+} release is typically mediated by ryanodine receptors (RyR). Accordingly, RyRs are found in drone photoreceptors, located on the SRCs and especially enriched in the parts of the SRCs that are closest to the microvilli.[94] This localization of RyRs is coordinated by and depends on the presence of actin filaments.[216]

These results strongly suggest that in drone photoreceptors two independent pathways for Ca^{2+} release exist, both of which seem to be either absent or of minor quantitative importance in flies. In drones, both types of Ca^{2+}-release mechanism are well functional and therefore well suited to produce a light-induced Ca^{2+} release. Light-induced Ca^{2+} release from intracellular stores has mainly been investigated by looking at the changes of the extracellular Ca^{2+} concentration. The rationale behind this approach is that Ca^{2+} released from intracellular stores will eventually be extruded into the extracellular space. Drone ommatidia do not have intraommatidial cavities, as the microvilli of all photoreceptor cells of one ommatidium are packed together to form a fused rhabdom.[217] The extracellular space in drone retinas is therefore very small, and Ca^{2+} uptake or release from the photoreceptor cells result in large changes of Ca_o. After onset of the light-stimulus, a small decrease in Ca_o occurs, consistent with a light-induced Ca^{2+} influx.[57] But then a strong increase of Ca_o occurs to values as twice as high as those found in dark adapted retinas, indicating extrusion of Ca^{2+} from the photoreceptor cells.[57] Importantly, this light-induced increase in Ca_o persists after chelating the extracellularly available Ca^{2+}.[218] The quantitative relationship between Ca^{2+} influx, $InsP_3$-induced and Ca^{2+} induced Ca^{2+} release has not yet been determined. This relationship is likely to depend on the light intensity of the stimulus, as low light stimulation only causes a decrease in Ca_o, but not the subsequent increase.[57] Both, $InsP_3$-induced and Ca^{2+}-induced Ca^{2+} release are reduced by applying the PIP_2-chelating drug neomycin to drone photoreceptors.[219] Neomycin also decreases the light-induced increase in Ca_i, further providing evidence that Ca^{2+} release from internal stores makes a substantial contribution to the light-induced rise of Ca_i.[219]

As in flies, Na^+/Ca^{2+} exchange has been shown to be an important mechanism of Ca^{2+} extrusion. The earliest evidence for the presence of Na^+/Ca^{2+} exchange was obtained by showing that the light-induced increase in Ca_o depends on extracellular Na^+.[57,220] Additionally, Ca^{2+} extrusion after a light stimulus was shown to correlate with a Na^+-dependent depolarization, indicative of a current generated by the exchangers.[220] Also, intracellular injection of Na^+ leads to a decrease in sensitivity, consistent with an increase in Ca_i mediated by Na^+/Ca^{2+} exchangers.[221]

The global rise in Ca^{2+} causes light adaptation in drone photoreceptors, very similar to what has been found in flies. Injecting Ca^{2+}-chelators into drone photoreceptors reduces the peak-to-plateau transition and diminishes the light-induced reduction in sensitivity.[136,221] Removing Ca_o similarly, albeit to a lesser extent, reduces the peak-to-plateau transition.[222] Conversely, injecting Ca^{2+} into drone photoreceptors produces smaller and faster responses to light flashes.[221] Light adaptation in drones has been shown to spread from the place of light absorption to neighboring regions of the cell.[223] As in flies,[133] this finding is consistent with a diffusible intracellular signal of adaptation, such as a rise in Ca_i. Thus, an increase in Ca_i is equally important for light adaptation in drones and in flies, however, the way in which the increase in Ca_i is brought about is quite different.

Do drone photoreceptor cells also make use of local Ca^{2+} signaling and the fast (positive and/or negative) feedback mechanisms that shape the bumps in *Drosophila*? Removing or reducing Ca_o leads to responses to short light flashes that are larger and have a longer duration; increasing Ca_o has the opposite effect.[221,224] These results show that Ca^{2+}-dependent negative feedback acts on the time scale of ca 50 ms, strongly suggesting a local action of Ca^{2+}. On the other hand, Ca_o was found not to influence the rising phase of the flash response.[224] Probably however, it is not possible to conclude from this finding that positive feedback due to Ca^{2+} influx is absent in drones, because in similar experiments in blowflies no dependence on Ca_o of the rising phase of the light response could be found either.[225] Possibly, under these experimental conditions (no voltage-clamp) other processes, such as voltage-gated channels, mask the reduced Ca^{2+}-dependent facilitation.

In deeply dark-adapted drone photoreceptor cells, a modest increase in Ca_i, brought about by very weak background illumination or by directly injecting Ca^{2+}, facilitates the light response.[226] Injecting EGTA, or reducing Ca_o prevents this facilitation.[226] This Ca^{2+}-dependent facilitation likely is an entirely different process than the very fast and local positive feedback that acts on the *Drosophila* phototransduction cascade, because it occurs on a considerably longer time scale and at largely different Ca^{2+} concentrations.

CONCLUSIONS AND OUTLOOK

In this review, I have highlighted the pivotal role of Ca^{2+} for the regulation of photoreceptor function. The evidence discussed strongly suggests that two different types of Ca^{2+} signals exist in fly photoreceptor cells, with largely differing properties and functions. Slow, global Ca^{2+} signals regulate the state of adaptation of these cells, while on the other hand the form of bumps is shaped by fast and very large Ca^{2+} signals confined to the microvilli. Remarkably, these two signals are caused by the same underlying event: Ca^{2+} influx through the light activated channels. The dichotomy of the signals arises by virtue of the unique anatomy of the rhabdomere, which is strongly compartmentalized into its tiny microvilli. There are however still major gaps in our understanding of Ca^{2+} signaling in fly photoreceptor cells. Most importantly, knowledge of cytosolic Ca^{2+} buffers, their molecular nature, their concentration and their biophysical characteristics is almost entirely lacking. Also, the diffusion coefficient of free Ca^{2+} has not yet been directly measured in fly photoreceptors. These data will be needed in order to quantitatively understand how, and to what extent, Ca^{2+} influx changes the calcium concentration in the rhabdomere, as well as in the cell body.

We are only starting to understand how Ca^{2+} influences the cellular response of photoreceptors at the molecular level. A large number of individual proteins have been identified that are regulated by Ca^{2+}, either directly or by binding to Ca^{2+}/calmodulin or by Ca^{2+}-dependent phosphorylation and dephosphorylations (Fig. 12). In the overwhelming majority of cases, interfering with these mechanisms produces flies that have abnormally prolonged termination phases of the light response, i.e., show defects in the negative regulation of the phototransduction. The PKC INAC and the unconventional myosin NINAC have been proposed to additionally play a role in light adaptation, i.e., in the regulation of the sensitivity of the phototransduction cascade. Also, two candidates for positive Ca^{2+}-dependent feedback have been identified, the channel-protein TRP and the PLCβ NORPA. However, the more interesting, albeit much harder, question how these different regulatory events play together to produce their macroscopic effect on the physiology of the photoreceptor cells has not yet been tackled. An advance of our knowledge in this direction will probably demand a more quantitative understanding of the Ca^{2+}-dependent processes and their consequences, as well as a detailed analysis of their time course, probably down to the millisecond range. This will allow to integrate the different pathways that have already been identified into a combined model of how Ca^{2+} brings about its regulation of fly phototransduction. The search for answers to these questions will no doubt increase not only our knowledge of fly phototransduction, but will also further our general understanding of how complex signaling cascades can be modulated and adapted to the prevailing environmental needs.

ACKNOWLEDGEMENT

I thank Drs. D. G. Stavenga, A. Huber and M. Postma for helpful discussions and critically reading the manuscript. M. Postma provided material for the preparation of Figures 8 to 11.

REFERENCES

1. Fasolato C, Innocenti B, Pozzan T. Receptor-activated Ca^{2+} influx: how many mechanisms for how many channels? Trends Pharmacol Sci 1994; 15:77-83.
2. Berridge M. Capacitative calcium entry. Biochem J 1995; 312:1-11.
3. Barritt GJ. Receptor-activated Ca^{2+} inflow in animal cells: a variety of pathways tailored to meet different intracellular Ca^{2+} signalling requirements. Biochem J 1999; 337:153-169.
4. Montell C, Rubin GM. Molecular characterization of the *Drosophila trp* locus: a putative integral membrane protein required for phototransduction. Neuron 1989; 2:1313-1323.
5. Phillips AM, Bull A, Kelly LE. Identification of a *Drosophila* gene encoding a calmodulin-binding protein with homology to the trp phototransduction gene. Neuron 1992; 8:631-642.
6. Putney JW Jr, McKay RR. Capacitative calcium entry channels. Bioessays 1999; 21:38-46.
7. Hofmann T, Schaefer M, Schultz G et al. Transient receptor potential channels as molecular substrates of receptor-mediated cation entry. J Mol Med 2000; 78:14-25.
8. Clapham DE, Runnels LW, Strübing C. The TRP ion channel family. Nat Rev Neurosci 2001; 2:387-396.
9. Montell C. Physiology, phylogeny, and functions of the TRP superfamily of cation channels. Sci STKE 2001; http://stke.sciencemag.org/cgi/content/full/OC_sigtrans;2001/90/re1.
10. Chyb S, Raghu P, Hardie RC. Polyunsaturated fatty acids activate the *Drosophila* light-sensitive channels TRP and TRPL. Nature 1999; 397:255-259.

11. Hofmann T, Obukhov AG, Schaefer M et al. Direct activation of human TRPC6 and TRPC3 channels by diacylglycerol. Nature 1999; 397:259-263.
12. Llinás R, Sugimori M, Silver RB. Microdomains of high calcium concentration in a presynaptic terminal. Science 1992; 256:677-679.
13. Roberts WM. Localization of calcium signals by a mobile calcium buffer in frog saccular hair cells. J Neurosci 1994; 14:3246-3262.
14. Naraghi M, Neher E. Linearized bu$^{ffered\ Ca2+}$ diffusion in microdomains and its implications for calculation of [Ca^{2+}] at the mouth of a calcium channel. J Neurosci 1997; 17:6961-6973.
15. Koch C, Zador A. The function of dendritic spines: devices subserving biochemical rather than electrical compartmentalization. J Neurosci 1993; 13:413-422.
16. Nimchinsky EA, Sabatini BL, Svoboda K. Structure and function of dendritic spines. Annu Rev Physiol 2002; 64:313-353.
17. Augustine GJ, Neher E. Neuronal Ca^{2+} signalling takes the local route. Curr Opin Neurobiol 1992; 2:302-307.
18. Bootman M, Lipp P, Berridge M. The organisation and functions of local Ca^{2+} signals. J Cell Sci 2001; 114:2213-2222.
19. Hardie RC. Functional organization of the fly retina. In: Ottoson D, ed. Progress in sensory physiology, Vol. 5. Berlin: Springer, 1985:1-79.
20. van Hateren JH. Photoreceptor optics, theory and practice. In: Stavenga DG, Hardie RC, eds. Facets of vision. Berlin: Springer, 1989:74-89.
21. Kirschfeld K, Franceschini N. Ein Mechanismus zur Steuerung des Lichtflusses in den Rhabdomeren des Komplexauges von Musca. Kybernetik 1969; 6:13-22.
22. Stavenga DG. Pseudopupils of compound eyes In: Autrum H, ed. Handbook of sensory physiology, Vol. VII/6A. Berlin: Springer, 1979:357-439.
23. Stavenga DG. Insect retinal pigments: spectral characteristics and physiological functions. Prog Ret Eye Res 1995; 15:231-259.
24. Howard J, Blakeslee B, Laughlin SB. The intracellular pupil mechanism and photoreceptor signal: noise ratios in the fly *Lucilia cuprina*. Proc R Soc Lond B Biol Sci 1987; 231:415-435.
25. Roebroek JGH, Stavenga DG. Insect pupil mechanisms. IV. Spectral characteristics and light intensity dependence in the blowfly, *Calliphora erythrocephala*. J Comp Physiol A 1990; 166:537-543.
26. Boschek CB. On the fine structure of the peripheral retina and lamina ganglionaris of the fly, *Musca domestica*. Z Zellforsch Mikrosk Anat 1971; 118:369-409.
27. Walz B. Calcium-sequestering smooth endoplasmic reticulum in retinula cells of the blowfly. J Ultrastruct Res 1982; 81:240-248.
28. Matsumoto-Suzuki E, Hirosawa K, Hotta Y. Structure of the subrhabdomeric cisternae in the photoreceptor cells of *Drosophila melanogaster*. J Neurocytol 1989; 18:87-93.
29. Hamdorf K, Hochstrate P, Höglund G et al. Light activation of the sodium pump in blowfly photoreceptors. J Comp Physiol A 1988; 162:285-300.
30. Laughlin SB, de Ruyter van Steveninck RR, Anderson JC. The metabolic cost of neural information. Nat Neurosci 1998; 1:36-41.
31. Trujillo-Cenóz O, Melamed J. Electron microscope observations on the peripheral and intermediate retinas of dipterans. In: Bernard CG, ed. The functional organization of the compound eye. Oxford: Pergamon, 1966:339-361.
32. Suzuki E, Katayama E, Hirosawa K. Structure of photoreceptive membranes of *Drosophila* compound eyes as studied by quick-freezing electron microscopy. J Electron Microsc (Tokyo) 1993; 42:178-184.
33. Postma M, Oberwinkler J, Stavenga DG. Does Ca^{2+} reach millimolar concentrations after single photon absorption in *Drosophila* photoreceptor microvilli? Biophys J 1999; 77:1811-1823.
34. Blest AD, Stowe S, Eddey W. A labile, Ca^{2+}-dependent cytoskeleton in rhabdomeral microvilli of blowflies. Cell Tissue Res 1982; 223:553-573.

35. Arikawa K, Hicks JL, Williams DS. Identification of actin filaments in the rhabdomeral microvilli of *Drosophila* photoreceptors. J Cell Biol 1990; 110:1993-1998.
36. Hamdorf K. The physiology of invertebrate visual pigments. In: Autrum H, ed. Handbook of sensory physiology, Vol. VII/6A. Berlin: Springer, 1979:145-224.
37. Porter JA, Yu M, Doberstein SK et al. Dependence of calmodulin localization in the retina on the NINAC unconventional myosin. Science 1993; 262:1038-1042.
38. Huber A, Sander P, Gobert A et al. The transient receptor potential protein (Trp), a putative store-operated Ca^{2+} channel essential for phosphoinositide-mdiated photoreception, forms a signaling complex with NorpA, InaC and InaD. EMBO J 1996; 15:7036-7045.
39. Devary O, Heichal O, Blumenfeld A et al. Coupling of photoexcited rhodopsin to inositol phospholipid hydrolysis in fly photoreceptors. Proc Natl Acad Sci USA 1987; 84:6939-6943.
40. Kirschfeld K. Discrete and graded receptor potentials in the compound eye of the fly (Musca). In: Bernhard CG, ed. The functional organization of the compound eye. Oxford: Pergamon, 1966:291-307.
41. Wu CF, Pak WL. Quantal basis of photoreceptor spectral sensitivity of *Drosophila melanogaster*. J Gen Physiol 1975; 66:149-168.
42. Hardie RC. Whole-cell recordings of the light induced current in dissociated *Drosophila* photoreceptors: evidence for feedback by calcium permeating the light-sensitive channels. Proc R Soc Lond B 1991; 245:203-210.
43. Henderson SR, Reuss H, Hardie RC. Single photon responses in *Drosophila* photoreceptors and their regulation by Ca^{2+}. J Physiol 2000; 524:179-194.
44. Martin AR. A further study of the statistical composition of the end-plate potential. J Physiol 1955; 130:114-122.
45. Minke B. *Drosophila* mutant with a transducer defect. Biophys Struct Mech 1977; 3:59-64.
46. Wu CF, Pak WL. Light-induced voltage noise in the photoreceptor of *Drosophila melanogaster*. J Gen Physiol 1978; 71:249-268.
47. Wong F, Knight BW. Adapting-bump model for eccentric cells of *Limulus*. J Gen Physiol 1980; 76:539-557.
48. Juusola M, Hardie RC. Light adaptation in *Drosophila* photoreceptors: I. Response dynamics and signaling efficiency at 25°C. J Gen Physiol 2001; 117:3-25.
49. Wong F, Knight BW, Dodge FA. Adapting bump model for ventral photoreceptors of *Limulus*. J Gen Physiol 1982; 79:1089-1113.
50. Ranganathan R, Harris GL, Stevens CF et al. A *Drosophila* mutant defective in extracellular calcium-dependent photoreceptor deactivation and rapid desensitization. Nature 1991; 354:230-232.
51. Hochstrate P, Hamdorf K. Microvillar components of light adaptation in blowflies. J Gen Physiol 1990; 95:891-910.
52. Scott K, Sun Y, Beckingham K et al. Calmodulin regulation of *Drosophila* light-activated channels and receptor function mediates termination of the light response in vivo. Cell 1997; 91:375-383.
53. Scott K, Zuker CS. Assembly of the *Drosophila* phototransduction cascade into a signalling complexshapes elementary responses. Nature 1998; 395:805-808.
54. Hardie RC. Phototransduction in *Drosophila melanogaster*. J Exp Biol 2001; 204:3403-3409.
55. Schraermeyer U, Polyanovsky A, Pivovarova N et al. Extracellular compartments of the blowfly eye: ionic content and topology. Vis Neurosci 1999; 16:461-474.
56. Sandler C, Kirschfeld K. Light intensity controls extracellular Ca^{2+} concentration in the blowfly retina. Naturwissenschaften 1988; 75:256-258.
57. Ziegler A, Walz B. Analysis of extracellular calcium and volume changes in the compound eye of the honeybee drone, *Apis mellifera*. J Comp Physiol A 1989; 165:697-709.
58. Muijser H. The receptor potential of retinular cells of the blowfly *Calliphora*: the role of sodium, potassium and calcium ions. J Comp Physiol 1979; 132:87-95.
59. Kirschfeld K, Vogt K. Calcium ions and pigment migration in fly photoreceptors. Naturwissenschaften 1980; 67:516-517.

60. Lo MC, Pak WL. Light-induced pigment migration in the retinula cells of *Drosophila melanogaster*. Comparison of wild type with ERG-defective mutants. J Gen Physiol 1981; 77:155-175.
61. Howard J. Calcium enables photoreceptor pigment migration in a mutant fly. J Exp Biol 1984; 113:471-475.
62. Hofstee CA, Stavenga DG. Calcium homeostasis in photoreceptor cells of *Drosophila mutant inaC* and *trp* studied with the pupil mechanism. Vis Neurosci 1996; 13:257-263.
63. Sandler C, Kirschfeld K. Light-induced extracellular calcium and sodium concentration changes in the retina of *Calliphora*: involvement in the mechanism of light adaptation. J Comp Physiol A 1991; 169:299-311.
64. Rom-Glas A, Sandler C, Kirschfeld K et al. The *nss* mutation or lanthanum inhibits light-induced Ca^{2+} influx into fly photoreceptors. J Gen Physiol 1992; 100:767-781.
65. Peretz A, Suss-Toby E, Rom-Glas A et al. The light response of *Drosophila* photoreceptors is accompanied by an increase in cellular calcium: effects of specific mutations. Neuron 1994; 12:1257-1267.
66. Hardie RC, Minke B. The trp gene is essential for a light-activated Ca^{2+} channel in *Drosophila* photoreceptors. Neuron 1992; 8:643-651.
67. Reuss H, Mojet MH, Chyb S et al. In vivo analysis of the *Drosophila* light-sensitive channels, TRP and TRPL. Neuron 1997; 19:1249-1259.
68. Peretz A, Sandler C, Kirschfeld K et al. Genetic dissection of light-induced Ca^{2+} influx into *Drosophila* photoreceptors. J Gen Physiol 1994; 104:1057-1077.
69. Ranganathan R, Bacskai BJ, Tsien RY et al. Cytosolic calcium transients: spatial localization and role in *Drosophila* photoreceptor cell function. Neuron 1994; 13:837-848.
70. Hardie RC. INDO-1 measurements of absolute resting and light-induced Ca^{2+} concentration in *Drosophila* photoreceptors. J Neurosci 1996; 16:2924-2933.
71. Hardie RC. A quantitative estimate of the maximum amount of light-induced Ca^{2+} release in *Drosophila* photoreceptors. J Photochem Photobiol B 1996; 35:83-89.
72. Niemeyer BA, Suzuki E, Scott K et al. The *Drosophila* light-activated conductance is composed of the two channels TRP and TRPL. Cell 1996; 85:651-659.
73. Hardie RC, Reuss H, Lansdell SJ et al. Functional equivalence of native light-sensitive channels in the *Drosophila* trp^{301} mutant and TRPL cation channels expressed in a stably transfected *Drosophila* cell line. Cell Calcium 1997; 21:431-440.
74. Xu XZ, Li HS, Guggino WB et al. Coassembly of TRP and TRPL produces a distinct store-operated conductance. Cell 1997; 89:1155-1164.
75. Paulsen R, Bähner M, Huber A. The PDZ assembled "transducisome" of microvillar photoreceptors: the TRP/TRPL problem. Pflügers Arch 2000; 439:R181-R183.
76. Hardie RC, Mojet MH. Magnesium-dependent block of the light-activated and *trp*-dependent conductance in *Drosophila* photoreceptors. J Neurophysiol 1995; 74:2590-2599.
77. Montell C. Visual transduction in *Drosophila*. Annu Rev Cell Dev Biol 1999; 15:231-268.
78. Xu XZ, Chien F, Butler A et al. TRPγ, a *Drosophila* TRP-related subunit, forms a regulated cation channel with TRPL. Neuron 2000; 26:647-657.
79. Gillo B, Chorna I, Cohen H et al. Coexpression of *Drosophila* TRP and TRP-like proteins in *Xenopus* oocytes reconstitutes capacitative Ca^{2+} entry. Proc Natl Acad Sci USA 1996; 93:14146-14151.
80. Li C, Geng C, Leung HT et al. INAF, a protein required for transient receptor potential Ca^{2+} channel function. Proc Natl Acad Sci USA 1999; 96:13474-13479.
81. Oberwinkler J, Stavenga DG. Calcium imaging demonstrates colocalization of calcium influx and extrusion in fly photoreceptors. Proc Natl Acad Sci USA 2000; 97:8578-8583.
82. Pollock JA, Assaf A, Peretz A et al. TRP, a protein essential for inositide-mediated Ca^{2+} influx is localized adjacent to the calcium stores in *Drosophila* photoreceptors. J Neurosci 1995; 15:3747-3760.

83. Polyanovsky AD, Hardie RC, Krause K. The TRP protein in *Drosophila* photoreceptors: an immunogold localization of the light-sensitive calcium-permeable channel. J Evol Biochem Physiol 1999; 35:506-515.
84. Berridge M. Inositol trisphosphate and calcium signalling. Nature 1993; 361:315-325.
85. Yoshioka T, Inoue H, Hotta Y. Absence of phosphatidylinositol phosphodiesterase in the head of a *Drosophila* visual mutant, *norpA* (no receptor potential A). J Biochem (Tokyo) 1985; 97:1251-1254.
86. Inoue H, Yoshioka T, Hotta Y. A genetic study of inositol trisphosphate involvement in phototransduction using *Drosophila* mutant. Biochem Biophys Res Commun 1985; 132:513-519.
87. Pearn MT, Randall LL, Shortridge RD et al. Molecular, biochemical, and electrophysiological characterization of *Drosophila norpA* mutants. J Biol Chem 1996; 271:4937-4945.
88. Cook B, Bar-Yaacov M, Cohen Ben-Ami H et al. Phospholipase C and termination of G-protein-mediated signalling in vivo. Nat Cell Biol 2000; 2:296-301.
89. Schneuwly S, Burg M, Lending C et al. Properties of photoreceptor-specific phospholipase C encoded by the *norpA* gene of *Drosophila melanogaster*. J Biol Chem 1991; 266:24314-24319.
90. Hasan G, Rosbash M. *Drosophila* homologs of two mammalian intracellular Ca^{2+}-release channels: identification and expression patterns of the inositol 1,4,5-triphosphate and the ryanodine receptor genes. Development 1992; 116:967-975.
91. Raghu P, Hasan G. The inositol 1,4,5-triphosphate receptor expression in *Drosophila* suggests a role for IP_3 signalling in muscle development and adult chemosensory functions. Dev Biol 1995; 171:564-577.
92. Acharya JK, Jalink K, Hardy RW et al. $InsP_3$ receptor is essential for growth and differentiation but not for vision in *Drosophila*. Neuron 1997; 18:881-887.
93. Raghu P, Colley NJ, Webel R et al. Normal phototransduction in *Drosophila* photoreceptors lacking an $InsP_3$ receptor gene. Mol Cell Neurosci 2000; 15:429-445.
94. Baumann O. Distribution of ryanodine receptor Ca^{2+} channels in insect photoreceptor cells. J Comp Neurol 2000; 421:347-361.
95. Hardie RC, Raghu P. Activation of heterologously expressed *Drosophila* TRPL channels: Ca^{2+} is not required and $InsP_3$ is not sufficient. Cell Calcium 1998; 24:153-163.
96. Sullivan KM, Scott K, Zuker CS et al. The ryanodine receptor is essential for larval development in *Drosophila melanogaster*. Proc Natl Acad Sci USA 2000; 97:5942-5947.
97. Arnon A, Cook B, Montell C et al. Calmodulin regulation of calcium stores in phototransduction of *Drosophila*. Science 1997; 275:1119-1121.
98. Arnon A, Cook B, Gillo B et al. Calmodulin regulation o light adaptation and store-operated dark current in *Drosophila* photoreceptors. Proc Natl AcadSci USA 1997; 94:5894-5899.
99. Cook B, Minke B. TRP and calcium stores in *Drosophila* phototransduction. Cell Calcium 1999; 25:161-171.
100.Chorna-Ornan I, Joel-Almagor T, Ben-Ami HC et al. A common mechanism underlies vertebrate calcium signaling and *Drosophila* phototransduction. J Neurosci 2001; 21:2622-2629.
101. Blaustein MP, Lederer WJ. Sodium/calcium exchange: its physiological implications. Physiol Rev 1999; 79:763-854.
102. Schwarz EM, Benzer S. Calx, a Na-Ca exchanger gene of *Drosophila melanogaster*. Proc Natl Acad Sci USA 1997; 94:10249-10254.
103. Haug-Collet K, Pearson B, Webel R et al. Cloning and characterization of a potassium-dependent sodium/calcium exchanger in *Drosophila*. J Cell Biol 1999; 147:659-670.
104. Hardie RC. Photolysis of caged Ca^{2+} facilitates and inactivates but does not directly excite light-sensitive channels in *Drosophila* photoreceptors. J Neurosci 1995; 15:889-902.
105. Hardie RC. Effects of intracellular Ca^{2+} chelation on the light response in *Drosophila* photoreceptors. J Comp Physiol A 1995; 177:707-721.
106. Hochstrate P. Electrogenic Na^{+}-Ca^{2+} exchange contributes to the light response of fly photoreceptors. Z Naturforsch 1991; 46c:451-460.

107. Gerster U. A quantitative estimate of flash-induced Ca^{2+}- and Na^+-influx and Na^+/Ca^{2+}-exchange in blowfly *Calliphora* photoreceptors. Vision Res 1997; 37:2477-2485.
108. Oberwinkler J, Stavenga DG. Light dependence of calcium and membrane potential measured in blowfly photoreceptors in vivo. J Gen Physiol 1998; 112:113-124.
109. Hochstrate P, Juse A. Intracellular free calcium concentration in the blowfly retina studied by Fura-2. Cell Calcium 1991; 12:695-712.
110. Bauer PJ, Schauf H, Schwarzer A et al. Direct evidence of Na^+/Ca^{2+} exchange in squid rhabdomeric membranes. Am J Physiol 1999; 276:C558-C565.
111. Meldolesi J, Pozzan T. The endoplasmic reticulum Ca^{2+} store: a view from the lumen. Trends Biochem Sci 1998; 23:10-14.
112. Blaustein MP, Golovina VA. Structural complexity and functional diversity of endoplasmic reticulum Ca^{2+} stores. Trends Neurosci 2001; 24:602-608.
113. Duchen MR. Mitochondria and calcium: from cell signalling to cell death. J Physiol 2000; 529:57-68.
114. Rizzuto R, Bernardi P, Pozzan T. Mitochondria as all-round players of the calcium game. J Physiol 2000; 529:37-47.
115. Magyar A, Bakos E, Varadi A. Structure and tissue-specific expression of the *Drosophila melanogaster* organellar-type Ca^{2+}-ATPase gene. Biochem J 1995; 310:757-763.
116. Periz G, Fortini ME. Ca^{2+}-ATPase function is required for intracellular trafficking of the Notch receptor in *Drosophila*. EMBO J 1999; 18:5983-5993.
117. Hardie RC. Excitation of *Drosophila* photoreceptors by BAPTA and ionomycin: evidence for capacitative Ca^{2+} entry? Cell Calcium 1996; 20:315-327.
118. Mojet MH, Tinbergen J, Stavenga DG. Receptor potential and light-induced mitochondrial activation in blowfly photoreceptor mutants. J Comp Physiol A 1991; 168:305-312.
119. Tsacopoulos M, Orkand RK, Coles JA et al. Oxygen uptake occurs faster than sodium pumping in bee retina after a light flash. Nature 1983; 301:604-606.
120. Fein A, Tsacopoulos M. Activation of mitochondrial oxidative metabolism by calcium ions in *Limulus* ventral photoreceptor. Nature 1988; 331:437-440.
121. Ballinger DG, Xue N, Harshman KD. A *Drosophila* photoreceptor cell-specific protein, calphotin, binds calcium and contains a leucine zipper. Proc Natl Acad Sci USA 1993; 90:1536-1540.
122. Martin JH, Benzer S, Rudnicka M et al. Calphotin: a *Drosophila* photoreceptor cell calcium-binding protein. Proc Natl Acad Sci USA 1993; 90:1531-1535.
123. Yang Y, Ballinger D. Mutations in *calphotin*, the gene encoding a *Drosophila* photoreceptor cell-specific calcium-binding protein, reveal roles in cellular morphogenesis and survival. Genetics 1994; 138:413-421.
124. Ukhanov KY. Ommochrome pigment granules: a calcium reservoir in the dipteran eyes. Comp Biochem Physiol 1991; 98A:9-16.
125. Schröder W, Frings D, Stieve H. Measuring calcium uptake and release by invertebrate photoreceptor cells by laser micromass spectroscopy. Scan Electron Microsc 1980; 11:647-652.
126. White RH, Michaud NA. Calcium as a component of ommochrome pigment granules in insect eyes. Comp Biochem Physiol 1980; 65A:239-242.
127. Laughlin SB, Hardie RC. Common strategies for light adaptation in the peripheral visual systems of fly and dragonfly. J Comp Physiol 1978; 128:319-340.
128. Oberwinkler J, Stavenga DG. Is the pupil mechanism of fly photoreceptors a useful calcium probe? In: Taddei-Ferretti C, Musio C, eds. From structure to information in sensory systems. Singapore: World Scientific, 1998:504-508.
129. Weckström M, Hardie RC, Laughlin SB. Voltage-activated potassium channels in blowfly photoreceptors and their role in light adaptation. J Physiol 1991; 440:635-657.
130. Hardie RC, Voss D, Pongs O et al. Novel potassium channels encoded by the *Shaker* locus in *Drosophila* photoreceptors. Neuron 1991; 6:477-486.
131. Hardie RC. Voltage-sensitive potassium channels in *Drosophila* photoreceptors. J Neurosci 1991; 11:3079-3095.

132. Laughlin SB, Weckström M. Fast and slow photoreceptors—a comparative study of the functional diversity of coding and conductance in the Diptera. J Comp Physiol A 1993; 172:593-609.
133. Minke B, Payne R. Spatial restriction of light adaptation and mutation-induced inactivation in fly photoreceptors. J Neurosci 1991; 11:900-909.
134. Peretz A, Abitbol I, Sobko A et al. A Ca^{2+}/calmodulin-dependent protein kinase modulates *Drosophila* photoreceptor K^{+} currents: a role in shaping the photoreceptor potential. J Neurosci 1998; 18:9153-9162.
135. Oberwinkler J, Stavenga DG. Calcium transients in the rhabdomeres of dark- and light-adaptd fly photoreceptor cells. J Neurosci 2000; 20:1701-1709.
136. Walz B, Zimmermann B, Seidl S. Intracellular Ca^{2+} concentration and latency of light-induced Ca^{2+} changes in photoreceptors of the honey bee drone. J Comp Physiol A 1994; 174:421-431.
137. Ukhanov K, Payne R. Light activated calcium release in *Limulus* ventral photoreceptors as revealed by laser confocal microscopy. Cell Calcium 1995; 18:301-313.
138. Payne R, Ukhanov K. Latencies of calcium elevation and depolarization in *Limulus* ventral photoreceptors injected with GDP-βS. J Photochem Photobiol B 1996; 35:91-95.
139. Brown JE, Rubin LJ. A direct demonstration that inositol-trisphosphate induces an increase in intracellular calcium in *Limulus* photoreceptors. Biochem Biophys Res Commun 1984; 125:1137-1142.
140. Fein A, Payne R, Corson DW et al. Photoreceptor excitation and adaptation by inositol 1,4,5-trisphosphate. Nature 1984; 311:157-160.
141. Nasi E, del Pilar Gomez M, Payne R. Phototransduction mechanisms in microvillar and ciliary photoreceptors of invertebrates. In: Stavenga DG, DeGrip WJ, Pugh EN jr, eds. Handbook of biological physics, Vol. 3: Molecular mechanisms in visual transduction. Amsterdam: Elsevier, 2000:389-448.
142. Hille B. Ionic channels of excitable membranes. 2nd ed. Sunderland, MA: Sinauer, 1992.
143. Hardie RC, Raghu P, Moore S et al. Calcium influx via TRP channels is required to maintain PIP_2 levels in *Drosophila* photoreceptors. Neuron 2001; 30:149-159.
144. Minke B, Hardie RC. Genetic dissection of *Drosophila* phototransduction. In: Stavenga DG, DeGrip WJ, Pugh EN jr, eds. Handbook of biological physics, Vol. 3: Molecular mechanisms in visual transduction. Amsterdam: Elsevier, 2000:449-525.
145. Porter JA, Minke B, Montell C. Calmodulin binding to *Drosophila* NinaC required for termination of phototransduction. EMBO J 1995; 14:4450-4459.
146. Yamanaka MK, Saugstad JA, Hanson-Painton O et al. Structure and expression of the *Drosophila* calmodulin gene. Nucleic Acids Res 1987; 15:335-3348.
147. Doyle KE, Kovalick GE, Lee E et al. *Drosophila melanogaster* contains a single calmodulin gene. Further structure and expression studies. J Mol Biol 1990; 213:599-605.
148. Maune JF, Klee CB, Beckinghm K. Ca^{2+} binding and conformational change in two series of point mutations to the individual Ca^{2+}-binding sites of calmodulin. J Biol Chm 1992; 267:5286-5295.
149. Martin SR, Maune JF, Beckingham K et al. Stopped-flow studies of calcium dissociation from calcium-binding-site mutants of *Drosophila melanogaster* calmodulin. Eur J Biochem 1992; 205:1107-1114.
150. Xu XZ,Wes PD, Chen H et al. Retinal targets for calmodulin include proteins implicated in synaptic transmission. J Biol Chem 1998; 273:31297-31307.
151. Hardie RC, Minke B. Spontaneous activation of light-snsitive channels in *Drosophila* photoreceptors. J Gen Physiol 1994; 103:389-407.
152. Hardie RC, Mnke B. Calcium-dependent inactivation of light-sensitive channels in *Drosophila* photoreceptors. J Gen Physiol 1994; 103:409-427.
153. Agam K, von Campenhausen M, Levy S et al. Metabolic stress reversibly activates the *Drosophila* light sensitive channels TRP and TRPL in vivo. J Neurosci 2000; 20:5748-5755.

154. Warr CG, Kelly LE. Identification and characterization of two distinct calmodulin-binding sites in the Trpl ion-channel protein of *Drosophila melanogaster*. Biochem J 1996; 314:497-503.
155. Chevesich J, Kreuz AJ, Montell C. Requirement for the PDZ domain protein, INAD, for localization of the TRP store-operated channel to a sigaling complex. Neuron 1997; 18:95-105.
156. Trost C, Marquart A, Zimmer S et al. Ca^{2+}-dependent interaction of the trpl cation channel and calmodulin. FEBS Lett 1999; 451:257-263.
157. Shieh BH, Zhu MY. Regulation of the TRP Ca^{2+} channel by INAD in *Drosophila* photoreceptors. Neuron 1996; 16:991-998.
158. Tsunoda S, Sierralta J, Sun Y et al. A multivalent PDZ-domain protein assembles signallin complexes in a G-protein-coupled cascade. Nature 1997; 388:243-249.
159. van Huizen R, Miller K, Chen DM et al. Two distantly positioned PDZ domains mediate multivalent INAD-phospholipase C interactions essential for G protein-coupled signaling. EMBO J 1998; 17:2285-2297.
160. Li HS, Montell C. TRP and the PDZ protein, INAD, form the core complex required for retention of the signalplex in *Drosophila* photoreceptor cells. J Cell Biol 2000; 150:1411-1422.
161. Tsunoda S, Sun Y, Suzuki E et al. Independent anchoring and assembly mechanisms of INAD signaling complexes in *Drosophila* photoreceptors. J Neurosci 2001; 21:150-158.
162. Xu XZ, Choudhury A, Li X et al. Coordination of an array of signaling proteins through homo- and heteromeric interactions between PDZ domains and target proteins. J Cell Biol 1998; 142:545-555.
163. Wes PD, Xu XZ, Li HS et al. Termination of phototransduction requires binding of the NINAC myosin III and the PDZ protein INAD. Nat Neurosci 1999; 2:447-453.
164. Shieh BH, Niemeyer B. A novel protein encoded by the InaD gene regulates recovery of visual transduction in *Drosophila*. Neuron 1995; 14:201-210.
165. Shieh BH, Zhu MY, Lee JK et al. Association of INAD with NORPA is essential for controlled activation and deactivation of *Drosophila* phototransduction in vivo. Proc Natl Acad Sci USA 1997; 94:12682-12687.
166. Adamski FM, Zhu MY, Bahiraei F et al. Interaction of eye protein kinase C and INAD in *Drosophila*. Localization of binding domains and electrophysiological characterization of a loss of association in transgenic flies. J Biol Chem 1998; 273:17713-17719.
167. Huber A, Sander P, Paulsen R. Phosphorylation of the InaD gene product, a photoreceptor membrane protein required for recovery of visual excitation. J Biol Chem 1996; 271:11710-11717.
168. Liu M, Parker LL, Wadzinski BE et al. Reversible phosphorylation of the signal transduction complex in *Drosophila* photoreceptors. J Biol Chem 2000; 275:12194-12199.
169. Hotta Y, Benzer S. Abnormal electroretinograms in visual mutants of *Drosophila*. Nature 1969; 222:354-356.
170. Pak W, Grossfield J, Arnold K. Mutants of the visual pathway of *Drosophila melanogaster*. Nature 1970; 227:518-520.
171. Toyoshima S, Matsumoto N, Wang P et al. Purification and partial amino acid sequences of phosphoinositide-specific phospholipase C of *Drosophila* eye. J Biol Chem 1990; 265:14842-14848.
172. Running Deer JL, Hurley JB, Yarfitz SL. G protein control of *Drosophila* photoreceptor phospholipase C. J Biol Chem 1995; 270:12623-12628.
173. Rack M, Xhonneux-Cremers B, Schraermeyer U et al. On the Ca^{2+}-dependence of inositol-phospholipid-specific phospholipase C of microvillar photoreceptors from *Sepia officinalis*. Exp Eye Res 1994; 58:659-664.
174. Richard EA, Ghosh S, Lowenstein JM et al. Ca^{2+}/calmodulin-binding peptides block phototransduction in *Limulus* ventral photoreceptors: evidence for direct inhibition of phospholipase C. Proc Natl Acad Sci USA 1997; 94:14095-14099.

175. Schaeffer E, Smith D, Mardon G et al. Isolation and characterization o two new *Drosophila* protein kinase C genes, including one specifically expressed in photoreceptor cells. Cell 1989; 57:403-412.
176. Smith DP, Ranganathan R, Hardy RW et al. Photoreceptor deactivation and retinal degeneration mediated by a photoreceptor-specific protein kinase C. Science 1991; 254:1478-1484.
177. Hardie RC, Peretz A, Suss-Toby E et al. Protein kinase C is required for light adaptation in *Drosophila* photoreceptors. Nature 1993; 363:634-637.
178. Huber A, Sander P, Bähner M et al. The TRP Ca^{2+} channel assembled in a signaling complex by the PDZ domain protein INAD is phosphorylated through the interaction with protein kinase C (ePKC). FEBS Lett 1998; 425:317-322.
179. Matsumoto H, Isono K, Pye Q et al. Gene encoding cytoskeletal proteins in *Drosophila* rhabdomeres. Proc Natl Acad Sci USA 1987; 84:985-989.
180. Montell C, Rubin GM. The *Drosophila* ninaC locus encodes two photoreceptor cell specific proteins with domains homologous to protein kinases and the myosin heavy chain head. Cell 1988; 52:757-772.
181. Hicks JL, Williams DS. Distribution of the myosin I-like ninaC proteins in the *Drosophila* retina and ultrastructural analysis of mutant phenotypes. J Cell Sci 1992; 101:247-254.
182. Porter JA, Hicks JL, Williams DS et al. Differential localizations of and requirements for the two *Drosophila* ninaC kinase/myosins in photoreceptor cells. J Cell Biol 1992; 116:683-693.
183. Li HS, Porter JA, Montell C. Requirement for the NINAC kinase/myosin for stable termination of the visual cascade. J Neurosci 1998; 18:9601-9606.
184. Porter JA, Montell C. Distinct roles of the *Drosophila* ninaC kinase and myosin domains revealed by systematic mutagenesis. J Cell Biol 1993; 122:601-612.
185. Hicks JL, Liu X, Williams DS. Role of the ninaC proteins in photoreceptor cell structure: ultrastructure of *ninaC* deletion mutants and binding to actin filaments. Cell Motil Cytoskeleton 1996; 35:367-379.
186. Hofstee CA, Henderson S, Hardie RC et al. Differential effects of ninaC proteins (p132 and p174) on light-activated currents and pupil mechanism in *Drosophila* photoreceptors. Vis Neurosci 1996; 13:897-906.
187. Chyb S, Hevers W, Forte M et al. Modulation of the light response by cAMP in *Drosophila* photoreceptors. J Neurosci 1999; 19:8799-8807.
188. Dolph PJ, Ranganathan R, Colley NJ et al. Arrestin function in inactivation of G protein-coupled receptor rhodopsin in vivo. Science 1993; 260:1910-1916.
189. Plangger A, Malicki D, Whitney M et al. Mechanism of arrestin 2 function in rhabdomeric photoreceptors. J Biol Chem 1994; 269:26969-26975.
190. Ranganathan R, Stevens CF. Arrestin binding determines the rate of inactivation of the G protein-coupled receptor rhodopsin in vivo. Cell 1995; 81:841-848.
191. Yamada T, Takeuchi Y, Komori N et al. A 49-kilodalton phosphoprotein in the *Drosophila* photoreceptor is an arrestin homolog. Science 1990; 248:483-486.
192. LeVine H 3rd, Smith DP, Whitney M et al. Isolation of a novel visual-system-specific arrestin: an in vivo substrate for light-dependent phosphorylation. Mech Dev 1990; 33:19-25.
193. Hyde D, Mecklenburg K, Pollock J et al. Twenty *Drosophila* visual system cDNA clones: one is a homolog of human arrestin. Proc Natl Acad Sci USA 1990; 87:1008-1012.
194. Smith DP, Shieh BH, Zuker CS. Isolation and structure of an arrestin gene from *Drosophila*. Proc Natl Acad Sci USA 1990; 87:1003-1007.
195. Matsumoto H, Yamada T. Phosrestins I and II: arrestin homologs which undergo differential light-induced phosphorylation in the *Drosophila* photoreceptor in vivo. Biochem Biophys Res Commun 1991; 177:1306-1312.
196. Bentrop J, Plangger A, Paulsen R. An arrestin homolog of blowfly photoreceptors stimulates visual-pigment phosphorylation by activating a membrane associated protein kinase. Eur J Biochem 1993; 216:67-73.

197. Alloway PG, Dolph PJ. A role for the light-dependent phosphorylation of visual arrestin. Proc Natl Acad Sci USA 1999; 96:6072-6077.
198. Matsumoto H, O'Tousa JE, Pak WL. Light-induced modification of *Drosophila* retinal polypeptides in vivo. Science 1982; 217:839-841.
199. Matsumoto H, Pak W. Light-induced phosphorylation of retina-specific polypeptides of *Drosophila* in vivo. Science 1984; 223:184-186.
200. Matsumoto H, Kurien BT, Takagi Y et al. Phosrestin I undergoes the earliest light-induced phosphorylation by a calcium/calmodulin-dependent protein kinase in *Drosophila* photoreceptors. Neuron 1994; 12:997-1010.
201. Kahn ES, Matsumoto H. Calcium/calmodulin-dependent kinase II phosphorylates *Drosophila* visual arrestin. J Neurochem 1997; 68:169-175.
202. Kahn ES, Kinum T, Tobin SL et al. Phosrestide-1, a peptide derived from the *Drosophila* photoreceptor protein phosrestin I, is a potent substrat for Ca^{2+}/calmodulin-dependent protein kinase II from rat brain. Comp Biochem Physiol B Biochem Mol Biol 1998; 119:739-746.
203. Ohsako S, Nishida Y, Ryo H et al. Molecular characterization and expression of the *Drosophila* Ca^{2+}/calmodulin-dependent protein kinase II ene. Identification of four forms of the enzyme generated from a single gene by alternative splicing. J Biol Chem 1993; 268:2052-2062.
204. Byk T, Bar-Yaacov M, Doza YN et al. Regulatory arrestin cycle secures the fideity and maintenance of the fly photoreceptor cell. Proc Natl Acad Sci USA 1993; 90:1907-1911.
205. Steele FR, Washburn T, Rieger R et al. *Drosophila* retinal degeneration C (*rdgC*) encodes a novel serine/threonine protein phosphatase. Cell 1992; 69:669-676.
206. Vinós J, Jalink K, Hardy RW et al. A G protein-coupled receptor phosphatase required for rhodopsin function. Science 1997; 277:687-690.
207. Lee SJ, Montell C. Regulation of the rodopsin protein posphatase, RDGC, through interaction with calmodulin. Neuron 2001; 32:1097-1106.
208. Vallet AM, Coles JA. A method for estimating the minimum visual stimulus that evokes a behavioural response in the drone, *Apis mellifera* male. Vision Res 1991; 31:1453-1455.
209. Coles JA, Vallet AM. Signal-to-noise atio at high light intensities in drone photoreceptors. Neurosci Res Suppl 1991; 15:S1-11.
210. Vallet AM, Coles JA. Is the membrane voltage amplifier of drone photoreceptors useful at physiological light intensities? J Comp Physiol A 1993; 173:163-168.
211. Perrelet A, Bader CR. Morphological evidence for calcium stores in photoreceptors of the honeybee drone retina. J Ultrastruct Res 1978; 63:237-243.
212. Baumann O, Walz B. Calcium- and inositol polyphosphate-sensitivity of the calcium-sequestering endoplasmic reticulum in the photoreceptor cells of the honeybee drone. J Comp Physiol A 1989; 165:627-636.
213. Baumann O, Walz B. Topography of Ca^{2+}-sequestering endoplasmic reticulum in photoreceptors and pigmented glial cells in the compound eye of the honeybee drone. Cell Tissue Res 1989; 255:511-522.
214. Bezprozvanny I, Watras J, Ehrlich B. Bell-shaped calcium-response curves of Ins(1,4,5)P_3-and calcium-gated channels from endoplasmic reticulum of cerebellum. Nature 1991; 351:751-754.
215. Walz B, Baumann O, Zimmermann B et al. Caffeine- and ryanodine-sensitive Ca^{2+}-induced Ca^{2+} release from the endoplasmic reticulum in honeybee photoreceptors. J Gen Physiol 1995; 105:537-567.
216. Baumann O. Disruption of actin filaments causes redistribution of ryanodine receptor Ca^{2+} channels in honeybee photoreceptor cells. Neurosci Lett 2001; 306:181-184.
217. Perrelet A. The fine structure of the retina of the honey bee drone. An electron microscopical study. Z Zellforsch Mikrosk Anat 1970; 108:530-562.
218. Ziegler A, Walz B. Evidence for light-induced release of Ca^{2+} from intracellular stores in bee photoreceptors. Neurosci Lett 1990; 111:87-91.

219. Walz B, Ukhanov K, Zimmermann B. Actions of neomycin on electrical light responses, Ca^{2+} release, and intracellular Ca^{2+} changes in photoreceptors of the honeybee drone. J Comp Physiol A 2000; 186:1019-1029.
220. Minke B, Tsacopoulos M. Light induced sodium dependent accumulation of calcium and potassium in the extracellular space of bee retina. Vision Res 1986; 26:679-690.
221. Bader C, Baumann F, Bertrand D. Role of intracellular calcium and sodium in light adaptation in the retina of the honey bee drone (*Apis mellifera*, L). J Gen Physiol 1976; 67:475-491.
222. Fulpius B, Baumann F. Effects of sodium, potassium, and calcium ions on slow and spike potentials in single photoreceptor cells. J Gen Physiol 1969; 53:541-561.
223. Bader CR, Baumann F, Bertrand D et al. Diffuse and local effects of light adaptation in photoreceptors of the honey bee drone. Vision Res 1982; 22:311-317.
224. Raggenbass M. Effects of extracellular calcium and of light adaptation on the response to dim light in honey bee drone photoreceptors. J Physiol 1983; 344:525-548.
225. Hochstrate P, Hamdorf K. The influence of extracellular calcium on the response of fly photoreceptors. J Comp Physiol A 1985; 156:53-64.
226. Walz B. Enhancement of sensitivity in photoreceptors of the honey been drone by light and by Ca^{2+}. J Comp Physiol A 1992; 170:605-613.

PHOTORECEPTOR DEGENERATION AND Ca^{2+} INFLUX THROUGH LIGHT-ACTIVATED CHANNELS OF *DROSOPHILA*

Chaoxian Geng and William L. Pak

ABSTRACT

We discuss in this chapter the role of Ca^{2+} homeostasis in maintaining the structural integrity of photoreceptor cells in *Drosophila*. Both insufficient and excessive amounts of Ca^{2+} in photoreceptor cells appear to lead to cell degeneration. Because one of the two classes of light-sensitive channels in *Drosophila* photoreceptors is highly Ca^{2+}-permeable, how well this class of channels functions can profoundly affect Ca^{2+} homeostasis. We will begin by reviewing *Drosophila* phototransduction, emphasizing what is known about the mechanism of activation of light-sensitive channels. We will then describe Ca^{2+} entry through light-sensitive channels and the presumed mechanisms by which too little and too much Ca^{2+} entry can both cause photoreceptor degeneration. We will conclude the chapter with discussions of two examples of mutations known to cause unregulated Ca^{2+} entry through light-sensitive channels, leading to massive photoreceptor degeneration.

INVERTEBRATE PHOTOTRANSDUCTION

Much has been learned about the phototransduction process in *Drosophila* in recent years (reviewed by, e.g., Pak[1]; Zuker[2]; Montell[3]; Minke and Hardie[4]). The phototransduction pathway is based on a phosphoinositide-mediated signaling cascade, in which photoexcited rhodopsin activates a phospholipase C_β (PLC) through the activation of a G protein, G_q. The central role PLC plays in this process is demonstrated by the fact that no light-induced response can be elicited from severely

Department of Biological Sciences, Purdue University, West Lafayette, IN 47907 , U.S.A.

affected mutants of the PLC-encoding norpA gene.[5,6] Activation of phospholipase C leads to the opening of two classes of phototransduction channels: TRP and TRPL channels, encoded, totally or in part, by the trp (transient receptor potential) and *trpl* (*trp-like*) genes,[7,8,9,10] respectively. The TRP channel is widely considered to be homomultimeric, consisting only of TRP as subunits. The subunit composition of the TRPL channel, however, is still under debate.[3,4] Evidence has been presented both for the model that the TRPL channel consists only of TRPL protein as subunits[11,12,13] and the model that it is heteromultimeric, consisting of both TRP and TRPL proteins as subunits.[14] However, a double mutant lacking both the TRP and TRPL proteins is totally unresponsive to light,[12,13] suggesting either that each channel is a homomultimer consisting only of TRP or TRPL or that the TRP and TRPL proteins constitute functionally indispensable portions of the respective channels.

Although the identity of these channels has been known for some time, the elucidation of the mechanisms of their activation and regulation proved elusive. In other phosphoinositide-mediated signaling systems, PLC_β, when activated, is known to hydrolyze the membrane phospholipid, phosphatidylinositol 4,5-bisphosphate (PIP_2), to generate two potential second messengers, inositol 1,4,5-trisphosphate (IP_3) and diacylglycerol (DAG) (review: Berridge[15]). IP_3 binds to the IP_3 receptor (IP_3R) to release Ca^{2+} from internal stores, while DAG is thought to activate protein kinase C. One of the earliest hypotheses to be put forth on *Drosophila* phototransduction was that IP_3 generated in the hydrolysis of PIP_2 by PLC binds to IP_3 receptors on internal stores and the ensuing depletion of internal Ca^{2+} stores signals the opening of the TRP channels on the plasma membrane (Minke and Selinger[16]). However, it was difficult to find definitive experimental support for the hypothesis. For example, depleting internal stores with thapsigargin, an endoplasmic Ca^{2+} ATPase inhibitor, does not activate light-sensitive channels in *Drosophila* photoreceptors (Ranganathan,[17] Hardie[18]). The most damaging piece of evidence against this hypothesis is that null or near-null mutations in the only known IP_3 receptor gene of *Drosophila* are found to have no effect on phototransduction (Acharya,[19] Raghu[20]). Unless there is another IP_3 receptor gene that has not yet been identified by the *Drosophila* Genome Project, the above finding appears to rule out participation of the IP_3 branch in *Drosophila* phototransduction. Moreover, Chyb[21] found that both TRP and TRPL channels could be activated by polyunsaturated fatty acids (PUFAs). Since PUFAs can be generated from DAG by DAG lipase, the above finding seems to suggest that the DAG branch, rather than the IP_3 branch, of the phosphoinositide cascade is utilized in Drosophila phototransduction.

The final chapter apparently has not been written on this story, however. Two recent papers challenged the idea that the DAG branch is involved in phototransduction. In one of these, Chorna-Ornan[22] reported that a newly discovered membrane-permeant IP_3 receptor antagonist, 2-APB (2-aminoethoxydiphenyl borate) was found to reversibly block responses of *Drosophila* photoreceptors to light, suggesting that the IP_3 branch may play a role in phototransduction after all. In the other paper, Agam[23] reported that metabolic stress, such as anoxia, mitochondrial uncoupling, and ATP depletion, reversibly activates TRP and TRPL channels.

They further pointed out that PUFAs are efficient mitochondrial uncouplers (Arslan[24]; Hermesh[25]) and suggested that the previously reported activation of TRP and TRPL channels by PUFAs may be an indirect result of mitochondrial uncoupling. It seems clear additional research is needed to sort out these conflicting findings.

Ca^{2+} ENTRY THROUGH TRP AND TRPL CHANNELS

Whole cell recordings of the light-induced current (LIC) from photoreceptors in dissociated ommatidia showed that in wild-type flies, in which both the TRP and TRPL channels are functional, the reversal potential of LIC is strongly dependent on extracellular concentration of Ca^{2+}, suggesting that the light-activated channels are highly Ca^{2+} permeable (P_{Ca}:P_{Na} of ~24:1) (Hardie[26]). Subsequently, Hardie and Minke[9] showed that the LIC in wild type is mediated by two distinct classes of light-activated channels, TRP and TRPL, and that the TRP channel is highly Ca^{+}-permeable (P_{Ca}:P_{Cs} of >100:1)(Reuss[13]), while the TRPL channel is nonspecifically cation selective but with a significant Ca^{2+} permeability (Peretz[27]; Hardie[18]), the P_{Ca}:P_{cs} ratio being about 4 (Reuss[13]). Ca^{2+} ions entering through these channels regulate the current through the two classes of channels differentially by feedback (Hardie[26]; Reuss[13]). The current through the TRP channel is facilitated first, causing a rapid increase in current influx, and then rapidly inhibited. The current through the TRPL channel is also inhibited but without the initial facilitation. Because only the TRP channel is initially facilitated under physiological conditions, most of the current entering the cell in response to light does so through the TRP channel (Reuss[13]).

MAGNITUDE OF Ca^{2+} ENTRY AND PHOTORECEPTOR DEGENERATION

Photoreceptors do not seem to tolerate well either insufficient or excessive amounts of Ca^{2+} in the cell. The mutations, *rdgA* and *Trp*^{P365}, have been shown recently to cause an excessive entry of Ca^{2+} into photoreceptors by rendering TRP channels constitutively active.[28,29] Cytotoxic effects of Ca^{2+} are well-documented,[30,31] and massive degeneration of photoreceptors ensues as a consequence of the constitutive activity. The degeneration caused by these mutations is among the most severe and earliest-onset reported for *Drosophila*. It can proceed in the absence of light, though light seems to further exacerbate the course of degeneration.[29] These mutations are discussed in some detail in the next sections.

Insufficient Ca^{2+} has been reported to cause light-dependent, late-onset photoreceptor degeneration by interfering with recycling of the rhodopsin inhibitor molecule, arrestin.[32] Following photoactivation of rhodopsin to metarhodopsin, inactivation of metarhodopsin begins almost immediately. As in vertebrates, inactivation of metarhodopsin involves arrestin binding.[33,34] Unlike in vertebrates, however, binding of *Drosophila* arrestins to metarhodopsin is not dependent on metarhodopsin phosphorylation.[35,36] The metarhodopsin-bound arrestin is then phosphorylated by Ca^{2+}-calmodulin dependent kinase II[37] to allow its release from metarhodopsin binding.[38] If arrestin is not phosphorylated or insufficiently phosphorylated, its release from metarhodopsin is inhibited, and metarhodopsin-arrestin complexes accumulate. [36] The complexes are then internalized by endocytosis, and this endocytosis appears to trigger apoptosis of photoreceptor cells by unknown mechanisms.[32,36] Thus, any mutation that substantially decreases Ca^{2+} entry through the light-activated channels is expected to cause degeneration through inefficient phosphorylation of arrestin.[32] It has long been known that many *Drosophila* mutations that affect phototransduction also cause photoreceptor degeneration.[39-45] Since these mutations affect the phototransduction process, many of them are expected to cause substantially reduced Ca^{2+} influx through the TRP channels. Light-dependent degeneration of a relatively slow time course observed in many of these mutants may have its origin in the above mechanism. Indeed, Alloway[32] showed that in two such mutants, *rdgB* and *norpA* (footnote 1), photoreceptor degeneration appears to result from the formation of stable metarhodopsin-arrestin complexes. In the next sections, we will discuss *trp* and inaF null and near-null mutants, in which most of the TRP channels are either absent or nonfunctional. It is possible that light-dependent, slowly progressing degeneration observed in these mutants also may have the same origin.

DEGENERATION CAUSED BY CONSTITUTIVE ACTIVITY OF THE TRP CHANNEL

In this section we discuss the mutants *rdgA* and Trp^{P365}, both of which display severe photoreceptor degeneration shown to be due to constitutive activity of the TRP channel.

rdgA

rdgA (*retinal degeneration A*) mutants were first isolated in 1970 by behavioral assays for non-phototaxis by Hotta and Benzer.[47] Shortly thereafter, additional *rdgA* mutant alleles were recovered in other laboratories using somewhat different methods (review: Pak[48]). In severely affected *rdgA* mutants, degeneration is well advanced already at eclosion (Johnson[49]; Matsumoto[50]). Although degeneration proceeds either in the presence or absence of light, illumination appears to accelerate the time course of degeneration (K. Isono and Pak, unpublished observation). Biochemical studies first provided evidence that rdgA mutants are deficient in the activity of diacylglycerol kinase (DGK),[51-53] an enzyme that catalyzes the phosphorylation of diacylglycerol (DAG) to convert it to phosphatidic acid (PA) in the first step of resynthesis of phosphatidylinositols. The rdgA gene was cloned by Masai[54] and shown to encode an eye-specific DGK.

The reason for the degeneration, however, remained obscure until Raghu[28] subjected *rdgA* mutants to whole cell, patch clamp recordings. The most significant finding to emerge from this study is that all *rdgA* photoreceptors on which recordings were made (n = 45) display a small (~50 pA), noisy, constitutive inward current, which is detectable immediately upon establishing the whole cell configuration. The following lines of evidence established that the TRP channels are primarily responsible for the constitutive current.

1) The current was blocked by La^{3+}, which is known to block TRP-mediated current (Hochstrate[55]; Suss[56]; Hardie and Minke[9]).
2) The I-V relationship, which plots current against the membrane voltage, closely resembled that of the TRP conductance.
3) The high frequency noise characteristics (power spectra) were indistinguishable between the constitutive current and the TRP-mediated current.

Eliminating the TRP channels in *rdgA* mutants by constructing *rdgA;;trp* double mutants rescued the severe early-onset degeneration. The rescue was specific to the elimination of the TRP channels because eliminating the TRPL channels by constructing *rdgA;;trpl* double mutants did not rescue the early degeneration phenotype. There was, however, also slower degeneration of photoreceptors in *rdgA;;trp* double mutants with a time course of 2-6 weeks, and this slow degeneration was still present when the TRPL channels were also eliminated, i.e., present in *rdgA;trpl;trp* triple mutants. This slow degeneration is also present in strong *trp* mutants (see next section) and may represent degeneration due to insufficient Ca^{2+} in the cell, mentioned in the previous section.

The above results suggested that the early-onset degeneration in *rdgA* is caused by an uncontrolled Ca^{2+} influx through the TRP channels. The problem with this interpretation was that the constitutive current was still present in *rdgA;;trp* double mutants, even though the early-onset degeneration phenotype had been essentially eliminated. Unlike in *rdgA*, however, the constitutive current in the double mutants was carried by the TRPL channels because (1) it was not blocked by La^{3+}, (2) its I-V relationship and noise power spectra were similar to those of TRPL channel currents, and (3) it was no longer present in *rdgA;trpl;trp* triple mutants, i.e., eliminating the TRPL channels eliminated the current. Since the TRPL channels also mediate substantial Ca^{2+} fluxes,[27,18] it would be difficult to explain the rescue of early-onset degeneration in *rdgA;;trp*, but not in *rdgA;trpl*, if uncontrolled influx of Ca^{2+} were responsible for the degeneration.

To resolve this paradox, these authors compared the developmental onsets of constitutive activity and early signs of degeneration at the pupal stage. They found that TRP channels become constitutively active somewhat earlier than the TRPL channels (~70 hr vs. 85-90 hr of the pupal stage). Electrophysiologically detectable early signs of degeneration seemed to occur at around 75-80 hr in *rdgA* mutants, coinciding with the first appearance of constitutive TRP channel activity in these mutants.

Raghu[28], therefore, suggested that there is a window of Ca^{2+} sensitivity for degeneration at ~75-85 hr of the pupal stage and that photoreceptors become resistant to Ca^{2+} past this window. Thus, they argued, while both the TRP and TRPL channels become constitutively active in *rdgA*, the constitutive activity of the TRP channels occurs somewhat earlier than the TRPL channels, and the uncontrolled Ca^{2+} through the TRP channels at this earlier time window is responsible for the early-onset photoreceptor degeneration in *rdgA*.

Chyb[21] had shown earlier that polyunsaturated fatty acids (PUFAs) could activate both TRP and TRPL channels in Drosophila photoreceptors. DAG can be metabolized via two potential pathways: (1) by DAG lipase to generate PUFAs or (2) by diacylglycerol kinase (DGK) to be converted to phosphatidic acid (PA). In *rdgA* mutants, conversion of DAG to PA is blocked, and, moreover, DAG levels are reported to be normal (Inoue[53]), suggesting enhanced metabolization of DAG via the DAG lipase pathway. Therefore, excess amounts of PUFAs would be generated resulting in constitutive activities of TRP and TRPL channels. Raghu[28] thus argued that their results with *rdgA* provided independent genetic support for the proposal that PUFAs may be messengers of excitation in *Drosophila* photoreceptors.

Trp^{P365}

Trp^{P365} is a newly identified mutation in the *trp* (transient receptor potential) gene that causes constitutive activity of the TRP channel.[29]

The first mutant isolated in the *trp* gene was a naturally occurring one that behaved as if blind in bright ambient light.[57,58] Subsequently, other recessive mutations of this gene were isolated through chemical mutagenesis.[59] Both ERG (electroretinogram: extracellularly recorded light-evoked mass response of the eye)[57] and intracellular[60] recordings revealed that the photoreceptor potential is much smaller than the wild-type potential and decays to baseline during a bright and/or prolonged light stimulus (Fig. 1D), thus the name *trp*. In contrast, in wild type a steady-state or maintained component is present throughout the duration of the stimulus (Fig. 1A). In addition to the electrophysiological phenotype, null or near-null *trp* mutants display late-onset, light-dependent photoreceptor degeneration (see Fig. 4).[61,62,63] This phenotype is discussed further below.

The trp gene was cloned independently by Montell and Rubin[7] and Wong[8]. As discussed previously (sections on Phototransduction and Ca^{2+} entry), the gene encodes the subunits of the highly Ca^{2+}-permeable class of light-sensitive channels, TRP. Protein (Western) blot analyses showed that the quantity of the TRP protein in null or near-null *trp* mutants is reduced to undetectable amounts. Thus, in trp null or near-null mutants, the amount of Ca^{2+} entering the photoreceptor cells in response to light is greatly reduced.[64,27,18] Insofar as insufficient Ca^{2+} in photoreceptor cells can be a cause of photoreceptor degeneration, the late-onset, slowly progressing degeneration seen in null or near-null *trp* mutants may be due to insufficient Ca^{2+} entry (see section on Magnitude of Ca^{2+} entry and photoreceptor degeneration).

The *Trp^{P365}* mutant was isolated in chemical mutagenesis. It has phenotypes very distinct from those of null and near-null *trp* mutants described above.[29] Its most conspicuous phenotype is the early-onset, very rapid, and massive photoreceptor degeneration (Fig. 2). The ERG is nearly absent in homozygotes, but the response remaining is not transient but lasts the entire duration of stimulus

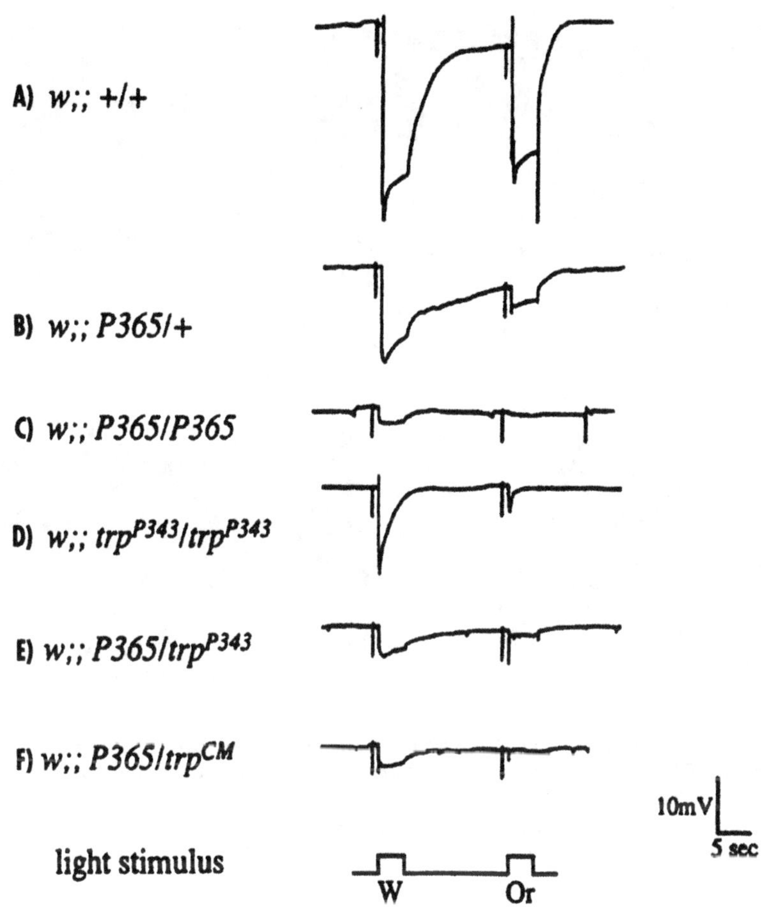

Figure 1. Representative electroretinograms (ERGs) of wild type, classical *trp* mutants, and mutants carrying Trp^{P365} allele, showing that Trp^{P365} causes a mutant phenotype distinct from that of the classical *trp* mutants. a) wild type, b) Trp^{P365} heterozygote, c) Trp^{P365} homozygote, d) *trp* null mutant, e) and f) flies carrying heteroallelic combinations of Trp^{P365} and *trp* null or near-null allele. Whereas classical *trp* mutants are charachterized by the fast decay of the photoreceptor potential during light stimulus, the maintained component is present throughout the duration of the stimulus in Trp^{P365}-carrying mutants. All flies were marked with the mutaiton w (*white*) to remove the red screening pigments in the eye, and all flies were 7 days post-eclosion at the time of recording. W, white light stimulus; Or, orange light stimulus. (Reproduced from ref. 29 with permission.)

(Fig. 1C). Moreover, the mutation does not seem to affect the amount of the TRP protein directly. Whatever decrease in the amount of the TRP protein observed appears to be attributable to photoreceptor degeneration. Finally, the mutation is semi-dominant for all these phenotypes in that the phenotypes are observed in heterozygotes as well as in homozygotes although phenotypes of heterozygotes are less severe than those of homozygotes (Figs. 1B, 1C, & 2).

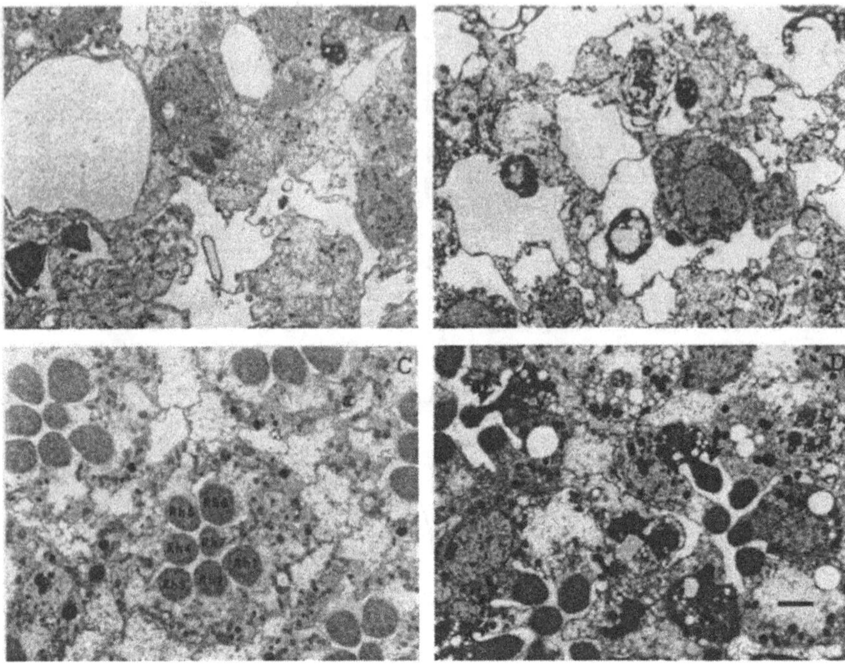

Figure 2. Electron micrographs showing severe photoreceptor degeneration caused by Trp^{P365} and exacerbation of degeneration by exposure to light. Transverse sections of retinas were obtained near the R7 and R8 rhabdomere bounderies from Trp^{P365} homozygotes raised in the dark A) or in 12-hr-light/12hr-dark cycles B) at 0 day post-eclosion, and Trp^{P365} hterozygotes raised in the dark C) or in 12-hr-light/12-hr-dark cycles D) at 2 days post-eclosion. All flies were raised at 25°C. Rh1,..., Rh7: Rhabdomeres of R1,..., R7 photoreceptors. Scale bar, 2 μm. (Reproduced with permission from ref. 29.)

Despite this very different set of phenotypes displayed by Trp^{P365} in comparison to all other known *trp* mutants, the following lines of evidence unequivocally established that mutation(s) in the *trp* gene is responsible for the Trp^{P365} phenotype:[29]

1) nucleotide substitutions were identified in the *trp* coding region of the Trp^{P365} mutant, which would cause alterations of amino acid sequence in the TRP protein;
2) a transgene containing the *trp* gene isolated from the Trp^{P365} mutant induced the mutant ERG phenotype in a wild-type background in a dose-dependent manner; and
3) a transgene containing the wild type *trp* gene partially rescued the mutant phenotype in the Trp^{P365} background in a dose-dependent manner.

Patch clamp whole-cell recordings were carried out on photoreceptors in dissociated ommatidia of Trp^{P365} in an attempt to determine the cause of degeneration.[29] The results showed that in Trp^{P365} homozygotes, but not in wild type, outwardly rectifying membrane currents were detected, as soon as the whole cell configuration was established, in response to membrane voltage steps in the absence of light stimulus. Applying La^{3+}, which blocks the TRP channel but not the TRPL channel (Hochstrate[55]; Suss[56]; Hardie and Minke[9]; Niemeyer[11]), abolished the currents. Thus, it appears that the TRP channel is constitutively active in *Trp*^{P365} mutants, and the consequent excess Ca^{2+} entry into the photoreceptors appears to be responsible for the observed early-onset, rapid degeneration.

Sequence analysis showed that *Trp*^{P365} carries four mutations that would alter the TRP protein sequence.[29] To determine which of these four might be responsible for the *Trp*^{P365} mutant phenotypes, transgenic flies, each carrying either one of the four mutations singly or three or more in various multiple combinations, were generated and tested for their ERG and degeneration phenotypes. Results showed that the primary cause of the *Trp*^{P365} mutant phenotypes is an alteration of the 550th amino acid residue, phenylalanine, to isoleucine (Hong et al, submitted). The TRP channel has some homology to voltage-gated Na^+ and Ca^{2+} channels,[10] and the Phe550 is located near the N-terminal end of the fifth transmembrane domain. The above results suggested that Phe550 might be important for the gating or regulation mechanism(s) of the TRP channel.

The mechanisms of the TRP channel activation and regulation are not understood (see Phototransduction section). We have hypothesized that additional, still unidentified proteins might be involved in the activation and regulation processes. One possible approach to identifying such proteins is to search for mutants with phenotypes similar to that of *trp* and then to identify the proteins corresponding to the mutants. The rationale for this approach is that any defect in the proteins involved in the activation or regulation of the TRP channel is likely to cause the TRP channel function to be defective. *inaF* mutants are among the first such mutants to be identified and characterized.[62]

The first *inaF* mutant was generated by P element insertional mutagenesis, and several other *inaF* alleles, including null alleles, were recovered subsequently through imprecise excision of the P element from the original *inaF* mutant. The receptor potential of a null mutant, *inaF*^{P106x}, decays to the baseline during bright light stimulus, in a manner very similar to that of null or near-null *trp* mutants. Furthermore, *inaF*^{P106x} displayed photoreceptor degeneration with an onset and kinetics approximating that of a near-null *trp* mutant (Fig. 4[62]).

The similarity of the ERG phenotype between *inaF*^{P106x} and null or near-null *trp* mutants suggested that the TRP channels are only marginally functional in the null *inaF* mutant as in near-null *trp* mutants. To compare the effects of *inaF* and *trp* mutations on the TRP channel function, the TRPL channels were genetically eliminated by introducing a null trpl mutation, *trpl*^{302}, into the *inaF*^{P106x} or *trp*^{P301} background, i.e., by constructing the *trpl*^{302};*trp*^{P301} and *inaF*^{P106x};*trpl*^{302} double mutants, where *trpl*^{302} and *inaF*^{P106x} are null alleles and *trp*^{P301} is a near-null allele. In *trpl*^{302}, which still has a full complement of TRP channels and used as a positive control, a light stimulus produces a robust response that lasts the entire duration of illumination (Fig. 3A). In *trpl*^{302};*trp*^{P301} and *inaF*^{P106x};*trpl*^{302} double mutants, on the other

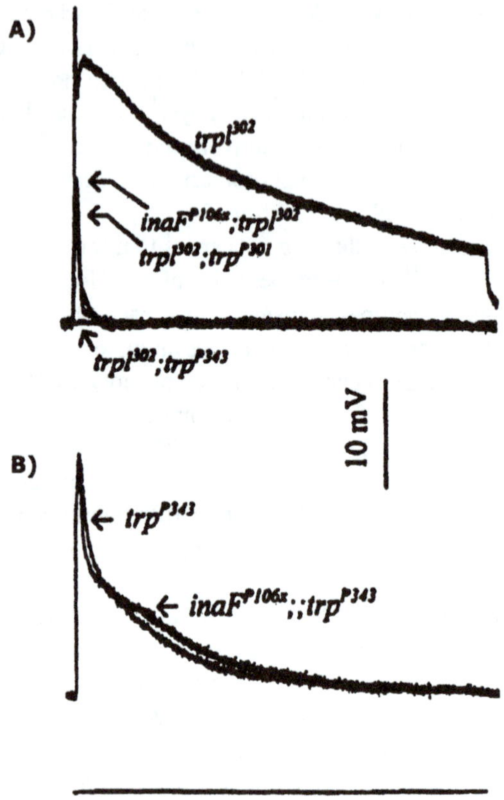

Figure 3. Effects of *inaF* and *trp* mutations on the photoreceptor responses generated through the TRP a) or TRPL b) channels. In a), the TRP channels were isolated using a *trpl* null mutation, and the effects of *inaF* null mutation and *trp* null and near-null mutations were assessed in appropriate double mutants. In b) the TRPL channels were isolated using a *trp* null mutation, and the effects if *inaF* null mutation were assessed in a *inaF;;trp* double mutant. Photoreceptor potentials were recorded intracellularly. All flies were marked with the mutation *w* (*white*) or *cn bw* (cinnabar brown) to remove the red screening pigments in the eye. The recordings were performed at 1 day post-eclosion, and the flies were dark-adapted for 3 min before each light stimulus. (Reproduced with permission from ref. 62).

hand, the responses are severely reduced in size and duration in a very similar manner (Fig. 3A). Thus, the effect of *inaF*^{P106x} on the TRP channel is very similar to that of *trp*,^{P301} which reduces the TRP concentration to undetectable levels. The effect is specific for the TRP channel because *inaF*^{P106x} has little or no effect on the TRPL channel (Fig. 3B). Western analysis revealed that the quantity of the TRP protein is drastically reduced in the *inaF*^{P106x} mutant to approximately 6% of that of wild type. This reduction appears to be specific for TRP because the reduction is not observed in three other retinal proteins Rh1, INAD and PLC that are known to function in the phototransduction pathway.[62] These results suggested that a part of the *inaF*^{P106x} phenotype might be due to the reduction in quantity of the TRP protein. It

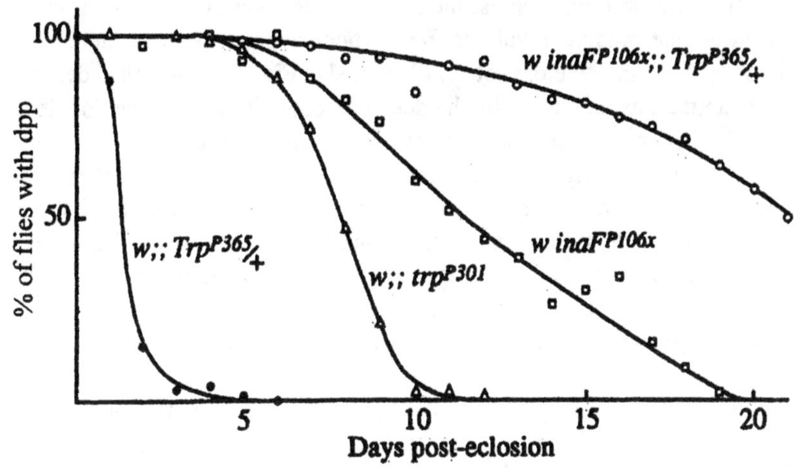

Figure 4. Time course of photoreceptor degeneration in *Trp*^{P365} hterozygotes, *trp* near-null mutants, and *inaF* null mutants, and mutual suppression of degeneration by *inaF* and Trp^{P365}. The degeneration time courses were determined by monitoring the disapearance of dpp in a sample popluation. All flies were raised under 12-hr-light/12-hr-dark illumination conditions. (Reproduced with permission from ref. 62).

should be noted, however, that in *trp*^{P301}, to which *inaF*^{P106x} is being compared, the TRP protein is reduced to an undetectable level in Western analysis.

Sequence analysis showed that the *inaF* gene encodes a novel protein INAF of 241 amino acid residues. INAF is devoid of any trans-membrane domains and contains no obvious domains or motifs. Database search has not resulted in the identification of any known proteins with significant sequence similarity to INAF.

The function of the INAF protein has not yet been determined. There are at least three viable hypotheses. One is that INAF is required for the gating of the TRP channels[62] and, without it, the TRP channels cannot open effectively upon light stimulus. The reduction in the quantity of the TRP protein may reflect secondary effects resulting from the absence of INAF. An alternative hypothesis is that INAF is required for maintaining the quantity of TRP protein either by enhancing the synthesis or preventing the degradation of the TRP protein. The apparent *trp*-like phenotype may reflect dearth of functional TRP protein. The third possibility is the combination of the above two, i.e., INAF is required for both the gating of the TRP channels and the maintenance of the quantity of the TRP protein. While experimental support can be found for all three, none is as yet definitive.

Regardless of the actual mechanism(s) involved, the net effect of *inaF* mutations is to make the TRP channels ineffective in mediating membrane currents. If the *Trp*^{P365} mutation causes the TRP channel to become constitutively active and a null or near-null *inaF* mutation renders the TRP channel ineffective, it seemed possible that the two mutations might compensate for each other's effects. Accordingly, the double mutant, *inaF*^{P106x};;*Trp*^{P365}/+, homozygous for *inaF*^{P106} and heterozygous for *Trp*^{P365} was constructed and tested for its phenotypes. The results showed that the two mutations, indeed, suppress the effects of each other. (Suppression was also present in double mutants involving *Trp*^{P365} homozygotes, but was more clearly seen in ones involving heterozygotes.) Illustrated in Fig. 4 are the retinal degeneration time courses, as monitored by the deep pseudopupil (dpp), of the double mutant, *inaF*^{P106x};;*Trp*^{P365}/+, and the parental stocks, *inaF*^{P106x} and *Trp*^{P365}/+, as well as *trp*^{P301}. The dpp is a superposed virtual image of a group of neighboring rhabdomere tips observed in a living fly using a dissecting microscope (Franceschini[65]). When the superposition fails, the dpp disappears. Thus, dpp is a sensitive indicator of not only the disappearance but also the disarrangement of the rhabdomeres resulting from degeneration. Because of its sensitivity, the degeneration time course determined by dpp generally tends to be faster than that determined by anatomical analyses. In *Trp*^{P365} heterozygotes, the dpp starts disappearing almost immediately after eclosion and is completely gone by 5 days post-eclosion. This fast dpp disappearance represents the rapid degeneration due to constitutive activity of the TRP channel. In *trp*^{P301} as well as *inaF*^{P106x}, the rapid degeneration is not present, but there is a slower disappearance of dpp that begins around 5-6 days post-eclosion. This slow disappearance represents the slowly progressing, late-onset degeneration normally seen in *trp* and *inaF* null and near-null mutants, and may be due to insufficient Ca^{2+} influx (see Section on Magnitude of Ca^{2+} entry). In the double mutant, *inaF*^{P106x};;*Trp*^{P365}/+, the rapid disappearance of dpp was no longer present, and even the slower dpp disappearance has greatly slowed in time course and improved in severity. We interpret these results to suggest that in the double mutant,

1) the early-onset degeneration due to *Trp*^{P365} is no longer present because of the restricted Ca^{2+} entry due to the presence of the *inaF*^{P106x} mutation, and
2) the late-onset degeneration is also much improved because *Trp*^{P365} mutant channels allow some Ca^{2+} to enter the cell even in the presence of *inaF*^{P106x}.

These results are consistent with the idea that the slowly progressing late-onset degeneration of trp and inaF null and near-null mutants may indeed be due to insufficient Ca^{2+} entry and suggest, furthermore, the importance of the INAF protein in TRP channel function and Ca^{2+} homoeostasis in photoreceptor cells.

CONCLUSION

Perturbed Ca^{2+} influx appears to be one of the major causes of photoreceptor degeneration in *Drosophila*. Excessive Ca^{2+} influx through the TRP channels causes severe, early-onset photoreceptor degeneration seen in such mutants as *rdgA* and *Trp*^{P365}. On the other hand, a drastic reduction in the TRP channel activity, as in the null or near-null trp and inaF mutants, causes photoreceptor degeneration of late-onset and slow progression.

In the double mutant, *inaF*^{P106x};;*Trp*^{P365}/+, the early-onset degeneration seen in *Trp*^{P365}/+ is eliminated and the late-onset, slowly progressing degeneration seen in *inaF*^{P106x} is greatly ameliorated. Because *Trp*^{P365} and *inaF*^{P106x} have opposing effects on Ca^{2+} influx, the rescue of the degeneration phenotypes may be due to restoration of appropriate Ca^{2+} influx through the TRP channels.

ACKNOWLEDGMENTS

Work done in the authors' laboratory was supported by a grant from the National Institutes of Health, National Eye Institute (EY 00033) to WLP. We thank Sara Dykes for help in preparation of the manuscript.

REFERENCES

1. Pak WL. *Drosophila* in vision research. Invest Ophthal Vis Sci 1995; 36:2340-2357.
2. Zucker CS. The biology of vision in *Drosophila*. Proc Natl Acad Sci USA 1996; 93:571-576.
3. Montell C. Visual transduction in *Drosophila*. Annu Rev Cell Dev Biol 1999; 15:231-268.
4. Minke B, Hardie RC. Genetic dissection of *Drosophila* Phototransduction. In: Stavenga DG, DeGrip WJ, Pugh Jr EN, eds. Handbook of Biological Physics, Vol. 3, 2000:449-525.
5. Pak WL, Ostroy SE, Deland MC et al. Photoreceptor mutant of *Drosophila*: Is protein involved in intermediate steps of phototransduction? Science 1976; 194:956-959.
6. Bloomquist BB, Shortridge RD, Schneuwly S et al. Isolation of a putative phospholipase C gene of *Drosophila*, *norpA*, and it role in phototransduction. Cell 1988; 54:723-733.
7. Montell C, Rubin GM. Molecular characterization of the *Drosophila trp* locus: A putative integral membrane protein required for phototransduction. Neuron 1989; 2:1313-1323.
8. Wong F, Schaefer EL, Roop BC et al. Proper function of the *Drosophila trp* gene product during pupal development is important for normal visual transduction in the adult. Neuron 1989; 3:81-94.
9. Hardie RC, Minke B. The *trp* gene is essential for a light activated Ca^{2+} channel in *Drosophila* photoreceptors. Neuron 1992; 8:643-651.
10. Phillips AM, Bull A, Kelly LE. Identification of a *Drosophila* gene encoding a calmodulin-binding protein with homology to the trp phototransduction gene. Neuron 1992; 8:631-642.
11. Niemeyer BA, Suzuki E, Scott K et al. The *Drosophila* light-activated conductance is composed of the two channels Trp and Trpl. Cell 1996; 85:651-659.
12. Scott K, Sun Y, Beckingham K et al. Calmodulin regulation of *Drosophila* light-activated channels and receptor function mediates termination of the light response in vivo. Cell 1997; 91:375-383.
13. Reuss H, Mojet MH, Chyb S et al. In vivo analysis of the Drosophila light-sensitive channels, TRP and TRPL. Neuron 1997; 19:1249-1259.
14. Xu X-ZS, Li H-S, Guggino WB et al. Coassembly of TRP and TRPL produces a distinct store-operated conductance. Cell 1997; 89:1155-1164.
15. Berridge MJ. Cell signalling—A tale of two messengers. Nature 1993; 365:388-389.
16. Minke B, Selinger Z. Inositol lipid pathway in fly photoreceptors: Excitation, calcium mobilization and retinal degeneration. In: Osborne NN, Chader GJ, eds. Progress in Retinal Research. Oxford: Pergamon, 1992:99-124.
17. Ranganathan R, Bacskai BJ, Tsein RY et al. Cytosolic calcium transients: spatial localization and role in *Drosophila* photoreceptor cell function. Neuron 1994; 13:837-848.
18. Hardie RC. Calcium signaling: setting store by calcium channels. Curr Biol 1996; 6:1371-1373.
19. Acharya JK, Jalink K, Hardy RW et al. $InsP_3$ receptor essential for growth and differentiation but not for vision in *Drosophila*. Neuron 1997; 18:881-887.
20. Raghu P, Colley NJ, Webel R et al. Normal phototransduction in *Drosophila* photoreceptors lacking an $InsP_3$ receptor gene. Mol and Cell Neurosci 2000; 15:429-445.
21. Chyb S, Raghu P, Hardie RC. Polyunsaturated fatty acids activate the *Drosophila* light-sensitive channels TRP and TRPL. Nature 1999; 397:255 259.
22. Chorna-Ornan I, Joel-Almagor T, Ben-Ami HC et al. A common mechanism underlies vertebrate calcium signaling and Drosophila phototransduction. J Neurosci 2001; 21:2622-2629.

23. Agam K, von Campenhausen M, Levy S et al. Metabolic stress reversibly activates the *Drosophila* light-sensitive channels TRP and TRPL in vivo. J Neurosci 2000; 20:5748-5755.
24. Arslan P, Corps AN, Hesketh TR et al. cis-unsaturated fatty acids uncouple mitochondria and stimulate glycolysis in intact lymphocytes. J Biochem 1984; 217:419-425.
25. Hermesh O, Kalderon B, Bar TJ. Mitochondria uncoupling by a long chain fatty acyl analogue. J Biol Chem 1998; 273:3937-3942.
26. Hardie RC. Whole-cell recordings of the light induced current in dissociated *Drosophila* photoreceptors: evidence for feedback by calcium permeating the light-sensitive channels. Proc Natl Acad Sci USA 1991; 245:203-210.
27. Peretz A, Sandler C, Kirschfeld K et al. Genetic dissection of light-induced Ca^{2+} influx into *Drosophila* photoreceptors. J Gen Physiol 1994; 104:1057-1077.
28. Raghu P, Usher K, Jonas S et al. Constitutive activity of the light-sensitive channels TRP and TRPL in the Drosophila diacylglycerol kinase mutant, rdgA. Neuron 2000; 26:169-179.
29. Yoon J, Ben-Ami HC, Hong YS et al. Novel mechanism of massive photoreceptor degeneration caused by mutations in the *trp* gene of *Drosophila*. J Neurosci 2000; 20:649-659.
30. Trump BF, Berezesky IK. The role of altered $[Ca^{2+}]_i$ regulation in apoptosis, oncosis, and necrosis. Biochim Biophys Acta 1996; 1313:173-178.
31. Lee JM, Zipfel GJ, Choi DW. The changing landscape of ischaemic brain injury mechanisms. Nature 1999; 399:A7-A14.
32. Alloway PG, Howard L, Dolph PJ. The formation of stable rhodopsin-arrestin complexes induces apoptosis and photoreceptor cell degeneration. Neuron 2000; 28:129-138.
33. Byk T, Bar Yaacov M, Doza YN et al. Regulatory arrestin cycle secures the fidelity and maintenance of the fly photoreceptor cell. Proc Natl Acad Sci USA 1993; 90:1907-1911.
34. Dolph PJ, Ranganathan R, Colley NJ et al. Arrestin function in inactivation of G protein-coupled receptor rhodopsin in vivo. Science 1993; 260:1910-1916.
35. Vinos J, Jalink K, Hardy RW et al. A G protein-coupled receptor phosphatase required for rhodopsin function. Science 1997; 277:687-690.
36. Kiselev A, Socolich M, Vinos J et al. A molecular pathway for light-dependent photoreceptor apoptosis in *Drosophila*. Neuron 1999; 28:139-152.
37. Kahn ES, Matsumoto H. Calcium/calmodulin-dependent kinase II phosphorylates *Drosophila* visual arrestin. J. Neurochem. 1997; 68:169-175.
38. Alloway PG, Dolph PJ. A role for the light-dependent phosphorylation of visual arrestin. Proc Natl Acad Sci USA 1999; 96:6072-6077.
39. Harris WA, Stark WS. Hereditary retinal degeneration of *Drosophila melanogaster*: A mutant defect associated with the phototransduction process. J Gen Physiol 1977; 69:261-291.
40. Stark WS, Sapp R. Retinal degeneration and photoreceptor maintenance in *Drosophila*: *rdgB* and its interaction with other mutants. Inherited and environmentally induced retinal degenerations: Allan R. Liss, 1989:467-489.
41. Steel F, O'Tousa JE. Rhodopsin activation causes retinal degeneration in *Drosophila rdgC* mutant. Neuron 1990; 4:883-890.
42. Smith DP, Ranganathan R, Hardy RW et al. Photoreceptor deactivation and retinal degeneration mediated by a photoreceptor-specific protein kinase C. Science 1991; 254:1478-1484.
43. Leonard DS, Bowman VD, Ready DF et al. Degeneration of photoreceptors in rhodopsin mutants of *Drosophila*. J Neurobiol 1992; 23:605-626.
44. Dolph PJ, Ranganathan R, Colley NJ et al. Arrestin function in inactivation of G protein-coupled receptor rhodopsin in vivo. Science 1993; 260:1910-1916.
45. Pak WL. Retinal degeneration mutants of *Drosophila*. In: Wright A, Jay B, eds. Modern Genetics: Molecular Genetics of Inherited Eye Disorders. Chur: Harwood Academic Publishers, 1994:29-52.
46. Vihtelic TS, Goebl M, Milligan S et al. Localization of *Drosophila* retinal degeneration B, a membrane-associated phosphatidylinositol transfer protein. J Cell Biol 1993; 122:1013-1022.
47. Hotta Y, Benzer S. Genetic dissection of the *Drosophila* nervous system by means of mosaics. Proc Natl Acad Sci USA 1970; 67:1156-1163.

48. Pak WL. Mutations affecting the vision of *Drosophila melanogaster*. In: King RC, ed. Handbook of Genetics, Vol. 3. New York: Plenum, 1975:703-733.
49. Johnson MA, Frayer KL, Stark WS. Characteristics of *rdgA*: Mutants with retinal degeneration in *Drosophila*. J Insect Physiol 1982; 28:233-242.
50. Matsumoto E, Hirosawa K, Takagawa K et al. Structure of retinular cells in a *Drosophila melanogaster* visual mutant, rdga, at early stages of degeneration. Cell Tissue Res. 1988; 252:293-300.
51. Yoshioka T, Inoue H, Hotta Y. Defective phospholipid metabolism in the retinular cell membrane of *norpA* (*no receptor potential*) visual transduction mutants of *Drosophila*. Biochem Biophys Res Com 1983; 111:567-573.
52. Yoshioka T, Inoue H, Hotta Y. Absence of diglyceride kinase activity in the photoreceptor cells of *Drosophila* mutant. Biochem Biophys Res Com 1984; 119:389-395.
53. Inoue H, Yoshioka T, Hotta Y. Diacylglycerol kinase defect in a *Drosophila* retinal degeneration mutant rdga. J Biol Chem 1989; 264:5996-6000.
54. Masai I, Okazaki A, Hosoya T et al. *Drosophila* retinal degeneration A gene encodes an eye-specific diacylglycerol kinase with cysteine-rich zinc-finger motifs and ankyrin repeats. Proc Natl Acad Sci USA 1993; 90:11157-11161.
55. Hochstrate P. Lanthanum mimics the *trp* photoreceptor mutant of *Drosophila* in the blowfly Calliphora. J Comp Physiol A 1989; 166:179-188.
56. Suss-Toby E, Selinger Z, Minke B. Lanthanum reduces the excitation efficiency in fly photoreceptors. J Gen Physiol 1991; 98:849-868.
57. Cosens DJ, Manning A. Abnormal retinogram from a *Drosophila* mutant. Nature 1969; 224:285-287.
58. Cosens D. Blindness in a *Drosophila* mutant. J Insect Physiol 1971; 17:285-302.
59. Pak WL. Study of photoreceptor function using *Drosophila* mutant. In: Breakfield X, ed. Neurogenetics: Genetic Approaches to the Nervous System. New York: Elsevier-North Holland, 1979:67-99.
60. Minke B, Wu C-F, Pak WL. Induction of photoreceptor voltage noise in the dark in *Drosophila* mutant. Nature 1975; 258:84-87.
61. Cosens DJ, Perry MM. The fine structure of the eye of a visual mutant, A-type, of *Drosophila melanogaster*. J Insect Physiol 1972; 18:1773-1786.
62. Li C, Geng C, Leung H-T et al. INAF, a protein required for transient receptor potential Ca^{2+} channel function. Proc Natl Acad Sci USA 1999; 96:13474-13479.
63. Pak WL. Molecular genetic studies of photoreceptor function using *Drosophila* mutant. In: Chader GJ, Farber D, eds. Molecular Biology of the Retina: Basic and Clinically Relevant Studies. New York: Wiley-Liss, 1991:1-32.
64. Peretz A, Suss-Toby E, Rom-Glas A et al. The light response of *Drosophila* photoreceptors is accompanied by an increase in cellular calcium: effects of specific mutations. Neuron 1994; 12:1257-1267.
65. Franceschini N. Pupil and pseudopupil in the compound eye of *Drosophila*. In: Wehner R, ed. Information Processing in the Visual System of Arthropods. New York: Springer-Verlag, 1972:75-82.

THE TRP CALCIUM CHANNEL AND RETINAL DEGENERATION

Baruch Minke

ABSTRACT

The *Drosophila* light activated channel TRP is the founding member of a large and diverse family of channel proteins that is conserved throughout evolution. These channels are Ca^{2+} permeable and have been implicated as important component of cellular Ca^{2+} homeostasis in neuronal and non-neuronal cells. The power of the molecular genetics of *Drosophila* has yielded several mutants in which constitutive activity of TRP leads to a rapid retinal degeneration in the dark. Metabolic stress activates rapidly and reversibly the TRP channels in the dark in a constitutive manner by a still unknown mechanism. The link of TRP gating to the metabolic state of the cell is shared also by mammalian homologues of TRP and makes cells expressing TRP extremely vulnerable to metabolic stress, a mechanism that may underlie retinal degeneration and neuronal cell death.

INTRODUCTION

The visual system of *Drosophila melanogaster*[1] has been the subject of genetic analysis through isolation of visual mutants[2] and application of molecular genetics combined with powerful electrophysiological and biochemical functional tests (for review see refs. 1-5). One of the key proteins that have been discovered due to the power of the *Drosophila* genetics is the protein designated by Minke Transient Receptor Potential (TRP) because of the unique phenotype of the *trp* mutant.[6] While genetic evidence has unequivocally demonstrated that phospholipase C is absolutely required for light excitation in *Drosophila* photoreceptors,[2,7] a major unresolved

Department of Physiology and the Kuhne Minerva Center for Studies of Visual Transduction, The Hebrew University-Hadassah Medical School Jerusalem 91120, Israel.

question of *Drosophila* phototransduction is the gating mechanism of light-sensitive channels TRP and TRP-like (TRPL) and the identity of the second messenger of excitation.[1]

Photoreceptor degeneration is a phenomenon, which appears in strong alleles of almost every phototransduction-defective mutant of *Drosophila*. Given the abundance of the signaling molecules in the microvilli (the signaling membranes of the photoreceptor cell) and the crucial roles of these molecules in the assembly and function of the phototransduction machinery, mutants defective in the signaling proteins are likely to show degeneration.[1] Because of the great interest in neuronal cell death in general and in human retinal degeneration, a search for mammalian homologues of *Drosophila* genes has been recently pursued with much effort. Indeed, mammalian homologues of the *Drosophila* genes *rdgB*,[8,9] *rdgC*,[10,11] *ninaC*[12] and *trp*[13] have been found in mammalian retina. These genes cause photoreceptor degeneration in *Drosophila* upon malfunction. However, the function of these molecules in vertebrate cells in not clear. Despite extensive investigation, relatively little is known about the molecular mechanisms underlying retinal degeneration due to mutations in the above genes also in invertebrate species.

It has been recently found that the TRP channel is involved in retinal degeneration of several *Drosophila* mutant in which constitutive activity of TRP seems to cause extremely rapid retinal degeneration.[14,15] Therefore, the phenomenon of constitutive activity of TRP and TRPL channels in the dark and the ensuing influx of Ca^{2+}, which possibly lead to retinal degeneration are the main topics of this review.

THE DROSOPHILA TRP AND TRPL CHANNELS

The Molecular Structure of Drosophila TRP and TRPL

The *trp* gene was cloned by Montell and Rubin[16] and by Wong and colleagues.[17] The molecular structure of the TRP protein shows multiple domains that are likely to play an important role in cellular functions.[3] TRPL was isolated using a screen to detect calmodulin binding proteins and shared overall 40% identity with TRP with much greater similarity (~ 70%) in the putative transmembrane regions.[18] The N and C-termini of TRP and TRPL both contain a number of recognizable motifs, but their function in TRP and TRPL is unknown (Fig. 1). Both TRP and TRPL have weak but significant homologies to known channel subunit of vertebrate voltage-gated Ca^{2+} channel (the dihydropyridine receptor). By analogy to voltage-gated K^+ channels and the cyclic nucleotide gated channels, both *trp* and *trpl* gene products represent subunits of putative tetrameric channels. Since null *trp* or *trpl* mutants both respond to light, each can clearly function without the other. However, heterologous co-expression studies and co-immunoprecipitation led to the suggestion that in wild type (WT) flies the light induced conductance was composed of TRP or TRPL homomers and TRP-TRPL heteromultimers.[19] Detailed in situ measurements of biophysical properties including ionic selectivity and single channel conductance

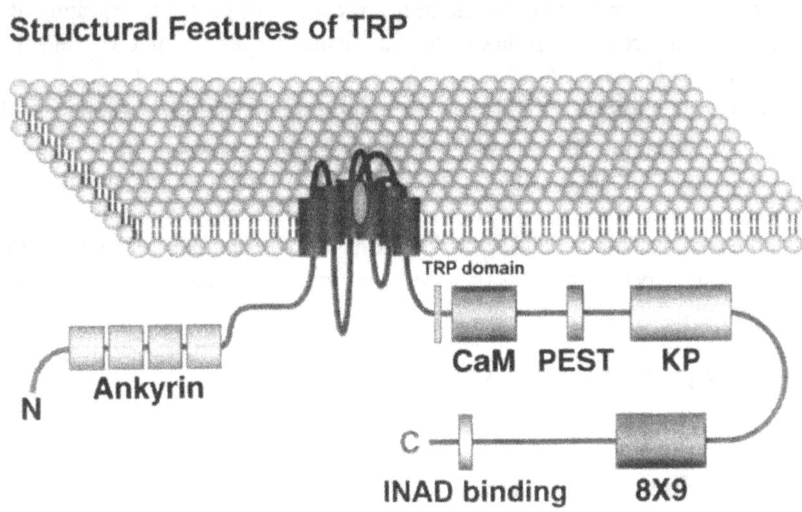

Figure 1. Putative domain structure and topology of *Drosophila* TRP. TRP contains six putative transmembrane helices S1-S6 and a putative pore region between S5 and S6, S3-S6 show sequence fragments identical to the equivalent regions of the dihydropyridine receptor (a vertebrate voltage-gated Ca^{2+} channel). TRP and all TRP homologue channels have 3-4 consensus ankyrin motifs towards the amino terminus. There is a highly conserved TRP domain with still unknown function and one putative calmodulin-binding site (CaM) in the TRP sequence. The TRP protein has a curious and unique hydrophilic sequence near the carboxyl terminus consisting of nine repeats of the sequence: DKDKKPG/AD (8x9), a PEST sequence, a proline rich region with the dipeptide KP, repeating 27 times (KP), an INAD binding domain at the end of the C-terminus. (From Minke and Selinger 1996, with permission from Current Opinion in Neurobiology)

questioned this conclusion.[20] A recent study has shown that there is additional subunit(s) called TRPγ,[21] which is highly enriched in the photoreceptor cells. The N terminal domain of TRPγ dominantly suppressed the TRPL dependent transient receptor potential on a *trp* mutant background suggesting that TRPL- TRPγ heteromultimers contribute to the photoresponse. Furthermore, TRPL and TRPγ coimmunoprecipitate, suggesting that physical interaction between these proteins forms heteromeric channels.[21]

TRP and TRPL Constitute the Light Sensitive Channels

Whole-cell voltage clamp recordings applied to isolated *Drosophila* ommatidia[22,23] makes it possible to examine the hypothesis that TRP is a Ca^{2+} channel/ transporter.[24] The experiments of Hardie and Minke revealed that the fundamental defect in the *trp* mutant is a change in the ionic selectivity of the light-sensitive conductance. Specifically, the relative Ca^{2+} permeability of the light-sensitive conductance in *trp* mutant was reduced by a factor ot ten,[20,25] along with a significant change in the relative pereability to different monovalent ions. The reduced Ca^{2+}

permeability in the trp mutant has also been corroborated by the demonstration of a reduced light-induced Ca^{2+} influx using Ca^{2+} indictor dyes[26,27] or Ca^{2+} selective microelectrodes.[28]

Zuker and colleagues [29] isolated a null mutant of the *trpl* gene. Under physiological conditions the *trpl* mutation has relatively small influence on the light response.[29,30] However, unlike in WT flies the response to light in *trpl* is completely abolished by La^{3+}, which specifically blocks TRP at low concentration[25,31,32] or in the double mutant combination (*trp;trpl*) indicate that TRP and TRPL channels make up all the light activated channels or are required for their activation. (for details see refs. 20,33, rev. 3,34,35).

THE TRP FAMILY OF CHANNEL PROTEINS

The "TRP-Homologue" Group

On the basis of similarity to *Drosophila* TRP and TRPL sequences, new mammalian TRP homologues have been cloned using database searches of expressed sequence tags (EST), RT-PCR or expression-cloning strategies. Seven major groups termed TRPC1-7 have been cloned and sequenced (for review see ref. 36). The TRPC homologues have been cloned from human, mouse, rat, rabbit, bovine and *Xenopus*. Six characteristic features of the *Drosophila* TRP and TRPL proteins have been found to be common to most members of the "TRP-homologue" group:[37] i) The predicted topology of six (but see ref. 38) transmembrane segments (S1-S6) including the typical pore region loop between transmembrane regions S5 and S6. ii) The charged residues in the putative S4 helix, which usually underlies voltage gating are replaced by noncharged residues. iii) Three to four ankyrin repeats are found at the N terminal. iv) A proline-rich sequence is found in the C terminal domain.[39] v) The TRP domain exists in all members of this group[13] vi) Calmodulin binding site is found in the C terminal domain (see Fig. 1).[40]

Birnbaumer and colleagues and Schultz and colleagues[39,41,42] have classified the vertebrate members of the "TRP-homologues" group into four subgroups according to their primary amino acid sequence. Type 1 includes all isoforms of TRPC1. Type 2 includes the TRPC2 homologues, which have the lowest similarity to the other groups. Type 3 includes TRPC3, TRPC6 and TRPC7 channel proteins. Type 4 includes TRPC4 and TRPC5 channels and has a higher similarity to the TRPC1 group relative to the other groups.

TRPC1

TRPC1 was the first mammalian homologue of TRP that was cloned independently by two laboratories using EST database of human fetal brain cDNA library. The expression pattern of TRPC1 made by several groups[43-47] show that TRPC1 is most abundantly expressed in the brain, heart, testes, ovary, bovine aortic endothelial cells and is barely detectable in the liver or adrenal gland. Interestingly, the rat orthologue changes its expression pattern during brain development[46] suggesting that it has a role in developmental signaling.

TRPC2

The human TRPC2, which was the first cloned TRPC2 gene,[43] is probably a pseudogene since several independent ESTs show mutations introducing early stop codons. A bovine homologue of this pseudogene, which is mainly expressed in the testes, spleen and liver was cloned and sequenced.[38] Later, a full-length 1072 amino acids (aa) TRPC2 isoform from mouse[48] and 1172 TRPC2 isoform from bovine (GenBank) were cloned. Thus TRPC2 is the longest of all vertebrate "TRP-homologue" group having longer cytoplasmic N-terminal domain. In contrast to the TRPC1, TRPC2 is tissue specific and in the mouse it was found only in the testes and vomeronasal organ (VNO). This organ plays a key role in the detection of pheromones, which are chemicals released by other rats and elicits stereotyped sexual behavior.[49] Turning TRPC2 gene into a pseudogene in humans fits well with the hypothesis that the vomeronasal organ is no longer functional in apes. TRPC2 shows only 25% to 30% identity to *Drosophila* TRP.

Recent studies revealed that $InsP_3$ is a second messenger of vomeronasal receptor neurons of snake.[50] This fact fits with TRPC2 as a channel, which is activated by the inositol-lipid signaling cascade.

A TRPC2 isoform is highly enriched in the sperm and seems to have an important function in the fertilization of mice.[51]

TRPC3, TRPC6 and TRPC7

This subgroup, especially the TRPC3 homologues have been thoroughly studied by a variety of functional tests and, therefore, significantly contribute to our understanding of TRP function and gating mechanism. The molecular structure of TRPC3 shows six features of the TRP family listed above. Human, mouse and rat orthologues of TRPC3 and TRPC6 have been cloned, as well as a mouse orthologue of TRPC7.

TRPC3 homologues are predominantly expressed in the brain and at much lower levels also in ovary, colon, small intestine, lung, prostate, placenta and testes.[41,52,53] TRPC6 expression is highest in the brain but it is also expressed in the lungs and ovaries. Interestingly, the development of tumors is associated with down regulation of TRPC6 isoform in a mouse model for an autocrine tumor.[54]

TRPC7 from mouse turned out to be very similar (81% identity and 89% similarity) to TRPC3 and also to TRPC6 (75% identity and 84% similarity) while only 33% identical (53% similar) to *Drosophila* TRP and TRPL. TRPC7 is mainly expressed in the heart, lung and eye and at a lower level in the brain, spleen and testes.[55]

TRPC4 and TRPC5

TRPC4 and TRPC5 channels are similar in structure to TRPC1. The rabbit and mouse orthologues of TRPC5 show 69% sequence identity to bovine TRPC4 and 41% identity to *Drosophila* TRP. The expression patterns of TRPC4 and TRPC5 are very different. TRPC5 mRNA is expressed predominantly in the brain,[56] while TRPC4 is expressed in the brain but also in the adrenal gland and at a much lower level in the heart, lung, liver, spleen kidney testes, thymus, aorta and uterus.[57] Of special interest are TRPC4 isoforms, which are expressed in vascular endothelial cells of various species (mouse, human and bovine).[58]

The "TRP-Related" Group

The "TRP-homologue" group has been discovered due to sequence homology to *Drosophila* TRP. Additional members of the TRP family with much smaller sequence homology to *Drosophila* TRP and TRPL have been cloned and sequenced. This group has been designated "TRP related" group.[37] The various members of the "TRP-related" group were found following studies aiming to explore specific sensory transduction pathways or specific genetic diseases. These pathways include olfaction and osmolarity in *C. elegans*,[59] defects in mechanosensory transduction in *C. elegans* and *Drosophila*[60] defects in pain mechanism in primary afferents of dorsal root ganglia,[61] Ca^{2+} transport in Ca^{2+} transporting epithelial cells,[62] polycystic kidney disease in cells expressing polycystin genes, *PKD*,[63] tumor suppression in the skin (rev. ref. 13) and mucolipidosis type IV in cells expressing defective mucolipin.[64,65] The different approaches resulted in a wealth of functional data about "TRP-related" members, in contrast to most mammalian members of the "TRP-homologue" group, whose functions in most of the native tissues are largely unknown. In contrast to mammals, none of the five *C. elegans* members of the "TRP-related" group or other 4 *C. elegans* members of the "TRP-homologue" sub family have been characterized at the cellular level, neither in the native cells nor in heterologous systems.[39]

In summary, the TRP family of channel proteins reveals a high structural diversity. This family has been divided on the basis of primary amino acid sequence into six subfamilies.[13,37] The founding member of this family, the *Drosophila* TRP shares structure homology with all subfamilies mainly at the region of the ion pore, strongly suggesting that the fundamental function of all members of this family is to constitute cation channels. Additional structural features that are common to most, but not all subfamilies are the ankyrin repeats at the N-terminal and the TRP domain at the C-terminal side of the transmembrane domain. In general, the diverse structural features of members of the TRP family strongly suggest that members of this family are involved in a wide range of cellular functions.

CONSTITUTIVE ACTIVITY OF THE TRP CHANNEL BY MUTATIONS AND BY METABOLIC STRESS MAY UNDERLIE RETINAL DEGENERATION

Mutations in the Transmembrane Domain of TRP Cause Rapid Photoreceptor Degeneration

Our understanding of TRP function and regulation has been augmented in the past by the availability of flies mutated in *trp*. However, especially informative should be flies that possess a TRP protein altered in its function. Unfortunately, the known *Drosophila trp* mutants either completely lack TRP protein, or have a highly reduced amount of functionally normal TRP. Furthermore, attempting to directly mutate the *trp* gene itself, by way of systematic site-directed mutagenesis, failed to yield any significant information on TRP function or structure. This is because mutated proteins were either unstable or not expressed at all in transgenic flies (leading to a full null phenotypes) or, alternately, the flies exhibited a completely wild-type phenotype, implying that those modifications had no phenotypic consequences (unpublished results). Recently, Pak and colleagues have identified a novel *trp* mutant (named Trp^{P365}), which genetically mapped to the *trp* locus and has three mutations in the transmembrane domain of TRP.[15] Trp^{P365} seems to be the first *trp* allele that expresses a modified (yet presumably active; see below) TRP protein. Phototransduction in Trp^{P365} mutant flies is aberrant, however instead of displaying the classical *trp* null phenotype (a receptor potential that declines to the dark level during illumination) the Trp^{P365} mutation leads to a total absence of light-response, with a constitutive current in the dark (see below). Strikingly, unlike any other *trp* alleles, these mutants also show extremely rapid retinal degeneration. Thus, the Trp^{P365} mutant provides us with a first and unique opportunity to characterize essential mechanisms underlying the physiological function of TRP and the TRP-dependent light-activated conductance.[15]

Functional Analysis at the Single Cell Level by Whole-Cell, Patch-Clamp Recordings Revealed Constitutive Activity of the TRP Channels

Whole-cell voltage clamp recordings were performed on R1-6 photoreceptors in dissociated ommatidia preparations of Trp^{P365} homozygotes, Trp^{P365} heterozygotes, and wild type. The light-induced current (LIC) of the heterozygote $Trp^{P365}/+$ was indistinguishable from that of wild type flies. This was not true for most of the ommatidia of the heteroallele Trp^{P365}/Trp^{CM}. In contrast, Trp^{P365}/Trp^{P365} homozygote did not respond to light of any intensity. Large fraction (20%-70%) of the Trp^{P365}/Trp^{CM} heterozygote ommatidia did not respond to light, like the Trp^{P365}/Trp^{P365} (Fig. 2A). The LICs of Trp^{P365}/Trp^{CM} flies revealed abnormal features such as highly reduced sensitivity to light and abnormally slow kinetics.

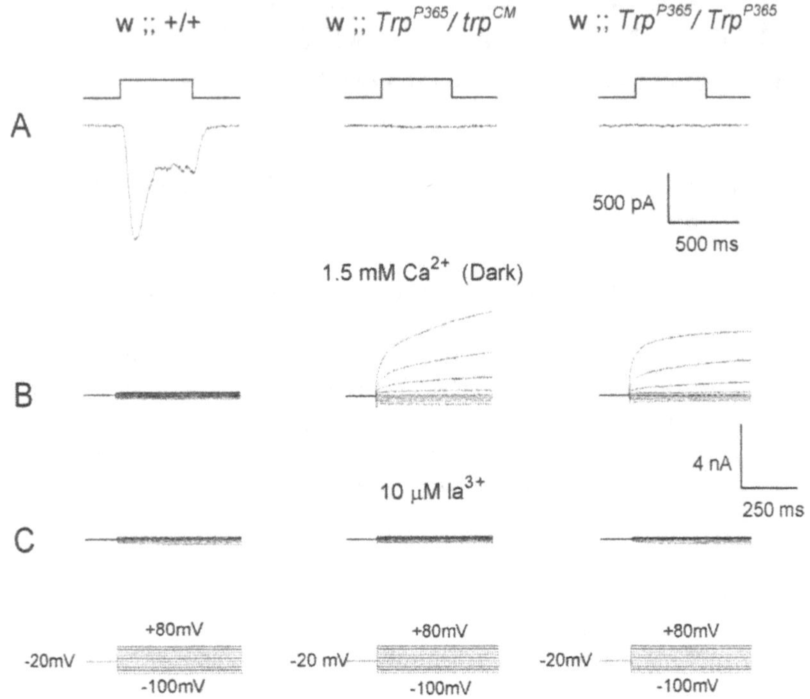

Figure 2. Single-cell functional analysis by whole-cell recordings showing constitutive activity of the TRP channel in the Trp^{P365}/Trp^{P365} and Trp^{P365}/trp^{CM} mutants. Panel A shows a typical light-induced current (LIC) of a wild type cell (left trace) in response to an orange stimulus (OG 590 sharp-cut filter, 1 log unit neutral density filter) and the absence of any responses in Trp^{P365}/trp^{CM} and Trp^{P365}/Trp^{P365} (middle and right traces). The duration of the orange light stimulus is indicated above each trace. Panels B and C compare families of current traces elicited by series of voltage steps from photoreceptors of wild type (left column), the light insensitive cells of Trp^{P365}/trp^{CM} (middle column), and Trp^{P365}/Trp^{P365} homozygotes (right column). For each experiment, a series of nine voltage steps was applied from a holding potential of –20 mV in 20 mV steps (bottom traces). (B) Membrane currents were recorded 30 s after establishing the whole-cell configuration with physiological concentrations (1.5 mM) of Ca^{2+} in the bath. In the mutants, the constitutive currents were recorded as soon as the whole-cell configuration was established. (C) Application of 10 μM La^{3+} to the bath suppressed the membrane currents. The currents shown were recorded 25 s after establishing the whole cell configuration. (From Yoon et al. 2000, Copyright 2000 by the Society for Neuroscience).

To investigate the reason for the inability of cells to respond to light, after the establishment of whole cell recordings we stepped the holding voltage to different membrane potentials in the range between +80 mV to -120 mV in steps of 20 mV in the dark. In WT cells no significant currents were observed. In contrast, in the non-responsive mutants, at positive membrane potentials large outward currents were recorded with strong outward rectification (Fig. 2B). The constitutive currents, which were recorded in a large fraction of Trp^{P365}/ Trp^{CM} heterozygote cells and in all

Trp^{P365}/Trp^{P365} cells, were very similar to the current, which reflects activation of the TRP channels of wild type cells during light. However, in sharp contrast to wild-type photoreceptors, the few Trp^{P365}/Trp^{P365} cells from which recordings could be made and in all light-insensitive Trp^{P365}/Trp^{CM} cells, voltage steps elicited outward currents in the dark from the moment the recording configuration was established.

The constitutive current of both homozygote and heterozygote Trp^{P365} cells had all the properties of the TRP-dependent current: In Ca^{2+} -containing medium large outward currents were observed in the homozygote and heterozygote Trp^{P365} cells, and these currents were very similar to these of WT cells under continuous light conditions. In addition, application of 10 μM La^{3+} to the extracellular medium blocked the TRP-dependent current and the constitutive currents in a very similar manner (Fig. 2C). Since, these currents are mediated by the TRP channel, the TRP channel in these photoreceptors appears to be already open at the time the recording configuration is being established in the nonresponding cells of the Trp^{P365}/Trp^{CM} heterozygote and in all Trp^{P365}/Trp^{P365} cells but not in WT ommatidia.

The above study thus shows that the Trp^{P365} mutation highly increased the probability of the TRP channels to open in the dark in correlation to the Trp^{P365} dosage. Strikingly, constitutively activity of TRP channels was recorded even in cells that do not show any sign of morphological degeneration (i.e., in Trp^{P365}/Trp^{CM} heterozygote cells), thus suggesting that the constitutive activation of the channels precedes the degeneration [15].

Constitutive Activity of the TRP and TRPL Channels in the *rdgA* Mutant Causes Rapid Retinal Degeneration

Mutations in the *Drosophila* retinal degeneration A (*rdgA*) gene causes rapid retinal degeneration as early as at the pupa stage.[66] Retinal degeneration in the *rdgA* mutant is of particular interest because it has been convincingly shown that retinal degeneration in the *rdgA* mutant is due to mutation in the structural gene, which encodes for eye-specific DAG-kinase.[67] The localization of the *rdgA* DAG-kinase in the Sub Microvillar Cisternae (SMC), en extention of the smooth endoplasmic reticulum, and the need for its product phosphatidic acid for resynthesis of phosphoinositides makes RDGA a key enzyme for synthesis of the PIP_2, which is the substrate for PLC activity. In a recent report it has been shown that the *rdgA* phenotype is rescued by the introduction of the eye specific DAG-kinase by germ line transformation.[68] Furthermore, construction of a double mutant *rdgA;inaC*, lacking both DAG-kinase and the eye specific PKC, which did not affect the induction of retinal degeneration, ruled out the possibility that persistent activation of PKC is responsible for retinal degeneration in the *rdgA* mutant. Based on previous determination of DAG in WT and *rdgA* flies showing that DAG content is not increased in *rdgA* flies,[69] the authors suggested that insufficient production of phosphatidic acid rather than excessive accumulation of DAG is responsible for retinal degeneration in *rdgA*. Hardie and colleagues have recently shown that the TRP and TRPL channels are constitutively active in *rdgA* photoreceptors. Strikingly, early degeneration and some responsiveness to light were rescued in *rdgA;trp* double mutants lacking TRP channels.[14] Interestingly, it has been recently shown that exogenous polyunsaturated fatty acids (PUFAs), a possible product of DAG, activate the TRP and TRPL channels in the dark.[70] Based on the effect of PUFAs on TRP, Hardie and colleagues have suggested that the constitutive activity of the TRP and TRPL channels in the *rdgA* mutant arise from accumulation of DAG. Accumulation of DAG was suggested to lead to excessive production of PUFA by DAG lipase, although this suggestion has not been supported experimentally.[14]

Metabolic Stress Constitutively Opens the TRP and TRPL Channels in the Dark at a Late Stage of the Cascade

Anoxia is known to rapidly and reversibly depolarize the photoreceptor cells of the fly in the dark, possibly via openings of the light activated channels. It is well established that Ca^{2+} influx into the photoreceptors takes place almost exclusively via the TRP and TRPL channels.[25,28] Therefore, Ca^{2+} influx was measured during anoxia to identify the specific component of the response to anoxia, which arises directly from activation of these channels. *Drosophila* mutant, which are defective in proteins crucial for the phototransduction cascade, were used to localize the transduction stage that underlies the effects of anoxia. Using Ca^{2+}-selective microelectrodes in WT flies, the well described reduction in extracellular Ca^{2+} concentration ($[Ca^{2+}]_{out}$) was measured during illumination (Fig. 3A, see also refs. 28,71). The reduction of $[Ca^{2+}]_{out}$ during illumination arises from Ca^{2+} influx into the photoreceptor cells due to openings of TRP and TRPL channels.[28] Application of anoxia indeed induced, after a delay, a reduction in $[Ca^{2+}]_{out}$ (Fig. 3A, bottom). This observation indicates that in response to anoxia the large depolarization phase and Ca^{2+} influx are due to openings of the light-sensitive channels. Fig. 3B shows that illumination of the blind mutant *norpA*^{P24}, in which light-activated PLC is missing[7,72] did not elicit any response to light as monitored by either voltage or $[Ca^{2+}]_{out}$ changes as expected (Fig. 3B). Application of anoxia in the dark induced a voltage response, similar to that of WT with an initial small phase and a subsequent larger phase. The calcium signal, which accompanied the larger phase of the voltage change, was similar in WT and the mutant except for a faster onset in the mutant and an overshoot after anoxia was turned off (Fig. 3). The effects of anoxia on the *ninaE*^{ora} mutant, which is an opsin Rh1 null mutant [73] and on the *Gαq1* mutant, which has a highly reduced level of light-activated G-protein[74] were measured. The effects of anoxia on these mutants were similar to those observed in WT flies. The results thus show that anoxia affects a late stage of the phototransduction cascade downstream to PLC activation.

Additional demonstration that anoxia opens both TRP and TRPL channels in the dark was obtained by measuring Ca^{2+} influx in the *trp* mutant, in which TRP is missing, and in the *trpl;trp* double mutant, which lacks both channels.[33] The ERG response to light of the *trp*^{P343} mutant revealed the typical decline towards baseline

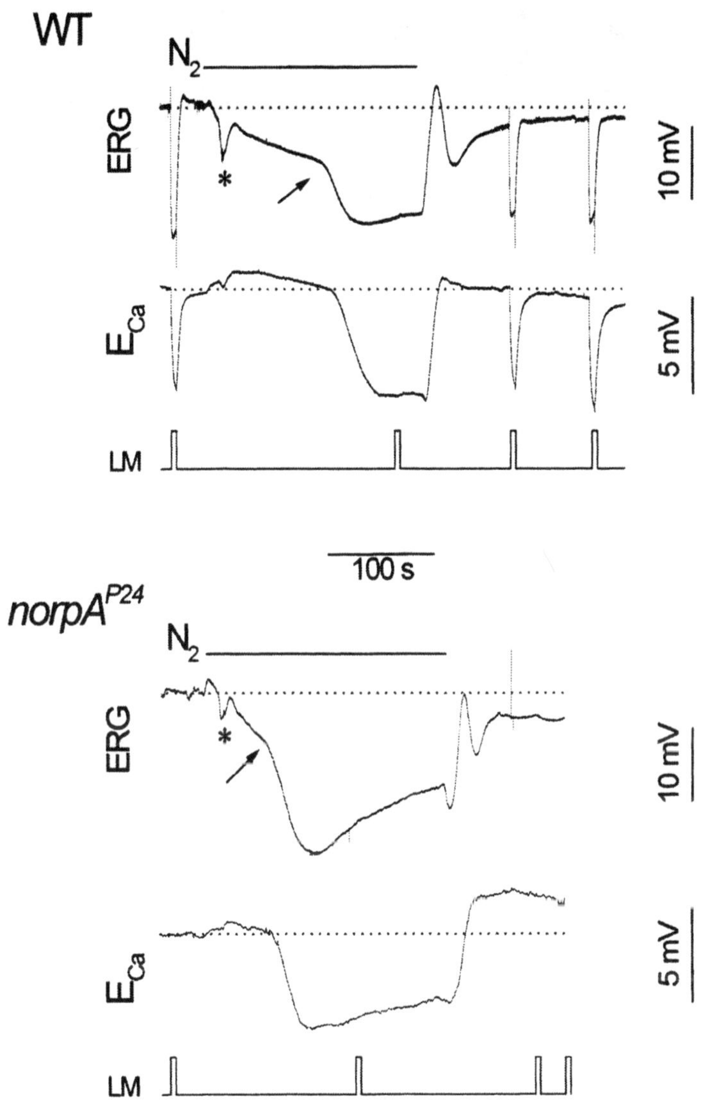

Figure 3. Anoxia activated the TRP and TRPL channels in both WT and PLC null mutant (*norpA*^{P24}) as monitored by Ca^{2+} influx. Extracellular voltage change (ERG, upper traces in both 3A and 3B) and potentiometric measurements with Ca^{2+}-selective microelectrode (E_{Ca}, lower traces in both 3A and 3B) in response to orange lights and anoxia in WT *Drosophila* and *norpA*^{P24} mutant. Note that there is no response to light in the *norpA*^{P24} mutant and the initial slow phase of the electrical response to anoxia is missing in the Ca^{2+} signals of both WT and the mutant. (From Agam et al. 2000, Copyright 2000 by the Society for Neuroscience).

during illumination while the Ca^{2+} signal was transient and relatively small, as previously reported (Fig. 4A).[28] Application of anoxia to the *trp*^{P343} mutant activated initially the slow and small phase of the voltage response (Fig. 4A). However,

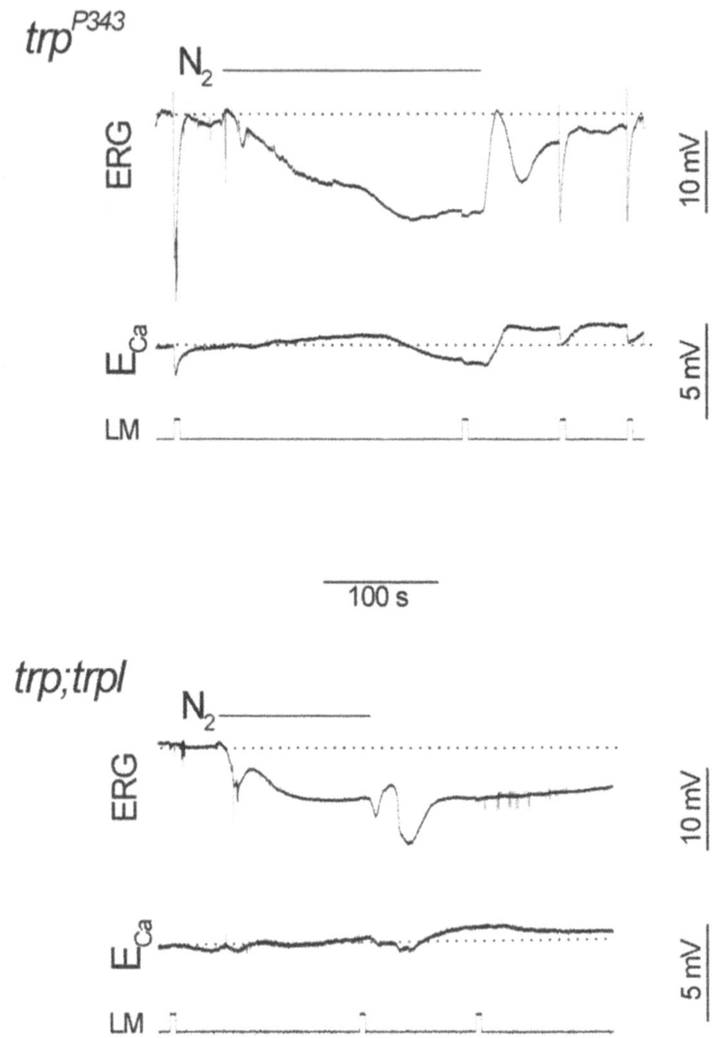

Figure 4. Genetic elimination of TRP resulted in a reduced Ca^{2+} influx through the remaining TRPL channels in response to anoxia, while elimination of both TRP and TRPL completely abolished both excitation and Ca^{2+} influx. The same paradigm of Fig. 3 was repeated in Fig. 4 except that the null *trp* mutant, *trp*^{P343} (3A) and the null double mutant *trpl*^{302};*trp*^{P343} (3B) were used. Note that in the *trp*^{P343} mutant the second phase of the electrical response to anoxia and the Ca^{2+} influx were relatively small but were maintained as long as anoxia was applied, in contrast to the transient responses to light. Also note that there is no response to light in the *trpl;trp* mutant and that the second phase of the electrical response to anoxia was absent. (From Agam et al. 2000, Copyright 2000 by the Society for Neuroscience)

in contrast to WT and the other mutants mentioned above, a significantly smaller amplitudes of the second faster phase of the voltage and Ca^{2+} signal were observed in *trp*^{P343} flies. The small Ca^{2+} signal in response to light and anoxia in the mutant reflects influx of Ca^{2+} into the photoreceptors via the TRPL channels.[28]

An interesting characteristic of the response to anoxia of the *trp*^{P343} mutant is the existence of a maintained component in both voltage and Ca^{2+} signals as long as anoxia was applied, as in WT (Fig. 4A). This maintained response to anoxia in the mutant is in sharp contrast to the transient nature of the response to continuous light.

To firmly establish that the Ca^{2+} influx in response to anoxia is due to activation of TRP and TRPL channels we measured $[Ca^{2+}]_{out}$ in response to anoxia in the double mutant *trpl*^{302};*trp*^{P343}. In this mutant the response to light is completely abolished (Fig. 4B).[33] Application of anoxia induced only the initial small voltage change with neither the subsequent larger voltage change nor any significant change in $[Ca^{2+}]_{out}$ during the initial 2 min of anoxia (Fig. 4).

The lack of virtually any Ca^{2+} influx in the *trpl;trp* double mutant in response to anoxia (Fig. 4) suggests that the mechanism which controls the opening of the TRP and TRPL channels is the target of anoxia.

Inhibition of Mitochondria Mimicked the Effects of Anoxia in vivo

Fly photoreceptor cells are known to have large number of mitochondria.[75] To investigate if the effects of anoxia in *Drosophila* retina are due to impaired function of the mitochondria, 2,4-dinitrophenol (DNP) and carbonyl cyanide *m*-chlorophenylhydrazone (CCCP) were applied to the intact eye. DNP and CCCP are known-uncouplers of the oxidative chain for ATP production in the mitochondria.[76] Application of DNP to the intact eyes of wild type *Drosophila* induced a negative voltage change and abolished light excitation in a manner similar to the larger phase of the response to anoxia. The effects of DNP were partially reversible ~20 min after the application. The effects of DNP on wild type flies were accelerated when combined with either anoxia or illumination, thus suggesting that light stimulation and all forms of metabolic stress were additive. These experiments further suggest that mitochondria uncouplers mimicked the effects of anoxia in wild type flies through effects on the TRP and TRPL channels. Whole cell recordings in single photoreceptor cells added more conclusive evidence to the in vivo studies.

Mitochondrial Uncouplers and Depletion of ATP Activate the TRP and TRPL Channels in the Dark in situ

Impairment of mitochondria function is expected to deplete ATP from the photoreceptor cells. To investigate more directly if depletion of ATP from the photoreceptor cells activates the TRP and TRPL channels, whole-cell patch clamp recordings in isolated ommatidia preparation was used.[22,23,25] When ATP and α-nicotinamide adenine dinucleotide (NAD) were omitted from the recording

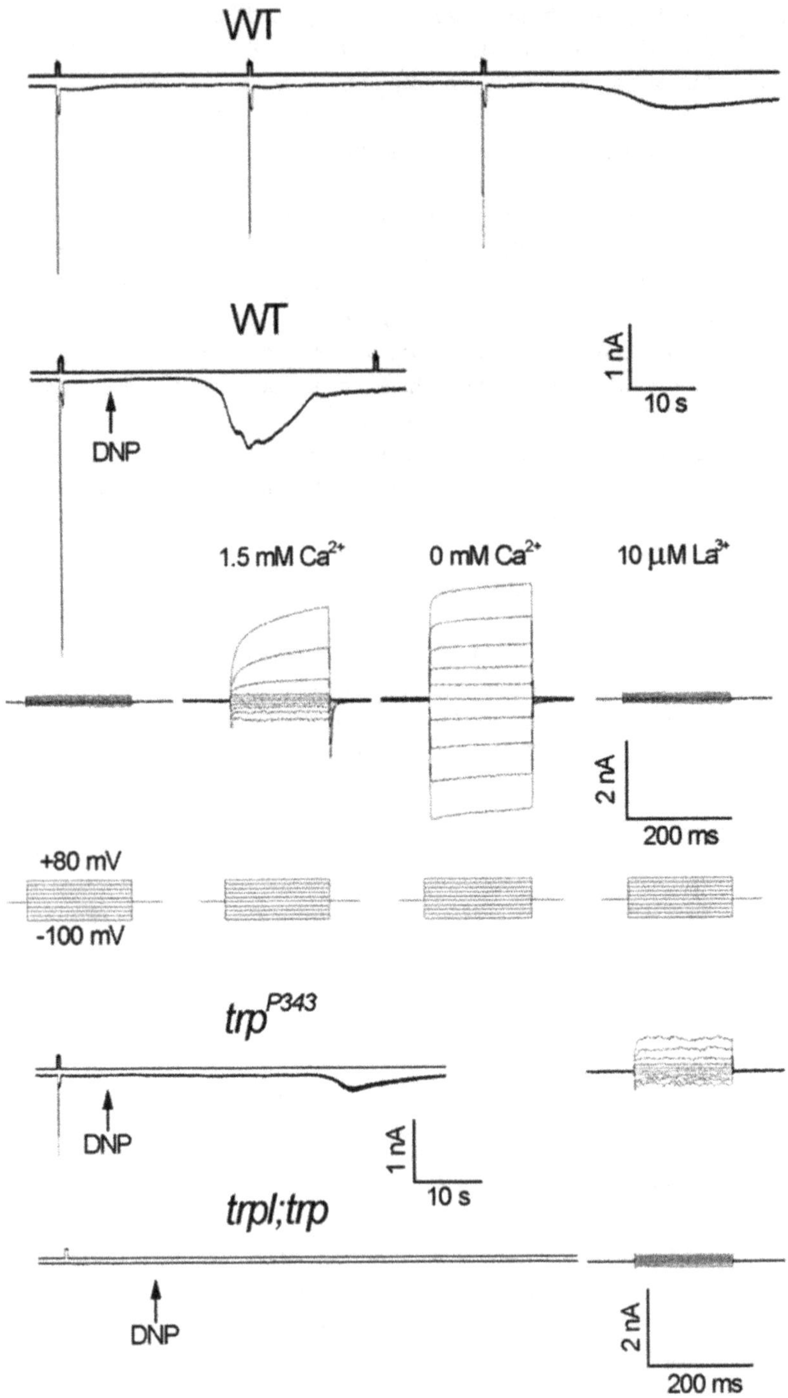

Figure 5. See figure legend on next page.

Figure 5. Single cell functional analysis by whole-cell recordings from newly eclosed flies, showing activation of the TRP and TRPL channels in WT and *trp*^{P343} mutant, respectively, but not in the *trpl;trp*^{P343} double mutant. Membrane currents are usually elicited by light in the present of ATP and NAD in the pipette. Omission of these agents, combined with either few light pulses (A) or with application of DNP to the bath (B) induced constitutive activation of the light sensitive channels in the dark as monitored by a sustained noisy inward current that had the characteristics of the TRP or TRPL-dependent current. None of these currents were observed in the double mutant *trpl;trp*^{P343} (E). A: Omission of ATP and NAD elicited a TRP dependent current in the dark after few responses to light. Fig. 5A shows the typical LIC of a wild type cell in response to orange lights (OG 590, attenuated by 1 log unit; left trace), followed by spontaneous inward current in the dark. The LM above all traces indicates the duration of the orange light stimulus. B,C: Application of DNP to the bath induced constitutive activation of the light sensitive channels. Membrane currents were recorded 30 s after establishing the whole-cell configuration with physiological concentrations (1.5 mM) of Ca^{2+} in the bath. (B) Application of DNP (0.1 mM) induced spontaneous dark inward current that abolished the response to light. (C) Dark membrane currents were elicited in a DNP treated cell by stepping the holding voltage in the range of –100 and +80 mV, the currents were suppressed when 10 μM La^{3+} was applied to the bath. (C, right). Currents obtained with 0 mM Ca^{2+} in the bath reveal the typical inward and outward rectifications of the TRP dependent current. D,E: The *trp* mutation reduced, and the double mutation *trpl;trp* abolished the dark current induced by DNP. Fig. 5D shows an inward small dark current induced by DNP and families of current traces elicited by series of voltage steps from photoreceptors of *trp*^{P343} (D) and *trpl;trp*^{P343} (E) during recordings with pipettes in which ATP and NAD were omitted and DNP was applied to the external medium. For each experiment, a series of nine voltage steps was applied from a holding potential of –20 mV in 20 mV steps (bottom traces in C). (From Agam et al. 2000, Copyright 2000 by the Society for Neuroscience)

pipette of WT cells, a few light pulses of medium intensity (orange light) induced an inward current in the dark, after a delay of ~100 s (Fig. 5A). Inclusion of ATP and NAD in the pipette prevented the induction of the inward dark current by repeated illumination, for at least 6 min. This observation suggests that although light pulses are known to reduce the ATP level in photoreceptor cells,[77] the supplement of exogenous ATP and NAD probably prevented depletion of ATP by illumination for at least 6 min. Application of 0.1 mM DNP to the bath (Fig. 5B) during recordings with pipettes without ATP and NAD induced the inward current in the dark in WT flies after a delay of only ~20 s (Fig. 5B). When DNP (0.1 mM) was included in the pipette solution (without ATP and NAD) the inward current was induced in less than 30s from the onset of whole cell recording.

In either *trp*^{P343} or in the double mutant *trpl*^{302};*trp*^{P343}, no dark current was elicited during prolonged recordings (<10 min) using pipettes without ATP and NAD. Application of 0.1 mM DNP either to the bath or into the pipette induced a noisy inward current of small amplitude in *trp* flies (Fig. 5D). In the double mutant *trpl*^{302};*trp*^{P343} DNP (applied either to the bath or in the pipette) or CCCP, without ATP and NAD in the pipette had no effect (Fig. 5E), even after incubation of 16 min.

The inward current of WT cells, which was induced in the dark following illumination or DNP application in the absence of ATP and NAD, had all the characteristics of the TRP dependent current.[78,79] In WT cells in zero external Ca^{2+} the current showed inward and outward rectification (Fig. 5C) while the amplitude of inward current at negative holding potentials was greatly reduced in the presence of Ca^{2+} in the medium (Fig. 5C). La^{3+}, a potent Ca^{2+} channel blocker that mimics the *trp* phenotype,[25,29,31,32,78] completely blocked the TRP-dependent current (Fig. 5C), as previously described.[78] In the *trp* mutant, after application of DNP the resulting inward current had all the characteristics of the TRPL-dependent current of *trp* mutant flies. It had small amplitude, high noise (Fig. 5C) and nearly absence of inward rectification even at 0 Ca^{2+} level (see ref. 79).[20] The effects of DNP were additive with light or absence of ATP and NAD, thus illumination accelerated the onset of the effect of DNP while inclusion of ATP and NAD in the pipette increased the latency of TRP channels activation to 2-4 min.

In summary, the results show that elimination of ATP and NAD from the recording pipette, combined with conditions that deplete ATP from the mitochondria opened the TRP and TRPL channels in the dark by affecting a late stage of the transduction cascade.

This study thus demonstrates that activation of TRP and TRPL in the dark results from depletion of ATP. The depletion of ATP probably directly activates the channels and not earlier stages of the transduction cascade because of the following reasons: i) Anoxia activated the TRP and TRPL channels at a stage downstream to PLC because anoxia induced Ca^{2+} influx in the PLC null mutant, $norpA^{P24}$. ii) Activation of earlier stages results in production of unitary events (quantum bumps)[70,80] while depletion of ATP was shown to produce a smooth response with channel noise (see also ref. 78). iii) Application of anoxia to the null *trp* mutant in vivo and application of DNP in situ produced maintained openings of the TRPL channels.[81]

The results suggest that the mitochondria are the primary target of anoxia leading to a reduction in cellular ATP. Indeed, previous studies on the large fly *Calliphora* have demonstrated that application of N_2 rapidly and reversibly inhibited a specific transient green fluorescence that normally arises from functional mitochondria in the eye.[82]

The observation that impaired mitochondia function leads to openings of TRP and TRPL channels has important implications on previous studies. In these studies activation of various TRP channels have been obtained by several means including oxidative stress[83] and application of polyunsaturated fatty acids (PUFAs).[70] The latter is of special interest because linoleic acid and several other long chain fatty acids have been shown to react as efficient uncouplers of mitochondria in a variety of cells.[84,85] The effect of DNP, therefore, suggest that PUFAs action on TRP and TRPL channels is indirect and result from their action as mitochondria uncouplers.[81]

The mechanism underlying activation of TRP and TRPL channels by depletion of ATP is not clear. However, the need to supply ATP in the dark at a high rate for proper function of the TRP and TRPL channels partially accounts for the well known, but unexplained, phenomenon of a high rate of oxygen consumption of insect retina in the dark.[86] The results, thus, suggest that a constitutive ATP dependent process operating at a late stage of the phototransduction cascade is required to keep the TRP and TRPL channels closed in the dark. This makes these channels extremely sensitive to anoxia and may lead to cell death under metabolic stress, not only in *Drosophila* but also in vertebrate cells, which express TRP channels.[83] The present study thus provides a novel target for metabolic stress with possible implications for brain damage since activation of TRP induces massive Ca^{2+} influx and TRP homologues are expressed in the brain.[36,39,52]

Taken together, the link of TRP gating to the metabolic state of cells opens up a new avenue of research that may explain many of the seemingly contradictory reports on pharmacological properties and mechanism of activation of TRP channels.[36,39]

Mammalian TRP Related Channels, Which are Sensitive to the Metabolic State of the Cell

The TRPM subfamily, which belongs to the TRP super-family is conserved through evolution and exist in *C. elegans* (CED-11, GON-2) *Drosophila* and humans. Members of this subfamily, share ~20% amino acid identity with the *Drosophila* TRP over ~325 residue region that includes S2-S6 transmembrane segments and the TRP domain.[13] The total length of this subfamily is 1000-2000 amino acids and it varies considerably because of large diversity of the C terminal domain. Members of the TRPM subfamily have a potential role in cell-cycle regulation. A very interesting member of this subfamily is the bi-functional channel protein designated PLC- interacting kinase (TRP-PLIK or LTRPC7, or TRPM7), which constitutes the first example of a channel with kinase activity.[87,88] Targeted deletion of LTRPC7 in DT-40 B cells was lethal, indicating that LTRPC7 has a fundamental and nonredundant role in cellular physiology. Heterologous expression of LTRPC7 in HEK-293 cells revealed a large outwardly rectifying current like the *Drosophila* TRPL channel. Further studies showed that the pore of the channel had a high affinity for and is permeant to both Ca^{2+} and Mg^{2+}. At the same time outward currents are inhibited with increasing Mg^{2+} concentration. The permeability of the channel to Mg^{2+} is reminiscent of a similar property of the *Drosophila* TRP channel.[25,89] Interestingly, Mg.ATP at mM concentration suppressed LTRPC7 channel activity. It turned out that Mg.ATP acted as physiological regulator of LTRPC7. Mg.ATPγS (but not Na.ATPγS) effectively and reversibly suppressed LTRPC7-dependent current suggesting that phosphorylation of the channel in Mg.ATP dependent manner directly regulates the channel activity. The studies on LTRPC7 indicate that this channel is an intracellular ligand-gated ion channel whose activation may be linked to cellular energy metabolism through its sensitivity to cytosolic Mg.ATP levels and that it effectively permeates both Ca^{2+} and Mg^{2+},[88]. It is thus striking that similar properties characterize the *Drosophila* TRP channel which is also permeable (and partially blocked) by Mg^{2+} [25,89] and shows extreme sensitivity to metabolic stress in vivo in a manner that depletion of ATP open the TRP channel in the dark while ATP suppresses channel opening.[81]

Additional important member of the TRPM subfamily, which is sensitive to the metabolic state of the cell is LTRPC2 (TRPM2) whose gating involves ADP-ribose (ADPR) and β-NAD. Heterologous expression of LTRPC2 in HEK-293 cells forms a current that was induced by application of 100 μM ADPR, showing a linear i-V relationship and reversing at 0 mV, thus, suggesting that this is a nonselective cation channel activated by ADPR.[90] The channel was permeable to both monovalent and divalent cations. Excised inside-out patch clamp recordings revealed that the channel is most likely directly opened by ADPR at concentration of 30 μM or higher showing no desensitization.[90] These studies show that ADPR can regulate directly the permeability of one member of the TRP family raising the possibility that the NUD9-H domain of the channel is involved in channel gating.

A more recent study has shown that both ADPR and β-NAD activate the LTRPC2 expressed in monocyte cell lines. Biophysical studies confirm LTRPC2 as Ca^{2+}-permeable nonselective cation channel. Interestingly, ATP strongly suppressed NAD and ADPR activated LTRPC2 channels.[91] Detailed experiments have demonstrated that LTRPC2 mediates Ca^{2+} influx into immunocytes where cellular ADPR and NAD directly activate LTRPC2 and enable Ca^{2+} influx. Importantly, NAD-induced activation was suppressed by ATP, suggesting that NAD activates LTRPC2 when ATP is depleted, while the activated LTRPC2 causes Ca^{2+} influx and apoptosis in these cells linking apoptosis with the metabolism of ADPR and NAD.[91]

CONCLUSION

The family of TRP channel proteins is unusually complex because of the large diversity of the structure and putative function of its members. Although the gating mechanism of any member of this large super-family is unknown, there are indications that in the *Drosophila* light activated channels and in some mammalian members, the gating of TRP is linked to the metabolic state of the cell. Although the function of the linkage between gating and metabolic stress is unknown it put the cell in a great hazard because TRP channels are permeable to Ca^{2+}. Accordingly, metabolic stress readily leads to toxic increase in cellular Ca^{2+} and cell death. Such a mechanism may underlie brain damage induced by ischemia, retinal degeneration in *Drosophila* and potentially degeneration of retinal neurons expressing TRPs.

ACKNOWLEDGMENTS

Experiments described in this paper were supported by the National Institutes of Health (EY 03529), the Israel Science Foundation (ISF), the German Israeli Foundation (GIF), the US-Israel Binational Science Foundation (BSF), the Minerva Foundation and the Moscona Foundation.

REFERENCES

1. Minke B, Hardie RC. Genetic dissection of *Drosophila* phototransduction. In: Stavenga DG, van der Hope DJN, Pugh E, eds. Molecular mechanisms in visual transduction. Elsevier: North Holland, 2000:449-525.
2. Pak WL. *Drosophila* in vision research. The Friedenwald Lecture. Invest Ophthalmol Vis Sci 1995; 36:2340-57.
3. Minke B, Selinger Z. The roles of *trp* and calcium in regulating photoreceptor function in *Drosophila*. Curr Opin Neurobiol 1996; 6:459-66.
4. Montell C. Visual transduction in *Drosophila*. Annu Rev Cell Dev Biol 1999; 15:231-68.
5. Ranganathan R, Malicki DM, Zuker CS. Signal transduction in *Drosophila* photoreceptors. Annu Rev Neurosci 1995; 18:283-317.
6. Minke B, Wu C, Pak WL. Induction of photoreceptor voltage noise in the dark in *Drosophila* mutant. Nature 1975; 258:84-7.
7. Bloomquist BT, Shortridge RD, Schneuwly S et al. Isolation of a putative phospholipase C gene of *Drosophila*, *norpA*, and its role in phototransduction. Cell 1988; 54:723-33.
8. Chang JT, Milligan S, Li Y et al. Mammalian homolog of *Drosophila* retinal degeneration B rescues the mutant fly phenotype. J Neurosci 1997; 17:5881-90.
9. Lev S, Hernandez J, Martinez R et al. Identification of a novel family of targets of PYK2 related to *Drosophila* retinal degeneration B (*rdgB*) protein. Mol Cell Biol 1999; 19:2278-88.
10. Ramulu P, Nathans J. Cellular and subcellular localization, N-terminal acylation, and calcium binding of *Caenorhabditis elegans* protein phosphatase with EF-hands. J Biol Chem 2001; 276:25127-35.
11. Sherman PM, Sun H, Macke JP et al. Identification and characterization of a conserved family of protein serine/threonine phosphatase homologous to *Drosophila* retinal degeneration C (*rdgC*). Proc Natl Acad Sci USA 1997; 94:11639-44.
12. Dose AC, Burnside B. Cloning and chromosomal localization of a human class III myosin. Genomics 2000; 67:333-42.
13. Montell C. Physiology, phylogeny, and functions of the TRP superfamily of cation channel. http://stke.sciencemag.org/cgi/content/full/OC_sigtrans2001/90/rel. 2001;1-17.
14. Raghu P, Usher K, Jonas S et al. Constitutive activity of the light-sensitive channels TRP and TRPL in the *Drosophila* diacylglycerol kinase mutant, *rdgA*. Neuron 2000; 26:169-79.
15. Yoon J, Cohen Ben-Ami H, Hong YS et al. Novel mechanism of massive photoreceptor degeneration caused by mutations in the *trp* gene of *Drosophila*. J Neurosci 2000; 20:649-59.
16. Montell C, Rubin GM. Molecular characterization of the *Drosophila trp* locus: a putative integral membrane protein required for phototransduction. Neuron 1989; 2:1313-23.
17. Wong F, Schaefer EL, Roop BC et al. Proper function of the *Drosophila trp* gene product during pupal development is important for normal visual transduction in the adult. Neuron 1989; 3:81-94.
18. Phillips AM, Bull A, Kelly LE. Identification of a *Drosophila* gene encoding a calmodulin-binding protein with homology to the *trp* phototransduction gene. Neuron 1992; 8:631-42.
19. Xu XZS, Li HS, Guggino WB et al. Coassembly of TRP and TRPL produces a distinct store-operated conductance. Cell 1997; 89:1155-64.
20. Reuss H, Mojet MH, Chyb S et al. In vivo analysis of the *Drosophila* light-sensitive channels, TRP and TRPL. Neuron 1997; 19:1249-59.
21. Xu XZ, Chien F, Butler A et al. TRPγ, a *Drosophila* TRP-related subunit, forms a regulated cation channel with TRPL. Neuron 2000; 26:647-57.
22. Hardie RC. Whole-cell recordings of the light induced current in dissociated *Drosophila* photoreceptors: evidence for feedback by calcium permeating the light-sensitive channels. Proc R Soc Lond B 1991; 245:203-10.
23. Ranganathan R, Harris GL, Stevens CF et al. A *Drosophila* mutant defective in extracellular calcium- dependent photoreceptor deactivation and rapid desensitization. Nature 1991; 354:230-2.

24. Minke B, Selinger Z. Inositol lipid pathway in fly photoreceptors: excitation, calcium mobilization and retinal degeneration. In: Osborne NA, Chader GJ, eds. Progress in retinal research. Oxford: Pergamon Press, 1991:99-124.
25. Hardie RC, Minke B. The *trp* gene is essential for a light-activated Ca^{2+} channel in *Drosophila* photoreceptors. Neuron 1992; 8:643-51.
26. Peretz A, Suss-Toby E, Rom-Glas A et al. The light response of *Drosophila* photoreceptors is accompanied by an increase in cellular calcium: effects of specific mutations. Neuron 1994; 12:1257-67.
27. Hardie RC. INDO-1 measurements of absolute resting and light-induced Ca^{2+} concentration in *Drosophila* photoreceptors. J Neurosci 1996; 16:2924-33.
28. Peretz A, Sandler C, Kirschfeld K et al. Genetic dissection of light-induced Ca^{2+} influx into *Drosophila* photoreceptors. J Gen Physiol 1994; 104:1057-77.
29. Niemeyer BA, Suzuki E, Scott K et al. The *Drosophila* light-activated conductance is composed of the two channels TRP and TRPL. Cell 1996; 85:651-9.
30. Leung HT, Geng C, Pak WL. Phenotypes of *trpl* mutants and interactions between the transient receptor potential (TRP) and TRP-like channels in *Drosophila*. J Neurosci 2000; 20:6797-803.
31. Hochstrate P. Lanthanum mimicks the trp photoreceptor mutant of *Drosophila* in the blowfly Calliphora J Comp Physiol A 1989; 166:179-87.
32. Suss Toby E, Selinger Z, Minke B. Lanthanum reduces the excitation efficiency in fly photoreceptors. J Gen Physiol 1991; 98:849-68.
33. Scott K, Sun Y, Beckingham K et al. Calmodulin regulation of *Drosophila* light-activated channels and receptor function mediates termination of the light response in vivo. Cell 1997; 91:375-83.
34. Scott K, Zuker C. TRP, TRPL and trouble in photoreceptor cells. Current Opinion in Neurobiology 1998; 8:383-8.
35. Montell C. New light on TRP and TRPL. Mol Pharmacol 1997; 52:755-63.
36. Putney-JW J, McKay RR. Capacitative calcium entry channels. Bioessays 1999; 21:38-46.
37. Minke B, Cook B. TRP channel proteins and signal transduction. Physiol Rev 2002; 82:429-272.
38. Wissenbach U, Schroth G, Philipp S et al. Structure and mRNA expression of a bovine *trp* homologue related to mammalian trp2 transcripts. FEBS Lett 1998; 429:61-6.
39. Harteneck C, Plant TD, Schultz G. From worm to man: three subfamilies of TRP channels. Trends Neurosci 2000; 23:159-66.
40. Tang J, Lin Y, Zhang Z et al. Identification of common binding sites for calmodulin and inositol 1,4,5-trisphosphate receptors on the carboxyl termini of Trp channels. J Biol Chem 2001; 276:21303-10.
41. Zhu X, Jiang MS, Peyton M et al. *trp*, a novel mammalian gene family essential for agonist- activated capacitative Ca^{2+} entry. Cell 1996; 85:661-71.
42. Birnbaumer L, Zhu X, Jiang MS et al. On the molecular basis and regulation of cellular capacitative calcium entry: Roles for Trp proteins. Proc Natl Acad Sci USA 1996; 93:15195-202.
43. Wes PD, Chevesich J, Jeromin A et al. TRPC1, a human homolog of a *Drosophila* store-operated channel. Proc Natl Acad Sci USA 1995; 92:9652-6.
44. Sinkins WG, Estacion M, Schilling WP. Functional expression of TrpC1: a human homologue of the *Drosophila* Trp channel. Biochem J 1998; 331:331-9.
45. Zhu X, Chu PB, Peyton M et al. Molecular cloning of a widely expressed human homologue for the *Drosophila trp* gene. FEBS Lett 1995; 373:193-8.
46. Funayama M, Goto K, Kondo H. Cloning and expression localization of cDNA for rat homolog of TRP protein, a possible store-operated Ca^{2+} channel. Brain Res Mol Brain Res 1996; 43:259-66.
47. Chang AS, Chang SM, Garcia RL et al. Concomitant and hormonally regulated expression of *trp* genes in bovine aortic endothelial cells. FEBS Lett 1997; 415:335-40.

48. Vannier B, Peyton M, Boulay G et al. Mouse trp2, the homologue of the human trpc2 pseudogene, encodes mTrp2, a store depletion-activated capacitative Ca^{2+} entry channel. Proc Natl Acad Sci USA 1999; 96:2060-4.
49. Liman ER, Corey DP, Dulac C. TRP2: a candidate transduction channel for mammalian pheromone sensory signaling. Proc Natl Acad Sci USA 1999; 96:5791-6.
50. Taniguchi M, Wang D, Halpern M. Chemosensitive conductance and inositol 1,4,5-trisphosphate-induced conductance in snake vomeronasal receptor neurons. Chem Senses 2000; 25:67-76.
51. Jungnickel MK, Marrero H, Birnbaumer L et al. Trp2 regulates entry of Ca^{2+} into mouse sperm triggered by egg ZP3. Nat Cell Biol 2001; 3:499-502.
52. Garcia RL, Schilling WP. Differential expression of mammalian TRP homologues across tissues and cell lines. Biochem Biophys Res Commun 1997; 239:279-83.
53. Mizuno N, Kitayama S, Saishin Y et al. Molecular cloning and characterization of rat *trp* homologues from brain. Brain Res Mol Brain Res 1999; 64:41-51.
54. Buess M, Engler O, Hirsch HH et al. Search for oncogenic regulators in an autocrine tumor model using differential display PCR: identification of novel candidate genes including the calcium channel mtrp6. Oncogene 1999; 18:1487-94.
55. Okada T, Inoue R, Yamazaki K et al. Molecular and functional characterization of a novel mouse transient receptor potential protein homologue TRP7. Ca^{2+}-permeable cation channel that is constitutively activated and enhanced by stimulation of G protein-coupled receptor. J Biol Chem 1999; 274:27359-70.
56. Philipp S, Hambrecht J, Braslavski L et al. A novel capacitative calcium entry channel expressed in excitable cells. EMBO J 1998; 17:4274-82.
57. Okada T, Shimizu S, Wakamori M et al. Molecular cloning and functional characterization of a novel receptor-activated TRP Ca^{2+} channel from mouse brain. J Biol Chem 1998; 273:10279-87.
58. Freichel M, Suh SH, Pfeifer A et al. Lack of endothelial store-operated Ca^{2+} current impairs agonist-dependent vasorelaxation in TRP4-/- mice. Neuron Cell Biology 2001; 3:121-7.
59. Colbert HA, Smith TL, Bargmann CI. OSM-9, a novel protein with structural similarity to channels, is required for olfaction, mechanosensation, and olfactory adaptation in *Caenorhabditis elegans*. J Neurosci 1997; 17:8259-69.
60. Walker RG, Willingham AT, Zuker CS. A *Drosophila* mechanosensory transduction channel. Science 2000; 287:2229-34.
61. Caterina MJ, Schumacher MA, Tominaga M et al. The capsaicin receptor: a heat-activated ion channel in the pain pathway. Neuron 1997; 389:816-24.
62. Hoenderop JG, van-der-Kemp AW, et al. Molecular identification of the apical Ca^{2+} channel in 1, 25-dihydroxyvitamin D3-responsive epithelia. J Biol Chem 1999; 274:8375-8.
63. Chen XZ, Vassilev PM, Basora N et al. Polycystin-L is a calcium-regulated cation channel permeable to calcium ions. Neuron 1999; 401:383-6.
64. Bargal R, Avidan N, Ben-Asher E et al. Identification of the gene causing mucolipidosis type IV. Nat Genet 2000; 26:118-23.
65. Sun M, Goldin E, Stahl S et al. Mucolipidosis type VI is caused by mutations in a gene encoding a novel transient receptor potential channel. Human Molecular Genetics 2000; 9:2471-8.
66. Hotta Y, Benzer S. Genetic dissection of the *Drosophila* nervous system by means of mosaics. Proc Natl Acad Sci USA 1970; 67:1156-63.
67. Masai I, Okazaki A, Hosoya T et al. *Drosophila* retinal degeneration A gene encodes an eye-specific diacylglycerol kinase with cysteine-rich zinc-finger motifs and ankyrin repeats. Proc Natl Acad Sci USA 1993; 90:11157-61.
68. Masai I, Suzuki E, Yoon CS et al. Immunolocalization of *Drosophila* eye-specific diacylgylcerol kinase, *rdgA*, which is essential for the maintenance of the photoreceptor. J Neurobiol 1997; 32:695-706.

69. Inoue H, Yoshioka T, Hotta Y. Diacylglycerol kinase defect in a *Drosophila* retinal degeneration mutant *rdgA*. J Biol Chem 1989; 264:5996-6000.
70. Chyb S, Raghu P, Hardie RC. Polyunsaturated fatty acids activate the *Drosophila* light-sensitive channels TRP and TRPL. Neuron 1999; 397:255-9.
71. Sandler C, Kirschfeld K. Light induced extracellular Ca^{2+} and Na^{+} concentration changes in the retinal of *Calliphora*: involvement in the mechanism of light adaptation. J Comp Physiol A 1991; 169:229-311.
72. Pearn MT, Randall LL, Shortridge RD et al. Molecular, biochemical, and electrophysiological characterization of *Drosohpila norpA* mutannts. J Biol Chem 1996; 271:4937-45.
73. O'Tousa JE, Leonard DS, Pak WL. Morphological defects in *oraJK84* photoreceptors caused by mutation in R1-6 opsin gene of *Drosophila*. J Neurogene. 1989; 6:41-52.
74. Scott K, Becker A, Sun Y et al. Gqa protein function *in vivo*: genetic dissection of its role in photoreceptor cell physiology. Neuron 1995; 15:919-27.
75. Boschek BC. On the fine structure of the peripheral retina and lamina ganglionaris of the fly, *Musca domestica* . Z Zellforsc. 1971; 118:369-409.
76. McLaughlin SG, Dilger JP. Transport of protons across membranes by weak acids. Physiol Rev 1980; 60:825-63.
77. Dimitracos SA, Tsacopoulos M. The recovery from a transient inhibition of the oxidative metabolism of the photoreceptors of the drone (*Apis mellifera*). J Exp Biol 1985; 119:165-81.
78. Hardie RC, Minke B. Spontaneous activation of light-sensitive channels in *Drosophila* photoreceptors. J Gen Physiol 1994; 103:389-407.
79. Hardie RC, Minke B. Calcium-dependent inactivation of light-sensitive channels in *Drosophila* photoreceptors. J Gen Physiol 1994; 103:409-27.
80. Scott K, Zuker CS. Assembly of the *Drosophila* phototransduction cascade into a signalling complex shapes elementary responses. Neuron 1998; 395:805-8.
81. Agam K, von-Campenhausen M, Levy S et al. Metabolic stress reversibly activates the *Drosophila* light-sensitive channels TRP and TRPL in vivo. J Neurosci 2000; 20:5748-55.
82. Stavenga DG, Tinbergen J. Light dependence of oxidative metabolism in fly compound eyes studied *in vivo* by micospectrophotometry. Naturwissenschaften 1983; 70,S.:618-20.
83. Balzer M, Lintschinger B, Groschner K. Evidence for a role of Trp proteins in the oxidative stress-induced membrane conductances of porcine aortic endothelial cells. Cardiovascular Research 1999; 42:543-9.
84. Hermesh O, Kalderon B, Bar TJ. Mitochondria uncoupling by a long chain fatty acyl analogue. J Biol Chem 1998; 273:3937-42.
85. Arslan P, Corps AN, Hesketh TR et al. cis-Unsaturated fatty acids uncouple mitochondria and stimulate glycolysis in intact lymphocytes. Biochem J 1984; 217:419-25.
86. Tsacopoulos M, Poitry S, Borsellino A. Diffusion and consumption of oxygen in the superfused retina of the drone (*Apis mellifera*) in darkness. J Gen Physiol 1981; 77:601-28.
87. Runnels LW, Yue L, Clapham DE. TRP-PLIK, a bifunctional protein with kinase and ion channel activities. Sciencexpress 2001; January 18.
88. Nadler MJ, Hermosura MC, Inabe K et al. LTRPC7 is a Mg.ATP-regulated divalent cation channel required for cell viability. Neuron 2001; 411:590-5.
89. Hardie RC, Mojet MH. Magnesium-dependent block of the light-activated and *trp*- dependent conductance in *Drosophila* photoreceptors. J Neurosci 1995; 74:2590-9.
90. Perraud AL, Fleig A, Dunn CA et al. ADP-ribose gating of the calcium-permeable LTRPC2 channel revealed by Nudix motif homology. Neuron 2001; 411:595-9.
91. Sano Y, Inamura K, Miyake A et al. Immunocyte Ca^{2+} influx system mediated by LTRPC2. Science 2001; 293:1327-30.

INDEX

www.ingramcontent.com/pod-product-compliance
Lightning Source LLC
LaVergne TN
LVHW081453200926
848397LV00077B/746
* 9 7 8 0 3 0 6 4 7 4 1 5 6 *